Chapman & Hall/CRC
Handbooks of Modern
Statistical Methods

Handbook of Missing Data Methodology

Edited by

Geert Molenberghs
Garrett Fitzmaurice
Michael G. Kenward
Anastasios Tsiatis
Geert Verbeke

CRC Press
Taylor & Francis Group
Boca Raton London New York

CRC Press is an imprint of the
Taylor & Francis Group, an **informa** business
A CHAPMAN & HALL BOOK

Chapman & Hall/CRC
Handbooks of Modern Statistical Methods

Series Editor

Garrett Fitzmaurice
Department of Biostatistics
Harvard School of Public Health
Boston, MA, U.S.A.

Aims and Scope

The objective of the series is to provide high-quality volumes covering the state-of-the-art in the theory and applications of statistical methodology. The books in the series are thoroughly edited and present comprehensive, coherent, and unified summaries of specific methodological topics from statistics. The chapters are written by the leading researchers in the field, and present a good balance of theory and application through a synthesis of the key methodological developments and examples and case studies using real data.

The scope of the series is wide, covering topics of statistical methodology that are well developed and find application in a range of scientific disciplines. The volumes are primarily of interest to researchers and graduate students from statistics and biostatistics, but also appeal to scientists from fields where the methodology is applied to real problems, including medical research, epidemiology and public health, engineering, biological science, environmental science, and the social sciences.

Published Titles

Handbook of Mixed Membership Models and Their Applications
Edited by Edoardo M. Airoldi, David M. Blei, Elena A. Erosheva, and Stephen E. Fienberg

Handbook of Markov Chain Monte Carlo
Edited by Steve Brooks, Andrew Gelman, Galin L. Jones, and Xiao-Li Meng

Longitudinal Data Analysis
Edited by Garrett Fitzmaurice, Marie Davidian, Geert Verbeke, and Geert Molenberghs

Handbook of Spatial Statistics
Edited by Alan E. Gelfand, Peter J. Diggle, Montserrat Fuentes, and Peter Guttorp

Handbook of Survival Analysis
Edited by John P. Klein, Hans C. van Houwelingen, Joseph G. Ibrahim, and Thomas H. Scheike

Handbook of Missing Data Methodology
Edited by Geert Molenberghs, Garrett Fitzmaurice, Michael G. Kenward, Anastasios Tsiatis, and Geert Verbeke

CRC Press
Taylor & Francis Group
6000 Broken Sound Parkway NW, Suite 300
Boca Raton, FL 33487-2742

First issued in paperback 2020

ISBN-13: 978-1-4398-5461-7 (hbk)
ISBN-13: 978-0-367-73929-4 (pbk)

Visit the Taylor & Francis Web site at
http://www.taylorandfrancis.com

and the CRC Press Web site at
http://www.crcpress.com

Dedication

To Conny, An, and Jasper

To Laura, Kieran, and Aidan

To Pirkko

To Marie

To Theo, Lien, Noor, and Aart

Contents

Preface

Missing data are a common and challenging problem that complicates the statistical analysis of data collected in almost every discipline. Since the 1990s there have been many important developments in statistical methodology for handling missing data. The goal of this handbook is to provide a comprehensive and up-to-date summary of many of the major advances. The book is intended to have a broad appeal. It should be of interest to all statisticians involved in the development of methodology or the application of missing data methods in empirical research.

The book is composed of 24 chapters, collected into a number of broad themes in the statistical literature on missing data methodology. Each part of the book begins with an introductory chapter that provides useful background material and an overview to set the stage for subsequent chapters. The first part begins by establishing notation and terminology, reviewing the general taxonomy of missing data mechanisms and their implications for analysis, and providing a historical perspective on early methods for handling missing data. The following three parts of the book focus on three alternative perspectives on estimation of parametric and semi-parametric models when data are missing: likelihood and Bayesian methods; semi-parametric methods, with particular emphasis on inverse probability weighting; and multiple imputation methods. These three parts synthesize major developments in methodology from the extensive statistical literature on parametric and semi-parametric models with missing data. Because inference about key parameters of interest generally requires untestable assumptions about the nature of the missingness process, the importance of sensitivity analysis is recognized. The next part of the book focuses on a range of approaches that share the broad aims of assessing the sensitivity of inferences to alternative assumptions about the missing data process. The final part of the book considers a number of special topics, including missing data in clinical trials and sample surveys and approaches to model diagnostics in the missing data setting.

In making our final selection of topics for this handbook, we focused on both established and emerging methodology for missing data that are advancing the field. Although our coverage of topics is quite broad, due to limitations of space, it is certainly not complete. For example, there is a related statistical literature on the broader concept of "coarsened" data (e.g., rounded, censored, or partially categorized data), of which missing data are a special case, that is not emphasized. In addition, we do not include the growing body of literature on causal inference as a missing data problem; other problems that can be put in the missing data framework, such as measurement error and record linkage, are only touched upon lightly. Nevertheless, we have tried to produce a handbook that is as comprehensive as possible, while also providing sufficient depth to set the scene for future research. We hope that this book provides the framework that will allow our readers to delve into research and practical applications of missing data methods.

Finally, we selected the chapter authors on the basis of their recognized expertise. We want to thank all of the contributors wholeheartedly for the high quality of the chapters they have written. We would also like to thank Rob Calver, Statistics Editor at Chapman & Hall/CRC Press, for his encouragement throughout all stages of this project.

Geert Molenberghs
Hasselt and Leuven, Belgium

Garrett Fitzmaurice
Boston, Massachusetts

Michael Kenward
London, United Kingdom

Butch Tsiatis
Raleigh, North Carolina

Geert Verbeke
Leuven, Belgium

Editors

Geert Molenberghs is a Professor of Biostatistics at Universiteit Hasselt and Katholieke Universiteit Leuven in Belgium. He earned a B.S. degree in mathematics (1988) and a Ph.D. in biostatistics (1993) from Universiteit Antwerpen. He has published on surrogate markers in clinical trials, and on categorical, longitudinal, and incomplete data. He was joint editor of *Applied Statistics* (2001–2004), co-editor of *Biometrics* (2007–2009), and co-editor of *Biostatistics* (2010–2015). He was president of the International Biometric Society (2004–2005), received the Guy Medal in Bronze from the Royal Statistical Society and the Myrto Lefkopoulou award from the Harvard School of Public Health. Geert Molenberghs is founding director of the Center for Statistics. He is also the director of the Interuniversity Institute for Biostatistics and statistical Bioinformatics (I-BioStat). Jointly with Geert Verbeke, Mike Kenward, Tomasz Burzykowski, Marc Buyse, and Marc Aerts, he has authored books on longitudinal and incomplete data, and on surrogate marker evaluation. Geert Molenberghs has received several Excellence in Continuing Education Awards of the American Statistical Association, for courses at Joint Statistical Meetings. Geert Molenberghs is a member of the Belgian Royal Academy of Medicine. He received accreditation as a professional statistician by the American Statistical Association (2011–2017).

Garrett Fitzmaurice is a Professor of Psychiatry (biostatistics) at the Harvard Medical School and a Professor in the Department of Biostatistics at the Harvard School of Public Health. He is a Fellow of the American Statistical Association and a member of the International Statistical Institute. He has served as Associate Editor for *Biometrics*, *Journal of the Royal Statistical Society, Series B*, and *Biostatistics* and as Statistics Editor for the journal *Nutrition*. His research and teaching interests are in methods for analyzing longitudinal and repeated measures data. A major focus of his methodological research has been on the development of statistical methods for analyzing repeated binary data and for handling the problem of attrition in longitudinal studies. Much of his collaborative research has concentrated on applications to mental health research, broadly defined. He co-authored the textbook *Applied Longitudinal Analysis*, 2nd edition (Wiley, 2011) and co-edited the handbook *Longitudinal Data Analysis* (Chapman & Hall/CRC Press, 2009).

Mike Kenward has been GSK Professor of Biostatistics at the London School of Hygiene and Tropical Medicine since 1999, with former positions at the Universities of Kent and Reading in the UK, and research institutes in the UK, Iceland, and Finland. His main research interests are in the analysis of longitudinal data, cross-over trials, small sample inference in REML, and the problem of missing data. He has co-authored three textbooks: *The Design and Analysis of Cross-Over Trials* with Byron Jones, *Missing Data in Clinical Studies* with Geert Molenberghs, and *Multiple Imputation and Its Application* with James Carpenter. He has been a co-editor of *Biometrics*, and is currently an associate editor of *Biostatistics*. During the past 25 years he has acted as a consultant in biostatistics, largely for the pharmaceutical industry, and he has given over 120 short courses worldwide on various topics in biostatistics, and has appeared as an expert witness in statistics in the U.S. Federal District Court of New York. He has recently completed the third Edition of *The Design and Analysis of Cross-Over Trials* with Byron Jones.

Anastasios A. Tsiatis is the Gertrude M. Cox Distinguished Professor of Statistics at

North Carolina State University. He earned a B.S. degree in mathematics (1970) from the Massachusetts Institute of Technology and a PhD in statistics (1974) from the University of California at Berkeley. His research has focused on developing statistical methods for the design and analysis of clinical trials, censored survival analysis, group sequential methods, surrogate markers, semi-parametric methods with missing and censored data, causal inference and dynamic treatment regimes and has been the major PhD advisor for more than 40 students working in these areas. He is Co-Editor of *Biostatistics* (2010–2015) and has authored a book, entitled: *Semiparametric Theory and Missing Data.* He is a Fellow of the American Statistical Association and the Institute of Mathematical Statistics, and is the recipient of the Spiegelman Award, the Snedecor Award, and the Princess Lilian Visiting Professorship in Belgium. He holds an honorary doctorate from Hasselt University in Belgium.

Geert Verbeke is a Professor in Biostatistics at Katholieke Universiteit Leuven and Universiteit Hasselt in Belgium. He earned a B.S. degree in mathematics (1989) from the Katholieke Universiteit Leuven, an M.S. in biostatistics (1992) from Universiteit Hasselt, and earned a Ph.D. in biostatistics (1995) from the Katholieke Universiteit Leuven. He has published extensively on various aspects of mixed models for longitudinal data analyses about which he co-authored and co-edited several textbooks (Springer Lecture Notes 1997; Springer Series in Statistics 2000 and 2005; Chapman & Hall/CRC 2009). He has held visiting positions at the Gerontology Research Center and Johns Hopkins University (Baltimore, MD), was International Program Chair for the International Biometric Conference in Montreal (2006), joint editor of the *Journal of the Royal Statistical Society, Series A* (2005–2008), and co-editor of *Biometrics* (2010–2012). He has served on a variety of committees of the International Biometric Society, is an elected Fellow of the American Statistical Association and an Elected Member of the International Statistical Institute. He was elected international representative on the Board of Directors of the American Statistical Association (2008–2010) and council member of the Royal Statistical Society (2013–2016). Geert Verbeke earned Excellence in Continuing Education Awards in 2002, 2004, 2008, and 2011 for short courses taught at the Joint Statistical Meetings of the American Statistical Association. He received the International Biometric Society Award for the best *Biometrics* paper in 2006 and received accreditation as professional statistician by the American Statistical Association (2010–2016).

List of Contributors

Thomas R. Belin	UCLA Jonathan and Karin Fielding School of Public Health Los Angeles, California
John B. Carlin	Murdoch Children's Research Institute & The University of Melbourne, Melbourne, Australia
James R. Carpenter	London School of Hygiene and Tropical Medicine & MRC Clinical Trials Unit at UCL London, United Kingdom
Michael J. Daniels	University of Texas at Austin Austin, Texas
Marie Davidian	North Carolina State University Raleigh, North Carolina
Garrett M. Fitzmaurice	Harvard University Boston, Massachusetts
Harvey Goldstein	University of Bristol Bristol, United Kingdom
Joseph W. Hogan	Brown University Providence, Rhode Island
Liangyuan Hu	Brown University Providence, Rhode Island
Michael G. Kenward	London School of Hygiene and Tropical Medicine London, United Kingdom
Craig Mallinckrodt	Eli Lilly & Company Indianapolis, Indiana
Geert Molenberghs	Hasselt University Hasselt, Belgium & University of Leuven Leuven, Belgium

Dimitris Rizopoulos	Erasmus University Medical Center Rotterdam, the Netherlands
Andrea Rotnitzky	Di Tella University Buenos Aires, Argentina
Juwon Song	Korea University Seoul, Korea
Anastasios A. Tsiatis	North Carolina State University Raleigh, North Carolina
Stef van Buuren	Netherlands Organisation for Applied Scientific Research TNO Leiden, the Netherlands & University of Utrecht Utrecht, the Netherlands
Stijn Vansteelandt	Ghent University Gent, Belgium
Geert Verbeke	University of Leuven Leuven, Belgium & Hasselt University Hasselt, Belgium
Ian R. White	MRC Biostatistics Unit Cambridge, United Kingdom

Part I

Preliminaries

1

Missing Data: Introduction and Statistical Preliminaries

Garrett M. Fitzmaurice
Harvard University, Boston, MA

Michael G. Kenward
London School of Hygiene and Tropical Medicine, London, UK

Geert Molenberghs, Geert Verbeke
Universiteit Hasselt & KU Leuven, Belgium

Anastasios A. Tsiatis
North Carolina State University, Raleigh, NC

CONTENTS

1.1 Introduction

Missing data are a ubiquitous problem that complicate the statistical analysis of data collected in almost every discipline. For example, the analysis of change is a fundamental component of many research endeavors that employ longitudinal designs. Although most longitudinal studies are designed to collect data on every individual (or unit) in the sample at each time of follow-up, many studies have missing observations at one or more occasions. Missing data are both a common and challenging problem for longitudinal studies. Indeed, missing data can be considered the rule, not the exception, in longitudinal studies in the health sciences. Frequently, study participants do not appear for a scheduled observation, or they may simply leave the study before its completion. When some observations are missing, the data are necessarily unbalanced over time in the sense that not all individuals have the same number of repeated measurements obtained at a common set of occasions. However, to distinguish missing data in a longitudinal study from other kinds of unbalanced data, such datasets are often referred to as being "incomplete." This distinction is important and emphasizes the fact that *intended* measurements on individuals could not be obtained.

With longitudinal studies problems of missing data are far more acute than in cross-sectional studies, because non-response can occur at any follow-up occasion. An individual's response can be missing at one follow-up time and then be measured at a later follow-up time. When missing values occur intermittently they yield a large number of distinct missingness patterns. At the same time, longitudinal studies often suffer from the problem of attrition or "dropout," that is, some individuals "drop out" or withdraw from the study before its intended completion. In either case, the term "missing data" is used to indicate that some intended measurements were not obtained. Because missing data are such a common problem in longitudinal studies, much of the statistical literature on methods for handling missing data has focused on their application to longitudinal study designs. For this reason, in this chapter we also concentrate attention on missing data arising in longitudinal study designs.

When the data collected from a study are incomplete, there are important implications for their analysis. It should be transparent that when a portion of the data are missing, there is necessarily a loss of information and a reduction in the precision with which key parameters of interest can be estimated. This reduction in precision is directly related to the amount of missing data and is influenced, to a certain extent, by the method of analysis. However, the potential for bias is usually of far greater concern. In certain circumstances, missing data can introduce bias and thereby lead to misleading inferences about the parameters of primary interest. It is this last feature, the potential for serious bias, that truly complicates the analysis of incomplete data. When there are missing data, the validity of any method of analysis will require that certain assumptions about the reasons why missing values occur, often referred to as the *missing data mechanism*, be tenable. Moreover, when data are missing inference about parameters of scientific interest generally requires untestable assumptions about the distribution of the missing data. As a result, a greater degree of caution is required in drawing conclusions from the analysis of incomplete data.

Put simply, the key issue is whether the reasons for missingness are related to the outcome of interest. When missingness is unrelated to the outcome, the impact of missing data is relatively benign and does not unduly complicate the analysis. On the other hand, when it is related to the outcome, somewhat greater care is required because there is potential for bias when individuals with missing data differ in important ways from those with complete data. Consequently, when data are incomplete, the reasons for any missingness must be carefully considered.

In this chapter, we introduce notation and terminology for missing data required for the remainder of the book. We do so within the context of missing data arising from longitudinal studies. We also review the general taxonomy of missing data mechanisms that was originally introduced by Rubin (1976). These missing data mechanisms differ in terms of assumptions about whether missingness is related to observed and unobserved responses. We mention the implications of these missing data mechanisms for analysis of data. We briefly review three broad classes of models for the missing data mechanism: selection models, pattern-mixture models, and shared-parameter models. Finally, we discuss some of the widely-used methods for handling missing data. These, and many other topics, are discussed in much greater detail in Chapter 2; specifically, Chapter 2 summarizes the historical developments in missing data methods and sets the scene for the remainder of the book.

1.2 Notation and Terminology

In this section, we introduce some notation and terminology for missing data. Because the problem of missing data is far more acute in longitudinal study designs, we present notation for incomplete longitudinal data that will be used extensively throughout the book. Specifically, we assume that for each of N individuals (or units), we intend to take n_i repeated measures of the response variable on the same individual. A subject with a *complete* set of responses has an $n_i \times 1$ response vector denoted by $\boldsymbol{Y}_i = (Y_{i1}, Y_{i2}, \ldots, Y_{in_i})'$. Regardless of the type of response (e.g., continuous, binary, or count data), Y_{ij} denotes the jth measurement for the ith subject (or unit) at time $t_{ij}, i = 1, \ldots, N,\ j = 1, \ldots, n_i$. In addition, associated with $\boldsymbol{Y}_i$ is an $n_i \times p$ matrix of covariates, X_i. In many, but not all, designed experiments where missingness arises, n_i is typically a constant, say $n_i = n$.

In what follows, we focus exclusively on missingness in the response. Although missingness can also arise in covariates, and raises similar considerations, unless stated otherwise, the covariates are assumed to be fully-observed for all individuals. Because we do not consider missingness in the covariates, we assume that any time-varying covariates are fixed by the study design. Also, there may be extraneous covariates that are distinct from X_i (e.g., auxiliary covariate that are predictive of missingness); to distinguish these from X_i, we let W_i denote the $n_i \times r$ matrix of extraneous covariates for the ith subject. We note that the inclusion of extraneous covariates often makes certain assumptions about the missingness process, assumptions that will be discussed in detail later, seem more plausible. In addition, certain methods for handling missing data (e.g., weighting methods and multiple imputation) can more easily incorporate extraneous covariates.

In the statistical literature on missing data, the $n_i \times 1$ vector of scheduled or *intended* responses (and the covariates) are referred to as the "complete data." This is the outcome vector that would have been recorded if there were no missing data. However, because of missingness, certain components of $\boldsymbol{Y}_i$ are not observed for at least some individuals. We let $\boldsymbol{R}_i$ be an $n_i \times 1$ vector of "response indicators," $\boldsymbol{R}_i = (R_{i1}, R_{i2}, \ldots, R_{in_i})'$, of the same length as $\boldsymbol{Y}_i$, with $R_{ij} = 1$ if Y_{ij} is observed and $R_{ij} = 0$ if Y_{ij} is missing. The stochastic process generating $\boldsymbol{R}_i$ is referred to as the missing data process. For the special case where missingness is restricted to "dropout" or attrition, such that a missing response at the jth occasion implies all subsequent responses are also missing, we can replace $\boldsymbol{R}_i$ by the scalar dropout indicator variable D_i, where

$$D_i = 1 + \sum_{j=1}^{n_i} R_{ij},$$

an indicator of the occasion at which dropout occurs. For a complete sequence, $D_i = n_i + 1$. Whether complete or incomplete, D_i indicates the length of the measurement sequence plus one. Dropout produces a particular *monotone* pattern of missingness. Missingness is said to be monotone when there is a permutation of the measurement occasions such that a measurement earlier in the permuted sequence is observed for at least those subjects who are observed at later measurements. Note that, for this definition to be meaningful, we need to have a balanced design in the sense of a common set of measurement occasions. All other patterns of missingness are referred to as *non-monotone.*

Given $\boldsymbol{R}_i$, the "complete data," $\boldsymbol{Y}_i = (Y_{i1}, \ldots, Y_{in_i})'$, can be partitioned into two components or subvectors $\boldsymbol{Y}_i^o$ and $\boldsymbol{Y}_i^m$, corresponding to those responses that are observed and missing, respectively. Here, $\boldsymbol{Y}_i^o$ denotes the vector of *observed* responses on the ith subject and contains those Y_{ij} for which $R_{ij} = 1$; $\boldsymbol{Y}_i^m$ denotes the complementary set of responses that are missing. These two components are usually referred to as the "observed data" and "missing data," respectively. This partition can differ from subject to subject, and the components are not necessarily temporally ordered in the sense that $\boldsymbol{Y}_i^o$ can contain components which are measured later than occasions at which components of $\boldsymbol{Y}_i^m$ ought to have been measured. Because the random vector $\boldsymbol{R}_i$ is recorded for all individuals, the "complete data" together with the "response indicators," $(\boldsymbol{Y}_i, \boldsymbol{R}_i)$, are referred to as the "full data." Except when all elements of $\boldsymbol{R}_i$ equal 1, the full data components are never jointly observed.

Note that one observes the measurements $\boldsymbol{Y}_i^o$ together with the response indicators $\boldsymbol{R}_i$. Confusion sometimes arises between the term "complete data" as introduced above and the term "complete case analysis." Whereas the former refers to the (hypothetical) dataset that would arise if there were no missing data, "complete case analysis" refers to an analysis based solely on those individuals with no missing data (i.e., omitting all subjects for which at least one component of $\boldsymbol{Y}_i$ is missing, the so-called incomplete cases).

When data are incomplete due to some stochastic mechanism we can write the full data density

$$f(\boldsymbol{Y}_i, \boldsymbol{R}_i | X_i, \boldsymbol{\gamma}, \boldsymbol{\psi}),$$

where X_i is the design matrix in the model for $\boldsymbol{Y}_i$, $\boldsymbol{\gamma}$ characterizes the model for $\boldsymbol{Y}_i$ given X_i and $\boldsymbol{\psi}$ characterizes the model for the response indicators $\boldsymbol{R}_i$. Often, we use $\boldsymbol{\gamma} = (\boldsymbol{\beta}', \boldsymbol{\alpha}')'$ and $\boldsymbol{\psi}$ to describe the data and missingness processes, respectively, where $\boldsymbol{\beta}$ is the parameter vector relating the mean of $\boldsymbol{Y}_i$ to X_i and $\boldsymbol{\alpha}$ assembles within-subject association parameters (e.g., covariances and/or variance components in linear models for the mean of $\boldsymbol{Y}_i$). In many applications, the main scientific interest is in making inferences about $\boldsymbol{\beta}$. In later sections, we consider likelihood-based methods for handling missing data that are based on parametric models for the data-generating and missing data mechanisms. When there are missing data, likelihood-based inferences are based on the so-called "observed-data likelihood," obtained by integrating the missing data $\boldsymbol{Y}_i^m$ out of the density:

$$\prod_{i=1}^{N} \int f(\boldsymbol{Y}_i, \boldsymbol{R}_i | X_i, \boldsymbol{\gamma}, \boldsymbol{\psi}) d\boldsymbol{Y}_i^m.$$

Therefore, for likelihood-based inferences to proceed, we require a parametric model for $f(\boldsymbol{Y}_i, \boldsymbol{R}_i | X_i, \boldsymbol{\gamma}, \boldsymbol{\psi})$. Such models can also form the basis of multiple imputation, another widely-used method for handling missing data that is discussed in detail in Part IV. Finally, semi-parametric approaches based on weighted estimating equations, e.g., inverse probability weighting (IPW) methods, provide another alternative to likelihood-based methods; semi-parametric models with missing data are discussed in detail in Part III.

1.3 Missing Data Mechanisms

To obtain valid inferences from incomplete data, we must consider the nature of the missing data mechanism. Ordinarily, the missing data mechanism is not under the control of study investigators; consequently, it is often not well understood. Instead, assumptions are made about the missing data mechanism, and the validity of the analysis depends on whether these assumptions hold for the data at hand.

Formally, the missing data mechanism describes the probability that a response is observed or missing. Specifically, it specifies a probability model for the distribution of the response indicators, $\boldsymbol{R}_i$, conditional on $\boldsymbol{Y}_i^o$, $\boldsymbol{Y}_i^m$, and X_i (and possibly W_i). A hierarchy of three different types of missing data mechanisms can be distinguished by considering how $\boldsymbol{R}_i$ is related to $\boldsymbol{Y}_i$ (and X_i): (i) *Missing Completely at Random* (MCAR); (ii) *Missing at Random* (MAR); and (iii) *Not Missing at Random* (NMAR). Rubin (1976) introduced this hierarchy of missing data mechanisms, and it is useful because the type of missing data mechanism determines the appropriateness of different methods of analyses.

1.3.1 Missing completely at random (MCAR)

Data are said to be missing completely at random (MCAR) when the probability that responses are missing is unrelated to either the specific values that, in principle, should have been obtained or the set of observed responses. That is, data are MCAR when $\boldsymbol{R}_i$ is independent of both $\boldsymbol{Y}_i^o$ and $\boldsymbol{Y}_i^m$, the observed and unobserved components of $\boldsymbol{Y}_i$, respectively. As such, missingness in $\boldsymbol{Y}_i$ is simply the result of a chance mechanism unrelated to observed or unobserved components of $\boldsymbol{Y}_i$. An example where partially missing longitudinal data can validly be assumed to be MCAR is the so-called "rotating panel" study design. In such a study design, commonly used in longitudinal surveys to reduce response burden, individuals rotate in and out of the study after providing a pre-determined number of repeated measures. However, the number and timing of the measurements is determined by study design and is not related to the vector of responses, $\boldsymbol{Y}_i$; the subset of unobserved responses on an individual can be considered "missing by design." Similarly, partially missing longitudinal data can generally be assumed to be MCAR when they are due to "administrative censoring" that occurs when subjects are entered in a staggered fashion and the study is terminated at a common scheduled date before all subjects have complete follow-up.

In the statistical literature, there does not appear to be universal agreement on whether the definition of MCAR also assumes no dependence of missingness on the covariates, X_i (and/or W_i if applicable; for ease of exposition, for the remainder of this section we assume there are no extraneous variables, W_i, present). Following Little (1995), we restrict the use of the term MCAR to the case where

$$\Pr(\boldsymbol{R}_i|\boldsymbol{Y}_i^o, \boldsymbol{Y}_i^m, X_i) = \Pr(\boldsymbol{R}_i). \tag{1.1}$$

When missingness depends on X_i, but is conditionally independent of $\boldsymbol{Y}_i$,

$$\Pr(\boldsymbol{R}_i|\boldsymbol{Y}_i^o, \boldsymbol{Y}_i^m, X_i) = \Pr(\boldsymbol{R}_i|X_i), \tag{1.2}$$

a subtle, but important, issue arises. The conditional independence of $\boldsymbol{Y}_i$ and $\boldsymbol{R}_i$, given X_i, may not hold when conditioning on only a subset of the covariates. Consequently, when an analysis is based on a subset of X_i that excludes a covariate predictive of $\boldsymbol{R}_i$, $\boldsymbol{Y}_i$ is no longer unrelated to $\boldsymbol{R}_i$. To avoid any potential ambiguity, Little (1995) suggests that MCAR be reserved for the case where there is no dependence of $\boldsymbol{R}_i$ on $\boldsymbol{Y}_i$ or X_i; when there is dependence on X_i alone, he suggests that the missing data mechanism be referred to as "covariate-dependent" missingness. However, we caution the reader that this subtle distinction is not universally adopted in the statistical literature on missing data.

The essential feature of MCAR is that the observed data can be thought of as a random sample of the complete data. Consequently, all moments, and even the joint distribution, of the observed data do not differ from the corresponding moments or joint distribution of the complete data. This has the following three implications. First, the so-called "completers" or "complete cases" (i.e., those subjects with no missing data) can be regarded as a random sample from the target population; moreover, any method of analysis that yields valid inferences in the absence of missing data will also yield valid, albeit inefficient, inferences when the analysis is restricted to the "completers." The latter is often referred to as a "complete case" analysis; we remind the reader not to confuse the terms *completers* and *complete case* analysis with the so-called *complete data*. Second, a similar result holds for subjects with any non-response pattern. The conditional distribution of $\boldsymbol{Y}_i^o$ for these subjects coincides with the distribution of the same components of $\boldsymbol{Y}_i$ in the target population. Third, the distribution of $\boldsymbol{Y}_i^m$ for subjects with any non-response pattern coincides with the distribution of the same components of $\boldsymbol{Y}_i$ for the "completers." Consequently, all available data can be used to obtain valid estimates of moments such as means, variances, and covariances.

An MCAR mechanism has important consequences for the analysis of data. In general, all methods for analysis that yield valid inferences in the absence of missing data will also yield valid inferences when the analysis is based on all available data, or even when it is restricted to the "completers." Finally, we note that the validity of MCAR can be checked empirically from the data at hand against the alternative of MAR, but only under the unverifiable assumption that the missing data mechanism is not NMAR. Specifically, we can compare and formally test equality of the distribution of the observed responses across patterns of missingness.

1.3.2 Missing at random (MAR)

In contrast to MCAR, data are said to be missing at random (MAR) when the probability that responses are missing depends on the set of observed responses, but is further unrelated to the specific missing values that, in principle, should have been obtained. Put another way, if subjects are stratified on the basis of similar values for the responses that have been observed, missingness is simply the result of a chance mechanism that does not depend on the values of the unobserved responses. In particular, data are MAR when $\boldsymbol{R}_i$ is conditionally independent of $\boldsymbol{Y}_i^m$, given $\boldsymbol{Y}_i^o$,

$$\Pr(\boldsymbol{R}_i|\boldsymbol{Y}_i^o, \boldsymbol{Y}_i^m, X_i) = \Pr(\boldsymbol{R}_i|\boldsymbol{Y}_i^o, X_i). \tag{1.3}$$

An example where longitudinal data are MAR arises when a study protocol requires that a subject be removed from the study as soon as the value of an outcome variable falls outside of a certain range of values. In that case, missingness in $\boldsymbol{Y}_i$ is under the control of the investigator and is related to observed components of $\boldsymbol{Y}_i$ only.

Because the missing data mechanism depends upon $\boldsymbol{Y}_i^o$, the distribution of $\boldsymbol{Y}_i$ in each of the distinct strata defined by the patterns of missingness is not the same as the distribution of $\boldsymbol{Y}_i$ in the target population. The MAR assumption has the following implications. First, the "completers" are a biased sample from the target population; consequently, an analysis restricted to the "completers" is not valid. Furthermore, the conditional distribution of $\boldsymbol{Y}_i^o$

for subjects with any non-response pattern does not coincide with the distribution of the same components of $\boldsymbol{Y}_i$ in the target population. Therefore, the sample means, variances, and covariances based on either the "completers" or the available data are biased estimates of the corresponding moments in the target population. With MAR, the observed data cannot be viewed as a random sample of the complete data.

However, the MAR assumption has the following important implication for the distribution of the missing data. The conditional distribution of an individual's missing values, $\boldsymbol{Y}_i^m$, given the observed values, $\boldsymbol{Y}_i^o$, is the same as the conditional distribution of the corresponding observations for the "completers," conditional on the "completers" having the same values as $\boldsymbol{Y}_i^o$. Put another way, upon stratification on values of $\boldsymbol{Y}_i^o$, the distribution of $\boldsymbol{Y}_i^m$ is the same as the distribution of the corresponding observations in the "completers" and also in the target population. When expressed this way, it should be apparent that the validity of the MAR assumption cannot be checked empirically from the data at hand, against NMAR, unless a very specific alternative model is assumed.

For likelihood-based inferences, the assumption that data are MAR implies that the likelihood contribution for the ith subject can be factored as

$$\begin{aligned} f(\boldsymbol{Y}_i^o, \boldsymbol{R}_i | X_i, \boldsymbol{\gamma}, \boldsymbol{\psi}) &= f(\boldsymbol{R}_i | \boldsymbol{Y}_i^o, X_i, \boldsymbol{\psi}) \times \int f(\boldsymbol{Y}_i^o, \boldsymbol{Y}_i^m | X_i, \boldsymbol{\gamma}) d\boldsymbol{Y}_i^m \\ &= f(\boldsymbol{R}_i | \boldsymbol{Y}_i^o, X_i, \boldsymbol{\psi}) f(\boldsymbol{Y}_i^o | X_i, \boldsymbol{\gamma}), \end{aligned} \tag{1.4}$$

because $f(\boldsymbol{R}_i | \boldsymbol{Y}_i^o, X_i, \boldsymbol{\psi})$ then does not depend on $\boldsymbol{Y}_i^m$ and consequently is independent of the integrator. Here, $\boldsymbol{\gamma} = (\boldsymbol{\beta}', \boldsymbol{\alpha}')'$ and $\boldsymbol{\psi}$ denotes parameters of the missing data mechanism. Thus, if $\boldsymbol{\gamma}$ and $\boldsymbol{\psi}$ are variation independent, the likelihood for $\boldsymbol{\gamma}$ is proportional to the likelihood obtained by ignoring the missing data mechanism. When $\boldsymbol{\gamma}$ and $\boldsymbol{\psi}$ are variation independent and the data are MAR, the missing data mechanism is referred to as "ignorable" (Rubin, 1976). Specifically, Rubin (1976) showed that likelihood-based inferences can be based on the likelihood ignoring the missing data mechanism, obtained by integrating the missing responses from the joint distribution, $f(\boldsymbol{Y}_i | X_i, \boldsymbol{\gamma})$:

$$L(\boldsymbol{\gamma} | \boldsymbol{Y}_i^o, X_i) = c \times \prod_{i=1}^{N} \int f(\boldsymbol{Y}_i^o, \boldsymbol{Y}_i^m | X_i, \boldsymbol{\gamma}) d\boldsymbol{Y}_i^m, \tag{1.5}$$

where c is a constant that does not depend on $\boldsymbol{\gamma}$.

Thus, when data are MAR, the missing values can be validly "predicted" or "extrapolated" using the observed data and a correct model for the joint distribution of $\boldsymbol{Y}_i$. Although with MAR, one does not require a model for $\Pr(\boldsymbol{R}_i | \boldsymbol{Y}_i^o, X_i)$, only a model for $\boldsymbol{Y}_i$ given X_i, the validity of the predictions of the missing values rests upon correct specification of the entire joint distribution of $\boldsymbol{Y}_i$. Because MCAR is a special case of MAR, the same is also true of MCAR; for this reason, MCAR and MAR are often referred to as *ignorable* mechanisms; the ignorability refers to the fact that once we establish that $\Pr(\boldsymbol{R}_i | \boldsymbol{Y}_i, X_i)$ does not depend on missing observations, we can ignore $\Pr(\boldsymbol{R}_i | \boldsymbol{Y}_i, X_i)$ and obtain a valid likelihood-based analysis provided we have correctly specified the model for $f(\boldsymbol{Y}_i | X_i)$.

Alternatively, methods of analysis that only require a model for the mean response (e.g., methods based on generalized estimating equations), but do not specify the joint distribution of the response vector, can be adapted to provide a valid analysis by explicitly modeling $\Pr(\boldsymbol{R}_i | \boldsymbol{Y}_i^o, X_i)$. When the data are MAR, the model for the mean response will generally not hold for the observed data. Consequently, the validity of generalized least square (GLS) and generalized estimating equations (GEE) methods (Liang and Zeger, 1986) applied to the available data is compromised. Recently, methods have been devised for making adjustments to the analysis by incorporating weights. The weights have to be estimated using a model for $\Pr(\boldsymbol{R}_i|\boldsymbol{Y}_i^o, X_i)$, hence the non-response model must be explicitly specified and estimated, although the joint distribution of the responses need not be. These weighting methods are discussed in greater detail in Part III (see Chapters 8 and 9).

In summary, the MAR assumption is far less restrictive on the missing data mechanism than MCAR and may be considered a more plausible assumption about missing data in many applications. When data are MAR, certain methods of analysis no longer provide valid estimates of parameters of interest without correctly specifying the joint distribution of the responses or correctly modeling the missing data process. Arguably, the MAR assumption should be the default assumption for the analysis of partially missing data unless there is a strong and compelling rationale to support the MCAR assumption.

1.3.3 Not missing at random (NMAR)

The third type of missing data mechanism is referred to as *not missing at random* (NMAR). In contrast to MAR, missing data are said to be NMAR when the probability that responses are missing is related to the specific values that should have been obtained, in addition to the ones actually obtained. That is, the conditional distribution of $\boldsymbol{R}_i$, given $\boldsymbol{Y}_i^o$, is related to $\boldsymbol{Y}_i^m$ and $\Pr(\boldsymbol{R}_i|\boldsymbol{Y}_i, X_i) = \Pr(\boldsymbol{R}_i|\boldsymbol{Y}_i^o, \boldsymbol{Y}_i^m, X_i)$ depends on at least some components of $\boldsymbol{Y}_i^m$. Alternatively, $\boldsymbol{R}_i$ may depend indirectly on $\boldsymbol{Y}_i^m$ via its dependence on unobserved random effects, say $\boldsymbol{b}_i$, with $\Pr(\boldsymbol{R}_i|\boldsymbol{Y}_i, \boldsymbol{b}_i, X_i) = \Pr(\boldsymbol{R}_i|\boldsymbol{b}_i, X_i)$.

An NMAR mechanism is often referred to as *non-ignorable* missingness because the missing data mechanism cannot be ignored when the goal is to make inferences about the distribution of the complete data. For example, in likelihood-based methods we can no longer factorize the individual likelihood contributions as in (1.4). In general, any valid inferential method under NMAR requires specification of a model for the missing data mechanism. We note that the term *informative* is sometimes used to describe missing data that are NMAR, especially for the monotone pattern of missingness due to dropout.

When data are NMAR, the model assumed for $\Pr(\boldsymbol{R}_i|\boldsymbol{Y}_i, X_i)$ is critical and must be included in the analysis. Moreover, the specific model chosen for $\Pr(\boldsymbol{R}_i|\boldsymbol{Y}_i, X_i)$ can drive the results of the analysis. It should be recognized that, short of tracking down the missing data, any assumptions made about the missingness process are wholly unverifiable from the data at hand. That is, the observed data provide no information that can either support or refute one NMAR mechanism over another. Recognizing that, without additional information, identification is driven by unverifiable assumptions, many authors have discussed the importance of conducting sensitivity analyses (e.g., Rosenbaum and Rubin, 1983, 1985; Nordheim, 1984; Little and Rubin, 1987, 2002; Laird, 1988; Vach and Blettner, 1995; Scharfstein, Rotnitzky, and Robins, 1999; Copas and Eguchi, 2001; Molenberghs, Kenward, and Goetghebeur, 2001; Verbeke et al., 2001). When missingness is thought to be NMAR, the sensitivity of inferences to a variety of plausible assumptions concerning the missingness process should be carefully assessed. Sensitivity analysis under different assumptions about missingness is the topic of Part V of the book.

Next, we consider the implications of NMAR for analysis. When data are NMAR, almost all standard methods of analysis are invalid. For example, standard likelihood-based methods that ignore the missing data mechanism yield biased estimates. To obtain valid estimators, *joint models* for the response vector and the missing data mechanism are required. In the following section we briefly review the literature on joint models and distinguish three model-based approaches: *selection*, *pattern-mixture*, and *shared-parameter* models. The main focus is on likelihood-based methods. However, we note that alternative methods for handling NMAR have been developed that do not require specification of the joint distribution of the responses; these weighting methods are discussed in detail in Part III (see Chapters 8 and 9).

1.4 Joint Models for Non-Ignorable Missingness

Joint models for $(\boldsymbol{Y}_i, \boldsymbol{R}_i)$ are often used to correct for non-ignorable non-response. Models for $f(\boldsymbol{Y}_i, \boldsymbol{R}_i|X_i, \gamma, \boldsymbol{\psi})$ can be specified in numerous ways, depending on how the joint distribution is factored. Three common factorizations of the joint distribution of the response and missing data mechanism give rise to selection, pattern-mixture, and shared-parameter models. In selection models, one uses a complete data model for the outcomes, and then the probability of non-response is modeled conditional on the possibly unobserved outcomes. That is, selection models specify the joint distribution of $\boldsymbol{R}_i$ and $\boldsymbol{Y}_i$ via models for the marginal distribution of $\boldsymbol{Y}_i$ and the conditional distribution of $\boldsymbol{R}_i$ given $\boldsymbol{Y}_i$:

$$f(\boldsymbol{Y}_i, \boldsymbol{R}_i|X_i, \gamma, \boldsymbol{\psi}) = f(\boldsymbol{Y}_i|X_i, \gamma) f(\boldsymbol{R}_i|\boldsymbol{Y}_i, X_i, \boldsymbol{\psi}).$$

For identifiability, the models are usually restricted in some way. With selection models identification comes from parametric assumptions about $f(\boldsymbol{Y}_i|X_i, \gamma)$ and unverifiable models for the dependence of the non-response probabilities on the unobserved outcomes.

In contrast, with pattern-mixture models, one uses a model for the conditional distribution of the outcomes given non-response patterns and then a model for non-response. Specifically, in pattern-mixture models, the joint distribution is specified in terms of models for the marginal distribution of $\boldsymbol{R}_i$ and the conditional distribution of $\boldsymbol{Y}_i$ given $\boldsymbol{R}_i$:

$$f(\boldsymbol{Y}_i, \boldsymbol{R}_i|X_i, \boldsymbol{\nu}, \boldsymbol{\delta}) = f(\boldsymbol{R}_i|X_i, \boldsymbol{\delta}) f(\boldsymbol{Y}_i|\boldsymbol{R}_i, X_i, \boldsymbol{\nu}),$$

where $\boldsymbol{\delta}$ and $\boldsymbol{\nu}$ denote parameters of the marginal and conditional densities, respectively. It should be immediately clear that the distribution of outcomes given patterns of non-response is not completely identifiable, because for all but the "completers" pattern, certain response variables are not observed. Hence, restrictions must be built into the model to ensure that there are links among the distributions of the outcomes conditional on the patterns of non-response (Little, 1993, 1994; Little and Wang, 1996).

Finally, in shared-parameter models, $\boldsymbol{Y}_i$ and $\boldsymbol{R}_i$ are assumed to depend on shared latent variables. Specifically, the model for $\boldsymbol{Y}_i$ is linked with a model for $\boldsymbol{R}_i$ through a vector of random effects that are shared between the complete data model and the model for the missing data mechanism. Conceptually, each individual is assumed to have subject-specific effects (random effects) that influence both the response, Y_{ij}, and the probability of missingness (R_{ij}). Shared-parameter models can be considered special cases of selection models for mixed effects models, where the probability of missing values is dependent on random effects:

$$f(\boldsymbol{Y}_i, \boldsymbol{R}_i, \boldsymbol{b}_i|X_i) = f(\boldsymbol{b}_i|X_i) f(\boldsymbol{Y}_i|\boldsymbol{b}_i, X_i) f(\boldsymbol{R}_i|X_i, \boldsymbol{Y}_i, \boldsymbol{b}_i).$$

In many applications of shared-parameter models it is assumed that $\boldsymbol{Y}_i$ and $\boldsymbol{R}_i$ are independent given the vector of random effects, $\boldsymbol{b}_i$,

$$\begin{aligned} f(\boldsymbol{Y}_i, \boldsymbol{R}_i, \boldsymbol{b}_i|X_i) &= f(\boldsymbol{b}_i|X_i) f(\boldsymbol{Y}_i|\boldsymbol{b}_i, X_i) f(\boldsymbol{R}_i|X_i, \boldsymbol{Y}_i, \boldsymbol{b}_i) \\ &= f(\boldsymbol{b}_i|X_i) f(\boldsymbol{Y}_i|\boldsymbol{b}_i, X_i) f(\boldsymbol{R}_i|X_i, \boldsymbol{b}_i). \end{aligned}$$

Because the vector of random effects, $\boldsymbol{b}_i$, is shared between the two models, marginally (after integrating over the distribution of the random effects) this induces correlation between $\boldsymbol{R}_i$ and the observed and unobserved components of $\boldsymbol{Y}_i$; consequently, the resulting missing data mechanism is NMAR although the underlying mechanism for the dependence of $\boldsymbol{R}_i$ on $\boldsymbol{Y}_i$ is not immediately transparent.

Although shared-parameter models are often treated as a separate category, for the remainder of this chapter we simply regard them as special cases of selection models where the dropout probabilities depend *indirectly* upon the unobserved responses, via their dependence on random effects (theoretically, we note that shared-parameter models can equally be represented as pattern-mixture models). In the following sections, we briefly review the early literature on selection and pattern-mixture models and highlight some of the potential advantages and disadvantages of each of the approaches.

1.4.1 Selection models

Some of the earlier work on methods for handling non-ignorable missingness was conducted by Wu and colleagues in the context of dropout in longitudinal studies. Wu and Carroll (1988) proposed a selection modeling approach used by many subsequent researchers that assumes the vector of continuous responses follow a simple linear random-effects model (Laird and Ware, 1982) and that the dropout process depends upon an individual's random effects (e.g., random intercept and slope). As noted earlier, models where the dropout probabilities depend indirectly upon the unobserved responses, via dependence on random effects, are referred to as *shared-parameter* models. An alternative selection modeling approach, based on earlier work in the univariate setting pioneered by econometricians (e.g., Heckman, 1976), was proposed by Diggle and Kenward (1994) who allowed the probability of non-response to depend directly on the unobserved outcome rather than on underlying random effects. Selection models, where the non-response probabilities depend directly or indirectly (via dependence on random effects) on unobserved responses, have also been extended to discrete longitudinal data (e.g., Baker, 1995; Fitzmaurice, Molenberghs, and Lipsitz, 1995; Fitzmaurice, Laird, and Zahner, 1996; Molenberghs, Kenward, and Lesaffre, 1997; Ten Have et al., 1998, 2000).

Finally, there is closely related work on selection models where the target of inference is the time-to-event (e.g., dropout time) distribution, rather than the distribution of the repeated measures. Schluchter (1992) and De Gruttola and Tu (1994) independently developed an alternative selection model which assumes the (transformed) event time and the repeated responses have a joint multivariate normal distribution, with an underlying random-effects structure; they proposed full likelihood approaches utilizing the EM algorithm (Dempster, Laird, and Rubin, 1977). Tsiatis, De Gruttola, and Wulfsohn (1995) extended this approach to permit a non-parametric time-to-event distribution with censoring. Note that the primary objective of this latter research is inferences about the time-to-event distribution and the ability of the repeated measures to capture covariate effects (e.g., treatment effects) on, for example, survival time. In contrast, the more usual focus of selection models for longitudinal data is on estimation of the mean time trend of the repeated measures, and its relation to covariates, regarding the dropout time or non-response patterns as a "nuisance" characteristic of the data.

1.4.2 Pattern-mixture models

Because of the complexity of model fitting in the selection modeling approach proposed by Wu and Carroll (1988), Wu and Bailey (1988, 1989) suggested approximate methods for inference about the time course of the vector of continuous responses. Their pattern-mixture modeling approach was based on method-of-moments type fitting of a linear model to the least squares slopes, conditional on dropout time, then averaging over the distribution of dropout time. Other than assuming dropouts occur at discrete times, this work made no distributional assumption on the event times (dropout times). Hogan and Laird (1997) extended this pattern-mixture model by permitting censored dropout times, as might arise when there are late entrants to a trial and interim analyses are performed. Follman and Wu (1995) generalized Wu and Bailey's (1988) conditional linear model to permit generalized linear models without any parametric assumption on the random effects. Other related work on pattern-mixture models is described in Rubin (1977), Glynn, Laird, and Rubin (1986), Mori, Woodworth, and Woolson (1992), Hedeker and Gibbons (1997), Ekholm and Skinner (1998), Molenberghs et al. (1999), Park and Lee (1999), Michiels, Molenberghs, and Lipsitz (1999), and Fitzmaurice and Laird (2000); see Little (2009) for a comprehensive review of both pattern-mixture and selection models.

There is an additional avenue of research on pattern-mixture models that can be distinguished. Little (1993, 1994) has considered pattern-mixture models that stratify the incomplete data by the pattern of missing values and formulate *distinct* models within each stratum. In these models, additional assumptions about the missing data mechanism have to be made to yield supplemental restrictions that identify the models. Interestingly, Little (1994) relies on a *selection* model for missingness to motivate a set of identifying restrictions for the pattern-mixture model. These identifying restrictions are unverifiable from the observed data; thus Little (1994) recommends conducting a sensitivity analysis for a range of plausible values. Thijs et al. (2002) provide a taxonomy for the various ways to conceive of and fit pattern-mixture models.

1.4.3 Contrasting selection and pattern-mixture models

Pattern-mixture and selection models each have their own distinct advantages and disadvantages. One key advantage of selection models is that they directly model the marginal distribution of the complete data, the usual target of inference. In addition, selection models often seem more intuitive to statisticians, as it is straightforward to formulate hypotheses about the non-response process. For example, using the taxonomy introduced in Section 1.3, it is straightforward to characterize the missing data mechanism within selection models. While assumptions about the non-response process are transparent in selection models, what is less clear is how these translate into assumptions about the distributions of the unobserved outcomes. Specifically, $f(\boldsymbol{Y}_i^m|\boldsymbol{Y}_i^o, \boldsymbol{R}_i, X_i)$, the distribution of the missing data given the observed data, cannot be obtained in closed form for parametric selection models. Furthermore, with selection models, identification comes from assumptions about $f(\boldsymbol{Y}_i|X_i, \gamma)$ and from postulating unverifiable models for the dependence of the non-response process on the unobserved outcomes. However, except in very simple cases, it can be very difficult to determine the identifying restrictions that must be placed on the model (Glonek, 1998; Bonetti, Cole, and Gelber, 1999). Moreover, once a parametric selection model has been postulated all parameters are identified by the observed data, including those for $f(\boldsymbol{Y}_i^m|\boldsymbol{Y}_i^o, \boldsymbol{R}_i, X_i)$; as a result, parametric selection models do not lend themselves well to sensitivity analysis because it is unclear how to incorporate sensitivity parameters. This has motivated much recent work on semi-parametric selection models (see Chapters 7–10). Semi-parametric selection models combine a non-parametric or semi-parametric model for $f(\boldsymbol{Y}_i|X_i)$ with a parametric model for $f(\boldsymbol{R}_i|\boldsymbol{Y}_i, X_i)$ but leave $f(\boldsymbol{Y}_i^m|\boldsymbol{Y}_i^o, \boldsymbol{R}_i, X_i)$ non-identifiable from the observed data. With semi-parametric selection models, sensitivity analysis is based on incorporating additional sensitivity parameters that make unverifiable assumptions about $f(\boldsymbol{Y}_i^m|\boldsymbol{Y}_i^o, \boldsymbol{R}_i, X_i)$; a truly expository guide to the theory of estimation for semi-parametric models with missing data can be found in the text by Tsiatis (2006). Finally, in general, selection models can be computationally demanding to fit.

In contrast, pattern-mixture models are often as easy to fit as standard models that assume non-response is ignorable. With pattern-mixture models, it is immediately clear that the distribution of the outcomes given patterns of non-response is not completely identifiable. Identification comes from postulating unverifiable links among the distributions of the outcomes conditional on the patterns of non-response. But, in contrast to selection models, it is relatively straightforward to determine the identifying restrictions that must be imposed. With pattern-mixture models the missing data mechanism can be derived via Bayes' rule,

$$f(\boldsymbol{R}_i|\boldsymbol{Y}_i, X_i) = \frac{f(\boldsymbol{R}_i|X_i)f(\boldsymbol{Y}_i|\boldsymbol{R}_i, X_i)}{f(\boldsymbol{Y}_i|X_i)},$$

although, in general, this is intractable. Pattern-mixture models do have one very important drawback that has, so far, limited their usefulness in many areas of application. The main drawback of pattern-mixture models is that the natural parameters of interest are not immediately available; they require marginalization of the distribution of outcomes over non-response patterns,

$$f(\boldsymbol{Y}_i|X_i) = \sum_R f(\boldsymbol{Y}_i|\boldsymbol{R}_i, X_i)f(\boldsymbol{R}_i|X_i).$$

For example, in the longitudinal setting, pattern-mixture models typically parameterize the mean of the longitudinal responses conditional upon covariates and the indicators of non-response patterns. As a result, the interpretation of the model parameters is unappealing due to the conditioning or stratification on non-response patterns. In most longitudinal studies the target of inference is the marginal distribution of the repeated measures (i.e., *not* conditional on non-response patterns). Although this marginal distribution can be obtained by averaging over the distribution of the non-response patterns (Wu and Bailey, 1989; Hogan and Laird, 1997; Fitzmaurice and Laird, 2000), the assumed form for the model (e.g., generalized linear model) for the conditional means no longer holds for the marginal means when a non-linear link function has been adopted.

There has been much recent research on selection and pattern-mixture models for handling missing data; this topic is discussed in greater detail in Part II (see Chapter 4). However, it is worth emphasizing that all models for handling non-ignorable non-response, whether a selection or pattern-mixture modeling approach is adopted, are fundamentally non-identifiable unless some arbitrary modeling assumptions are imposed. That is, inference is possible only once some unverifiable assumptions about the non-response process or the distributions of the missing responses have been made. As a result, somewhat greater care is required in drawing any conclusions from the analysis. In general, it is important to conduct sensitivity analyses, with results reported for analyses conducted under a range of plausible assumptions about non-ignorable non-response. Sensitivity analysis under different assumptions about missingness is the topic of Part V of the book.

1.5 Methods for Handling Missing Data

In this final section, we very briefly review three commonly used methods for handling missing data: (i) likelihood and Bayesian methods; (ii) weighting methods; and (iii) multiple imputation methods. In a very general way, these three approaches are interrelated in the sense that they "impute" or fill-in certain values for the missing data; the difference is that in (iii) the imputation is explicit; whereas in (i) and (ii), it is more implicit. We also discuss the assumptions about missingness required for each of the methods to yield valid inferences in most practical settings.

1.5.1 Likelihood and Bayesian methods

In general, likelihood-based methods for handling missing data assume a parametric model for the complete data and, in cases where missingness is assumed to be non-ignorable, a parametric model for the missing data mechanism. With missing data, likelihood inference is based on the *observed-data likelihood*,

$$L(\gamma, \psi | \boldsymbol{Y}_i^o, \boldsymbol{R}_i, X_i) = c \times \prod_{i=1}^{N} \int f(\boldsymbol{Y}_i, \boldsymbol{R}_i | X_i, \gamma, \psi) d\boldsymbol{Y}_i^m,$$

where c is a factor that does not depend on (γ, ψ). The maximum likelihood estimate (MLE) of γ (and ψ) is the value that maximizes $L((\gamma, \psi | \boldsymbol{Y}_i^o, \boldsymbol{R}_i, X_i)$; in large-sample samples, the MLE has an approximate normal distribution with covariance matrix given by the inverse of the observed information matrix. Standard errors based on the inverse of the observed information matrix properly account for the fact that some of the data are missing.

The *observed-data likelihood* function is also central to Bayesian inference. In the Bayesian approach, the parameters (γ, ψ) are assigned a prior distribution and inference for γ, (and ψ) is based on its posterior distribution, obtained by multiplying the *observed-data likelihood* by the prior for (γ, ψ). Point estimates of (γ, ψ) can be obtained as summary measures for the posterior distribution, e.g., the posterior mean or posterior mode. Uncertainty about the point estimates can be expressed in terms of the posterior standard deviations or by "credible intervals" based on percentiles of the posterior distribution. Ordinarily, calculation of these intervals involves draws from the posterior distribution via Markov chain Monte Carlo (MCMC) simulation. Because Bayesian inference for (γ, ψ) is based on its exact posterior distribution given a particular choice of prior distribution, inference with small samples is likely to be relatively sensitive to the choice of prior. However, the Bayesian approach with quite dispersed priors can often yield better small sample inferences than those obtained by relying on large sample approximations. Finally, the Bayesian approach lends itself well to sensitivity analysis where assumptions about the distribution of the missing data, and uncertainty about those assumptions, can be formulated in terms of the prior.

Regardless of whether the missing data mechanism is ignored or modeled, likelihood-based methods can be thought to fall under the general heading of imputation. For example, when missingness is assumed to be ignorable, likelihood-based methods are effectively imputing the missing values by modeling and estimating parameters for $f(\boldsymbol{Y}_i | X_i, \gamma)$. When missingness is ignorable, likelihood-based methods are based solely on the marginal distribution of the observed data and ML estimates are obtained by maximizing (1.5), where the likelihood contribution for the ith individual is $f(\boldsymbol{Y}_i^o | X_i, \gamma)$. In a certain sense, the missing values are validly predicted by the observed data via the model for the conditional mean, $E(\boldsymbol{Y}_i^m|\boldsymbol{Y}_i^o, X_i, \gamma)$ and the model for the covariance. This form of "imputation" becomes more transparent when a particular implementation of ML, known as the *EM algorithm* (Dempster, Laird, and Rubin, 1977) is adopted. In the EM algorithm, a two-step iterative algorithm alternates between filling-in missing values with their conditional means, given the observed responses and parameter estimates from the previous iteration (the expectation or E-step), and maximizing the likelihood for the resulting "complete data" (the maximization or M-step).

For example, if the responses are assumed to have a multivariate normal distribution, then "predictions" of the missing values in the E-step of the EM algorithm are based on the conditional mean of $\boldsymbol{Y}_i^m$, given $\boldsymbol{Y}_i^o$,

$$E(\boldsymbol{Y}_i^m|\boldsymbol{Y}_i^o, \gamma) = \boldsymbol{\mu}_i^m + \Sigma_i^{mo}\Sigma_i^{o-1}(\boldsymbol{Y}_i^o - \boldsymbol{\mu}_i^o),$$

where $\boldsymbol{\mu}_i^m$ and $\boldsymbol{\mu}_i^o$ denote those components of the mean response vector corresponding to $\boldsymbol{Y}_i^m$ and $\boldsymbol{Y}_i^o$, and Σ^o and Σ_i^{mo} denote those components of the covariance matrix corresponding to the covariance among the elements of $\boldsymbol{Y}_i^o$ and the covariance between $\boldsymbol{Y}_i^m$ and $\boldsymbol{Y}_i^o$.

Thus, when missingness is ignorable, likelihood-based inference does not require specification of the missing data mechanism, but does require full distributional assumptions about $\boldsymbol{Y}_i$ (given X_i). Furthermore, the entire model for $f(\boldsymbol{Y}_i|X_i, \gamma)$ must be correctly specified. In the example above, any misspecification of the model for the covariance will, in general, yield biased estimates of the mean response trend. When missingness is non-ignorable, a joint model (e.g., selection or pattern-mixture model) is required and inferences tend to be exquisitely sensitive to unverifiable model assumptions; as a result, sensitivity analysis is warranted.

1.5.2 Weighting methods

An alternative approach for handling missing data is to weight the observed data in some appropriate way. In weighting methods, the under-representation of certain response profiles in the observed data is taken into account and corrected. A variety of different weighting methods that adjust for missing data have been proposed. These approaches are often called propensity weighted or inverse probability weighted (IPW) methods and are the main topic of Chapters 8 and 9.

In weighting methods, the underlying idea is to base estimation on the observed responses but weight them to account for the probability of non-response. Under MAR, the propensities for non-response can be estimated as a function of the observed responses and also as a function of the covariates and any additional variables that are thought likely to predict non-response. Moreover, dependence on unobserved responses can also be incorporated in these methods (Rotnitzky and Robins, 1997).

Weighting methods are especially simple to apply in the case of monotone missingness due to dropout. When there is dropout, we can replace the vector of response indicators, $\boldsymbol{R}_i$, with a simple dropout indicator variable, D_i, with $D_i = k$ if the ith individuals drops out between the $(k-1)$th and kth occasion. In that case, the required weights can be obtained sequentially as the product of the propensities for dropout,

$$w_{ij} = (1 - \pi_{i1}) \times (1 - \pi_{i2}) \times \cdots \times (1 - \pi_{ij}),$$

where w_{ij} denotes the probability that the ith subject is still in the study at the jth occasion and $\pi_{ik} = \Pr(D_i = k|D_i \geq k)$ can be estimated from those remaining at the $(k-1)$th occasion, given the recorded history of all available data up to the $(k-1)$th occasion. Then, given estimated weights, say $\widehat{w}_{ij}$ (based on the $\widehat{\pi}_{ij}$'s), a weighted analysis can be performed where the available data at the jth occasion are weighted by $\widehat{w}_{ij}^{-1}$. For example, the generalized estimating equations (GEE) approach can be adapted to handle data that are MAR by making adjustments to the analysis for the propensities for dropout, or for patterns of missingness more generally (Robins, Rotnitzky, and Zhao, 1995).

The intuition behind the weighting methods is that each subject's contribution to the weighted analysis is replicated w_{ij}^{-1} times, to count once for herself and $(w_{ij}^{-1} - 1)$ times for those subjects with the same history of prior responses and covariates, but who dropped out. Thus, weighting methods can also be thought of as methods for imputation where the observed responses are counted more than once (specifically, w_{ij}^{-1} times).

Inverse probability weighted methods have a long and extensive history of use in statistics. They were first proposed in the sample survey literature (Horvitz and Thompson, 1952), where the weights are known and based on the survey design. In contrast to sample surveys, here the weights are not ordinarily known, but must be estimated from the observed data (e.g., using a repeated sequence of logistic regressions for the π_{ij}'s). Therefore, the variance of inverse probability weighted estimators must also account for estimation of the weights. In general, weighting methods are valid provided the model that produces the estimated w_{ij}'s is correctly specified.

In general, weighting methods are best suited to the setting of monotone missing data. Although in principle weighting methods can be applied to non-monotone missing data patterns, in practice the methods are conceptually more difficult to formulate and are certainly not so straightforward to implement:

> Not much is available for the analysis of semi-parametric models of longitudinal studies with intermittent non-response. One key difficulty is that realistic models for the missingness mechanism are not obvious. As argued in Robins and Gill (1997) and Vaansteelandt, Rotnitzky and Robins (2007), the [coarsened at random] CAR assumption with non-monotone data patterns is hard to interpret and rarely realistic ... More investigation into realistic, easy to interpret models for intermittent non-response is certainly needed.
>
> Rotnitzky (2009)

Finally, because IPW methods require a correctly specified model for the weights, this has motivated research on IPW methods that incorporate an "augmentation term"; in general, the augmentation term is a function of the observed data chosen so that it has conditional mean of zero given the complete data. In particular, there is an important class of inverse probability weighted augmented (IPWA) estimators known as "doubly robust" estimators that relax the assumption that the model for π_{ij} has been correctly specified, but require additional assumptions on the model for $f(\boldsymbol{Y}_i|X_i)$. Doubly robust methods specify two models, one for the missingness probabilities and another for the distribution of the complete data. The appealing property of doubly robust methods is that they yield estimators that are consistent when *either*, but not necessarily both, the model for missingness *or* the model for the distribution of the complete data has been correctly specified; these estimators are discussed in detail in Chapter 9.

1.5.3 Multiple imputation methods

Finally, methods that *explicitly* impute values for the missing data are widely used in practice. The basic idea behind imputation is very simple: substitute or fill-in the values that were not recorded with imputed values. One of the chief attractions of imputation methods is that, once a filled-in data set has been constructed, standard methods for complete data can be applied. However, methods that rely on just a single imputation, creating only a single filled-in dataset, fail to acknowledge the uncertainty inherent in the imputation of the unobserved responses. Multiple imputation circumvents this difficulty. In multiple imputation the missing values are replaced by a set of M plausible values, thereby acknowledging the uncertainty about what values to impute for the missing responses. The M filled-in datasets produce M different sets of parameter estimates and their standard errors. Using simple rules developed by Rubin (1987), these are then appropriately combined to provide a single estimate of the parameters of interest, together with standard errors that reflect the uncertainty inherent in the imputation of the unobserved responses. Typically, a small number of imputations, e.g., $10 \leq M \leq 25$, is sufficient to obtain realistic estimates of the sampling variability. Imputation methods are discussed in greater detail in Part IV of the book.

Although the main idea behind multiple imputation is very simple; what is less clear-cut is how to produce the imputed values for the missing responses. Many imputation methods aim at drawing values of $\boldsymbol{Y}_i^m$ from the conditional distribution of the missing responses given the observed responses, $f(\boldsymbol{Y}_i^m|\boldsymbol{Y}_i^o, X_i)$. For example, in a longitudinal study with the monotone missing data patterns produced by dropouts, it is relatively straightforward to impute missing values by drawing values of $\boldsymbol{Y}_i^m$ from $f(\boldsymbol{Y}_i^m|\boldsymbol{Y}_i^o, X_i)$ in a sequential manner. A variety of imputation methods can be used to draw values from $f(\boldsymbol{Y}_i^m|\boldsymbol{Y}_i^o, X_i)$; two commonly used methods are propensity score and predictive mean matching (see Chapters 11 and 12 for a more detailed discussion of different imputation methods.) When missing values are imputed from $f(\boldsymbol{Y}_i^m|\boldsymbol{Y}_i^o, X_i)$, regardless of the particular imputation method adopted, subsequent analyses of the observed and imputed data are valid when missingness is MAR (or MCAR).

1.6 Concluding Remarks

For many longitudinal studies, missing data are the rule, not the exception, and pose a major challenge for data analysis. In this chapter we have focused on issues related to missing data primarily within the context of longitudinal study designs. We have chosen to focus on longitudinal studies because problems of missing data are far more acute in that setting and also because much of the recent statistical literature on missing data has been devoted to longitudinal settings. Since the seminal paper on missing data by Rubin (1976), this topic has been a fertile area for methodological research. There now exists an extensive body of literature on likelihood-based, semi-parametric, and multiple imputation methods for handling missing data. Much of this literature is synthesized in Chapters 3–14. In the extant statistical literature on missing data there is an important distinction between models that assume missing data are MAR and models that assume the missing data are NMAR. We have summarized some of the widely used methods for handling missing values when the data are MAR, including the use of likelihood-based methods that ignore the missing data mechanism and semi-parametric weighting methods that directly model the missing data mechanism. In addition, we have described three classes of models for NMAR data: selection, pattern-mixture, and shared-parameter models. Models for NMAR data are somewhat problematic in the sense that they rest on inherently unverifiable assumptions. As a result, there is increasing consensus that the role of these models should be confined to sensitivity analysis. Part V of this book is devoted to recent advances in conducting sensitivity analyses to incorporate reasonable departures from MAR. As noted at the outset of this chapter, missing data is a ubiquitous problem that complicates the analysis in many important ways. When data are missing, one of the key challenges is that we cannot usually verify the assumptions required to analyze the data. As a result, a recurring theme in much of the more recent statistical literature is the importance of sensitivity analysis.

References

Baker, S. G. (1995). Marginal regression for repeated binary data with outcome subject to non-ignorable non-response. *Biometrics* **51,** 1042–1052.

Bonetti, M., Cole, B. F., and Gelber, R. D. (1999). A method-of-moments estimation procedure for categorical quality-of-life data with non-ignorable missingness. *Journal of the American Statistical Association* **94,** 1025–1034.

Copas, J. and Eguchi, S. (2001). Local sensitivity approximations for selectivity bias. *Journal of the Royal Statistical Society, Series B* **63,** 871–895.

De Gruttola, V. and Tu, X. M. (1994). Modeling progression of CD4 lymphocyte count and its relationship to survival time. *Biometrics* **50,** 1003–1014.

Dempster, A. P., Laird, N. M., and Rubin, D. B. (1977). Maximum likelihood with incomplete data via the EM algorithm. *Journal of the Royal Statistical Society, Series B* **39,** 1–38.

Diggle, P. J. and Kenward, M. G. (1994). Informative dropout in longitudinal data analysis (with discussion). *Applied Statistics* **43,** 49–93.

Ekholm, A. and Skinner, C. (1998). The Muscatine children's obesity data reanalysed using pattern-mixture models. *Applied Statistics* **47,** 251–263.

Fitzmaurice, G. M. and Laird, N. M. (2000). Generalized linear mixture models for handling non-ignorable dropouts in longitudinal studies. *Biostatistics* **1,** 141–156.

Fitzmaurice, G. M., Laird, N. M., and Zahner, G. E. P. (1996). Multivariate logistic models for incomplete binary responses. *Journal of the American Statistical Association* **91,** 99–108.

Fitzmaurice, G. M., Molenberghs, G., and Lipsitz, S. R. (1995). Regression models for longitudinal binary responses with informative drop-outs. *Journal of the Royal Statistical Society, Series B* **57,** 691–704.

Follman, D. and Wu, M. (1995). An approximate generalized linear model with random effects for informative missing data. *Biometrics* **51,** 151–168.

Glonek, G. F. V. (1998). On identifiability in models for incomplete binary data. *Statistics and Probability Letters* **41,** 191–197.

Glynn, R. J., Laird, N. M., and Rubin, D.B. (1986). Selection modelling versus mixture modelling with non-ignorable non-response. In: *Drawing Inferences from Self-Selected Samples.* H. Wainer, ed. New York: Springer-Verlag.

Heckman, J. (1976). The common structure of statistical models of truncation, sample selection and limited dependent variables, and a simple estimator for such models. *Annals of Economic Social Measurements* **5,** 475–492.

Hedeker, D. and Gibbons, R. D. (1997). Application of random-effects pattern-mixture models for missing data in longitudinal studies. *Psychological Methods* **2,** 64–78.

Hogan, J. W. and Laird, N. M. (1997). Mixture models for the joint distribution of repeated measures and event times. *Statistics in Medicine* **16,** 239–258.

Horvitz, D. G. and Thompson, D. J. (1952). A generalization of sampling without replacement from a finite universe. *Journal of the American Statistical Association* **47,** 663–685.

Laird, N. M. (1988). Missing data in longitudinal studies. *Statistics in Medicine* **7,** 305–315.

Laird, N.M. and Ware, J.H. (1982). Random effects models for longitudinal data. *Biometrics* **38,** 963–974.

Liang, K.-Y. and Zeger, S. L. (1986). Longitudinal data analysis using generalized linear models. *Biometrika* **73,** 13–22.

Little, R. J. A. (1993). Pattern-mixture models for multivariate incomplete data. *Journal of the American Statistical Association* **88,** 125–34.

Little, R. J. A. (1994). A class of pattern-mixture models for normal incomplete data. *Biometrika* **81,** 471–483.

Little, R. J. A. (1995). Modelling the drop-out mechanism in repeated-measures studies. *Journal of the American Statistical Association* **90,** 1112–1121.

Little, R. J. A. (2009). Selection and pattern-mixture models. In Fitzmaurice, G., Davidian, M., Molenberghs, G., and Verbeke, G. (editors), *Longitudinal Data Analysis.* Handbooks of Modern Statistical Methods, pp. 409–431. New York: Chapman & Hall/CRC.

Little, R. J. A. and Rubin, D. B. (1987). *Statistical Analysis with Missing Data.* New York: Wiley.

Little, R. J. A. and Rubin, D. B. (2002). *Statistical Analysis with Missing Data,* 2nd ed. New York: Wiley.

Little, R. J. A. and Wang, Y. (1996). Pattern-mixture models for multivariate incomplete data with covariates. *Biometrics* **52,** 98–111.

Michiels, B., Molenberghs, G., and Lipsitz, S. R. (1999). A pattern-mixture odds ratio model for incomplete categorical data. *Communications in Statistics: Theory and Methods* **28,** 2843–2869.

Molenberghs, G., Goetghebeur, E. J. T., Lipsitz, S. R., and Kenward, M. G. (1999). Nonrandom missingness in categorical data: Strengths and limitations. *American Statistician* **53,** 110–118.

Molenberghs, G., Kenward, M. G., and Goetghebeur, E. (2001). Sensitivity analysis for incomplete contingency tables: The Slovenian plebiscite case. *Applied Statistics* **50,** 15–29.

Molenberghs, G., Kenward, M. G., and Lesaffre, E. (1997). The analysis of longitudinal ordinal data with nonrandom drop-out. *Biometrika* **84,** 33–44.

Molenberghs, G., Michiels, B., Kenward, M. G., and Diggle, P. J. (1999). Missing data mechanisms and pattern-mixture models. *Statistica Neerlandica* **52,** 153–161.

Mori, M., Woodworth, G. G., and Woolson, R. F. (1992). Application of empirical Bayes inference to estimation of rate of change in the presence of informative right censoring. *Statistics in Medicine* **11,** 621–631.

Nordheim, E. V. (1984). Inference from nonrandomly missing categorical data: An example from a genetic study in Turner's syndrome. *Journal of the American Statistical Association* **79,** 772–780.

Park, T. and Lee, S. L. (1999). Simple pattern-mixture models for longitudinal data with missing observations: analysis of urinary incontinence data. *Statistics in Medicine* **18,** 2933–2941.

Robins, J. M. and Gill, R. D. (1997). Non-response models for the analysis of non-monotone ignorable missing data. *Statistics in Medicine* **16,** 39–56.

Robins, J. M., Rotnitzky, A., and Zhao, L. P. (1995). Analysis of semiparametric regression models for repeated outcomes in the presence of missing data. *Journal of the American Statistical Association* **90,** 106–121.

Rosenbaum, P. R. and Rubin, D. B. (1983). Assessing sensitivity to an unobserved binary covariate in an observational study with binary outcome. *Journal of the Royal Statistical Society, Series B* **45,** 212–218.

Rosenbaum, P. R. and Rubin, D. B. (1985). The bias due to incomplete matching. *Biometrics* **41,** 103–116.

Rotnitzky, A. (2009). Inverse probability weighted methods. In Fitzmaurice, G., Davidian, M., Molenberghs, G., and Verbeke, G. (editors), *Longitudinal Data Analysis.* Handbooks of Modern Statistical Methods, pp. 453–476. New York: Chapman & Hall/CRC.

Rotnitzky, A. and Robins, J. M. (1997). Analysis of semi-parametric regression models with non-ignorable non-response. *Statistics in Medicine* **16,** 81–102.

Rubin, D. B. (1976). Inference and missing data. *Biometrika* **63,** 581–592.

Rubin, D. B. (1977). Formalizing subjective notions about the effect of nonrespondents in sample surveys. *Journal of the American Statistical Association* **72,** 538–543.

Rubin, D. B. (1987). *Multiple Imputation for Nonresponse in Surveys.* New York: Wiley.

Scharfstein, D., Rotnitzky, A., and Robins, J. M. (1999). Adjusting for non-ignorable drop-out using semiparametric non-response models (with discussion). *Journal of the American Statistical Association* **94,** 1096–1146.

Schluchter, M. D. (1992). Methods for the analysis of informatively censored longitudinal data. *Statistics in Medicine* **11,** 1861–1870.

Ten Have, T. R., Kunselman, A. R., Pulkstenis, E. P., and Landis, J. R. (1998). Mixed effects logistic regression models for longitudinal binary response data with informative drop-out. *Biometrics* **54,** 367–383.

Ten Have, T. R., Miller, M. E., Reboussin, B. A., and James, M. M. (2000). Mixed effects logistic regression models for longitudinal ordinal functional response data with multiple-cause drop-out from the longitudinal study of aging. *Biometrics* **56,** 279–287.

Thijs, H., Molenberghs, G., Michiels, B., Verbeke, G., and Curran, D. (2002). Strategies to fit pattern-mixture models. *Biostatistics* **3,** 245–265.

Tsiatis, A. A. (2006). *Semiparametric Theory and Missing Data.* New York: Springer.

Tsiatis, A. A., De Gruttola, V., and Wulfsohn, M. S. (1995). Modeling the relationship of survival to longitudinal data measured with error: Applications to survival and CD4 counts in patients with AIDS. *Journal of the American Statistical Association* **90,** 27–37.

Vach, W. and Blettner, M. (1995). Logistic regression with incompletely observed categorical covariates: Investigating the sensitivity against violations of the missing at random assumption. *Statistics in Medicine* **12,** 1315–1330.

Vansteelandt, S., Rotnitzky, A., and Robins, J. M. (2007). Estimation of regression models for the mean of repeated outcomes under non-ignorable non-monotone non-response. *Biometrika* **94,** 841–860.

Verbeke, G., Molenberghs, G., Thijs, H., Lesaffre, E., and Kenward, M. G. (2001). Sensitivity analysis for non-random dropout: A local influence approach. *Biometrics* **57,** 7–14.

Wu, M. C. and Bailey, K. R. (1988). Analyzing changes in the presence of informative right censoring caused by death and withdrawal. *Statistics in Medicine* **7,** 337–346.

Wu, M. C. and Bailey, K. R. (1989). Estimation and comparison of changes in the presence of informative right censoring: Conditional linear model. *Biometrics* **45,** 939–955.

Wu, M. C. and Carroll, R. J. (1988). Estimation and comparison of changes in the presence of informative right censoring by modeling the censoring process. *Biometrics* **44,** 175–188.

2

Development of Methods for the Analysis of Partially Observed Data and Critique of ad hoc Methods

James R. Carpenter

London School of Hygiene and Tropical Medicine and MRC Clinical Trials Unit at UCL, London, UK

Michael G. Kenward

London School of Hygiene and Tropical Medicine, UK

CONTENTS

2.1 Introduction

It is inevitable that empirical research will result in missing data—that is, information that researchers intended, but were unable, to collect. Thus, the issues raised by missing data are as old as empirical research itself. However, statisticians' awareness of these issues, and the approaches they have taken to address them, have changed and developed steadily since the 1950s. Advances in computing power and associated software mean that many of these developments are now readily available to quantitative researchers, who are at risk of being overwhelmed by the choices available.

The aim of this chapter is to sketch the development of research on the analysis of partially observed data, discussing the evolution of the various approaches, and how they relate to each other. As part of this, we will review a number of what we term *ad hoc* approaches, highlighting their inferential shortcomings compared to the more principled methods that are now available. Such *ad hoc* methods have been used extensively in the past but will not generally be covered elsewhere in this Handbook.

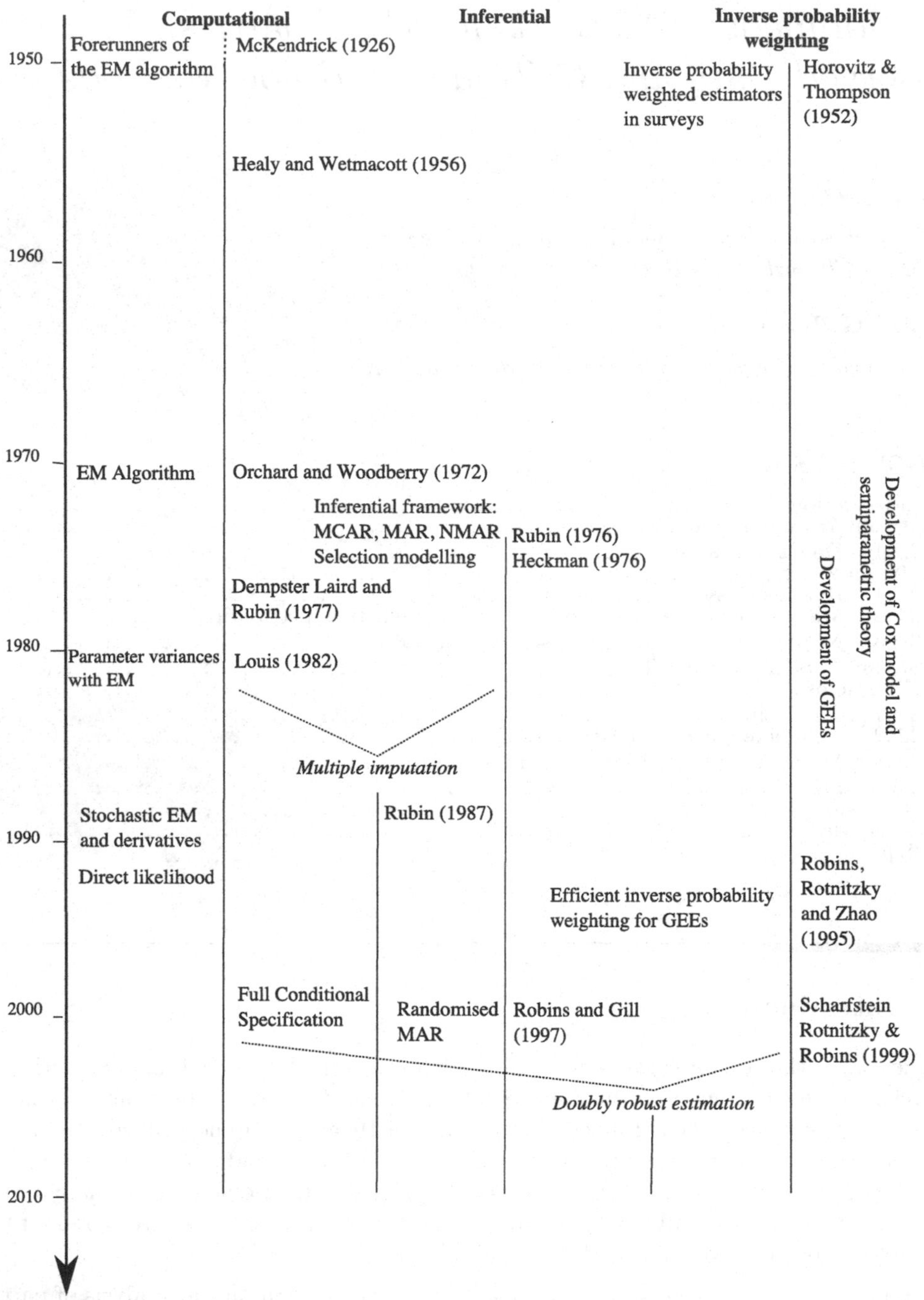

FIGURE 2.1
Missing data methodology timeline.

Figure 2.1 gives an overview of the development of research in missing data, separating the ideas into three broad streams: computational, inferential (including parametric selection modelling), and inverse probability weighting (semi-parametric).

While no taxonomy is wholly satisfactory, it nevertheless highlights the different ways of tackling the issues raised by missing data which researchers have pursued. In Sections 2.2–2.4 we give a more detailed description of how the ideas have evolved, and in particular how the streams have interacted and cross-fertilized each other. Inevitably, this has been coloured by our own experiences, and we apologize to those whose contributions we have inadvertently overlooked.

Section 2.5 comprises the second part of the chapter, where we review and critique what we refer to as *ad hoc* methods for the analysis of partially observed data. While this is not a universally pejorative term, we will see that these methods are characterized by simple methods for fixing up a dataset with missing data—typically lacking a statistically principled motivation—so that the originally intended analysis can be performed. Finally, in Section 2.6 we summarize the main points from this chapter.

2.2 Stream 1: Computational

When statistical computing was in its infancy, many analyses were only feasible because of the carefully planned balance in the dataset. Missing data meant the available data for analysis were unbalanced, thus complicating the planned analysis and in some instances rendering it infeasible. Early work on the problem was therefore largely computational.

The earliest reference to missing data we are aware of is McKendrick (1926). This is reprinted with a commentary in Kotz and Johnson (1997). In the context of fitting stochastic epidemic models McKendrick introduced a form of the so-called *Expectation-Maximisation (EM)* algorithm. Soon after this, in the 1930s, the problem of missing data in the analysis of designed experiments was addressed. Much effort was put into designing agricultural experiments in such a way that analyses remained tractable in the pre-computer age. Missing data from plots disturb the necessary balance of these designs, and so attempts to deal with the problem centered around filling in the missing plots with suitable values, and then using the standard analysis on the completed dataset. For most designs this process needed to be iterated. It turns out, again, that these methods are special cases of the general EM algorithm. The first example of this that we have seen is Allan and Wishart (1930), with several authors extending the idea to increasingly complex designs (Yates 1933, Yates and Hale 1939, Cornish 1940ab). Around the same time, the use of analysis of covariance for the same purpose was noted (Nair 1940). Healy and Westmacott (1956) set out the general form of the algorithm for designed experiments. The essentials of the algorithm are the following:

1. estimate the missing data given *current* parameter values;
2. use the observed and current estimates of the missing data to estimate updated parameter values, and
3. iterate till convergence.

The above algorithm has been increasingly formalised over the years, as described in Little and Rubin (2002, Ch. 8), with 'individual observations' being replaced by 'sufficient statistics' and then, more generally, by the log-likelihood. Orchard and Woodbury (1972) described the general applicability of the underlying idea. The approach was formalized in its full generality, and given the name Expectation-Maximisation (EM) algorithm by Dempster, Laird, and Rubin (1977). In this, step (1) in the algorithm above is the expectation step, and step (2) is the maximization step. These authors also showed that, under quite general conditions, each iteration of the algorithm increases the log-likelihood. They also provide a number of examples, and show that the algorithm extends well beyond classic missing datasettings to models with mixtures and latent variables, for example.

The approach has spawned a series of developments further exploring the properties of the EM algorithm. Under regularity conditions the algorithm will converge linearly and it can be shown, see, for example, McLachlan and Krishnan (2008), that the rate is given by the largest eigenvalue of the information matrix ratio

$$\mathbf{J}_C^{-1}\mathbf{J}_{M|O}$$

for $\mathbf{J}_{M|O}$ the conditional expected information matrix for the missing data given the observed, and $\mathbf{J}_C$ the expected information matrix for the complete data (observed and missing). This measures the proportion of information that remains following the loss of the missing data.

The maximisation step can be computationally awkward and a range of modifications have been developed to accommodate this. For example in the Generalized EM (GEM) algorithm the full maximisation step is replaced by one that simply increases the value of the likelihood. Sometimes a conditional maximisation step is available as an alternative, giving the Expectation Conditional Maximization Expectation (ECME) algorithm. A range of such modifications are available, we refer to Little and Rubin (2002, Ch. 8) and McLachlan and Krishnan (2008) for further details. Stochastic versions of the E step have been used when the expectation is computationally awkward, producing the so-called Monte Carlo EM (MCEM) algorithm, see, for example, Tanner and Wong (1987). A more radical stochastic version, the so-called Stochastic EM (SEM) algorithm, replaces iterative maximisation of the likelihood by a Markov chain Monte Carlo approach, in which a chain of parameter estimates is generated with a stationary distribution whose mode corresponds to the maximum likelihood estimates. In its basic form this can be used only in very restricted settings, such as simple generalized linear and mixture models. The method was introduced by Celeux and Diebolt (1985), with comments on its properties and limitations by Wang and Robins (1998), Nielsen (2000), and Robins and Wang (2000).

Although the EM algorithm does not provide an estimate of precision as a byproduct, various methods have been proposed for this. These include the bootstrap, which can be rather demanding computationally, and methods for obtaining the second derivatives of the log likelihood at the obtained maximum. An elegant approach was proposed by Louis (1982), who re-expresses the missing information in terms of full data quantities, and this provides a useful computational approach (where the expectations may, as usual, be estimated by simulation if necessary).

The EM algorithm remains an important tool for obtaining maximum likelihood estimators when data are missing, and, importantly, for other apparently different settings, such as latent variables models, which it transpires can be formulated as missing data problems.

2.2.1 Direct likelihood

With increasing computational power, there has been increased interest in maximising the likelihood directly. Under the important assumption of ignorability (for a definition, see

Chapter 3) units with missing outcomes in a regression model contribute nothing to the log-likelihood, and can therefore be omitted. Likewise with longitudinal data, information only comes from the marginal distribution of the observed responses, which can typically be derived relatively simply.

However, for missing covariates in a regression model (again assuming ignorability) computation is more complicated. For example, consider a linear regression model of Y on binary X. If, for unit i, X_i is missing, then to compute the likelihood of the regression parameters we must specify a marginal model for X_i, say $\pi(X;\phi)$. Then the likelihood contribution for unit i is:

$$L(Y_i|X_i;\theta;\phi) = \sum_{j=1,2} \{L(Y_i|X_i = j;\theta)\pi(X_i = j;\phi)\}\,. \tag{2.1}$$

While the resulting computations are straightforward in this case, they are less so with continuous covariates, and become rapidly more awkward as the number of covariates (and hence the complexity of the missing data pattern) increases. Direct likelihood has been widely used in structural equation modelling (for example, Skrondal and Rabe-Hesketh 2004, Ch. 6). It is the central estimation engine, employing the EM algorithm, in the commercial latent variable modelling program M*plus* (https://www.statmodel.com/).

Equation (2.1) highlights the close connection between direct likelihood and the EM algorithm, because the right-hand side is simply the expectation of the likelihood over the distribution of the missing data. Let $\boldsymbol{Y}_i$ be the vector of observations on unit i, and partition this into missing and observed parts, $\boldsymbol{Y}_i^{\mathrm{m}}, \boldsymbol{Y}_i^{\mathrm{o}}$. Developing this line of reasoning, it can be shown that the score statistic for the observed data on unit i is equal to the expectation of the full data score statistic over the conditional distribution of the missing data given the observed, that is:

$$\boldsymbol{s}(\boldsymbol{Y}_i^{\mathrm{o}};\theta) = \int \boldsymbol{s}(\boldsymbol{Y}_i;\theta) f(\boldsymbol{Y}_i^{\mathrm{m}}|\boldsymbol{Y}_i^{\mathrm{o}};\eta)\, d\boldsymbol{Y}_i^{\mathrm{m}}. \tag{2.2}$$

Thus, if we can estimate η, (typically from a model of the partially observed variables given the fully observed ones) and calculate the expectation, we can solve the resulting score equation for the maximum likelihood estimate of θ.

In practice, this will typically be analytically awkward. As usual, though, expectation on the right-hand side of (2.2) can be estimated using simulation. Specifically, suppose we are able to draw

$$\boldsymbol{Y}_1^{\mathrm{m}}, \boldsymbol{Y}_2^{\mathrm{m}}, \ldots, \boldsymbol{Y}_K^{\mathrm{m}} \quad i.i.d. \;\; f(\boldsymbol{Y}^{\mathrm{m}} \mid \boldsymbol{Y}^{\mathrm{o}};\eta).$$

Then, at any value of θ, the right-hand side of (2.2) can be estimated by

$$\frac{1}{K}\sum_{k=1}^{K} \boldsymbol{s}(\boldsymbol{Y}^{\mathrm{m}}{}_k, \boldsymbol{Y}^{\mathrm{o}};\theta).$$

We can then use a numerical algorithm to solve for the maximum likelihood estimate. For a review of this, and related approaches, see Clayton et al. (1998). A variant of this method is known as the *mean score* method (Reilly and Pepe 1995), which has been shown to have close links to the inverse probability weighting approaches discussed below in Section 2.4 and the subject of Part III.

Now, suppose that we consider again the same setup, but reverse the order of expectation and subsequent solution of score equation (i.e., obtaining the maximum likelihood estimate). This would lead us to:

1. draw $\boldsymbol{Y}_1^{\mathrm{m}}, \boldsymbol{Y}_2^{\mathrm{m}}, \ldots, \boldsymbol{Y}_K^{\mathrm{m}} \;\; i.i.d. \;\; f(\boldsymbol{Y}^{\mathrm{m}}|\boldsymbol{Y}^{\mathrm{o}};\eta)$;

2. solve each of $\boldsymbol{s}(\boldsymbol{Y}_k^{\mathrm{m}}, \boldsymbol{Y}^{\mathrm{o}}; \theta)$ for $\widehat{\theta}_k$, $k = 1, \ldots, K$, and

3. average the resulting parameter estimates, giving

$$\widehat{\theta} = \frac{1}{K} \sum_{k=1}^{K} \widehat{\theta}_k.$$

This is essentially the multiple imputation (MI) algorithm introduced by Rubin, e.g., Rubin (1978), and set out in full detail in Rubin (1987). This approach is the subject of Part IV of this Handbook. The attraction of MI is its practicality and wide applicability. It is practical because, in order to use it we need only (i) our substantive model and (ii) an appropriate imputation model for the missing data given the observed. It is widely applicable because, having imputed the data and fitted the model to each imputed dataset, Rubin's rules for combining the results from fitting the substantive model to each imputed dataset for final inference, are simple to apply yet remarkably robust across a wide range of applications. Although originally introduced in a sample survey setting, the use of MI has subsequently spread to most areas of quantitative research as elaborated on in Chapters 11–14 of this Handbook and in the later chapters of Carpenter and Kenward (2013).

2.3 Stream 2: Inferential

2.3.1 Rubin's classes

We have already noted that early work on missing data was primarily algorithmic. The computational tools discussed above, at least in their simpler forms as originally introduced, rely on ignorability of the missing data mechanism. The wider question of the consequences of non-trivial proportions of missing data for subsequent inference was relatively neglected until the seminal paper by Rubin (1976). This set out a typology for assumptions about the reasons for missing data, and sketched their implications for analysis and inference.

The concepts of data missing completely at random (MCAR), missing at random (MAR) and not missing at random (NMAR) set out in that paper, and developed in Chapter 3 have been at the heart of many subsequent developments, as witnessed by the many thousands of citations of the original paper.

The typology makes clear that under an MAR mechanism the analysis can proceed without requiring the explicit formulation of a model (i.e., *ignoring*) for the the missingness mechanism. This is extremely useful in practice, but a strong assumption. It also makes clear that MAR is not a property of a whole dataset *per se*, but an assumption we make for a particular analysis. Since the MAR assumption cannot be checked from the data at hand, it is often important to explore the robustness of inferences to departures from MAR. Originally, Rubin defined these concepts from a Bayesian perspective, i.e., conditional on the data at hand. This makes them specific to the observed data and, as Rubin (1976) shows, a common data generating mechanism may give rise to datasets that belong to the MCAR/MAR or NMAR missingness classes. However, under the frequentist paradigm, these classes *are* a property of the missingness mechanism and must hold for all the datasets that can be generated from it (see, Diggle and Kenward 1994, for example).

As the next section shows, an alternative stream of development has focussed on semi-

TABLE 2.1
Some missingness patterns with three variables.

Missingness Pattern	Variable X	Y	Z
1.	$\surd$	$\surd$	$\surd$
2.	·	$\surd$	$\surd$
3.	$\surd$	·	$\surd$
4.	$\surd$	$\surd$	·

parametric methods and inverse probability weighting. Such an approach relies on estimating a model for the missingness mechanism, and this can be very complex for general missing data patterns. This has led to a debate in the literature about the plausibility and appropriateness of the MAR/ignorability assumption in applications.

For example, Robins and Gill (1997) point out that if we have non-monotone missingness, and a common missingness mechanism, then that mechanism must often be either MCAR or NMAR. To illustrate, Table 2.1 shows a non-monotone missingness pattern for three variables. If we have a common ignorable missingness mechanism, then it cannot depend on any of the three variables, X, Y, and Z, and must be missing completely at random. The alternative is that it depends on one or more of X, Y, and Z. However, since all these variables have missing data, the mechanism must then be NMAR. Or—which may be plausible in certain contexts—we have to say that different subsets of the dataset which have different missingness patterns, which are collectively ignorable. For example, in Table 2.1 patterns (1, 2) may be MAR (with mechanism dependent on Y, Z) while patterns (3, 4) be MAR (with mechanism dependent on X). Of course, even if data are strictly NMAR, an analysis assuming MAR may be preferable to an analysis assuming MCAR. We return to this point below in Section 2.3.2.

Building on this, Robins and Gill (1997) introduced a new mechanism, *Randomised Monotone Missingness (RMM)*, a subclass of MAR, and argue that this is the only plausible non-monotone MAR mechanism that is not MCAR. They show (Gill and Robins 1996) that there exist mechanisms that are MAR but not RMM, but that for a computer to generate data under such a mechanism, it requires knowledge of one or more variables which are then 'concealed' later in the process. This is essentially equivalent to saying—as we did above—that there are different MAR missingness mechanisms applying in different portions of the data but that the variable indicating the portion is concealed. This may happen, for example, when certain sets of variables, which increase the risk that individuals may be identified, are withheld when survey data are released to researchers. Nevertheless, due to the precise manner in which this must occur in order for the missingness mechanism to be MAR but not RMM, Robins and Gill argue that 'natural missing data processes that are not representable as RMM processes will be [NMAR].' Related issues for non-monotone longitudinal data are considered by Vansteelandt, Rotnitzky, and Robins (2007).

2.3.2 Missing at random and the conditional predictive distribution

It is important to realize that in making an assumption about the missingness mechanism, we implicitly make a corresponding assumption about the distribution of the missing data, given the observed data. We call this the *Conditional Predictive Distribution*. Carpenter and Kenward (2013, pp. 12–21) discuss this in more detail; we now give an indicative example.

Consider two variables, Y_1 and Y_2, where Y_1 has missing values, and R is defined as

$$R_i = \begin{cases} 1 \text{ if } Y_{i1} \text{ is observed, and} \\ 0 \text{ if } Y_{i1} \text{ is missing.} \end{cases} \tag{2.3}$$

Then values of Y_1 are MAR if $\Pr(R_i = 1|Y_{i1}, Y_{i2}) = \Pr(R_i = 1|Y_{i2})$. To see what this means for the distribution of Y_{i1} given Y_{i2}, notice that

$$\begin{aligned} \Pr(Y_{i1}|Y_{i2}, R_i = 1) &= \frac{\Pr(Y_{i1}, Y_{i2}, R_i = 1)}{\Pr(Y_{i2}, R_i = 1)} \\ &= \frac{\Pr(R_i = 1|Y_{i1}, Y_{i2}) \Pr(Y_{i1}, Y_{i2})}{\Pr(R_i = 1|Y_{i2}) \Pr(Y_{i2})} \\ &= \Pr(Y_{i1}|Y_{i2}). \end{aligned} \tag{2.4}$$

Thus, under MAR, the distribution of $Y_{i1}|Y_{i2}$ is the same whether or not Y_{i1} is observed. Were $\Pr(Y_{i1}|Y_{i2}) \neq \Pr(Y_{i1}|Y_{i2}, R_i)$, the missingness mechanism would be NMAR, and the extent of the differences between $\Pr(Y_{i1}|Y_{i2}, R_i = 1)$ and $\Pr(Y_{i1}|Y_{i2}, R_i = 0)$ would be reflected in the strength of the corresponding NMAR selection mechanism.

Looking at (2.4) we see the result holds if we replace Y_1, Y_2, and R by—respectively—sets of partially observed variables, fully observed variables and the corresponding response indicators. Viewing the MAR assumption through the conditional framework (2.4), we see that it is a very natural starting point for an analysis of partially observed data, even if—from a strict selection mechanism perspective—it is unlikely to hold precisely. This is why in applications precise discussions over the MAR selection mechanism are usually of secondary interest to including appropriate (relatively) fully observed variables which improve our estimate of the distribution of the missing data given the observed.

2.3.3 Not missing at random and selection models

The assumption underlying MAR is key because it effectively means that the observed data are sufficient to break the dependence between the probability of a variable being missing and the particular value it takes. It is this dependence that lies behind so-called selection bias. As an assumption it may not hold, and then the missing data mechanism *cannot* be ignored to produce valid inferences. To address this problem, Heckman (1976) introduced the so-called *selection model.* This can be expressed in several ways, but these all share, in some form or another, dependence between the *unobserved* data and the probability of missingness. Heckman introduced this model to correct for selection bias that is due to just such a dependence, and showed how such a model can be fitted to the data using a two-stage regression procedure. Heckman's paper has led to a broad stream of subsequent development across several disciplines, including econometrics, the social and life sciences. Likelihood methods of estimation for this class of models are now widespread. A development of the model for dropout in longitudinal data, in which the missingness component takes the form of the discrete hazard model, is known as the Diggle and Kenward model (Diggle and Kenward 1994, Diggle 2005). Selection models for NMAR settings are the subject of Section 4.2.

Although these models have become important tools in econometrics and the social sciences, their use remains controversial. This is because the estimated model parameters and subsequent inferences are very sensitive to untestable assumptions. Little (1986) was one of the first to highlight this, and it has subsequently been explored extensively, see, for example, Little and Rubin (2002, Sec. 15.4). The special structure of these models means that they are not regular in the usual sense, and convergence to asymptotic properties is non-standard (Rotnitzky et al. 2000). For these reasons, in the life sciences at least, such selection models tend not to be used in primary analyses. Instead, their role is more typically as part for sensitivity analyses.

More broadly we can think of the use of the Heckman model and its many generalizations and developments as an approach to missing data that is built on ever more complex models to handle the potential associations between unobserved data and the probability of missingness. The aim is to correct for the bias that this may cause if ignored, and so arrive at single 'corrected' and hence valid inferences. However, because any inference depends strongly on untestable assumptions it can be argued that a better approach is to (i) explicitly state the assumptions, (ii) draw valid inferences under these assumptions, and then (iii) review the range of inferences consistent with the observed data within the constraints provided by the substantive setting.

2.4 Stream 3: Semi-Parametric

The third significant stream in missing data methodology is based on inverse probability weighting, the subject of Part III of this Handbook. This approach was first introduced by Horvitz and Thompson (1952) for estimation in the presence of sample weights as commonly occur in a survey setting. The key idea can be illustrated using a sample mean. Suppose we sample data $\{Y_{i1}\}_{i=1,\ldots,N}$ from a population with common mean, θ, which we wish to estimate. Then solving

$$\sum_{i=1}^{N}(Y_{i1} - \theta) = 0 \tag{2.5}$$

gives a consistent estimator of θ because the expectation of (2.5) is zero.

Now suppose some of the Y_{i1} are missing, and define the missing data indicator R_i as in (2.3). Let $\mathrm{E}[R_i] = \pi_i$. Then, while solving

$$\sum_{i=1}^{N} R_i(Y_{i1} - \theta) = 0$$

will give a biased estimate of θ unless data are MCAR, solving the inverse probability weighted equation

$$\sum_{i=1}^{N} \frac{R_i}{\pi_i}(Y_{i1} - \theta) = 0 \tag{2.6}$$

gives a consistent estimator. This follows because the expectation of (2.6) is zero, when we take expectations first over R and then over Y.

Now suppose we wish to estimate β in the regression of Y_{i2} on Y_{i1},

$$Y_{i2} = \beta Y_{i1} + e_i, \quad e_i \ \ i.i.d. \ N(0, \sigma^2).$$

Assume that values of Y_{i1} are MAR dependent on Y_{i2}, so that π_i can be estimated from a logistic regression of R_i on Y_{i2}. Then solving the inverse probability weighted equation

$$\sum_{i=1}^{N} \frac{R_i}{\pi_i}(Y_{i2} - \beta Y_{i1})Y_{i1} = 0 \tag{2.7}$$

gives a consistent estimator of β. Again, this follows because taking expectations, first over R, then $Y_{i2}|Y_{i1}$, gives zero. In practice the π_i also need to be estimated, but this is not difficult, in principle, under MAR using, for example, logistic regression.

However, the drawback of (2.7) is that all the units with Y_{i1} missing (but Y_{i2} observed) are omitted. This means that the estimate of β from (2.7) is inefficient relative to one obtained using multiple imputation. To improve the efficiency, Scharfstein, Rotnitzky, and Robins (1999) proposed bringing in this information by augmenting the estimating equation with a term whose expectation is zero, and which can be calculated from the observed data. This leads to the use of so-called Augmented Inverse Probability Weighted (AIPW) methods. In this simple example, it turns out the most efficient choice for this additional term is

$$\sum_{i=1}^{N} \left\{ \frac{R_i}{\pi_i}[(Y_{i2} - \beta Y_{i1})Y_{i1}] + \left(1 - \frac{R_i}{\pi_i}\right) E_{Y_{i1}|Y_{i2}}[(Y_{i2} - \beta Y_{i1})Y_{i1}] \right\} = 0. \tag{2.8}$$

In other words, for records where Y_{i1} is missing, we add to (2.7) the expectation of the estimating equation, $(Y_{i2} - \beta Y_{i1})Y_{i1}$, over the distribution of the missing data given the observed.

As before, we can take a series of conditional expectations to show that (2.8) has expectation zero, and hence the estimator is consistent. Moreover, the estimator is efficient if the conditional distribution of $Y_{i2}|Y_{i1}$ is specified correctly.

Estimators following from score equations like (2.8) have an additional property. If either (i) the model for R_i is correct, regardless of the model for the distribution of $[Y_{i2}|Y_{i1}]$, *or* the model for the distribution of $[Y_{i2}|Y_{i1}]$ is correct, regardless of the model for R_i, then we get consistent estimates of β. This property is known as *double robustness.* Vansteelandt, Carpenter, and Kenward (2010) give an accessible introduction to these developments.

We have only considered a very simple example here, but the idea of augmenting the inverse probability weighted estimating equations in this way applies quite generally, allowing efficient, doubly robust estimation from generalized estimating equations when data are missing. For an accessible overview of the theory see Tsiatis (2006). This makes the inverse probability-based approach comparable, in terms of efficiency, with approaches based around parametric approaches such as multiple imputation. It is important to note that both approaches require a model for the distribution of the missing data given the observed; the difference is that the AIPW approach has the potential to confer a degree of robustness to mis-specification of this model, by drawing on a model for the probability of observing the data.

The extent to which this gain is realised in practice has been the subject of lively debate; see, for example, Kang and Schafer (2007). Further, the need to estimate the weights, and so explicitly specify the missingness model, has fueled the discussion of the plausibility of MAR in non-monotone missingness patterns mentioned above. It turns out that the robustness gains are limited if the weights are poorly estimated, especially if some records have relatively large weights. This has in turn led to additional work on increasing the reliability, or stability, of the estimated weights. See Tsiatis, Davidian, and Cao (2011), for example. This continues to be an area of active research.

We conclude by noting that, because the various methods for imputing data all involve a model for the missing data given the observed, AIPW approaches show how these, too, may be made more robust. This has been explored by Bang and Robins (2005), and taken up in the context of multiple imputation by Daniel and Kenward (2012), see also Carpenter and Kenward (2013, Ch. 12).

2.5 Critique of *Ad Hoc* Methods

We now turn from our historical sketch to discuss several *ad hoc* methods for the analysis of partially observed data. We consider these in the light of the more principled developments discussed above. The term *ad hoc* is not universally pejorative, but it does indicate that the methods described below have been primarily motivated by their convenience, rather than considered methodological development.

It is useful to separate two issues. First, does the chosen method lead to consistent estimators for the model of interest (hereafter to be termed the *substantive model*)? This is much easier to achieve for simple univariate marginal quantities, such as a mean or a percentile, than essentially multivariate quantities such as regression coefficients, measures of association, and so on. The latter is much more of a challenge because the method must recover the required dependencies among the variables of interest, as well as their marginal behaviour. Second, does the method lead to appropriate measures of precision? A common feature of these methods is the construction of a 'complete' dataset, typically indexed by units in rows and variables in columns, in which there are no missing data. This allows conventional methods of analysis, such as regression, to be applied directly, at least in a computational sense. It does not follow that all aspects of such analyses will necessarily be valid. This construction can be done through *reduction*, by deleting partially observed units in so-called complete-records analysis, or *augmentation*, in which missing data are replaced by *imputed* values, in so-called single imputation methods. In the following, in addition to complete records analysis, we consider two such examples of single imputation: mean imputation and last observation carried forward (LOCF).

One method that has been commonly used in the past and does not fit exactly into this reduction/augmentation framework is the *missing indicator method*, which is used for incomplete *categorical* variables, and is most commonly seen in the analysis of observational datasets where there may be many incomplete categorical variables to deal with. In this method, a new category corresponding to 'missing' is created for each partially observed categorical variable. The problem with this approach is that the units allocated to this new category will be a mixture of cases from the other categories, with the particular mixture depending on the missing data mechanism. For example, under MCAR the proportions of the actual categories represented in the 'missing' category will match those in the sampled population. Other mechanisms will lead to different proportions. Greenland and Finkle (1995) provide a thorough critique of this method, and recommend that it be avoided. They show that such augmented categorical variables will, in general, cause either bias when used either as direct exposure variables, or as covariates for adjustment. There is another rather special role for such missing indicator variables where their use *can* be justified. This is for the special case of missing baseline covariates in clinical trials. We refer to White and Thompson (2005) Carpenter and Kenward (2008, Sec. 2.4) for details.

TABLE 2.2
Isolde Trial. Number of patients attending follow-up visits, by treatment group.

Visit	Placebo Arm	FP Arm
Baseline	376	374
6 months	298	288
12 months	269	241
18 months	246	222
24 months	235	194
30 months	216	174
36 months	168	141

TABLE 2.3
Isolde Trial. Adjusted odds ratios for withdrawal.

Variable	Odds Ratio	(95% CI)	p
Exacerbation rate (# /year)	1.51	(1.35,1.69)	< 0.001
BMI (kg/m^2)	0.95	(0.92,0.99)	0.025
FEV$_1$ slope (ml/year)	0.98	(0.96,0.99)	0.003
Age (years)	1.03	(1.01,1.06)	0.011
Sex (Male vs Female)	1.51	(0.99,2.32)	0.057

2.5.1 The Isolde trial

To illustrate the *ad hoc* approaches to be considered in the following, we consider data from the *Isolde* trial (Burge et al. 2000). In this trial, 751 patients with chronic obstructive pulmonary disease (COPD) were randomised to receive either 50 mg/day of fluticasone propionate (FP) or an identical placebo. Patients were followed up for 3 years, and their FEV$_1$ (litres, ℓ) was recorded every 3 months, although here we only use the 6 monthly measures. Interest focuses on how patients respond to treatment over time; especially in any treatment by time interaction.

As Table 2.2 indicates, only 45% of FP patients completed, compared with 38% of placebo patients. Of these, many had interim missing values. To identify key predictors of withdrawal, we carried out a logistic regression of the probability of withdrawal on the available baseline variables together with the post-randomisation exacerbation rate and rate of change in FEV$_1$. Table 2.3 shows the results after excluding variables with p-values exceeding 0.06. The effect of age, sex, and BMI are all in line with expectations from other studies. After adjusting for these, it is clear that the response to treatment is a key predictor of patient withdrawal. In particular, discussions with the trialists suggested that high exacerbation rates were probably acting as a direct trigger for withdrawal.

2.5.2 Complete records analysis

One of the simplest and most common methods of handling missing data is to include in the analysis only those units with complete records for those variables in the substantive model. Virtually all implementations of regression analysis in commercial software will do this automatically for all units that do not have complete data on all variables included in

TABLE 2.4
Isolde trial. Complete record analysis: t-test of treatment effect 3 years after randomisation.

Group	# Pat.	Mean FEV_1 (ℓ)	s.d.
Active treatment (FP)	168	1.33	0.46
Placebo	141	1.30	0.49
$t = 0.48$, 307 d.f., $p = 0.63$			
95% CI for difference $(-0.08, 0.13)$			

the model. Such analyses will be valid, if potentially inefficient, under MCAR. There are other circumstances in the regression setting when such 'completers' analyses are also valid, depending on the mechanism underlying the missingness and the model of interest. For a thorough exploration using causal diagrams of such cases we refer to Daniel et al. (2012) and Carpenter and Kenward (2013, Sec. 1.6).

We now turn to our illustrative case of a longitudinal clinical trial. If the missingness mechanism here is not MCAR, complete record analysis will typically not lead to valid inferences. One exception to this is the special case when, conditional on observed baseline covariates, including treatment group, missingness is not associated with any of the outcomes, whether observed or not provided that the analysis conditions on these covariates. (Bell et al. 2013). This is sometimes called "covariate-dependent MAR" (CD-MAR). In many settings this is not particularly plausible, but there are exceptions to this. From the results in Table 2.3 it is unlikely to be the case here, where there is evidence of strong dependence on post-baseline data.

Suppose we wish to estimate the effect of treatment at 3 years. Table 2.4 shows the results of a t-test to estimate this, using data from complete records only.

There does not appear to be a treatment effect. However, of the 705 patients randomised, only 309 (44%) completed and are included in this analysis. It is clearly possible that the analysis of the full dataset, had we obtained it, would have produced quite different conclusions.

Suppose that the end of follow-up had been reached in a trial, and that we wish to estimate the treatment effect half way through the follow up period. A complete record analysis would only include data from patients who went on to complete the trial. An *available record* analysis (sometimes called an *all observed data* analysis) includes data from all patients who have not withdrawn from the trial at the half way point. If the missingness mechanism is MCAR or CD-MAR, both the complete record analysis and the available record analysis will give sensible answers. In this record, the available record analysis would be preferable, as it will include more patients and so give more precise estimates of treatment effects. However, if the missingness mechanism is not MCAR or CD-MAR, then neither method is sensible. Only if data were MCAR or CD-MAR in the early part of the trial and not MCAR or CD-MAR later would observed record analysis be sensible but complete record analysis not sensible. This is very unlikely in practice, as we have seen in the *Isolde* trial.

2.5.3 Marginal and conditional mean imputation

In keeping with the *ad hoc* methods considered in this section, we again consider a technique for producing a dataset which has no missing values, and which is then amenable to standard

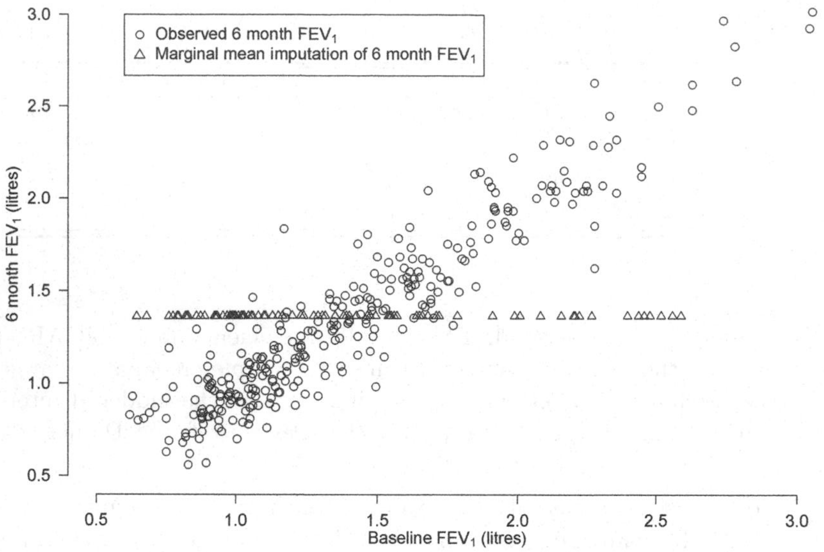

FIGURE 2.2
Isolde trial. Placebo arm: plot of baseline FEV_1 against 6 month FEV_1 with missing 6 month FEV_1's imputed by the marginal mean.

complete data methods of analysis, such as simple regression. Missing values are here filled-in using predicted *mean* values, and the approaches differ according to what is conditioned on in constructing the means. Using the terminology introduced above, we call these replaced values *single imputed values.*

Marginal mean imputation, as its name suggests, ignores other variables (Little and Rubin 2002, Sec. 4.2.1). Missing values are imputed using the average of the observed values for that variable. It is also sometimes referred to as *simple mean* imputation or just *mean* imputation.

Clearly, marginal mean imputation is problematic for categorical variables, where the 'average category' has no meaning. However, the problems go far beyond this. First, as marginal mean imputation ignores all the other variables in the dataset, its use reduces, and so misrepresents, the actual associations in the dataset. Second, in common with all single imputation methods, if the 'completed' dataset is analysed as though all the observations are genuine, then precision will be over-estimated. Data are being 'made up.' This over-estimation is the more marked because the missing data would never actually equal their marginal mean. It should be noted in passing however, that in some special settings, such as particular estimators from sample surveys, bespoke variance estimators have been constructed that reflect the single imputation method used.

We again use the *Isolde* trial to illustrate the methods. Consider the FEV_1 response 6 months after randomisation for the 375 patients in the placebo group. Eighty seven have a missing response. The mean FEV_1 of the remaining 288 is 1.36 litres. Marginal mean imputation sets each of the missing values equal to 1.36.

Figure 2.2 shows, for the 375 placebo patients, a plot of baseline FEV_1 against 6 month FEV_1. The 87 patients with marginal mean imputed values are shown with a '△.' The

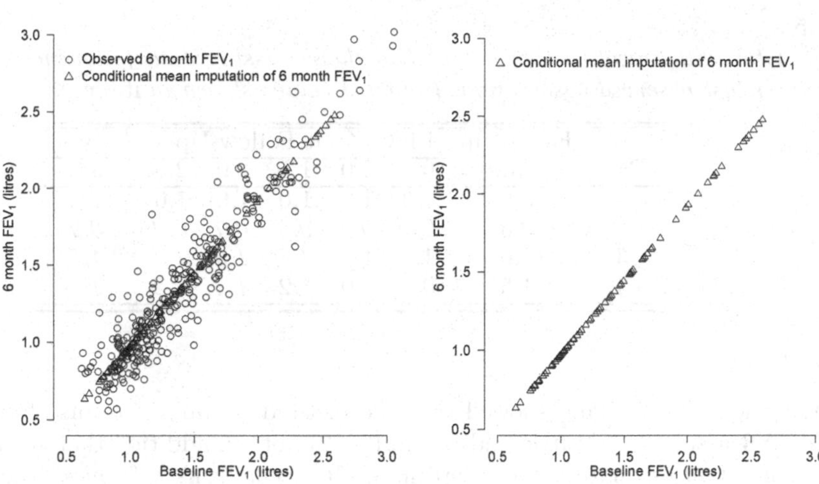

FIGURE 2.3

Isolde trial. Placebo arm: plots of baseline FEV_1 against 6 month FEV_1 with missing 6 month FEV_1's imputed by the conditional mean (2.10). Left panel: Observed and imputed data; right panel: imputed data only.

shortcomings of marginal mean imputation are immediately obvious. Unless a patient's baseline FEV_1 is close to the mean baseline FEV_1, the marginal mean is very unlikely to be close to the unobserved value.

We now consider *conditional mean imputation*, sometimes called *Buck's Method* (Buck 1960, Little and Rubin 2002). In the simplest setting, suppose we have one fully observed variable, x, linearly related to the variable with missing data, y. Using the observed pairs, (x_i, y_i), $i \in (1, \ldots, n_1)$, fit the regression of y on x:

$$E(Y_i) = \alpha + \beta x_i, \tag{2.9}$$

obtaining estimates $(\widehat{\alpha}, \widehat{\beta})$ of (α, β). Then, for the missing y_i's, $i \in (n_1 + 1, \ldots, n)$, impute them as $y_i = \widehat{\alpha} + \widehat{\beta} x_i$.

Consider again the baseline (denoted x) and 6 month (denoted y) FEV_1 measurements for the 375 placebo patients. Fitting (2.9) to the 288 patients with both values observed gives

$$E(Y_i) = 0.024 + 0.947 \times x_i. \tag{2.10}$$

We can use this to calculate the conditional mean imputation for each patient with missing 6 month FEV_1. For example, a patient with baseline 0.645 litres is imputed a 6 month value of $0.024 + 0.947 \times 0.645 = 0.635$ litres.

Figure 2.3 shows the results of using (2.10) for the 88 placebo patients with missing 6 month FEV_1. Conditional mean imputed values are shown with a '$\triangle$'. It is clear that the conditional imputations are much more plausible than the marginal imputations. However, as the right panel indicates, they are much less variable than the observed data. Thus, regarding the conditional mean imputations as 'observed data' and using them in an analysis will generally lead to underestimated standard errors, and p-values.

More generally, in settings where the regressions are all linear, such as with the multivariate

TABLE 2.5
Isolde trial. After withdrawal, patients have had their missing data following withdrawal imputed by their last observed observation, (imputed values shown in italics*).*

Pat.	Follow-up (years)	$FEV_1(\ell)$ at Follow-Up Visit (yrs) 0.5	1.0	1.5	2.0	2.5	3.0
1	3.0	1.3	1.2	1.0	1.0	1.0	1.1
2	0.5	0.7	*0.7*	*0.7*	*0.7*	*0.7*	*0.7*
3	1.0	1.7	1.5	*1.5*	*1.5*	*1.5*	*1.5*
4	1.5	0.9	1.0	1.2	*1.2*	*1.2*	*1.2*

normal distribution, Buck (1960) showed that the method is valid in terms of estimation under the MCAR assumption. Little and Rubin (2002) note in addition that the approach can produce consistent estimators under certain MAR mechanisms. Of course, the problem with the subsequent measures of precision remain, unless specific steps are taken to adjust the variances produced by the simple analyses of the 'completed' datasets.

2.5.4 Last observation carried forward

Consider a study with longitudinal follow up, such as the *Isolde* trial. Suppose that units, often individuals, withdraw (or drop out, or are lost) over the course of the follow up period. After a unit is lost to follow-up, the subsequent data are missing. Consider a particular measurement of interest. Suppose that missing values of this measurement on a given unit are *imputed* (i.e., replaced) by the last observed value of the measurement from that same unit. This is called *Last Observation Carried Forward* (LOCF); see, for example, Heyting, Tolboom, and Essers (1992). Using LOCF gives a dataset with no missing values, to which the analysis method intended for the fully observed data can be directly applied. We say the missing values have been *imputed using LOCF*, and refer to the assumption that a missing unit's responses are equal to their last observed response as the *LOCF assumption.*

We now apply this to the *Isolde* trial. In Table 2.5 we illustrate LOCF. Follow-up data are shown for four patients. The first completed the trial. The subsequent 3 have had their missing data imputed using LOCF (values shown in italics).

To illustrate the use of LOCF, we return to the *Isolde* trial. Missing responses for a patient following withdrawal are imputed by their last observed observation, apart from the 134 who withdrew before the first follow-up visit. Figure 2.4 shows the mean FEV_1 at each follow-up visit, by treatment group, using (i) all available data at each follow-up visit and (ii) LOCF to impute the missing data. The LOCF imputed means are similar for the FP arm, but markedly lower for the placebo arm. The exception is the last visit, where LOCF gives a higher mean for the FP arm. Table 2.6 shows a t-test for treatment effect using the LOCF imputed data. In contrast to Table 2.4, the estimated treatment effect is now significant at the 5% level.

However, this raises a number of questions. Under which missing data mechanisms is LOCF sensible, and are these plausible here? We need to be confident about the answers to these before concluding a treatment effect actually exists.

LOCF has been a widely used technique for dealing with dropout in clinical trials, where it has even been extended to the carrying forward of baseline values (so-called BOCF). However, in the light of severe criticism (e.g., Lavori 1992, Carpenter et al. 2004, Carpenter

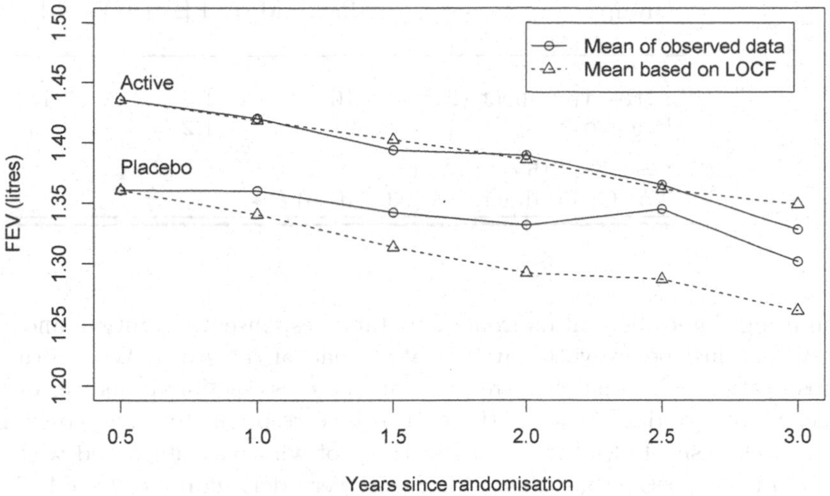

FIGURE 2.4
Isolde trial. Mean FEV_1 *(ℓ) at each follow-up visit, by treatment arm. Solid line, means calculated using all available data at each visit. Broken line, means calculated after imputing missing data using LOCF. Note that 134 patients with no readings after baseline are omitted.*

and Kenward 2008, Molnar, Hutton, and Fergusson 2008, Siddiqui, Hung, and O'Neill 2009, Kenward and Molenberghs 2009, Mallinckrodt, 2013) it is now falling out of favour. There are several strands to this criticism.

First, as usually invoked, it is a *single imputation* method, in which no account is taken of the imputation process. That is, the subsequent analysis gives these imputed responses the same status as actual observed responses. This is unsatisfactory, as a single value is being used as an estimate of a *distribution*. Hence, the subsequent measures of precision are biased downwards.

Second, leaving aside issues with precision, it is not self-evident that the subsequent estimator represents a meaningful or useful treatment comparison. It is straightforward to show that the LOCF procedure leads to biased "on treatment" estimators under both MCAR and MAR mechanisms (e.g., Molenberghs et al. 2004). Indeed, only under very specific, restrictive and unrealistic assumptions can an LOCF analysis be shown to be valid, and this corresponds to an "intention to treat" (ITT) estimator. Kenward and Molenberghs (2009) show that these assumptions imply a particular NMAR mechanism, in which exchangeability is required between the last observed and missing observation that is the target of the analysis, often the final one. This means that we effectively assume that the distribution around the marginal (i.e., treatment group) mean stays the same for patients who dropout, and we are prepared to accept that, for each patient who drops out, their condition has stabilised *before* their last observation, and the distribution of their responses does not change *at all* for the remainder of the study. For example, there is no evidence of this sort of behaviour in the *Isolde* trial. Under this strong assumption, the patient's last

TABLE 2.6
Isolde trial. LOCF imputed data: t-test of treatment effect 3 years after randomisation.

Group	# Pat.	Mean FEV_1 (ℓ)	s.d.
Active treatment (FP)	316	1.35	0.47
Placebo	301	1.26	0.48
$t = 2.28$, 615 d.f., $p = 0.02$			
95% CI for difference $(0.01, 0.16)$ ℓ			

observation is a genuine observation from their stable response distribution, and could have equally been seen just before withdrawal as at the end of the study. We can therefore use this last observation as the patient's response in the cross-sectional analysis of treatment effect at the end of the trial follow-up. However, this corresponds to a very counter-intuitive missingness mechanism. Indeed it is hard to think of why patients would withdraw *after* they had stabilised unless either the protocol were very demanding or they had no expectation that their condition would change whether they were in or out of the trial—so that the distribution of their responses would not change *at all* for the remainder of the study.

Third, defenders of LOCF sometimes argue that it leads to *conservative* estimates of treatment effects. However, it is easy to show that this cannot be true in general (Molenberghs and Thijs 2004). Rather, the direction of the bias depends on the (unknown) true treatment effect, the missing value mechanism and, most importantly, the target of the estimator (e.g., on-treatment or ITT).

Fourth, in general, LOCF leads to bias even when a complete record analysis is sensible (Molenberghs and Thijs 2004). If investigators or regulators have strong prior beliefs about the relationship between missing and observed responses, a far better way to allow for these is through an appropriate sensitivity analysis (see Part V of this Handbook, for example).

Fifth, another point sometimes made in favour of LOCF is that if there is no treatment effect it preserves the 'Type I error' (i.e., the chance of finding a statistically significant treatment effect when none in fact exists) at 5%. Although in a limited sense this is true (if both groups have identical distributions of response *and* withdrawal) the problem lies rather under the alternative hypothesis. There are many possible patterns of treatment effect and withdrawal for which the power of the LOCF test is the *same* as the test size. Further, merely maintaining the test size is not sufficient to justify a test procedure: if it were we could use the throw of a 20-sided die to calculate a test statistic with perfect nominal 5% type I error! Clearly we also need to consider the behaviour of the statistic under the range of alternative hypotheses. In this the LOCF test falls down badly, as it is unable to detect a wide range of actual treatment effects (Carpenter et al. 2004).

In summary, we would argue that if we really wish to 'carry forward' information after withdrawal, then the appropriate distribution should be carried forward, not the observation. This is not difficult to do, and will give valid inference much more generally than carrying forward the last observation. Examples of this can be seen later in Chapter 19 of this Handbook.

As LOCF is neither valid under general assumptions nor based on statistical principles, it is not a sensible method, and should not be used. It is therefore unfortunate that Wood, White, and Thompson (2004) found that LOCF is commonly used as a sensitivity analysis when the principal analysis is complete records. In effect, LOCF is actually just an analysis of each patient's last observed value (so-called Last Observation Analysis, LOA). If LOA is really of interest then by definition the last observed measurement needs to be analysed, but in this setting it is equally obvious that the time to this *event* must also be relevant, yet this is almost never considered in such analyses. When estimating treatment effects at the end of a trial, though, LOA is not useful, as it may well reflect misleading transient effects. Although there has been some confusion on this point (Shao and Zhong 2003), seeing LOCF in this light helps expose its lack of credibility (Carpenter et al. 2004). It is definitely not a sensitivity analysis in the sense described in Part V of this Handbook. Finally, Pocock (1996) reinforces Heyting, Tolboom, and Essers (1992), noting that "it is doubtful whether this [LOCF] actually answers a scientifically relevant question."

2.6 Discussion

The historic overview in Sections 2.2–2.4 show how the main threads that make up current missing data methodology have their origins in very early statistical developments, in particular the EM type algorithms that were developed for designed experiments, inverse probability weighting in sample surveys and single imputation methods. While the early developments were largely computational, the landmark papers of Rubin (1976) and Heckman (1976) on the inferential and modelling issues surrounding the handling of missing data triggered a major surge in research and development that continues into the present, with many themes and strands. This has been accompanied by the revolution in computing power that has enabled analyses to be undertaken that could have only been addressed from a conceptual perspective when the subject was in its early stages. One notable point of evolution is the absorption of the missing datasetting into broader classes of problem such as causal inference (van der Laan and Robins 2003, Pearl 2009) and data-based sampling (Molenberghs et al. 2004). It would now be difficult, if not impossible, for any individual to have a thorough knowledge of the entire corpus of the current missing data literature. The following chapters in this Handbook span much of this literature and so set out to provide the reader with the opportunity to pursue whichever of these main threads are of relevance and interest and to obtain a sense of contrast among them.

The *ad hoc* methods discussed in the second part of this chapter are increasingly taking on a historical rôle. They largely reflect the earlier methodology that was necessarily simplistic in nature given the limitations of available computation. Apart from the restriction to complete record analyses the remaining methods are based on various forms of single imputation. The single imputation methods reviewed here are not the only ones, but all have similar issues. Little and Rubin (2002) mention several others. For example, methods such as hot deck imputation are based on filling in missing values from 'matching' subjects, where an appropriate matching criterion is used. Almost all single imputation techniques suffer from the following limitations:

1. The performance of imputation techniques is unreliable. Situations where they do work are difficult to distinguish from situations were they prove misleading. Imputation often requires *ad hoc* adjustments to obtain satisfactory results. The methods fail to provide simple, valid estimators of precision.

2. In addition, most methods require the MCAR assumption to hold and some even go beyond MCAR in their requirements.

The main advantage, shared with complete case analysis, is that complete data software can be used. But without appropriate adjustment, the resulting estimates of precision will be wrong, even if the estimators themselves are consistent. As the remainder of this book illustrates, methodological advances, coupled with increasing computing power, have largely removed the rationale for using *ad hoc* methods. Rather, as the research streams shown in Figure 2.1 continue to develop and cross-fertilize, the challenge is to identify the appropriate methodology for the analysis at hand. The remainder of this Handbook sets out to address this.

Acknowledgments

We are grateful to Astra-Zeneca for permission to use the data from the clinical trial to assess the efficacy and safety of budesonide on patients with chronic asthma.

References

Allan, F.E. and Wishart, J. (1930). A method of estimating the yield of a missing plot in field experimental work. *Journal of Agricultural Science* **20**, 399–406.

Bang, H. and Robins, J. M. (2005). Doubly robust estimation in missing data and causal inference models. *Biometrics* **61**, 962-973.

Bell, M.L., Kenward, M.G., Fairclough, D.L., and Horton, N.J. (2013). Differential dropout and bias in RCTs: when it matters and when it may not. *British Medical Journal* **346**, e8668.

Buck, S.F. (1960). A method of estimation for missing values in multivariate data suitable for use with an electronic computer. *Journal of the Royal Statistical Society, Series B* **22**, 302–306.

Burge, P.S., Calverley, P.M.A., Jones, P.W., Spencer, S., Anderson, J.A., and Maslen, T.K. (2000). Randomised, double blind, placebo controlled study of fluticasone propionate in patients with moderate to severe chronic obstructive pulmonary disease: the Isolde trial. *British Medical Journal* **320**, 1297–1303.

Carpenter, J., Kenward, M., Evans, S., and White, I. (2004) Letter to the editor: Last observation carry forward and last observation analysis, by J. Shao and B. Zhong, *Statistics in Medicine*, 2003, **22**, 2429–2441. *Statistics in Medicine* **23**, 3241–3244.

Carpenter, J.R. and Kenward, M.G. (2008) *Missing Data in Clinical Trials: A Practical Guide.* UK National Health Service, National Co-ordinating Centre for Research on Methodology.

Carpenter, J.R. and Kenward, M.G. (2013). *Multiple Imputation and Its Application.* Chichester: John Wiley & Sons.

Celeux, G., and Diebolt, J. (1985). The SEM algorithm: A probabilistic teacher algorithm derived from the EM algorithm for the mixture problem. *Computational Statistics* **2**, 73–82.

Clayton, D., Spiegelhalter, D., Dunn, G., and Pickles, A. (1998) Analysis of longitudinal binary data from multi-phase sampling (with discussion). *Journal of the Royal Statistical Society, Series B* **60**, 71–87.

Cornish, E.A. (1940a). The estimation of missing values in incomplete randomized block experiments. *Annals of Eugenics, London* **10**, 112–118.

Cornish, E.A. (1940b). The estimation of missing values in quasi-factorial designs. *Annals of Eugenics, London* **10**, 137–143.

Daniel, R.M. and Kenward, M.G. (2012). A method for increasing the robustness of multiple imputation. *Computational Statistics and Data Analysis* **56**, 1624–1643.

Daniel, R.M., Kenward, M.G., Cousens, S., de Stavola, B. (2012). Using directed acyclic graphs to guide analysis in missing data problems. *Statistical Methods in Medical Research* **21**, 243–256.

Dempster, A.P., Laird, N.M., and Rubin, D.B. (1977). Maximum likelihood from incomplete data via the EM algorithm (with discussion). *Journal of the Royal Statistical Society, Series B* **39**, 1–38.

Diggle, P.J. (2005). Diggle-Kenward model for dropouts. In: P. Armitage and T. Colton (Eds.), *Encyclopaedia of Biostatistics (2nd Ed.).* New York: Wiley, pp. 1160–1161.

Diggle, P.J. and Kenward, M.G. (1994). Informative dropout in longitudinal data analysis (with discussion). *Applied Statistics* **43**, 49–94.

Gill, R. and Robins, J. (1996). Missing at random from an algorithmic viewpoint. In *Proceedings of the First Seattle Symposium on Survival Analysis.*

Greenland, S. and Finkle, W.D. (1995). A critical look at methods for handling missing covariates in epidemiologic regression analyses. *American Journal of Epidemiology* **142**, 1255–1264.

Healy, M.J.R. and Westmacott, M. (1956). Missing values in experiments analyzed on automatic computers. *Applied Statistics* **5**, 203–206.

Heckman, J. (1976). The common structure of statistical models of truncation, sample selection and limited dependent variables and a simple estimator for such models. *Annals of Economic and Social Measurement* **5**, 475–492.

Heyting, A., Tolboom, J., and Essers, J. (1992). Statistical handling of drop-outs in longitudinal clinical trials. *Statistics in Medicine* **11**, 2043–2061.

Horvitz, D.G. and Thompson, D.J. (1952). A generalisation of sampling without replacement from a finite universe. *Journal of the American Statistical Association* **47**, 663–685.

Kang, J.D.Y. and Schafer, J.L. (2007). Demystifying double robustness: A comparison of alternative strategies for estimating a population mean from incomplete data (with discussion). *Statistical Science* **22**, 523–539.

Kenward, M.G. and Molenberghs, G. (2009). Last observation carried forward: A crystal ball? *Journal of Biopharmaceutical Statistics* **19**, 872–888.

Kotz, S. and Johnson, N.L. (Eds.) (1997). *Breakthroughs in Statistics: Volume III.* New York: Springer.

Louis, T. A. (1982). Finding the observed information matrix when using the EM algorithm. *Journal of the Royal Statistical Society, Series B*, **44**, 226–233.

Lavori, P.W. (1992). Clinical trials in psychiatry: should protocol deviation censor patient data. *Neuropsychopharmacology* **6**, 39–48.

Little, R.J.A. (1986). A note about models for selectivity bias. *Econometrika* **53,** 1469–1474

Little, R.J.A. and Rubin, D.B. (2002). *Statistical Analysis With Missing Data (2nd Ed.).* Chichester: John Wiley & Sons.

Louis, T. (1982). Finding the observed information matrix when using the EM algorithm. *Journal of the Royal Statistical Society, Series B* **44**, 226–233.

Mallinckrodt, C.H. (2013) *Preventing and Treating Missing Data in Longitudinal Clinical Trials.* Cambridge University Press.

McKendrick, A.,G. (1926). Application of mathematics to medical problems. *Proceedings of the Edinburgh Mathematical Society* **44**, 98–130.

McLachlan, G.J. and Krishnan, T. (2008). *The EM Algorithm and Its Extensions (2nd Ed.).* New York: John Wiley & Sons.

Molenberghs, G., Thijs, H., Jansen, I., Beunkens, C., Kenward, M.G., Mallinkrodt, C., and Carroll, R.J. (2004). Analyzing incomplete longitudinal clinical trial data. *Biostatistics* **5**, 445–464.

Molnar, F.J., Hutton B., and Fergusson, D. (2008). Does analysis using "last observation carried forward" introduce bias in dementia research? *Canadian Medical Association Journal* **179**, 751–753.

Nair, N.R. (1940). The application of covariance technique to field experiments with missing or mixed up-plots. *Sankhya* **4**, 581–588.

Nielsen, F.E. (2000). The stochastic EM algorithm: estimation and asymptotic results. *Bernoulli* **6**, 457–489.

Orchard, T. and Woodbury, M. (1972) A missing information principle: theory and applications. In *Proceedings of the Sixth Berkeley Symposium on Mathematics, Statistics and Probability, Volume 1,* L.M.L. Cam, J. Neyman, and E.L. Scott (Eds.), pp. 697–715. Berkeley: University of California Press.

Pearl, J. (2009). *Causality (2nd Ed.).* Cambridge University Press.

Pocock, S.J. (1996). Clinical trials: A statistician's perspective. In *Advances in Biometry* P. Armitage and H.A. David (2nd Ed.), pp. 405–421. London: John Wiley & Sons.

Reilly, M. and Pepe, M. (1995). A mean score method for missing and auxiliary covariate data in regression models. *Biometrika* **82**, 299–314.

Robins, J.M. and Gill, R. (1997). Non-response models for the analysis of non-monotone ignorable missing data. *Statistics in Medicine* **16**, 39–56.

Robins, J.M., Rotnitzky, A., and Zhao, L.P. (1995). Analysis of semiparametric regression models for repeated outcomes in the presence of missing data. *Journal of the American Statistical Association* **90**, 106–121.

Robins, J. and Wang, N. (2000). Inference for imputation estimators. *Biometrika* **87**, 113–124.

Rotnitzky, A., Cox, D.R., Bottai, M., and Robins, J.M. (2000). Likelihood-based inference with a singular information matrix. *Bernouilli* **6**, 243–284.

Rubin, D.B. (1976). Inference and missing data. *Biometrika* **63**, 581–592.

Rubin, D.B. (1978) Multiple imputations in sample surveys: A phenomenological Bayesian approach to nonresponse. *Proceedings of the Survey Research Methods Section of the American Statistical Association*, 20–34.

Rubin, D.B. (1987) *Multiple Imputation for Nonresponse in Surveys.* New York: John Wiley & Sons.

Scharfstein, D.O. Rotnizky, A., and Robins, J.M. (1999). Adjusting for nonignorable drop-out using semiparametric nonresponse models (with discussion). *Journal of the American Statistical Association* **94**, 1096–1146.

Shao, J. and Zhong, B. (2003). Last observation carry-forward and last observation analysis. *Statistics in Medicine* **22**, 3241–3244.

Siddiqui, O., Hung, H.M.J., and O'Neill, R. (2009). MMRM vs. LOCF: a comprehensive comparison based on simulation study and 25 NDA datasets, *Journal of Biopharmaceutical Statistics* **19**, 227–246.

Skrondal, A. and Rabe-Hesketh, S. (2004) *Generalized Latent Variable Modelling.* Boca Raton: Chapman & Hall/CRC.

Tanner, M.A. and Wong, W.H. (1987). An application of imputation to an estimation problem in grouped lifetime analysis. *Technometrics* **29**, 23–32.

Tsiatis, A.A. (2006). *Semiparametric Theory and Missing Data.* New York: Springer.

Tsiatis, A.A., Davidian, M. and Cao, W. (2011). Improved doubly robust estimation when data are monotonely coarsened, with application to longitudinal studies with drop-out. *Biometrics* **67**, 536–545.

van der Laan, M. and Robins, J.M. (2003). *Unified Methods for Censored Longitudinal Data and Causality.* New York: Springer.

Vansteelandt, S., Carpenter, J., and Kenward, M.G. (2010). Analysis of Incomplete Data Using Inverse Probability Weighting and Doubly Robust Estimators. *Methodology* **6**, 37–48.

Vansteelandt, S., Rotnitzky, A., and Robins, J.M. (2007). Estimation of regression models for the mean of repeated outcomes under nonignorable nonmonotone nonresponse. *Biometrika* **94**, 841–860.

Wang, N. and Robins, J. (1998). Large-sample theory for parametric multiple imputation procedures. *Biometrika* **85**, 935–948.

White, I.R. and Thompson, S.G. (2005). Adjusting for partially missing baseline measurements in randomized trials. *Statistics in Medicine* **24**, 993–1007.

Wood, A.M., White, I.R., and Thompson, S.G. (2004). Are missing outcome data adequately handled? a review of published randomized controlled trials in major medical journals. *Clinical Trials* **1**, 368–376.

Yates, F. (1933). The analysis of replicated experiments where the field plots are incomplete. *The Empire Journal of Experimental Agriculture* **20**, 129–142.

Yates, F. and Hale, R.W. (1939) Latin squares when two or more rows, columns or treatments are missing. *Supplement to the Journal of the Royal Statistical Society* **6**, 132–139.

Part II

Likelihood and Bayesian Methods

3

Likelihood and Bayesian Methods: Introduction and Overview

Michael G. Kenward

London School of Hygiene and Tropical Medicine, London, UK

Geert Molenberghs

Universiteit Hasselt & KU Leuven, Belgium

Geert Verbeke

KU Leuven & Universiteit Hasselt, Belgium

CONTENTS

3.1 Likelihood and Bayesian Inference and Ignorability

In Chapter 1, key concepts were set out that are relevant throughout this volume. First, missing data mechanisms were considered: missing completely at random (MCAR), missing at random (MAR), and not missing at random (NMAR). Second, the choices of model framework to simultaneously model the outcome and missing-data were described: selection models, pattern-mixture models, and shared-parameter models. Third, the major routes of inference were reviewed: likelihood and Bayesian inference, methods based on inverse probability weighting, and multiple imputation. The latter classification also applies to Parts II, III, and IV of this volume.

In this part, we are concerned with the likelihood and Bayesian routes. A key concept, with both of these, is ignorability, as already discussed in Section 1.3.2. Likelihood and Bayesian inferences rest upon the specification of the full joint distribution of the outcomes and missing-data mechanism, regardless of which framework is chosen. However, the missing-data mechanism does not always need to be specified. To see this, consider the following. The full data likelihood contribution for unit i takes the form

$$L^*(\boldsymbol{\theta}, \boldsymbol{\psi}|X_i, \boldsymbol{Y}_i, \boldsymbol{R}_i) \propto f(\boldsymbol{Y}_i, \boldsymbol{R}_i|X_i, \boldsymbol{\theta}, \boldsymbol{\psi}).$$

Because inference is based on what is observed, the full data likelihood L^* is replaced by the observed data likelihood L:

$$L(\boldsymbol{\theta}, \boldsymbol{\psi}|X_i, \boldsymbol{Y}_i^o, \boldsymbol{R}_i) \propto f(\boldsymbol{Y}_i^o, \boldsymbol{R}_i|X_i, \boldsymbol{\theta}, \boldsymbol{\psi}) \tag{3.1}$$

with

$$\begin{aligned} f(\boldsymbol{Y}_i^o, \boldsymbol{R}_i|\boldsymbol{\theta}, \boldsymbol{\psi}) &= \int f(\boldsymbol{Y}_i, \boldsymbol{R}_i|X_i, \boldsymbol{\theta}, \boldsymbol{\psi}) d\boldsymbol{Y}_i^m \\ &= \int f(\boldsymbol{Y}_i^o, \boldsymbol{Y}_i^m|X_i, \boldsymbol{\theta}) f(\boldsymbol{R}_i|\boldsymbol{Y}_i^o, \boldsymbol{Y}_i^m, \boldsymbol{\psi}) d\boldsymbol{Y}_i^m. \end{aligned} \tag{3.2}$$

Under MAR, we obtain

$$\begin{aligned} f(\boldsymbol{Y}_i^o, \boldsymbol{R}_i|\boldsymbol{\theta}, \boldsymbol{\psi}) &= \int f(\boldsymbol{Y}_i^o, \boldsymbol{Y}_i^m|X_i, \boldsymbol{\theta}) f(\boldsymbol{R}_i|\boldsymbol{Y}_i^o, \boldsymbol{\psi}) d\boldsymbol{Y}_i^m \\ &= f(\boldsymbol{Y}_i^o|X_i, \boldsymbol{\theta}) f(\boldsymbol{R}_i|\boldsymbol{Y}_i^o, \boldsymbol{\psi}). \end{aligned} \tag{3.3}$$

The likelihood then factors into two components. While this seems a trivial result at first sight, the crux is that the first factor depends on $\boldsymbol{\theta}$ only, while the second one is a function of $\boldsymbol{\psi}$ only. Factoring also the left-hand side of (3.3), this can be written as:

$$f(\boldsymbol{Y}_i^o|X_i, \boldsymbol{\theta}, \boldsymbol{\psi}) f(\boldsymbol{R}_i|\boldsymbol{Y}_i^o, \boldsymbol{\theta}, \boldsymbol{\psi}) = f(\boldsymbol{Y}_i^o|X_i, \boldsymbol{\theta}) f(\boldsymbol{R}_i|\boldsymbol{Y}_i^o, \boldsymbol{\psi}).$$

Therefore, if, further, $\boldsymbol{\theta}$ and $\boldsymbol{\psi}$ are disjoint (also termed: variationally independent) in the sense that the parameter space of the full vector $(\boldsymbol{\theta}', \boldsymbol{\psi}')'$ is the product of the parameter spaces of $\boldsymbol{\theta}$ and $\boldsymbol{\psi}$, then inference can be based solely on the marginal observed data density. This is the so-called *separability condition.* A formal derivation is given in Rubin (1976). The practical implication is that, essentially an ignorable likelihood or ignorable Bayesian analysis is computationally as simple as the corresponding analysis in a non-missing data context.

A few remarks apply. First, with a likelihood analysis, the observed information matrix should be used rather than the expected one (Kenward and Molenberghs 1998), even though the discrepancies are usually minor. Second, ignoring the missing data mechanism assumes there is no scientific interest attached to this. When this is untrue, the analyst can, in a straightforward way, fit appropriate models to the missing data indicators. Third, regardless of the appeal of an ignorable analysis, NMAR can almost never be ruled out as a mechanism, and therefore one should also consider the possible impact of such mechanisms. In Chapters 4 and 5 such models are explored in the likelihood and Bayesian frameworks, respectively. Fourth, a particular NMAR model can never provide a definitive analysis, because of the necessary uncertainty about what is unobserved. Even when data are balanced by design, recording an incomplete version of it may induce imbalance. This, in turn, may lead to inferences that are more dependent on correctly specified modeling assumptions than is the case with balanced data. This is one of the main reasons why much research has been devoted to sensitivity analysis, the topic of Part V. The same issue has also stimulated important work in semi-parametric methods for missing data; these are studied in Part III. Fifth, while the interpretation of an MAR mechanism and ignorability is relatively straightforward in a monotone setting, this need not be the case in some non-monotone settings (Robins and Gill 1997, Molenberghs *et al* 2008). In such settings, the predictive model of what is unobserved given what is observed will necessarily be pattern-specific, however this is not an issue for all inferential targets. As stated above, should the need arise, an analysis in which the outcomes and the missing-data mechanism are jointly modeled is available (Chapters 4–5).

3.2 Joint Models

The specific but rapidly growing field of joint models for longitudinal and time-to-event data is the topic of Chapter 6. Such models have strong connections with those that arise in the missing-data literature. For example, much of the work done in the area uses shared-parameter models, partly because a missingness process, especially when confined to dropout, can be seen as a discrete version of a survival process and the use of a latent process to link the two is particularly convenient. As will be seen in Chapter 6, joint modeling may place the main focus on the time-to-event process, the longitudinal process, or both. Of course, when the time-to-event process refers to dropout, either in continuous or in discrete time, inferences are usually, though not always, directed more towards the longitudinal process than to that describing dropout.

References

Kenward, M.G. and Molenberghs, G. (1998). Likelihood based frequentist inference when data are missing at random. *Statistical Science* **13**, 236–247.

Molenberghs, G., Beunckens, C., Sotto, C., and Kenward, M.G. (2008). Every missing not at random model has got a missing at random counterpart with equal fit. *Journal of the Royal Statistical Society, Series B* **70**, 371–388.

Robins, J.M. and Gill, R. (1997). Non-response models for the analysis of non-monotone ignorable missing data. *Statistics in Medicine* **16**, 39–56.

Rubin, D. B. (1976). Inference and missing data. *Biometrika* **63,** 581–592.

4

A Perspective and Historical Overview on Selection, Pattern-Mixture and Shared Parameter Models

Michael G. Kenward

London School of Hygiene and Tropical Medicine, London, UK

Geert Molenberghs

Universiteit Hasselt & KU Leuven, Belgium

CONTENTS

4.1 Introduction

In this chapter we build on the introduction to parametric modelling presented in Section 4 of Chapter 1. As well as providing a historical overview, much of the material presented here serves as an underpinning to the parametric analyses presented in later chapters. Our overarching framework is likelihood. We have seen in Chapter 3 that, in the likelihood setting, the missing at random (MAR) assumption, together with parameter separability, is a sufficient condition for ignorability of the missing data mechanism. One implication of this is that, under the MAR assumption, likelihood-based analyses can proceed in a conventional way, and so fall into the much broader category of model construction in a general sense. The main implications of missing data then lie with models that incorporate NMAR mechanisms and the focus in this chapter is on such settings. In one form or another, such models jointly represent both the outcome and the missing value process. In practice, such models tend to be used as part of a sensitivity analysis in which they parameterize departures from the MAR assumption (see for example Part V of this volume). A key to their success therefore lies in their ability to represent such departures in a substantively meaningful and transparent way, and while there has been much methodological development of such models, not all reflects this important need.

NMAR mechanisms can be introduced into the model for the data in both a *direct* and an *indirect* way. We can motivate two such broad approaches by considering two alternative ways of defining MAR in the setting of attrition or dropout in a longitudinal design with n common times of measurement. Suppose that we have only baseline covariates (X) that are fully observed. Let $\boldsymbol{Y}$ be the vector of n repeated measurements from a given individual. If dropout occurs between measurement times t and $t+1$, the missing data $\boldsymbol{Y}^m$ corresponds to the final unobserved $n-t$ measurements.

1. *Direct:* under MAR the missing data mechanism does not depend on $\boldsymbol{Y}^m$, given the inclusion of the covariates and observed (previous) outcomes $\boldsymbol{Y}^o$,

$$\Pr(D = t+1 \mid \boldsymbol{Y}^o, X),$$

for D the time of dropout. Non-randomness of the missing data mechanism is then governed by additional dependence of this mechanism on $\boldsymbol{Y}^m$, as expressed through the mechanism itself,

$$\Pr(D = t+1 \mid \boldsymbol{Y}^o, \boldsymbol{Y}^m, X).$$

2. *Indirect:* an alternative, equivalent, definition of MAR can be expressed in terms of the conditional distribution of the unobserved data given the observed,

$$f(\boldsymbol{Y}^m \mid \boldsymbol{Y}^o, X, D),$$

which, under MAR, does not depend on D. In words, under MAR, the future statistical behaviour, given the past, is the same for those who remain and those who drop out. The introduction of non-random missingness is therefore accomplished by allowing this conditional distribution to differ for different values of D, and the difference among these conditional distributions determines the nature of the non-randomness of the missing value mechanism.

More succinctly, we can express these two approaches as different decompositions of the same joint distribution:

$$f(\boldsymbol{Y}, D \mid X) = f(\boldsymbol{Y} \mid X)\Pr(D \mid \boldsymbol{Y}, X) = \Pr(D \mid X) f(\boldsymbol{Y} \mid D, X). \tag{4.1}$$

Although it is clear that either decomposition determines the other, the important distinction between the two approaches is the way in which departures from MAR are represented by the relevant parameters. Typically, an accessible and interpretable form for the parameters in one approach will not translate equivalently to the other. In practice the choice among such models depends on the appropriateness of the representation of non-randomness for the particular substantive setting.

The first approach corresponds to the class of *selection* models and, given the explicit formulation of the missing data mechanism in such models, it is perhaps unsurprising that these were the first to be introduced and developed. Selection models form the subject of the next section. Developments based on the second approach correspond to so-called *pattern-mixture* models, which appeared in the literature somewhat later. At first sight, the pattern-mixture approach may seem intuitively less appealing, but it turns out that the viewpoint is valuable for making explicit the assumptions underlying particular NMAR models in terms of separating out components of the model that can, and cannot, be estimated from the data without untestable assumptions. As a consequence such models have an important role in sensitivity analysis. We consider pattern-mixture models in Section 4.3.

These two formulations are based on modelling observables, the outcomes and the covariates; the only unobservables are the missing data. By introducing, in addition, unobserved latent variables, a much richer family of models can be constructed that jointly represents the outcome and missing value processes. The two processes can be linked through common (or dependent) latent variables, and such models are known as *shared-parameter* models in the missing datasetting. These are considered in Section 4.4. Such an approach is common in the structural equation framework. Indeed, it is not difficult to formulate over-arching models which have all three forms (selection, pattern-mixture, and shared-parameter) as special cases. The price of such generality is the increased number of assumptions that need to hold for valid inferences to be made. Indeed, in more complex examples it may not even be obvious what assumptions are required for MAR to hold, for example, or what is the implied form of the corresponding selection model, and whether this is plausible. For this reason selection and pattern-mixture models still have an important role in NMAR modelling for missing data.

One emphasis in this chapter will be on illustrating the special features of NMAR models that do not necessarily correspond to intuitions learned from conventional modelling under MCAR and MAR. For this, some very simple examples will be used, which allow these special aspects to stand out clearly for scrutiny. It is partly such features that support the broad recommendation that such NMAR models are best suited to sensitivity analyses rather than as the bases of definitive analyses.

4.2 Selection Models

4.2.1 The Heckman model

Much of the early development of, and debate about, selection models appeared in the econometrics literature. Indeed the term selection model was coined by Heckman (1976) who introduced an example of a Tobit model for a potentially missing continuous outcome. This model combines a marginal Gaussian regression model for the response, as might be used in the absence of missing data, with a probit model for the missing-data mechanism, that is

$$\Pr(R = 0 \mid Y = y) = \Phi(\psi_0 + \psi_1 y) \tag{4.2}$$

for suitably chosen parameters ψ_0 and ψ_1, $\Phi(\cdot)$ the Gaussian cumulative distribution function, and R the missing data indicator (1 if the observation is taken, 0 if it is missing). We note in passing that Heckman (1976) originally derived this model through a latent Gaussian process.

To avoid some of the complications of direct likelihood maximization, a two-stage estimation procedure was proposed by Heckman (1976) for this type of model. The use of this form of Tobit model and associated two-stage procedure was the subject of considerable debate in the econometrics literature, much of it focusing on the issues of identifiability and sensitivity (e.g., Amemiya 1984, Little 1986). The same issues arise whenever such models are used, irrespective of the substantive setting.

This basic structure underlies the simplest forms of selection model that have been proposed in a wide range of settings. A suitable response model, such as the multivariate Gaussian, is combined with a binary regression model for missingness. With dropout in a longitudinal setting this latter component takes the form of a discrete hazard model; at each time point the occurrence of dropout is regressed on previous and current values of the response as well as covariates. We now explore such models in more detail.

4.2.2 Models for categorical data

Although modelling a continuous response is generally more straightforward than a categorical response, we begin by considering the latter case. By reducing the problem to a very simple setting, that of two repeated binary responses, we are able to illustrate some key points in a very explicit way. With both measurements (Y_1, Y_2) observed, a subject will generate one of four responses, with associated probabilities π_{ij}:

	Time 2		
Time 1	0	1	Total
0	π_{00}	π_{01}	$\pi_{0\cdot}$
1	π_{10}	π_{11}	$\pi_{1\cdot}$
Total	$\pi_{\cdot 0}$	$\pi_{\cdot 1}$	1

These probabilities can be parameterized in many different ways, for example using marginal or conditional representations. For the moment we are not concerned with the particular parameterization chosen and will work with the joint probabilities, noting that three of these determine the fourth. It is assumed that missing values are restricted to a monotone pattern, i.e., only the second can be missing. Use $R = 0$ to indicate that the second is missing, and $R = 1$ if observed:

$$\Pr(R = 1 \mid Y_1 = s, Y_2 = t) = \phi_{st}, \quad s = 0, 1;\ t = 0, 1.$$

There are four probabilities, making a total of seven degrees-of-freedom for the model that combines the response and missingness components. If we could observe the data from those who dropout, the full data could be represented as a $2 \times 2 \times 2$ contingency table, classified by the time 1 and 2 outcomes and missingness status. The model described saturates the degrees-of-freedom in this table. In practice, two tables of data are observed; a 2×2 table from the completers and a 2×1 table of time 1 outcomes from the dropouts, making a total of five degrees-of-freedom. Clearly some parameters in the full model cannot be identified, and if a model is to be estimated from the observed data some constraints must be applied. This reflects the information lost with the dropouts. The MCAR and MAR cases correspond to the simple constraints

$$\begin{aligned} \text{MCAR} &: \quad \phi_{st} = \phi, \\ \text{MAR} &: \quad \phi_{st} = \phi_s. \end{aligned}$$

Allowing ϕ_{st} to depend on t makes the dropout non-random. We can relate this dropout model in a simple way to the logistic regression models that have often been used in the selection model framework (e.g., Greenlees, Reece and Zieschang 1982, Diggle and Kenward 1994, Baker 1995, Fitzmaurice, Laird and Zahner 1996, Fitzmaurice, Heath and Clifford 1996, Molenberghs, Kenward and Lesaffre 1997),

$$\text{logit}(\phi_{st}) = \psi_0 + \psi_1 Y_1 + \psi_2 Y_2. \tag{4.3}$$

Note that this does not saturate the dropout model, the introduction of the interaction term in Y_1Y_2 would be required for this. Non-randomness is determined by ψ_2 which can be expressed as the log odds-ratio

$$\psi_2 = \ln\left\{\frac{\phi_{11}(1-\phi_{10})}{(1-\phi_{11})\phi_{10}}\right\}.$$

To highlight the issues that can arise when using such a seemingly simple model, we use an illustrative example that has an ordinal outcome: the Fluvoxamine study. These data originate from a multicentre study involving 315 patients that were treated for psychiatric symptoms described as possibly resulting from a dysregulation of serotonin in the brain. Analyses of these data have been discussed in several publications, see for example Section 2.2 of Molenberghs and Kenward (2007) and the references given there. After recruitment to the study, the patients were assessed at four visits. The therapeutic effect and the extent of side effects were scored at each visit on an ordinal scale. The side effect response is coded as (1) none; (2) not interfering with functionality; (3) interfering significantly with functionality; (4) side effects surpasses the therapeutic effect. Similarly, the effect of therapy is recorded on a four point ordinal scale: (1) no improvement or worsening; (2) minimal improvement; (3) moderate improvement; and (4) important improvement. Thus a side effect occurs if new symptoms occur, while there is therapeutic effect if old symptoms disappear. At least one measurement was obtained from 299 subjects, including 242 subjects with a complete set of measurements (completers). There is also baseline covariate information on each subject: sex, age, initial severity (scale 1 to 7) and duration of actual illness. The aim of the analysis is to assess dependence of the two end points, therapeutic and side-effect, on baseline covariate values and to assess changes with time.

To begin with, to match the simple model setting introduced above, we restrict ourselves to the data from visits 1 and 4 and reduce the ordinal outcome to a binary one (absence/presence of side effects), by assigning absence of side effects to category 0, and presence (outcomes 2, 3, and 4) to category 1. We then take the data on the first and fourth visits, for the 299 subjects with an observation on the first visit. Of these, 75 have missing values on the fourth visit. These data can be summarized as follows.

Completers				Dropouts
	Visit 4			
Visit 1	0	1	Total	Total
0	89	13	102	26
1	57	65	122	49
Total	146	78	224	75

Examples of the model described above can then be fit to these data. To concentrate on issues associated with the dropout mechanism we will use the saturated response model with three parameters. The random dropout model, defined by $\phi_{st} = \phi_s$ or, equivalently, in terms of the logistic dropout model

$$\text{logit}(\phi_{st}) = \psi_0 + \psi_1 Y_1,$$

saturates the observed data and so fits the observed counts exactly. But, because it is notionally a model for all the data, whether observed or not, the fitted model predicts, or imputes, the behaviour of the outcome of the dropouts on the fourth visit by distributing the dropouts among the 0 and 1 outcomes, giving the counts in the first line of Table 4.1. This

TABLE 4.1
Fluvoxamine Study. Predicted counts for the dropouts (dichotomized side-effects, visits 1 and 4). The definition of the models is given in the text.

	Outcome				Marginal Success Probabilities		
Model	(0,0)	(0,1)	(1,0)	(1,1)	Visit 1	Visit 4	Odds Ratios
MAR	22.7	3.3	22.9	26.1	0.65	0.39	7.81
NMAR(2)	25.5	0.5	42.4	6.6	0.65	0.09	6.10
NMAR(-2)	12.5	13.5	5.2	43.8	0.65	0.76	6.70
NMAR(+)	21.6	4.4	19.1	29.9	0.65	0.46	7.96
NMAR(-)	24.2	1.76	31.3	17.7	0.65	0.26	7.19

table also contains summaries of the fitted response model in terms of the estimated marginal success probabilities for each of the visits and the odds-ratio $\phi_{00}\phi_{11}/\phi_{01}\phi_{10}$. We would like to compare these results with those obtained from a non-random dropout model. The data contain no information on ψ_2 in (4.3), so some additional structure must be imposed. One possibility is to select a range of values for ψ_2 in a simple sensitivity analysis. Maximum likelihood estimates and imputed values are presented in Table 4.1 for two values: 2 and -2; the models being labelled NMAR(2) and NMAR(-2) in the table. It can be seen that with NMAR(2) nearly all the dropouts are being assigned to category 0 in visit 4, in contrast to NMAR(-2) where the majority are assigned to category 1. This qualitative difference is to be expected from the signs of ψ_2. The marginal probability in visit 1 is unaffected by the choice of dropout model, which follows directly from the use of the saturated response model. The second marginal probability is highly dependent on the choice of ψ_2 and it can be seen that there are limits to the range of values of this parameter, depending on the proportion of dropouts assigned to each category in visit 4. By comparison, the odds-ratio is not greatly affected. Another approach to identifying the non-random dropout model is to impose constraints on the parameters, such as assuming that the coefficients of Y_1 and Y_2 are equal in (4.3). The imposition of such constraints, often implicitly, is usually the route taken when such selection models are applied in more complicated and realistic settings. We consider two such possibilities here, one in which the two parameters are equal $\psi_1 = \psi_2$ labelled NMAR(+) and one in which they are equal, but of opposite sign, $\psi_1 = -\psi_2$, labelled NMAR(-). The corresponding full models each have five parameters and are estimable from the data. The results from the first, NMAR(+), are very similar to those from the MAR model, while those of the second tends towards NMAR(2). In summary we have fitted several NMAR models by imposing additional information or constraints on the original unidentifiable model. Each of these models saturates the observed data and so fits exactly. Therefore we cannot distinguish between the models using their fit. In this very limited example we have no reason to chose between the different constraints applied and as a consequence the conclusions to be drawn would depend on what were believed to be a plausible range of assignments of dropout to category on the fourth visit.

Consider now the extension in this setting to any pattern of missing values, in other words, we may have measurements missing on the first occasion but not on the second and vice versa. This introduces a further set of partially classified outcomes. This apparently simple situation becomes surprisingly complex when we start to fit non-random models. It has received much attention in the literature, (Fay 1986; Baker and Laird 1988; Baker, Rosenberger and Desimonium 1992; Baker 1994, Park and Brown 1994, Baker 1995, Molenberghs et al. 1999) probably because it illustrates very well the range of issues and problems that can arise when using non-random models with categorical data. As well as the issues seen above concerning non-identifiability of parameters and models fitting the observed data identically yet with entirely different behaviour at the complete data level, we have the following examples that can occur with maximum likelihood estimators from non-random models (Molenberghs et al. 1999).

- Solutions can be non-unique, or be outside the parameter space. This may be true for predicted complete data cell counts as well as for model parameters.
- Examples exist that saturate the degrees-of-freedom yet have a positive deviance.
- Models that saturate the degrees-of-freedom can yield boundary solutions in both the predicted complete data counts and some of the parameter estimates. Without constraints built in to ensure non-negative solutions, invalid solutions can be found.
- Non-unique solutions can be obtained, in particular when the two variables (time 1 and 2) are independent. When both variables are only weakly associated, this implies that the likelihood surface becomes very flat. Similar effects are seen with non-random models for continuous data.

We do not pursue this setup further, other than by adding that we should expect to see such issues arising when non-random dropout models are fitted in more complicated and realistic settings, for example with covariates and more time points, but recognising that in such settings it becomes very much harder, if not impossible, to study their occurrence systematically.

Returning to the Fluvoxamine data we now illustrate the use of a selection model for both four-category ordinal responses from the first three visits. First an appropriate model is chosen for the responses, in this case we use a 3-way Dale model in which a proportional odds model represents the marginal response at each of the three visits and dependence among these three responses is introduced through global odds-ratios (Molenberghs and Lesaffre 1994). The logistic regression form (4.3) is used to model dropout. Similar approaches have been used for binary data which combine various models for the repeated binary response with a logistic dropout model (Fitzmaurice, Laird and Zahner 1996; Fitzmaurice, Heath and Clifford 1996, Baker 1995). In this example the Y_i are ordinal outcomes, so considerable structure is being applied to the dropout relationship, that is, the dropout model is far from saturated and consequently all the parameters are identified. This is in contrast to the simple binary setting described earlier. Other terms could in principle be added to the dropout model, for example allowing time dependence of the parameters, or explicit dependence of dropout on other covariates. It does not necessarily follow however that the data will contain sufficient information to estimate such additional terms. General rules are difficult to formulate, but experience suggests that commonly information on the structure in the dropout model will be limited.

Previous analyses of the Fluvoxamine data have shown that the therapeutic effects are largely independent of baseline covariates, while for side effects, there is non-negligible dependence of response on age, sex and there are interactions between visit and severity and between visit and duration (Molenberghs and Lesaffre, 1994; Kenward, Lesaffre and Molenberghs 1994). Using maximum likelihood, these marginal models have been fitted together with the logistic regression model (4.3). Full sets of parameter estimates are given in Molenberghs et al. (1997); here, we concentrate on the dropout part of the fitted models. The estimated parameters are given in Table 4.2. Within the framework of the model, we can test

TABLE 4.2
Fluvoxamine Study. Parameter estimates (standard errors) for the dropout model parameters.

Parameter	Side Effects	Therapeutic Effect
ψ_0	−4.26 (0.48)	−2.00 (0.48)
ψ_1	0.18 (0.45)	0.77 (0.19)
ψ_2	1.08 (0.54)	−1.11 (0.42)

TABLE 4.3
Fluvoxamine Study. Test statistics (p-values) for the comparison of (1) NMAR versus MAR and (2) MAR versus MCAR.

	Side Effects		Therapeutic Effect	
Test	NMAR—MAR	MAR—MCAR	NMAR—MAR	MAR —MCAR
Wald	4.02 (0.045)	38.91 (< 0.001)	6.98 (0.008)	3.98 (0.046)
LR	4.26 (0.039)	39.99 (< 0.001)	6.94 (0.008)	4.03 (0.044)
Score	4.42 (0.040)	45.91 (< 0.001)	9.31 (0.002)	4.02 (0.045)

for the different classes of dropout (MCAR, MAR, NMAR) using tests on the parameters governing dropout, that is $\psi_1 = \psi_2 = 0$ corresponds to MCAR and $\psi_1 \neq 0, \psi_2 = 0$ to MAR. For these tests, conventional statistics with asymptotic null χ^2 distributions, such as the likelihood ratio, Wald, and score, can be used. Test results are displayed in Table 4.3. For the side effects there is a large dependence of dropout on previous measurement, a result that could have been shown using standard logistic regression. There is at most marginal evidence for a non-zero ψ_2. The full fitted dropout model can be written

$$\text{logit}\{\Pr(\text{dropout})\} = -4.26 + 0.18Y_p + 1.08Y_c.$$

for Y_p and Y_c the previous and current measurements, respectively. It is often helpful to express such models in terms of the increment and sum of the successive measurements. Standard errors of the estimated parameters have been added in square brackets:

$$\text{logit}\{\Pr(\text{dropout})\} = -4.26 + 0.63[0.08](Y_p + Y_c) + 0.45[0.49](Y_c - Y_p).$$

It can be seen that the estimated probability of dropout increases greatly with larger number of side-effects. The corresponding standard error is comparatively small. Although the coefficient of the increment does not appear negligible in terms of its absolute size, in the light of its standard error it cannot be said to be significantly different from zero, reflecting the lack of information in the data on this term. The fitted Dale model indicates high association between successive measurements, and the results here point to a high dependence of dropout on level on side-effects. Given the large association among responses, we see that this level can be measured in terms of a single side effect measurement, as in the original dropout model, or in terms of the sum, as in the reparameterized model. Because the addition of ψ_2 to the model has minimal effect, parameter estimates and imputed missing values are very similar under the MAR and NMAR models. Given the clear lack of fit of the MCAR model compared with the MAR, it is not surprising that the same quantities differ appreciably between these two models (Kenward et al. 1994).

A different picture of dropout emerges from the therapeutic response in which the NMAR model fits the observed data appreciably better than the MAR, which in turn is only marginally better than the MCAR. The fitted dropout model is

$$\text{logit}\{\text{P(dropout)}\} = -2.00 + 0.77Y_p - 1.11Y_c,$$

and in terms of increment and sum,

$$\text{logit}\{\text{P(dropout)}\} = -2.00 - 0.17[0.17](Y_p + Y_c) - 0.94[0.28](Y_c - Y_p). \tag{4.4}$$

From the model the dependence on dropout appears to largely involve the change in response between two visits, the probability of dropout decreasing when there is a favorable change in therapeutic effect and increasing in comparison by a small amount only when there is little therapeutic effect. The introduction of the NMAR component of the dropout model makes a significant contribution to the fit of the model, and also leads to non-trivial changes in the resulting parameters estimates for the response model and imputed missing values (Molenberghs, Kenward and Lesaffre, 1997).

While the interpretations from the fit of the models to the side-effects and therapeutic response may appear plausible, and appear to suggest, for example that an MAR process operates with side-effects, in contrast to the NMAR structure seen for the therapeutic response, such conclusions must be treated very cautiously indeed. The use of this particular dropout model makes implicit assumptions about the behaviour of the missing data that can only be assessed using external information, should such information exist. Another choice of model could reverse these conclusions. This does not imply that nothing has been learnt however: the contribution of ψ_2 to the fit of the model with the therapeutic response indicates the presence of structure in the data not accommodated by the MAR model. It is the origins of this structure that cannot necessarily be inferred from the observed data. In this sense these non-random dropout models can be viewed as a way of assessing the assumption of ignorability of the chosen response model. The NMAR dropout model provides one possible alternative hypothesis. Absence of evidence for NMAR dropout within the given model framework does not imply that the dropout is ignorable: the model may be inappropriate or there may be insufficient information on the relevant parameters. On the other hand, clear evidence for NMAR dropout implies some inadequacy in the original MAR model, and if the introduction of non-random dropout changes the conclusions drawn then this inadequacy needs to be addressed.

4.2.3 Models for continuous data

Similar models to those seen above for categorical data can be constructed in a fairly obvious way for a continuous response, combining the multivariate Gaussian linear model with a suitable missingness model. This is particularly straightforward for dropout or monotone missingness. Diggle and Kenward (1994) (see also Diggle 1998) introduced a logistic dropout model, essentially a discrete hazard model, in a longitudinal development of the model of Greenlees et al. (1982). This was done as follows, assuming that any covariates are complete and external (or exogenous), so models can be expressed conditionally on these without added complication. Let $\boldsymbol{H}_{ij} = (Y_{i1}, \dots, Y_{i,j-1})$ denote the observed history of subject i up to time $t_{i;j-1}$. The Diggle and Kenward model for the dropout process allows the probability of dropout between occasions $j-1$ and j (i.e., $D_i = j$) to depend on the history $\boldsymbol{H}_{ij}$ and the possibly unobserved outcome Y_{ij}, but not on future outcomes Y_{ik}, $k > j$. We call this latter property *Non Future Dependence*, and we will see that this has an important part to play in the development of pattern-mixture models below in Section 4.3. These conditional probabilities

$$P(D_i = j | D_i \geq j, \boldsymbol{H}_{ij}, Y_{ij}, \boldsymbol{\psi})$$

can then be used to construct the probability of dropout for each time interval:

$$\begin{aligned} P(D_i = j | \boldsymbol{Y}_i, \boldsymbol{\psi}) &= P(D_i = j | \boldsymbol{H}_{ij}, Y_{ij}, \boldsymbol{\psi}) \qquad (4.5) \\ &= \begin{cases} P(D_i = j | D_i \geq j, \boldsymbol{H}_{ij}, Y_{ij}, \boldsymbol{\psi}) & j = 2 \\ P(D_i = j | D_i \geq j, \boldsymbol{H}_{ij}, Y_{ij}, \boldsymbol{\psi}) & \\ \quad \times \prod_{k=2}^{j-1} [1 - P(D_i = k | D_i \geq k, \boldsymbol{H}_{ik}, Y_{ik}, \boldsymbol{\psi})] & j = 3, \ldots, n_i \\ \prod_{k=2}^{n_i} [1 - P(D_i = k | D_i \geq k, \boldsymbol{H}_{ik}, Y_{ik}, \boldsymbol{\psi})] & j = n_i + 1. \end{cases} \end{aligned}$$

A commonly used, and simple, version of such a dropout model is

$$\text{logit}\,[P(D_i = j \mid D_i \geq j, \boldsymbol{H}_{ij}, y_{ij}, \boldsymbol{\psi})] = \psi_0 \; + \; \psi_1 Y_{i;j-1} \; + \; \psi_2 Y_{i;j}. \qquad (4.6)$$

The special cases of the model corresponding to MAR and MCAR are obtained from setting $\psi_2 = 0$ or $\psi_1 = \psi_2 = 0$, respectively. In the first case, dropout is no longer allowed to depend on the current measurement, and in the second case, dropout is independent of the outcome altogether. These properties make the model convenient for some forms of sensitivity analysis, see for example Chapter 16.

In principle, this model is easily generalized by including the complete history $\boldsymbol{H}_{ij} = (y_{i1}, \ldots, y_{i;j-1})$, and/or covariates, and allowing interactions with time. While it is entirely plausible in some settings that extensions are needed, especially changes in dependence over time, in practice there will typically be very little information available to estimate such parameters and an appropriate compromise will be needed. Moreover, it is very unlikely that one single missingness model will apply to all subjects or units, and again, the model must be regarded as a compromise which perhaps represents a single dominant process that is most likely to be relevant for the substantive setting. Note as well that, strictly speaking, one could allow dropout at a specific occasion to be related to all future responses as well, whether observed or not. However, in many settings this is very counter-intuitive. Such dependence can occur when the dropout mechanism depends on unobserved subject-specific latent variables and if this is thought likely then the shared-parameter approach of Section 4.4 might be more appropriate.

In the Diggle and Kenward model, the dropout model (4.5) or an extension of this, is combined with a conventional multivariate Gaussian model for the measurement process:

$$\boldsymbol{Y}_i \sim N(X_i \boldsymbol{\beta}, V_i), \;\; (i = 1, \ldots, N). \qquad (4.7)$$

for X_i the covariate matrix for the ith subject. Such models are widely used for modelling hierarchical and longitudinal continuous outcomes, and have a large associated literature, see for example Verbeke and Molenberghs (2000). We have seen that *under MAR*, (e.g., $\psi_2 = 0$) the missing value mechanism is ignorable, and model (4.7) can be fitted directly to the data using conventional tools and subsequence inferences made in the usual way, noting only that the observed, not the expected, information matrix must be used for obtaining measures of precision (Kenward and Molenberghs 1999).

Under NMAR however, that is, when there is dependence of the missing value mechanism on the potentially unobserved data, the marginal likelihood requires integration over the missing data:

$$\begin{aligned} f(\boldsymbol{Y}_i^{\mathrm{O}}, D_i|\boldsymbol{\theta},\boldsymbol{\psi}) &= \int f(\boldsymbol{Y}_i, d_i|\boldsymbol{\theta},\boldsymbol{\psi})\, d\boldsymbol{Y}^m \\ &= \int f(\boldsymbol{Y}_i|\boldsymbol{\theta}) f(D_i|\boldsymbol{Y}_i,\boldsymbol{\psi})\, d\boldsymbol{Y}^m. \end{aligned} \tag{4.8}$$

One advantage of the Non Future Dependence assumption in the Diggle and Kenward model is the reduction of this integral, which at first sight appears multivariate, to a univariate integral involving only Y_{ij} for $D_i = j$. This integral has the same basic structure as the one that occurs with the commonly used mixed logistic regression model with a Gaussian random effect in the linear predictor, and numerical tools for the latter problem can be applied to the former. In fact a range of numerical techniques have been used for this integral. Diggle and Kenward (1994) originally used the Nelder–Mead simplex algorithm (Nelder and Mead 1965) to optimize the integrated likelihood. However, such an approach lacks flexibility, is inefficient for high-dimensional problems, and does not exploit the well-known routines that are implemented for the two separate components of the model. A range of alternative numerical approaches exists. Yun, Lee, and Kenward (1997) use the Laplace approximation for the Diggle and Kenward model, while for some combinations of response and dropout model, the EM algorithm can be used and this does allow separate maximisation of response and dropout, as Molenberghs, Kenward, and Lesaffre (1997) did for ordinal data (see also Molenberghs and Kenward 2007). Thus, this exploits the relatively simple structure for the complete-data models, but integration is still required in the expectation step of the algorithm. Markov chain Monte Carlo methods have also been used, both in a fully Bayesian framework (e.g., Carpenter, Pocock, and Lam 2002) and in hybrid methods such as the stochastic EM algorithm (e.g., Gad 1999).

The Diggle and Kenward model has subsequently been extended to the non-monotone setting: Troxel, Harrington, and Lipsitz (1988) and Troxel, Lipsitz, and Harrington, (1988) use an antedependence covariance structure for the response with full likelihood and pseudo-likelihood, respectively, while Gad and Ahmed (2005, 2007) apply the stochastic EM algorithm with full likelihood.

As in the categorical case we use a very simple example with two time points to illustrate the issues that can arise when using such models in practice. Our example is taken from Diggle and Kenward (1994) and has been re-analysed on several occasions, in particular see Kenward (1998), Molenberghs and Kenward (2007), and Chapter 17 of this book. It concerns a study on the infectious disease mastitis in dairy cows. Data were available of the milk yields in thousands of liters of 107 dairy cows from a single herd in 2 consecutive years: Y_{ij} $(i = 1, \ldots, 107; j = 1, 2)$. In the first year, all animals were supposedly free of mastitis; in the second year, 27 became infected. Mastitis typically reduces milk yield, and the question of scientific interest is whether the probability of occurrence of mastitis is related to the yield that would have been observed had mastitis not occurred. So here, occurrence of mastitis corresponds to dropout.

A simple bivariate Gaussian linear model will be used to represent the marginal milk yield in the 2 years, the yield that would be, or was, observed in the absence of mastitis:

$$\begin{bmatrix} Y_1 \\ Y_2 \end{bmatrix} = N\left(\begin{bmatrix} \mu \\ \mu + \Delta \end{bmatrix}, \begin{bmatrix} \sigma_1^2 & \rho\sigma_1\sigma_2 \\ \rho\sigma_1\sigma_2 & \sigma_2^2 \end{bmatrix}\right).$$

Note that the parameter Δ represents the change in average yield between the 2 years. The

TABLE 4.4
Mastitis in Dairy Cattle. Maximum likelihood estimates (standard errors) of random and nonrandom dropout models under 5 scenarios: (a) the original model with a Gaussian model for the measurements; (b) the same model with observations #4 and #5 removed; (c) a t_{25} distribution for the measurements; (d) a t_{10} distribution; (e) a t_2 distribution.

Par.	(a) Gauss.	(b) Gauss.(−2)	(c) t_{25}	t_{10}	t_2
		Missing at Random			
Δ	0.72(0.11)	0.64(0.09)	0.69(0.10)	0.67(0.09)	0.61(0.08)
ψ_1	0.27(0.25)	0.22(0.25)	0.27(0.24)	0.27(0.24)	0.27(0.24)
−2 l.l.	280.02	237.94	275.54	271.22	267.87
		Not Missing at Random			
Δ	0.33(0.14)	0.40(0.18)	0.35(0.13)	0.38(0.14)	0.54(0.11)
ψ_1	2.25(0.77)	1.61(1.13)	2.11(0.78)	1.84(0.82)	0.80(0.66)
ψ_2	−2.54(0.83)	−1.66(1.29)	−2.33(0.88)	−1.96(0.95)	−0.65(0.73)
−2 l.l.	274.91	237.86	271.77	269.12	266.79

probability of mastitis is assumed to follow the logistic regression model:

$$\text{logit}\{\Pr(\text{dropout})\} = \psi_0 + \psi_1 Y_1 + \psi_2 Y_2. \quad (4.9)$$

The combined response/dropout model was fitted to the milk yields by maximum likelihood using a generic function maximization routine. In addition, the MAR model ($\psi_2 = 0$) was fitted. This latter is equivalent to fitting separately the Gaussian linear model for the milk yields and the logistic regression model for the occurrence of mastitis. These fits produced the parameter estimates as displayed in Table 4.4, standard errors and minimized value of twice the negative log-likelihood.

It is tempting to compare the two minus twice the log-likelihood values to provide a likelihood ratio test for the null hypothesis $\psi_2 = 0$, or equivalently for non-randomness. The difference here is $280.02 - 274.91 = 5.01$, with a tail probability from the χ^2_1 equals 0.02. This is suggestive of a real difference. However, there are two main reasons to be careful in giving such a test result the usual interpretation. First, information on parameters in such selection models does not accrue at the usual rate with increasing numbers of subjects, and this means that, in finite samples, estimates of error, and asymptotic-based tests like the likelihood-ratio can have behaviour that is far from the nominal (Rotnitzky et al. 2000; Jansen et al. 2006). For example, the Wald test that corresponds to the above likelihood-ratio test is $(-2.53)^2/0.83 = 9.35$, with corresponding χ^2_1 probability of 0.002. The discrepancy between the results of the two tests is large, suggesting strongly that the usual asymptotic approximation is not holding here. Such behaviour is confirmed more generally by the simulation results in Jansen et al. (2006). The second reason to be careful with such tests, is the fact that the ability to differentiate between the two models is based wholly on modelling assumptions that cannot be assessed from the data at hand. Indeed, as is shown in Molenberghs et al. (2008), for any given NMAR model, an MAR counterpart can be constructed that has exactly the same fit to the observed data. We return to this second point below.

First we consider a little more closely the nature of the dropout model implied by the NMAR fit. The dropout model estimated is:

$$\text{logit}\{\Pr(\text{dropout})\} = 0.37 + 2.25 Y_1 - 2.54 Y_2. \quad (4.10)$$

Some insight into this fitted model can be obtained by rewriting it in terms of the milk yield totals $(Y_1 + Y_2)$ and increments $(Y_2 - Y_1)$:

$$\text{logit}\{\Pr(\text{dropout})\} = 0.37 - 0.15(Y_1 + Y_2) - 2.40(Y_2 - Y_1). \tag{4.11}$$

The probability of mastitis increases with larger negative increments; that is, those animals who showed (or would have shown) a greater decrease in yield over the 2 years have a higher probability of getting mastitis. With highly correlated repeated measurements we can expect such behaviour in Diggle and Kenward type models. If, given the model assumptions, there is a strong dependence of dropout on the simple level of the measurement at the next occasion, then, because of the dependence among the measurements, the previous measurement(s) can act as proxy for this, and conditional dependence in the logistic regression on the next measurement will be small. So if the likelihood does show a non-trivial difference in fit between the MAR and NMAR models, the dependence on the current measurement must take a form that is largely independent of the previous average level, that is the increment. If, with longer sequences, the dropout model included earlier measurements than just the previous, then earlier *increments* may act as a proxy, and remove much of the apparent difference between the MAR and NMAR fits. As a final point on the nature of the estimated dropout model in such NMAR models, the log-likelihood surface for the dropout parameters is typically awkward in shape, with a strong ridge, or even multi-modality. This again reflects the fact that information on the dropout parameters arises in a non-standard way in such settings. The log-likelihood surface (figure not shown; see Molenberghs et al. 2001) around the maximum for ψ_1 and ψ_2 is strongly ridged in the direction of the difference of these two. Finally we note that the choice of model does have a non-trivial impact on the regression parameters in the Gaussian part of the model: the NMAR dropout model predicts a smaller average increment in yield (Δ), with larger second year variance and smaller correlation caused by greater negative imputed differences between yields.

The type of numerical issues observed here have also been noted elsewhere for this class of model. For example, Freedman and Sekhon (2010) present a detailed investigation for the original Heckman model (4.2), which uses the probit model for the missing value process in a cross-sectional setting. They provide a careful study of identifiability in such models and explore methods of estimation, concluding that Heckman's two-stage procedure should be avoided, and also identify numerical problems surrounding the use of full maximum likelihood, such as multiple and spurious maxima.

We now return to the second reason for caution in interpreting the results from such NMAR selection models. We first take a closer look at the raw data and the predictive behavior of the Gaussian NMAR model. Under an NMAR model, the predicted, or imputed, value of a missing observation is given by the ratio of expectations:

$$\widehat{\boldsymbol{Y}}^{\text{m}} = \frac{E_{\boldsymbol{Y}^{\text{m}}|\boldsymbol{Y}^{\text{o}}}\left[\boldsymbol{Y}^{\text{m}}\Pr(\boldsymbol{R} \mid \boldsymbol{Y}^{\text{o}}, \boldsymbol{Y}^{\text{m}})\right]}{E_{\boldsymbol{Y}^{\text{m}}|\boldsymbol{Y}^{\text{o}}}\left[\Pr(\boldsymbol{R} \mid \boldsymbol{Y}^{\text{o}}, \boldsymbol{Y}^{\text{m}})\right]}. \tag{4.12}$$

Recall that the fitted dropout model (4.10) implies that the probability of mastitis increases with decreasing values of the increment $Y_2 - Y_1$. We therefore plot the 27 imputed values of this quantity together with the 80 observed increments against the first year yield Y_1. This is presented in Figure 4.1, in which the imputed values are indicated with triangles and the observed values with crosses. Note how the imputed values are almost linear in Y_1: This is a well-known property of the ratio (4.12) within this range of observations. The imputed values are all negative, in contrast to the observed increments, which are nearly all positive. With animals of this age, one would normally expect an increase in yield between

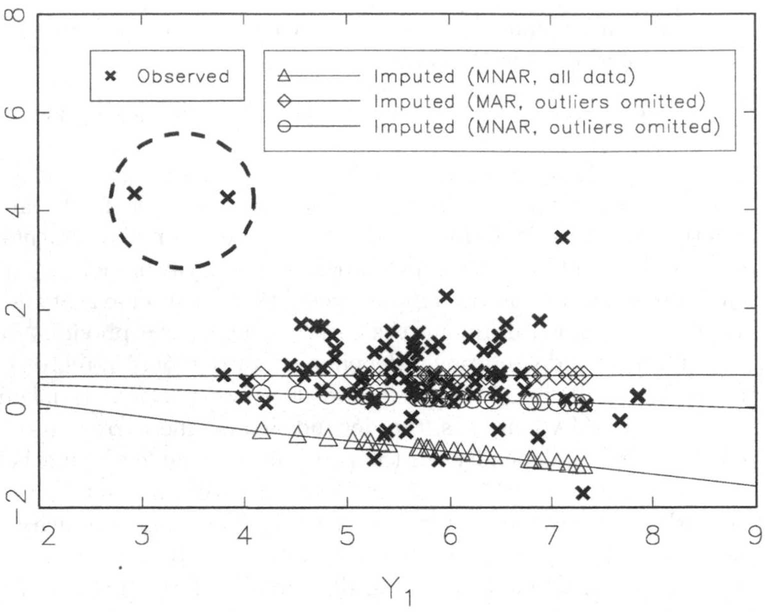

FIGURE 4.1
Mastitis in Dairy Cattle. Plot of observed and imputed year 2 − year 1 yield differences against year 1 yield. Two outlying points are circled.

the 2 years. The dropout model is imposing very atypical behavior on these animals and this corresponds to the statistical significance of the NMAR component of the model (ψ_2).

Another feature of this plot is the pair of outlying observed points circled in the top left-hand corner. These two animals have the lowest and third lowest yields in the first year, but moderately large yields in the second, leading to the largest positive increments. In a well-husbanded dairy herd, one would expect approximately Gaussian joint milk yields, and these two then represent outliers. It is likely that there is some anomaly, possibly illness, leading to their relatively low yields in the first year. One can conjecture that these two animals are the cause of the structure identified by the Gaussian NMAR model. Under the joint Gaussian assumption, the NMAR model essentially "fills in" the missing data to produce a complete Gaussian distribution. To counterbalance the effect of these two extreme positive increments, the dropout model predicts negative increments for the mastitic cows, leading to the results observed. As a check on this conjecture, we omit these two animals from the dataset and refit the MAR and NMAR Gaussian models. The resulting estimates are also presented in Table 4.4.

The difference in log-likelihoods is very small indeed and the NMAR model now shows no improvement in fit over MAR. The estimates of the dropout parameters, although still moderately large in an absolute sense, are of the same size as their standard errors. In the absence of the two anomalous animals, the structure identified earlier in terms of the NMAR dropout model no longer exists. The increments imputed by the fitted model are also plotted in Figure 4.1, indicated by circles. Although still lying among the lower region of the observed increments, these are now all positive and lie close to the increments imputed by the MAR model (diamonds). Thus, we have a plausible representation of the data in terms of joint Gaussian milk yields, two pairs of outlying yields and no requirement for an NMAR dropout process.

The two key assumptions underlying the non-random selection model are, first, the form chosen for the relationship between dropout (or missingness) probability and partially observed outcome and, second, the distribution of the response or, more precisely, the conditional distribution of the possibly unobserved outcome given the observed outcome. In the current setting for the first assumption, if there is dependence of mastitis occurrence on yield, experience with logistic regression tells us that the exact form of the link function in this relationship is unlikely to be critical. In terms of the role of the model in the conclusions reached, we therefore consider the second assumption, the distribution of the response.

All the data from the first year are available, and a normal probability plot of these (not shown here) does not show great departures from the Gaussian assumption. Leaving this distribution unchanged, we therefore examine the effect of changing the conditional distribution of Y_2 given Y_1. One simple and obvious choice is to consider a heavy-tailed distribution, and for this, we use the translated and scaled t_m-distribution with density:

$$f(Y_2 \mid Y_1) = \{\sigma\sqrt{m}B(1/2, m/2)\}^{-1} \left\{1 + \frac{1}{m}\left(\frac{Y_2 - \mu_{2|1}}{\sigma}\right)^2\right\}^{-(m+1)/2},$$

where

$$\mu_{2|1} = \mu + \Delta + \frac{\rho\sigma_2(Y_1 - \mu)}{\sigma_1}$$

is the conditional mean of $Y_2 \mid Y_1$. The corresponding conditional variance is

$$\frac{m}{m-2}\sigma^2.$$

Relevant parameter estimates from the fits of both MAR and NMAR models are presented in Table 4.4 as well for three values of m: 2, 10, and 25. Smaller values of m correspond to greater kurtosis and, as m becomes large, the model approaches the Gaussian one used in the previous section. It can be seen from the results for the NMAR model in Table 4.4, that as the kurtosis increases the estimate of ψ_2 decreases. Also, the maximized likelihoods of the MAR and NMAR models converge. With 10 and 2 degrees-of-freedom, there is no evidence at all to support the inclusion of ψ_2 in the model; that is, the MAR model provides as good a description of the observed data as the NMAR, in contrast to the Gaussian-based conclusions. Further, as m decreases, the estimated yearly increment in milk yield Δ from the NMAR model increases to the value estimated under the MAR model. In most applications of outcome-based selection models, it will be quantities of this type that will be of prime interest, and it is clearly seen in this example how the dropout model can have a crucial influence on the estimate of this. Comparing the values of the deviance from the t-based model with those from the original Gaussian model, we also see that the former with $m = 10$ or 2 produces a slightly better fit, although no meaning can be attached to the statistical significance of the difference in these likelihood values.

The results observed here are consistent with those from the deletion analysis. The two outlying pairs of measurements identified earlier are not inconsistent with the heavy-tailed t-distribution; so it would require no "filling in" and hence no evidence for non-randomness in the dropout process under the second model. In conclusion, if we consider the data with outliers included, we have two models that effectively fit equally well to the observed data. The first assumes a joint Gaussian distribution for the responses and a NMAR dropout model. The second assumes a Gaussian distribution for the first observation and a conditional t_m-distribution (with small m) for the second given the first, with no requirement for a NMAR dropout component. Each provides a different explanation for what has been observed, with quite a different biological interpretation. In likelihood terms, the second model fits a little better than the first, but a key feature of such dropout models is that the distinction between them should not be based on the observed data likelihood alone. It is always possible to specify models with identical maximized observed data likelihoods that differ with respect to the unobserved data and dropout mechanism and such models can have very different implications for the underlying mechanism generating the data (Molenberghs et al. 2008). Finally, the most plausible explanation for the observed data is that the pairs of milk yields have joint Gaussian distributions, with no need for an NMAR dropout component, and that two animals are associated with anomalous pairs of yields.

Although we have illustrated features of a likelihood-based analysis for a typical continuous outcome NMAR selection model using a particularly simple example, all the issues observed arise in more general and complex settings. For these however, it can be less obvious how the assumptions made in the NMAR model influence the conclusions drawn. In practice therefore such models have found their main use as part of sensitivity analyses — the subject of the chapters in Part V of this book.

4.3 Pattern-Mixture Models

4.3.1 Introduction

We recall from Section 4.1 that we can summarize the pattern-mixture decomposition as follows, for a general missing value process $\boldsymbol{R}$ and covariates X:

$$f(\boldsymbol{Y},\boldsymbol{R} \mid X) = \Pr(\boldsymbol{R} \mid X) f(\boldsymbol{Y} \mid \boldsymbol{R}, X).$$

The right-hand component $\Pr(\boldsymbol{R} \mid X)$ can be estimated directly from the observed data, provided that the sample is sufficiently large to provide information on less common patterns $\boldsymbol{R}$. Pattern-mixture models are rarely used in practice for arbitrary patterns of missingness, because of the proliferation of potential patterns, which obviously does not happen when missingness is restricted to dropout. We note in passing that arbitrary missingness patterns can also be an issue in selection models: there, the appropriate formulation of a missingness model is the issue. It follows that it is the choice, and estimation of, distributions $f(\boldsymbol{Y} \mid \boldsymbol{R}, X)$ that raise most of the issues with pattern-mixture models. Typically it is very clear what components of these distributions can, and cannot, be estimated without additional external assumptions. To illustrate this point we begin with a very simple example of a pattern-mixture decomposition, and follow this with a paradox. Early references to pattern-mixture models include Rubin (1977), who even mentioned the concept of a sensitivity analysis, Glynn, Laird, and Rubin (1986), and Little and Rubin (2002). Important early development was provided by Little (1993, 1994, 1995).

We consider a trivariate Gaussian outcome, with dropout, where D_i can take values 1 and 2 for dropouts and 3 for completers. A pattern-mixture model implies a different distribution for each time of dropout. We can write

$$\boldsymbol{Y}_i \mid D_i = d \quad \sim \quad N\{\boldsymbol{\mu}(d), \Sigma(d)\}, \tag{4.13}$$

where

$$\boldsymbol{\mu}_d = \begin{bmatrix} \mu_1(d) \\ \mu_2(d) \\ \mu_3(d) \end{bmatrix} \quad \text{and} \quad \Sigma_d = \begin{bmatrix} \sigma_{11}(d) & \sigma_{21}(d) & \sigma_{31}(d) \\ \sigma_{21}(d) & \sigma_{22}(d) & \sigma_{32}(d) \\ \sigma_{31}(d) & \sigma_{32}(d) & \sigma_{33}(d) \end{bmatrix},$$

for $d = 1, 2, 3$. Let $P(d) = \pi_d = \Pr(D_i = d \mid \boldsymbol{\psi})$, then the marginal distribution of the response is a mixture of normals with overall mean

$$\boldsymbol{\mu} = \sum_{d=1}^{3} \pi_d \boldsymbol{\mu}_d.$$

The variance for an estimator of $\boldsymbol{\mu}$ can be derived by application of the delta method, given we have variances and covariances for the estimated component means $\boldsymbol{\mu}_d$.

As noted above, the π_d are simply estimated from the observed proportions in each dropout group. However, of the parameters in the outcome part of the model, only 16 of the 27 can be identified from the data without making further assumptions. These 16 comprise all 9 parameters from the completers ($d = 3$) plus those from the following two sub-models. For $d = 2$,

$$N\left(\begin{bmatrix} \mu_1(2) \\ \mu_2(2) \end{bmatrix}; \begin{bmatrix} \sigma_{11}(2) & \sigma_{12}(2) \\ \sigma_{21}(2) & \sigma_{22}(2) \end{bmatrix} \right),$$

is identified, and for $d = 1$,

$$N\left(\mu_1(1); \sigma_{11}(1)\right)$$

contains the estimable parameters. This is a *saturated* pattern-mixture model and the above representation makes it very clear what information each dropout group provides and, consequently, the assumptions that need to be made if we are to predict the behavior of the unobserved responses, and so obtain marginal models for the response. If the three sets of parameters $\boldsymbol{\mu}_d$ are simply equated, with the same holding for the corresponding variance components, then this implies that dropout is MCAR. Progress can be made with less stringent restrictions however, and we return to this below in Section 4.3.2. From a practical perspective the way in which this external information is introduced needs to be guided by the context, and we shall see various examples of this later in this section, and also in later chapters, particularly Chapters 16, 18, 19, and 22.

An obvious consideration is the relationship between the assumptions required to make a pattern-mixture model identified, and those seen earlier for a selection model. Little (1995) and Little and Wang (1996) consider the restrictions implied by a selection dropout model in the pattern-mixture framework. For example, with two time points and a Gaussian response, Little proposes a general form of dropout model:

$$\Pr(\text{dropout} \mid \boldsymbol{Y}) = g(Y_1 + \lambda Y_2), \tag{4.14}$$

with the function $g(\cdot)$ left unspecified. We have seen above that in a selection modeling context, (4.14) is often assumed to have a logistic form, as in the Diggle and Kenward model for example. This relationship implies that the conditional distribution of Y_1 given $Y_1 + \lambda Y_2$ is the same for those who drop out and those who do not. With this restriction and given λ, the parameters of the full distribution of the dropouts is identified. Indeed, as Little and Wang (1996) recognize, this forms the basis of one approach to sensitivity analysis based on selection models, with λ as a sensitivity parameter. Our main interest now, however, is to compare the sources of identifiability in the pattern-mixture and selection models. In the former, the information comes from the assumption that the dropout probability is some function of a linear combination of the two observations with known coefficients. In the latter, it comes from the shape of the assumed conditional distribution of Y_2 given Y_1 (typically Gaussian), together with the functional form of the dropout probability. The difference is highlighted if we consider a sensitivity analysis for the selection model that varies λ in the same way as with the pattern-mixture model. Such sensitivity analysis is much less convincing because the data can, through the likelihood, distinguish between the fit associated with different values of λ. Some observations are also made on this point in Chapter 19. Therefore, identifiability problems in the selection context tend to be masked; there are always unidentified parameters, although a related "problem" seems absent in the selection model. This apparent paradox has been noted by Glynn, Laird, and Rubin (1986). It is assumed again that there are two measurements with dropout, i.e., Y_1 is always observed and Y_2 is either observed ($d = 2$) or missing ($d = 1$). The notation is simplified further by suppressing dependence on parameters and additionally adopting the following definitions:

$$\begin{aligned} g(d|Y_1, Y_2) &:= f(d|Y_1, Y_2), \\ p(d) &:= f(d), \\ f_d(Y_1, Y_2) &:= f(Y_1, Y_2|d). \end{aligned}$$

Equating the selection model and pattern-mixture model factorizations yields

$$\begin{aligned} f(Y_1, Y_2)g(d = 2|Y_1, Y_2) &= f_2(Y_1, Y_2)p(d = 2), \\ f(Y_1, Y_2)g(d = 1|Y_1, Y_2) &= f_1(Y_1, Y_2)p(d = 1). \end{aligned}$$

Since we have only two patterns, this obviously simplifies further to

$$\begin{aligned} f(Y_1, Y_2)g(Y_1, Y_2) &= f_2(Y_1, Y_2)p, \\ f(Y_1, Y_2)[1 - g(Y_1, Y_2)] &= f_1(Y_1, Y_2)[1 - p], \end{aligned}$$

of which the ratio produces

$$f_1(Y_1, Y_2) = \frac{1 - g(Y_1, Y_2)}{g(Y_1, Y_2)} \frac{p}{1 - p} f_2(Y_1, Y_2).$$

All selection model factors are identified, as are the pattern-mixture quantities on the right-hand side. However, the left-hand side is not entirely identifiable. We can further separate the identifiable from the unidentifiable quantities:

$$f_1(Y_2|Y_1) = f_2(Y_2|Y_1) \frac{1 - g(Y_1, Y_2)}{g(Y_1, Y_2)} \frac{p}{1 - p} \frac{f_2(Y_1)}{f_1(Y_1)}. \quad (4.15)$$

In other words, the conditional distribution of the second measurement given the first one, *in the incomplete first pattern*, about which there is no information in the data, is identified by equating it to its counterpart from the complete pattern, modulated via the ratio of the "prior" and "posterior" odds for dropout [$p/(1-p)$ and $g(Y_1, Y_2)/(1-g(Y_1, Y_2))$, respectively] and via the ratio of the densities for the first measurement.

Thus, although an identified selection model is seemingly less arbitrary than a pattern-mixture model, it incorporates *implicit* restrictions. Indeed, it is precisely these that are used in (4.15) to identify the component for which there is no information.

We can distinguish two main routes to identifying pattern-mixture models. The first, which is arguably less used in practice, uses outcome models that are sufficiently constrained so that they can be identified within the different dropout patterns. The main class of these is the so-called *Random Coefficient Pattern-Mixture* model which is based on the Gaussian linear mixed model. The different patterns constitute an additional categorical grouping variable, and interactions between this and the other linear model parameters imply different models in the different patterns (Hogan and Laird 1997a, 1997b). Such a model can be fit using standard tools, and so has the advantage of convenience. The problem is that the constraints required to make the model estimable in all the different dropout patterns implies the use of polynomial extrapolation, which may be hard to justify from a substantive perspective. See for example the discussion in Demirtas and Schafer (2003) and Molenberghs and Kenward (2007, Section 16.4). The second, much more widely used route, uses the broad idea of identifying restrictions, and we consider this now.

4.3.2 Identifying restrictions

Rather than build models that can be estimated within the incomplete patterns and then extrapolate, one can instead "borrow" the unidentifiable distributional information from other patterns. This leads to a class of methods that fall under the general heading of identifying restrictions. These methods differ according to the particular rules used to borrow the information. Perhaps the simplest example of this is the one in which the unidentifiable conditional distributions are all borrowed from the completers (Little 1993, 1994a), sometimes called complete case restrictions. Alternatively, one can set equal all distributions, estimable or not, that condition on the same set of observations. The implied restrictions are called the available case restrictions and it has been shown that these correspond to MAR in the pattern-mixture framework, for monotone missing value patterns (Molenberghs et al. 1998). Nearly all development of pattern-mixture models has been done within the monotone setting, in practice usually associated with dropout in longitudinal studies, and we restrict ourselves to this in the following. To simplify notation we drop the unit index i. It follows that there are $t = 1, \ldots, n = T$ dropout patterns where the dropout indicator, is $D = t + 1$. The indices j for measurement occasions and t for dropout patterns assume the same values, but using both simplifies notation.

For pattern t, the complete data density, is given by

$$f_t(Y_1, \ldots, Y_{\mathrm{T}}) = f_t(Y_1, \ldots, Y_t) f_t(Y_{t+1}, \ldots, Y_{\mathrm{T}} | Y_1, \ldots, Y_t). \tag{4.16}$$

The first factor is clearly identified from the observed data, while the second factor is not. It is assumed that the first factor is known or, more realistically, modeled using the observed data. Then, identifying restrictions are applied in order to identify the second component.

Although, in principle, completely arbitrary restrictions can be used by means of any valid density function over the appropriate support, strategies that imply links back to the observed data are likely to have more practical relevance. One can base identification on all patterns for which a given component, Y_s say, is identified. A general expression for this is

$$f_t(Y_s | Y_1, \ldots Y_{s-1}) = \sum_{j=s}^{T} \omega_{sj} f_j(Y_s | Y_1, \ldots Y_{s-1}), \quad s = t+1, \ldots, T. \tag{4.17}$$

We will use $\boldsymbol{\omega}_s$ as shorthand for the set of ω_{sj}'s used, the components of which are typically positive. Every $\boldsymbol{\omega}_s$ that sums to one provides a valid identification scheme.

Let us incorporate (4.17) into (4.16):

$$f_t(Y_1,\ldots,Y_{\mathrm{T}}) = f_t(Y_1,\ldots,Y_t) \prod_{s=0}^{T-t-1} \left[\sum_{j=T-s}^{T} \omega_{\mathrm{T}-s,j} f_j(Y_{\mathrm{T}-s}|Y_1,\ldots,Y_{\mathrm{T}-s-1}) \right]. \tag{4.18}$$

We will consider three special but important cases. Little (1993) proposes CCMV (complete case missing values) which uses the following identification:

$$f_t(Y_s|Y_1,\ldots Y_{s-1}) = f_{\mathrm{T}}(Y_s|Y_1,\ldots Y_{s-1}), \quad s = t+1,\ldots,T, \tag{4.19}$$

corresponding to $\omega_{sT} = 1$ and all others zero. In other words, information which is unavailable is always borrowed from the completers. Alternatively, the nearest identified pattern can be used:

$$f_t(Y_s|Y_1,\ldots Y_{s-1}) = f_s(Y_s|Y_1,\ldots Y_{s-1}), \qquad s = t+1,\ldots,T, \tag{4.20}$$

corresponding to $\omega_{ss} = 1$ and all others zero. We will refer to these restrictions as *neighboring case missing values.*

The third special case is ACMV, mentioned briefly above. The corresponding $\boldsymbol{\omega}_s$ vectors can be shown (Molenberghs et al. 1998) to have components:

$$\omega_{sj} = \frac{\alpha_j f_j(Y_1,\ldots,Y_{s-1})}{\sum_{\ell=s}^{T} \alpha_\ell f_\ell(Y_1,\ldots,Y_{s-1})}, \tag{4.21}$$

$(j = s,\ldots,T)$ where α_j is the fraction of observations in pattern j (Molenberghs et al. 1998).

This MAR–ACMV link connects the selection and pattern-mixture families. It is of further interest to consider specific sub-families of the NMAR family. In the context of selection models for longitudinal data, one typically restricts attention to a class of mechanisms where dropout may depend on the current, possibly unobserved, measurement, but not on future measurements. The entire class of such models will be termed missing non-future dependent (MNFD). Although they are natural and easy to consider in a selection model situation, there exist important examples of mechanisms that do not satisfy MNFD, such as a broad class of shared-parameter models (see Section 4.4).

Kenward, Molenberghs, and Thijs (2003) have shown there is a counterpart to MNFD in the pattern-mixture context. The conditional probability of pattern t in the MNFD selection models obviously satisfies

$$f(d = t|Y_1,\ldots,Y_T) = f(d = t|Y_1,\ldots,Y_{t+1}). \tag{4.22}$$

Within the PMM framework, we define non-future dependent missing value restrictions (NFMV) as follows:

$$\begin{aligned} \forall t \geq 2, \forall j < t-1 &\quad : \\ f(Y_t|Y_1,\ldots,Y_{t-1},d=j) &= f(Y_t|Y_1,\ldots,Y_{t-1},d \geq t-1). \end{aligned} \tag{4.23}$$

NFMV is not a single set of restrictions, but rather leaves one conditional distribution per incomplete pattern unidentified:

$$f(Y_{t+1}|Y_1,\ldots,Y_t,d=t). \tag{4.24}$$

SEM	:	MCAR	$\subset$	MAR	$\subset$	MNFD	$\subset$	general NMAR
		$\updownarrow$		$\updownarrow$		$\updownarrow$		$\updownarrow$
PMM	:	MCAR	$\subset$	ACMV	$\subset$	NFMV	$\subset$	general NMAR
					$\cap$	$\neq$	$\cup$	
						interior		

FIGURE 4.2
Relationship between nested families within the selection model (SEM) and pattern-mixture model (PMM) families. MCAR: missing completely at random; MAR: missing at random; NMAR: missing not at random; MNFD: missing non-future dependence; ACMV: available-case missing values; NFMV: non-future missing values; interior: restrictions based on a combination of the information available for other patterns. The $\subset$ *symbol here indicates "is a special case of." The* $\updownarrow$ *symbol indicates correspondence between a class of SEM models and a class of PMM models.*

In other words, the distribution of the "current" unobserved measurement, given the previous ones, is unconstrained. Note that (4.23) excludes such mechanisms as CCMV and NCMV. Kenward, Molenberghs, and Thijs (2003) have shown that, for longitudinal data with dropouts, MNFD and NFMV are equivalent.

For pattern t, the complete data density is given by

$$\begin{aligned} f_t(Y_1,\ldots,Y_{\mathrm{T}}) &= f_t(Y_1,\ldots,Y_t)f_t(Y_{t+1}|Y_1,\ldots,Y_t) \\ &\quad \times f_t(Y_{t+2},\ldots,Y_T|Y_1,\ldots,Y_{t+1}). \end{aligned} \tag{4.25}$$

It is assumed that the first factor is known or, more realistically, modeled using the observed data. Then, identifying restrictions are applied to identify the second and third components. First, from the data, estimate $f_t(Y_1,\ldots,Y_t)$. Second, the user has full freedom to choose

$$f_t(Y_{t+1}|Y_1,\ldots,Y_t). \tag{4.26}$$

Substantive considerations could be used to identify this density. Alternatively, a family of densities might be considered by way of sensitivity analysis. Third, using (4.23), the densities $f_t(Y_j|Y_1,\ldots,Y_{j-1})$, $(j \geq t+2)$ are identified. This identification involves not only the patterns for which Y_j is observed, but also the pattern for which Y_j is the current and hence the first unobserved measurement. An overview of the connection between selection and pattern-mixture models is given in Figure 4.2.

Two obvious mechanisms, within the MNFD family but outside MAR, are NFD1 (NFD standing for "non-future dependent"), i.e., choose (4.26) according to CCMV, and NFD2, i.e., choose (4.26) according to NCMV. NFD1 and NFD2 are strictly different from CCMV and NCMV. For further details see Kenward, Molenberghs, and Thijs (2003).

The type of identifying restrictions so far considered might be termed "within-group." They make no explicit reference to other structures in the data, particularly covariates. Another class of restrictions, which has proved very useful in the longitudinal trial setting, instead equates conditional distributions from *different* treatment groups with the aim of representing subjects who deviate from the treatment regime set out in the protocol. Suppose that there are now G groups, and we distinguish the joint distributions from the different groups by the superscript g:

$$f^g(Y_1,\ldots,Y_{\mathrm{T}}),\ \ g = 1,\ldots,G.$$

TABLE 4.5
Alzheimer Study. Sample size per treatment arm and drop-out pattern.

Pattern	1	2	3	4	5	6	7
Treatment 1	4	5	16	3	9	6	71
Treatment 2	4	9	7	6	3	5	81
Treatment 3	12	4	15	9	5	3	67

Suppose that a subject drops out in group g at visit t, and moves to another treatment, g' say. Then in some situations it might be sensible to replace the non-estimable conditional distribution

$$f^g(Y_{t+1}|Y_1,\ldots,Y_t,d=t).$$

by one that is estimable using information from those who continue in group g':

$$f^{g'}(Y_{t+1}|Y_1,\ldots,Y_t,d>t).$$

It can be seen that this is simply another form of identification. Whether this would be sensible in any given setting depends very much on the context, especially the goal of the analysis, the nature of the outcome measurement and the actions of the interventions. Many variations on this basic theme are possible, however, which can all be considered as alternative ways of identifying the non-estimable conditional distributions from the observed data. Analyses based on this broad approach have been developed by Little and Yau (1996), Kenward and Carpenter (2009) and Carpenter, Roger, and Kenward (2013), and are described in Chapter 19 of this volume.

Estimation and inference for such "identified" pattern-mixture models follow similar routes however, whether within-group as seen earlier or using between-group borrowing as just described. Multiple imputation provides a particularly convenient, although not essential, tool for this; see for example Thijs et al. (2002). The principal application of pattern-mixture models lies in the construction of non-random models for use in sensitivity analysis. They have a range of roles for this, and these will be explored in several later chapters in Part V of this volume.

4.3.3 An example

To illustrate the use of identifying restrictions in pattern-mixture models, we use data from a three-armed clinical trial involving patients with Alzheimer's disease (Reisberg et al. 1987), conducted by 50 investigators in 8 countries. The outcome is a dementia score, ranging from 0 to 43. Treatment arm 1 is placebo, with 114 patients, while arms 2, with 115 patients, and 3, with 115 patients, involve active compounds. Of the patient population, 56.4% are female. Measurements are taken at baseline, at weeks 1, 2 and then every two weeks until week 12. In agreement with the protocol, we will analyze change versus baseline. This outcome is sufficiently close to normality, unlike the raw score.

Attrition over time is fairly steady for each treatment arm. The sample size per drop-out pattern and per treatment arm is displayed in Table 4.5. In each of the arms, about 40% drop out before the end of the study. Unfortunately, very little is known about the reasons for drop-out, in this particular study. While such information is generally important, one also needs to be able to analyse incomplete data in the absence of such knowledge.

In Verbeke and Molenberghs (2000), a conventional linear mixed model was fitted to the

TABLE 4.6
Alzheimer Study. Inference for treatment contrasts. For the contrasts, parameter estimates and standard errors, in parentheses, are reported.

Pat.	Cont.	ACMV	CCMV	NCMV	NFD1	NFD2
			Stratified Analysis			
1	1	9.27(6.42)	5.84(5.16)	-4.19(6.27)	4.90(8.29)	5.44(6.52)
	2	−8.19(6.58)	−6.92(6.15)	2.56(5.12)	−7.78(7.62)	−4.48(7.76)
2	1	2.78(4.75)	−0.00(2.90)	−4.43(3.54)	0.61(4.88)	−1.49(4.07)
	2	−3.57(4.53)	−5.08(3.92)	−1.37(4.12)	−6.48(5.22)	−4.54(5.46)
3	1	6.78(4.20)	6.95(2.66)	0.10(2.40)	4.18(2.64)	0.18(3.65)
	2	−1.75(2.76)	−3.44(2.12)	0.83(2.14)	−2.66(2.29)	−0.10(2.20)
4	1	11.05(3.21)	10.87(2.85)	6.59(3.09)	9.65(3.56)	9.97(2.90)
	2	−3.84(4.09)	−6.55(3.88)	−3.23(4.09)	−6.84(3.78)	−4.30(4.24)
5	1	0.15(5.71)	−2.05(6.29)	−5.60(6.46)	−3.02(5.92)	−6.13(6.42)
	2	−0.74(3.99)	−0.87(4.51)	0.92(4.68)	−0.53(4.24)	1.05(4.57)
6	1	14.16(3.75)	12.91(3.71)	13.44(3.72)	13.28(3.82)	12.72(3.79)
	2	−5.24(3.48)	−4.74(3.69)	−4.95(3.79)	−4.71(3.63)	−4.77(3.70)
7	1	−0.99(0.85)	−0.99(0.85)	−0.99(0.85)	−0.99(0.85)	−0.99(0.85)
	2	1.68(0.88)	1.68(0.88)	1.68(0.88)	1.68(0.88)	1.68(0.88)
F		2.45	2.96	1.76	1.92	1.77
p		0.0024	0.0002	0.0407	0.0225	0.0413
			Marginal Analysis			
	1	1.97(1.05)	1.47(0.87)	−0.48(0.85)	1.05(1.04)	0.37(0.96)
	2	−0.24(0.81)	−0.56(0.86)	0.91(0.77)	−0.59(1.01)	0.19(0.84)
F		2.15	1.23	0.66	0.52	0.19
p		0.1362	0.3047	0.5208	0.6043	0.8276

Pat., pattern; Cont., contrast; ACMV, available-case missing values; CCMV, complete-case missing values; NCMV, neighbouring-case missing values; NFD1, Case 4; NFD2, Case 5.

outcomes, in which the variance structure was modeled by means of a random subject effect, an exponential serial correlation process and measurement error. The fixed effects considered in the model were, apart from treatment effect, those of age, time, investigator and country, as well as 2- and 3-way interactions. From an initial model-selection exercise, only main effects of age, time, time^2 and treatment group were retained. Scientific interest is in the effect of treatment. Since there are three arms, we consider two treatment contrasts of the experimental arms versus the standard arm. Our focus here will be on estimates and standard errors for these contrasts, as well as on tests for the null hypothesis of no treatment effect.

First, pattern-specific linear mixed models are fitted to the data. The fixed-effects structure comprises treatment indicators, time and its square, and age. The covariance structure is captured by means of a random subject effect, an exponential serial correlation process and measurement error. Second, five instances of identifying restrictions are applied: (a) CCMV, as in (4.19), (b) NCMV, as in (4.20), and (c) ACMV, as in (4.17)–(4.18), and (d) and (e) the NFD counterparts to CCMV, and NCMV, respectively, labeled NFD1 and NFD2 in Section 4.3.2. Third, from these choices, the conditional distributions of the unobserved outcomes, given the observed ones, are constructed. Fourth, inferences are drawn based on multiple imputation; five imputations are drawn from these conditionals and the resulting multiply imputed datasets are analyzed, using the same pattern-mixture model as was applied to the incomplete data. For further details on multiple imputation we refer to Part IV of this volume.

Given the nature of the model, the natural parameters are pattern-specific. When there is also interest in effects marginalized over patterns, some extra steps are needed. Let $\theta_{\ell t}$ represent the treatment contrasts $\ell = 1, 2$ in pattern $t = 1, \ldots, 7$, and let π_t be the proportion of patients in pattern t, easily derived from Table 4.5. Then the marginal contrasts are

$$\theta_\ell = \sum_{t=1}^{7} \theta_{\ell t}\pi_t, \qquad \ell = 1, 2. \tag{4.27}$$

The marginalized within-imputation variance is obtained using the delta method:

$$\widetilde{W} = (A_1|A_2)\left(\begin{array}{c|c} W & 0 \\ \hline 0 & \mathrm{var}(\pi_t) \end{array}\right)\left(\begin{array}{c} A_1' \\ \hline A_2' \end{array}\right), \tag{4.28}$$

where

$$A_1 = \frac{\partial(\theta_1, \theta_2)}{\partial(\theta_{11}, \theta_{12}, \theta_{21}, \ldots, \theta_{72})}, \qquad A_2 = \frac{\partial(\theta_1, \theta_2)}{\partial(\pi_1, \ldots, \pi_7)}.$$

Similarly, the between-imputation variance is

$$\widetilde{B} = A_2 B A_2'. \tag{4.29}$$

Expressions (4.28) and (4.29) are combined to produce the total marginal variance $\widetilde{V}$. These quantities can be used for precision estimation and hypothesis testing.

The results of our analysis are reported in Table 4.6. The marginal treatment effect assessments are all non-significant. However, all stratified treatment assessments produce significant p values, although to various levels of strength. Strong evidence is obtained from the available-case missing values model. Of course, the complete-case missing values analysis provides even stronger evidence, but this assumption may be unrealistic, since even patterns with few observations are completed using the set of completers, corresponding to pattern 7. Both of the other non-future missing values mechanisms, corresponding to the NFD restrictions, where drop-out does not depend on future unobserved values, provide mild evidence for treatment effect. Importantly, we are in a position to consider which patterns are responsible for an identified treatment effect. Note that the contrasts are nowhere near significant in the complete pattern 7, while patterns 4 and 6 seem to contribute to the effect, consistently across patterns. The first contrast of pattern 3 is significant only under complete-case missing values, perhaps explaining why this strategy yields the most significant result.

Figure 4.3 graphically summarizes the fit of these models for the first treatment arm; very similar displays for the other arms have been omitted. Clearly, the chosen identifying restrictions have a strong impact, especially for the patterns with earlier drop-out. Of course, from Table 4.5 it is clear that the earlier patterns are rather sparsely filled. It is striking to see that the missing non-future dependence patterns are not all grouped together. An important and perhaps counterintuitive feature is that the fitted averages depend on the identification method chosen, even at time points prior to drop-out. The reason for this is that, after imputation, a parametric model is fitted to the completed sequences as a whole, as opposed to, for example, change point models with change point at the time of drop-out. Hence, the smoothing induced by the parametric model applies across the entire sequence, before and after drop-out.

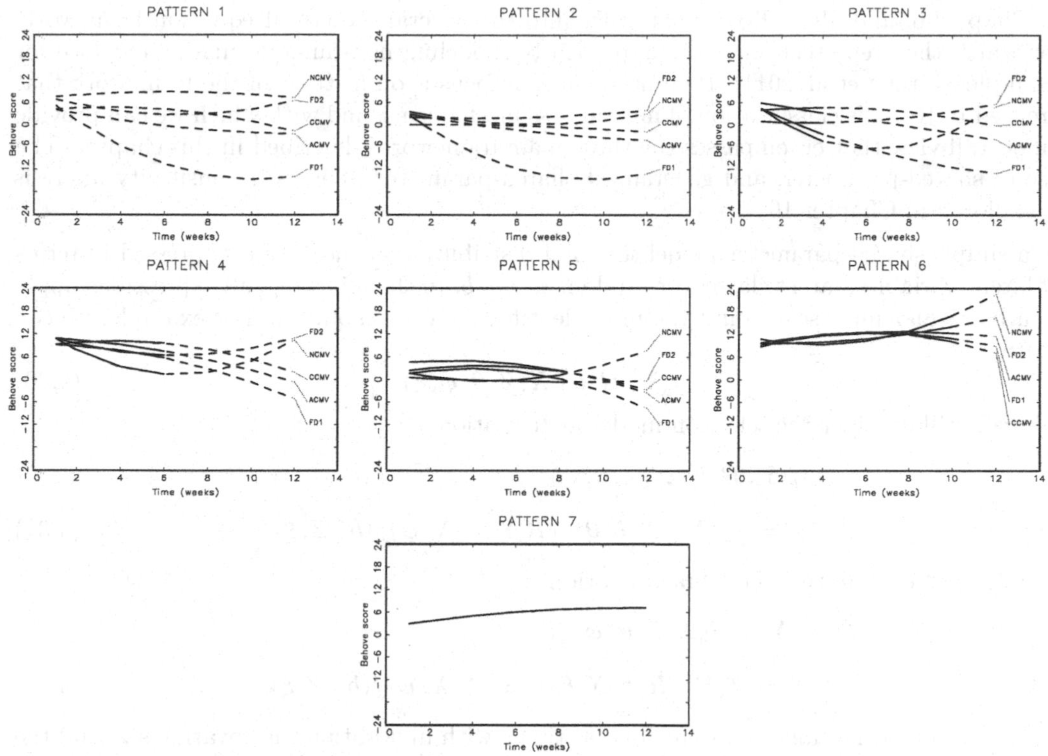

FIGURE 4.3
Alzheimer Study. Pattern-mixture models. Fitted average profiles for each of the five identification strategies. Treatment arm 1. ACMV, available-case missing values; CCMV, complete-case missing values; NCMV, neighbouring-case missing values; NFD1, Case 4; NFD2, Case 5.

4.4 Shared-Parameter Models

4.4.1 The simple shared-parameter framework

We have seen above that much of the interest in the selection and pattern-mixture frameworks lies in the ways in which they allow the formulation of non-random missingness models. In the former case this is achieved directly in terms of the missing data mechanism itself, and in the latter case, indirectly, through the posited distribution of the missing data. In both cases the models are expressed in terms of directly (or potentially) observable data. In contrast to these approaches, our third framework instead uses unobservable latent variables to link missingness with potentially missing data. Such models are very closely related to, and could even be considered a special case of, so-called *joint models*, the topic of Chapter 6, and also fall very naturally into the general structural equation framework, for which they represent the main approach to modeling non-random missingness (see for example Muthén et al. 2011). Here our principal focus is on features of the framework that are tied to Rubin's classification of missingness mechanisms, and with which we can provide an overall view that encompasses the three main frameworks described in this chapter. The use of shared-parameter, and generalized shared-parameter, models in sensitivity analysis is explored in Chapter 16.

In a simple shared-parameter model the joint distribution of the data is expressed in terms of latent variables, latent classes, or random effects $\boldsymbol{b}$. In the following, all expressions apply to a particular unit, so we omit the unit identifier from the notation. For example, we can write

$$f(\boldsymbol{Y}, \boldsymbol{R}, \boldsymbol{b} \mid X, Z, \boldsymbol{\theta}, \boldsymbol{\psi}, \boldsymbol{\xi}). \tag{4.30}$$

We can still consider the selection-model factorization

$$\begin{aligned} & f(\boldsymbol{Y}, \boldsymbol{R}, \boldsymbol{b} \mid X, Z, \boldsymbol{\theta}, \boldsymbol{\psi}) \\ = \; & f(\boldsymbol{Y} \mid X, \boldsymbol{b}, \boldsymbol{\theta}) f(\boldsymbol{R} | \boldsymbol{Y}, \boldsymbol{b}, X, \boldsymbol{\psi}) f(\boldsymbol{b} \mid Z, \boldsymbol{\xi}) \end{aligned} \tag{4.31}$$

and the pattern-mixture model factorization

$$\begin{aligned} & f(\boldsymbol{Y}, \boldsymbol{R}, \boldsymbol{b} | X, Z, \boldsymbol{\theta}, \boldsymbol{\psi}, \boldsymbol{\xi}) \\ = \; & f(\boldsymbol{Y} \mid \boldsymbol{R}, \boldsymbol{b}, X, \boldsymbol{\theta}) f(\boldsymbol{R} \mid \boldsymbol{b}, X, \boldsymbol{\psi}) f(\boldsymbol{b} \mid Z, \boldsymbol{\xi}). \end{aligned} \tag{4.32}$$

The notation is the same as in previous sections, with in addition the covariates Z and the parameters $\boldsymbol{\xi}$, which describe the random-effects distribution. Little (1995) refers to such decompositions as random-coefficient selection and pattern-mixture models, respectively.

Important early references to such models are Wu and Carroll (1988) and Wu and Bailey (1988, 1989). Wu and Carroll (1988) proposed such a model for what they termed informative right censoring. For a continuous response, Wu and Carroll (1988) suggested using a conventional Gaussian random-coefficient model combined with an appropriate model for time to dropout, such as proportional hazards, logistic, or probit regression. The combination of probit and Gaussian response allows explicit solution of the integral and was used in their application.

In a slightly different approach to modelling dropout time as a continuous variable in the random effects setting, Schluchter (1992) and DeGruttola and Tu (1994) proposed joint multivariate Gaussian distributions for the latent variable(s) of the response process and a variable representing time to dropout. The correlation between these variables induces dependence between dropout and response. To permit more realistic distributions for dropout time Schluchter proposes that dropout time itself should be some monotone transformation of the corresponding Gaussian variable. The use of a joint Gaussian representation does simplify computational problems associated with the likelihood. There are clear links here with the Tobit model which was used to introduce the selection framework in Section 4.2, and this is made explicit by Cowles, Carlin, and Connett (1996) who use a number of correlated latent variables to represent various aspects of an individual's behaviour, such as compliance and attendance at scheduled visits. Models of this type handle non-monotone missingness quite conveniently. There are many ways in which such models can be extended and generalized.

An important simplification arises when $\boldsymbol{Y}$ and $\boldsymbol{R}$ are assumed to be independent, conditional on the random effects $\boldsymbol{b}$. We then obtain the shared-parameter decomposition or, set out in full,

$$\begin{aligned} & f(\boldsymbol{Y}, \boldsymbol{R}, \boldsymbol{b} \mid X, Z, \boldsymbol{\theta}, \boldsymbol{\psi}, \boldsymbol{\xi}) \\ = \; & f(\boldsymbol{Y} \mid X, \boldsymbol{b}, \boldsymbol{\theta}) f(\boldsymbol{R} \mid \boldsymbol{b}, X, \boldsymbol{\psi}) f(\boldsymbol{b} \mid Z, \boldsymbol{\xi}). \end{aligned} \tag{4.33}$$

This route was followed by Follman and Wu (1995). Note that, when $\boldsymbol{b}$ is assumed to be discrete, a latent-class or mixture model follows.

It follows from the general form of a shared-parameter model, in particular the presence of the unobserved $\boldsymbol{b}$ in both the outcome and missing value models, that the resulting missing value mechanism is non-random, but the form of this mechanism is not immediately obvious in most cases.

From (4.33) it follows that the missing-value mechanism takes the form:

$$f(\boldsymbol{R} \mid \boldsymbol{Y}, X, Z, \boldsymbol{\theta}, \boldsymbol{\psi}, \boldsymbol{\xi}) = \frac{\int f(\boldsymbol{Y} \mid X, \boldsymbol{b}, \boldsymbol{\theta}) f(\boldsymbol{R} \mid \boldsymbol{b}, \boldsymbol{\psi}) f(\boldsymbol{b} \mid Z, \boldsymbol{\xi}) d\boldsymbol{b}}{\int\int f(\boldsymbol{Y} \mid X, \boldsymbol{b}, \boldsymbol{\theta}) f(\boldsymbol{R} \mid \boldsymbol{b}, \boldsymbol{\psi}) f(\boldsymbol{b} \mid Z, \boldsymbol{\xi}) d\boldsymbol{b} d\boldsymbol{r}.}$$

Obviously, given the dependence on $\boldsymbol{Y}$ as a whole, and therefore in particular on $\boldsymbol{Y}^m$, this represents, in general, an NMAR mechanism.

4.4.2 The generalized shared-parameter model

Following Creemers et al. (2010, 2011) we now consider an extension of the SPM model, which these authors term the *Generalized Shared-parameter Model (GSPM)*. In the GSPM it is assumed that there is a *set* of random-effects vectors $\boldsymbol{b} = (\boldsymbol{g}, \boldsymbol{h}, \boldsymbol{j}, \boldsymbol{k}, \boldsymbol{\ell}, \boldsymbol{m}, \boldsymbol{q})$, characterized by the components of the full-density factorization to which they apply, in the following way:

$$\begin{aligned} & f(\boldsymbol{Y}, \boldsymbol{R} | \boldsymbol{g}, \boldsymbol{h}, \boldsymbol{j}, \boldsymbol{k}, \boldsymbol{\ell}, \boldsymbol{m}, \boldsymbol{q}, \boldsymbol{\theta}, \boldsymbol{\psi}) \\ = \; & f(\boldsymbol{Y}^{\mathrm{o}} \mid \boldsymbol{g}, \boldsymbol{h}, \boldsymbol{j}, \boldsymbol{\ell}, \boldsymbol{\theta}) f(\boldsymbol{Y}^{\mathrm{m}} \mid \boldsymbol{Y}^{\mathrm{o}}, \boldsymbol{g}, \boldsymbol{h}, \boldsymbol{k}, \boldsymbol{m}, \boldsymbol{\theta}) \\ & \times f(\boldsymbol{R} \mid \boldsymbol{g}, \boldsymbol{j}, \boldsymbol{k}, \boldsymbol{q}, \boldsymbol{\psi}). \end{aligned} \tag{4.34}$$

This is the most general shared-parameter model that can be constructed in the sense that $\boldsymbol{g}$ is common to all three factors, $\boldsymbol{h}$, $\boldsymbol{j}$, and $\boldsymbol{k}$ are shared between a pair of factors, and $\boldsymbol{\ell}$, $\boldsymbol{m}$, and $\boldsymbol{q}$ are restricted to a single factor. The random effect $\boldsymbol{m}$ is not identifiable, because it merely describes the missing data. The same holds for $\boldsymbol{k}$ because it is aliased with $\boldsymbol{q}$, of which one is used twice, the other one in a single factor only. However, the occurrence of $\boldsymbol{k}$ in the middle factor does not separate it from $\boldsymbol{q}$, because the middle factor is unidentifiable. The same applies to $\boldsymbol{j}$ and $\boldsymbol{g}$, which are not separable. As a consequence these are of use only in the context of sensitivity analysis (see Chapter 16). In its full generality this model may appear rather contrived. However, as Creemers et al. (2011) stress, the objective of the model is not to provide a practical data analytic tool, but rather to provide the most general form of SPM from which substantively appropriate models follow as sub-classes. So, depending on the setting, one may choose to either retain all the sets of random effects or to omit some. The conventional SPM (4.33) is obtained in an obvious way by omitting all vectors except $\boldsymbol{g}$, i.e., $\boldsymbol{b} = \boldsymbol{g}$.

From the form of (4.34) it can appear at first sight that two different distributions are being assumed for the outcome vector, that is, the observed and missing components are divorced from each other. However, it can be seen that this is not the case because $\boldsymbol{g}$ and $\boldsymbol{h}$ still tie both factors together. The impact of $\boldsymbol{j}$, $\boldsymbol{k}$, $\boldsymbol{\ell}$, and $\boldsymbol{m}$ is to modify an individual's latent process in terms of missingness. In other words, the most general model assumes that observed and missing components are governed in part by common processes and partly by separate processes.

Turning to the longitudinal setting, we are now in a position to establish conditions on the GSPM under which, on the one hand, MAR holds, and on the other, missingness does not depend on future, unobserved measurements. This allows us to present Rubin's missing-data taxonomy in a symmetric way for all three frameworks: selection, pattern-mixture, and shared-parameter. It is assumed that we are conditioning on all covariates, which are completely observed, and so the following statements are made conditionally on these. In other settings with partially observed covariates, these definitions would have to be extended in an obvious way. Thus, we have, for MCAR

$$\begin{aligned}
\text{SM} &: & f(\boldsymbol{R} \mid \boldsymbol{Y}, \boldsymbol{\psi}) &= f(\boldsymbol{R} \mid \boldsymbol{\psi}),\\
\text{PMM} &: & f(\boldsymbol{Y} \mid \boldsymbol{R}, \boldsymbol{\theta}) &= f(\boldsymbol{Y} \mid \boldsymbol{\theta}),\\
\text{SPM} &: & f(\boldsymbol{Y}, \boldsymbol{R} \mid \boldsymbol{b}, \boldsymbol{\theta}, \boldsymbol{\psi}) &= f(\boldsymbol{Y} \mid \boldsymbol{\theta}) f(\boldsymbol{R} \mid \boldsymbol{\psi}),\\
\text{GSPM} &: & & f(\boldsymbol{Y}^{\text{o}} \mid \boldsymbol{g}, \boldsymbol{h}, \boldsymbol{j}, \boldsymbol{\ell}, \boldsymbol{\theta}) f(\boldsymbol{Y}^{\text{m}} \mid \boldsymbol{Y}^{\text{o}}, \boldsymbol{g}, \boldsymbol{h}, \boldsymbol{k}, \boldsymbol{m}, \boldsymbol{\theta}) f(\boldsymbol{R} \mid \boldsymbol{g}, \boldsymbol{j}, \boldsymbol{k}, \boldsymbol{q}, \boldsymbol{\psi})\\
&& &= f(\boldsymbol{Y}^{\text{o}} \mid \boldsymbol{h}, \boldsymbol{\ell}, \boldsymbol{\theta}) f(\boldsymbol{Y}^{\text{m}} \mid \boldsymbol{Y}^{\text{o}}, \boldsymbol{h}, \boldsymbol{m}, \boldsymbol{\theta}) f(\boldsymbol{R} \mid \boldsymbol{q}, \boldsymbol{\psi}).
\end{aligned}$$

We have seen that the mechanism is MAR when missingness depends on the observed outcomes and observed covariates but not, in addition, on the unobserved outcomes. In the various frameworks, MAR holds if and only if

$$\begin{aligned}
\text{SM} &: f(\boldsymbol{R} \mid \boldsymbol{Y}, \boldsymbol{\psi}) = f(\boldsymbol{R} \mid \boldsymbol{Y}^{\text{o}}, \boldsymbol{\psi}),\\
\text{PMM} &: f(\boldsymbol{Y}^{\text{m}} \mid \boldsymbol{Y}^{\text{o}}, \boldsymbol{R}, \boldsymbol{\theta}) = f(\boldsymbol{Y}^{\text{m}} \mid \boldsymbol{Y}^{\text{o}}, \boldsymbol{\theta}), \qquad (4.35)\\
\text{SPM} &: \int f(\boldsymbol{Y}^{\text{o}}, \boldsymbol{R}, \boldsymbol{b}) \left\{ f(\boldsymbol{Y}^{\text{m}} \mid \boldsymbol{Y}^{\text{o}}, \boldsymbol{b}) - f(\boldsymbol{Y}^{\text{m}} \mid \boldsymbol{Y}^{\text{o}}) \right\} d\boldsymbol{b} = 0,\\
\text{GSPM} &: \frac{\int f(\boldsymbol{Y}^{\text{o}} \mid \boldsymbol{g}, \boldsymbol{h}, \boldsymbol{j}) f(\boldsymbol{Y}^{\text{m}} \mid \boldsymbol{Y}^{\text{o}}, \boldsymbol{g}, \boldsymbol{h}, \boldsymbol{k}) f(\boldsymbol{R} \mid \boldsymbol{g}, \boldsymbol{j}, \boldsymbol{k}) f(\boldsymbol{b})\, d\boldsymbol{b}}{\int f(\boldsymbol{Y}^{\text{o}} \mid \boldsymbol{g}, \boldsymbol{j}) f(\boldsymbol{R} \mid \boldsymbol{g}, \boldsymbol{j}) f(\boldsymbol{b})\, d\boldsymbol{b}}\\
&= \frac{\int f(\boldsymbol{Y}^{\text{o}} \mid \boldsymbol{g}, \boldsymbol{h}) f(\boldsymbol{Y}^{\text{m}} \mid \boldsymbol{Y}^{\text{o}}, \boldsymbol{g}, \boldsymbol{h}) f(\boldsymbol{b})\, d\boldsymbol{b}}{f(\boldsymbol{Y}^{\text{o}})}.
\end{aligned}$$

The PMM model in (4.35) states that the conditional distribution of the unobserved outcome $\boldsymbol{Y}^{\text{m}}$ given the observed outcome $\boldsymbol{Y}^{\text{o}}$ does not depend on the missingness indicator. These results are derived in Creemers et al. (2011). These authors also define the following MAR sub-class:

$$f(\boldsymbol{Y}^{\text{o}} \mid \boldsymbol{j}, \boldsymbol{\ell}) f(\boldsymbol{Y}^{\text{m}} \mid \boldsymbol{Y}^{\text{o}}, \boldsymbol{m}) f(\boldsymbol{R} \mid \boldsymbol{j}, \boldsymbol{q}), \qquad (4.36)$$

where the random effects $\boldsymbol{g}, \boldsymbol{h}$ and $\boldsymbol{k}$ vanish. Although this subclass does not contain all MAR models, it has intuitive appeal. Examples of models that satisfy GSPM (4.35) without belonging to (4.36) are described in Creemers et al. (2011).

Within the simple SPM framework we have only the following trivial result: the corresponding MAR sub-class containing only MAR models in this case is

$$f(\boldsymbol{Y}^{\text{m}} \mid \boldsymbol{Y}^{\text{o}}, \boldsymbol{b}) = f(\boldsymbol{Y}^{\text{m}} \mid \boldsymbol{Y}^{\text{o}}). \qquad (4.37)$$

TABLE 4.7
Congestive Heart Failure Study. Potential outcomes.

Observation Time Point	Echocardiogram Performed	Outcome
at end of study	no	missing
	yes	measurement
before end of study	no	missing
	yes	measurement

4.4.3 An example: Congestive heart failure

As a relatively simple application of a shared-parameter model we consider a study that has been conducted to evaluate the safety of a new agent in patients with congestive heart failure (CHF). The analyses closely follow those in Kenward and Rosenkranz (2011).

The primary goal was to investigate the effect of treatment on cardiac function, determined by echocardiography. Randomization was stratified by the patient's baseline CHF status (NYHA class I, II or III, American Heart Association, 1994) and whether the patient used a specific background therapy or not. An echocardiogram was to be taken at baseline and week 52 (end of study, EOS). Patients who discontinued early were to provide an echocardiogram performed at the time of discontinuation, to substitute for the missing end of study measurement. Some data will be missing because of patients' refusal to undergo the post-baseline measurement procedure, whether at the time of withdrawal or at the study end. The potential outcomes of this study are displayed in Table 4.7.

The probability of missingness may depend on whether a patient withdraws early or completes the study. Time to withdrawal and the measured endpoint may be statistically dependent as well. This may happen, for example, when the condition of a patient tends to deteriorate over time in the absence of treatment or even under standard therapy. We assume that there is a random variable Y of primary interest, a missingness indicator R, the observation time T, and some random effect, or latent variable b reflecting the unknown properties of an individual patient. We use an index $1 \leq i \leq n_i$ to refer to the repetitions pertaining to patient i. We set $R_i = 1$ if data for Y_i are observed and zero when they are missing.

Let T_i^0 denote the time of the observed repeat measurement and $z_i \in \{0, 1\}$ the treatment indicator of patient i. For some maximum observation time $t_{\max} > 0$, let $\delta_i = 1$ if $T_i^0 < t_{\max}$ and zero otherwise be the censoring indicator, and $T_i = \min(t_{\max}, T_i^0)$. We assume that $[\log(T_i^0) \mid b_i]$ is Gaussian with mean

$$\mu_{T_i} = \mu_T + \theta_T z_i + \tau_T b_i,$$

and standard deviation σ_T. For later reference we define

$$\tilde{t}_i = \frac{\log t_i - \mu_{T_i}}{\sigma_T}.$$

Next, let $Y_i(t)$ denote a continuous measurement at time point t, where $t = 0$ corresponds to baseline. We assume that the distribution of $Y_i(t)$ conditional on b_i and $T_i = t$ is $N(\mu_{Y_i}(t), \sigma_Y^2)$ with

$$\mu_{Y_i}(t_i, b_i, z_i) = \mu_Y + \tau_Y b_i + (\theta_Y z_i + \gamma_Y) \log t_i.$$

It follows that the value of Y tends to increase over time in the absence of the experimental treatment for $\gamma_Y > 0$, or to decrease for $\gamma_Y < 0$. This term in the model accounts for a condition that tends to deteriorate over time (like CHF). A treatment effect θ_Y may slow down or reverse the deterioration of the disease. The latent variable b leads to a lower or higher mean of Y. We define the standardized variable

$$\tilde{Y}_i(t) = \frac{Y_i(t) - \mu_{Y_i}(t)}{\sigma_Y}.$$

We now consider the model for the response indicator $R_i(t_i)$. Let

$$r_i = r(T_i, b_i, z_i) = \Pr[R_i(t_i) = 1 \mid T_i, b_i, z_i].$$

We assume a logistic model for r where the latent variable b acts on the intercept while the treatment effect θ_R affects the slope of the change of the logit of r over time:

$$\text{logit}(r_i) = \mu_R + \tau_R b_i + (\theta_R z_i + \gamma_R)\log t_i.$$

For a positive γ_R, the chance of a missing observation increases over time, while it decreases if $\gamma_R < 0$.

These components of the model can then be combined to produce the overall likelihood. Let $\xi = (\mu_T, \sigma_T, \theta_T, \tau_T, \mu_Y, \sigma_Y, \gamma_Y, \theta_Y, \tau_Y, \mu_R, \gamma_R, \theta_R, \tau_R)$ be the parameter vector to be estimated. The likelihood can then be written

$$\ell(\mathbf{Y}, \mathbf{R}, \mathbf{T}; \xi) = \prod_{i=1}^{n} \int \phi\{\tilde{y}_i(0)\}[r_i\phi\{\tilde{Y}_i(t_i)\}]^{R_i}(1-r_i)^{1-R_i}\phi(\tilde{T}_i)^{\delta_i} \times [1 - \Phi(\tilde{T}_i)]^{1-\delta_i}\phi(b_i)\, db_i, \quad (4.38)$$

where Φ is the standard normal distribution function and ϕ the corresponding probability density function.

This model implies a missingness mechanism that is NMAR, since Y and R are generally correlated. This is true even if $\tau_T = \tau_Y = \tau_R = 0$ is assumed since the distribution of Y and R may depend on the observation time t. Independence of Y and R holds only if $\gamma_Y = 0$ or $\gamma_R = 0$ in addition. Now, γ_R is identifiable given that R is observable for all patients. This is not the case for the parameters defining the distribution of Y, that is, γ_Y, μ_Y, θ_Y are generally not identifiable without further assumptions because some Y's may be missing. However, under the assumption that Y and R are assumed conditionally independent given (T, b), these parameters can be estimated from the observed Y's and are therefore identifiable. Because $\tau_T = 0$ or $\tau_Y = 0$ implies that the distribution of T does not depend on Y, μ_T and θ_T are identifiable.

For the full model, R and T are not directly dependent on the post-baseline measurement of Y, but only indirectly through the latent variable b. Under the non-verifiable assumptions made about the distribution of b and the likewise non-verifiable conditional independence assumption all model parameters are in principle identifiable.

The likelihood (4.38) can be maximized and parameter estimates obtained using standard statistical software for function optimization like, for example, the NLMIXED procedure from SAS. This procedure approximates the integral in (4.38) using adaptive Gaussian quadrature in every iteration step of a gradient-based optimization procedure.

We now fit this model to the dataset from the CHF study. Data from 271 patients with baseline echocardiography measurements were available. We present in Table 4.8 the missingness pattern for one response variable: the left ventricular ejection fraction (LVEF). About 10%

TABLE 4.8
Congestive Heart Failure Study. Missing value and discontinuation pattern.

	Control ($z = 0$)			Experimental ($z = 1$)		
	At EOS ($\delta = 0$)	Before EOS ($\delta = 1$)	Total	At EOS ($\delta = 0$)	Before EOS ($\delta = 1$)	Total
Missing ($m = 0$)	8	7	15	9	5	14
Observed ($m = 1$)	116	8	124	110	8	118
Total	124	15	139	119	13	132

TABLE 4.9
Congestive Heart Failure Study. Estimates and standard errors for the parameters of the distribution of Y. (*) stands for $\tau_T = \tau_Y = \tau_R = 0$, "—" for the full model; p-value refers to $H_0 : \theta_Y = 0$ and $-2\ell\ell$ denotes twice the negative log-likelihood.

τ_R	μ_Y	θ_Y	γ_Y	σ_Y	τ_Y	p-value ($-2\ell\ell$)
(*)	30.78	0.219	0.176	5.503	–	0.17
	(0.334)	(0.160)	(0.135)	(0.172)	–	(3590.1)
0.0	30.81	0.018	0.350	2.149	5.165	0.83
	(0.340)	(0.085)	(0.061)	(0.098)	(0.246)	(3293.1)
—	30.81	0.022	0.363	2.145	5.186	0.80
	(0.341)	(0.085)	(0.061)	(0.098)	(0.246)	(3273.3)

of the subjects in each group had a missing post-baseline assessment. In addition, about 10% withdrew from the study before 54 weeks. Hence, the probability of being missing, or of early withdrawal, does not seem to be associated with treatment, at least in a simple way.

A comparison of the post-baseline values including 242 patients with both a baseline and a post-baseline LVEF by analysis of covariance with baseline LVEF and treatment as covariates does not result in a significant difference between the groups ($p = 0.99$).

First it is assumed that T, Y, and R do not depend on b, that is, $\tau_T = \tau_Y = \tau_R = 0$. This implies that Y and R are conditionally independent given T, which is completely observed. The parameter estimates and standard errors for this case are contained in the first row of Table 4.9. As might be expected, there is a strong dependency on the probability of missingness of the observation time, reflected in a point estimate of γ_R of -2.1 (s.e. $= 0.47$) and $p < 0.0001$ (not shown).

Next the full model is fit. The full models were fitted both with and without τ_R, to assess its impact on the likelihood. The results are displayed in Table 4.9. For $\tau_R = 0$, the estimates of the parameters of the Y process, compared with those obtained from the previous analysis, differ considerably. First, the estimate for the treatment effect became much smaller and the corresponding p-value increased to $p = 0.83$. The time dependence of Y is now obvious from the estimate of γ_Y. The estimate of σ_Y decreases by 50%, a consequence of some of the variability being redistributed to τ_Y. The parameter estimates related to T and R changed only trivially (not shown). The value of minus twice the log-likelihood decreased from 3590.1 from the previous analysis, to 3293.1. For the full model we obtain $\widehat{\tau}_R = 1.037$ and -2 log-likelihood decreases further to 3273.3.

In this example, the principal treatment estimate changes very little, with the associated p-values remaining far from significant. The treatment effects on T and R also remained non-significant, which is consistent with the patterns observed in Table 4.9. The parameter estimates changed to a negligible degree.

This analysis rests heavily on untestable modelling assumptions, especially, but not only, those concerning the distributional shape of the random variables. In the light of this, the authors of the original analyses (Kenward and Rosenkranz, 2011) conducted sensitivity analyses as an adjunct to the results presented here. Such analyses are explored in more detail in Part V.

4.5 Concluding Remarks

In this chapter, we have explored and developed the classes of parametric model that simultaneously capture the statistical behaviour of both the outcome data and the missing data process. The aim has been to provide a foundation for the many extensions and applications of parametric models that appear in later chapters. We have deliberately kept the exposition and illustrative examples as simple as possible while allowing, at the same time, the key features of such models to emerge. In this way, we hope that the issues that are special to such models, and which do not always coincide with intuitions gained from conventional statistical modeling where data are not missing, are not obscured by the inevitable complexities that accompany such applications in most real settings. While it is not difficult to construct a very general framework that contains as special cases the three classes considered here, selection, pattern-mixture and shared parameter models, this does not imply that the classification explored here has no value. As stressed at the beginning of this chapter, the importance of the models considered here lies in their role in accommodating potentially nonrandom missing data mechanisms in sensitivity analyses. As such, their value lies in the way in which assumptions expressed in terms of the substantive setting can be translated into components, or parameters, of these models. Such translations can be very different in the different classes, and it would be expected that the different routes available will be of markedly different value in different applications. Hence it is important that each of the three classes is considered in its own right.

While there are many technicalities that surround the use of these models in practice, especially algorithms for obtaining maximum likelihood estimates, we have focused rather on the structure of the models and their interpretation, especially the way in which Rubin's classification of missing data mechanisms emerges in the three classes. There is a large literature on the computational aspects, which are less interesting from an inferential perspective, and for those who would like to pursue this further, appropriate references can be found in the bibliography.

We have seen repeatedly in the models considered in this chapter how model identification rests on untestable assumptions. All statistical analyses that provide non-trivial inferences rely on some assumptions that the data under analysis cannot support; the missing dataset-ting is special in the degree to which such reliance is made. This has been pointed out on many occasions. For a good cross-section of the arguments see, for example, the discussions to Diggle and Kenward (1994) and Scharfstein et al. (1999). There is, as a consequence, a broad agreement that their most appropriate role is as part sensitivity analyses, as explored in the chapters in Part V of this volume.

Acknowledgments

The authors gratefully acknowledge support from IAP research Network P7/06 of the Belgian Government (Belgian Science Policy).

References

Amemiya, T. (1984) Tobit models: a survey. *Journal of Econometrics* **24**, 3–61.

American Heart Association. (1994). *Nomenclature and Criteria for Diagnosis of Diseases of the Heart and Great Vessels. 9th ed.* Boston, Mass: Little, Brown & Co, 253–256.

Baker, S. (1994). Composite linear models for incomplete data. *Statistics in Medicine* **13**, 609–622.

Baker, S. (1995). Marginal regression for repeated binary data with outcome subject to non-ignorable non-response. *Biometrics* **51**, 1042–1052.

Baker, S. and Laird, N. (1988). Regression analysis for categorical variables with outcome subject to nonignorable nonresponse. *Journal of the American Statistical Association* **83**, 62–69.

Baker, S., Rosenberger, W., and DerSimonian, R. (1992). Closed-form estimates for missing counts in two-way contingency tables. *Statistics in Medicine* **11**, 643–657.

Carpenter, J., Pocock, S., and Lamm, C. (2002). Coping with missing values in clinical trials: a model based approach applied to asthma trials. *Statistics in Medicine* **21**, 1043–1066.

Carpenter, J.R., Roger, J.H., and Kenward, M.G. (2013). Analysis of longitudinal trials with protocol deviations: A framework for relevant, accessible assumptions and inference via multiple imputation. *Journal of Biopharmaceutical Statistics* **23**, 1352–1371.

Cowles, M., Carlin, B., and Connett, J. (1996). Bayesian tobit modeling of longitudinal ordinal clinical trial compliance data with nonignorable missingness. *Journal of the American Statistical Association* **91**, 86–98.

Creemers, A., Hens, N., Aerts, M., Molenberghs, G., Verbeke, G., and Kenward M.G. (2010). A sensitivity analysis for shared-parameter models. *Biometrical Journal* **52**, 111–125.

Creemers, A., Hens, N., Aerts, M., Molenberghs, G., Verbeke, G., and Kenward M.G. (2011). Generalized shared-parameter models and missingness at random. *Statistical Modelling* **11**, 279–310.

DeGruttola, V. and Tu, X.M. (1994). Modelling progression of CD4 lymphocyte count and its relationship to survival time. *Biometrics* **50**, 1003–1014.

Demirtas, H. and Schafer, J. (2003). On the performance of random-coefficient pattern-mixture models for non-ignorable drop-out. *Statistics in Medicine* **22**, 2553–2575.

Diggle, P.J. (1998). Diggle–Kenward model for dropouts. In: Armitage, P. and Colton, T. (Eds.) *Encyclopaedia of Biostatistics*, New York: Wiley, pp. 1160–1161.

Diggle, P.J. and Kenward, M.G.(1994). Informative dropout in longitudinal data analysis (with discussion). *Journal of the Royal Statistical Society, Series C* **43**, 49–94.

Fay, R. (1986). Causal models for patterns of nonresponse. *Journal of the American Statistical Association* **81**, 354–365.

Fitzmaurice, G., Heath, G., and Clifford, P. (1996). Logistic regression models for binary data panel data with attrition. *Journal of the Royal Statistical Society, Series C* **159**, 249–264.

Fitzmaurice, G., Laird, N., and Zahner, G. (1996). Multivariate logistic models for incomplete binary response. *Journal of the American Statistical Association* **91**, 99–108.

Freedman, D.A. and Sekhon, J.S. (2010). Endogeneity in probit response models. *Political Analysis* **18**, 138–150.

Follman, D. and Wu, M. (1995). An approximate generalized linear model with random effects for informative missing data. *Biometrics* **51**, 151–168.

Gad, A.M. (1999). Fitting selection models to longitudinal data with dropout using the Stochastic EM algorithm. Unpublished PhD thesis, School of Mathematics and Statistics, University of Kent and Canterbury, UK.

Gad, A.M. and Ahmed, A.S. (2005). Analysis of longitudinal data with intermittent missing values using the stochastic EM algorithm. *Computational Statistics and Data Analysis* **50**, 2702–2714

Gad, A.M. and Ahmed, A.S. (2007). Sensitivity analysis of longitudinal data with intermittent missing values. *Statistical Methodology* **4**, 217–226.

Glynn, R., Laird, N., and Rubin, D.B. (1986). Selection modelling versus mixture modelling with nonignorable nonresponse. In: *Drawing Inferences from Self-selected Samples*, Wainer, H. (Ed.) New-York: Springer-Verlag, 115-142

Greenlees J.S., Reece, W.R., and Zieschang, K.D. (1982). Imputation of missing values when the probability of response depends on the variable being imputed. *Journal of the American Statistical Association* **77**, 2251–2261.

Heckman, J. (1976). The common structure of statistical models of truncation, sample selection and limited dependent variables and a simple estimator for such models. *Annals of Economic and Social Measurement* **5**, 475–492.

Hogan, J. and Laird, N.M. (1997a). Mixture models for joint distribution of repeated measures and event times. *Statistics in Medicine* **16**, 239–257.

Hogan, J. and Laird, N.M. (1997b). Model-based approaches to analysing incomplete longitudinal and failure time data. *Statistics in Medicine* **16**, 259–272.

Jansen, I., Hens, N., Molenberghs, G., Aerts, M., Verbeke, G., and Kenward, M.G. (2006). The nature of sensitivity in missing not at random models. *Computational Statistics and Data Analysis* **50**, 830–858.

Kenward, M.G. (1998). Selection models for repeated measurements with nonrandom dropout: An illustration of sensitivity. *Statistics in Medicine* **17**, 2723–2732.

Kenward M.G. and Carpenter J.R. (2009). Multiple imputation. In: *Longitudinal Data Analysis*, G. Fitzmaurice et al. (Eds.) London:Chapman & Hall/CRC.

Kenward, M.G, Lesaffre, E., and Molenberghs, G. (1994). An application of maximum likelihood and generalized estimating equations to the analysis of ordinal data from a longitudinal study with cases missing at random. *Biometrics* **50**, 945–953.

Kenward, M.G. and Molenberghs, G. (1999). Parametric models for incomplete continuous and categorical longitudinal studies data. *Statistical Methods in Medical Research* **8**, 51–83.

Kenward, M.G, Molenberghs, G., and Thijs, H. (2003). Pattern-mixture models with proper time dependence. *Biometrika* **90**, 53–71.

Kenward, M.G. and Rosenkranz, G. (2011). Joint modelling of outcome, observation time and missingness. *Journal of Biopharmaceutical Statistics*, **21**, 252–262.

Little, R.J.A. (1986). A note about models for selectivity bias. *Econometrika* **53,** 1469–1474.

Little, R.J.A. (1993). Pattern-mixture models for multivariate incomplete data. *Journal of the American Statistical Association* **88**, 125–134.

Little, R.J.A. (1994). A class of pattern-mixture models for multivariate incomplete data. *Biometrika* **81**, 471–483.

Little, R.J.A. (1995). Modeling the dropout mechanism in repeated-measures studies. *Journal of the American Statistical Association* **90**, 1112–1121.

Little, R.J.A. and Rubin, D.B. (2002). *Statistical Analysis with Missing Data (Second Edition).* Chichester: Wiley.

Little, R.J.A. and Wang, Y. (1996). Pattern-mixture models for multivariate incomplete data with covariates. *Biometrics* **52,** 98–111.

Little, R.J.A. and Yau, L. (1996). Intent-to-treat analysis for longitudinal studies with drop-outs. *Biometrics* **52**, 1324–1333.

Molenberghs, G., Beunckens, C., Sotto, C., and Kenward, M.G. (2008). Every missing not at random model has got a missing at random counterpart with equal fit. *Journal of the Royal Statistical Society, Series B* **70**, 371–388.

Molenberghs, G. and Kenward, M.G. (2007). *Missing Data in Clinical Studies.* Chichester: Wiley.

Molenberghs, G., Beunckens, C., Sotto, C., and Kenward, M.G. (2008). Every missing not at random model has got a missing at random counterpart with equal fit. *Journal of the Royal Statistical Society, Series B* **70**, 371–388.

Molenberghs, G., Goetghebeur, E., Lipsitz, S., and Kenward, M.G. (1999). Non-random missingness in categorical data: strengths and limitations. *American Statistician* **52,** 110–118.

Molenberghs, G. and Lesaffre, E. (1994). Marginal modelling of correlated ordinal data using a multivariate Plackett distribution. *Journal of the American Statistical Association* **89**, 633–644

Molenberghs, G., Kenward, M.G., and Lesaffre, E. (1997). The analysis of longitudinal ordinal data with non-random dropout. *Biometrika* **84**, 33–44.

Molenberghs, G., Michiels, B., Kenward, M.G., and Diggle, P.J. (1998). Missing data mechanisms and pattern-mixture models. *Statistica Nederlandica* **52**, 153–161.

Molenberghs, G. and Verbeke, G. (2005). *Models for Discrete Longitudinal Data.* New York: Springer.

Molenberghs, G., Verbeke, G., Thijs, H., Lesaffre, E., and Kenward, M.G. (2001). Mastitis in dairy cattle: Local influence to assess sensitivity of the dropout process. *Computational Statistics and Data Analysis* **37**, 93–113.

Muthén, B., Asparouhov, T., Hunter, A., and Leuchter, A. (2011). Growth modeling with non-ignorable dropout: Alternative analyses of the STAR*D antidepressant trial. *Psychological Methods* **16**, 17–33.

Nelder, J.A. and Mead, R. (1965). A simplex method for function minimisation. *The Computer Journal* **7**, 303–313.

Park, T. and Brown, M. (1994). Models for categorical data with nonignorable nonresponse. *Journal of the American Statistical Association* **89,** 44–51.

Reisberg, B., Borenstein, J., Salob, S.P., Ferris, S.H., Franssen, E., and Georgotas, A. (1987). Behavioral symptoms in Alzheimer's disease: Phenomenology and treatment. *Journal of Clinical Psychiatry* **48**, 9–13.

Rotnitzky, A., Cox, D.R., Bottai, M., and Robins, J.M. (2000). Likelihood-based inference with a singular information matrix. *Bernouilli* **6**, 243–284.

Rubin, D. B. (1977). Formalizing subjective notions about the effect of nonrespondents in sample surveys. *Journal of the American Statistical Association* **72**, 538–543.

Scharfstein, D. O., Rotnizky, A. and Robins, J. M. (1999). Adjusting for nonignorable drop-out using semiparametric nonresponse models (with discussion). *Journal of the American Statistical Association* **94**, 1096–1146.

Schluchter, M. (1992). Methods for the analysis of informatively censored longitudinal data. *Statistics in Medicine* **11**, 1861–1870

Thijs, H., Molenberghs, G., Michiels, B., Verbeke, G., and Curran, D. (2002). Strategies to fit pattern-mixture models. *Biostatistics* **3**, 245–265.

Troxel, A., Harrington, D., and Lipsitz, S. (1988). Analysis of longitudinal data with non-ignorable non-monotone missing values. *Journal of the Royal Statistical Society, Series C* **47**, 425–438.

Troxel, A., Lipsitz, S., and Harrington, D. (1988). Marginal models for the analysis of longitudinal data with nonignorable non-monotone missing data. *Biometrika* **85**, 661–673.

Verbeke, G. and Molenberghs, G. (2000). *Linear Mixed Models for Longitudinal Data.* New York: Springer–Verlag.

Wu, M. C. and Bailey, K.R. (1988). Analysing changes in the presence of informative right censoring caused by death and withdrawal. *Statistics in Medicine* **7**, 337–346.

Wu, M. and Bailey, K. (1989). Estimation and comparison of changes in the presence of informative right censoring: Conditional linear model. *Biometrics* **45**, 939–955.

Wu, M. and Carroll, R. (1988). Estimation and comparison of changes in the presence of informative right censoring by modelling the censoring process. *Biometrics* **44,** 175–188.

Yun, S., Lee, Y., and Kenward, M. G. (2007). Using hierarchical likelihood for missing data problems. *Biometrika* **94,** 905–919.

5

Bayesian Methods for Incomplete Data

Michael J. Daniels

University of Texas at Austin, TX

Joseph W. Hogan

Brown University, Providence, RI

CONTENTS

5.1 Introduction

Concepts for Bayesian inference for incomplete data began to be formalized in the mid-1970s. Bayesian inference provides a powerful and appropriate framework for the analysis of incomplete data. Inherent in models and drawing inference in the presence of missing data is a lack of identifiability. Functionals of the distribution of the full data are generally not identifiable without uncheckable (from the data) assumptions. These can include parametric assumptions about the full data model and/or specific assumptions about the mechanism of missingness (discussed in considerable detail in Section 5.5). The Bayesian approach provides a principled way to account for uncertainty about the missingness and the lack of identifiability via prior distributions. Common approaches in the literature which result in identification of the full data response (e.g., parametric selection models) tend to ignore this uncertainty. Moment-based approaches (Scharfstein et al., 1999) vary parameters not identified by the data, but do not have a formal way to account for the underlying uncertainty of such parameters in the final inference. Given that we account for uncertainty in parameters identified by the data, it would seem unsatisfactory to allow for no uncertainty for parameters that are *not* identified by the data.

The chapter will provide an overview of key concepts and introduce some new ideas. The overall structure of the chapter is as follows. For the reader less familiar with Bayesian inference, Section 5.2 provides a quick primer on relevant concepts and demonstrates the importance of using a Bayesian approach for incomplete data. In Section 5.3, we stress the importance of correct model specification and introduce essential properties for assessing fit and selecting models. We also introduce new approaches to assess model fit. Several recent papers have been written that focus on model selection and model checking for incomplete data (Gelman et al., 2005; Chatterjee and Daniels, 2013; Wang and Daniels, 2011; Daniels et al., 2012; Mason et al., 2012).

Bayesian ignorability (Rubin 1976) and its connection to multiple imputation (including with auxiliary covariates) are explored in Section 5.4 including the introduction of a new modeling approach for Bayesian ignorability with auxiliary covariates. The concept of ignorability is not a part of moment-based methods as they do not specify a full probability model for the distribution of the full data response.

Section 5.5 explores Bayesian non-ignorability. Related work in the literature includes finding parameters in the model for which the data provide no information and, in a Bayesian setting, specifying an informative prior. Such parameters are called sensitivity parameters and will be formally defined in this section. Some of the most important literature includes Rubin (1977), who in the context of mixture models and nonrespondents in sample surveys, found sensitivity parameters and proposed an approach to construct subjective priors. Identifying restrictions are another popular approach to addressing non-ignorability. Little (1994) discusses restrictions to identify all the parameters in mixture models without impacting the fit of the observed data in the context of a bivariate normal model with monotone missingness. He also introduces a sensitivity parameter on which a sensitivity analysis could be done or an informative prior constructed. Little and Wang (1996) extend his earlier work to more complex settings. Additional work on identifying restrictions in mixture models and their explicit connections to selection models can be found in Thijs et al. (2002) and Kenward, Molenberghs, and Thijs (2003). Although Bayesian inference is not explicitly discussed in these last two papers, performing a Bayesian analysis would be straightforward (Wang et al., 2011; Wang and Daniels, 2011).

For Bayesian inference, it is typical that the posterior distribution is not available analytically. In Section 5.6, we outline general computational issues, including data augmentation and algorithms to perform Bayesian inference in the setting of incomplete data. The chapter concludes with a section describing how Bayesian non-parametric inference can be, and has been, used for incomplete data (Section 5.7) and a short section that points out important open problems (Section 5.8).

Before delving into the chapter, we point out several monographs on missing data that have a substantial (or complete) Bayesian focus. Rubin (1987) introduces and explains the theory and practice of multiple imputation in the context of surveys. Little and Rubin (1987, 2002) (first and second editions) focus on model-based approaches to handle missing data in more general settings. Schafer (1997) explores Bayesian ignorable missingness (for more details, see Section 5.4) in the context of multivariate normal models, log-linear models for categorical data, and general location models (GLOMs). Most recently, Daniels and Hogan (2008) present a comprehensive Bayesian treatment of ignorable and non-ignorable missingness in the setting of longitudinal data.

We will not address Bayesian approaches for missing covariates in this chapter (see e.g., Huang, Chen, and Ibrahim, 2005). We will also not explicitly address causal inference with counterfactuals (Rubin, 1978), for which identifiability issues exist similar to those discussed in this chapter.

5.2 Primer on Bayesian Inference and Its Importance in Incomplete Data

Bayesian inference is based on a posterior distribution, which combines a prior distribution with a data model (and the data) through the likelihood. Consider vector-valued responses, $\boldsymbol{Y}_i : i = 1, \ldots, N$. Consider a probability model for the data, $p(\boldsymbol{y}_i|\boldsymbol{\theta})$ parameterized by $\boldsymbol{\theta}$. The corresponding likelihood, $L(\boldsymbol{\theta}|\boldsymbol{y})$ is given as

$$L(\boldsymbol{\theta}|\boldsymbol{y}) = \prod_{i=1}^{N} p(\boldsymbol{y}_i|\boldsymbol{\theta}).$$

Assume prior information on $\boldsymbol{\theta}$ is collected through a prior distribution, $p(\boldsymbol{\theta})$. Then the posterior distribution, $p(\boldsymbol{\theta}|\boldsymbol{y})$ is given as

$$p(\boldsymbol{\theta}|\boldsymbol{y}) = \frac{L(\boldsymbol{\theta}|\boldsymbol{y})p(\boldsymbol{\theta})}{\int L(\boldsymbol{\theta}|\boldsymbol{y})p(\boldsymbol{\theta})}. \tag{5.1}$$

Note that the posterior is proportional to the numerator. This is very important for computations as the denominator is often a high-dimensional intractable integral and sample-based approaches to evaluate the posterior are based on the numerator alone. We discuss relevant Bayesian computations in Section 5.6. We will discuss the importance of specification of the probability model for the data in Section 5.3.

5.2.1 Complete data concepts and prior specification

A variety of summaries of the posterior are used for "point estimation." These can be justified based on loss functions (Carlin and Louis, 2008). The most common are the posterior mean, $E(\boldsymbol{\theta}|\boldsymbol{y})$ or the posterior median, $\tilde{\theta}_j$ which satisfies

$$\int_{-\infty}^{\tilde{\theta}_j} p(\theta_j|\boldsymbol{y})d\theta_j = 0.5$$

and is based on the marginal posterior of θ_j. Uncertainty is often characterized using $100(1-\alpha)\%$ credible intervals that can be computed using $\alpha/2$ and $1-\alpha/2$ percentiles of the posterior. Inference about parameters in terms of hypotheses can be quantified by computing relevant quantities from the posterior. For example, the strength of evidence for the hypothesis, $H_0 : \theta_j = 0$ could be quantified as $P(\theta_j > 0|\boldsymbol{y})$, i.e., the posterior probability that $\theta_j > 0$.

Priors are classified in a variety of ways. The first is tied to closed-form posteriors and computations, i.e., (conditionally) conjugate priors. Other classifications differentiate how informative the prior is, e.g., informative *versus* diffuse/default/non-informative.

Conditionally conjugate priors are such that the full conditional distribution of the parameter has the same distributional form as the prior. This can be most easily seen with a simple example:

$$\begin{aligned} Y_i|\mu, \sigma &\sim N(\mu, \sigma) : i = 1, \ldots, N \\ \mu &\sim N(a, b) \\ \sigma &\sim IG(c, d). \end{aligned}$$

The full conditional distribution of μ, $p(\mu|\sigma, \boldsymbol{y})$, will be a normal distribution and the full conditional distribution of σ, $p(\sigma|\mu, \boldsymbol{y})$, will be an inverse gamma (IG) distribution. However, the marginal posterior of μ, $p(\mu|\boldsymbol{y})$, will not be normal.

The reason such priors are useful is that sampling-based approaches for the posterior distribution typically involve sampling from these full conditional distributions. Further details can be found in Section 5.6.

In most situations, there are parameters for which the data can be used to provide most of the information and reliance on the prior is not necessary. In such situations, it is desirable to use a diffuse or default prior for such parameters. There is an extensive literature on such priors from Laplace (flat priors) (Laplace, 1878) to Jeffreys (1961) to the reference priors of (Berger and) Bernardo (Bernardo, 1979; Berger and Bernardo, 1992) among many others. Often these diffuse priors are improper (they are not integrable) and in such cases, it is necessary to verify that the posterior distribution is proper (i.e., the denominator in (5.1) is finite); sampling-based approaches to compute the posterior do not diagnose the problem (Hobert and Casella, 1996). To avoid this checking, diffuse proper prior equivalents are often used. We discuss some of the typical default choices next.

For mean (or regression) parameters, a normal prior with large variance is the typical default choice. For variances, see Gelman (2006) for a nice review. His recommendation for random-effects variances are flat (uniform) priors on the standard deviations. If a proper prior is needed, either the flat prior can be truncated or a half normal prior can be used. Uniform shrinkage priors (Daniels, 1999) are another proper choice. For covariance matrices, the default choice is the inverse Wishart prior (which is the conditionally conjugate prior for the covariance matrix in a multivariate normal model) for computational reasons. However, it is not straightforward to specify the scale matrix and there is a lower bound on the degrees-of-freedom (which limits how non-informative the prior can be in small samples). For more discussion of these problems, see e.g., Daniels and Kass (1999).

5.2.2 The prior distribution and incomplete data

The aforementioned discussion focuses on key concepts regarding Bayesian inference for complete data. We now explore its power for incomplete data, starting with the prior distribution. The prior distribution is an integral component of Bayesian inference with incomplete data. We demonstrate this with a simple example.

Consider univariate responses $Y_i : i = 1, \ldots, N$. Suppose we only observe a subset of the N responses. Define a random variable R_i which is an indicator of whether Y_i is observed ($R_i = 1$) or missing ($R_i = 0$). We assume the following model for the response, conditional on R_i,

$$Y_i|R_i = r \sim N(\mu_r, \sigma^2).$$

We define the unconditional marginal mean, $E(Y) \equiv \mu$. We want to draw inference about this population mean, μ. Note that we can write μ as a mixture over r,

$$\mu = \mu_0\theta + \mu_1(1 - \theta)$$

where $\mu_0 = E[Y|R=0]$, $\mu_1 = E[Y|R=1]$, and $\theta = P(R=0)$. Clearly, the data contain no information on μ_0. We further explore the impact of the prior on μ_0 on posterior inference on μ in what follows. For ease of exposition and without loss of generality, we only place priors on μ_1 and μ_0 and assume θ is known; specifically, we place a $N(a_0, \tau_0)$ prior on μ_0 and a $N(a_1, \tau_1)$ prior on μ_1. The posterior distribution of μ will be a mixture of normals with expectation,

$$E(\mu|\boldsymbol{y}_{\text{obs}}, \boldsymbol{r}) = \theta a_0 + (1-\theta)\left(B\bar{y}_1 + (1-B)a_1\right),$$

where $N_1 = \sum_{i=1}^{N} R_i$, $\bar{y}_1 = \sum_{i:R_i=1} y_i/N_1$, and $B = \frac{\tau_1}{\tau_1 + \sigma^2/N_1}$. So clearly, the prior on μ_0 is very influential. As $N \to \infty$ with θ fixed, the prior on μ_1 is swamped as $B \to 1$, but not the prior on μ_0. Thus, the prior mean a_0 carries a weight of $\theta = P(R=0) > 0$ in the posterior mean of μ. In addition, the posterior variance is

$$\text{Var}(\mu|\boldsymbol{y}_{\text{obs}}, \boldsymbol{r}) = \theta^2\tau_0 + (1-\theta)^2 \left(\frac{1}{\frac{1}{\tau_1} + \frac{\sum_i R_i}{\sigma^2}}\right).$$

Note that as $N \to \infty$ with θ fixed, the posterior variance goes to $\theta^2\tau_0$ (not zero) since the first term is $O(1)$. Thus, the posterior standard deviation is bounded below by $\theta\sqrt{\tau_0}$ and is proportional to the probability of missingness, θ. However, a simple assumption, like MAR, would specify $p(\mu_0|\mu_1) = I\{\mu_0 = \mu_1\}$, an extremely informative, degenerate prior.

From the above example, it is clear that informative priors are needed for inference with incomplete data. A key component when using an informative prior is eliciting it. In regression settings, it is recommended to elicit observables rather than the regression coefficients themselves (Bedrick et al., 1996). Some recent examples in the context of missing data include Daniels and Hogan (2008, Chapter 10, Case study 2) and Lee et al. (2012) who elicit probabilities to obtain a prior for an odds ratio, Scharfstein et al. (2006) and Wang et al. (2011) who elicit relative risks to obtain a prior for an odds ratio, and Paddock and Ebener (2009) who use a carefully constructed questionnaire to elicit expert opinions about change over time.

Besides the natural approach of using prior distributions to address uncertainty about missingness in the Bayesian approach, other advantages include simple approaches to assess model fit based on the posterior predictive distribution (which we discuss in Section 5.3), the connection to multiple imputation (which is explored in Section 5.4), and computations using data augmentation (Section 5.6).

5.3 Importance of Correct Model Specification

For model-based inference with missing data, there is always an underlying joint model for the missingness and the responses, $p(\boldsymbol{y}, \boldsymbol{r}|\boldsymbol{\omega})$. In what follows, we define the full data as $(\boldsymbol{y}, \boldsymbol{r})$ and the full data response as $\boldsymbol{y}$. A convenient way to understand identifiability of full data parameters in missing data is to factor the joint distribution in the following way:

$$p(\boldsymbol{y}, \boldsymbol{r}|\boldsymbol{\theta}) = p(\boldsymbol{y}_{\text{obs}}, \boldsymbol{r}|\boldsymbol{\theta}_O)p(\boldsymbol{y}_{\text{mis}}|\boldsymbol{y}_{\text{obs}}, \boldsymbol{r}, \boldsymbol{\theta}_E), \tag{5.2}$$

where $\boldsymbol{\theta} = (\boldsymbol{\theta}_O, \boldsymbol{\theta}_E)$. We call this the *extrapolation factorization* (Daniels and Hogan, 2008). The first component is the observed data model and the second component is the *extrapolation distribution*; in the context of mixture models, Little previously called the second component the *missing-value distribution* (Little, 1993). Clearly, there is no information in the data itself about the second component. Thus uncheckable assumptions are needed for inference about the full data response model,

$$p(\boldsymbol{y}|\gamma(\boldsymbol{\theta})) = \int p(\boldsymbol{y}, \boldsymbol{r}|\boldsymbol{\theta})dF(\boldsymbol{r}|\boldsymbol{\theta}),$$

where $\boldsymbol{\gamma}(\boldsymbol{\theta})$ are the parameters of the full data response model that are a function of the parameters, $\boldsymbol{\theta}$, in the full data models.

In Bayesian inference for missing data, we need to specify a probability model for the distribution of the full data. It is typical to first use a criteria to choose the "best" among several models and then to assess the "absolute" fit of the selected model. In the rest of this section, we first introduce a property which should be possessed by all criteria that either select among models or assess model fit and then discuss various approaches to do both. We wrap up this section by briefly discussing the importance of the dependence specification in longitudinal incomplete data.

From the extrapolation factorization (5.2), it is obvious that data-based selection among several models should only be based on the fit of the data to the first component, $p(\boldsymbol{y}_{\text{obs}}, \boldsymbol{r}|\boldsymbol{\theta}_O)$, the induced observed-data distribution. As such, criteria for model selection should satisfy the following property:

Property: Invariance to the extrapolation distribution: Two models for the full data with the same model specification for the observed data, $p(\boldsymbol{y}_{\text{obs}}, \boldsymbol{r}; \boldsymbol{\theta}_O)$ and same prior, $p(\boldsymbol{\theta}_O)$ should give the same value for the Bayesian model selection criterion (Daniels et al., 2012).

Some criteria proposed in the literature satisfy this property and some do not. Criteria based on the observed data likelihood, e.g., Bayes factors, conditional predictive ordinates, and the deviance information criterion (DIC) obviously satisfy the property. Posterior predictive criteria based on replicated observed data, proposed in Daniels et al. (2012) also satisfy this property. However, versions of the DIC that are not based on the observed data likelihood do not (Celeux et al., 2006; Mason et al, 2012). We provide details on replicated observed data in what follows.

Methods to assess model fit should also be invariant to the extrapolation distribution. A natural approach for model checking in the Bayesian setting is posterior predictive checks. Within this paradigm, there is a choice of using replicated full data response checks (Gelman et al., 2005) versus replicated observed data checks. However, only the latter satisfies the property stated above in terms of being invariant to the extrapolation distribution. Replicated observed data correspond to sampling from

$$p(\boldsymbol{y}^{\text{rep}}, \boldsymbol{r}^{\text{rep}}|\boldsymbol{y}_{\text{obs}}, \boldsymbol{r}) = \int p(\boldsymbol{y}^{\text{rep}}, \boldsymbol{r}^{\text{rep}}|\boldsymbol{\theta})p(\boldsymbol{\theta}|\boldsymbol{y}_{\text{obs}}, \boldsymbol{r})d\boldsymbol{\theta},$$

from which the observed replicated data, $\boldsymbol{y}_{\text{obs}}^{\text{rep}}$ is formed as

$$\boldsymbol{y}_{\text{obs}}^{\text{rep}} = \{y_i^{\text{rep}} : r_i^{\text{rep}} = 1\}.$$

Replicated full data responses (i.e., replicated complete data) correspond to sampling from

$$p(\boldsymbol{y}^{\text{rep}}|\boldsymbol{y}_{\text{obs}}, \boldsymbol{r}) = \int p(\boldsymbol{y}^{\text{rep}}, \boldsymbol{r}^{\text{rep}}|\boldsymbol{\theta})p(\boldsymbol{\theta}|\boldsymbol{y}_{\text{obs}}, \boldsymbol{r})d\boldsymbol{r}^{\text{rep}}d\boldsymbol{\theta}.$$

The latter can be sampled in ignorable models (to be defined formally in Section 5.4) since it is not a function of the replicated observed data indicators, ($\boldsymbol{r}^{\text{rep}}$) but is *not* invariant to the extrapolation distribution. For example, posterior predictive checks based on replicated complete data for different non-ignorable models which provide identical fits to the observed data will result in different values for the posterior predictive checks (Chatterjee and Daniels, 2013). There is also the issue of the power supplied by the different methods for checking as the replicated observed data only uses part of the replicated response data (i.e., those where $r^{\text{rep}} = 1$). However, the replicated complete data fills in the missing responses based on the assumed model. For further discussion, see Chatterjee and Daniels (2013).

In the context of correlated data (e.g., longitudinal data), specification of the dependence structure is essential. In complete data, mis-specification often results in only a loss of efficiency for estimating mean parameters, but not (asymptotic) bias. However, in the context of missing data, mis-specification of the dependence structure leads to bias as well. For further discussion and details, see Chapter 6 in Daniels and Hogan (2008).

In a full likelihood (or Bayesian) framework, categorizing missing data as ignorable *versus* non-ignorable is most natural. We start with Bayesian ignorability next.

5.4 Multiple Imputation and Bayesian Ignorability with and without Auxiliary Covariates

In this section, we first formally define Bayesian ignorability and show its connection to multiple imputation. We then define Bayesian ignorability in the context of auxiliary covariates, which are covariates that are not to be included in the full-data response model, $p(\boldsymbol{y}|\boldsymbol{x};\boldsymbol{\gamma})$. Finally, we discuss the complications associated with auxiliary covariates for probability models in Bayesian inference and a proposed solution. In most of this section, we will explicitly include covariates of interest in the full data model, $\boldsymbol{x}$ for clarity.

5.4.1 Bayesian ignorability

The full data model, $p(\boldsymbol{y},\boldsymbol{r}|\boldsymbol{x};\boldsymbol{\theta})$ can be factored as $p(\boldsymbol{y}|\boldsymbol{x};\boldsymbol{\gamma}(\boldsymbol{\theta}))p(\boldsymbol{r}|\boldsymbol{y},\boldsymbol{x};\boldsymbol{\psi}(\boldsymbol{\theta}))$, the product of the full data response model $p(\boldsymbol{y}|\boldsymbol{x};\boldsymbol{\gamma}(\boldsymbol{\theta}))$ and the missing data mechanism (mdm), $p(\boldsymbol{r}|\boldsymbol{y},\boldsymbol{x};\boldsymbol{\psi}(\boldsymbol{\theta}))$; this is often known as a selection model factorization and is helpful to understand ignorable missingness. We will suppress the dependence of $(\boldsymbol{\gamma},\boldsymbol{\psi})$ on $\boldsymbol{\theta}$ in what follows. Ignorable missingness in Bayesian inference holds under the following three conditions:

1. MAR missingness.
2. Distinct parameters: The set of parameters $\boldsymbol{\gamma}$ and $\boldsymbol{\psi}$ where $\boldsymbol{\gamma}$ parameterizes the full data response model, $p(\boldsymbol{y}|\boldsymbol{x};\boldsymbol{\psi})$ and $\boldsymbol{\psi}$ the missing data mechanism, $p(\boldsymbol{r}|\boldsymbol{y},\boldsymbol{x};\boldsymbol{\psi})$ are distinct.
3. A priori independence: The prior for the parameters of the full data model, $\boldsymbol{\theta}$, factors as follows, $p(\boldsymbol{\theta}) = p(\boldsymbol{\gamma},\boldsymbol{\psi}) = p(\boldsymbol{\gamma})p(\boldsymbol{\psi})$

Under Bayesian ignorability, the posterior distribution of $\boldsymbol{\theta}$ can be factored as

$$p(\boldsymbol{\theta}|\boldsymbol{y}_{\text{obs}},\boldsymbol{r},\boldsymbol{x}) = p(\boldsymbol{\gamma}|\boldsymbol{y}_{\text{obs}},\boldsymbol{x})p(\boldsymbol{\psi}|\boldsymbol{y}_{\text{obs}},\boldsymbol{r},\boldsymbol{x}).$$

So inference on the full data response model parameters γ only requires specification of the full data response model, $p(\boldsymbol{y}|\boldsymbol{x};\gamma)$ and the prior on γ since $p(\gamma|\boldsymbol{y}_{\text{obs}},\boldsymbol{x}) \propto p(\boldsymbol{y}_{\text{obs}}|\boldsymbol{x};\gamma)p(\gamma)$.

The extrapolation distribution, $p(\boldsymbol{y}_{\text{mis}}|\boldsymbol{y}_{\text{obs}},\boldsymbol{r})$ is identified under ignorability. This can be seen (ignoring $\boldsymbol{x}$) as

$$\begin{aligned}
p(\boldsymbol{y}_{\text{mis}}|\boldsymbol{y}_{\text{obs}},\boldsymbol{r}) &= \frac{p(\boldsymbol{r}|\boldsymbol{y}_{\text{mis}},\boldsymbol{y}_{\text{obs}})p(\boldsymbol{y}_{\text{mis}},\boldsymbol{y}_{\text{obs}})}{p(\boldsymbol{y}_{\text{obs}},\boldsymbol{r})} \\
&= \frac{p(\boldsymbol{r}|\boldsymbol{y}_{\text{obs}})p(\boldsymbol{y}_{\text{mis}}|\boldsymbol{y}_{\text{obs}})p(\boldsymbol{y}_{\text{obs}})}{p(\boldsymbol{y}_{\text{obs}},\boldsymbol{r})} \\
&= \frac{p(\boldsymbol{r}|\boldsymbol{y}_{\text{obs}})p(\boldsymbol{y}_{\text{mis}}|\boldsymbol{y}_{\text{obs}})p(\boldsymbol{y}_{\text{obs}})}{p(r|\boldsymbol{y}_{\text{obs}})p(\boldsymbol{y}_{\text{obs}})} \\
&= p(\boldsymbol{y}_{\text{mis}}|\boldsymbol{y}_{\text{obs}}).
\end{aligned}$$

So, for example, in the setting of a bivariate response, $\boldsymbol{y} = (y_1, y_2)$ and only missingness in y_2 (denoted by $R = 0$),

$$\begin{aligned}
p(y_2|y_1, r = 0) &= p(y_2|y_1) \\
&= p(y_2|y_1, r = 1).
\end{aligned}$$

5.4.2 Connection to multiple imputation

For Bayesian ignorability, the posterior of interest is the observed data response posterior (with the missing responses integrated out),

$$p(\gamma|\boldsymbol{y}_{\text{obs}},\boldsymbol{x}) = \int p(\gamma,\boldsymbol{y}_{\text{mis}}|\boldsymbol{y}_{\text{obs}},\boldsymbol{x})d\boldsymbol{y}_{\text{mis}}.$$

Alternatively, the missing data could be directly imputed from the extrapolation distribution, which is equal to $p(\boldsymbol{y}_{\text{mis}}|\boldsymbol{y}_{\text{obs}},\boldsymbol{x})$ as shown above. This distribution is the basis for the imputation model and is derived directly from the full data response model. As has been pointed out previously, Bayesian ignorability is the limiting case of multiply imputing a finite number of datasets and then correcting the variance (Rubin, 1987). As such, multiple imputation is inherently an "approximation" to a fully Bayesian procedure and such imputation has been termed Bayesian proper imputation (Schafer, 1997). However, in the presence of auxiliary covariates, that are needed for imputing the missing responses, but not for inclusion in the final model for inference, several issues arise, which we discuss in what follows. For further discussion on multiple imputation, see Part IV of this book.

5.4.3 Bayesian ignorability with auxiliary covariates

Auxiliary covariates, $\boldsymbol{w}$ are covariates that have been collected (often at baseline) that are not of interest for the primary analysis (especially in randomized clinical trials), but may be predictive of missingness (and the response). In this setting, we need to jointly model $p(\boldsymbol{y},\boldsymbol{r},\boldsymbol{w}|\boldsymbol{\theta}_v)$. It may be the case that MAR holds conditionally, but not unconditionally, on auxiliary covariates (sometimes called auxiliary variable MAR or A-MAR). That is,

$$p(\boldsymbol{r}|\boldsymbol{y}_{\text{obs}},\boldsymbol{y}_{\text{mis}},\boldsymbol{x};\psi) \neq p(\boldsymbol{r}|\boldsymbol{y}_{\text{obs}},\boldsymbol{y}'_{\text{mis}},\boldsymbol{x};\psi)$$

for $\boldsymbol{y}_{\text{mis}} \neq \boldsymbol{y}'_{\text{mis}}$. However,

$$p(\boldsymbol{r}|\boldsymbol{w}, \boldsymbol{y}_{\text{obs}}, \boldsymbol{y}_{\text{mis}}, \boldsymbol{x}; \boldsymbol{\psi}) = p(\boldsymbol{r}|\boldsymbol{w}, \boldsymbol{y}_{\text{obs}}, \boldsymbol{x}; \boldsymbol{\psi})$$

for $\boldsymbol{y}_{\text{mis}} \neq \boldsymbol{y}'_{\text{mis}}$. As a result, we now need to jointly model $p(\boldsymbol{y}, \boldsymbol{r}, \boldsymbol{w}|\boldsymbol{x})$.

In this case, the ignorability definition is modified as follows,

1. the missingness is A-MAR
2. distinct parameters: r (γ_w, ν) and $\boldsymbol{\psi}$ where γ_w parameterizes the full data response model conditional on auxiliary covariates, $p(\boldsymbol{y}|\boldsymbol{w}, \boldsymbol{x}; \gamma_w)$, $\boldsymbol{\psi}$ the missing data mechanism, $p(\boldsymbol{r}|\boldsymbol{y}, \boldsymbol{w}, \boldsymbol{x}; \boldsymbol{\psi})$ and $\boldsymbol{\nu}$ the distribution of the auxiliary covariates, $p(\boldsymbol{w}|\boldsymbol{x}; \boldsymbol{\nu})$.
3. a priori independence: $p(\gamma_w) = p(\gamma_w, \boldsymbol{\psi}, \nu) = p(\gamma_w, \boldsymbol{\nu})p(\boldsymbol{\psi})$.

Similar to the case without auxiliary covariates, the posterior distribution factors,

$$\begin{aligned}
&\int p(\boldsymbol{y}, \boldsymbol{r}, \boldsymbol{w}|\boldsymbol{x}; \boldsymbol{\theta}_w) d\boldsymbol{y}_{\text{mis}} \\
&\quad = \int p(\boldsymbol{r}|\boldsymbol{y}_{\text{mis}}, \boldsymbol{y}_{\text{obs}}, \boldsymbol{w}, \boldsymbol{x}; \boldsymbol{\theta}_w) p(\boldsymbol{y}_{\text{mis}}, \boldsymbol{y}_{\text{obs}}, \boldsymbol{w}|\boldsymbol{x}; \boldsymbol{\theta}_w) d\boldsymbol{y}_{\text{mis}} \\
&\quad \overset{\text{A-MAR}}{=} p(\boldsymbol{r}|\boldsymbol{y}_{\text{obs}}, \boldsymbol{w}, \boldsymbol{x}; \boldsymbol{\theta}_w) \int p(\boldsymbol{y}_{\text{mis}}, \boldsymbol{y}_{\text{obs}}, \boldsymbol{w}|\boldsymbol{x}; \boldsymbol{\theta}_w) d\boldsymbol{y}_{\text{mis}} \\
&\quad \overset{\text{cond 2}}{=} p(\boldsymbol{r}|\boldsymbol{y}_{\text{obs}}, \boldsymbol{w}, \boldsymbol{x}; \boldsymbol{\psi}) p(\boldsymbol{y}_{\text{obs}}, \boldsymbol{w}|\boldsymbol{x}, \gamma_w, \boldsymbol{\nu}).
\end{aligned}$$

If the prior for $\boldsymbol{\theta}_w$ factors as in condition 3, the posterior factors as

$$p(\boldsymbol{\theta}_w|\boldsymbol{y}_{\text{obs}}, \boldsymbol{r}, \boldsymbol{x}) = p(\gamma_w, \boldsymbol{\nu}|\boldsymbol{y}_{\text{obs}}, \boldsymbol{w}, \boldsymbol{x}) p(\boldsymbol{\psi}|\boldsymbol{y}_{\text{obs}}, \boldsymbol{w}, \boldsymbol{x}, \boldsymbol{r}).$$

However, to ultimately draw inferences about $p(\boldsymbol{y}|\boldsymbol{x}; \gamma(\gamma_w, \boldsymbol{\nu}))$, we need to integrate over $\boldsymbol{w}$,

$$p(\boldsymbol{y}|\boldsymbol{x}; \gamma(\boldsymbol{\theta})) = \int p(\boldsymbol{y}|\boldsymbol{w}, \boldsymbol{x}; \gamma_w) p(\boldsymbol{w}|\boldsymbol{x}; \boldsymbol{\nu}) d\boldsymbol{w}. \tag{5.3}$$

We discuss the complications that arise with this in the next subsection.

5.4.4 Fully Bayesian modeling with auxiliary covariates

In the setting of auxiliary covariates above, one typically constructs an *imputation model*, $p^\star(\boldsymbol{y}|\boldsymbol{x}, \boldsymbol{w}; \gamma_w)$ that is used to impute missing data under an assumption of A-MAR. The *inference model*, $p(\boldsymbol{y}|\boldsymbol{x}; \gamma)$ is the model one would like to use for inference (as discussed above). Also, as discussed above, it seems clear that (5.3) should hold, i.e.,

$$p^\star(\boldsymbol{y}|\boldsymbol{w}; \gamma_w) = p(\boldsymbol{y}|\boldsymbol{w}; \gamma_w). \tag{5.4}$$

When this equality does not hold, the imputation has been called *uncongenial* (Meng, 1994).

Up until recently, it was common that imputation models were built separately from inference models (by different individuals, with those ultimately doing inference not necessarily even being aware of the form of the imputation model); as such, the relationship in (5.4) generally did not hold. Now, it is typical that the same investigator (research group) constructs both models, but it is still often the case that the two models are uncongenial.

In terms of coherent probability models, it is necessary for the two models to be *congenial*, but there are complications regarding interpretability of the inference model and the need to address the distribution of the auxiliary covariates (which is essentially a potentially high-dimensional nuisance parameter). In the context of longitudinal data, we refer the reader to Daniels et al. (2013) for the details of a solution that involves Bayesian shrinkage priors (to estimate the distribution of the auxiliary covariates) and marginalized models (Heagerty, 2002; Roy and Daniels, 2008) to allow for simple interpretation of the full data response model.

5.5 Bayesian Non-Ignorability

When ignorability does not hold (even with auxiliary covariates), missingness is called *non-ignorable*; this typically occurs from a violation of condition 1, i.e., NMAR, but also happens via violations of conditions 2 and 3. The latter violation typically arises when making an assumption of MAR in mixture models (an example is provided later in this section) and is referred to as *non-ignorable MAR*.

In this setting, the entire full data model, $p(\boldsymbol{y}, \boldsymbol{r}|\boldsymbol{\theta})$ needs to be (correctly) specified, unlike in the ignorable case. In addition, any valid approach obviously needs to identify, either explicitly or implicitly, the extrapolation distribution. This can be done in several ways:

1. (unverifiable) assumptions about the missingness, e.g., with identifying restrictions.
2. modeling assumptions.
3. an informative prior for $\boldsymbol{\theta}_E$.

A powerful asset of the Bayesian approach is to have the ability to characterize uncertainty about the extrapolation distribution (equivalently, about the missing data mechanism) through an informative, but non-degenerate, prior. Before discussing the approaches listed above to identify the extrapolation distribution, we will formally define sensitivity parameters that are essential in drawing inference for incomplete data.

Recall the extrapolation factorization, which factors the distribution of the full data as

$$p(\boldsymbol{y}, \boldsymbol{r}|\boldsymbol{\theta}) = p(\boldsymbol{y}_{\text{obs}}, \boldsymbol{r}|\boldsymbol{\theta}_O)p(\boldsymbol{y}_{\text{mis}}|\boldsymbol{y}_{\text{obs}}, \boldsymbol{r}, \boldsymbol{\theta}_E), \tag{5.5}$$

where $\boldsymbol{\theta} = (\boldsymbol{\theta}_O, \boldsymbol{\theta}_E)$. As mentioned earlier, the data provide no information on the extrapolation distribution, $p(\boldsymbol{y}_{\text{mis}}|\boldsymbol{y}_{\text{obs}}, \boldsymbol{r}, \boldsymbol{\theta}_E)$. Thus, sensitivity parameters should be some functional of $\boldsymbol{\theta}_E$. A formal definition (Daniels and Hogan, 2008) is given as follows. The parameter $\boldsymbol{\xi}_S$ is said to be a *sensitivity parameter*, if there exists of reparameterization of the parameters of the full data model, $\boldsymbol{\xi}(\boldsymbol{\theta}) = (\boldsymbol{\xi}_S, \boldsymbol{\xi}_M)$ such that the following three conditions hold:

1. $\boldsymbol{\xi}_S$ is a non-constant function of $\boldsymbol{\theta}_E$.
2. the observed data likelihood is constant as a function of $\boldsymbol{\xi}_S$
3. at a fixed value of $\boldsymbol{\xi}_S$, the observed data likelihood is a non-constant function of $\boldsymbol{\xi}_M$.

For further discussion of sensitivity parameters and an alternative (but equivalent) definition, see Chapter 18 in this volume. We start by explaining why mixture models are a straightforward approach to model non-ignorable missing data and identify sensitivity parameters.

5.5.1 Why mixture models?

Mixture models are advantageous for non-ignorable missingness, since they can be parameterized in terms of the extrapolation factorization, and make explicit what is identified (or not) from the observed data. This can be seen in the simple example of a bivariate normal mixture model given below. In general, it can be seen by writing the mixture model as

$$p(\boldsymbol{y}|\boldsymbol{r}) = p(\boldsymbol{y}|\boldsymbol{r})p(\boldsymbol{r}) = p(\boldsymbol{y}_{\text{mis}}|\boldsymbol{y}_{\text{obs}}, \boldsymbol{r})p(\boldsymbol{y}_{\text{obs}}|\boldsymbol{r})p(\boldsymbol{r}).$$

The first component is the extrapolation distribution and the product of the last two is the distribution of the observed data. The connection of the mixture model factorization with the extrapolation factorization provides a natural basis for finding sensitivity parameters and specifying informative priors. The latter is essential in appropriately quantifying uncertainty about the missingness process in final inference. Settings where all the parameters are identified do not allow for this, which we discuss later in the section.

To illustrate these we start out by introducing a simple example of a mixture model with a bivariate longitudinal response where only the second component can be missing ($R = 0$):

$$\begin{aligned}
p(y_1|R = r) &\sim N(\mu^{(r)}, \sigma^{(r)}) : r = 0, 1 \\
p(y_2|y_1, R = r) &\sim N(\beta_0^{(r)} + \beta_1^{(r)} y_1, \tau^{(r)}) : r = 0, 1. \\
R &\sim Ber(\phi)
\end{aligned}$$

Here, $\boldsymbol{\theta}_O = (\mu^{(0)}, \mu^{(1)}, \sigma^{(0)}, \sigma^{(1)}, \beta_0^{(1)}, \beta_1^{(1)}, , \tau^{(1)})$ and $\boldsymbol{\theta}_E = (\beta_0^{(0)}, \beta_1^{(0)}, \tau^{(0)})$. So, functions of $\boldsymbol{\theta}_E$ are candidates for sensitivity parameters on which an informative prior can be specified.

Note that approaches 1 and 2 above, in fact, are special cases of 3, in that they correspond to very informative, degenerate (or point mass) priors. In particular, they assume that

$$p(\boldsymbol{\theta}_E|\boldsymbol{\theta}_o) = I\{\boldsymbol{\theta}_E = g(\boldsymbol{\theta}_O)\},$$

where g is a fixed and known function. We will see this in our discussion of identifying restrictions and modeling assumptions next. Also, we discuss non-parametric estimation of the observed data model, $p(\boldsymbol{y}_{\text{obs}}|\boldsymbol{r})p(\boldsymbol{r})$ in Section 5.7.

5.5.2 Identifying restrictions

One way to identify the unidentified parameters in mixture models for incomplete longitudinal data is via identifying restrictions. A popular choice is the available case missing value restriction (ACMV), which in the context of monotone missingness is the same as MAR (Molenberghs et al., 1998). In the bivariate mixture model from the previous subsection, the restriction sets $p(y_2|y_1, R = 0) = p(y_2|y_1, R = 1)$. This identifies the full set of non-identified parameters, $\boldsymbol{\theta}_E$ in the following way,

$$\begin{aligned}
\beta_0^{(0)} &= \beta_0^{(1)} \equiv g_0(\boldsymbol{\theta}_O), \\
\beta_1^{(0)} &= \beta_1^{(1)} \equiv g_1(\boldsymbol{\theta}_O), \\
\tau^{(0)} &= \tau^{(1)} \equiv g_2(\boldsymbol{\theta}_O).
\end{aligned}$$

For Bayesian inference, clearly these are just degenerate priors, corresponding to $p(\boldsymbol{\theta}_E|\boldsymbol{\theta}_o) = I\{\boldsymbol{\theta}_E = g(\boldsymbol{\theta}_O)\}$ where $g(\boldsymbol{\theta}_O) = (g_0(\boldsymbol{\theta}_O), g_1(\boldsymbol{\theta}_O), g_2(\boldsymbol{\theta}_O))$. Note that this is a case of non-ignorable MAR. We can see this by noting that a particular functional of the full data response model, $E[Y_1]$, is a function of $\mu^{(j)} : j = 0, 1$ and ϕ,

$$E[Y_1] = \mu^{(1)}\phi + \mu^{(0)}(1 - \phi)$$

and the missing data mechanism, $P(R = 1|Y_1 = y_1)$ is a function of $\mu^{(j)} : j = 0, 1$ and ϕ (see Example 5.9 in Daniels and Hogan, 2008). Clearly the parameters of the missing data mechanism and the full data response model are not distinct and thus the missingness is not ignorable. For characterizing uncertainty about the missingness, identifying restrictions are less than satisfactory as outlined above. However, they can provide a centering point for the construction of informative priors for the sensitivity parameters.

There are also various NMAR identifying restrictions that have been proposed. We introduce these in the context of a trivariate longitudinal response that has monotone missingness. Define $S = \sum_{j=1}^{3} R_j$ where $R_j = I\{Y_j \text{ is observed}\}$. Consider the following mixture model (based on S),

$$\begin{aligned} p(y_1|S = s) &\sim N(\mu^{(s)}, \sigma^{(s)}) \\ p(y_2|y_1, S = s) &\sim N(\beta_{01}^{(s)} + \beta_{11}^{(s)} y_1, \tau_1^{(s)}) \\ p(y_3|y_2, y_1, S = s) &\sim N(\beta_{02}^{(s)} + \beta_{12}^{(s)} y_1 + \beta_{22}^{(s)} y_2, \tau_2^{(s)}) \\ S &\sim Mult(\boldsymbol{\phi}). \end{aligned}$$

Clearly, the distributions, $p(y_2|y_1, S = 1)$, $p(y_3|y_2, y_1, S = 1)$, and $p(y_3|y_2, y_1, S = 2)$, and their corresponding parameters, are not identified by the observed data. In this model, the ACMV restriction identifies these distributions/parameters via the following restrictions:

$$\begin{aligned} p(y_2|y_1, S = 1) &= p(y_2|y_1, S \geq 2) \\ p(y_3|y_2, y_1, S = 1) &= p(y_3|y_2, y_1, S = 3) \\ p(y_3|y_2, y_1, S = 2) &= p(y_3|y_2, y_1, S = 3). \end{aligned}$$

The neighboring case missing value restriction (NCMV), identifies them via equating the parameters in a pattern that is not identified to the nearest neighbor identified pattern:

$$\begin{aligned} p(y_2|y_1, S = 1) &= p(y_2|y_1, S = 2) \\ p(y_3|y_2, y_1, S = 1) &= p(y_3|y_2, y_1, S = 3) \\ p(y_3|y_2, y_1, S = 2) &= p(y_3|y_2, y_1, S = 3). \end{aligned}$$

The non-future dependent missing value restriction (NFMV) (Kenward et al, 2003), partially identifies them via:

$$p(y_3|y_2, y_1, S = 1) = p(y_3|y_2, y_1, S \geq 2). \tag{5.6}$$

Implicit in the right-hand side of (5.6) is one unidentified distribution, $p(y_3|y_2, y_1, S = 2)$; there is also an unidentified distribution in the pattern $S = 1$, $p(y_2|y_1, S = 1)$. This restriction reduces the dimension of the non-identified parameter space and in general, with each additional pattern, adds *one* additional unidentified distribution.

The name for the restriction is related to the fact that the missing data mechanism depends on the past responses and the current response, but not the future responses. So for example,

$$p(R_j = 1|R_{j-1} = 1, y_1, \ldots, y_n) = p(R_j = 1|R_{j-1} = 1, y_1, \ldots, y_{j-1}, y_j).$$

The non-future dependent missing value restriction greatly reduces the number of potential sensitivity parameters. See Wang and Daniels (2011) for a suggested Bayesian approach with sensitivity analysis for using this restriction in a more general setting. For the exact specification of these restrictions for $n > 3$, we refer the reader to Thijs et al. (2002) and Kenward et al. (2003).

5.5.3 Identification by modeling assumptions

We provide a few examples of identification via modeling assumptions in what follows. Some of these approaches allow for sensitivity parameters, but some do not. As an example of modeling assumptions that allow for sensitivity parameters, consider a normal mixture model with n possible dropout times (and monotone missingness) and patterns based on these dropout times (or equivalent, the number of observed responses); more precisely, define S to be the number of observed responses (which takes values $\{1, \ldots, n\}$). Identification of the covariance matrix within each pattern, Σ_s might be done by assuming it is constant across patterns, $\Sigma_s = \Sigma$ for $s = 1, \ldots, n$; this corresponds to a degenerate prior on the covariance parameters as discussed above. This assumption is partially testable. For example, for patterns $k \geq j$, the variance, σ_{jj} is identifiable and the equality across the patterns is testable. However, for patterns $k < j$ it is not. An alternative would be to assume the unidentified Modified Choleski parameters (Pourahmadi, 1999; Daniels and Pourahmadi, 2002) in a pattern are equal to their corresponding identified counterparts (e.g., set to their values under MAR). More specifically, consider

$$p(y_j|y_{j-1}, \ldots, y_1, S = s) = N(\beta_0^{(s)} + \sum_{k=1}^{j-1} \beta_k^{(s)} y_k, \tau^{(s)}).$$

We then set $\beta_k^{(s)} = \beta_k^{(\geq s)}$ for $j < s$ and $\tau^{(s)} = \tau^{(\geq s)}$ which are the corresponding coefficients and variance for the conditional distribution, $p(y_j|y_{j-1}, \ldots, y_1, S \geq j)$ given below

$$p(y_j|y_{j-1}, \ldots, y_1, S \geq j) = N(\beta_0^{(\geq j)} + \sum_{k=1}^{j-1} \beta_k^{(\geq j)} y_k, \tau^{(\geq j)}).$$

However, here we focus our discussion on the mean. Within each pattern, we assume the mean changes linearly in time. Without loss of generality, assume times $\{1, \ldots, n\}$. Define S to be the number of observed responses. For example, for $S = j$ (pattern j),

$$Y_{ik}|S = j \sim N(\beta_0^{(j)} + \beta_1^{(j)} k, \Sigma).$$

As long as $j > 1$, $(\beta_0^{(j)}, \beta_1^{(j)})$ will be identified based on the linearity assumption. In terms of parameters of the extrapolation factorization, there is an implicit very informative prior. Clearly, the slope after time j is not identified (since no data beyond time j in pattern j). So the slope should be more precisely written as $\beta_1^{(j)} kI\{j \geq k\} + \beta_1^{(j)\star} kI\{j < k\}$. As a result, implicitly, there is the following conditional prior on $\beta_1^{(j)\star}$, $p(\beta_1^{(j)\star}|\beta_1^{(j)}) = I\{\beta_1^{(j)\star} = \beta_1^{(j)}\}$.

Other approaches that identify the extrapolation distribution via parametric modeling assumptions do not allow for sensitivity parameters (see Chapter 18 of this volume). These include parametric selection models (Diggle and Kenward, 1994, Cowles et al., 1996, Nandram and Choi, 2010) and parametric shared-parameters models (Wu and Carroll, 1989; Li et al., 2007; Yuan and Yin, 2010).

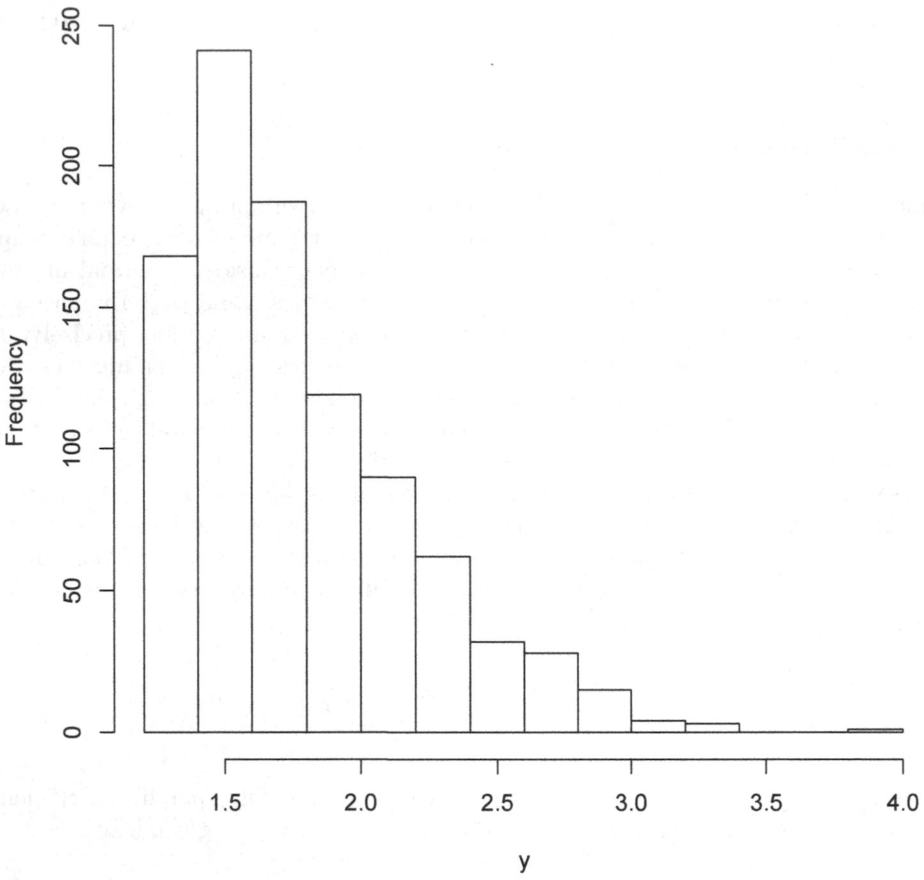

FIGURE 5.1
Histogram of observed data response.

We provide a simple cross-sectional example to understand this approach to identifiability. Consider the model

$$\begin{aligned} g(P(R_i = 0|y_i, \psi)) &= \psi_0 + \psi_1 y_i \\ Y_i|\mu, \sigma &\sim N(\mu, \sigma). \end{aligned}$$

Suppose the observed data look like that given in Figure 5.1, i.e., a normal distribution with the left tail missing. Since we assumed the full data response is normal, this indicates small values of y are unlikely to be observed. As such, ψ_1 will be negative. We discuss ways to avoid identification and allow sensitivity parameters using Bayesian non-parametrics in Section 5.7. Further discussion of the identification can be found in Chapter 18 of this volume.

Informative priors can be used on the parameters, like ψ_1 in these models, though they are identified. As an example, consider the following more general parametric selection model for longitudinal data with covariates and monotone missingness,

$$\begin{aligned} g(P(R_{ij}=0|R_{ij-1}=1,\boldsymbol{y},\boldsymbol{\psi})) &= \psi_0+\psi_1 y_{j-1}+\psi_2 y_j \\ Y_i|\gamma &\sim N(\boldsymbol{x}_i\boldsymbol{\beta},\Sigma). \end{aligned}$$

A seeming sensitivity parameter in the above would be ψ_2 since y_j is not observed when $R_j = 0$. However, this parameters will generally be identified as discussed above (see also the discussion of Diggle and Kenward, 1994; Daniels and Hogan, 2008 Chapter 8). An informative prior can be placed on this parameter, but the prior and the posterior will typically be very different (see the OASIS case study in Chapter 10 of Daniels and Hogan, 2008). Further development can be found in Chapter 18 of this volume.

5.5.4 Some challenges for mixture models

We have advocated mixture models as the best approach to handle non-ignorable missingness. However, there are several challenging issues that occur in such models. We describe a few next.

A general mixture model defines the mixture based on each different pattern of missingness, defined by $\boldsymbol{r}$. For p responses, always observing the first value, and monotone missingness, there will be p patterns as in previous examples. Obviously, even in this case, if p is large, it can be problematic to fit the model conditional on pattern. In a longitudinal setting, Bayesian approaches have been proposed to reduce the number of patterns by grouping the "dropout" times in a data-dependent way (Roy and Daniels, 2008). For continuous dropout time, Hogan, Lin, and Herman (2004), Daniels and Hogan (2008, case study 3), Su and Hogan (2010), and Su (2012) propose varying coefficient models; the latter three, Bayesian varying coefficient models. We provide some details next.

Define U to be the dropout time. Now specify the mixture model (conditional on dropout time),

$$\begin{aligned} Y_{ij}|\boldsymbol{\beta}_i,U &\sim N(\beta_0(u)+b_{0i}+(\beta_1(u)+b_{1i})t_j,\sigma^2), \\ \boldsymbol{\beta}_i &\sim N(0,\boldsymbol{\theta}(u)), \\ U &\sim F_{\boldsymbol{\theta}}. \end{aligned}$$

The distribution, $F_{\boldsymbol{\theta}}$ can be specified non-parametrically; see Section 5.7 for details. The regression coefficients, $\boldsymbol{\beta}(u)$ need to be modeled as a function of dropout time, u. Penalized splines (Ruppert et al., 2003) are a convenient and simple choice to do this. Ultimate inference is typically based on

$$E[\boldsymbol{Y}] = \int E[\boldsymbol{Y}|U=u]dF(u).$$

Interpretation can be problematic for non-linear link functions for $Y|U$. This can be overcome using marginalized models; see Su (2012) for details. As specified, there are no sensitivity parameters. However, similar to linear trends in finite mixture models, there is a sensitivity parameter, Δ, that is implicitly set to zero,

$$E[Y_{ij}|U] = \beta_0(u)+\beta_1(u)t_j+\Delta(t_j-u)_+.$$

Priors can be placed on Δ.

However, for monotone missingness that is not ordered or non-monotone missingness, it is unclear how to form these patterns. Little (1993) defined patterns based on specific "clusters" of missingness. However, it is often not straightforward to form these sets. Lin et al. (2004) proposed using latent class models to form sets (or classes). Conditional on being in one of these sets, any remaining missingness is typically assumed ignorable (though it is possible to weaken this assumption). For example, if patterns are defined based on dropout times, which is common, and there is intermittent missing, the default assumption would be the intermittent missingness is ignorable *conditional* on dropout time. This is a special case of what has been termed *partial ignorability* and was formalized in Harel and Schafer (2009), but has been implicitly assumed in previous models (e.g., Hogan and Laird, 1997). In general, simultaneously dealing with intermittent missingness and dropout missingness in Bayesian (and frequentist) inference in appropriate and flexible ways is an open and active research area.

5.6 General Computational Issues with the Posterior Distribution

For most models for incomplete data, under ignorability or non-ignorability, the posterior distribution of the parameters is sampled using a Markov chain Monte Carlo (MCMC) algorithm. In particular, Gibbs sampling is often the specific algorithm chosen, sampling sequentially from the full conditional distribution of each parameter (or blocks of parameters). Consider a parameter vector $\boldsymbol{\theta}$ with p components, $\theta_j : j = 1, \ldots, p$ and data $\boldsymbol{y}$. The idea is to sample from the full conditional distribution of each parameter. For example, at iteration k, the algorithm would proceed as follows:

$$\begin{aligned}
\text{Sample } \theta_1^{(k)} &\sim p(\theta_1 | \theta_2^{(k-1)}, \ldots, \theta_p^{(k-1)}, \boldsymbol{y}), \\
\text{Sample } \theta_2^{(k)} &\sim p(\theta_2 | \theta_1^{(k)}, \theta_3^{(k-1)}, \ldots, \theta_p^{(k-1)}, \boldsymbol{y}), \\
&\vdots \\
\text{Sample } \theta_p^{(k)} &\sim p(\theta_p | \theta_1^{(k)}, \ldots, \theta_{p-1}^{(k)}, \boldsymbol{y}).
\end{aligned}$$

This can also be done in blocks. For example, sampling (θ_1, θ_2) from their joint full conditional distribution, as opposed to sampling them individually as above. As an example in the context of a specific model, assume the full data response follows a multivariate normal distribution with mean $\boldsymbol{\mu}$ and covariance matrix $\boldsymbol{\Sigma}$, the algorithm, at iteration k would sample

$$\begin{aligned}
\boldsymbol{\mu}^{(k)} &\sim p(\boldsymbol{\mu} | \boldsymbol{\Sigma}^{(k-1)}, \boldsymbol{y}), \\
\boldsymbol{\Sigma}^{(k)} &\sim p(\boldsymbol{\Sigma} | \boldsymbol{\mu}^{(k)}, \boldsymbol{y}).
\end{aligned}$$

Here, there are two blocks: the vector of mean parameters, $\boldsymbol{\mu}$ and the covariance matrix, $\boldsymbol{\Sigma}$. If conditionally conjugate priors are used for $\boldsymbol{\mu}$ and $\boldsymbol{\Sigma}$ then the respective full conditional distributions will be multivariate normal and inverse Wishart. If conditionally conjugate priors are not used (or are not available), the full conditional distribution is not available in closed form and various approaches can be used to sample from it including Metropolis-Hastings algorithm (Hastings, 1970) or slice sampling (Neal, 2003), among other approaches. We provide some details on the Metropolis-Hasting (M-H) algorithm next.

The M-H algorithm samples from a candidate distribution and then accepts the candidate value with a probability specified such that the resulting sample is from the distribution of interest. The two most common variations are a random-walk M-H algorithm, which perturbs the current value via a mean zero candidate distribution or to actually sample from an approximation to the distribution of interest. Define $p(\cdot)$ to be the true distribution and $q(\cdot|\cdot)$ to be the candidate distribution. At iteration k, we sample $\theta^{(k)} \sim q(\cdot|\theta^{(k-1)})$, and accept it with probability given by

$$\min\left\{1, \frac{p(\theta^{(k)})q(\theta^{(k-1)}|\theta^{(k)})}{p(\theta^{(k-1)})q(\theta^{(k)}|\theta^{(k-1)})}\right\}.$$

For the random walk M-H algorithm, $q(\cdot|\theta^{(k-1)})$ is often set to a normal (or t-distribution) with mean $\theta^{(k-1)}$ and variance τ^2. The variance τ^2 should be fixed such that the probability of acceptance is around 30% (Gelman et al., 1997). Another option is to sample from an approximation to $p(\cdot)$, e.g., using a Laplace approximation (Tierney and Kadane, 1986). In this case, $q(\cdot|\theta)$ often does not depend on θ, i.e., $q(\cdot|\theta) = \widehat{p}(\cdot)$. In this situation, acceptance close to 1.0 is desired.

Additional issues arise in incomplete data. Consider the multivariate normal model above in the setting of Bayesian ignorability. We denote the observed responses as $\boldsymbol{y}_{\text{obs}}$. Replacing $\boldsymbol{y}$ with $\boldsymbol{y}_{\text{obs}}$ in the above, the full conditional distribution of Σ will no longer be available in closed form (i.e., not inverse Wishart). This unnecessarily complicates the computations. In general, conditionally conjugate priors (specified on the full data response model parameters) for the full data response posterior are often not conditionally conjugate for the full conditional distributions from the observed data response posterior. This problem can be avoided by using data augmentation (Hobert, 2010). For data augmentation, the missing data, $\boldsymbol{y}_{\text{mis}}$ is sampled at every iteration along with the parameters. As such, sampling from the full conditional distribution of the parameters will be the same as if we had no missing data. This corresponds to explicitly doing the imputation of the missing data at each iteration (as opposed to implicitly doing it by having the missing data integrated out and working with the observed data posterior as discussed in Section 5.4); so we are still obtaining a sample of the parameters from the observed data posterior as desired. At iteration k, we would sample sequentially

$$\begin{aligned}
\boldsymbol{y}_{\text{mis}}^{(k)} &\sim p(\boldsymbol{y}_{\text{mis}}|\boldsymbol{\mu}^{(k-1)}, \boldsymbol{\Sigma}^{(k-1)}, \boldsymbol{y}_{\text{obs}}),\\
\boldsymbol{\mu}^{(k)} &\sim p(\boldsymbol{\mu}|\boldsymbol{\Sigma}^{(k-1)}, \boldsymbol{y}_{\text{mis}}^{(k)}, \boldsymbol{y}_{\text{obs}}),\\
\boldsymbol{\Sigma}^{(k)} &\sim p(\boldsymbol{\Sigma}|\boldsymbol{\mu}^{(k)}, \boldsymbol{y}_{\text{mis}}^{(k)}, \boldsymbol{y}_{\text{obs}}).
\end{aligned}$$

By doing this at each iteration, the full conditional distribution of the parameters is now based on the full data, $\boldsymbol{y}^{(k)} = (\boldsymbol{y}_{\text{mis}}^{(k)}, \boldsymbol{y}_{\text{obs}})$ and the conditional conjugacy is preserved, i.e., the full conditionals for $\boldsymbol{\mu}$ and $\boldsymbol{\Sigma}$ are multivariate normal and inverse Wishart, respectively. In addition, that data augmentation step itself is just a (multivariate) normal.

Regarding data augmentation, similar computational advantages occur in models for Bayesian non-ignorability. Consider the following parametric selection model presented earlier,

$$g(P(R_{ij} = 0|R_{ij-1} = 1, \boldsymbol{y}, \boldsymbol{\psi})) = \psi_0 + \psi_1 y_{j-1}, \psi_2 y_j \tag{5.7}$$

$$Y_i|\gamma \sim N(\boldsymbol{x}_i\boldsymbol{\beta}, \boldsymbol{\Sigma}). \tag{5.8}$$

where $j = 1, \ldots, n$ and $\boldsymbol{\gamma} = (\boldsymbol{\beta}, \boldsymbol{\Sigma})$. Under monotone missingness, we assume individual i

has $n_i - 1$ observed responses. Then, the observed data posterior is proportional to

$$\prod_i \int \left\{ \prod_{j=1}^{n_i-1} p(r_{ij} = 1 | r_{j-1} = 1, y_{i,j-1}, y_{ij}, \boldsymbol{\psi}) p(y_{ij} | y_{i1}, \ldots, y_{i,j-1}; \boldsymbol{\gamma}) \right\}$$
$$\times p(r_{i,n_i} = 0 | r_{i,n_i-1} = 1, y_{i,n_i-1}, y_{i,n_i}; \boldsymbol{\psi}) p(y_{i,n_i} | y_{i1}, \ldots, y_{i,n_i-1}; \boldsymbol{\gamma})$$
$$\times dy_{i,n_i} p(\boldsymbol{\psi}, \boldsymbol{\gamma}). \tag{5.9}$$

This creates several problems. First, the observed data likelihood is not available in closed form, i.e., the integral in (5.9) is not available in closed form (conditional on the observed data, $(\boldsymbol{y}_{\text{obs}}, \boldsymbol{r})$). Second, none of the full conditional distributions of the parameters are available in closed form. However, with data augmentation, and thus, conditional on the filled-in missing data (here, $(y_{i,n_i}, \ldots, y_{in})$), sampling proceeds on a binary data model with no missingness (5.7) and a multivariate normal model with no missingness (5.8).

Similar advantages occur with shared parameter models. Consider the following simple shared parameter model,

$$g(P(R_{ij} = 0 | R_{ij-1} = 1, \boldsymbol{y}, \boldsymbol{\beta}_i, \boldsymbol{\psi})) = \psi_0 + \psi_1 b_i, \tag{5.10}$$
$$Y_i | \boldsymbol{\gamma}, b_i \sim N(\boldsymbol{x}_i \boldsymbol{\beta} + b_i, \sigma^2 I), \tag{5.11}$$
$$b_i | \tau^2 \sim N(0, \tau^2). \tag{5.12}$$

where $j = 1, \ldots, n$ and $\boldsymbol{\gamma} = (\boldsymbol{\beta}, \sigma^2)$. As with the selection model, if we assume monotone missingness and that individual i has $n_i - 1$ observed responses. Then, the observed data posterior is proportional to

$$\prod_i \int \int \{ \prod_{j=1}^{n_i-1} p(r_{ij} = 1 | r_{j-1} = 1, b_i, \boldsymbol{\psi}) p(y_{ij} | y_{i1}, \ldots, y_{i,j-1}; \boldsymbol{\gamma}, b_i) \}$$
$$\times p(r_{i,n_i} = 0 | r_{i,n_i-1} = 1, b_i; \boldsymbol{\psi}) p(y_{i,n_i} | y_{i1}, \ldots, y_{i,n_i-1}, b_i; \boldsymbol{\gamma}) p(b_i | \tau^2)$$
$$db_i dy_{i,n_i} p(\boldsymbol{\psi}, \boldsymbol{\gamma}, \tau^2). \tag{5.13}$$

As with the selection model, the observed data likelihood is intractable and the full conditional distribution of the parameters is not available in closed form. However, by sampling from the full conditional distribution b_i and data augmenting the missing y's $(y_{i,n_i}, \ldots, y_{in})$), sampling then proceeds as a Bayesian logistic regression model (with no missingness) and a multivariate normal regression with a random effect.

For mixture models, it is often the case that the parameters of the observed data distribution are given priors directly. In this setting, data augmentation is often not needed. As an example, we revisit the following mixture models under monotone missingness,

$$Y_j | y_1, \ldots, y_{j-1}, S = k \sim N \left(\beta_0^{(k)} + \sum_{k=1}^{j-1} \beta_l^{(k)} y_k, \tau^{(k)} \right).$$

For $j \leq k$, the parameters above are identified. With this specification, priors are typically put directly on $\boldsymbol{\beta}$ and τ (for $j \leq k$, i.e., observed data model parameters) and data augmentation (for $S = k$, augmenting $(Y_{k+1}, \ldots, Y_n)$) provides no computational advantages.

Data augmentation is closely related to the EM algorithm (Dempster, Laird, and Rubin, 1977). Both are used when the full data likelihood (posterior) is easier to work with than the observed data likelihood (posterior). Situations where the data augmentation step of the MCMC is available in closed form correspond to situations where the E-step is available in closed form in the EM algorithm.

Software is available to sample from many Bayesian models for missing data. The most popular is WinBUGS (Lunn et al., 2012). The advantage of this software is that the user only needs to specify the model and the priors (i.e., not derive the full conditional distributions). However, care must be taken to ensure proper mixing and convergence of the MCMC algorithm (Gelman et al., 2010). Most of the worked examples in Daniels and Hogan (2008) use this software to sample from the posterior distribution.

5.7 Non-Parametric Bayesian Inference

In principle, to make inference about the full data model, one can estimate the observed data model using Bayesian non-parametric methods, but not the extrapolation distribution. A recent paper by Wang et al. (2011) used this approach for the case of longitudinal binary data with non-ignorable dropout. In this situation, the joint distribution of the binary response data and the observed data indicators is a multinomial distribution. However, the problem is that as the number of observation times increases, the multinomial distribution grows in dimension and will not be well estimated using a saturated multinomial model due to sparse (or empty) cells. A strategy to address this is to put a Dirichlet prior on the multinomial cell probabilities centered on a parsimonious (and appropriate) model (with unknown parameters) with an unknown precision parameter. More specifically, let $\boldsymbol{W}$ follow a saturated multinomial distribution with a conjugate Dirichlet prior,

$$\begin{aligned} \boldsymbol{W} &\sim \text{Mult}(\boldsymbol{\theta}), \\ \boldsymbol{\theta} &\sim \text{Dir}(\boldsymbol{\pi}(\boldsymbol{\alpha}), \eta), \end{aligned}$$

where $E[\boldsymbol{\theta}] = \boldsymbol{\pi}(\boldsymbol{\alpha})$ and $\text{Var}[\boldsymbol{\theta}] \propto \eta$. $\boldsymbol{\pi}$ can be specified as a parsimonious shrinkage target (based on a log linear model) and the variance parameters, η can be given a uniform shrinkage prior (Daniels, 1999), which is proper; for its derivation in this setting, see Daniels, Wang, and Marcus (2013). For a vector of continuous responses, mixtures of Dirichlet processes (MacEachern and Mueller, 1998) can be used.

It is difficult to be effectively non-parametric with selection models and shared-parameter models for continuous responses beyond univariate settings. Scharfstein et al. (2003), in the setting of a univariate response, used a Dirichlet process (DP) prior for the full data response distribution and parametric logistic model for the missing data mechanism as follows,

$$\begin{aligned} g(P(R_i = 0 | y_i, \boldsymbol{\psi})) &= \psi_0 + \psi_1 y_i \\ Y_i | F &\sim F \\ F &\sim DP(\alpha G_0), \end{aligned}$$

where G_0 is specified as a normal distribution with mean μ and variance σ^2. In this model, ψ_1 is not identified and is a sensitivity parameter and as such,

$$p(\psi_1 | \psi_0, F, \boldsymbol{y}_{\text{obs}}, \boldsymbol{r}) = p(\psi_1 | \psi_0, F);$$

for further discussion and clarification, see Chapter 18 in this volume. So being non-parametric for the full data response model allows for a sensitivity parameter in the missing data mechanism here. In longitudinal settings, it is still possible to be non-parametric (as above) with the full data response model. However, it is difficult to be appropriately non-parametric in the missing data mechanism when the full data response is continuous. Define

$\bar{\boldsymbol{y}}_{ij} = (y_{i1}, \ldots, y_{i,j-1})$. Then, the missing data mechanism can be written as

$$g(P(R_{ij} = 0|\bar{\boldsymbol{y}}_{ij}, \boldsymbol{\psi})) = h(\bar{\boldsymbol{y}}_{ij}; \boldsymbol{\psi}_0) + \psi_1 y_{ij},$$

in which the potentially high-dimensional function, $h_j(\bar{\boldsymbol{y}}_{ij}; \boldsymbol{\psi}_0)$ needs to be estimated non-parametrically which is difficult. Similar problems occur with shared parameter models. For further discussion, see Chapter 8 in Daniels and Hogan (2008).

Being completely non-parametric for the observed data model in mixture models is more feasible. This can be done by specifying saturated multinomial distributions (as discussed above) with shrinkage for categorical data or by specifying Dirichlet process mixture of models (Escobar and West, 1995) for continuous data. Identification of the full data response model can then proceed by specification of an informative prior for the extrapolation distribution (Wang et al., 2011) or via identifying restrictions (Linero and Daniels, 2013; working paper). Various authors have specified parts of models non-parametrically to provide some robustness to parametric assumptions. In the setting of Bayesian ignorable inference with auxiliary covariates (i.e., Bayesian proper multiple imputation), Daniels, Wang and Marcus (2013) estimated the marginal distribution of the auxiliary covariates non-parametrically using the saturated multinomial approach with shrinkage described at the beginning of this section. In the setting of mixture models with continuous time dropout, Su (2012) specified the marginal distribution of dropout time using non-parametric Bayesian methods. In the context of selection models, Chen and Ibrahim (2006) use splines in the missing data mechanism and response model; in mixture models with continuous time dropout, Daniels and Hogan (2008), Su and Hogan (2010), and Su (2012) use splines to allow mean parameters to vary with dropout time. Non-parametric random effect in shared-parameter models (along with splines) have been implemented in Rizopoulos and Ghosh (2011). However, although these approaches are more robust to model specification than a full parametric model, most still do not allow for sensitivity parameters. In addition, in these settings, it is then essential to assess model fit (as discussed in Section 5.3) as the fit of the full data model to the observed data can be "lost" in the full data model specification.

5.8 Concluding Remarks and Future Directions

In this chapter, we have reviewed and introduced new Bayesian approaches for inference with incomplete data, including appropriate ways to do model selection and assess model fit, coherent Bayesian inference in the presence of auxiliary covariates, the connection of common approaches to model non-ignorable missingness and priors, general computational strategies, including data augmentation (and its connection to multiple imputation), and the use of Bayesian non-parametrics. Further insight into informative priors, sensitivity analysis, and identification in Bayesian inference will be examined and discussed in Chapter 18.

Most Bayesian approaches for incomplete data in the literature, and in this chapter, focus implicitly on mean regression. However, the impact of covariates on quantiles can also be of inferential interest. Yuan and Yin (2010) discuss an approach for Bayesian quantile regression under non-ignorable missingness using shared parameter models that is fully identified and does not allow sensitivity parameters. There is much development still to be done for non-ignorable missingness and quantile regression, including the development of approaches that allow for sensitivity parameters.

We discussed issues of comparison of Bayesian models with incomplete data in Section 5.3.

Adhering to the principles discussed there in the setting of trying to compare ignorable and non-ignorable models is an open problem. The reason for this is that in ignorable models, we do not explicitly specify a model for the missingness.

Bayesian models for ignorability with auxiliary covariates (as discussed in Section 5.4) also need further development, including extensions to allow for departures from ignorability (using sensitivity parameters) and allowing for all types of auxiliary covariates (e.g., continuous and/or time-varying). Such work is ongoing.

As discussed in Section 5.5, fully model-based approaches to handle both dropout and intermittent missingness need further development. As discussed earlier, intermittent missingness in mixture models is often assumed to be missing at random, given the dropout time. How to specify alternative assumptions in the context of mixture models is difficult. In addition, elicitation of priors on sensitivity parameters in the context of incomplete data is another area which requires further development, often in the context of specific applications. The key idea here is providing an easy to understand parameterization on which to elicit expert information.

In causal inference models, counterfactuals are often thought of as missing data as mentioned earlier and explored in Chapter 18. But, in addition, there are also issues with missingness in the data that is potentially observable. Bayesian methods for these settings need further development as many current approaches do not provide a principled way to address the missingness.

Many methods and application papers have appeared in the literature which address Bayesian approaches for missingness. For the interested reader, some of these that have been published since 2000 include Daniels and Hogan (2000), Nandram and Choi (2002), Scharfstein et al. (2003), Nandram and Choi (2004), Kaciroti et al. (2006), Zhang and Heitjan (2007), Kaciroti et al. (2008), Su and Hogan (2008), Paddock and Ebener (2009), Xie (2009), Yang et al. (2010), Wang et al. (2011), and Lee, Hogan, and Hitsman (2012).

References

Bedrick, E.J., Christensen, R., and Johnson, W. (1996). A new perspective on priors for generalized linear models. *Journal of the American Statistical Association* **91**, 1450–1460.

Berger, J.O. and Bernardo, J.M. (1992). On the development of the reference prior method. In: J.O. Berger, J.M. Bernardo, A.P. Dawid, and A.F.M. Smith (Eds.) *Bayesian Statistics 4*, pp. 35–60. Oxford: Oxford University Press.

Bernardo, J.M. (1979). Reference posterior distributions for Bayesian inference. *Journal of the Royal Statistical Society, Series B* **41**, 113–147.

Carlin, B.P. and Louis, T.A. (2008). *Bayesian Methods for Data Analysis.* Boca Raton: CRC Press.

Celeux, G., Forbes, F., Robert, C.P. and Titterington, D.M. (2006). Deviance information criteria for missing data models. *Bayesian Analysis* **1**, 651–674.

Chatterjee, A. and Daniels, M.J. (2013). An exploration of posterior predictive assessment for incomplete longitudinal data. *Working paper.*

Chen, Q. and Ibrahim, J.G. (2006). Semiparametric models for missing covariate and response data in regression models. *Biometrics* **62**, 177–184.

Cowles, M.K., Carlin, B.P., and Connett, J.E. (1996). Bayesian tobit modeling of longitudinal ordinal clinical trial compliance data with non-ignorable missingness. *Journal of the American Statistical Association* **91**, 86–98.

Daniels, M.J. (1999). A prior for the variance in hierarchical models. *The Canadian Journal of Statistics / La Revue Canadienne de Statistique* **27**, 567–578.

Daniels, M.J., Chatterjee, A. and Wang, C. (2012). Bayesian model selection for incomplete longitudinal data using the posterior predictive distribution. *Biometrics* **68**, 1055–1063.

Daniels, M.J. and Hogan, J.W. (2000). Reparameterizing the pattern mixture model for sensitivity analyses under informative dropout. *Biometrics* **56**, 1241–1248.

Daniels, M.J. and Hogan, J.W. (2008). *Missing Data in Longitudinal Studies: Strategies for Bayesian Modeling and Sensitivity Analysis.* Boca Raton: CRC Press.

Daniels, M.J. and Kass, R.E. (1999). Nonconjugate Bayesian estimation of covariance matrices and its use in hierarchical models. *Journal of the American Statistical Association* **94**, 254–1263.

Daniels, M.J. and Pourahmadi, M. (2002). Bayesian analysis of covariance matrices and dynamic models for longitudinal data. *Biometrika* **89**, 553–566.

Daniels, M.J., Wang, C. and Marcus, B. (2013). Fully Bayesian inference under ignorable missingness in the presence of auxiliary covariates. *Submitted for publication.*

Dempster, A.P., Laird, N.M., and Rubin, D.B. (1977). Maximum likelihood from incomplete data via the EM algorithm. *Journal of the Royal Statistical Society. Series B* **39**, 1–38.

Diggle, P.J. and Kenward, M.G. (1994). Informative drop-out in longitudinal data analysis (Disc: P73-93). *Applied Statistics* **43**, 49–73.

Escobar, M.D. and West, M. (1995). Bayesian density estimation and inference using mixtures. *Journal of the American Statistical Association* **90**, 577–588.

Gelman, A. (2006). Prior distributions for variance parameters in hierarchical models. *Bayesian Analysis* **1**, 515–533.

Gelman, A., Brooks, S., Jones, G., and Meng, X.-L. (2010). *Handbook of Markov Chain Monte Carlo.* Boca Raton: Chapman & Hall/CRC.

Gelman, A., Gilks, W.R., and Roberts, G.O. (1997). Weak convergence and optimal scaling of random walk metropolis algorithms. *Annals of Applied Probability* **7**, 110–120.

Gelman, A., Van Mechelen, I., Verbeke, G., Heitjan, D.F., and Meulders, M. (2005). Multiple imputation for model checking: Completed-data plots with missing latent data. *Biometrics* **61**, 74–85.

Harel, O. and Schafer, J.L. (2009). Partial and latent ignorability in missing-data problems. *Biometrika* **96**, 37–50.

Hastings, W.K. (1970). Monte Carlo sampling methods using Markov chains and their applications. *Biometrika* **57**, 97–109.

Heagerty, P.J. (2002). Marginalized transition models and likelihood inference for longitudinal categorical data. *Biometrics* **58**, 342–351.

Hobert, J.P. (2010). The data augmentation algorithm: Theory and methodology. In: S. Brooks, A. Gelman, G. Jones, and X.-L. Meng (Eds.). *Handbook of Markov Chain Monte Carlo.* Boca Raton: Chapman & Hall/CRC Press.

Hobert, J.P. and Casella, G. (1996). The effect of improper priors on Gibbs sampling in hierarchical linear mixed models. *Journal of the American Statistical Association* **91**, 1461–1473.

Hogan, J.W. and Laird, N.M. (1997). Mixture models for the joint distribution of repeated measures and event times. *Statistics in Medicine* **16**, 239–257.

Hogan, J.W., Lin, X., and Hernan, B. (2004). Mixtures of varying coefficient models for longitudinal data with discrete or continuous non-ignorable dropout. *Biometrics* **60**, 854–864.

Huang, L., Chen, M.H., and Ibrahim, J.G. (2005). Bayesian analysis for generalized linear models with non-ignorably missing covariates. *Biometrics* **61**, 767–780.

Jeffreys, H. (1961). *Theory of Probability (3rd ed.).* Oxford: Oxford University Press.

Kaciroti, N.A., Raghunathan, T.E., Schork, M.A., and Clark, N.M. (2008). A Bayesian model for longitudinal count data with non-ignorable dropout. *Applied Statistics* **57**, 521–534.

Kaciroti, N.A., Raghunathan, T.E., Schork, M.A., Clark, N.M., and Gong, M. (2006). A Bayesian approach for clustered longitudinal ordinal outcome with non-ignorable missing data. *Journal of the American Statistical Association* **101**, 435–446.

Kenward, M.G., Molenberghs, G., and Thijs, H. (2003). Pattern-mixture models with proper time dependence. *Biometrika* **90**, 53–71.

Laplace, P.S. (1878). *Œuvres Completes de Laplace.* Paris: Gauthier-Villars.

Lee, J., Hogan, J.W., and Hitsman, B. (2012). Sensitivity analysis and informative priors for longitudinal binary data with outcome dependent dropout. *Submitted for publication.*

Li, J., Yang, X., Wu, Y., and Shoptaw, S. (2007). A random-effects Markov transition model for Poisson-distributed repeated measures with non-ignorable missing values. *Statistics in Medicine* **26**, 2519–2532.

Lin, H., McCulloch, C.E., and Rosenheck, R.A. (2004). Latent pattern mixture models for informative intermittent missing data in longitudinal studies. *Biometrics* **60**, 295–305.

Linero, A. and Daniels, M.J. (2013). A Bayesian non-parametric approach to monotone missing data in longitudinal studies with informative missingness. *Technical report.*

Little, R.J.A. (1993). Pattern-mixture models for multivariate incomplete data. *Journal of the American Statistical Association* **88**, 125–134.

Little, R.J.A. (1994). A class of pattern-mixture models for normal incomplete data. *Biometrika* **81**, 471–483.

Little, R.J.A. and Rubin, D.B. (1987). *Statistical Analysis with Missing Data (1st Ed.).* New York: John Wiley & Sons.

Little, R.J.A. and Rubin, D.B. (2002). *Statistical Analysis with Missing Data (2nd Ed.).* New York: John Wiley & Sons.

Little, R.J.A. and Wang, Y. (1996). Pattern-mixture models for multivariate incomplete data with covariates. *Biometrics* **52**, 98–111.

Lunn, D., Jackson, C., Best, N., Thomas, A., and Spiegelhalter, D. (2012). *The BUGS Book.* Boca Raton: Chapman & Hall/CRC Press.

MacEachern, S.N. and Müller, P. (1998). Estimating mixture of Dirichlet process models. *Journal of Computational and Graphical Statistics* **7**, 223–238.

Mason, A., Richardson, S., and Best, N. (2012). Two-pronged strategy for using DIC to compare selection models with non-ignorable missing responses. *Bayesian Analysis* **7**, 109–146.

Meng, X.-L. (1994). Multiple-imputation inferences with uncongenial sources of input. *Statistical Science* **9**, 538–558.

Molenberghs, G. Michiels, B., Kenward, M.G., and Diggle, P.J. (1998). Monotone missing data and pattern-mixture models. *Statistica Neerlandica* **52**, 153–161.

Nandram, B. and Choi, J.W. (2002). Hierarchical Bayesian nonresponse models for binary data from small areas with uncertainty about ignorability. *Journal of the American Statistical Association* **97**, 381–388.

Nandram, B. and Choi, J.W. (2004). Nonparametric Bayesian analysis of a proportion for a small area under non-ignorable nonresponse. *Journal of Nonparametric Statistics* **16**, 821–839.

Nandram, B. and Choi, J.W. (2010). A Bayesian analysis of body mass index data from small domains under non-ignorable nonresponse and selection. *Journal of the American Statistical Association* **105**, 120–135.

Neal, R.M. (2003). Slice sampling. *The Annals of Statistics* **31**, 705–767.

Paddock, S.M. and Ebener, P. (2009). Subjective prior distributions for modeling longitudinal continuous outcomes with non-ignorable dropout. *Statistics in Medicine* **28**, 659–678.

Pourahmadi, M. (1999). Joint mean-covariance models with applications to longitudinal data: unconstrained parameterisation. *Biometrika* **86**, 677–690.

Rizopoulos, D. and Ghosh, P. (2011). A Bayesian semi-parametric multivariate joint model for multiple longitudinal outcomes and a time-to-event. *Statistics in Medicine* **30**, 1366–1380.

Roy, J. and Daniels, M.J. (2008). A general class of pattern mixture models for non-ignorable dropout with many possible dropout times. *Biometrics* **64**, 538–545.

Rubin, D.B. (1976). Inference and missing data. *Biometrika* **63**, 581–590.

Rubin, D.B. (1977). Formalizing subjective notions about the effect of nonrespondents in sample surveys. *Journal of the American Statistical Association* **72**, 538–543.

Rubin, D.B. (1978). Bayesian inference for causal effects: The role of randomization. *The Annals of Statistics* **6**, 34–58.

Rubin, D.B. (1987). *Multiple Imputation for Nonresponse in Surveys.* New York: John Wiley & Sons.

Ruppert, D., Wand, M.P., and Carroll, R.J. (2003). *Semiparametric Regression.* Cambridge: Cambridge Univeristy Press.

Schafer, J.L. (1997). *Analysis of Incomplete Multivariate Data.* Boca Raton: Chapman & Hall/CRC.

Scharfstein, D.O., Daniels, M.J., and Robins, J.M. (2003). Incorporating prior beliefs about selection bias into the analysis of randomized trials with missing outcomes. *Biostatistics* **4**, 495–512.

Scharfstein, D.O., Halloran, M.E., Chu, H., and Daniels, M.J. (2006). On estimation of vaccine efficacy using validation samples with selection bias. *Biostatistics* **7**, 615–629.

Scharfstein, D.O., Rotnitzky, A., and Robins, J.M. (1999). Adjusting for non-ignorable drop-out using semi-parametric nonresponse models. *Journal of the American Statistical Association* **94**, 1096–1120.

Su, L. (2012). A marginalized conditional linear model for longitudinal binary data when informative dropout occurs in continuous time. *Biostatistics* **13**, 355–368.

Su, L. and Hogan, J.W. (2008). Bayesian semi-parametric regression for longitudinal binary processes with missing data. *Statistics in Medicine* **27**, 3247–3268.

Su, L. and Hogan, J.W. (2010). Varying-coefficient models for longitudinal processes with continuous-time informative dropout. *Biostatistics* **11**, 93–110.

Thijs, H., Molenberghs, G., Michiels, B., Verbeke, G., and Curran, D. (2002). Strategies to fit pattern-mixture models. *Biostatistics* **3**, 245–265.

Tierney, L. and Kadane, J.B. (1986). Accurate approximations for posterior moments and marginal densities. *Journal of the American Statistical Association* **81**, 82–86.

Wang, C. and Daniels, M.J. (2011). A note on MAR, identifying restrictions, model comparison, and sensitivity analysis in pattern mixture models with and without covariates for incomplete data. *Biometrics* **67**, 810–818 (Correction in 2012, Vol. 68, p. 994).

Wang, C., Daniels, M.J., Scharfstein, D.O., and Land, S. (2011). A Bayesian shrinkage model for incomplete longitudinal binary data with application to the breast cancer prevention trial. *Journal of the American Statistical Association* **105**, 1333–1346.

Wu, M.C. and Carroll, R.J. (1988). Estimation and comparison of changes in the presence of informative right censoring by modeling the censoring process. *Biometrics* **44**, 175–188 (Correction in 1989, Vol. 45, p. 1347; 1991, Vol. 47, p. 357).

Xie, H. (2009). Bayesian inference from incomplete longitudinal data: A simple method to quantify sensitivity to non-ignorable dropout. *Statistics in Medicine* **28**, 2725–2747.

Yang, Y., Halloran, M.E., Daniels, M.J., Longini, I.M., Burke, D.S., and Cummings, D.A.T. (2010). Modeling competing infectious pathogens from a Bayesian perspective: Application to influenza studies with incomplete laboratory results. *Journal of the American Statistical Association* **105**, 1310–1322.

Yuan, Y. and Yin, G. (2010). Bayesian quantile regression for longitudinal studies with non-ignorable missing data. *Biometrics* **66**, 105–114.

Zhang, J. and Heitjan, D.F. (2007). Impact of non-ignorable coarsening on Bayesian inference. *Biostatistics* **8**, 722–743.

6

Joint Modeling of Longitudinal and Time-to-Event Data

Dimitris Rizopoulos
Erasmus University Medical Center, the Netherlands

CONTENTS

6.1 Introduction

It has been long recognized that missing data are the norm rather than the exception when it comes to the analysis of real data. In this chapter we focus on follow-up studies and on the statistical analysis of longitudinal outcomes with missing data. Due to the wide use of longitudinal studies in many different disciplines, there has been a lot of research in the literature on extensions of selection and pattern-mixture models, which can be considered as the two traditional modeling frameworks for handling missing data (Little and Rubin, 2002; Molenberghs and Kenward, 2007), to the longitudinal setting; see for instance, Verbeke and Molenberghs, (2000), Fitzmaurice et al. (2004), and references therein. These models are applied in non-random missing datasettings, i.e., when the probability of missingness may depend on unobserved longitudinal responses, and require defining the joint distribution of the longitudinal and dropout processes.

The majority of the extensions of selection and pattern-mixture models to longitudinal data have focused on standard designs assuming a fixed set of time points at which patients are expected to provide measurements. Nonetheless, in reality, patients often do not adhere to the posited study schedule and may skip visits and dropout from the study at random time points. Even though in many of those occasions information on the exact dropout time is available, the typical convention in selection and pattern mixture modeling has been to ignore this feature and coerce measurements back to discrete follow-up times. Another alternative that makes better use of the available data is to acknowledge that dropout occurs in continuous time and consider it as a time-to-event outcome. Thus, in this context such an approach will entail defining an appropriate model for the joint distribution of a longitudinal and a survival outcome. A lot of research in this type of joint models has been done primarily in the survival analysis context. In particular, when interest is on testing for the association between patient-generated longitudinal outcomes (e.g., biomarkers) and the risk for an event (e.g., death, development of a disease or dropout from the study), traditional approaches for analyzing time-to-event data, such as the partial likelihood for the Cox proportional hazards model, are not applicable. This is due to the fact that these standard approaches require that time-dependent covariates are external; that is, the value of the covariate at any time point t is not affected by the occurrence of an event at time point u, with $t > u$ (Kalbfleisch and Prentice, 2002, Section 6.3). Often, the time-dependent covariates encountered in longitudinal studies do not satisfy this condition, because they are the output of a stochastic process generated by the subject, which is directly related to the failure mechanism. To accurately measure the strength of the association between the endogenous time-dependent covariate and the risk for an event, a model for the joint distribution of the longitudinal and survival outcomes is required. Linking back to the missing data context, a joint model for the longitudinal and time-to-dropout outcomes is required in order to make valid inferences for the longitudinal process.

The aim of this chapter is to introduce a joint modeling framework for longitudinal and time-to-event data, focusing on dropout. Even though there are various ways to build joint models by appropriately defining the joint distribution of a longitudinal response and a time-to-event (Verbeke and Davidian, 2008), here we will focus on the class of joint models that utilize random effects to explain the associations between the two processes (Wu and Carroll, 1988; Wu and Bailey, 1989; Follman and Wu, 1995; Wulfsohn and Tsiatis, 1997; Henderson et al., 2000; Rizopoulos, 2012a). We present the key assumptions behind these models, explain their features and refer to estimation approaches. Special attention is given to the posited association structure between the two outcomes for which we present several types of parameterizations. The appropriate choice for this parameterization is of particular importance in the missing data context because it can considerably affect inferences. We illustrate the use of joint models of this type in a case study that considers 467 patients with advanced human immunodeficiency virus infection during antiretroviral treatment who had failed or were intolerant of zidovudine therapy. The main aim of this study was to compare the efficacy and safety of two alternative antiretroviral drugs, namely didanosine (ddI) and zalcitabine (ddC). Patients were randomly assigned to receive either ddI or ddC, and CD4 cell counts were recorded at study entry, where randomization took place, as well as at 2, 6, 12, and 18 months thereafter. By the end of follow-up, 188 patients had died, resulting in about 60% censoring. More details regarding the design of this study can be found in Abrams et al. (1994). Our aim here is to explore both the relationship between the CD4 cell count and the risk for death, and to investigate how inferences for the average longitudinal evolutions of the CD4 cell count are affected by the occurrence of events.

6.2 Joint Modeling Framework

The basic idea behind joint models is to couple a survival model for the continuous time-to-dropout process with a mixed-effects model for the longitudinal outcome, with the two models glued together with a set of random effects. A graphical representation is given in Figure 6.1, where, at each time point we associate the true underlying level of the longitudinal outcome (bottom panel) with the risk for dropout (top panel). To formally introduce

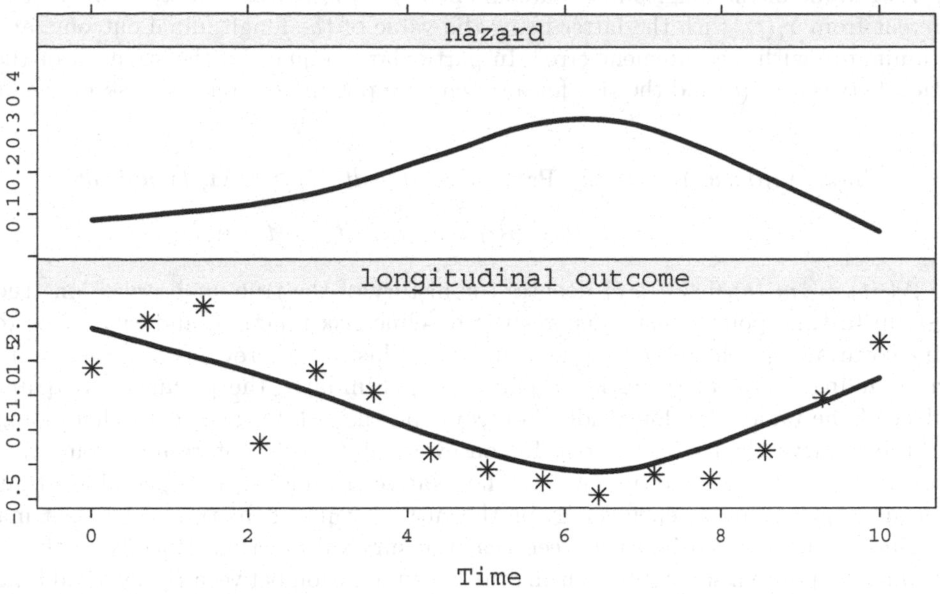

FIGURE 6.1
Intuitive representation of joint models. In the top panel the solid line illustrates how the hazard function evolves over time. In the bottom panel the solid line represents the joint model approximation.

this type of models, we let $Y_i(t)$ denote the observed value of the longitudinal outcome at time point t for the ith subject. Discontinuation of the data collection may occur for various reasons possibly related to the longitudinal outcome process. We distinguish between two types of events, namely we let T_i denote the time to outcome-dependent dropout, and C_i, the time to outcome-independent dropout (i.e., censoring). For example, in our motivating case study, dropout due to AIDS-related problems is considered outcome-dependent dropout, whereas dropout due to relocation away from the study center as censoring. The observed data are $Y_{ij} = \{Y_i(t_{ij}), j = 1, \ldots, n_i\}$, the longitudinal responses at the specific occasions t_{ij} at which measurements were taken, $U_i = \min(T_i, C_i)$ the observed event time, and $\delta_i = I(T_i \leq C_i)$ the corresponding event indicator.

For the sake of simplicity, we focus on normally distributed longitudinal outcomes and use a linear mixed-effects model to describe the subject-specific longitudinal trajectories, namely

$$\left\{\begin{array}{lcl} \boldsymbol{Y}_i(t) & = & m_i(t) + \varepsilon_i(t), \\ m_i(t) & = & \boldsymbol{x}_i'(t)\boldsymbol{\beta} + \boldsymbol{z}_i'(t)\boldsymbol{b}_i, \\ \boldsymbol{b}_i & \sim & \mathcal{N}(\boldsymbol{0}, \boldsymbol{D}), \quad \varepsilon_i(t) \sim \mathcal{N}(0, \sigma^2). \end{array}\right. \tag{6.1}$$

In the specification of the model we explicitly note that the design vectors $\boldsymbol{x}_i(t)$ for the fixed effects $\boldsymbol{\beta}$, and $\boldsymbol{z}_i(t)$ for the random effects $\boldsymbol{b}_i$, as well as the error terms $\varepsilon_i(t)$ are time-dependent. Moreover, we assume that the error terms are mutually independent, independent of the random effects, and normally distributed with mean zero and variance σ^2.

We assume that the risk for outcome-dependent dropout is associated with the *true* and *unobserved* value of the longitudinal outcome denoted by the term $m_i(t)$. Note that $m_i(t)$ is different from $Y_i(t)$, with the latter being the value of the longitudinal outcome at time t contaminated with measurement error. In particular, to quantify the strength of the association between $m_i(t)$ and the risk for an event, we postulate a relative risk model of the form

$$\begin{aligned} h_i(t \mid \mathcal{M}_i(t), \boldsymbol{w}_i) &= \lim_{dt \to 0} \Pr\{t \le T_i < t + dt \mid T_i \ge t, \mathcal{M}_i(t), \boldsymbol{w}_i\}/dt \\ &= h_0(t) \exp\{\boldsymbol{\gamma}' \boldsymbol{w}_i + \alpha m_i(t)\}, \quad t > 0, \end{aligned} \tag{6.2}$$

where $\mathcal{M}_i(t) = \{m_i(s), 0 \le s < t\}$ denotes the history of the true unobserved longitudinal process up to time point t, $h_0(\cdot)$ denotes the baseline risk function, and $\boldsymbol{w}_i$ is a vector of baseline covariates (such as the treatment indicator, history of previous diseases, etc.) with a corresponding vector of regression coefficients $\boldsymbol{\gamma}$. Similarly, the parameter α quantifies the effect of the underlying longitudinal outcome on the risk for an event. That is, $\exp(\alpha)$ equals the relative change in the risk for an event at time t that results from one unit increase in $m_i(t)$ at the same time point. The relative risk model (6.2) postulates that the risk for an event at time t depends only on the current value of the time-dependent marker $m_i(t)$. Nevertheless, it can be easily seen that the survival function depends on the whole longitudinal response history $\mathcal{M}_i(t)$, using the known relation between the survival function and the cumulative hazard function, i.e.,

$$\begin{aligned} \mathcal{S}_i(t \mid \mathcal{M}_i(t), \boldsymbol{w}_i) &= \Pr\big(T_i > t \mid \mathcal{M}_i(t), \boldsymbol{w}_i\big) \\ &= \exp\bigg(-\int_0^t h_0(s) \exp\{\boldsymbol{\gamma}' \boldsymbol{w}_i + \alpha m_i(s)\}\, ds\bigg). \end{aligned} \tag{6.3}$$

Therefore, for accurate estimation of $\mathcal{S}_i(t)$ it is important to obtain a good estimate of $\mathcal{M}_i(t)$. This entails considering an elaborate specification of the time structure in $\boldsymbol{x}_i(t)$ and $\boldsymbol{z}_i(t)$, and possibly interaction terms between the postulated time structure and baseline covariates. For instance, in applications in which subjects show highly non-linear longitudinal trajectories, it is advisable to consider flexible representations for $\boldsymbol{x}_i(t)$ and $\boldsymbol{z}_i(t)$ using a possibly high-dimensional vector of functions of time t, expressed in terms of high-order polynomials or splines. Compared to polynomials, splines are usually preferred due to their local nature and better numerical properties (Ruppert et al., 2003). In the joint modeling framework several authors have considered spline-based approaches to model flexibly the subject-specific longitudinal profiles. For example, Rizopoulos et al. (2009) and Brown et al. (2005) have utilized B-splines with multidimensional random effects, Ding and Wang (2008) proposed the use of B-splines with a single multiplicative random effect, Rizopoulos and Ghosh (2011) have considered natural cubic splines. An alternative approach to model highly nonlinear shapes of subject-specific evolutions is to incorporate in the linear mixed model an additional stochastic term that aims to capture the remaining serial correlation in the observed measurements not captured by the random effects. In this framework the linear mixed model takes the form,

$$Y_i(t) = m_i(t) + u_i(t) + \varepsilon_i(t), \tag{6.4}$$

where $u_i(t)$ is a mean-zero stochastic process, independent of $\boldsymbol{b}_i$ and $\varepsilon_i(t)$, and $m_i(t)$ has the same mixed-effects model structure as in (6.1). Joint models with such terms have been considered by Wang and Taylor (2001), who postulated an integrated Ornstein-Uhlenbeck process, and by Henderson et al. (2000), who considered a latent stationary Gaussian process shared by both the longitudinal and event processes. The choice between the two approaches is to a large extent a philosophical issue dictated in part by the analyst's belief about the "true" underlying biological mechanism that generates the data. In particular, model (6.1) posits that the trajectory followed by a subject throughout time is dictated by the time-independent random effects $\boldsymbol{b}_i$ alone. This implies that the shape of the longitudinal profile of each subject is an inherent characteristic of this subject that is constant in time. On the other hand, model (6.4), which includes both a serial correlation stochastic term and random effects attempts to more precisely capture the features of the longitudinal trajectory by allowing the subject-specific trends to vary over time. We should note that because both the random effects and a stochastic serial correlation process attempt to appropriately model the marginal correlation in the data, there is a conflict for information between the two approaches. For instance, if we have postulated a linear mixed model with a random-intercepts and random-slopes structure and there is excess serial correlation that the random effects have not adequately captured, then extending this model by either including the serial correlation term $u_i(t)$ or by considering a more elaborate random-effects structure (using e.g., splines in the design matrix $\boldsymbol{Z}_i$) could produce practically indistinguishable fits to the data. Therefore, even though it might seem intuitively appealing to use both elaborate random-effects and serial correlation structures, it is advisable to opt for either of those. Computationally, the random-effects approach is easier to implement in practice because it only requires an appropriate specification of the random-effects design matrix $\boldsymbol{Z}_i$. In what follows, we will focus on the random-effects approach.

To complete the specification of (6.2), we need to discuss the choice for the baseline risk function $h_0(\cdot)$. In standard survival analysis it is customary to leave $h_0(\cdot)$ completely unspecified in order to avoid the impact of misspecifying the distribution of survival times. However, within the joint modeling framework it turns out that following such a route may lead to underestimation of the standard errors of the parameter estimates (Hsieh et al., 2006). To avoid such problems we will need to explicitly define $h_0(\cdot)$. A standard option is to use a risk function corresponding to a known parametric distribution, such as the Weibull, the log-normal and the Gamma. Alternatively and even more preferably, we can opt for a parametric but flexible specification of the baseline risk function. Several approaches have been proposed in the literature to flexibly model the baseline risk function. For instance, Whittemore and Killer (1986) used step-functions and linear splines to obtain a non-parametric estimate of the risk function, Rosenberg (1995) utilized a B-splines approximation, and Herndon and Harrell (1996) used restricted cubic splines. Two simple options that often work quite satisfactorily in practice are the piecewise-constant and regression splines approaches.

6.2.1 Estimation

The main estimation methods that have been proposed for joint models are (semi-parametric) maximum likelihood (Henderson et al., 2000; Hsieh et al., 2006; Wulfsohn and Tsiatis, 1997) and Bayes using MCMC techniques (Brown and Ibrahim, 2003; Chi and Ibrahim, 2006; Wang and Taylor, 2001; Xu and Zeger, 2001; Rizopoulos and Ghosh, 2011). Moreover, Tsiatis and Davidian (2001) have proposed a conditional score approach in which the random effects are treated as nuisance parameters, and they developed a set of unbiased estimating equations that yield consistent and asymptotically normal estimators. Here we give the basics of the maximum likelihood method for joint models as one of the more traditional approaches.

Maximum likelihood estimation for joint models is based on the maximization of the log-likelihood corresponding to the joint distribution of the time-to-event and longitudinal outcomes $\{U_i, \delta_i, \boldsymbol{Y}_i\}$. To define this joint distribution we will assume that the vector of time-independent random effects $\boldsymbol{b}_i$ is shared by both the longitudinal and survival processes. This means that these random effects account for both the association between the longitudinal and event outcomes, and the correlation between the repeated measurements in the longitudinal process (conditional independence). Formally, we have that,

$$\begin{aligned} p(U_i, \delta_i, \boldsymbol{y}_i \mid \boldsymbol{b}_i; \boldsymbol{\theta}) &= p(U_i, \delta_i \mid \boldsymbol{b}_i; \boldsymbol{\theta})\, p(\boldsymbol{y}_i \mid \boldsymbol{b}_i; \boldsymbol{\theta}), \qquad (6.5)\\ p(\boldsymbol{y}_i \mid \boldsymbol{b}_i; \boldsymbol{\theta}) &= \prod_j p\{y_i(t_{ij}) \mid \boldsymbol{b}_i; \boldsymbol{\theta}\}, \qquad (6.6) \end{aligned}$$

where $\boldsymbol{\theta}$ denotes the parameter vector, $\boldsymbol{y}_i$ is the $n_i \times 1$ vector of longitudinal responses of the ith subject, and $p(\cdot)$ denotes an appropriate probability density function. Due to the fact that the survival and longitudinal submodels share the same random effects, joint models of this type are also known as shared-parameter models. Under the modeling assumptions presented in the previous section, and the above conditional independence assumptions the joint log-likelihood contribution for the ith subject can be formulated as

$$\begin{aligned} &\log p(U_i, \delta_i, \boldsymbol{y}_i; \boldsymbol{\theta})\\ &= \log \int p(U_i, \delta_i \mid \boldsymbol{b}_i; \boldsymbol{\theta}) \Big[\prod_j p\{y_i(t_{ij}) \mid \boldsymbol{b}_i; \boldsymbol{\theta}\}\Big] p(\boldsymbol{b}_i; \boldsymbol{\theta})\, d\boldsymbol{b}_i, \qquad (6.7) \end{aligned}$$

where the likelihood of the survival part is written as

$$p(U_i, \delta_i \mid \boldsymbol{b}_i; \theta) = \{h_i(U_i \mid \mathcal{M}_i(U_i); \boldsymbol{\theta})\}^{\delta_i} \mathcal{S}_i(U_i \mid \mathcal{M}_i(U_i); \boldsymbol{\theta}), \qquad (6.8)$$

with $h_i(\cdot)$ given by (6.2), the survival function by (6.3), $p\{y_i(t_{ij}) \mid \boldsymbol{b}_i; \boldsymbol{\theta}\}$ denotes the univariate normal density for the longitudinal responses, and $p(\boldsymbol{b}_i; \boldsymbol{\theta})$ is the multivariate normal density for the random effects.

Maximization of the log-likelihood function corresponding to (6.7) with respect to $\boldsymbol{\theta}$ is a computationally challenging task. This is mainly because both the integral with respect to the random effects in (6.7), and the integral in the definition of the survival function (6.3) do not have an analytical solution, except in very special cases. Standard numerical integration techniques such as Gaussian quadrature and Monte Carlo have been successfully applied in the joint modeling framework (Henderson et al., 2000; Song et al., 2002; Wulfsohn and Tsiatis, 1997). Furthermore, Rizopoulos (2012b) and Rizopoulos et al. (2009) have recently discussed improvements on the standard Gaussian quadrature for joint models and the use of Laplace approximations, respectively. These methods can be especially useful in high-dimensional random-effects settings (e.g., when splines are used in the random-effects design matrix). For the maximization of the approximated log-likelihood the EM algorithm has traditionally been used in which the random effects are treated as 'missing data'. The main motivation for using this algorithm is the closed-form M step updates for certain parameters of the joint model. However, a serious drawback of the EM algorithm is its linear convergence rate that results in slow convergence especially near the maximum. Nonetheless, Rizopoulos et al. (2009) have noted that a direct maximization of the observed data log-likelihood, using for instance, a quasi-Newton algorithm (Lange, 2004), requires computations very similar to the EM algorithm. Therefore, hybrid optimization approaches that start with EM (which is more stable far from the optimum), and then continue with direct maximization using quasi-Newton (which has faster convergence rate near the maximum) can be easily employed.

6.3 Missing Data Mechanism

To understand how joint models handle missing data we need to derive the corresponding missing data mechanism, which is the conditional distribution of the time-to-dropout given the complete vector of longitudinal responses $(\boldsymbol{y}_i^o, \boldsymbol{y}_i^m)$, where the observed part $\boldsymbol{y}_i^o = \{y_i(t_{ij}) : t_{ij} < T_i, j = 1, \ldots, n_i\}$ consists of all observed longitudinal measurements of the ith subject before the event time, and the missing part $\boldsymbol{y}_i^m = \{y_i(t_{ij}) : t_{ij} \geq T_i, j = 1, \ldots, n_i'\}$ contains the longitudinal measurements that would have been taken until the end of the study, had the event not occurred. Under the conditional independence assumptions (6.5) and (6.6), the dropout mechanism implied by a joint model takes the form:

$$\begin{aligned} p(T_i \mid \boldsymbol{y}_i^o, \boldsymbol{y}_i^m; \boldsymbol{\theta}) &= \int p(T_i, \boldsymbol{b}_i \mid \boldsymbol{y}_i^o, \boldsymbol{y}_i^m; \boldsymbol{\theta}) \, d\boldsymbol{b}_i \\ &= \int p(T_i \mid \boldsymbol{b}_i, \boldsymbol{y}_i^o, \boldsymbol{y}_i^m; \boldsymbol{\theta}) \, p(\boldsymbol{b}_i \mid \boldsymbol{y}_i^o, \boldsymbol{y}_i^m; \boldsymbol{\theta}) \, d\boldsymbol{b}_i \\ &= \int p(T_i \mid \boldsymbol{b}_i; \boldsymbol{\theta}) \, p(\boldsymbol{b}_i \mid \boldsymbol{y}_i^o, \boldsymbol{y}_i^m; \boldsymbol{\theta}) \, d\boldsymbol{b}_i. \end{aligned} \tag{6.9}$$

We observe that the time-to-dropout depends on $\boldsymbol{Y}_i^m$ through the posterior distribution of the random effects $p(\boldsymbol{b}_i \mid \boldsymbol{y}_i^o, \boldsymbol{y}_i^m; \boldsymbol{\theta})$, which means that joint models correspond to a NMAR missing data mechanism (Little and Rubin, 2002; Molenberghs and Kenward, 2007). A closer inspection of (6.9) reveals that the key component behind the attrition mechanism in joint models is the random effects $\boldsymbol{b}_i$. More specifically, and as we have already seen, under the joint model the survival and longitudinal sub-models *share* the same random effects. Due to this feature joint models belong to the class of shared-parameter models (Wu and Carroll, 1988; Wu and Bailey, 1989; Follman and Wu, 1995; Vonesh et al., 2006). Under a simple random-effects structure (i.e., random intercepts and random slopes), this missing data mechanism implies that subjects which show steep increases in their longitudinal profiles may be more (or less) likely to drop out than subjects who exhibit more stable trajectories. A relevant issue here is the connection between the association parameter α and the type of missing data mechanism. In particular, a null value for α corresponds to a MCAR missing data mechanism (Little and Rubin, 2002; Molenberghs and Kenward, 2007), because once conditioned upon available covariates, the dropout process does not depend on either missing or observed longitudinal responses. Moreover, since under $\alpha = 0$ the parameters in the two sub-models are distinct, the joint probability of the dropout and longitudinal processes can be factorized as follows:

$$\begin{aligned} p(U_i, \delta_i, \boldsymbol{y}_i; \boldsymbol{\theta}) &= p(U_i, \delta_i; \boldsymbol{\theta}_t) \, p(\boldsymbol{y}_i; \boldsymbol{\theta}_y, \boldsymbol{\theta}_b) \\ &= p(U_i, \delta_i; \boldsymbol{\theta}_t) \int p(\boldsymbol{y}_i \mid \boldsymbol{b}_i; \boldsymbol{\theta}_y) p(\boldsymbol{b}_i; \boldsymbol{\theta}_b) \, d\boldsymbol{b}_i, \end{aligned}$$

where $\boldsymbol{\theta} = (\boldsymbol{\theta}_t', \boldsymbol{\theta}_y', \boldsymbol{\theta}_b')'$ denotes the full parameter vector, with $\boldsymbol{\theta}_t$ denoting the parameters for the event time outcome, $\boldsymbol{\theta}_y$ the parameters for the longitudinal outcomes and $\boldsymbol{\theta}_b$ the unique parameters of the random-effects covariance matrix. This implies that the parameters in the two sub-models can be estimated separately. Nonetheless, since we have adopted a full likelihood approach, the estimated parameters derived from maximizing the log-likelihood of the longitudinal process $\ell(\boldsymbol{\theta}_y) = \sum_i \log p(\boldsymbol{y}_i; \boldsymbol{\theta}_y)$, will also be valid under a MAR missing data mechanism, i.e., under the hypothesis that dropout depends on the observed responses only. Thus, while strictly speaking $\alpha = 0$ corresponds to a MCAR mechanism, the parameter estimates that we will obtain will still be valid under MAR. As a side note, we should also mention that, in practice, discontinuation of the data collection for the longitudinal process also occurs when a subject leaves the study because of censoring. However, in the formulation of the likelihood function of joint models we have assumed that the censoring mechanism may depend on the observed history of longitudinal responses and covariates, but is independent of future longitudinal outcomes. Hence, under this assumption, censoring corresponds to a MAR missing data mechanism.

An additional nice feature of the shared-parameter models framework is that these models can very easily handle both intermittent missingness and attrition. To see how this is achieved we write the observed data log-likelihood under the complete data model $\{\boldsymbol{y}_i^o, \boldsymbol{y}_i^m\}$ for the longitudinal outcome:

$$\begin{aligned}
\ell(\boldsymbol{\theta}) &= \sum_{i=1}^n \log \int p(U_i, \delta_i, \boldsymbol{y}_i^o, \boldsymbol{y}_i^m; \boldsymbol{\theta})\, d\boldsymbol{y}_i^m \\
&= \sum_{i=1}^n \log \int \int p(U_i, \delta_i, \boldsymbol{y}_i^o, \boldsymbol{y}_i^m \mid \boldsymbol{b}_i; \boldsymbol{\theta})\, p(\boldsymbol{b}_i; \boldsymbol{\theta})\, d\boldsymbol{y}_i^m d\boldsymbol{b}_i \\
&= \sum_{i=1}^n \log \int p(U_i, \delta_i \mid \boldsymbol{b}_i; \boldsymbol{\theta}) \left\{ \int p(\boldsymbol{y}_i^o, \boldsymbol{y}_i^m \mid \boldsymbol{b}_i; \boldsymbol{\theta})\, d\boldsymbol{y}_i^m \right\} p(\boldsymbol{b}_i; \boldsymbol{\theta})\, d\boldsymbol{b}_i \\
&= \sum_{i=1}^n \log \int p(U_i, \delta_i \mid \boldsymbol{b}_i; \boldsymbol{\theta})\, p(\boldsymbol{y}_i^o \mid \boldsymbol{b}_i; \boldsymbol{\theta})\, p(\boldsymbol{b}_i; \boldsymbol{\theta})\, d\boldsymbol{b}_i.
\end{aligned}$$

Under the first conditional independence assumption (6.5) we obtain that the missing longitudinal responses $\boldsymbol{y}_i^m$ are only involved in the density of the longitudinal sub-model. Moreover, under the second conditional independence assumption (6.6) the longitudinal responses conditionally on the random effects are independent of each other, and therefore, the integral with respect to $\boldsymbol{y}_i^m$ is easily dropped. Thus, even if some subjects have intermittently missing responses, the likelihood of a joint model is easily obtained without requiring integration with respect to the missing responses. This is in contrast to other approaches frequently used for handling missing data, such as, e.g., selection and pattern mixture models (Troxel et al., 1998; Jansen et al., 2006).

6.4 Parameterizations for the Association Structure

As it has been stressed in earlier chapters of this handbook, inferences in non-random dropout settings can be highly sensitive to modeling assumptions, and in particular to the assumptions regarding the association structure between the unobserved longitudinal outcomes and the dropout process (Copas and Li, 1997; Molenberghs et al., 2008). The key component in shared-parameter models that drive this association are the random effects. Motivated by this fact, several papers have investigated how a misspecification of the random-effects distribution may affect the ensuing inferences. In particular, Song et al. (2002) proposed a flexible model for this distribution based on the class of smooth densities studied by Gallant and Nychka (1987), and Tsiatis and Davidian (2001) proposed a semi-parametric estimating equations approach that provides valid inferences without requiring to specify the random-effects distribution. However, simulation findings of these authors suggested that parameter estimates and standard errors were rather robust to misspecification. This feature has later been theoretically corroborated by Rizopoulos et al. (2008) and Huang et al. (2009), who showed that, as the number of repeated measurements per subject n_i increases, misspecification of the random-effects distribution has a minimal effect on parameter estimates and standard errors.

Another key component of shared-parameter models that has been studied less in the literature but nonetheless may prove important with respect to sensitivity of inferences to modeling assumptions, is the form of the association structure between the two processes. In the standard joint model introduced in Section 6.2, we assumed that the current underlying value of the longitudinal process $m_i(t)$ is associated with the risk for an event at the same time point t, with parameter α measuring the strength of this association. Even though this is an intuitively very appealing parameterization with a clear interpretation for α, it is not realistic to expect that this parameterization will always be appropriate in expressing the correct relationship between the two processes. In the following, we focus on this issue, and present several alternative parameterizations that extend the standard parameterization (6.2) in different ways. These different parameterizations can be seen as special cases of the following general formulation of the association structure between the longitudinal outcome and the risk for dropout:

$$h_i(t) = h_0(t)\exp\bigl[\boldsymbol{\gamma}'\boldsymbol{w}_{i1} + f\{m_i(t), \boldsymbol{b}_i, \boldsymbol{w}_{i2}; \boldsymbol{\alpha}\}\bigr], \tag{6.10}$$

where $f(\cdot)$ is a function of the true level of the marker $m_i(\cdot)$, of the random effects $\boldsymbol{b}_i$ and extra covariates $\boldsymbol{w}_{i2}$. Under this general formulation $\boldsymbol{\alpha}$ can potentially denote a vector of association parameters.

6.4.1 Interaction and lagged effects

The standard parameterization (6.2) assumes that the risk for dropout at time t depends on the underlying value of the longitudinal outcome at the same time point, and in addition that the strength of this association between the two outcomes is the same in all different subgroups in the population. However, at many occasions it may be more reasonable to assume that the underlying level of the longitudinal outcome at a specific time point is related to the risk for an event at a later time point, and moreover that the strength of this association may be different for different subgroups of subjects. A straightforward extension to handle situations in which these two assumptions may not hold is to allow for lagged effects and include in the linear predictor of the relative risk model interaction terms of the marker with baseline covariates of interest, i.e.,

$$h_i(t) = h_0(t)\exp\bigl[\boldsymbol{\gamma}'\boldsymbol{w}_{i1} + \boldsymbol{\alpha}'\{\boldsymbol{w}_{i2} \times m_i(t-c)_+\}\bigr], \tag{6.11}$$

where $(t-c)_+ = t-c$ when $t > c$ and 0 otherwise, $\boldsymbol{w}_{i1}$ is used to accommodate the direct effects of baseline covariates to the risk for an event, and $\boldsymbol{w}_{i2}$ contains interaction terms that expand the association of $m_i(t-c)_+$ in different subgroups in the data. When the value c is not equal to zero, we postulate that the risk at time t depends on the true value of the longitudinal marker at time $t-c$. This can be treated either as a constant fixed a priori or as a parameter to be estimated from the data. At first glance the incorporation of lagged effects suggests that dropout depends only on observed longitudinal responses, which eventually corresponds to a missing at random mechanism. Nevertheless, we should stress that the risk depends on the true value of the trajectory at time $t-c$ and not the observed one. Since $m_i(t-c)_+$ depends on the random effects, parameterization (6.11) will still correspond to a missing not at random mechanism.

6.4.2 Time-dependent slopes parameterization

Under the linear mixed model we postulate that the longitudinal outcome of each patient follows a trajectory in time. It is therefore reasonable to consider parameterizations that allow the risk for an event to depend not only on the current value of the trajectory but also on other features. A parameterization of this type has been considered by Ye et al. (2008) who postulated a joint model in which the risk depends on both the current true value of the trajectory and the slope of the true trajectory at time t. More specifically, the relative risk survival sub-model for the time-to-dropout takes the form,

$$h_i(t) = h_0(t) \exp\{\boldsymbol{\gamma}' \boldsymbol{w}_i + \alpha_1 m_i(t) + \alpha_2 (dm_i(t)/dt)\}, \tag{6.12}$$

where

$$\frac{d}{dt} m_i(t) = \frac{d}{dt} \{\boldsymbol{x}_i'(t)\boldsymbol{\beta} + \boldsymbol{z}_i'(t)\boldsymbol{b}_i\}.$$

The interpretation of parameter α_1 remains the same as in the standard parameterization (6.2). Parameter α_2 measures how strongly the slope of the true longitudinal trajectory at time t is associated with the risk for an event at the same time point, provided that $m_i(t)$ remains constant. This parameterization could capture situations where, for instance, at a specific time point two patients show similar true marker levels, but they may differ in the rate of change of the marker.

6.4.3 Cumulative-effects parameterization

A common characteristic of the previous parameterizations is that they assume that the risk for an event at a specific time depends on features of the longitudinal trajectory at only a single time point. That is, from the whole history of the true marker levels $\mathcal{M}_i(t) = \{m_i(s), 0 \leq s < t\}$, the risk at time t is typically assumed to depend on either the marker level at the same time point $m_i(t)$ or at a previous time point $m_i(t-c)$, if lagged effects are considered as in Section 6.4.1. However, several authors have argued that this assumption is not always realistic, and in many cases we may benefit by allowing the risk to depend on a more elaborate function of the longitudinal marker history (Sylvestre and Abrahamowicz, 2009; Hauptmann et al., 2000; Vacek, 1997).

One approach to account for the cumulative effect of the longitudinal outcome up to time point t is to include in the linear predictor of the relative risk sub-model the integral of the longitudinal trajectory. More specifically, the survival sub-model takes the form

$$h_i(t) = h_0(t) \exp\Big\{\boldsymbol{\gamma}' \boldsymbol{w}_i + \alpha \int_0^t m_i(s)\, ds\Big\}, \tag{6.13}$$

where for any particular time point t, α measures the strength of the association between the risk for an event at time point t and the area under the longitudinal trajectory up to the same time t, where the area under the longitudinal trajectory is regarded as a suitable summary of the whole trajectory.

A restriction of this parameterization is that it assigns the same weight to all past values of the marker, which may not be reasonable in some situations. A straightforward extension to account for this issue is to adjust the integrand and multiply the $m_i(t)$ with an appropriately chosen weight function that places different weights at different time points:

$$h_i(t) = h_0(t) \exp\Big\{\boldsymbol{\gamma}' \boldsymbol{w}_i + \alpha \int_0^t \varpi(t-s) m_i(s)\, ds\Big\}, \tag{6.14}$$

where $\varpi(\cdot)$ denotes the weight function. A desirable property of $\varpi(\cdot)$ would be to assign smaller weights to points further away in the past. One possible family of functions with this property are probability density functions of known parametric distributions, such as the normal, the Student's t and the logistic. The scale parameter in these densities and also the degrees-of-freedom parameter in the Student's-t density can be utilized to tune the relative weights of more recent marker values compared to older ones.

6.4.4 Random-effects parameterization

Another type of parameterization that is frequently used in joint models includes in the linear predictor of the risk model only the random effects of the longitudinal sub-model, i.e.,

$$h_i(t) = h_0(t) \exp(\boldsymbol{\gamma}' \boldsymbol{w}_i + \boldsymbol{\alpha}' \boldsymbol{b}_i), \tag{6.15}$$

where $\boldsymbol{\alpha}$ denotes a vector of association parameters each one measuring the association between the corresponding random effect and the hazard for an event. This parameterization is more meaningful when a simple random-intercepts and random-slopes structure is assumed for the longitudinal sub-model, in which case, the random effects express subject-specific deviations from the average intercept and average slope. Under this setting this parameterization postulates that patients who have a lower/higher level for the longitudinal outcome at baseline (i.e., intercept) or who show a steeper increase/decrease in their longitudinal trajectories (i.e., slope) are more likely to experience the event (i.e., drop out). This parameterization shares also similarities with the time-dependent slopes parameterization presented in Section 6.4.2. In particular, under a longitudinal sub-model with a random-intercepts and random-slopes structure, i.e.,

$$y_i(t) = \beta_0 + \beta_1 t + b_{i0} + b_{i1} t + \varepsilon_i(t),$$

the relative risk sub-model for the event process under the time-dependent slopes parameterization (6.12) with $\alpha_1 = 0$ takes the form

$$h_i(t) = h_0(t) \exp\{\boldsymbol{\gamma}' \boldsymbol{w}_i + \alpha_2(\beta_1 + b_{i1})\},$$

whereas the relative risk sub-model under the same longitudinal sub-model, but with parameterization (6.15) becomes

$$h_i(t) = h_0(t) \exp(\boldsymbol{\gamma}' \boldsymbol{w}_i + \alpha_1 b_{i0} + \alpha_2 b_{i1}).$$

If we set $\alpha_1 = 0$ in the latter formulation of the relative risk sub-model, we observe that this model also postulates that the risk depends only on the random-slopes component of the linear mixed model.

A computational advantage of parameterization (6.15) is that it is time-independent, and therefore leads to a closed-form solution (under certain baseline risk functions) for the integral in the definition of the survival function (6.3). This facilitates computations because we do not have to numerically approximate this integral. However, an important disadvantage of (6.15) emerges when an elaborate formulation of the subject-specific mean structure of longitudinal sub-model is assumed. In particular, when polynomials or splines are used to capture nonlinear subject-specific evolutions, the random effects do not have a straightforward interpretation, which in turn complicates the interpretation of the association parameters $\boldsymbol{\alpha}$.

6.5 Analysis of the AIDS Data

We return to the AIDS dataset introduced in Section 6.1 for which we are interested in comparing the average longitudinal evolutions of CD4 cell counts for the two treatment groups, and also in investigating the association structure between CD4 and the hazard of death. The CD4 cell counts are known to exhibit rightly skewed distributions, and therefore, for the remainder of this analysis we will work with the square root of the CD4 count. Figure 6.2 shows the subject-specific longitudinal trajectories of the patients in the two treatment groups. We observe considerable variability in the shapes of these trajectories,

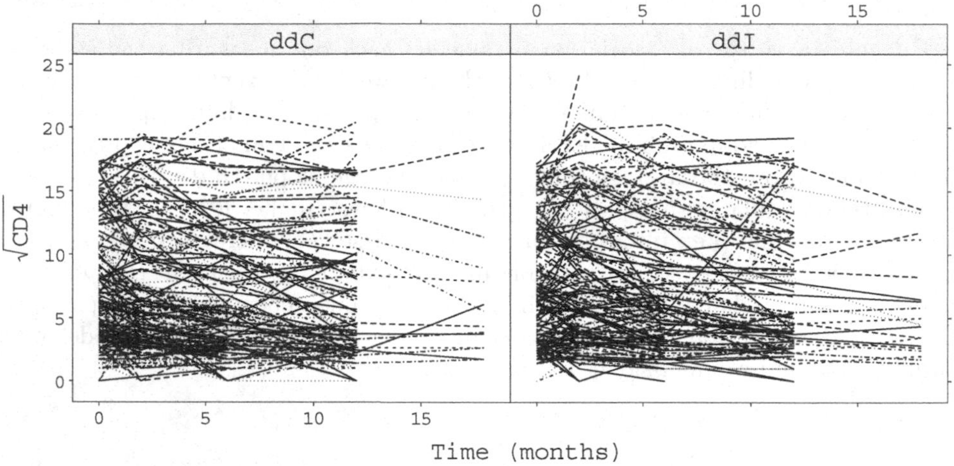

FIGURE 6.2
AIDS Data. Subject-specific longitudinal trajectories of CD4 cell counts for the two treatment groups.

but there are no apparent differences between the two groups.

Traditionally, in a missing data context we start with a MAR analysis, which corresponds to fitting an appropriate linear mixed-effects model for the square root of the CD4 count. In particular, taking advantage of the randomization set-up of the study, we fit the linear mixed model

$$\begin{aligned} Y_i(t) &= m_i(t) + \varepsilon_i(t) \\ &= \beta_0 + \beta_1 t + \beta_2\{t \times \texttt{ddI}_i\} + b_{i0} + b_{i1}t + \varepsilon_i(t), \end{aligned}$$

where in the fixed-effects part we include the main effect of time and the interaction of treatment with time but not the main effect of treatment, and in the random-effects design matrix we include an intercept and a time term. The results of this analysis are presented in Table 6.1. From the corresponding t-test for the coefficient β_2 of the interaction term we do not have any evidence for a difference in the average longitudinal evolutions in the two groups. To gain more confidence in the results obtained from the simple linear mixed model fit, we continue our analysis by allowing for nonrandom dropout. In particular, we assume that the risk for dropout at any particular time point t may depend on treatment, and on

TABLE 6.1
AIDS Data. Parameter estimates, standard errors and p-values under the linear mixed and joint models. D_{ij} denote the ij-element of the covariance matrix for the random effects.

	Mixed Model			Joint Model		
	Value	Std.Err	p-value	Value	Std.Err	p-value
β_0	7.19	0.22	< 0.0001	7.22	0.22	< 0.0001
β_1	−0.16	0.03	< 0.0001	−0.19	0.02	< 0.0001
β_2	0.03	0.03	0.3415	0.01	0.03	0.7014
$\log(\sigma)$	0.56	0.03		0.55	0.03	
D_{11}	21.07	1.34		21.01	1.50	
D_{12}	−0.12	0.1		−0.04	0.07	
D_{22}	0.03	0.02		0.03	0.01	
γ				0.33	0.16	0.0324
α_1				−0.29	0.04	< 0.0001

the true underlying profile of the CD4 cell count as estimated from the longitudinal model, i.e.,

$$h_i(t) = h_0(t) \exp\{\gamma \mathtt{ddI}_i + \alpha_1 m_i(t)\},$$

where the baseline risk function is assumed piecewise-constant

$$h_0(t) = \sum_{q=1}^{Q} \xi_q I(v_{q-1} < t \leq v_q),$$

with six internal knots placed at equally spaced percentiles of the observed event times. The results of the joint modeling analysis, presented in Table 6.1, also do not provide any indication for a difference between the two treatment groups. For the survival process we observe that the square root CD4 cell count is strongly related to the risk for death, with a unit decrease in the marker corresponding to a $\exp(-\alpha_1) = 1.33$-fold increase in the risk for death (95% CI: 1.24; 1.43).

Since in both the MAR and NMAR analyses we reached the same conclusion, one could opt stopping and reporting these results. Nevertheless, as it has been mentioned in Section 6.4 and stressed in earlier chapters, inferences in missing datasettings can be rather sensitive to modeling assumptions. The only viable alternative in this case is to perform a sensitivity analysis. To proceed with such an analysis, we should decide between staying within the shared-parameter models framework or moving to other types of NMAR models, such as selection or pattern-mixture models. The majority of the literature has in fact focused on sensitivity analysis for the latter two classes of models (Ibrahim and Molenberghs, 2009; Molenberghs and Kenward, 2007; Little and Rubin, 2002; Little, 1995; Diggle and Kenward, 1994; Little, 1994, 1993) with very little work done in the shared-parameter models framework (Creemers et al., 2010; Tsonaka et al., 2010; Tsonaka et al., 2009). Because our focus here is on joint models we opt staying within this modeling framework. Following the ideas presented in Section 6.4, we perform a sensitivity analysis by considering alternative parameterizations for the association structure between the longitudinal responses and the risk for dropout. More specifically, we refit the joint model with the same linear mixed model for

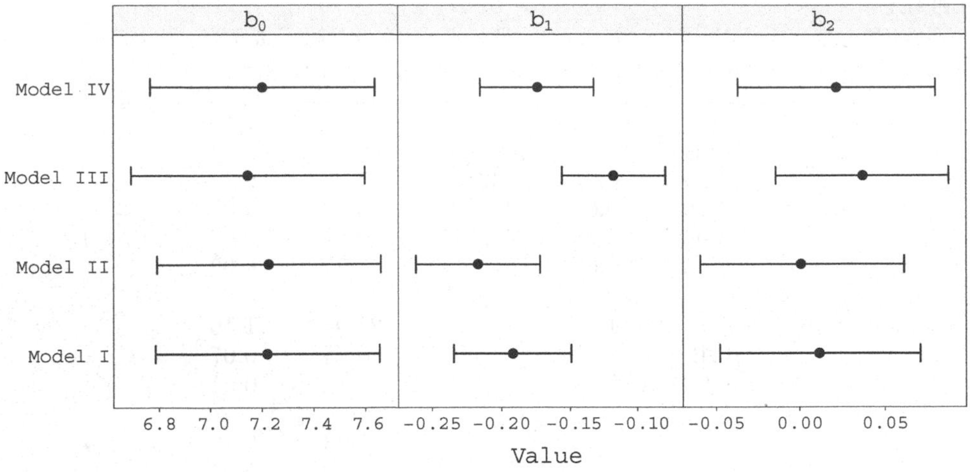

FIGURE 6.3
AIDS Data. Parameter estimates (points) and 95% confidences intervals (error bars) for the fixed-effects coefficients of the linear mixed model under the four parameterizations.

the longitudinal outcome, and the following four survival sub-models for the event process:

$$\begin{aligned}
\text{(I)}\quad h_i(t) &= h_0(t)\exp\{\gamma \mathtt{ddI}_i + \alpha_1 m_i(t)\},\\
\text{(II)}\quad h_i(t) &= h_0(t)\exp\{\gamma \mathtt{ddI}_i + \alpha_1 m_i(t) + \alpha_2 (dm_i(t)/dt)\},\\
\text{(III)}\quad h_i(t) &= h_0(t)\exp\{\gamma \mathtt{ddI}_i + \alpha_3 b_{i1}\},\\
\text{(IV)}\quad h_i(t) &= h_0(t)\exp\Big\{\gamma \mathtt{ddI}_i + \alpha_4 \int_0^t m_i(s)\, ds\Big\},
\end{aligned}$$

where $dm_i(t)/dt = \beta_1 + \beta_2 \mathtt{ddI}_i + b_{i1}$, and

$$\int_0^t m_i(s)\, ds = (\beta_0 + b_{i0})t + \frac{(\beta_1 + b_{i1})}{2} t^2 + \frac{\beta_2}{2}\{t^2 \times \mathtt{ddI}_i\}.$$

Relative risks Models (I) and (II) assume that the risk for dropout at time t depends on the true level of the square root CD4 cell count at the same time point, and on both the true level and the slope of the true trajectory at t, respectively. Model (III) posits that the risk only depends on the random effect b_{i1} denoting the deviation of the subject-specific slope from the average slope β_1. Finally, Model (IV) postulates that the risk is associated with the area under the longitudinal trajectory of each patient up to the same time t. The estimated regression coefficients and the corresponding 95% confidence intervals for the longitudinal sub-model are depicted in Figure 6.3. We can clearly observe that the slope coefficients β_1 and β_2 are rather sensitive to the chosen association structure. In particular, the time effect for treatment group ddC is estimated to be the lowest under parameterization (II) and highest under parameterization (III), with the confidence intervals between the two not overlapping. In addition, under parameterization (III) we observe the largest difference in magnitude in the slopes of the two groups. As expected, these differences also have a direct influence on the shapes of the average longitudinal evolutions of the two groups illustrated in Figure 6.4.

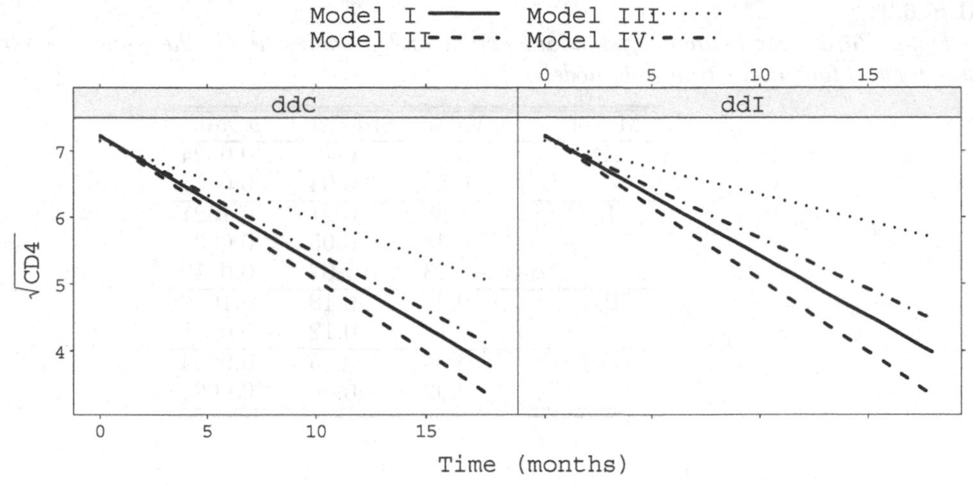

FIGURE 6.4
AIDS Data. Average longitudinal evolutions for the square root CD4 cell count under the four parameterizations.

The parameter estimates and standard errors of the regression coefficients of the four survival sub-models are shown in Table 6.2. The results suggest that not only the current value of the square root CD4 cell count, represented by term $m_i(t)$, but also other features of the patients' longitudinal trajectories are strongly associated with the risk for death. Moreover, we observe noticeable differences in the magnitude of the treatment effect, which is found marginally significant under parameterizations (I) and (IV) and non-significant under the other two; however, note that the interpretation of parameter γ is not exactly the same under the four models because it measures the treatment effect conditional on different aspects of the longitudinal process.

6.6 Discussion and Extensions

Joint modeling of longitudinal and time-to-event data is one of the most rapidly evolving areas of current biostatistics research, with several extensions of the standard joint model that we have presented here already proposed in the literature. These include, among others, the consideration of categorical longitudinal outcomes (Faucett et al., 1998; Pulkstenis et al., 1998) or more generally of multiple longitudinal outcomes (Brown et al., 2005; Rizopoulos and Ghosh, 2011), replacing the relative risk model by an accelerated failure time model (Tseng et al., 2005), and associating the two outcomes via latent classes instead of random effects (Proust-Lima et al., 2009; Lin et al., 2002).

A nice feature of these models is that they explicitly distinguish between random and non-random dropout, treating the former as censoring and the latter as "true" events. This can be further extended to multiple failure times (Elashoff et al., 2008; Williamson et al., 2008) thereby allowing for different non-random dropout processes (i.e., different reasons for nonrandom dropout) to be associated with the longitudinal outcome. Moreover,

TABLE 6.2
AIDS Data. Parameter estimates, standard errors and p-values under the joint modeling analysis for the four event time sub-models.

Model		Value	Std.Err	p-value
(I)	γ	0.33	0.16	0.0324
	α_1	−0.29	0.04	< 0.0001
(II)	γ	0.34	0.21	0.1021
	α_1	−0.34	0.05	< 0.0001
	α_2	−3.33	1.07	0.0019
(III)	γ	0.29	0.18	0.1076
	α_3	0.73	0.12	< 0.0001
(IV)	γ	0.30	0.15	0.0464
	α_4	−0.02	0.00	< 0.0001

we have focused our sensitivity analysis presented in Section 6.5 on the parameterization describing the association structure between the longitudinal responses and the time-to-dropout. Nevertheless, in a broader sensitivity analysis other modeling assumption can be explored as well, such as the form of the baseline hazard function or the consideration of an accelerated failure time model instead of a relative risk model.

Finally, even though joint models enjoy a wide range of applications in many different statistical fields, they have not yet found their rightful place in the toolbox of modern applied statisticians mainly due to the fact that they are rather computationally intensive to fit. The main difficulty arises from the requirement for numerical integration with respect to the random effects, and the lack of software. The R package **JM** (freely available from the CRAN Web site at: `http://cran.r-project.org/`) has been developed to fill this gap to some extent. **JM** has a user-friendly interface to fit joint models and also provides several supporting functions that extract or calculate various quantities based on the fitted model (e.g., residuals, fitted values, empirical Bayes estimates, various plots, and others). At `http://wiki.r-project.org/rwiki/doku.php?id=packages:cran:jm`, more information can be found. See also Rizopoulos (2010).

References

Abrams, D., Goldman, A., Launer, C., et al. (1994). Comparative trial of didanosine and zalcitabine in patients with human immunodeficiency virus infection who are intolerant of or have failed zidovudine therapy. *New England Journal of Medicine* **330**, 657–662.

Brown, E. and Ibrahim, J. (2003). A Bayesian semiparametric joint hierarchical model for longitudinal and survival data. *Biometrics* **59**, 221–228.

Brown, E., Ibrahim, J., and DeGruttola, V. (2005). A flexible B-spline model for multiple longitudinal biomarkers and survival. *Biometrics* **61**, 64–73.

Chi, Y.-Y. and Ibrahim, J. (2006). Joint models for multivariate longitudinal and multivariate survival data. *Biometrics* **62**, 432–445.

Copas, J. and Li, H. (1997). Inference for non-random samples (with discussion). *Journal of the Royal Statistical Society, Series B* **59**, 55–95.

Creemers, A., Hens, N., Aerts, M., Molenberghs, G., Verbeke, G., and Kenward, M. (2010). A sensitivity analysis for shared-parameter models for incomplete longitudinal data. *Biometrical Journal* **52**, 111–125.

Diggle, P. and Kenward, M. (1994). Informative dropout in longitudinal data analysis (with discussion). *Journal of the Royal Statistical Society, Series C* **43**, 49–93.

Ding, J. and Wang, J.-L. (2008). Modeling longitudinal data with nonparametric multiplicative random effects jointly with survival data. *Biometrics* **64**, 546–556.

Elashoff, R., Li, G., and Li, N. (2008). A joint model for longitudinal measurements and survival data in the presence of multiple failure types. *Biometrics* **64**, 762–771.

Faucett, C., Schenker, N., and Elashoff, R. (1998). Analysis of censored survival data with intermittently observed time-dependent binary covariates. *Journal of the American Statistical Association* **93**, 427–437.

Fitzmaurice, G., Laird, N., and Ware, J. (2004). *Applied Longitudinal Analysis.* Hoboken: Wiley.

Follmann, D. and Wu, M. (1995). An approximate generalized linear model with random effects for informative missing data. *Biometrics* **51**, 151–168.

Gallant, A. and Nychka, D. (1987). Seminonparametric maximum likelihood estimation. *Econometrica* **55**, 363–390.

Hauptmann, M., Wellmann, J., Lubin, J., Rosenberg, P., and Kreienbrock, L. (2000). Analysis of exposure-time-response relationships using a spline weight function. *Biometrics* **56**, 1105–1108.

Henderson, R., Diggle, P., and Dobson, A. (2000). Joint modelling of longitudinal measurements and event time data. *Biostatistics* **1**, 465–480.

Herndon, J. and Harrell, F. (1996). The restricted cubic spline hazard model. *Communications in Statistics – Theory and Methods* **19**, 639–663.

Hsieh, F., Tseng, Y.-K., and Wang, J.-L. (2006). Joint modeling of survival and longitudinal data: likelihood approach revisited. *Biometrics* **62**, 1037–1043.

Huang, X., Stefanski, L., and Davidian, M. (2009). Latent-model robustness in joint models for a primary endpoint and a longitudinal process. *Biometrics* **64**, 719–727.

Ibrahim, J. and Molenberghs, G. (2009). Missing data methods in longitudinal studies: A review. *Test* **18**, 1–43.

Jansen, I., Hens, N., Molenberghs, G., Aerts, M., Verbeke, G., and Kenward, M. (2006). The nature of sensitivity in monotone missing not at random models. *Computational Statistics & Data Analysis* **50**, 830–858.

Kalbfleisch, J. and Prentice, R. (2002). *The Statistical Analysis of Failure Time Data*, 2nd edition. New York: Wiley.

Lange, K. (2004). *Optimization.* New York: Springer-Verlag.

Lin, H., Turnbull, B., McCulloch, C., and Slate, E. (2002) Laten class models for joint analysis of longitudinal biomarker and event process data: Application to longitudinal prostate-specific antigen readings and prostate cancer. *Journal of the American Statistical Association* **97**, 53–65.

Little, R. (1993). Pattern-mixture models for multivariate incomplete data. *Journal of the American Statistical Association* **88**, 125–134.

Little, R. (1994). A class of pattern-mixture models for normal incomplete data. *Biometrika* **81**, 471–483.

Little, R. (1995). Modeling the drop-out mechanism in repeated-measures studies. *Journal of the American Statistical Association* **90**, 1112–1121.

Little, R. and Rubin, D. (2002). *Statistical Analysis with Missing Data*, 2nd edition. New York: Wiley.

Molenberghs, G., Beunckens, C., Sotto, C., and Kenward, M. (2008). Every missingness not at random model has a missingness at random counterpart with equal fit. *Journal of the Royal Statistical Society, Series B* **70**, 371–388.

Molenberghs, G. and Kenward, M. (2007). *Missing Data in Clinical Studies.* New York: Wiley.

Proust-Lima, C., Joly, P., Dartigues, J. F., and Jacqmin-Gadda, H. (2009) Joint modelling of multivariate longitudinal outcomes and a time-to-event: A nonlinear latent class approach. *Computational Statistics and Data Analysis* **53**, 1142–1154.

Pulkstenis, E., Ten Have, T., and Landis, R. (1998). Model for the analysis of binary longitudinal pain data subject to informative dropout through remedication. *Journal of the American Statistical Association* **93**, 438–450.

Rizopoulos, D. (2010). JM: An R package for the joint modelling of longitudinal and time-to-event data. *Journal of Statistical Software* **35** (9), 1–33.

Rizopoulos, D. (2012a). *Joint Models for Longitudinal and Time-to-Event Data with Applications in R.* Boca Raton: Chapman & Hall/CRC.

Rizopoulos, D. (2012b). Fast fitting of joint models for longitudinal and event time data using a pseudo-adaptive Gaussian quadrature rule. *Computational Statistics & Data Analysis* **56**, 491–501.

Rizopoulos, D. and Ghosh, P. (2011). A Bayesian semiparametric multivariate joint model for multiple longitudinal outcomes and a time-to-event. *Statistics in Medicine* **30**, 1366–1380.

Rizopoulos, D., Verbeke, G., and Lesaffre, E. (2009). Fully exponential Laplace approximations for the joint modelling of survival and longitudinal data. *Journal of the Royal Statistical Society, Series B* **71**, 637–654.

Rizopoulos, D., Verbeke, G., and Molenberghs, G. (2008). Shared parameter models under random effects misspecification. *Biometrika* **95**, 63–74.

Rosenberg, P. (1995). Hazard function estimation using B-splines. *Biometrics* **51**, 874–887.

Ruppert, D., Wand, M., and Carroll, R. (2003). *Semiparametric Regression.* Cambridge: Cambridge University Press, Cambridge.

Song, X., Davidian, M., and Tsiatis, A. (2002). A semiparametric likelihood approach to joint modeling of longitudinal and time-to-event data. *Biometrics* **58**, 742–753.

Sylvestre, M.-P. and Abrahamowicz, M. (2009). Flexible modeling of the cumulative effects of time-dependent exposures on the hazard. *Statistics in Medicine* **28**, 3437–3453.

Troxel, A., Harrington, D., and Lipsitz, S. (1998). Analysis of longitudinal data with non-ignorable non-monotone missing values. *Journal of the Royal Statistical Society, Series C* **74**, 425–438.

Tseng, Y.-K., Hsieh, F., and Wang, J.-L. (2005) Joint modelling of accelerated failure time and longitudinal data. *Biometrika* **92**, 587–603.

Tsiatis, A. and Davidian, M. (2001). A semiparametric estimator for the proportional hazards model with longitudinal covariates measured with error. *Biometrika* **88**, 447–458.

Tsonaka, R., Rizopoulos, D., Verbeke, G., and Lesaffre, E. (2010). Nonignorable models for intermittently missing categorical longitudinal responses. *Biometrics* **66**, 834–844.

Tsonaka, R., Verbeke, G., and Lesaffre, E. (2009). A semi-parametric shared parameter model to handle non-monotone non-ignorable missingness. *Biometrics* **65**, 81–87.

Vacek, P. (1997). Assessing the effect of intensity when exposure varies over time. *Statistics in Medicine* **16**, 505–513.

Verbeke, G. and Davidian, M. (2008). Joint models for longitudinal data: Introduction and overview. In *Handbook on Longitudinal Data Analysis* by Fitzmaurice, G., Davidian, M., Verbeke, G. and Molenberghs, G. Boca Raton: Chapman & Hall/CRC.

Verbeke, G. and Molenberghs, G. (2000). *Linear Mixed Models for Longitudinal Data.* New York: Springer-Verlag.

Vonesh, E., Greene, T., and Schluchter, M. (2006). Shared parameter models for the joint analysis of longitudinal data and event times. *Statistics in Medicine* **25**, 143–163.

Wang, Y. and Taylor, J. (2001). Jointly modeling longitudinal and event time data with application to acquired immunodeficiency syndrome. *Journal of the American Statistical Association* **96**, 895–905.

Whittemore, A. and Killer, J. (1986). Survival estimation using splines. *Biometrics* **42**, 495–506.

Williamson, P., Kolamunnage-Dona, R., Philipson, P., and Marson, A. (2008). Joint modelling of longitudinal and competing risks data. *Statistics in Medicine* **27**, 6426–6438.

Wu, M. and Bailey, K. (1989). Estimation and comparison of changes in the presence of informative right censoring: conditional linear model. *Biometrics* **45**, 939–955.

Wu, M. and Carroll, R. (1988). Estimation and comparison of changes in the presence of informative right censoring by modeling the censoring process. *Biometrics* **44**, 175–188.

Wulfsohn, M. and Tsiatis, A. (1997). A joint model for survival and longitudinal data measured with error. *Biometrics* **53**, 330–339.

Xu, J. and Zeger, S. (2001). Joint analysis of longitudinal data comprising repeated measures and times to events. *Applied Statistics* **50**, 375–387.

Ye, W., Lin, X., and Taylor, J. (2008). Semiparametric modeling of longitudinal measurements and time-to-event data-a two stage regression calibration approach. *Biometrics* **64**, 1238–1246.

Part III

Semi-Parametric Methods

7

Semiparametric Methods: Introduction and Overview

Garrett M. Fitzmaurice

Harvard University, Boston, MA

CONTENTS

7.1 Introduction

Parametric methods for handling missing data have been well developed since the early 1970s. Much of the statistical literature on parametric models for the data-generating and missing data mechanisms was summarized in the landmark book on missing data by Little and Rubin (1987); see Little and Rubin (2002) for an updated survey that incorporates more recent developments. A potential limitation of these methods is that inferences about parameters of scientific interest depend heavily on parametric assumptions about other nuisance parameters that are of much less interest. For example, parametric selection models directly model the marginal distribution of the complete data, $f(\boldsymbol{Y}_i|X_i, \gamma)$, and a common target of inference is $E(\boldsymbol{Y}_i|X_i, \boldsymbol{\beta})$, where $\gamma = (\boldsymbol{\beta}', \boldsymbol{\alpha}')'$, $\boldsymbol{\beta}$ is the parameter vector of interest relating the mean of $\boldsymbol{Y}_i$ to X_i, and $\boldsymbol{\alpha}$ is the vector of nuisance parameters for the within-subject association (e.g., covariances and/or variance components parameters). Consistency of the estimator of $\boldsymbol{\beta}$ requires correct specification of the entire parametric model for the complete data, $f(\boldsymbol{Y}_i|X_i, \gamma)$; identification comes from assumptions about $f(\boldsymbol{Y}_i|X_i, \gamma)$ and from postulating unverifiable models for the dependence of the non-response process on the unobserved outcomes. Similarly, valid inferences from pattern-mixture and shared-parameter models also require correct specification of joint parametric models for the response vector and the missing data mechanism.

During the last 20 years there has been much research on semi-parametric models that avoid assumptions about aspects of $f(\boldsymbol{Y}_i|X_i)$ that are not of scientific interest. The key feature of semi-parametric models for incomplete data is that they combine a non-parametric or semi-parametric model for $f(\boldsymbol{Y}_i|X_i)$ with a parametric (or sometimes semi-parametric) model for $f(\boldsymbol{R}_i|\boldsymbol{Y}_i, X_i)$. For example, semi-parametric models might combine a logistic regression model for $f(\boldsymbol{R}_i|\boldsymbol{Y}_i, X_i)$ with a suitable regression model for $E(\boldsymbol{Y}_i|X_i)$ but without making any additional distributional assumptions for $f(\boldsymbol{Y}_i|X_i)$. Because semi-parametric models are less restrictive than parametric models, inferences based on them may be more robust. That is, by avoiding assumptions about aspects of $f(\boldsymbol{Y}_i|X_i)$ that are not of scientific interest, inferences about $\boldsymbol{\beta}$ are robust to misspecification of parametric models for these secondary features of the data. Importantly, semi-parametric models leave the so-called "extrapolation distribution" (Daniels and Hogan, 2008), $f(\boldsymbol{Y}_i^m|\boldsymbol{Y}_i^o, \boldsymbol{R}_i, X_i)$, non-identifiable from the observed data. With semi-parametric models sensitivity analysis is ordinarily based on incorporating additional sensitivity parameters that make unverifiable assumptions about $f(\boldsymbol{Y}_i^m|\boldsymbol{Y}_i^o, \boldsymbol{R}_i, X_i)$.

There is a growing literature on semi-parametric models with missing data and the methodology is constantly evolving in important ways; there is probably no better source for the theory of estimation for semi-parametric models with missing data than the excellent text by Tsiatis (2006).The focus of this part of the book is on three major developments in methodology: (i) inverse probability weighting, (ii) double robust estimators, and (iii) pseudo-likelihood methods.

7.2 Inverse Probability Weighting

A general methodology for constructing estimators of regression parameters, $\boldsymbol{\beta}$, in semi-parametric models with missing data was developed in a series of seminal papers by Robins and Rotnitzky (and colleagues); see Robins and Rotnitzky (1992) and Robins, Rotnitzky and Zhao (1994, 1995). Specifically, Robins and Rotnitzky suggested the use of inverse probability weighted (IPW) estimating functions for constructing regression estimators in missing datasettings. Interestingly, the IPW methodology was first proposed in the sample survey literature (Horvitz and Thompson, 1952), where the weights are known and based on the survey design. In the missing datasetting, however, the appropriate weights are not ordinarily known, but must be estimated from the observed data (e.g., using a repeated sequence of logistic regression models).

In a simple version of the IPW method applied to the longitudinal datasetting, for the case where missingness is restricted to dropout, a single weight, say w_i, is estimated only for the study completers. The weight for each individual denotes the inverse probability of remaining in the study until the last intended measurement occasion; that is, $w_i = \{\Pr(D_i = n+1)\}^{-1}$. This weight can be calculated sequentially as the inverse of the product of the conditional probabilities of remaining in the study at each occasion:

$$\begin{aligned} w_i &= \{\Pr(D_i = n+1)\}^{-1} \\ &= \{\Pr(D_i > 1|D_i \geq 1) \times \Pr(D_i > 2|D_i \geq 2) \times \cdots \times \Pr(D_i > n|D_i \geq n)\}^{-1} \\ &= (\pi_{i1} \times \pi_{i2} \times \cdots \times \pi_{in})^{-1}, \end{aligned}$$

where $\pi_{ik} = \Pr(D_i > k|D_i \geq k)$ is conditional on $\boldsymbol{Y}_i$ and X_i (and perhaps also on auxiliary variables, say W_i). Note, to ensure that w_i is identifiable from the observed data, additional assumptions about the dependence of π_{ik} on $\boldsymbol{Y}_i$, X_i, and W_i are required.

Given the estimated weight, $\widehat{w}_i$, for each individual a weighted complete-case analysis can then be performed. For example, the generalized estimating equations (GEE) approach (Liang and Zeger, 1986) can be adapted to handle missing data by making adjustments to the analysis for the probability of remaining in the study. One simple version of this approach is to use an inverse probability weighted GEE to analyze the data from the study completers. In this complete-case IPW GEE each subject's contribution to the analysis is weighted by $\widehat{w}_i$, thereby providing valid estimates of the mean response trends.

The basic intuition behind this weighting method is that each subject's contribution to the weighted complete-case analysis is replicated w_i times, in order to count once for herself and (w_i-1) times for those subjects with the same responses and covariates who do not complete the study. For example, a subject given a weight of 4 has a probability of completing the study of 0.25 (or $\frac{1}{w_i} = 0.25$). As a result, in a complete-case analysis, data from this subject should count once for herself and 3 times for those subjects who do not complete the study. In general, the weights correct for the under-representation of certain response profiles in the observed data due to dropout. That is, the study completers are reweighted so that they resemble a sample from the full population. We note that the weighted GEE described above is a very simple special case of a general class of weighted estimators. In particular, Robins et al. (1995) discuss more general weighted estimating equations for longitudinal data and the construction of "semi-parametric efficient" weighted estimating equations, a generalization known as "augmented" IPW methods.

The type of IPW complete-case analysis described above is valid if the model that provides the estimated w_i is correct. In general, weighting methods yield consistent estimators provided that the non-response model is not misspecified. In the missing datasetting, the w_i are not ordinarily known but must be estimated from the observed data. However, under the assumption that data are MAR, the π_{ij}'s are assumed to be a function only of the *observed* covariates and the *observed* responses prior to dropout. The π_{ij}'s can also depend on *observed* auxiliary variables, say W_i, that are thought to be predictive of dropout. Finally, dependence on unobserved responses can also be incorporated into IPW methods (Rotnitzky and Robins, 1997, Rotnitzky, Robins, and Scharfstein, 1998); in particular, Scharfstein, Rotnitzky, and Robins (1999), Robins, Rotnitzky, and Scharfstein (2000), and Rotnitzky et al. (2000) discuss the use of NMAR models within the context of sensitivity analysis.

Chapter 8 provides a comprehensive survey of IPW methods for constructing estimators of $\boldsymbol{\beta}$ in semi-parametric models for $f(\boldsymbol{Y}_i|X_i)$. In particular, the chapter considers the large sample properties of estimators of $\boldsymbol{\beta}$ via their so-called "influence functions" and discusses strategies for deriving a class of estimators that are both robust and efficient when data are missing. The chapter describes the geometry of influence functions and the correspondence between estimators and influence functions for semi-parametric models. This geometric perspective on semi-parametric theory is used to characterize the class of "augmented" IPW complete-case estimators of $\boldsymbol{\beta}$ when data are missing.

7.3 Double Robustness

In the previous section we described the key features of IPW methods for missing data. Recall that in the simple complete-case IPW method the weight,

$$w_i = \{\Pr(D_i = n+1)\}^{-1} = (\pi_{i1} \times \pi_{i2} \times \cdots \times \pi_{in})^{-1},$$

is the inverse probability of remaining in the study until the last intended measurement occasion. For the simple complete-case IPW method it is readily apparent that there is information in the available data that is ignored by focusing exclusively on the complete cases. This has provided the impetus for research on IPW methods that incorporate an "augmentation term." By incorporating an augmentation term that extracts information in the available data, the inverse probability weighted augmented (IPWA) methodology yields estimators of $\boldsymbol{\beta}$ that can be discernibly more efficient than their non-augmented IPW counterparts. In addition, there are special types of IPWA estimators known as "doubly robust" estimators that have very appealing properties (Robins, 2000; Robins and Rotnitzky, 2001); these estimators are the main focus of Chapter 9.

The IPWA estimating functions provide a general method for constructing estimators of $\boldsymbol{\beta}$. The essential feature of the IPWA estimating function is that it is composed of two terms, the usual IPW term and an augmentation term. The basis for the IPW term is a complete data unbiased estimating function for $\boldsymbol{\beta}$; whereas the basis for the augmentation term is some function of the observed data chosen so that it has conditional mean of zero given the complete data. In general, the addition of the augmentation term to the IPW term increases the efficiency of the resulting estimator of $\boldsymbol{\beta}$.

Although the augmentation term may be used to gain efficiency, there is an important class of IPWA estimators that have desirable robustness properties. The key feature of so-called "doubly robust" IPWA estimators is that they relax the assumption that the model for the missingness probabilities, π_{ij}, has been correctly specified, albeit requiring additional assumptions on the model for $f(\boldsymbol{Y}_i|X_i)$. That is, doubly robust methods require the specification of two models, one for the missingness probabilities and another for the distribution of the incomplete data. For example, for doubly robust estimation of $\boldsymbol{\beta}$ the IPW GEE can be augmented by a function of the observed data. When this augmentation term is selected and modeled correctly according to the distribution of the complete data, the resulting estimator of $\boldsymbol{\beta}$ is consistent even if the model for missingness is misspecified. On the other hand, if the model for missingness is correctly specified, the augmentation term no longer needs to be correctly specified to yield consistent estimators of $\boldsymbol{\beta}$ (Scharfstein, Rotnitzky, and Robins, 1999, see Section 3.2 of Rejoinder; Robins and Rotnitzky, 2001; van der Laan and Robins, 2003; Bang and Robins, 2005; also see Lipsitz, Ibrahim, and Zhao, 1999; Lunceford and Davidian, 2004).

Thus, with doubly robust estimators, by construction, an unbiased estimating function for $\boldsymbol{\beta}$ is obtained if *either* the model for the incomplete data *or* the model for the missingness mechanism has been correctly specified, *but not necessarily both.* These estimators are said to be "doubly robust" in the sense of providing double protection against model misspecification. In general, when both models are correctly specified, the doubly robust estimator is more efficient than the non-augmented IPW estimator; the gain in efficiency arises from the doubly robust estimator's reliance on a model for the incomplete data. Thus, by capitalizing on both robustness and efficiency, doubly robust methods provide a very appealing approach for analyzing incomplete data. Chapter 9 provides an extensive and comprehensive review of the literature on doubly robust estimation as it applies to the analysis of incomplete data, with a focus on the problem of incomplete outcomes in cross-sectional and longitudinal studies. Taking an historical perspective on the development of doubly robust methods, the chapter reviews a number of recently proposed strategies for modeling the missingness process and the incomplete data with the goal of constructing doubly robust estimators with improved finite-sample behavior and asymptotic efficiency. In particular, the chapter highlights how the finite-sample behavior and asymptotic efficiency of doubly robust estimators are heavily influenced by the choice of estimators for the missingness process and the incomplete data. The chapter concludes with a discussion of data-driven model selection strategies to guide the choice of models for the missingness process and the incomplete data.

7.4 Pseudo-Likelihood

In earlier sections, we discussed estimators for semi-parametric models with missing data that are based on estimating functions. One of the key motivating factors for the development of these semi-parametric methods is avoidance of assumptions about aspects of $f(\boldsymbol{Y}_i|X_i)$ that are not of primary scientific interest. An alternative, but closely related, approach is to base estimation on pseudo-likelihood (PL) functions (Gong and Samaniego, 1982; Arnold and Strauss, 1991; Liang and Self, 1996). An important feature of pseudo-likelihood functions is that they approximate true likelihood functions but are far simpler to evaluate and maximize. Specifically, pseudo-likelihood functions are most commonly constructed by multiplying a sequence of lower dimensional likelihoods.

For example, a widely used version of pseudo-likelihood is the pairwise likelihood function (Cox and Reid, 2004) formed by replacing $f(\boldsymbol{Y}_i|X_i, \boldsymbol{\gamma})$ with the product of all pairwise densities, $f(Y_{ij}, Y_{ik}|X_i, \boldsymbol{\gamma}^*)$ for $1 \leq j < k \leq n_i$, over the set of all possible pairs within the response vector $\boldsymbol{Y}_i$. The parameter $\boldsymbol{\gamma}^*$ is usually a sub-vector of $\boldsymbol{\gamma}$. The ith individual's contribution to the pairwise pseudo-likelihood is

$$\mathcal{P}L_i(\boldsymbol{\gamma}^*) = \prod_{1 \leq j < k \leq n_i} f(Y_{ij}, Y_{ik}|X_i, \boldsymbol{\gamma}^*);$$

the PL estimate, $\widehat{\boldsymbol{\gamma}}^*$, maximizes the log pseudo-likelihood and is obtained by setting the partial derivative of the log pseudo-likelihood with respect to $\boldsymbol{\gamma}^*$ to zero and solving for $\widehat{\boldsymbol{\gamma}}^*$. Because $\boldsymbol{\gamma}^*$ is typically a sub-vector of $\boldsymbol{\gamma}$, the PL approach avoids assumptions about many higher-order moments that are not of scientific interest. By modeling only the means and pairwise associations, PL inference is robust to misspecification of parametric models for higher-order moments. However, because of the working independence assumption among the likelihood terms forming the pairwise pseudo-likelihood, standard errors must be based on the empirical or so-called "sandwich" variance estimator (Huber, 1967; White, 1982; Royall, 1986).

Although we have focused on the pairwise likelihood above, in their most general form, PL functions can be constructed by compounding any sequence of lower dimensional likelihoods and the choice of component likelihood terms, often determined by the context, will have an influence on the accuracy and efficiency of PL inference. For example, another widely used version of PL is the pairwise *conditional* likelihood function, formed by replacing $f(\boldsymbol{Y}_i|X_i, \boldsymbol{\gamma})$ with the product of all pairwise *conditional* densities, $f(Y_{ij}|Y_{ik}, X_i, \boldsymbol{\gamma}^*)$ for $1 \leq j < k \leq n_i$; see, for example, Molenberghs and Verbeke (2005) in the context of longitudinal studies. An alternative version of PL is based on compounding the *full conditional* densities, $f(Y_{ij}|\boldsymbol{Y}_{i(-j)}, X_i, \boldsymbol{\gamma}^*)$ for $j = 1, \ldots, n_i$, where $\boldsymbol{Y}_{i(-j)}$ denotes the random vector with Y_{ij} omitted. We note that when the pseudo-likelihood function is restricted to component terms that are marginal and/or conditional densities, it is usually referred to as a *composite likelihood* function; see Varin, Reid and Firth (2011) for an excellent survey of recent developments in the theory and application of composite likelihoods.

The motivation for PL estimation is best illustrated by the following example that involves no missing data. Consider the case where $\boldsymbol{Y}_i$ is a vector of binary responses and the main target of inference is $E(\boldsymbol{Y}_i|X_i, \boldsymbol{\beta})$. Maximum likelihood (ML) estimation of $\boldsymbol{\beta}$ requires complete specification of the joint distribution of $f(\boldsymbol{Y}_i|X_i, \boldsymbol{\gamma})$, where $\boldsymbol{\gamma} = (\boldsymbol{\beta}', \boldsymbol{\alpha}')'$ and $\boldsymbol{\alpha}$ is a vector of parameters for higher-order moments of $\boldsymbol{Y}_i$. Whereas $\boldsymbol{\beta}$ parameterizes the mean of the vector of binary responses, $\boldsymbol{\alpha}$ parameterizes the joint probabilities of pairs of responses, triples of responses, quadruples, and so on. The joint distribution, $f(\boldsymbol{Y}_i|X_i, \gamma)$, is multinomial and many alternative multinomial models have been proposed that differ in terms of how the higher-order moments are parameterized. For example, using the elegant Bahadur (1961) expansion, Kupper and Haseman (1978) and Altham (1978) propose multinomial models where the within-subject association is parameterized in terms of correlations. Dale (1984), McCullagh and Nelder (1989), Lipsitz, Laird, and Harrington (1990), Liang, Zeger, and Qaqish (1992), Becker and Balagtas (1993), Molenberghs and Lesaffre (1994), Lang and Agresti (1994), Glonek and McCullagh (1995), and others propose models where the higher-order moments are parameterized in terms of marginal odds ratios. An alternative approach is to parameterize the within-subject association in terms of conditional associations, leading to so-called "mixed-parameter" models (Fitzmaurice and Laird, 1993; Glonek, 1996; Molenberghs and Ritter, 1996). Although numerous multinomial models for regression analysis of repeated binary responses have been proposed, ML estimation is impractical in many applications and is hampered by at least three main factors. First, with the exception of the Bahadur expansion model, there are no simple expressions for evaluating the joint probabilities in terms of the model parameters. This makes maximization of the likelihood particularly difficult. Second, even with current advances in computing, these models are computationally demanding to fit except when the number of binary responses is relatively small. Finally, many of these models are not robust to misspecification of the higher-order moments.

Because there is no convenient and tractable specification of $f(\boldsymbol{Y}_i|X_i, \gamma)$ when $\boldsymbol{Y}_i$ is a vector of binary responses, it is advantageous to use a method of estimation that avoids assumptions about many aspects of $f(\boldsymbol{Y}_i|X_i)$ that are not of primary scientific interest. Pseudo-likelihood methods yield estimators of $\boldsymbol{\beta}$ without making full distributional assumptions and provide a remarkably convenient alternative to ML estimation. For example, PL estimation can be based on the pairwise pseudo-likelihood; the resulting PL estimator is consistent provided the models for the mean and pairwise association have been correctly specified.

Next, we briefly consider the use of PL methods with missing data. When the primary target of inference is $E(\boldsymbol{Y}_i|X_i, \boldsymbol{\beta})$ and data are incomplete, estimation of $\boldsymbol{\beta}$ can be based, for example, on a pairwise pseudo-likelihood formed by replacing the full data density $f(\boldsymbol{Y}_i, \boldsymbol{R}_i|X_i, \gamma, \psi)$ with the product of all pairwise densities, $f(Y_{ij}, Y_{ik}, R_{ij}, R_{ik}|X_i, \gamma^*, \psi) = f(Y_{ij}, Y_{ik}|X_i, \gamma^*) f(R_{ij}, R_{ik}|Y_{ij}, Y_{ik}, X_i, \psi)$ for $1 \leq j < k \leq n_i$. The resulting pseudo-likelihood requires only partial specification of the distribution of the complete data and missingness indicators at pairs of occasions. Moreover, $\gamma^* = (\boldsymbol{\beta}', \boldsymbol{\alpha}^{*\prime})'$, where $\boldsymbol{\alpha}^*$ is the sub-vector of $\boldsymbol{\alpha}$ corresponding to associations among pairs of responses only, thereby avoiding assumptions about many higher-order moments. An alternative approach for handling missing data is to base inference on a weighted version of the pairwise pseudo-likelihood for the complete data. That is, inference for γ^* is based on the so-called pseudo-score vector,

$$S(\gamma^*) = \sum_{i=1}^{N} \frac{\partial}{\partial \gamma^*} \log \mathcal{P}L_i(\gamma^*),$$

but with each individual's contribution to $S(\gamma^*)$ weighted by the inverse probability of missingness. The latter approach, discussed in detail in Chapter 10, leads to a class of IPW estimators.

In general, the pseudo-likelihood approach shares many of the appealing properties of the estimating equations approach discussed in Section 7.2. Because $\boldsymbol{\alpha}^*$ is a sub-vector of $\boldsymbol{\alpha}$, the pseudo-likelihood approach avoids making assumptions about many higher-order moments that are not of scientific interest. As a result, maximization of the pseudo-likelihood yields estimators of $\boldsymbol{\beta}$ (and of $\boldsymbol{\alpha}^*$) that are robust to misspecification of parametric models for these higher-order moments. This is similar in spirit to the robustness of the estimating equations approach. However, when data are missing, pseudo-likelihood methods must take into account the missingness mechanism in order to yield valid inferences based on the observed data. Chapter 10 provides a comprehensive survey of the theory of pseudo-likelihood for incomplete data. The chapter describes how inverse probability weighting (IPW) and ideas of double robustness can also be incorporated in pseudo-likelihood methods. Specifically, pseudo-likelihood estimating equations are developed in which each subject's contribution is weighted by the inverse probability, either of being fully observed, or of being observed up to a certain time. The resulting estimators are robust against misspecification of the likelihood function for the complete data when missingness is MAR, provided the inverse probability weights have been correctly specified. Doubly robust PL estimators are also developed.

7.5 Concluding Remarks

The early statistical literature on methods for missing data focused heavily on the development of parametric models for both the data-generating and missing data mechanisms. Many of these models for missing data can now be considered classical and the theory underlying parametric models for missing data is well-established. Recognizing that the primary target of inference is often on a relatively low-dimensional parametric component (e.g., the mean response, and its dependence on covariates), there has been much recent research on semi-parametric methods that avoid strong assumptions about nuisance characteristics of the data. During the last two decades there has been very significant progress in research on the construction and properties of estimators of regression parameters in semi-parametric models for missing data. Regardless of whether they are based on estimating functions or objective functions defined by pseudo-likelihoods, estimators of semi-parametric models have robustness properties that make them especially appealing in many scientific applications. An intriguing avenue of research in semi-parametric methods is the recent advances in constructing doubly robust estimators that maintain good properties under certain kinds of model misspecification. Despite the many advances in semi-parametric methods with missing data over the past two decades, it is somewhat unfortunate that their implementation in statistical software has lagged painfully far behind these developments in theory. This situation needs to be quickly remedied if the tremendous promise of these new methods for handling missing data is to be realized in scientific applications.

References

Altham, P.M.E. (1978). Two generalizations of the binomial distribution. *Applied Statistics* **27,** 162–167.

Arnold, B.C. and Strauss, D. (1991). Pseudolikelihood estimation: Some examples. *Sankhya: The Indian Journal of Statistics, Series B* **53,** 233–243.

Bahadur, R.R. (1961). A representation of the joint distribution of responses to n dichotomous items. In: *Studies in Item Analysis and Prediction,*, H. Solomon (Ed.). Stanford Mathematical Studies in the Social Sciences VI. Stanford, CA: Stanford University Press.

Bang, H. and Robins, J.M. (2005). Doubly robust estimation in missing data and causal inference models. *Biometrics* **61,** 962–973.

Becker, M.P. and Balagtas, C.C. (1993). Marginal modeling of binary cross-over data. *Biometrics* **49,** 997–1009.

Cox, D. and Reid, N. (2004). A note on pseudolikelihood constructed from marginal densities. *Biometrika* **91,** 729–737.

Dale, J.R. (1984). Local versus global association for bivariate ordered responses. *Biometrika* **71,** 507–514.

Daniels, M.J. and Hogan, J.W. (2008). *Missing Data in Longitudinal Studies: Strategies for Bayesian Modeling and Sensitivity Analysis.* New York: Chapman & Hall/CRC.

Fitzmaurice, G.M. and Laird, N.M. (1993). A likelihood-based method for analysing longitudinal binary responses. *Biometrika* **80,** 141–151.

Glonek, G.F.V. (1996). A class of regression models for multivariate categorical responses. *Biometrika* **83,** 15–28.

Glonek, G.F.V. and McCullagh, P. (1995). Multivariate logistic models. *Journal of the Royal Statistical Society, Series B* **57,** 533–546.

Gong, G., and Samaniego, F. (1982). Pseudo maximum likelihood estimation: Theory and applications. *Annals of Statistics* **9,** 861–869.

Horvitz, D.G. and Thompson, D.J. (1952). A generalization of sampling without replacement from a finite universe. *Journal of the American Statistical Association* **47,** 663–685.

Huber, P.J. (1967) The behavior of maximum likelihood estimates under nonstandard conditions. In *Proceedings of the Fifth Berkeley Symposium on Mathematical Statistics and Probability,* Vol. 1, pp. 221–233. Berkeley: University of California Press.

Kupper, L.L. and Haseman, J.K. (1978). The use of a correlated binomial model for the analysis of certain toxicological experiments. *Biometrics* **34,** 69–76.

Lang, J.B. and Agresti, A. (1994). Simultaneous modeling joint and marginal distributions of multivariate categorical responses. *Journal of the American Statistical Association* **89,** 625–632.

Liang, K.-Y. and Self, S.G. (1996). On the asymptotic behavior of the pseudolikelihood ratio test statistic. *Journal of the Royal Statistical Society, Series B* **58,** 785–796.

Liang, K.-Y. and Zeger, S.L. (1986). Longitudinal data analysis using generalized linear models. *Biometrika* **73,** 13–22.

Liang, K.-Y., Zeger, S.L., and Qaqish, B. (1992). Multivariate regression analyses for categorical data (with discussion). *Journal of the Royal Statistical Society, Series B* **54,** 2–24.

Lipsitz, S. R., Ibrahim, J.G., and Zhao, L.P. (1999). A weighted estimating equation for missing covariate data with properties similar to maximum likelihood. *Journal of the American Statistical Association* **94,** 1147–1160.

Lipsitz, S.R., Laird, N.M., and Harrington, D.P. (1990). Maximum likelihood regression methods for paired binary data. *Statistics in Medicine* **9,** 1417–1425.

Little, R.J. A. and Rubin, D.B. (1987). *Statistical Analysis with Missing Data, (1st ed.).* New York: John Wiley & Sons.

Little, R.J.A. and Rubin, D.B. (2002). *Statistical Analysis with Missing Data, (2nd ed.).* New York: John Wiley & Sons.

Lunceford, J.K. and Davidian, M. (2004). Stratification and weighting via the propensity score in estimation of causal treatment effects: A comparative study. *Statistics in Medicine* **23,** 2937–2960.

Molenberghs, G. and Lesaffre, E. (1994). Marginal modelling of correlated ordinal data using a multivariate Plackett distribution. *Journal of the American Statistical Association* **89,** 633–644.

Molenberghs, G. and Ritter, L. (1996). Likelihood and quasi-likelihood based methods for analyzing multivariate categorical data, with the association between outcome of interest. *Biometrics* **52,** 1121–1133.

Molenberghs, G. and Verbeke, G. (2005). *Models for Discrete Longitudinal Data.* New York: Springer.

McCullagh, P. and Nelder, J. (1989). *Generalized Linear Models, (2nd ed.)* London: Chapman and Hall/CRC Press.

Robins, J.M. (2000). Robust estimation in sequentially ignorable missing data and causal inference models. *Proceedings of the 1999 Joint Statistical Meetings.*

Robins, J.M., and Rotnitzky, A. (1992). Recovery of information and adjustment for dependent censoring using surrogate markers. In: N.P. Jewell, K. Dietz, and V. Farewell (Eds.).*AIDS Epidemiology: Methodological Issues.* Boston: Birkhäuser.

Robins, J.M. and Rotnitzky, A. (2001). Discussion of "Inference for semi-parametric models: some questions and an answer," by P.J. Bickel and J. Kwon, *Statistica Sinica* **11,** 920–936.

Robins, J.M., Rotnitzky, A., and Scharfstein, D. (2000). Sensitivity analysis for selection bias and unmeasured confounding in missing data and causal inference models. *Statistical Models in Epidemiology, the Environment, and Clinical Trials. The IMA Volumes in Mathematics and its Applications* **116**, 1—94.

Robins, J.M., Rotnitzky, A., and Zhao, L.P. (1994). Estimation of regression coefficients when some of the regressors are not always observed. *Journal of the American Statistical Association* **89,** 846–866.

Robins, J.M., Rotnitzky, A., and Zhao, L. P. (1995). Analysis of semi-parametric regression models for repeated outcomes in the presence of missing data. *Journal of the American Statistical Association* **90,** 106–121.

Rotnitzky, A. and Robins, J.M. (1997). Analysis of semi-parametric regression models with non-ignorable non-response. *Statistics in Medicine* **16,** 81–102.

Rotnitzky, A., Robins, J.M., and Scharfstein, D. (1998). Semiparametric regression for repeated outcomes with nonignorable nonresponse. *Journal of the American Statistical Association* **93,** 1321–1339.

Rotnitzky, A., Scharfstein, D., Su, T.L., and Robins, J. M. (2000). Methods for conducting sensitivity analysis of trials with possibly non-ignorable competing causes of censoring. *Biometrics* **57,** 111–121.

Royall, R.M. (1986). Model robust confidence intervals using maximum likelihood estimators. *International Statistical Review* **54,** 221–226.

Scharfstein, D.O., Rotnitzky, A., and Robins, J.M. (1999). Adjusting for nonignorable drop-out using semi-parametric nonresponse models. *Journal of the American Statistical Association* **94,** 1096–1120 (with Rejoinder, 1135–1146).

Tsiatis, A.A. (2006). *Semiparametric Theory and Missing Data.* New York: Springer.

van der Laan, M.J. and Robins, J.M. (2003). *Unified Methods for Censored Longitudinal Data and Causality.* New York: Springer.

Varin, C., Reid, N., and Firth, D. (2011). An overview of composite likelihood methods. *Statistica Sinica* **21,** 5–42.

White, H. (1982). Maximum likelihood estimation under mis-specified models. *Econometrica* **50**, 1–26.

8

Missing Data Methods: A Semi-Parametric Perspective

Anastasios A. Tsiatis

North Carolina State University, Raleigh, NC

Marie Davidian

North Carolina State University, Raleigh, NC

CONTENTS

8.1 Introduction

As discussed in other chapters in this book, when the mechanism governing missingness is that of missing at random (MAR), three main approaches to inference on parameters describing the complete data using the observed data have been advocated. Broadly speaking, these may be characterized as likelihood methods, imputation methods, and methods involving the use of so-called inverse probability weighting. Likelihood, and imputation methods generally are predicated on the assumption of a parametric statistical model for the complete data; that is, the class of probability densities that is believed to contain the true density generating the data may be described using a finite number of parameters. One of the advantages of these approaches is that estimators for the model parameters may be derived without the need to model explicitly the missing data process.

Over the past several decades, there has been increasing interest in statistical models that are less restrictive than parametric models. In particular, considerable research has focused on semi-parametric statistical models, in which densities in the class of densities thought to contain the true density of the complete data are described by both finite-dimensional parametric and infinite-dimensional components. Because these models involve a larger class of densities, inferences based on them may be more robust than those based on fully parametric models. Without the complication of missing data, an extensive theory has been developed for inference under semi-parametric models; a rigorous treatment is given by Bickel et al. (1993), and a more heuristic account is provided by Tsiatis (2006).

When data are missing, however, under the assumption of a semi-parametric model for the complete data, likelihood and imputation methods become theoretically challenging and often difficult to implement. Accordingly, alternative approaches are of interest.

Inverse probability weighted complete-case (IPWCC) estimators were first introduced in the context of sampling by Horvitz and Thompson (1952) and have been adapted to and used in missing data problems for some time. In contrast to likelihood and imputation approaches, these methods require modeling of the missing data mechanism. IPWCC estimators and a generalization of them known as augmented inverse-probability weighted complete-case (AIPWCC) estimators are often viewed as ad hoc alternatives to likelihood and imputation methods that are not model-based and hence are criticized on the basis that the assumptions underlying them (that would be clearly stated through a model) cannot be checked.

In a landmark paper (Robins, Rotnitzky, and Zhao, 1994), the authors demonstrated that, in fact, these estimators arise directly from the application of semi-parametric theory to the problem of inference on parameters of interest in a semi-parametric statistical model for complete data when some data are missing. In particular, Robins, Rotnitzky, and Zhao (1994) showed how estimators for parameters in semi-parametric models when there are no missing data may be used to derive estimators for these parameters when there are missing data, and that these estimators have exactly the form of AIPWCC estimators. Thus, the theory makes explicit that AIPWCC estimators, which include IPWCC estimators as a special case, are not ad hoc but come about from a rigorous theory based on semi-parametric statistical models and hence are in fact model-based. Indeed, because the statistical models underlying them involve a larger class of densities, some analysts may find these estimators preferable to those based on parametric models on grounds of robustness.

In this chapter, we provide an overview of these developments. We begin by defining formally semi-parametric models and the concept of an influence function. We then review semi-parametric theory in the case of no missing data and demonstrate how influence functions may be derived and the correspondence between estimators and influence functions. The remainder of the chapter systematically presents the steps involved in using semi-parametric theory to derive estimators for parameters of interest when some data are missing and shows how this leads to the class of AIPWCC estimators.

Throughout, we draw heavily on material in Davidian, Tsiatis, and Leon (2005) and Tsiatis (2006). Readers wishing further details on the developments presented here should consult these references.

The convention in this book is to refer to the data that would be observed if there were no missing data, as the *complete data*, with the term *full data* reserved for the complete data together with indicator variables for the missing components. In the literature on semi-parametric theory for missing data problems, it is conventional to refer to the complete data instead as the full data when discussing the influence functions that are central to this theory. Because this nomenclature is widespread, in Sections 8.5 and beyond, we adopt it to circumvent confusion for readers consulting this literature.

8.2 Semi-Parametric Models and Influence Functions

Problems in statistical inference are formalized by placing them in the framework of a statistical model. That is, data are envisioned as realizations of random vectors $\boldsymbol{V}_1, \ldots, \boldsymbol{V}_N$, where $\boldsymbol{V}_i$ represents the data collected on the ith individual in a sample of n individuals from some population of interest. For example, $\boldsymbol{V}_i = (\boldsymbol{X}_i', Y_i)'$, where Y_i denotes a scalar response, $\boldsymbol{X}_i$ a vector of covariates, and interest focuses on characterizing the relationship between response and covariates. As is conventional, assume that $\boldsymbol{V}_1, \ldots, \boldsymbol{V}_N$ are independent and identically distributed (i.i.d.).

A statistical model is a class of densities believed to have generated the data. Thus, $\boldsymbol{V}_i$ is assumed to have a density within the class of densities dictated by the model. The statistical model may be indexed by a parameter, $\boldsymbol{\theta}$, that characterizes the density in some meaningful way. Specifically, we write a statistical model as the class of densities

$$\mathcal{P} = \{f_{\boldsymbol{V}}(\boldsymbol{v}, \boldsymbol{\theta}), \boldsymbol{\theta} \in \Omega\}, \tag{8.1}$$

where for each $\boldsymbol{\theta} \in \Omega$, $f_{\boldsymbol{V}}(\boldsymbol{v}, \boldsymbol{\theta})$ denotes a uniquely identified density of $\boldsymbol{V}$. We will be more precise about the nature of $\boldsymbol{\theta}$ shortly. We take densities to be defined with respect to a dominating measure $\nu(\boldsymbol{v})$ that is generally a counting measure for elements of $\boldsymbol{V}$ that are discrete and Lebesgue measure for elements that are continuous.

It is often the case that inference focuses on only part of $\boldsymbol{\theta}$; however, $\boldsymbol{\theta}$ is necessary to describe fully the statistical model (8.1). The most familiar form of a statistical model is that where $\boldsymbol{\theta}$ is a finite-dimensional $(p \times 1)$ parameter, in which case (8.1) is referred to as a *parametric model.* For example, in the regression context discussed above with $\boldsymbol{V} = (\boldsymbol{X}', Y)'$, one might assume that $Y = \mu(\boldsymbol{X}, \boldsymbol{\beta}) + \varepsilon$, where $\mu(\boldsymbol{X}, \boldsymbol{\beta})$ is a known function of $\boldsymbol{X}$, linear or nonlinear in the unknown q-dimensional parameter $\boldsymbol{\beta}$; and ε is independent of $\boldsymbol{X}$ and has a normal distribution with mean zero and variance σ^2. Here, $\boldsymbol{\beta}$ is of interest for describing the relationship between Y and $\boldsymbol{X}$, and σ^2 is a nuisance parameter. Thus, assumed densities for this problem may be described by $\boldsymbol{\theta} = (\boldsymbol{\beta}', \sigma^2)'$. In general, this suggests partitioning $\boldsymbol{\theta}$ as $(\boldsymbol{\beta}', \boldsymbol{\eta}')'$, where $\boldsymbol{\beta}$ is the q-dimensional parameter on which inference focuses, and $\boldsymbol{\eta}$ $(r \times 1)$ is a nuisance parameter that, with $\boldsymbol{\beta}$, characterizes the assumed statistical model.

Remark. Strictly speaking, the foregoing model is not a finite-dimensional parametric model, as the marginal distribution of $\boldsymbol{X}$ is left unspecified. Only the conditional distribution of Y given $\boldsymbol{X}$ is fully specified parametrically. In the regression context, however, only this distribution is of interest, and, accordingly it is conventional to refer to this as a parametric model.

A drawback of parametric models is the potential for compromised inference on $\boldsymbol{\beta}$ in the event that the assumed class (8.1) is too restrictive to represent the true data generating mechanism. A popular approach that acknowledges concern over such sensitivity to parametric assumptions is to take a semi-parametric perspective. A *semi-parametric model* may involve both parametric and non-parametric components. The non-parametric component represents features on which the analyst is unwilling or unable to make parametric assumptions, and interest may focus on a parametric component or on some functional of the non-parametric components, as we now describe.

In the regression setting, for example, instead of adopting the full normal-based, homoscedastic model above, one might specify only that $E(Y|\boldsymbol{X}) = \mu(\boldsymbol{X}, \boldsymbol{\beta})$ and place no other restrictions on the class (8.1). In this case, inference still focuses on $\boldsymbol{\beta}$, but now one may view the nuisance portion of $\boldsymbol{\theta}$, $\boldsymbol{\eta}$, as infinite-dimensional. Under these conditions, we may write $\boldsymbol{\theta}$ in (8.1) as $\boldsymbol{\theta}(\cdot) = \{\boldsymbol{\beta}', \boldsymbol{\eta}(\cdot)\}$, where $\boldsymbol{\beta}$ is $(q \times 1)$ and $\boldsymbol{\eta}(\cdot)$ is infinite-dimensional, to emphasize that all other aspects of the densities beyond the form of $E(Y|\boldsymbol{X})$ are left unspecified. In Section 1.2 of Tsiatis (2006), this construction is formalized for this semi-parametric model.

In general, then, in a semi-parametric model, allowing the nuisance parameter $\boldsymbol{\eta}(\cdot)$ to be infinite-dimensional places fewer probabilistic constraints on the data generating mechanism relative to fully parametric models. Accordingly, inferences on $\boldsymbol{\beta}$ under the semi-parametric model should enjoy greater applicability and robustness than those based on a parametric model.

An alternative formulation of a semi-parametric model is to index densities in $\mathcal{P}$ by $\boldsymbol{\theta}(\cdot)$ and identify the parameter of interest directly as a functional $\boldsymbol{\beta}\{\boldsymbol{\theta}(\cdot)\}$. A special case is where the analyst may wish to place no restrictions on the class (8.1), in which case $\boldsymbol{\theta}(\cdot)$ are positive functions of $\boldsymbol{v}$ such that $\int \boldsymbol{\theta}(\boldsymbol{v})\, d\nu(\boldsymbol{v}) = 1$. This is the particular case of a semi-parametric model referred to as a *non-parametric model.* For example, if $\boldsymbol{V} = (\boldsymbol{X}', Y)$, interest may focus on $E(Y)$, and the parameter of interest is thus $\boldsymbol{\beta}\{\boldsymbol{\theta}(\cdot)\} = \int y\, \boldsymbol{\theta}(\boldsymbol{x}, y)\, d\nu(\boldsymbol{x}, y)$ (so $q = 1$). See Tsiatis (2006, Section 1.2) for further discussion.

Henceforth, for a parametric or semi-parametric model where $\boldsymbol{\theta}$ is partitioned as a finite-dimensional parameter $\boldsymbol{\beta}$ and a finite- or infinite-dimensional parameter $\boldsymbol{\eta}$, we sometimes write densities in $\mathcal{P}$ as $f_{\boldsymbol{V}}(\boldsymbol{v}, \boldsymbol{\beta}, \boldsymbol{\eta})$ to highlight dependence on $\boldsymbol{\beta}$ and $\boldsymbol{\eta}$.

Because a semi-parametric model leaves features of the class of assumed densities unspecified, it represents a larger class of densities than those under a parametric model. Accordingly, in the theoretical developments that follow, we assume that a specified semi-parametric model contains the true density generating the data. That is, in (8.1), we assume that the true density $f_{\boldsymbol{V}0}(\boldsymbol{v})$ of $\boldsymbol{V}$ may be written as $f_{\boldsymbol{V}}\{\boldsymbol{v}, \boldsymbol{\theta}_0(\cdot)\}$ for some $\boldsymbol{\theta}_0(\cdot) \in \Omega$.

Given a specified parametric or semi-parametric statistical model (8.1), the goal is to use the data $\boldsymbol{V}_1, \ldots, \boldsymbol{V}_N$ to estimate the parameter $\boldsymbol{\beta}$ of interest. Most reasonable estimators for the parameter $\boldsymbol{\beta}$ are *regular and asymptotically linear.* Regularity is a technical condition that rules out "pathological" estimators with undesirable local properties (Newey, 1990), such as the "superefficient" estimator of Hodges (e.g., Casella and Berger, 2002, p. 515). Generically, an estimator $\widehat{\boldsymbol{\beta}}$ for $\boldsymbol{\beta}$ $(q \times 1)$ in a parametric or semi-parametric statistical model for random vector $\boldsymbol{V}$ based on i.i.d. data $\boldsymbol{V}_i$, $i = 1, \ldots, N$, is asymptotically linear if it satisfies, for a q-dimensional function $\boldsymbol{\varphi}(\boldsymbol{V})$,

$$N^{1/2}(\widehat{\boldsymbol{\beta}} - \boldsymbol{\beta}_0) = N^{-1/2} \sum_{i=1}^{n} \boldsymbol{\varphi}(\boldsymbol{V}_i) + o_p(1), \tag{8.2}$$

where $\boldsymbol{\beta}_0$ is the true value of $\boldsymbol{\beta}$ generating the data, $o_p(1)$ is a remainder term that converges in probability to zero, $E\{\boldsymbol{\varphi}(\boldsymbol{V})\} = \boldsymbol{0}$, $E\{\boldsymbol{\varphi}'(\boldsymbol{V})\boldsymbol{\varphi}(\boldsymbol{V})\} < \infty$, and expectation is with respect to the true distribution of $\boldsymbol{V}$. The function $\boldsymbol{\varphi}(\boldsymbol{V})$ is referred to as the *influence function* of $\widehat{\boldsymbol{\beta}}$, as to a first order $\boldsymbol{\varphi}(\boldsymbol{V})$ is the influence of a single observation on $\widehat{\boldsymbol{\beta}}$ in the sense given in Casella and Berger (2002, Section 10.6.4).

An estimator that is both regular and asymptotically linear (RAL) with influence function $\boldsymbol{\varphi}(\boldsymbol{V})$ is consistent and asymptotically normal with asymptotic covariance matrix $E\{\boldsymbol{\varphi}(\boldsymbol{V})\boldsymbol{\varphi}'(\boldsymbol{V})\}$. For RAL estimators, there exists an influence function $\boldsymbol{\varphi}^{\mathit{eff}}(\boldsymbol{V})$ such that $E\{\boldsymbol{\varphi}(\boldsymbol{V})\boldsymbol{\varphi}'(\boldsymbol{V})\} - E\{\boldsymbol{\varphi}^{\mathit{eff}}(\boldsymbol{V})\boldsymbol{\varphi}^{\mathit{eff}'}(\boldsymbol{V})\}$ is non-negative definite for any influence function $\boldsymbol{\varphi}(\boldsymbol{V})$; $\boldsymbol{\varphi}^{\mathit{eff}}(\boldsymbol{V})$ is referred to as the *efficient influence function* and the corresponding estimator as the *efficient estimator.*

An important example of a RAL estimator is that of an m-estimator, which we now review. This particular case highlights the correspondence between estimators and influence functions.

8.3 Review of m-Estimators

Throughout this chapter, we consider the two illustrative examples introduced in the preceding section. To facilitate our discussion of m-estimators, we now present these examples more formally.

Example 1. As in Section 8.2, the complete data consist of $\boldsymbol{V}_i = (\boldsymbol{X}_i', Y_i)'$, $i = 1, \ldots, N$, and no assumptions are made regarding the distribution of $\boldsymbol{V}$ (the non-parametric model). Interest focuses on estimation of $\beta = E(Y)$, so that $q = 1$. Here, the obvious estimator for β is $\widehat{\beta} = N^{-1} \sum_{i=1}^{N} Y_i$, which is the solution to

$$\sum_{i=1}^{N} (Y_i - \beta) = 0.$$

Example 2. Consider again the complete data $\boldsymbol{V}_1, \ldots, \boldsymbol{V}_N$, $\boldsymbol{V}_i = (\boldsymbol{X}_i', Y_i)'$, and suppose that we adopt the semi-parametric model in which it is assumed that

$$E(Y|\boldsymbol{X}) = \mu(\boldsymbol{X}, \boldsymbol{\beta})$$

for $\mu(\boldsymbol{X}, \boldsymbol{\beta})$ a known function of $\boldsymbol{X}$, linear or non-linear in the q-dimensional parameter $\boldsymbol{\beta}$, with no other assumptions on the class of densities generating the complete data. Under these conditions, a standard approach to estimation of $\boldsymbol{\beta}$ is to solve in $\boldsymbol{\beta}$

$$\sum_{i=1}^{N} \boldsymbol{A}(\boldsymbol{X}_i, \boldsymbol{\beta}) \{Y_i - \mu(\boldsymbol{X}_i, \boldsymbol{\beta})\} = \boldsymbol{0}. \tag{8.3}$$

In (8.3), $\boldsymbol{A}(\boldsymbol{X}, \boldsymbol{\beta})$ $(q \times 1)$ is an arbitrary function of $\boldsymbol{X}$ and $\boldsymbol{\beta}$, and (8.3) is an example of a generalized estimating equation (GEE; Liang and Zeger, 1986). Specifically, if we take $\boldsymbol{A}(\boldsymbol{X}, \boldsymbol{\beta}) = \partial/\partial\boldsymbol{\beta}\{\mu(\boldsymbol{X}, \boldsymbol{\beta})\}$, the q-dimensional vector of partial derivatives of $\mu(\boldsymbol{X}, \boldsymbol{\beta})$ with respect to the q elements of $\boldsymbol{\beta}$, then the solution $\widehat{\boldsymbol{\beta}}$ to (8.3) is the least squares estimator for $\boldsymbol{\beta}$.

In both of these examples, it is possible to deduce an estimator for the parameter $\boldsymbol{\beta}$ via solution to an *estimating equation* of the form

$$\sum_{i=1}^{N} \boldsymbol{m}(\boldsymbol{V}_i, \widehat{\boldsymbol{\beta}}) = \boldsymbol{0}. \tag{8.4}$$

In Example 1, $m(\boldsymbol{V}, \beta) = Y - \beta$, while $\boldsymbol{m}(\boldsymbol{V}, \boldsymbol{\beta}) = \boldsymbol{A}(\boldsymbol{X}, \boldsymbol{\beta})\{Y - \mu(\boldsymbol{X}, \boldsymbol{\beta})\}$ in Example 2. The resulting solutions to such estimating equations are *m-estimators*, as follows.

Generically, an m-estimator is defined by a q-dimensional function $\boldsymbol{m}(\boldsymbol{V}, \beta)$ of, in our context, $\boldsymbol{V}$ and $\boldsymbol{\beta}$, referred to as an *estimating function*, such that

$$E_{\boldsymbol{\beta},\boldsymbol{\eta}}\{\boldsymbol{m}(\boldsymbol{V}, \boldsymbol{\beta})\} = \boldsymbol{0},$$

for all $\boldsymbol{\beta}$ and $\boldsymbol{\eta}$, where $\boldsymbol{\eta}$ may be infinite-dimensional as in a semi-parametric model. An estimating function with this property is said to be *unbiased.* Here, $E_{\boldsymbol{\beta},\boldsymbol{\eta}}(\cdot)$ denotes expectation with respect to the distribution indexed by $\boldsymbol{\beta}$ and $\boldsymbol{\eta}$ such that, for any function $\boldsymbol{h}(\boldsymbol{V},\boldsymbol{\beta})$, $E_{\boldsymbol{\beta},\boldsymbol{\eta}}\{\boldsymbol{h}(\boldsymbol{V},\boldsymbol{\beta})\} = \int \boldsymbol{h}(\boldsymbol{v},\boldsymbol{\beta}) f_{\boldsymbol{V}}(\boldsymbol{v},\boldsymbol{\beta},\boldsymbol{\eta}) d\nu(\boldsymbol{v})$, $E_{\boldsymbol{\beta},\boldsymbol{\eta}}\{\boldsymbol{m}'(\boldsymbol{V},\boldsymbol{\beta})\boldsymbol{m}(\boldsymbol{V},\boldsymbol{\beta})\} < \infty$; and $E_{\boldsymbol{\beta},\boldsymbol{\eta}}\{\boldsymbol{m}(\boldsymbol{V},\boldsymbol{\beta})\boldsymbol{m}'(\boldsymbol{V},\boldsymbol{\beta})\}$ is positive definite for all $(\boldsymbol{\beta}',\boldsymbol{\eta}')' \in \Omega$. Based on i.i.d. data $\boldsymbol{V}_1,\ldots,\boldsymbol{V}_N$, the m-estimator $\widehat{\boldsymbol{\beta}}$ is defined as the solution (assuming it exists) to (8.4).

There is a close connection between estimating functions $\boldsymbol{m}(\boldsymbol{V},\boldsymbol{\beta})$ and influence functions $\boldsymbol{\varphi}(\boldsymbol{V})$. To see this, consider the influence function for the m-estimator. The influence function for $\widehat{\boldsymbol{\beta}}$ is found via the expansion

$$\mathbf{0} = \sum_{i=1}^{N} \boldsymbol{m}(\boldsymbol{V}_i,\widehat{\boldsymbol{\beta}}) = \sum_{i=1}^{N} \boldsymbol{m}(\boldsymbol{V}_i,\boldsymbol{\beta}_0) + \left\{ \sum_{i=1}^{N} \frac{\partial \boldsymbol{m}(\boldsymbol{V}_i,\boldsymbol{\beta}^*)}{\partial \boldsymbol{\beta}'} \right\} (\widehat{\boldsymbol{\beta}} - \boldsymbol{\beta}_0),$$

where $\partial/\partial\boldsymbol{\beta}'\{\boldsymbol{m}(\boldsymbol{V},\boldsymbol{\beta}^*)\}$ is the $(q \times q)$ matrix of partial derivatives of the q elements of $\boldsymbol{m}(\boldsymbol{V},\boldsymbol{\beta})$ with (j,j') element $\partial/\partial\beta_{j'}\{m_j(\boldsymbol{V},\boldsymbol{\beta})\}$, $\beta_{j'}$ is the j'th element of $\boldsymbol{\beta}$, and $m_j(\boldsymbol{V},\boldsymbol{\beta})$ is the jth element of $\boldsymbol{m}(\boldsymbol{V},\beta)$, evaluated at $\boldsymbol{\beta}^*$; and $\boldsymbol{\beta}^*$ is an intermediate value between $\widehat{\boldsymbol{\beta}}$ and $\boldsymbol{\beta}_0$.

Remark. More precisely, the intermediate value theorem applies separately to each of the q components of $\boldsymbol{m}(\boldsymbol{V},\boldsymbol{\beta})$, and therefore the value $\boldsymbol{\beta}^*$ is row dependent. However, all such row dependent $\boldsymbol{\beta}^*$ are intermediate values between $\widehat{\boldsymbol{\beta}}$ and $\boldsymbol{\beta}_0$ and hence converge in probability to $\boldsymbol{\beta}_0$, which allows the subsequent development to follow.

Under suitable regularity conditions

$$\left\{ N^{-1} \sum_{i=1}^{N} \frac{\partial \boldsymbol{m}(\boldsymbol{V}_i,\boldsymbol{\beta}^*)}{\partial \boldsymbol{\beta}'} \right\}^{-1} \xrightarrow{p} \left[E\left\{ \frac{\partial \boldsymbol{m}(\boldsymbol{V},\boldsymbol{\beta}_0)}{\partial \boldsymbol{\beta}'} \right\} \right]^{-1}, \tag{8.5}$$

where the notation on the right-hand side of (8.5) indicates the matrix of partial derivatives evaluated at $\boldsymbol{\beta}_0$. Therefore,

$$\begin{aligned} N^{1/2}(\widehat{\boldsymbol{\beta}} - \boldsymbol{\beta}_0) &= -\left\{ N^{-1} \sum_{i=1}^{N} \frac{\partial \boldsymbol{m}(\boldsymbol{V}_i,\boldsymbol{\beta}^*)}{\partial \boldsymbol{\beta}'} \right\}^{-1} \left\{ N^{-1/2} \sum_{i=1}^{N} \boldsymbol{m}(\boldsymbol{V}_i,\boldsymbol{\beta}_0) \right\} \\ &= -\left[E\left\{ \frac{\partial \boldsymbol{m}(\boldsymbol{V}_i,\boldsymbol{\beta}_0)}{\partial \boldsymbol{\beta}'} \right\} \right]^{-1} \left\{ N^{-1/2} \sum_{i=1}^{N} \boldsymbol{m}(\boldsymbol{V}_i,\boldsymbol{\beta}_0) \right\} + o_p(1). \end{aligned}$$

Because, by definition, $E\{\boldsymbol{m}(\boldsymbol{V},\boldsymbol{\beta}_0)\} = \mathbf{0}$, we immediately deduce that the influence function of $\widehat{\boldsymbol{\beta}}$ is given by

$$\boldsymbol{\varphi}(\boldsymbol{V}) = -\left[E\left\{ \frac{\partial \boldsymbol{m}(\boldsymbol{V},\boldsymbol{\beta}_0)}{\partial \boldsymbol{\beta}'} \right\} \right]^{-1} \boldsymbol{m}(\boldsymbol{V},\boldsymbol{\beta}_0); \tag{8.6}$$

that is, the influence function is a constant (matrix) times the estimating function. It follows that $N^{1/2}(\widehat{\boldsymbol{\beta}} - \boldsymbol{\beta}_0)$ converges in distribution to a random vector distributed as

$$N\left(\mathbf{0}, \left[E\left\{ \frac{\partial \boldsymbol{m}(\boldsymbol{V},\boldsymbol{\beta}_0)}{\partial \boldsymbol{\beta}'} \right\} \right]^{-1} \mathrm{Cov}\{\boldsymbol{m}(\boldsymbol{V},\boldsymbol{\beta}_0)\} \left[E\left\{ \frac{\partial \boldsymbol{m}(\boldsymbol{V},\boldsymbol{\beta}_0)}{\partial \boldsymbol{\beta}'} \right\} \right]^{-1'} \right),$$

where

$$\text{Cov}\{\boldsymbol{m}(\boldsymbol{V},\boldsymbol{\beta}_0)\} = E\{\boldsymbol{m}(\boldsymbol{V},\boldsymbol{\beta}_0)m'(\boldsymbol{V},\boldsymbol{\beta}_0)\}.$$

From the above, the asymptotic covariance matrix of an m-estimator may be estimated by the sandwich estimator given by

$$\left[\widehat{E}\left\{\frac{\partial \boldsymbol{m}(\boldsymbol{V},\boldsymbol{\beta}_0)}{\partial \boldsymbol{\beta}'}\right\}\right]^{-1} \widehat{\text{Cov}}\left\{\boldsymbol{m}(\boldsymbol{V},\boldsymbol{\beta}_0)\right\} \left[\widehat{E}\left\{\frac{\partial \boldsymbol{m}(\boldsymbol{V},\boldsymbol{\beta}_0)}{\partial \boldsymbol{\beta}'}\right\}\right]^{-1'}, \tag{8.7}$$

where

$$\widehat{E}\left\{\frac{\partial \boldsymbol{m}(\boldsymbol{V},\boldsymbol{\beta}_0)}{\partial \boldsymbol{\beta}'}\right\} = N^{-1}\sum_{i=1}^{N}\frac{\partial \boldsymbol{m}(\boldsymbol{V}_i,\widehat{\boldsymbol{\beta}})}{\partial \boldsymbol{\beta}'},$$

and

$$\widehat{\text{Cov}}\{\boldsymbol{m}(\boldsymbol{V},\boldsymbol{\beta}_0)\} = N^{-1}\sum_{i=1}^{N}\boldsymbol{m}(\boldsymbol{V}_i,\widehat{\boldsymbol{\beta}})\boldsymbol{m}'(\boldsymbol{V}_i,\widehat{\boldsymbol{\beta}}).$$

As indicated by (8.2) and exemplified by the derivation here, there is a relationship between influence functions and consistent and asymptotically normal estimators; thus, by identifying influence functions, one may deduce corresponding such estimators. In the next section, we describe the geometry of influence functions for RAL estimators for $\boldsymbol{\beta}$ for parametric and semi-parametric models and how influence functions may be derived.

In both examples above, the estimating function depends only on the parameter of interest, $\boldsymbol{\beta}$, so that an estimator for $\boldsymbol{\beta}$ may be obtained without the need to estimate additional nuisance parameters. In other settings, this might not be the case. For instance, in Example 2, if the semi-parametric model is modified to include the additional assumption that $\text{var}(Y|\boldsymbol{X}) = v(\boldsymbol{X},\boldsymbol{\delta})$ for some known variance function depending on $(r \times 1)$ parameter $\boldsymbol{\delta}$, but interest is still focused on $\boldsymbol{\beta}$, then $\boldsymbol{\delta}$ is a component of the nuisance parameter $\boldsymbol{\eta}(\cdot)$. In this situation, one would solve in both $\boldsymbol{\beta}$ and $\boldsymbol{\delta}$ a modified version of (8.3),

$$\sum_{i=1}^{N}\boldsymbol{A}(\boldsymbol{X}_i,\boldsymbol{\beta})\{v(\boldsymbol{X}_i,\boldsymbol{\delta})\}^{-1}\{Y_i - \mu(\boldsymbol{X}_i,\boldsymbol{\beta})\} = \boldsymbol{0},$$

jointly with a separate estimating equation for estimation of $\boldsymbol{\delta}$. By "stacking" the estimating equations and identifying the $(q + r \times 1)$ estimating function for $\boldsymbol{\beta}$ and $\boldsymbol{\delta}$ jointly, via an argument analogous to that above, one could derive the influence function for the associated estimator $(\widehat{\boldsymbol{\beta}}', \widehat{\boldsymbol{\delta}}')'$ for $(\boldsymbol{\beta}', \boldsymbol{\delta}')'$, in which case the influence function for $\widehat{\boldsymbol{\beta}}$ would be its first q elements; see Tsiatis (2006, Section 3.2).

It is worth noting that the maximum likelihood estimator for $\boldsymbol{\beta}$ in a parametric model is an m-estimator, where the estimating function is the usual score function; see, for example, Stefanski and Boos (2002).

8.4 Influence Functions and Semi-Parametric Theory

Because of the relationship between influence functions and RAL estimators discussed above, influence functions are the essential element underlying semi-parametric theory. We first present an argument that justifies more formally this correspondence.

Correspondence between influence functions and RAL estimators. It may be shown by contradiction that an asymptotically linear estimator, i.e., satisfying (8.2), has a unique (almost surely) influence function $\varphi(\boldsymbol{V})$, say. If not, there would exist another influence function $\tilde{\varphi}(\boldsymbol{V})$ with $E\{\tilde{\varphi}(\boldsymbol{V})\} = \mathbf{0}$ also satisfying (8.2). If (8.2) holds for both $\varphi(\boldsymbol{V})$ and $\tilde{\varphi}(\boldsymbol{V})$, then it must be that $\boldsymbol{C} = N^{-1/2}\sum_{i=1}^{N}\{\varphi(\boldsymbol{V}_i) - \tilde{\varphi}(\boldsymbol{V}_i)\} = o_p(1)$. Because the $\boldsymbol{V}_i$ are i.i.d., it also follows that $\boldsymbol{C}$ converges in distribution to a mean-zero normal random vector with covariance matrix $\Sigma = E[\{\varphi(\boldsymbol{V}) - \tilde{\varphi}(\boldsymbol{V})\}\{\varphi(\boldsymbol{V}) - \tilde{\varphi}(\boldsymbol{V})\}']$. As this limiting distribution is $o_p(1)$, it must be that $\Sigma = 0$, so that $\varphi(\boldsymbol{V}) = \tilde{\varphi}(\boldsymbol{V})$ almost surely.

The foregoing result demonstrates that working with influence functions is equivalent to working with estimators. Semi-parametric theory is predicated on geometric representation of all influence functions for $\boldsymbol{\beta}$ in a given semi-parametric model. To lay the groundwork for understanding this geometric perspective, we consider a fully parametric model such that densities in the class $\mathcal{P}$ may be written as $f_{\boldsymbol{V}}(\boldsymbol{v}, \boldsymbol{\theta})$, where $\boldsymbol{\theta} = (\boldsymbol{\beta}', \boldsymbol{\eta}')'$ for $\boldsymbol{\beta}$ $(q \times 1)$ and $\boldsymbol{\eta}$ $(r \times 1)$, $p = q + r$. We first present familiar results for maximum likelihood estimation of the parameter of interest, $\boldsymbol{\beta}$.

Maximum likelihood estimator in a parametric model. Let $\boldsymbol{S}_{\boldsymbol{\theta}}(\boldsymbol{V}, \boldsymbol{\theta}) = \{\boldsymbol{S}'_{\boldsymbol{\beta}}(\boldsymbol{V}, \boldsymbol{\theta}), \boldsymbol{S}'_{\boldsymbol{\eta}}(\boldsymbol{V}, \boldsymbol{\theta})\}' = [\partial/\partial\boldsymbol{\beta}'\{\log f_{\boldsymbol{V}}(\boldsymbol{V}, \boldsymbol{\beta}, \boldsymbol{\eta})\}, \partial/\partial\boldsymbol{\eta}'\{\log f_{\boldsymbol{V}}(\boldsymbol{V}, \boldsymbol{\beta}, \boldsymbol{\eta})\}]'$, and let $\boldsymbol{\theta}_0 = (\boldsymbol{\beta}'_0, \boldsymbol{\eta}'_0)'$ be the true value of $\boldsymbol{\theta}$. The score vector for $\boldsymbol{\theta}$ is then given by $\boldsymbol{S}_{\boldsymbol{\theta}}(\boldsymbol{V}) = \{\boldsymbol{S}'_{\boldsymbol{\beta}}(\boldsymbol{V}), \boldsymbol{S}'_{\boldsymbol{\eta}}(\boldsymbol{V})\}' = \boldsymbol{S}_{\boldsymbol{\theta}}(\boldsymbol{V}, \boldsymbol{\theta}_0)$. We have $E\{\boldsymbol{S}_{\boldsymbol{\theta}}(\boldsymbol{V})\} = \mathbf{0}$, and the information matrix is

$$\mathcal{I} = E\{\boldsymbol{S}_{\boldsymbol{\theta}}(\boldsymbol{V})\boldsymbol{S}'_{\boldsymbol{\theta}}(\boldsymbol{V})\} = \begin{pmatrix} \mathcal{I}_{\boldsymbol{\beta\beta}} & \mathcal{I}_{\boldsymbol{\beta\eta}} \\ \mathcal{I}'_{\boldsymbol{\beta\eta}} & \mathcal{I}_{\boldsymbol{\eta\eta}} \end{pmatrix},$$

where expectation is with respect to the true density $f_{\boldsymbol{V}}(\boldsymbol{v}, \boldsymbol{\beta}_0, \boldsymbol{\eta}_0)$. Writing $\widehat{\boldsymbol{\theta}} = (\widehat{\boldsymbol{\beta}}', \widehat{\boldsymbol{\eta}}')'$ to denote the maximum likelihood estimator for $\boldsymbol{\theta}$ found by maximizing $\sum_{i=1}^{n} \log f_{\boldsymbol{V}}(\boldsymbol{v}_i, \boldsymbol{\beta}, \boldsymbol{\eta})$, under "regularity conditions" (e.g., Bickel et al., 1993, Section 2.4), it is well-known that

$$N^{1/2}(\widehat{\boldsymbol{\beta}} - \boldsymbol{\beta}_0) = N^{-1/2}\sum_{i=1}^{N} \varphi^{\mathit{eff}}(\boldsymbol{V}_i) + o_p(1),$$

$$\varphi^{\mathit{eff}}(\boldsymbol{V}) = \mathcal{I}^{-1}_{\boldsymbol{\beta\beta\bullet\eta}}\{\boldsymbol{S}_{\boldsymbol{\beta}}(\boldsymbol{V}) - \mathcal{I}_{\boldsymbol{\beta\eta}}\mathcal{I}^{-1}_{\boldsymbol{\eta\eta}}\boldsymbol{S}_{\boldsymbol{\eta}}(\boldsymbol{V})\}, \quad \mathcal{I}_{\boldsymbol{\beta\beta\bullet\eta}} = \mathcal{I}_{\boldsymbol{\beta\beta}} - \mathcal{I}_{\boldsymbol{\beta\eta}}\mathcal{I}^{-1}_{\boldsymbol{\eta\eta}}\mathcal{I}'_{\boldsymbol{\beta\eta}},$$

so that $E\{\varphi^{\mathit{eff}}(\boldsymbol{V})\} = \mathbf{0}$, and $\widehat{\boldsymbol{\beta}}$ is RAL with influence function $\varphi^{\mathit{eff}}(\boldsymbol{V})$. As $\boldsymbol{S}^{\mathit{eff}}(\boldsymbol{V}) = \boldsymbol{S}_{\boldsymbol{\beta}}(\boldsymbol{V}) - \mathcal{I}_{\boldsymbol{\beta\eta}}\mathcal{I}^{-1}_{\boldsymbol{\eta\eta}}\boldsymbol{S}_{\boldsymbol{\eta}}(\boldsymbol{V})$ has covariance matrix $\mathcal{I}_{\boldsymbol{\beta\beta\bullet\eta}}$, $\widehat{\boldsymbol{\beta}}$ is consistent and asymptotically normal with asymptotic covariance matrix $E\{\varphi^{\mathit{eff}}(\boldsymbol{V})\varphi^{\mathit{eff}\prime}(\boldsymbol{V})\} = \mathcal{I}^{-1}_{\boldsymbol{\beta\beta\bullet\eta}}$. This quantity is the well-known Cramèr–Rao lower bound, the "smallest" possible covariance matrix in the sense of non-negative definiteness for (regular) estimators for $\boldsymbol{\beta}$. This implies that $\widehat{\boldsymbol{\beta}}$ is the efficient estimator, and, accordingly, $\boldsymbol{S}^{\mathit{eff}}(\boldsymbol{V})$ is called the *efficient score*. Consequently, it is clear that $\varphi^{\mathit{eff}}(\boldsymbol{V})$ is the efficient influence function.

We now review the geometric construction underlying semi-parametric theory; our presentation is non-rigorous and covers only the essential elements. We then return to the parametric model and place it in this geometric context.

A fundamental concept is that of a Hilbert space. We first provide some basic definitions and results and then indicate why these are essential to the geometric perspective.

Hilbert space. A *Hilbert space* $\mathcal{H}$ is a linear vector space, so that $a\boldsymbol{h}_1 + b\boldsymbol{h}_2 \in \mathcal{H}$ for $\boldsymbol{h}_1, \boldsymbol{h}_2 \in \mathcal{H}$ and any real a, b, equipped with an inner product; see Luenberger (1969, Chapter 3). Influence functions based on data $\boldsymbol{V}$ corresponding to a q-dimensional parameter $\boldsymbol{\beta}$ in a statistical model are elements in the Hilbert space $\mathcal{H}$ of all q-dimensional, mean-zero random functions $\boldsymbol{h}(\boldsymbol{V})$ with $E\{\boldsymbol{h}'(\boldsymbol{V})\boldsymbol{h}(\boldsymbol{V})\} < \infty$, inner product $E\{\boldsymbol{h}_1'(\boldsymbol{V})\boldsymbol{h}_2(\boldsymbol{V})\}$ for $\boldsymbol{h}_1, \boldsymbol{h}_2 \in \mathcal{H}$, and corresponding norm $\|\boldsymbol{h}\| = [E\{\boldsymbol{h}'(\boldsymbol{V})\boldsymbol{h}(\boldsymbol{V})\}]^{1/2}$ measuring "distance" from the origin, $\boldsymbol{h} \equiv \boldsymbol{0}$. Thus, underlying the geometric perspective of semi-parametric theory is this view of influence functions as elements of the Hilbert space $\mathcal{H}$, which allows the geometry of Hilbert spaces to be exploited to obtain results on the nature of influence functions in parametric and semi-parametric models.

Before using this framework in this manner, we review some key Hilbert space results. For any closed linear subspace M of $\mathcal{H}$, denote the set of all elements of $\mathcal{H}$ *orthogonal* to those in M by $M^{\perp}$. (A closed linear subspace is one that contains all of its limit points.) That is, if $\boldsymbol{h}_1 \in M$ and $\boldsymbol{h}_2 \in M^{\perp}$, the inner product of $\boldsymbol{h}_1$ and $\boldsymbol{h}_2$ is zero. Then $M^{\perp}$ is also a linear subspace of $\mathcal{H}$. The *direct sum* of two linear subspaces M and N, $M \oplus N$, is such that every element in $M \oplus N$ has a unique representation of the form $\boldsymbol{m} + \boldsymbol{n}$ for $\boldsymbol{m} \in M$, $\boldsymbol{n} \in N$. Intuitively, then, $\mathcal{H} = M \oplus M^{\perp}$.

A concept key to the geometric perspective is that of a *projection.* For $\boldsymbol{h} \in \mathcal{H}$, the projection of $\boldsymbol{h}$ onto a closed linear subspace M of $\mathcal{H}$ is the element in M, denoted by $\boldsymbol{\Pi}(\boldsymbol{h}|M)$, such that $\|\boldsymbol{h} - \boldsymbol{\Pi}(\boldsymbol{h}|M)\| < \|\boldsymbol{h} - \boldsymbol{m}\|$ for all $\boldsymbol{m} \in M$. The Projection Theorem states that the projection exists, is unique, and is such that the *residual* $\boldsymbol{h} - \boldsymbol{\Pi}(\boldsymbol{h}|M)$ is orthogonal to all $\boldsymbol{m} \in M$ (Luenberger, 1969, Section 3.3).

See Chapter 2 of Tsiatis (2006) for a more detailed account and a graphical depiction of the notion of a projection.

Armed with these definitions, we are now in a position to examine how they may be used to identify influence functions and the efficient influence function. The rationale is as follows. RAL estimators may be compared via their asymptotic covariance matrices. As noted in Section 8.2, the asymptotic covariance matrix of an RAL estimator is equal to the covariance matrix of its influence function, so that comparisons may be made via the covariance matrices of influence functions. In the case $q = 1$, by the definition of Hilbert space above, in the Hilbert space $\mathcal{H}$ in which influence functions lie, distance to the origin ($h \equiv \boldsymbol{0}$) of any $h \in \mathcal{H}$ is equal to the variance of h; for $q > 1$, this may be generalized, as discussed in Section 3.4 of Tsiatis (2006). Accordingly, the search for the efficient estimator with smallest covariance matrix in the sense of non-negative definiteness is equivalent to the search for the influence function in $\mathcal{H}$ having smallest distance from the origin.

We first return to the case of a parametric model.

Geometric perspective on the parametric model. Here, $\boldsymbol{\theta}$ may be partitioned as $\boldsymbol{\theta} = (\boldsymbol{\beta}', \boldsymbol{\eta}')'$. Let Λ be the linear subspace of $\mathcal{H}$ consisting of all linear combinations of $\boldsymbol{S}_{\boldsymbol{\eta}}(\boldsymbol{V})$ of the form $B\boldsymbol{S}_{\boldsymbol{\eta}}(\boldsymbol{V})$, where B is a constant $(q \times r)$ matrix. That is, $\Lambda = \{B\boldsymbol{S}_{\boldsymbol{\eta}}(\boldsymbol{V})$ for all $(q \times r)$ constant matrices $B\}$ is the linear subspace of $\mathcal{H}$ spanned by $\boldsymbol{S}_{\boldsymbol{\eta}}(\boldsymbol{V})$. Because Λ depends on the score for nuisance parameters, it is referred to as the *nuisance tangent space.*

A key result that we do not demonstrate here is that all influence functions for RAL estimators for $\boldsymbol{\beta}$ may be shown to lie in the subspace $\Lambda^{\perp}$ orthogonal to Λ. Thus, an influence function $\boldsymbol{\varphi}(\boldsymbol{V})$ is orthogonal to the nuisance tangent space Λ. Moreover, $\boldsymbol{\varphi}(\boldsymbol{V})$ must also satisfy the property that $E\{\boldsymbol{\varphi}(\boldsymbol{V})\boldsymbol{S}_{\boldsymbol{\beta}}'(\boldsymbol{V})\} = I_q$, where I_q is the $(q \times q)$ identity matrix. See Tsiatis (2006, Chapter 3) for details.

As we noted earlier, there is a close connection between influence functions and unbiased estimating functions. It is clear that an estimating function can be defined up to a multiplicative constant. That is, if $\boldsymbol{m}(\boldsymbol{V},\boldsymbol{\beta})$ is an unbiased estimating function, then $B\boldsymbol{m}(\boldsymbol{V},\boldsymbol{\beta})$, where B is a constant $(q \times q)$ matrix, is also an unbiased estimating function, and both lead to the same m-estimator for $\boldsymbol{\beta}$ with the same influence function. Thus, one approach to identifying estimating functions for a particular model with $\boldsymbol{\theta} = (\boldsymbol{\beta}', \boldsymbol{\eta}')'$ is to characterize the form of elements in $\Lambda^{\perp}$ directly.

Other representations are possible. For general $f_{\boldsymbol{V}}(\boldsymbol{v},\boldsymbol{\theta})$, the *tangent space* Γ is the linear subspace of $\mathcal{H}$ spanned by the entire score vector $\boldsymbol{S}_{\boldsymbol{\theta}}(\boldsymbol{V})$; i.e., $\Gamma = \{B\boldsymbol{S}_{\theta}(\boldsymbol{V})$ for all $(q \times p)$ $B\}$. We state the following two key results without proof; see Tsiatis (2006, Chapter 3) and Appendix A.1 of Davidian et al. (2005) for details.

Result 1. *Representation of influence functions.* All influence functions for RAL estimators for $\boldsymbol{\beta}$ may be represented as $\boldsymbol{\varphi}(\boldsymbol{V}) = \tilde{\boldsymbol{\varphi}}(\boldsymbol{V}) + \boldsymbol{\psi}(\boldsymbol{V})$, where $\tilde{\boldsymbol{\varphi}}(\boldsymbol{V})$ is any influence function and $\boldsymbol{\psi}(\boldsymbol{V}) \in \Gamma^{\perp}$, the subspace of $\mathcal{H}$ orthogonal to Γ.

Result 1 implies that, by identifying any influence function and $\Gamma^{\perp}$, all influence functions may be characterized. Result 1 also leads to a useful characterization of the efficient influence function $\boldsymbol{\varphi}^{\mathit{eff}}(\boldsymbol{V})$, which, as discussed in Section 8.2, must be such that $E\{\boldsymbol{\varphi}(\boldsymbol{V})\boldsymbol{\varphi}'(\boldsymbol{V})\} - E\{\boldsymbol{\varphi}^{\mathit{eff}}(\boldsymbol{V})\boldsymbol{\varphi}^{\mathit{eff}\prime}(\boldsymbol{V})\}$ is non-negative definite for all influence functions $\boldsymbol{\varphi}(\boldsymbol{V})$. By Result 1, because for arbitrary $\boldsymbol{\varphi}(\boldsymbol{V})$, $\boldsymbol{\varphi}^{\mathit{eff}}(\boldsymbol{V}) = \boldsymbol{\varphi}(\boldsymbol{V}) - \boldsymbol{\psi}(\boldsymbol{V})$ for some $\boldsymbol{\psi} \in \Gamma^{\perp}$, if $E\{\boldsymbol{\varphi}^{\mathit{eff}}(\boldsymbol{V})\boldsymbol{\varphi}^{\mathit{eff}\prime}(\boldsymbol{V})\} = E[\{\boldsymbol{\varphi}(\boldsymbol{V}) - \boldsymbol{\psi}(\boldsymbol{V})\}\{\boldsymbol{\varphi}(\boldsymbol{V}) - \boldsymbol{\psi})\boldsymbol{V})\}']$ is to be as small as possible, it must be that $\boldsymbol{\psi}(\boldsymbol{V}) = \boldsymbol{\Pi}(\boldsymbol{\varphi}|\Gamma^{\perp})(\boldsymbol{V})$. This leads to the following.

Result 2. *Representation of the efficient influence function.* $\boldsymbol{\varphi}^{\mathit{eff}}(\boldsymbol{V})$ may be represented as $\boldsymbol{\varphi}(\boldsymbol{V}) - \boldsymbol{\Pi}(\boldsymbol{\varphi}|\Gamma^{\perp})(\boldsymbol{V})$ for any influence function $\boldsymbol{\varphi}(\boldsymbol{V})$.

In the case of a parametric model, with $\boldsymbol{\theta} = (\boldsymbol{\beta}', \boldsymbol{\eta}')'$, the form of the efficient influence function may be identified explicitly from this perspective. Define the *efficient score* to be the residual of the score vector for $\boldsymbol{\beta}$ evaluated at the truth after projecting it onto the nuisance tangent space; that is, $\boldsymbol{S}^{\mathit{eff}}(\boldsymbol{V}) = \boldsymbol{S}_{\boldsymbol{\beta}}(\boldsymbol{V}) - \boldsymbol{\Pi}(\boldsymbol{S}_{\boldsymbol{\beta}}|\Lambda)(\boldsymbol{V})$. From Result 2, the unique efficient influence function, $\boldsymbol{\varphi}^{\mathit{eff}}(\boldsymbol{V})$, is also the only influence function that belongs to the tangent space Γ; that is, is a linear combination of $\boldsymbol{S}_{\boldsymbol{\beta}}(\boldsymbol{V})$ and $\boldsymbol{S}_{\boldsymbol{\eta}}(\boldsymbol{V})$. As noted above, all influence functions $\boldsymbol{\varphi}(\boldsymbol{V})$ must be

(i) orthogonal to the nuisance tangent space Λ, and

(ii) such that $E\{\boldsymbol{\varphi}(\boldsymbol{V})\boldsymbol{S}_{\boldsymbol{\beta}}'(\boldsymbol{V})\} = E\{\boldsymbol{\varphi}(\boldsymbol{V})\boldsymbol{S}^{\mathit{eff}\prime}(\boldsymbol{V})\} = I_q$.

By construction, the efficient score is an element of Γ and orthogonal to Λ. Hence, an appropriately scaled version of $\boldsymbol{S}^{\mathit{eff}}(\boldsymbol{V})$ given by $[E\{\boldsymbol{S}^{\mathit{eff}}(\boldsymbol{V})\boldsymbol{S}^{\mathit{eff}\prime}(\boldsymbol{V})\}]^{-1}\boldsymbol{S}^{\mathit{eff}}(\boldsymbol{V})$ satisfies (i) and (ii) and is an element of Γ; thus, it is the unique efficient influence function $\boldsymbol{\varphi}^{\mathit{eff}}(\boldsymbol{V})$. See Appendix A.1 of Davidian et al. (2005) for details.

We may relate this to the maximum likelihood results given earlier in this section. By the definition of a projection, $\boldsymbol{\Pi}(\boldsymbol{S}_{\boldsymbol{\beta}}|\Lambda) \in \Lambda$ is the unique element $B_0\boldsymbol{S}_{\boldsymbol{\eta}} \in \Lambda$ such that $E[\{\boldsymbol{S}_{\boldsymbol{\beta}}(\boldsymbol{V}) - B_0\boldsymbol{S}_{\boldsymbol{\eta}}(\boldsymbol{V})\}B\boldsymbol{S}_{\boldsymbol{\eta}}(\boldsymbol{V})] = \boldsymbol{0}$ for all B $(q \times r)$. This is the same as requiring that $E[\{\boldsymbol{S}_{\boldsymbol{\beta}}(\boldsymbol{V}) - B_0\boldsymbol{S}_{\boldsymbol{\eta}}(\boldsymbol{V})\}\boldsymbol{S}_{\boldsymbol{\eta}}'(\boldsymbol{V})] = \boldsymbol{0}$ $(q \times r)$, which implies that $B_0 = \mathcal{I}_{\boldsymbol{\beta}\boldsymbol{\beta}}\mathcal{I}_{\boldsymbol{\eta}\boldsymbol{\eta}}^{-1}$. Thus, $\boldsymbol{\Pi}(\boldsymbol{S}_{\boldsymbol{\beta}}|\Lambda)(\boldsymbol{V}) = \mathcal{I}_{\boldsymbol{\beta}\boldsymbol{\beta}}\mathcal{I}_{\boldsymbol{\eta}\boldsymbol{\eta}}^{-1}\boldsymbol{S}_{\boldsymbol{\eta}}(\boldsymbol{V})$ and $\boldsymbol{S}^{\mathit{eff}}(\boldsymbol{V}) = \boldsymbol{S}_{\boldsymbol{\beta}}(\boldsymbol{V}) - \mathcal{I}_{\boldsymbol{\beta}\boldsymbol{\eta}}\mathcal{I}_{\boldsymbol{\eta}\boldsymbol{\eta}}^{-1}\boldsymbol{S}_{\boldsymbol{\eta}}(\boldsymbol{V})$, which of course is the same result derived earlier.

Although the preceding development illustrates key aspects of the geometric construction, for a parametric model, it is usually unnecessary to exploit this approach to identify the efficient estimator and influence functions. However, for a semi-parametric model, this is less straightforward. Fortunately, the geometric perspective may be generalized to this case and provides a framework for identifying influence functions, as we now discuss.

Geometric perspective on the semi-parametric model. To generalize the results for parametric models to the semi-parametric model setting, a key concept required is that of a *parametric sub-model*. A parametric sub-model is a parametric model contained in the semi-parametric model that contains the truth.

First consider the most general formulation of a semi-parametric model, with $\mathcal{P}$ consisting of densities $f_V\{v, \boldsymbol{\theta}(\cdot)\}$ and functional of interest $\boldsymbol{\beta}\{\boldsymbol{\theta}(\cdot)\}$. As in Section 8.2, there is a true $\boldsymbol{\theta}_0(\cdot)$ such that $f_{V0}(\boldsymbol{v}) = f_V\{\boldsymbol{v}, \boldsymbol{\theta}_0(\cdot)\} \in \mathcal{P}$ is the density generating the data. Then a parametric sub-model is the class of all densities $\mathcal{P}_{\boldsymbol{\xi}}$ characterized by a finite-dimensional parameter $\boldsymbol{\xi}$ such that $\mathcal{P}_{\boldsymbol{\xi}} \subset \mathcal{P}$ and the true density $f_{V0}(\boldsymbol{v}) = f_V\{\boldsymbol{v}, \boldsymbol{\theta}_0(\cdot)\} = f_V(\boldsymbol{v}, \boldsymbol{\xi}_0) \in \mathcal{P}_{\boldsymbol{\xi}}$, where the dimension r of $\boldsymbol{\xi}$ varies according to the particular choice of sub-model. That is, there exists a density identified by the parameter $\boldsymbol{\xi}_0$ within the parameter space of the parametric sub-model such that $f_{V0}(\boldsymbol{v}) = f_V(\boldsymbol{v}, \boldsymbol{\xi}_0)$.

The key implication is that an estimator is an RAL estimator for $\boldsymbol{\beta}$ under the semi-parametric model if it is an estimator under every parametric sub-model. Thus, the class of RAL estimators for $\boldsymbol{\beta}$ for the semi-parametric model must be contained in the class of estimators for a parametric sub-model. Accordingly, any influence function for an estimator for $\boldsymbol{\beta}$ under the semi-parametric model must be an influence function for a parametric sub-model. We may then appeal to the foregoing results for parametric models as follows. If $\Gamma_{\boldsymbol{\xi}}$ is the tangent space for a given sub-model $f_V(\boldsymbol{v}, \boldsymbol{\xi})$ with $\boldsymbol{S}_{\boldsymbol{\xi}}(\boldsymbol{V}, \boldsymbol{\xi}) = \partial/\partial\boldsymbol{\xi}\{\log f_V(\boldsymbol{V}, \boldsymbol{\xi})\}$, Result 1 implies that the corresponding influence functions for estimators for $\boldsymbol{\beta}$ may be represented as as $\boldsymbol{\varphi}(\boldsymbol{V}) = \tilde{\boldsymbol{\varphi}}(\boldsymbol{V}) + \boldsymbol{\gamma}(\boldsymbol{V})$. Here, $\tilde{\boldsymbol{\varphi}}(\boldsymbol{V})$ is any influence function for estimation of $\boldsymbol{\xi}$ in the parametric sub-model, and $\boldsymbol{\gamma}(\boldsymbol{V}) \in \Gamma_{\boldsymbol{\xi}}^{\perp}$. If we define Γ to be the mean-square closure of all parametric sub-model tangent spaces, i.e., $\Gamma = \{\boldsymbol{h} \in \mathcal{H}$ such that there exists a sequence of parametric sub-models $\mathcal{P}_{\boldsymbol{\xi}_j}$ with $\|\boldsymbol{h}(\boldsymbol{V}) - B_j \boldsymbol{S}_{\boldsymbol{\xi}_j}(\boldsymbol{V}, \boldsymbol{\xi}_{0j})\|^2 \to 0$ as $j \to \infty\}$, where B_j are $(q \times r_j)$ constant matrices, then it may be shown that Result 1 holds for semi-parametric model influence functions. That is, all influence functions $\boldsymbol{\varphi}(\boldsymbol{V})$ for estimators for $\boldsymbol{\beta}$ in the semi-parametric model may be represented as $\tilde{\boldsymbol{\varphi}}(\boldsymbol{V}) + \boldsymbol{\psi}(\boldsymbol{V})$, where $\tilde{\boldsymbol{\varphi}}(\boldsymbol{V})$ is any semi-parametric model influence function and $\boldsymbol{\psi}(\boldsymbol{V}) \in \Gamma^{\perp}$. In addition, Result 2 holds, so that, as in the parametric case, the efficient estimator for $\boldsymbol{\beta}$ has influence function $\boldsymbol{\varphi}^{\mathit{eff}}(\boldsymbol{V})$ that may be represented as $\boldsymbol{\varphi}^{\mathit{eff}}(\boldsymbol{V}) = \boldsymbol{\varphi}(\boldsymbol{V}) - \boldsymbol{\Pi}(\boldsymbol{\varphi}|\Gamma^{\perp})(\boldsymbol{V})$ for any semi-parametric model influence function $\boldsymbol{\varphi}(\boldsymbol{V})$.

In Example 1, we considered a particular case of the above formulation, the non-parametric model with interest focused on estimating $\beta = E(Y)$. In Section 4.4 of Tsiatis (2006), it is shown that the tangent space Γ for the non-parametric model is the entire Hilbert space $\mathcal{H}$. Therefore, the orthogonal complement $\Gamma^{\perp}$ is the zero vector. Because $Y - \beta$ is the influence function for the RAL estimator $\widehat{\beta} = N^{-1}\sum_{i=1}^{N} Y_i$ for β, by the results above, this is the unique and, hence, efficient RAL estimator for β.

In the case where $\boldsymbol{\theta}(\cdot) = \{\boldsymbol{\beta}, \boldsymbol{\eta}(\cdot)\}$ for $\boldsymbol{\beta}$ $(q \times 1)$, a related development is possible. Here, we may write densities in $\mathcal{P}$ as $f_V\{\boldsymbol{v}, \boldsymbol{\beta}, \boldsymbol{\eta}(\cdot)\}$ with true values $\boldsymbol{\beta}_0, \boldsymbol{\eta}_0(\cdot)$ such that the true density is $f_{V0}(\boldsymbol{v}) = f_V\{v, \boldsymbol{\beta}_0, \boldsymbol{\eta}_0(\cdot)\} \in \mathcal{P}$. Write $\boldsymbol{S}_{\boldsymbol{\beta}}\{\boldsymbol{V}, \boldsymbol{\beta}, \boldsymbol{\eta}(\cdot)\} = \partial/\partial\boldsymbol{\beta}[f_V\{\boldsymbol{v}, \boldsymbol{\beta}, \boldsymbol{\eta}(\cdot)\}]$, and let $\boldsymbol{S}_{\boldsymbol{\beta}}(\boldsymbol{V}) = \boldsymbol{S}_{\boldsymbol{\beta}}\{\boldsymbol{V}, \boldsymbol{\beta}_0, \boldsymbol{\eta}_0(\cdot)\}$. A parametric sub-model $\mathcal{P}_{\boldsymbol{\beta},\boldsymbol{\xi}}$ is the class of all densities characterized by $\boldsymbol{\beta}$ and finite-dimensional $\boldsymbol{\xi}$ such that $\mathcal{P}_{\boldsymbol{\beta},\boldsymbol{\xi}} \subset \mathcal{P}$ and $f_V\{\boldsymbol{v}, \boldsymbol{\beta}_0, \boldsymbol{\eta}_0(\cdot)\} = f_V(\boldsymbol{v}, \boldsymbol{\beta}_0, \boldsymbol{\xi}_0) \in \mathcal{P}_{\boldsymbol{\beta},\boldsymbol{\xi}}$. Because it is a parametric model, a parametric sub-model has a corresponding nuisance tangent space. Moreover, as above, because the class of estimators for $\boldsymbol{\beta}$ for the semi-parametric model must be contained in the class of estimators for a parametric sub-model, influence functions for estimators for $\boldsymbol{\beta}$ for the semi-parametric model must lie in a space orthogonal to all parametric sub-model nuisance tangent spaces. Thus, defining the semi-parametric model nuisance tangent space Λ as the mean-square closure of all parametric sub-model nuisance tangent spaces, it may be shown that all influence functions for the semi-parametric model lie in $\Lambda^{\perp}$. Such influence functions $\boldsymbol{\varphi}(\boldsymbol{V})$ must then also be such that $E\{\boldsymbol{\varphi}(\boldsymbol{V})\boldsymbol{S}'_{\boldsymbol{\beta}}(\boldsymbol{V})\} = \boldsymbol{I}_q$.

The semi-parametric model efficient score $\boldsymbol{S}^{\mathit{eff}}(\boldsymbol{V})$ is $\boldsymbol{S}_{\boldsymbol{\beta}}(\boldsymbol{V}) - \boldsymbol{\Pi}(\boldsymbol{S}_{\boldsymbol{\beta}}|\Lambda)(\boldsymbol{V})$ with efficient influence function

$$\boldsymbol{\varphi}^{\mathit{eff}}(\boldsymbol{V}) = \{\boldsymbol{S}^{\mathit{eff}}(\boldsymbol{V})\boldsymbol{S}^{\mathit{eff}'}(\boldsymbol{V})\}^{-1}\boldsymbol{S}^{\mathit{eff}}(\boldsymbol{V}),$$

analogous to the parametric model case. The covariance matrix of $\boldsymbol{\varphi}^{\mathit{eff}}(\boldsymbol{V})$, $E\{\boldsymbol{\varphi}^{\mathit{eff}}(\boldsymbol{V})\boldsymbol{\varphi}^{\mathit{eff}'}(\boldsymbol{V})\} = [E\{\boldsymbol{S}^{\mathit{eff}}(\boldsymbol{V})\boldsymbol{S}^{\mathit{eff}'}(\boldsymbol{V})\}]^{-1}$, achieves the so-called *semi-parametric efficiency bound*; i.e., the supremum over all parametric sub-models of the Cramèr–Rao lower bounds for $\boldsymbol{\beta}$.

In Section 4.5 of Tsiatis (2006), it is shown that elements of the orthogonal complement of the nuisance tangent space $\Lambda^{\perp}$ for the semi-parametric regression of Example 2 above, are given by $\boldsymbol{A}(\boldsymbol{X})\{Y - \mu(\boldsymbol{X}, \boldsymbol{\beta}_0)\}$ for arbitrary q-dimensional function $\boldsymbol{A}(\boldsymbol{X})$. Thus, using these as estimating functions leads us to GEEs for such models. It is also shown that the corresponding efficient score $\boldsymbol{S}^{\mathit{eff}}(\boldsymbol{V}) = \boldsymbol{S}_{\boldsymbol{\beta}}(\boldsymbol{V}) - \boldsymbol{\Pi}(\boldsymbol{S}_{\boldsymbol{\beta}}|\Lambda)(\boldsymbol{V})$ is given by choosing $\boldsymbol{A}(\boldsymbol{X}) = \boldsymbol{D}'(\boldsymbol{X})V^{-1}(\boldsymbol{X})$, where $\boldsymbol{D}(\boldsymbol{X}) = \partial/\partial\boldsymbol{\beta}'\{\mu(\boldsymbol{X}, \boldsymbol{\beta})\}|_{\boldsymbol{\beta}=\boldsymbol{\beta}_0}$, and $V(\boldsymbol{X}) = \text{var}(Y|\boldsymbol{X})$. This is the optimal GEE as derived by Liang and Zeger (1986).

We next outline arguments on how this geometric perspective for deriving influence functions and estimating functions for RAL estimators for $\boldsymbol{\beta}$ may be used when the data are missing at random (MAR). As prelude to this discussion, we introduce notation for the missing data problem and provide a heuristic overview.

8.5 Missing Data and Inverse Probability Weighting of Complete Cases

Our development thus far has assumed that $\boldsymbol{V}$ is observed on all individuals in the sample; that is, we observe $\boldsymbol{V}_i$, $i = 1, \ldots, N$. We now tackle the situation where some elements of $\boldsymbol{V}$ may be missing for some individuals in the sample, either by design or happenstance.

To represent this, we adopt the following notation. Assume that $\boldsymbol{V}$ may be partitioned as $(\boldsymbol{V}^{(1)'}, \ldots, \boldsymbol{V}^{(K)'})'$, where each $\boldsymbol{V}^{(j)}$, $j = 1, \ldots, K$, is a possibly vector valued subset of $\boldsymbol{V}$ such that, for each individual, $\boldsymbol{V}^{(j)}$ is either fully observed or is missing entirely. Define the K-dimensional vector $\boldsymbol{R} = (R_1, \ldots, R_K)$ of missing data indicators, where, for each $j = 1, \ldots, K$, $R_j = 1$ if $\boldsymbol{V}^{(j)}$ is observed and $R_j = 0$ if it is missing. Define the many-to-one function $\boldsymbol{G}_{\boldsymbol{R}}(\boldsymbol{V})$ that "picks off" the elements of $\boldsymbol{V}$ for which the corresponding elements of the vector $\boldsymbol{R}$ are equal to one. For example, if $K = 3$, so $\boldsymbol{V} = (\boldsymbol{V}^{(1)'}, \boldsymbol{V}^{(2)'}, \boldsymbol{V}^{(3)'})'$, and $\boldsymbol{R} = (1, 0, 1)$, then $\boldsymbol{G}_{\boldsymbol{R}}(\boldsymbol{V}) = (\boldsymbol{V}^{(1)'}, \boldsymbol{V}^{(3)'})'$. If none of the elements of $\boldsymbol{V}$ are missing, then $\boldsymbol{R} = \boldsymbol{1}$, where $\boldsymbol{1}$ is a K-dimensional vector of ones, and $\boldsymbol{G}_{\boldsymbol{1}}(\boldsymbol{V}) = \boldsymbol{V}$.

With this notation, the complete data are $\boldsymbol{V}_i$, $i = 1, \ldots, N$; the full data are $(\boldsymbol{R}_i, \boldsymbol{V}_i)$,

$i = 1, \ldots, N$; and the observed data are $\{\boldsymbol{R}_i, \boldsymbol{G}_{\boldsymbol{R}_i}(\boldsymbol{V}_i)\}, i = 1, \ldots, n$. We may then express the usual taxonomy of missing data mechanisms as follows. Under the assumption of missing completely at random (MCAR) $P(\boldsymbol{R} = \boldsymbol{r}|\boldsymbol{V}) = \varpi(\boldsymbol{r})$; i.e., the probability of missingness is independent of $\boldsymbol{V}$. MAR, of which MCAR is a special case, assumes that $P(\boldsymbol{R} = \boldsymbol{r}|\boldsymbol{V}) = \varpi\{\boldsymbol{r}, \boldsymbol{G}_{\boldsymbol{r}}(\boldsymbol{V})\}$, so that the probability of missingness only depends on the data that are observed. If missingness is not at random (NMAR), then missingness depends on data that are not observed. We restrict attention henceforth only to identifiable missing mechanisms (MCAR, MAR).

Although some data are missing, the focus remains on estimating the parameter of interest $\boldsymbol{\beta}$ in the statistical model for the complete data. The missingness mechanism is not of direct interest; however, it must be taken into account in some way, as inference can be based only on the observed data $\{\boldsymbol{R}_i, \boldsymbol{G}_{\boldsymbol{R}_i}(\boldsymbol{V}_i)\}, i = 1, \ldots, n$. As a prelude to our subsequent derivation using semi-parametric theory, we now present an introduction to IPWCC and AIPWCC estimators for $\boldsymbol{\beta}$.

The basic intuition behind IPWCC estimators is as follows. With complete data $\boldsymbol{V}_1, \ldots, \boldsymbol{V}_N$, assume that an m-estimator for $\boldsymbol{\beta}$ may be obtained as the solution to the estimating equation

$$\sum_{i=1}^{N} \boldsymbol{m}(\boldsymbol{V}_i, \boldsymbol{\beta}) = \boldsymbol{0},$$

where $\boldsymbol{m}(\boldsymbol{V}, \boldsymbol{\beta})$ is an unbiased estimating function; i.e., $E_{\boldsymbol{\beta},\boldsymbol{\eta}}\{\boldsymbol{m}(\boldsymbol{V}, \boldsymbol{\beta})\} = \boldsymbol{0}$ for all $\boldsymbol{\beta}$ and $\boldsymbol{\eta}$. However, when some data are missing, complete data are observed only for individuals i for whom $\boldsymbol{R}_i = \boldsymbol{1}$. For any randomly chosen individual in the population, the probability of observing the complete data in the presence of missingness is given by $P(\boldsymbol{R} = \boldsymbol{1}|\boldsymbol{V}) = \varpi(\boldsymbol{1}, \boldsymbol{V})$. Therefore, any individual in the sample with complete data $\boldsymbol{V}$ fully observed may be thought of as representing $1/\varpi(\boldsymbol{1}, \boldsymbol{V})$ individuals chosen at random from the population, some of whom may have missing data. This suggests estimating $\boldsymbol{\beta}$ by solving the estimating equation

$$\sum_{i=1}^{N} \frac{I(\boldsymbol{R}_i = \boldsymbol{1})\boldsymbol{m}(\boldsymbol{V}_i, \boldsymbol{\beta})}{\varpi(\boldsymbol{1}, \boldsymbol{V}_i)} = \boldsymbol{0}.$$

Here, the estimating function that would be used if complete data were available is "inverse weighted" by the probability of observing complete data, and only "complete cases" with $\boldsymbol{V}$ fully observed (i.e., $\boldsymbol{R} = \boldsymbol{1}$) contribute. The resulting IPWCC estimator is in fact an m-estimator with unbiased estimating function $I(\boldsymbol{R} = \boldsymbol{1})\boldsymbol{m}(\boldsymbol{V}, \boldsymbol{\beta})/\varpi(\boldsymbol{1}, \boldsymbol{V})$. That this is an unbiased estimating function follows from the fact that

$$\begin{aligned} E_{\boldsymbol{\beta},\boldsymbol{\eta}}\left\{\frac{I(\boldsymbol{R} = \boldsymbol{1})\boldsymbol{m}(\boldsymbol{V}, \boldsymbol{\beta})}{\varpi(\boldsymbol{1}, \boldsymbol{V})}\right\} &= E_{\boldsymbol{\beta},\boldsymbol{\eta}}\left[E\left\{\left.\frac{I(\boldsymbol{R} = \boldsymbol{1})\boldsymbol{m}(\boldsymbol{V}, \boldsymbol{\beta})}{\varpi(\boldsymbol{1}, \boldsymbol{V})}\right| \boldsymbol{V}\right\}\right] \\ &= E_{\boldsymbol{\beta},\boldsymbol{\eta}}\left\{\frac{\boldsymbol{m}(\boldsymbol{V}, \boldsymbol{\beta})}{\varpi(\boldsymbol{1}, \boldsymbol{V})} P(\boldsymbol{R} = \boldsymbol{1}|\boldsymbol{V})\right\} \\ &= E_{\boldsymbol{\beta},\boldsymbol{\eta}}\{\boldsymbol{m}(\boldsymbol{V}, \boldsymbol{\beta})\} = \boldsymbol{0}. \end{aligned}$$

It is important to note that, in order for $P(\boldsymbol{R} = \boldsymbol{1}|\boldsymbol{V})/\varpi(\boldsymbol{1}, \boldsymbol{V}) = 1$, it is necessary that $P(\boldsymbol{R} = \boldsymbol{1}|\boldsymbol{V})$ be strictly positive for all realizations of $\boldsymbol{V}$. Also, to ensure that $P(\boldsymbol{R} = \boldsymbol{1}|\boldsymbol{V})$ is identifiable from the observed data, some additional assumptions are required on the missing mechanism, such as MAR.

We illustrate with the two examples introduced previously.

Example 1. Here, the complete data are $\boldsymbol{V} = (\boldsymbol{X}', Y)'$, and interest is in estimating $\beta = E(Y)$, where no assumptions are made regarding the distribution of $\boldsymbol{V}$ (the non-parametric model). We demonstrated in Section 8.3 that, without missing data, the unbiased estimating function $(Y - \beta)$ leads to the m-estimator $N^{-1}\sum_{i=1}^{N} Y_i$.

Assume that, while $\boldsymbol{X}$ is observed for all individuals in the sample, Y may be missing for some of them. Here, there are two levels of missingness; that is, writing $\boldsymbol{V} = (\boldsymbol{V}^{(1)'}, \boldsymbol{V}^{(2)})'$, where $\boldsymbol{V}^{(1)} = \boldsymbol{X}$, $\boldsymbol{V}^{(2)} = Y$, the vector $\boldsymbol{R}$ is either equal to $\mathbf{1} = (1,1)$ if Y is observed or $(1,0)$ if Y is missing. Thus, $\boldsymbol{G}_{(1,1)}(\boldsymbol{V}) = (\boldsymbol{X}', Y)'$, and $\boldsymbol{G}_{(1,0)}(\boldsymbol{V}) = \boldsymbol{X}$. This simple missingness structure allows an alternative representation of the missingness by defining a single missingness indicator $R = I(\boldsymbol{R} = \mathbf{1})$ taking on values 0 or 1 according to whether Y is missing or not. If we make the MAR assumption, then it follows that $P(R = 1|\boldsymbol{X}, Y) = 1 - P(R = 0|\boldsymbol{X}, Y) = 1 - P(R = 0|\boldsymbol{X}) = P(R = 1|\boldsymbol{X})$. Denote $P(R = 1|\boldsymbol{X})$ as $\varpi(\boldsymbol{X})$.

An IPWCC estimator $\widehat{\beta}$ for β is then obtained readily as the solution to

$$\sum_{i=1}^{N} \frac{R_i(Y_i - \beta)}{\varpi(\boldsymbol{X}_i)} = 0,$$

which yields

$$\widehat{\beta} = \frac{\sum_{i=1}^{N} R_i Y_i / \varpi(\boldsymbol{X}_i)}{\sum_{i=1}^{N} R_i / \varpi(\boldsymbol{X}_i)}.$$

Using the theory of semi-parametrics, we show in subsequent sections that all RAL estimators for β for this example may be written as the solution to an estimating equation of the form

$$\sum_{i=1}^{N} \left[\frac{R_i(Y_i - \beta)}{\varpi(\boldsymbol{X}_i)} + \frac{\{R_i - \varpi(\boldsymbol{X}_i)\}}{\varpi(\boldsymbol{X}_i)} \phi(\boldsymbol{X}_i) \right] = 0 \tag{8.8}$$

for any arbitrary function $\phi(\cdot)$ of $\boldsymbol{X}$. It is straightforward to show by a conditioning argument that the summand in (8.8) is an unbiased estimating function. The second term in the summand of (8.8), $\{R - \varpi(\boldsymbol{X})\}\phi(\boldsymbol{X})/\varpi(\boldsymbol{X})$, is referred to as an *augmentation term* and the resulting estimator for β as an augmented inverse probability weighted complete case (AIPWCC) estimator. The augmentation term can be shown to enhance the efficiency of the resulting estimator relative to that of the IPWCC estimator through judicious choice of $\phi(\boldsymbol{X})$.

In the sequel, via semi-parametric theory arguments, we show that the optimal choice is $\phi(\boldsymbol{X}) = -E(Y|\boldsymbol{X})$. Of course, this optimal choice depends on the conditional expectation of Y given $\boldsymbol{X}$, which is unknown and for the non-parametric model can be any arbitrary function of $\boldsymbol{X}$. In practice, as estimating $E(Y|\boldsymbol{X})$ non-parametrically is generally difficult and may lead to unstable estimators, it is customary to posit a finite-dimensional parametric model $\phi(\boldsymbol{X}, \boldsymbol{\psi})$, say, for $E(Y|\boldsymbol{X})$. Although it is likely that $\phi(\boldsymbol{X}, \boldsymbol{\psi})$ is misspecified in the sense that there is no value $\boldsymbol{\psi}_0$, say, for which it is equal to $E(Y|\boldsymbol{X})$, one may nonetheless use appropriate methods to obtain an estimator $\widehat{\boldsymbol{\psi}}$ using the $(\boldsymbol{X}_i', Y_i)'$ for $\{i : R_i = 1\}$. An AIWCC estimator for β would then be found as the solution to

$$\sum_{i=1}^{N} \left[\frac{R_i(Y_i - \beta)}{\varpi(\boldsymbol{X}_i)} + \frac{\{R_i - \varpi(\boldsymbol{X}_i)\}}{\varpi(\boldsymbol{X}_i)} \phi(\boldsymbol{X}_i, \widehat{\boldsymbol{\psi}}) \right] = 0.$$

As long as $P(R = 1|\boldsymbol{X}) = \varpi(\boldsymbol{X})$, the resulting estimator is a consistent and asymptotically normal RAL estimator for β regardless of whether the posited model $\phi(\boldsymbol{X}, \boldsymbol{\psi})$ is correctly specified and will be efficient (i.e., will have smallest asymptotic covariance matrix (in the sense of non-negative definiteness) among the class of RAL estimators) if the posited model is correctly specified.

In most cases, unless the missingness is by design, the missingness probability is also unknown. Accordingly, it is standard to postulate a model $\varpi(\boldsymbol{X}, \gamma)$ for $P(R = 1|\boldsymbol{X})$, depending on a finite-dimensional parameter γ; because R is a binary indicator, a logistic regression model is often convenient for this purpose. The parameter γ may be estimated by maximum likelihood using the data $(R_i, \boldsymbol{X}_i), i = 1, \ldots, n$. Specifically, the estimator $\widehat{\gamma}$ may be obtained by maximizing the likelihood

$$\prod_{i=1}^{N} \varpi(\boldsymbol{X}_i, \gamma)^{R_i} \{1 - \varpi(\boldsymbol{X}_i, \gamma)\}^{(1-R_i)}$$

in γ. The resulting AIPWCC estimator is then the solution to the estimating equation

$$\sum_{i=1}^{N} \left[\frac{R_i(Y_i - \beta)}{\varpi(\boldsymbol{X}_i, \widehat{\gamma})} + \frac{\{R_i - \varpi(\boldsymbol{X}_i, \widehat{\gamma})\}}{\varpi(\boldsymbol{X}_i, \widehat{\gamma})} \phi(\boldsymbol{X}_i, \widehat{\psi}) \right] = 0.$$

An interesting property of the above estimator is that of *double robustness.* That is, the estimator for β is a consistent, asymptotically normal RAL estimator if either the model $\varpi(\boldsymbol{X}, \gamma)$ for $P(R = 1|\boldsymbol{X})$ or the model $\phi(\boldsymbol{X}, \psi)$ for $E(Y|\boldsymbol{X})$ is correctly specified. This is discussed in detail in Chapter 9.

Example 2. Consider again the semi-parametric regression model where $\boldsymbol{V} = (\boldsymbol{X}', Y)'$, and it is assumed only that

$$E(Y|\boldsymbol{X}) = \mu(\boldsymbol{X}, \boldsymbol{\beta}).$$

The parameter $\boldsymbol{\beta}$ can be estimated by an m-estimator with unbiased estimating function

$$\boldsymbol{A}(\boldsymbol{X})\{Y - \mu(\boldsymbol{X}, \boldsymbol{\beta})\};$$

that is, by the estimator solving a GEE.

To illustrate a situation in which the MAR assumption is satisfied by design, consider the following two stage scheme. Suppose the vector of covariates $\boldsymbol{X}$ is partitioned as $\boldsymbol{X} = (\boldsymbol{X}^{(1)'}, \boldsymbol{X}^{(2)'})'$, where $\boldsymbol{X}^{(1)}$ are variables that are relatively inexpensive to collect, whereas $\boldsymbol{X}^{(2)}$ are expensive to collect. For example, $\boldsymbol{X}^{(2)}$ may contain genetic markers that are expensive to process, and $\boldsymbol{X}^{(1)}$ may involve demographic variables such as age, race, sex, and so on. In this setting, on grounds of cost, a decision may be made to collect the response variable Y and the inexpensive covariates $\boldsymbol{X}^{(1)}$ on all individuals in the sample but collect the expensive covariates $\boldsymbol{X}^{(2)}$ on only a subset of individuals. Moreover, the probability of collecting $\boldsymbol{X}^{(2)}$ may be taken to depend on the values of Y and $\boldsymbol{X}^{(1)}$. This might be the case if, say, it is desired to overrepresent some values of Y and $\boldsymbol{X}^{(1)}$ in the subset of subjects on whom complete data are collected.

Reordering the complete data as $\boldsymbol{V} = (Y, \boldsymbol{X}^{(1)'}, \boldsymbol{X}^{(2)'})'$, write $\boldsymbol{V} = (\boldsymbol{V}^{(1)'}, \boldsymbol{V}^{(2)'})'$, where $\boldsymbol{V}^{(1)} = (Y_i, \boldsymbol{X}^{(1)'})'$ is observed for all individuals while $\boldsymbol{V}^{(2)} = \boldsymbol{X}^{(2)}$ may be missing for some individuals. Thus, $\boldsymbol{G}_{(1,1)}(\boldsymbol{V}) = (Y, \boldsymbol{X}^{(1)'}, \boldsymbol{X}^{(2)'})'$ and $\boldsymbol{G}_{(1,0)}(\boldsymbol{V}) = \boldsymbol{X}^{(2)}$. As in the previous example, $\boldsymbol{R}$ is either equal to $\boldsymbol{1} = (1, 1)$ or $(1, 0)$, so that there are two levels of missingness. Accordingly, again define the binary indicator $R = I(\boldsymbol{R} = \boldsymbol{1})$.

This is an example of missing data by design. To implement such a design, we would collect

$(Y_i, \boldsymbol{X}_i^{(1)'})'$ for all patients $i = 1, \ldots, n$, as well as blood samples that may be used to obtain the expensive genetic markers. For each patient i, generate R_i at random taking value 1 or 0 with probability $\varpi(Y_i, \boldsymbol{X}_i^{(1)})$ and $1 - \varpi(Y_i, \boldsymbol{X}_i^{(1)})$, respectively, where $0 < \varpi(y, \boldsymbol{x}^{(1)}) \leq 1$ is a known function of the response $Y = y$ and the covariates $\boldsymbol{X}^{(1)} = \boldsymbol{x}^{(1)}$, chosen by the investigator. If $R_i = 1$, then the blood sample from patient i is processed and the genetic markers $\boldsymbol{X}_i^{(2)}$ obtained; otherwise, the genetic markers are not ascertained and are hence missing.

An IPWCC estimator for $\boldsymbol{\beta}$ may be obtained readily as the solution to the estimating equation

$$\sum_{i=1}^{N} \frac{R_i \boldsymbol{A}(\boldsymbol{X}_i)\{Y_i - \mu(\boldsymbol{X}_i, \boldsymbol{\beta})\}}{\varpi(Y_i, \boldsymbol{X}_i^{(1)})} = \boldsymbol{0}$$

for arbitrary $\boldsymbol{A}(\boldsymbol{X})$. Note that this is an m-estimator with unbiased estimating function

$$\frac{R\boldsymbol{A}(\boldsymbol{X})\{Y - \mu(\boldsymbol{X}, \boldsymbol{\beta})\}}{\varpi(Y, \boldsymbol{X}^{(1)})}.$$

As in Example 1, one can use the theory of semi-parametrics to show that all RAL estimators for $\boldsymbol{\beta}$ may be obtained as the solution to

$$\sum_{i=1}^{N} \left(\frac{R_i \boldsymbol{A}(\boldsymbol{X}_i)\{Y_i - \mu(\boldsymbol{X}_i, \boldsymbol{\beta})\}}{\varpi(Y_i, \boldsymbol{X}_i^{(1)})} + \left\{ \frac{R_i - \varpi(Y_i, \boldsymbol{X}_i^{(1)})}{\varpi(Y_i, \boldsymbol{X}_i^{(1)})} \right\} L(Y_i, \boldsymbol{X}_i^{(1)}) \right) = \boldsymbol{0} \tag{8.9}$$

for arbitrary $\boldsymbol{A}(\boldsymbol{X})$ and $L(Y, \boldsymbol{X}^{(1)})$. The IPWCC estimator results from taking $L(Y_i, \boldsymbol{X}_i^{(1)}) \equiv 0$. In general, the resulting AIPWCC estimator is a consistent, asymptotically normal RAL estimator for $\boldsymbol{\beta}$ regardless of the choice of $\boldsymbol{A}(\boldsymbol{X})$ and $L(Y, \boldsymbol{X}^{(1)})$; however, the choice of these functions will affect the efficiency of the estimator. We discuss this issue further in subsequent sections. Note that, because the missingness mechanism is by design here, there is no concern regarding the plausibility of the MAR assumption, nor is there concern over misspecification of the missingness probability.

In the two examples above, we were able to derive estimators for $\boldsymbol{\beta}$ with observed data that are MAR by using IPWCC or AIPWCC estimating equations based on complete-data influence (estimating) functions. For a general semi-parametric model, the pioneering contribution of Robins, Rotnitzky, and Zhao (1994) was to characterize the class of all observed-data influence functions when data are MAR, including the efficient influence function, and to demonstrate that observed-data influence functions may be expressed in terms of full-data influence functions. Because, for many popular semi-parametric models, the form of full-data influence functions is known or straightforwardly derived, this provides an attractive basis for identifying estimators when data are MAR. We now review some of the elements of semi-parametric theory as it applies to missing data.

8.6 Density and Likelihood of Missing Data

For definiteness, we focus on the situation where the objective is to estimate the parameter $\boldsymbol{\beta}$ in a model $f_{\boldsymbol{V}}(\boldsymbol{v}, \boldsymbol{\theta})$ for the complete data, where $\boldsymbol{\theta} = (\boldsymbol{\beta}, \boldsymbol{\eta})$ may be either finite-dimensional (parametric model) or infinite-dimensional (semi-parametric model), when some data are missing. Thus, the goal is to estimate $\boldsymbol{\beta}$ based on the observed i.i.d. data $\{\boldsymbol{R}_i, \boldsymbol{G}_{\boldsymbol{R}_i}(\boldsymbol{V}_i)\}$, $i = 1, \ldots, N$. We assume that the missingness mechanism is MAR, so that $P(\boldsymbol{R} = \boldsymbol{r}|\boldsymbol{V}) = \varpi\{\boldsymbol{r}, \boldsymbol{G}_{\boldsymbol{r}}(\boldsymbol{V})\}$, where, for the time being, we take $\varpi\{\boldsymbol{r}, \boldsymbol{G}_{\boldsymbol{r}}(\boldsymbol{V})\}$ to be known.

In order to derive the underlying semi-parametric theory, we first derive the likelihood of the observed data in terms of $\boldsymbol{\beta}$ and other nuisance parameters. Consider the full data; that is, the unobservable i.i.d. random vectors

$$(\boldsymbol{R}_i, \boldsymbol{V}_i), \quad i = 1, \ldots, N;$$

these are unobservable because, when $\boldsymbol{R}_i = \boldsymbol{r}$, $\boldsymbol{r} \neq \boldsymbol{1}$, we are only able to observe $\boldsymbol{G}_{\boldsymbol{r}}(\boldsymbol{V}_i)$ and not $\boldsymbol{V}_i$ itself.

Because the observed data $\{\boldsymbol{R}, \boldsymbol{G}_{\boldsymbol{R}}(\boldsymbol{V})\}$ are a known function of the full data $(\boldsymbol{R}, \boldsymbol{V})$, the density of the observed data is induced by that of the full data. The density of the full data and the corresponding likelihood, in terms of the parameters $\boldsymbol{\beta}$ and $\boldsymbol{\eta}$, are given by

$$f_{\boldsymbol{R},\boldsymbol{V}}(\boldsymbol{r}, \boldsymbol{v}, \boldsymbol{\beta}, \boldsymbol{\eta}) = P(\boldsymbol{R} = \boldsymbol{r}|\boldsymbol{V} = \boldsymbol{v}) f_{\boldsymbol{V}}(\boldsymbol{v}, \boldsymbol{\beta}, \boldsymbol{\eta}).$$

That is, the density and likelihood of the full data are determined from the density and likelihood of $\boldsymbol{V}$, $f_{\boldsymbol{V}}(\boldsymbol{v}, \boldsymbol{\beta}, \boldsymbol{\eta})$, and the density and likelihood for the missingness mechanism (i.e., the probability of missingness given $\boldsymbol{V}$). The density and likelihood of the observed data may then be written as

$$\begin{aligned} f_{\boldsymbol{R},\boldsymbol{G}_{\boldsymbol{R}}(\boldsymbol{v})}(\boldsymbol{r}, \boldsymbol{g}_{\boldsymbol{r}}, \boldsymbol{\beta}, \boldsymbol{\eta}) &= \int_{\{\boldsymbol{v}:\boldsymbol{G}_{\boldsymbol{r}}(\boldsymbol{v}) = \boldsymbol{g}_{\boldsymbol{r}}\}} f_{\boldsymbol{R},\boldsymbol{V}}(\boldsymbol{r}, \boldsymbol{v}, \boldsymbol{\beta}, \boldsymbol{\eta})\, d\nu(\boldsymbol{v}) \\ &= \int_{\{\boldsymbol{v}:\boldsymbol{G}_{\boldsymbol{r}}(\boldsymbol{v}) = \boldsymbol{g}_{\boldsymbol{r}}\}} P(\boldsymbol{R} = \boldsymbol{r}|\boldsymbol{V} = \boldsymbol{v}) f_{\boldsymbol{V}}(\boldsymbol{v}, \boldsymbol{\beta}, \boldsymbol{\eta})\, d\nu(\boldsymbol{v}), \end{aligned} \tag{8.10}$$

where $\boldsymbol{g}_{\boldsymbol{r}}$ is used to denote a realization of $\boldsymbol{G}_{\boldsymbol{R}}(\boldsymbol{V})$ when $\boldsymbol{R} = \boldsymbol{r}$. Viewed as a function of $\boldsymbol{\beta}$ and $\boldsymbol{\eta}$, the likelihood in (8.10) allows for any missingness mechanism, including NMAR. When the data are MAR, (8.10) becomes

$$\begin{aligned} f_{\boldsymbol{R},\boldsymbol{G}_{\boldsymbol{R}}(\boldsymbol{V})}(\boldsymbol{r}, \boldsymbol{g}_{\boldsymbol{r}}, \boldsymbol{\beta}, \boldsymbol{\eta}) &= \int_{\{\boldsymbol{v}:\boldsymbol{G}_{\boldsymbol{r}}(\boldsymbol{v}) = \boldsymbol{g}_{\boldsymbol{r}}\}} \varpi(\boldsymbol{r}, \boldsymbol{g}_{\boldsymbol{r}}) f_{\boldsymbol{V}}(\boldsymbol{v}, \boldsymbol{\beta}, \boldsymbol{\eta}) d\nu(\boldsymbol{v}) \\ &= \varpi(\boldsymbol{r}, \boldsymbol{g}_{\boldsymbol{r}}) \int_{\{\boldsymbol{v}:\boldsymbol{G}_{\boldsymbol{r}}(\boldsymbol{v}) = \boldsymbol{g}_{\boldsymbol{r}}\}} f_{\boldsymbol{V}}(\boldsymbol{v}, \boldsymbol{\beta}, \boldsymbol{\eta}) d\nu(\boldsymbol{v}). \end{aligned} \tag{8.11}$$

An important feature is that, under MAR, the observed data likelihood (8.11) is the product of two separate terms, one involving the missingness mechanism and one involving the parameters $\boldsymbol{\beta}$ and $\boldsymbol{\eta}$. See Part II for detailed discussion of the role of this feature in likelihood-based inference.

We are now in a position to describe the geometry of influence functions of RAL estimators for $\boldsymbol{\beta}$ when data are MAR.

8.7 Geometry of Semi-Parametric Missing Data Models

The key to deriving the class of influence functions for RAL estimators for $\boldsymbol{\beta}$ and the corresponding geometry when data are missing is to build on the theory of influence functions for estimators for $\boldsymbol{\beta}$ and its geometry had the data not been missing. That is, having characterized the class of influence functions for estimators for $\boldsymbol{\beta}$ based on the complete data $\boldsymbol{V}$, we demonstrate how the class of influence functions for estimators for $\boldsymbol{\beta}$ based on the observed data $\{\boldsymbol{R}, \boldsymbol{G_R}(\boldsymbol{V})\}$ may be derived.

Recall from Section 8.4 that influence functions based on the complete data $\boldsymbol{V}$ are elements of the Hilbert space of all q-dimensional, mean-zero functions $\boldsymbol{h}(\boldsymbol{V})$ with $E\{\boldsymbol{h}'(\boldsymbol{V})\boldsymbol{h}(\boldsymbol{V})\} < \infty$ equipped with the covariance inner product. As we noted at the end of Section 8.1, it is conventional in the literature on semi-parametric missing data methods to refer to this Hilbert space as the *full-data* Hilbert space (rather than the "complete-data" Hilbert space) and to refer to $\boldsymbol{V}$ as the full (rather than complete) data. To maintain consistency with the literature, we will henceforth adopt this terminology and denote this Hilbert space by $\mathcal{H}^F$.

In what follows, $\mathcal{H}^F$ is to be distinguished from the *observed-data* Hilbert space of all q-dimensional, mean-zero functions $\boldsymbol{h}\{\boldsymbol{R}, \boldsymbol{G_R}(\boldsymbol{V})\}$ with $E[\boldsymbol{h}'\{\boldsymbol{R}, \boldsymbol{G_R}(\boldsymbol{V})\}\boldsymbol{h}\{\boldsymbol{R}, \boldsymbol{G_R}(\boldsymbol{V})\} < \infty$ equipped with the covariance inner product, which we denote by $\mathcal{H}$ (without the superscript F).

Under MAR, the observed-data likelihood for a single observation is given by (8.11), and hence the log-likelihood for a single observation is given by

$$\log \varpi(\boldsymbol{r}, \boldsymbol{g_r}) + \log \int_{\{\boldsymbol{v}:\boldsymbol{G_r}(\boldsymbol{v})=\boldsymbol{g_r}\}} f_{\boldsymbol{V}}(\boldsymbol{v}, \boldsymbol{\beta}, \boldsymbol{\eta})\, d\nu(\boldsymbol{v}). \tag{8.12}$$

The observed-data likelihood and log-likelihood are functions of the parameters $\boldsymbol{\beta}$ and $\boldsymbol{\eta}$.

Because influence functions of RAL estimators for $\boldsymbol{\beta}$ must lie in the space orthogonal to the nuisance tangent space, and because unbiased estimating functions can often be derived by considering elements of this space, deriving the orthogonal complement of the observed-data nuisance tangent space plays an important role in deriving observed-data estimators. To this end, we now consider how the nuisance tangent space and its orthogonal complement are derived when data are MAR. In so doing, we make the connection to the full-data nuisance tangent space and its orthogonal complement, which allows us to show how estimators with missing data can be obtained using full-data estimators.

8.7.1 Observed-data nuisance tangent space

The observed-data nuisance tangent space $\Lambda_{\boldsymbol{\eta}}$ is defined as the mean-square closure of parametric sub-model nuisance tangent spaces associated with the nuisance parameter $\boldsymbol{\eta}$. Therefore, we begin by considering the parametric sub-model for the full data $\boldsymbol{V}$ given by $f_{\boldsymbol{V}}(\boldsymbol{v}, \boldsymbol{\beta}, \boldsymbol{\xi})$ for $\boldsymbol{\xi}$ $(r \times 1)$ and computing the corresponding observed-data score vector. Let $\boldsymbol{S}^F_{\boldsymbol{\xi}}(\boldsymbol{V}, \boldsymbol{\beta}, \boldsymbol{\xi}) = \partial/\partial\boldsymbol{\xi}\{\log f_{\boldsymbol{V}}(\boldsymbol{V}, \boldsymbol{\beta}, \boldsymbol{\xi})\}$, and denote the parametric sub-model full-data score vector by $\boldsymbol{S}^F_{\boldsymbol{\xi}}(\boldsymbol{V}) = \boldsymbol{S}^F_{\boldsymbol{\xi}}(\boldsymbol{V}, \beta_0, \boldsymbol{\xi}_0)$.

Lemma 1. The parametric sub-model observed-data score vector with respect to $\boldsymbol{\xi}$ is given by

$$\boldsymbol{S}_{\boldsymbol{\xi}}(\boldsymbol{r}, \boldsymbol{g}_{\boldsymbol{r}}) = E\{\boldsymbol{S}^F_{\boldsymbol{\xi}}(\boldsymbol{V})|\boldsymbol{G}_{\boldsymbol{r}}(\boldsymbol{V}) = \boldsymbol{g}_{\boldsymbol{r}}\}. \tag{8.13}$$

Proof. The part of the log-likelihood for the observed data that involves $\boldsymbol{\xi}$ is, from (8.12),

$$\log\left\{\int_{\{\boldsymbol{v}:\boldsymbol{G}_{\boldsymbol{r}}(\boldsymbol{v})=\boldsymbol{g}_{\boldsymbol{r}}\}} f_{\boldsymbol{V}}(\boldsymbol{v}, \boldsymbol{\beta}, \boldsymbol{\xi})\, d\nu(\boldsymbol{v})\right\}.$$

The score vector with respect to $\boldsymbol{\xi}$ is

$$\begin{aligned}\boldsymbol{S}_{\boldsymbol{\xi}}(\boldsymbol{r}, \boldsymbol{g}_{\boldsymbol{r}}) &= \frac{\partial}{\partial \boldsymbol{\xi}}\left[\log\left\{\int_{\{\boldsymbol{v}:\boldsymbol{G}_{\boldsymbol{r}}(\boldsymbol{v})=\boldsymbol{g}_{\boldsymbol{r}}\}} f_{\boldsymbol{V}}(\boldsymbol{v}, \boldsymbol{\beta}, \boldsymbol{\xi})\, d\nu(\boldsymbol{v})\right\}\right]\Bigg|_{\boldsymbol{\beta}=\boldsymbol{\beta}_0, \boldsymbol{\xi}=\boldsymbol{\xi}_0} \\ &= \frac{\displaystyle\int_{\{\boldsymbol{v}:\boldsymbol{G}_{\boldsymbol{r}}(\boldsymbol{v})=\boldsymbol{g}_{\boldsymbol{r}}\}} \partial/\partial\boldsymbol{\xi}\{f_{\boldsymbol{V}}(\boldsymbol{v}, \boldsymbol{\beta}, \boldsymbol{\xi})\}|_{\boldsymbol{\beta}=\boldsymbol{\beta}_0, \boldsymbol{\xi}=\boldsymbol{\xi}_0}\, d\nu(\boldsymbol{v})}{\displaystyle\int_{\{\boldsymbol{v}:\boldsymbol{G}_{\boldsymbol{r}}(\boldsymbol{v})=\boldsymbol{g}_{\boldsymbol{r}}\}} f_{\boldsymbol{V}}(\boldsymbol{v}, \boldsymbol{\beta}_0, \boldsymbol{\xi}_0)\, d\nu(\boldsymbol{v})}. \end{aligned} \tag{8.14}$$

Dividing and multiplying by $f_{\boldsymbol{V}}(\boldsymbol{v}, \boldsymbol{\beta}_0, \boldsymbol{\xi}_0)$ in the integral of the numerator of (8.14) yields

$$\boldsymbol{S}_{\boldsymbol{\xi}}(\boldsymbol{r}, \boldsymbol{g}_{\boldsymbol{r}}) = \frac{\displaystyle\int_{\{\boldsymbol{v}:\boldsymbol{G}_{\boldsymbol{r}}(\boldsymbol{v})=\boldsymbol{g}_{\boldsymbol{r}}\}} \boldsymbol{S}^F_{\boldsymbol{\xi}}(\boldsymbol{v}) f_{\boldsymbol{V}}(\boldsymbol{v}, \boldsymbol{\beta}_0, \boldsymbol{\xi}_0)\, d\nu(\boldsymbol{v})}{\displaystyle\int_{\{\boldsymbol{v}:\boldsymbol{G}_{\boldsymbol{r}}(\boldsymbol{v})=\boldsymbol{g}_{\boldsymbol{r}}\}} f_{\boldsymbol{V}}(\boldsymbol{v}, \boldsymbol{\beta}_0, \boldsymbol{\xi}_0)\, d\nu(\boldsymbol{v})}.$$

Hence, (8.13) follows.

Remark. Equation (8.13) involves only the conditional probability distribution of $\boldsymbol{V}$ given $\boldsymbol{G}_{\boldsymbol{r}}(\boldsymbol{V})$ for a fixed value of $\boldsymbol{r}$ and should be contrasted with

$$E\{\boldsymbol{S}^F_{\boldsymbol{\xi}}(\boldsymbol{V})|\boldsymbol{R} = \boldsymbol{r}, \boldsymbol{G}_{\boldsymbol{R}}(\boldsymbol{V}) = \boldsymbol{g}_{\boldsymbol{r}}\}. \tag{8.15}$$

The conditional expectations in (8.15) and (8.13) are not equal in general. However, they are equal under the assumption of MAR, as shown in the following lemma.

Lemma 2. When the missingness mechanism is MAR,

$$\begin{aligned}\boldsymbol{S}_{\boldsymbol{\xi}}(\boldsymbol{r}, \boldsymbol{g}_{\boldsymbol{r}}) &= E\{\boldsymbol{S}^F_{\boldsymbol{\xi}}(\boldsymbol{V})|\boldsymbol{G}_{\boldsymbol{r}}(\boldsymbol{V}) = \boldsymbol{g}_{\boldsymbol{r}}\} \\ &= E\{\boldsymbol{S}^F_{\boldsymbol{\xi}}(\boldsymbol{V})|\boldsymbol{R} = \boldsymbol{r}, \boldsymbol{G}_{\boldsymbol{R}}(\boldsymbol{V}) = \boldsymbol{g}_{\boldsymbol{r}}\}.\end{aligned} \tag{8.16}$$

Proof. Equation (8.16) follows if

$$f_{\boldsymbol{V}|\boldsymbol{R},\boldsymbol{G}_{\boldsymbol{R}}(\boldsymbol{V})}(\boldsymbol{v}|\boldsymbol{r}, \boldsymbol{g}_{\boldsymbol{r}}) = f_{\boldsymbol{V}|\boldsymbol{G}_{\boldsymbol{r}}(\boldsymbol{V})}(\boldsymbol{v}|\boldsymbol{g}_{\boldsymbol{r}}), \tag{8.17}$$

The equality in (8.17) is true because, when $\boldsymbol{G_r}(\boldsymbol{v}) = \boldsymbol{g_r}$,

$$
\begin{aligned}
f_{\boldsymbol{V}|\boldsymbol{R},\boldsymbol{G_R}(\boldsymbol{V})}(\boldsymbol{v}|\boldsymbol{r},\boldsymbol{g_r}) &= \frac{f_{\boldsymbol{R},\boldsymbol{V}}(\boldsymbol{r},\boldsymbol{v})}{\displaystyle\int_{\{\boldsymbol{u}:\boldsymbol{G_r}(\boldsymbol{u})=\boldsymbol{g_r}\}} f_{\boldsymbol{R},V}(\boldsymbol{r},\boldsymbol{u})\,d\nu(\boldsymbol{u})} \\
&= \frac{f_{\boldsymbol{R}|\boldsymbol{V}}(\boldsymbol{r}|\boldsymbol{v})f_{\boldsymbol{V}}(\boldsymbol{v})}{\displaystyle\int_{\{\boldsymbol{u}:\boldsymbol{G_r}(\boldsymbol{u})=\boldsymbol{g_r}\}} f_{\boldsymbol{R}|\boldsymbol{V}}(\boldsymbol{r}|\boldsymbol{u})f_{\boldsymbol{V}}(\boldsymbol{u})\,d\nu(\boldsymbol{u})} \\
&= \frac{\varpi(\boldsymbol{r},\boldsymbol{g_r})f_{\boldsymbol{V}}(\boldsymbol{v})}{\varpi(\boldsymbol{r},\boldsymbol{g_r})\displaystyle\int_{\{\boldsymbol{u}:\boldsymbol{G_r}(\boldsymbol{u})=\boldsymbol{g_r}\}} f_{\boldsymbol{V}}(\boldsymbol{u})\,d\nu(\boldsymbol{u})} \qquad (8.18) \\
&= \frac{f_{\boldsymbol{V}}(\boldsymbol{v})}{\displaystyle\int_{\{\boldsymbol{u}:\boldsymbol{G_r}(\boldsymbol{v})=\boldsymbol{g_r}\}} f_{\boldsymbol{V}}(\boldsymbol{u})\,d\nu(\boldsymbol{u})} = f_{\boldsymbol{V}|\boldsymbol{G_r}(\boldsymbol{V})}(\boldsymbol{v}|\boldsymbol{g_r}).
\end{aligned}
$$

The last equality holds because $\varpi(\boldsymbol{r},\boldsymbol{g_r})$ cancels in the numerator and denominator of (8.18), which follows because of the MAR assumption.

The implication of Lemma 2 is that, when the missingness mechanism is MAR, the corresponding observed-data nuisance score vector for the parametric sub-model $f_{\boldsymbol{V}}(\boldsymbol{v},\boldsymbol{\beta},\boldsymbol{\xi})$ is equivalently

$$\boldsymbol{S_\xi}\{\boldsymbol{R},\boldsymbol{G_R}(\boldsymbol{V})\} = E\{\boldsymbol{S^F_\xi}(\boldsymbol{V})|\boldsymbol{R},\boldsymbol{G_R}(\boldsymbol{V})\}. \qquad (8.19)$$

This is useful in defining the observed-data nuisance tangent space, as we now demonstrate.

Theorem 1. The observed-data nuisance tangent space Λ_η (i.e., the mean-square closure of parametric sub-model nuisance tangent spaces spanned by $\boldsymbol{S_\xi}\{\boldsymbol{R},\boldsymbol{G_R}(\boldsymbol{V})\}$) is the space of elements

$$\Lambda_\eta = \left[E\left\{\boldsymbol{\alpha}^F(\boldsymbol{V})|\boldsymbol{R},\boldsymbol{G_R}(\boldsymbol{V})\right\} \text{ for all } \boldsymbol{\alpha}^F \in \Lambda^F\right], \qquad (8.20)$$

where Λ^F denotes the full-data nuisance tangent space.

Proof. By Lemma 2, we have from (8.19) that the linear subspace within $\mathcal{H}$ spanned by the parametric sub-model score vector $\boldsymbol{S_\xi}\{\boldsymbol{R},\boldsymbol{G_R}(\boldsymbol{V})\}$ is

$$
\begin{aligned}
&[BE\{\boldsymbol{S^F_\xi}(\boldsymbol{V})|\boldsymbol{R},\boldsymbol{G_R}(\boldsymbol{V})\} \text{ for all } B\ (q\times r)\}] \\
&= [E\{B\boldsymbol{S^F_\xi}(\boldsymbol{V})|\boldsymbol{R},\boldsymbol{G_R}(\boldsymbol{V})\} \text{ for all } B\ (q\times r)].
\end{aligned}
$$

The linear subspace Λ_η consisting of elements $BE\{\boldsymbol{S^F_\xi}(\boldsymbol{V})|\boldsymbol{R},\boldsymbol{G_R}(\boldsymbol{V})\}$ for some parametric sub-model or a limit (as $j\to\infty$) of elements $B_jE\{\boldsymbol{S^F_{\xi_j}}(\boldsymbol{V})|\boldsymbol{R},\boldsymbol{G_R}(\boldsymbol{V})\}$ for B_j $(q\times r_j)$, a sequence of parametric sub-models and conformable matrices, is the same as the space consisting of elements $E\{B\boldsymbol{S^F_\xi}(\boldsymbol{V})|\boldsymbol{R},\boldsymbol{G_R}(\boldsymbol{V})\}$ or limits of elements $E\{B_j\boldsymbol{S^F_{\xi_j}}(\boldsymbol{V})|\boldsymbol{R},\boldsymbol{G_R}(\boldsymbol{V})\}$. But the space of elements $B\boldsymbol{S^F_\xi}(\boldsymbol{V})$ or limits of elements $B_j\boldsymbol{S^F_{\xi_j}}(\boldsymbol{V})$ for parametric sub-models is precisely the definition of the full-data nuisance tangent space Λ^F. Consequently, the space Λ_η can be characterized by (8.20).

8.7.2 Orthogonal complement of the nuisance tangent space

As in the case with no missing data, influence functions for observed-data RAL estimators for $\boldsymbol{\beta}$ lie in the space orthogonal to Λ_η. Accordingly, we now characterize $\Lambda_\eta^\perp$.

Lemma 3. The space $\Lambda_\eta^\perp$ consists of all elements $\boldsymbol{h}\{\boldsymbol{R}, \boldsymbol{G_R}(\boldsymbol{V})\}$, $\boldsymbol{h} \in \mathcal{H}$, such that

$$E[\boldsymbol{h}\{\boldsymbol{R}, \boldsymbol{G_R}(\boldsymbol{V})\}|\boldsymbol{V}] \in \Lambda^{F\perp}, \tag{8.21}$$

where $\Lambda^{F\perp}$ is the space orthogonal to the full-data nuisance tangent space.

Proof. By definition, $\Lambda_\eta^\perp$ consists of all $(q \times 1)$ mean-zero functions $\boldsymbol{h}\{\boldsymbol{R}, \boldsymbol{G_R}(\boldsymbol{V})\}$ in $\mathcal{H}$ that are orthogonal to Λ_η. By Theorem 1, this is the set of elements $\boldsymbol{h} \in \mathcal{H}$ such that

$$E[\boldsymbol{h}'\{\boldsymbol{R}, \boldsymbol{G_R}(\boldsymbol{V})\}E\{\boldsymbol{\alpha}^F(\boldsymbol{V})|\boldsymbol{R}, \boldsymbol{G_R}(\boldsymbol{V})\}] = 0 \text{ for all } \boldsymbol{\alpha}^F \in \Lambda^F. \tag{8.22}$$

It may be shown that elements of the form (8.22) satisfy (8.21) by a repeated conditioning argument, namely,

$$\begin{aligned} 0 &= E\left(E[\boldsymbol{h}'\{\boldsymbol{R}, \boldsymbol{G_R}(\boldsymbol{V})\}\boldsymbol{\alpha}^F(\boldsymbol{V})|\boldsymbol{R}, \boldsymbol{G_R}(\boldsymbol{V})]\right) \\ &= E[\boldsymbol{h}'\{\boldsymbol{R}, \boldsymbol{G_R}(\boldsymbol{V})\}\boldsymbol{\alpha}^F(\boldsymbol{V})] = E\left(E[\boldsymbol{h}'\{\boldsymbol{R}, \boldsymbol{G_R}(\boldsymbol{V})\}\boldsymbol{\alpha}^F(\boldsymbol{V})|\boldsymbol{V}]\right) \\ &= E\left(E[\boldsymbol{h}'\{\boldsymbol{R}, \boldsymbol{G_R}(\boldsymbol{V})\}|\boldsymbol{V}]\ \boldsymbol{\alpha}^F(\boldsymbol{V})\right) \text{ for all } \boldsymbol{\alpha}^F(\boldsymbol{V}) \in \Lambda^F. \end{aligned} \tag{8.23}$$

Because (8.23) implies that $\boldsymbol{h} \in \mathcal{H}$ belongs to $\Lambda_\eta^\perp$ if and only if $E[\boldsymbol{h}\{\boldsymbol{R}, \boldsymbol{G_R}(\boldsymbol{V})\}|\boldsymbol{V}]$ is orthogonal to every element $\boldsymbol{\alpha}^F \in \Lambda^F$, (8.21) follows.

Lemma 3 provides a general characterization of elements of $\Lambda_\eta^\perp$, the space in which observed-data influence functions lie. To determine more precisely the form of elements of $\Lambda_\eta^\perp$ that suggests estimating functions for estimators for $\boldsymbol{\beta}$ based on the observed data, it is useful to introduce the notion of a (linear) mapping, or operator, from one Hilbert space to another. We summarize the salient features here; see Luenberger (1969, Chapter 6) for more detail.

A mapping $\mathcal{K}$ is a function that maps each element of some linear space to an element of another linear space. In our context, these spaces are Hilbert spaces. More concretely, if $\mathcal{H}^{(1)}$ and $\mathcal{H}^{(2)}$ are two Hilbert spaces, then the mapping $\mathcal{K} : \mathcal{H}^{(1)} \to \mathcal{H}^{(2)}$ is such that, for any $\boldsymbol{h} \in \mathcal{H}^{(1)}$, $\mathcal{K}(\boldsymbol{h}) \in \mathcal{H}^{(2)}$. For a linear mapping, $\mathcal{K}(a\boldsymbol{h}_1 + b\boldsymbol{h}_2) = a\mathcal{K}(\boldsymbol{h}_1) + b\mathcal{K}(\boldsymbol{h}_2)$ for any two elements $\boldsymbol{h}_1, \boldsymbol{h}_2 \in \mathcal{H}^{(1)}$ and any scalar constants a and b. A many-to-one mapping is such that more than one element $\boldsymbol{h} \in \mathcal{H}^{(1)}$ maps to the same element in $\mathcal{H}^{(2)}$.

For our purposes, it is useful to define the many-to-one mapping $\mathcal{K} : \mathcal{H} \to \mathcal{H}^F$ given by

$$\mathcal{K}(\boldsymbol{h}) = E[\boldsymbol{h}\{\boldsymbol{R}, \boldsymbol{G_R}(\boldsymbol{V})\}|\boldsymbol{V}] \tag{8.24}$$

for $\boldsymbol{h} \in \mathcal{H}$. Because conditional expectation is a linear operation, $\mathcal{K}$ defined in (8.24) is a linear mapping.

The inverse mapping $\mathcal{K}^{-1}(\boldsymbol{h}^F)$ for any $\boldsymbol{h}^F$ is the set of all elements $\boldsymbol{h} \in \mathcal{H}$ such that $\mathcal{K}(\boldsymbol{h}) = \boldsymbol{h}^F$. Thus, we denote by $\mathcal{K}^{-1}(\Lambda^{F\perp})$ the space of all elements $\boldsymbol{h} \in \mathcal{H}$ such that $\mathcal{K}(\boldsymbol{h}) \in \Lambda^{F\perp}$. Then, by Lemma 3, the space $\Lambda_\eta^\perp$ may be defined as

$$\Lambda_\eta^\perp = \mathcal{K}^{-1}(\Lambda^{F\perp}).$$

Because $\mathcal{K}$ is a linear operator and $\Lambda^{F\perp}$ is a linear subspace of $\mathcal{H}^F$, it is straightforward that $\mathcal{K}^{-1}(\Lambda^{F\perp})$ is a linear subspace of $\mathcal{H}$.

To move toward a description of the form of observed-data influence functions, consider $\Lambda_\eta^\perp = \mathcal{K}^{-1}(\Lambda^{F\perp})$ element by element. In what follows, we refer to elements in the space $\Lambda^{F\perp}$ using the notation $\varphi^{*F}(\boldsymbol{V})$, which is to be contrasted with our notation for full-data influence functions, $\varphi^F(\boldsymbol{V})$ (without the asterisk). Full-data influence functions belong to the space $\Lambda^{F\perp}$, but, as discussed in Section 8.4, in order to be influence functions, they must also satisfy $E\{\varphi^F(\boldsymbol{V})\boldsymbol{S}^{\mathit{eff}\,F\,'}(\boldsymbol{V})\} = I_q$, where $\boldsymbol{S}^{\mathit{eff}\,F}(\boldsymbol{V})$ is the full-data efficient score. Thus, for any $\varphi^{*F} \in \Lambda^{F\perp}$, we may construct an influence function that equals $\varphi^{*F}(\boldsymbol{V})$ up to a multiplicative constant as

$$\varphi^F(\boldsymbol{V}) = [E\{\varphi^{*F}(\boldsymbol{V})\boldsymbol{S}^{\mathit{eff}\,F'}(\boldsymbol{V})\}]^{-1}\varphi^{*F}(\boldsymbol{V}).$$

Lemma 4. For any $\varphi^{*F} \in \Lambda^{F\perp}$, let $\mathcal{K}^{-1}(\varphi^{*F})$ denote the space of elements $\tilde{\boldsymbol{h}}\{\boldsymbol{R}, \boldsymbol{G}_{\boldsymbol{R}}(\boldsymbol{V})\}$ in $\mathcal{H}$ such that

$$\mathcal{K}[\tilde{\boldsymbol{h}}\{\boldsymbol{R}, \boldsymbol{G}_{\boldsymbol{R}}(\boldsymbol{V})\}] = E[\tilde{\boldsymbol{h}}\{\boldsymbol{R}, \boldsymbol{G}_{\boldsymbol{R}}(\boldsymbol{V})\}|\boldsymbol{V}] = \varphi^{*F}(\boldsymbol{V}).$$

Let Λ_2 be the linear subspace of $\mathcal{H}$ consisting of elements $\boldsymbol{L}_2\{\boldsymbol{R}, \boldsymbol{G}_{\boldsymbol{R}}(\boldsymbol{V})\}$ such that

$$E[\boldsymbol{L}_2\{\boldsymbol{R}, \boldsymbol{G}_{\boldsymbol{R}}(\boldsymbol{V})\}|\boldsymbol{V}] = \boldsymbol{0};$$

that is, $\Lambda_2 = \mathcal{K}^{-1}(\boldsymbol{0})$. Then, for any element $\boldsymbol{h}\{\boldsymbol{R}, \boldsymbol{G}_{\boldsymbol{R}}(\boldsymbol{V})\}$ such that $\mathcal{K}(\boldsymbol{h}) = \varphi^{*F}(\boldsymbol{V})$,

$$\mathcal{K}^{-1}(\varphi^{*F}) = \boldsymbol{h}\{\boldsymbol{R}, \boldsymbol{G}_{\boldsymbol{R}}(\boldsymbol{V})\} + \Lambda_2, \tag{8.25}$$

where the right-hand side of (8.25) denotes the space with elements $\boldsymbol{h}\{\boldsymbol{R}, \boldsymbol{G}_{\boldsymbol{R}}(\boldsymbol{V})\} + \boldsymbol{L}_2\{\boldsymbol{R}, \boldsymbol{G}_{\boldsymbol{R}}(\boldsymbol{V})\}$ for all $\boldsymbol{L}_2 \in \Lambda_2$.

Proof. If $\tilde{\boldsymbol{h}}\{\boldsymbol{R}, \boldsymbol{G}_{\boldsymbol{R}}(\boldsymbol{V})\}$ is an element of the space $\boldsymbol{h}\{\boldsymbol{R}, \boldsymbol{G}_{\boldsymbol{R}}(\boldsymbol{V})\}+\Lambda_2$, then $\tilde{\boldsymbol{h}}\{\boldsymbol{R}, \boldsymbol{G}_{\boldsymbol{R}}(\boldsymbol{V})\} = \boldsymbol{h}\{\boldsymbol{R}, \boldsymbol{G}_{\boldsymbol{R}}(\boldsymbol{V})\} + \boldsymbol{L}_2\{\mathcal{R}, \boldsymbol{G}_{\boldsymbol{R}}(\boldsymbol{V})\}$ for some $\boldsymbol{L}_2 \in \Lambda_2$. It follows that

$$\begin{aligned} E[\tilde{\boldsymbol{h}}\{\boldsymbol{R}, \boldsymbol{G}_{\boldsymbol{R}}(\boldsymbol{V})\}|\boldsymbol{V}] &= E[h\{\boldsymbol{R}, \boldsymbol{G}_{\boldsymbol{R}}(\boldsymbol{V})\}|\boldsymbol{V}] + E[\boldsymbol{L}_2\{\boldsymbol{R}, \boldsymbol{G}_{\boldsymbol{R}}(\boldsymbol{V})\}|\boldsymbol{V}] \\ &= \varphi^{*F}(\boldsymbol{V}) + \boldsymbol{0} = \varphi^{*F}(\boldsymbol{V}). \end{aligned}$$

Conversely, if $E[\tilde{\boldsymbol{h}}\{\boldsymbol{R}, \boldsymbol{G}_{\boldsymbol{R}}(\boldsymbol{V})\}|\boldsymbol{V}] = \varphi^{*F}(\boldsymbol{V})$, then

$$\tilde{\boldsymbol{h}}\{\boldsymbol{R}, \boldsymbol{G}_{\boldsymbol{R}}(\boldsymbol{V})\} = \boldsymbol{h}\{\boldsymbol{R}, \boldsymbol{G}_{\boldsymbol{R}}(\boldsymbol{V})\} + [\tilde{\boldsymbol{h}}\{\boldsymbol{R}, \boldsymbol{G}_{\boldsymbol{R}}(\boldsymbol{V})\} - \boldsymbol{h}\{\boldsymbol{R}, \boldsymbol{G}_{\boldsymbol{R}}(\boldsymbol{V})\}],$$

where clearly $(\tilde{\boldsymbol{h}} - \boldsymbol{h}) \in \Lambda_2$.

The implication of Lemma 4 is that, in order to deduce $\Lambda_\eta^\perp = \mathcal{K}^{-1}(\Lambda^{F\perp})$, we must, for each $\varphi^{*F} \in \Lambda^{F\perp}$,

(i) identify one element $\boldsymbol{h}\{\boldsymbol{R}, \boldsymbol{G}_{\boldsymbol{R}}(\boldsymbol{V})\}$ in $\mathcal{H}$ such that $E[\boldsymbol{h}\{\boldsymbol{R}, \boldsymbol{G}_{\boldsymbol{R}}(\boldsymbol{V})\}|\boldsymbol{V}] = \varphi^{*F}(\boldsymbol{V})$, and

(ii) find $\Lambda_2 = \mathcal{K}^{-1}(\boldsymbol{0})$.

In the following theorem, the precise form of the elements in the space $\Lambda_\eta^\perp$ is derived by following these two steps.

Theorem 2. Assume that

$$E\{I(\boldsymbol{R}=1)|\boldsymbol{V}\} = \varpi(1,\boldsymbol{V}) > 0 \quad \text{for all } \boldsymbol{V} \text{ (a.e.)}, \tag{8.26}$$

where $\varpi(1,\boldsymbol{V}) = P\{\boldsymbol{R}=1|G_1(\boldsymbol{V})\} = P(\boldsymbol{R}=1|\boldsymbol{V})$, as $\boldsymbol{R}=1$ indicates that $\boldsymbol{V}$ is completely observed. Under (8.26), $\Lambda_\eta^\perp$ consists of all elements that can be written as

$$\frac{I(\boldsymbol{R}=1)\varphi^{*F}(\boldsymbol{V})}{\varpi(1,\boldsymbol{V})} + \tag{8.27}$$

$$\frac{I(\boldsymbol{R}=1)}{\varpi(1,\boldsymbol{V})}\left[\sum_{r\neq 1}\varpi\{r,\boldsymbol{G_r}(\boldsymbol{V})\}\boldsymbol{L_{2r}}\{\boldsymbol{G_r}(\boldsymbol{V})\}\right] - \sum_{r\neq 1} I(\boldsymbol{R}=r)\boldsymbol{L_{2r}}\{\boldsymbol{G_r}(\boldsymbol{V})\},$$

where, for $r \neq 1$, $\boldsymbol{L_{2r}}\{\boldsymbol{G_r}(\boldsymbol{V})\}$ is an arbitrary $(q \times 1)$ function of $\boldsymbol{G_r}(\boldsymbol{V})$, and $\varphi^{*F}(\boldsymbol{V})$ is an arbitrary element of $\Lambda^{F\perp}$.

Proof. We first address step (i) above and identify $\boldsymbol{h} \in \mathcal{H}$ such that $E[\boldsymbol{h}\{\boldsymbol{R},\boldsymbol{G_R}(\boldsymbol{V})\}|\boldsymbol{V}] = \varphi^{*F}(\boldsymbol{V})$ for $\varphi^{*F} \in \Lambda^{F\perp}$. Take $\boldsymbol{h}\{\boldsymbol{R},\boldsymbol{G_R}(\boldsymbol{V})\}$ to be

$$\frac{I(\boldsymbol{R}=1)\varphi^{*F}(\boldsymbol{V})}{\varpi(1,\boldsymbol{V})}.$$

From Section 8.5, this has the form of an IPWCC estimating function, and is clearly a function of the observed data. In fact,

$$E\left\{\frac{I(\boldsymbol{R}=1)\varphi^{*F}(\boldsymbol{V})}{\varpi(1,\boldsymbol{V})}\middle|\boldsymbol{V}\right\} = \frac{\varphi^{*F}(\boldsymbol{V})}{\varpi(1,\boldsymbol{V})}E\{I(\boldsymbol{R}=1)|\boldsymbol{V}\} = \varphi^{*F}(\boldsymbol{V}),$$

where assumption (8.26) ensures that no division of 0 by 0 can take place.
Hence, we have identified $\boldsymbol{h} \in \mathcal{H}$ satisfying (i). Following Lemma 4, $\Lambda_\eta^\perp = \mathcal{K}^{-1}(\Lambda^{F\perp})$ may be written as the direct sum of two linear subspaces of $\mathcal{H}$ given by

$$\Lambda_\eta^\perp = \Lambda_{IPWCC} \oplus \Lambda_2, \tag{8.28}$$

where Λ_{IPWCC} on the right-hand side of (8.28) is the linear subspace of $\mathcal{H}$ with elements

$$\frac{I(\boldsymbol{R}=1)\varphi^{*F}(\boldsymbol{V})}{\varpi(1,\boldsymbol{V})} \text{ for } \varphi^{*F} \in \Lambda^{F\perp}.$$

That is, $\Lambda_\eta^\perp$ is the linear subspace of $\mathcal{H}$

$$\Lambda_\eta^\perp = \left\{\frac{I(\boldsymbol{R}=1)\varphi^{*F}(\boldsymbol{V})}{\varpi(1,\boldsymbol{V})} + \boldsymbol{L}_2\{\boldsymbol{R},\boldsymbol{G_R}(\boldsymbol{V})\} : \varphi^{*F} \in \Lambda^{F\perp},\right.$$
$$\left. \boldsymbol{L}_2 \in \Lambda_2;\ \text{i.e., } E[\boldsymbol{L}_2\{\boldsymbol{R},\boldsymbol{G_R}(\boldsymbol{V})\}|\boldsymbol{V}] = \boldsymbol{0}\right\}. \tag{8.29}$$

To complete the proof, we address (ii) and derive the linear space $\Lambda_2 = \mathcal{K}^{-1}(\boldsymbol{0})$. Because $\boldsymbol{R}$ is discrete, any $\boldsymbol{h} \in \mathcal{H}$ may be expressed as

$$\boldsymbol{h}\{\boldsymbol{R},\boldsymbol{G_R}(\boldsymbol{V})\} = I(\boldsymbol{R}=1)\boldsymbol{h}_1(\boldsymbol{V}) + \sum_{r\neq 1} I(\boldsymbol{R}=r)\boldsymbol{h_r}\{\boldsymbol{G_r}(\boldsymbol{V})\},$$

where $\boldsymbol{h}_1(\boldsymbol{V})$ denotes an arbitrary $(q \times 1)$ function of $\boldsymbol{V}$, and $\boldsymbol{h_r}\{\boldsymbol{G_r}(\boldsymbol{V})\}$ denotes an arbitrary $(q \times 1)$ function of $\boldsymbol{G_r(V)}$. The space $\Lambda_2 \subset \mathcal{H}$ has elements $\boldsymbol{L}_2\{\boldsymbol{R}, \boldsymbol{G_R(V)}\}$ satisfying

$$E[\boldsymbol{L}_2\{\boldsymbol{R}, \boldsymbol{G_R(V)}\}|\boldsymbol{V}] = \boldsymbol{0},$$

which hence may be written equivalently

$$\begin{aligned} \boldsymbol{0} &= E\Big[I(\boldsymbol{R}=\boldsymbol{1})\boldsymbol{L}_{21}(\boldsymbol{V}) + \sum_{\boldsymbol{r}\neq\boldsymbol{1}} I(\boldsymbol{R}=r)\boldsymbol{L}_{2r}\{\boldsymbol{G_r(V)}\} \mid \boldsymbol{V}\Big] \\ &= \varpi(\boldsymbol{1},\boldsymbol{V})\boldsymbol{L}_{21}(\boldsymbol{V}) + \sum_{\boldsymbol{r}\neq\boldsymbol{1}} \varpi\{\boldsymbol{r}, \boldsymbol{G_r(V)}\}\boldsymbol{L}_{2r}\{\boldsymbol{G_r(V)}\} \end{aligned} \tag{8.30}$$

for arbitrary q-dimensional functions $\boldsymbol{L}_{2r}\{\boldsymbol{G_r(V)}\}$. Thus, an arbitrary element $\boldsymbol{L}_2 \in \Lambda_2$ may be characterized by defining any set of such functions and taking

$$\boldsymbol{L}_{21}(\boldsymbol{V}) = -\{\varpi(\boldsymbol{1},\boldsymbol{V})\}^{-1} \sum_{\boldsymbol{r}\neq\boldsymbol{1}} \varpi\{\boldsymbol{r}, \boldsymbol{G_r(V)}\}\boldsymbol{L}_{2r}\{\boldsymbol{G_r(V)}\},$$

satisfying (8.30). Here again, (8.26) is needed to guarantee no division by zero.

Hence, for any $\boldsymbol{L}_{2r}\{\boldsymbol{G_r(V)}\}$, $\boldsymbol{r} \neq \boldsymbol{1}$, we can define a typical element of Λ_2 as

$$\begin{aligned} &\frac{I(\boldsymbol{R}=\boldsymbol{1})}{\varpi(\boldsymbol{1},\boldsymbol{V})}\left[\sum_{\boldsymbol{r}\neq\boldsymbol{1}} \varpi\{\boldsymbol{r}, \boldsymbol{G_r(V)}\}\boldsymbol{L}_{2r}\{\boldsymbol{G_r(V)}\}\right] \\ &\qquad - \sum_{\boldsymbol{r}\neq\boldsymbol{1}} I(\boldsymbol{R}=\boldsymbol{r})\boldsymbol{L}_{2r}\{\boldsymbol{G_r(V)}\}. \end{aligned} \tag{8.31}$$

Combining (8.29) and (8.31), the space $\Lambda_\eta^\perp$ may be explicitly defined as that consisting of all elements given by (8.27).

Armed with this representation of elements contained in the orthogonal complement of the observed-data nuisance tangent space $\Lambda_\eta^\perp$, we now consider how it may be used to identify RAL estimators for $\boldsymbol{\beta}$ based on the observed data. As discussed previously, estimators are often derived by deducing unbiased estimating functions associated with elements that are orthogonal to the nuisance tangent space.

Unbiased estimating functions leading to RAL estimators for $\boldsymbol{\beta}$ when there are no missing data, which we have denoted as $\boldsymbol{m}(\boldsymbol{V}, \boldsymbol{\beta})$, are associated with elements of $\Lambda^{F\perp}$. Because the first space in the direct sum in (8.28), Λ_{IPWCC}, consists of inverse probability weighted complete-case elements of $\Lambda^{F\perp}$, this suggests that observed-data estimators for $\boldsymbol{\beta}$ can be constructed based on inverse weighted full-data estimating functions. This may be accomplished in several ways.

The second space in the direct sum in (8.28), Λ_2, is referred to as the *augmentation space*, with elements of the form (8.31). Taking the $(q \times 1)$ functions $\boldsymbol{L}_{2r}\{\boldsymbol{G_r(V)}\} \equiv \boldsymbol{0}$, $\boldsymbol{r} \neq \boldsymbol{1}$, leads to IPWCC estimators, of which we presented examples in Section 8.5. In particular, given a full-data estimating function $\boldsymbol{m}(\boldsymbol{V}, \boldsymbol{\beta})$, the corresponding observed data estimating function is $I(\boldsymbol{R} = \boldsymbol{1})\boldsymbol{m}(\boldsymbol{V}, \boldsymbol{\beta})/\varpi(\boldsymbol{1}, \boldsymbol{V})$, and the IPWCC observed-data estimator for $\boldsymbol{\beta}$ is the solution to the estimating equation

$$\sum_{i=1}^N \frac{I(\boldsymbol{R}_i = \boldsymbol{1})\boldsymbol{m}(\boldsymbol{V}_i, \boldsymbol{\beta})}{\varpi(\boldsymbol{1}, \boldsymbol{V}_i)} = \boldsymbol{0}.$$

Other specifications for $\boldsymbol{L}_{2r}\{\boldsymbol{G_r(V)}\}$, $\boldsymbol{r} \neq \boldsymbol{1}$, incorporate non-zero elements of the augmentation space Λ_2, and lead to AIPWCC estimators for $\boldsymbol{\beta}$, as exemplified in Section 8.5. As discussed there, such estimators can yield increased efficiency over IPWCC estimators and, in some cases, enjoy the property of double robustness.

Specifically, given a full-data estimating function $\boldsymbol{m}(\boldsymbol{V}, \boldsymbol{\beta})$ and the form of a typical element of Λ_2 in (8.31) for arbitrary $\boldsymbol{L}_{2r}\{\boldsymbol{G}_r(\boldsymbol{V})\}$, $\boldsymbol{r} \neq \mathbf{1}$, an AIPWCC observed-data estimator for $\boldsymbol{\beta}$ is the solution to an estimating equation having the general form

$$\sum_{i=1}^{N} \left(\frac{I(\boldsymbol{R}_i = \mathbf{1})\boldsymbol{m}(\boldsymbol{V}_i, \boldsymbol{\beta})}{\varpi(\mathbf{1}, \boldsymbol{V}_i)} + \frac{I(\boldsymbol{R}_i = \mathbf{1})}{\varpi(\mathbf{1}, \boldsymbol{V}_i)} \left[\sum_{\boldsymbol{r} \neq \mathbf{1}} \varpi\{\boldsymbol{r}, \boldsymbol{G}_{\boldsymbol{r}}(\boldsymbol{V}_i)\} \boldsymbol{L}_{2\boldsymbol{r}}\{\boldsymbol{G}_{\boldsymbol{r}}(\boldsymbol{V}_i)\} \right] - \sum_{\boldsymbol{r} \neq \mathbf{1}} I(\boldsymbol{R}_i = \boldsymbol{r}) \boldsymbol{L}_{2\boldsymbol{r}}\{\boldsymbol{G}_{\boldsymbol{r}}(\boldsymbol{V}_i)\} \right) = \mathbf{0}. \tag{8.32}$$

The foregoing development shows that semi-parametric theory leads to RAL estimators for $\boldsymbol{\beta}$ based on the observed data when missingness is MAR that are solutions to the AIPWCC estimating equations of the form (8.32). To construct such an estimator, one must specify a full-data estimating function $\boldsymbol{m}(\boldsymbol{V}, \boldsymbol{\beta})$ and functions $\boldsymbol{L}_{2\boldsymbol{r}}\{\boldsymbol{G}_{\boldsymbol{r}}(\boldsymbol{V})\}$ for $\boldsymbol{r} \neq \mathbf{1}$. As we discuss in the next section, these choices have implications for the efficiency of the resulting estimator for $\boldsymbol{\beta}$.

Before we address efficiency, it is useful to demonstrate explicitly this formulation for two common missingness structures.

8.7.3 Augmentation space Λ_2 with two levels of missingness

Examples of two levels of missingness were given in Section 8.5. In particular, with $\boldsymbol{V} = (\boldsymbol{V}^{(1)'}, \boldsymbol{V}^{(2)'})'$, $\boldsymbol{V}^{(1)}$ is always observed whereas $\boldsymbol{V}^{(2)}$ is missing for some individuals, so that the possible values of $\boldsymbol{R}$ are $\boldsymbol{r} = (1,1)$ and $(1,0)$. Letting $R = I(\boldsymbol{R} = \mathbf{1})$ as in that section, $I(\boldsymbol{R} \neq \mathbf{1}) = I\{\boldsymbol{R} = (1,0)\} = (1 - R)$, $\boldsymbol{G}_{(1,0)}(V) = \boldsymbol{V}^{(1)}$, and $\varpi\{(1,1), \boldsymbol{V}\} = 1 - \varpi\{(1,0), \boldsymbol{V}^{(1)}\} = \varpi(\boldsymbol{V}^{(1)})$, say, after some algebra, a typical element in Λ_2 given in (8.31) may be written as

$$\left\{ \frac{R - \varpi(\boldsymbol{V}^{(1)})}{\varpi(\boldsymbol{V}^{(1)})} \right\} \boldsymbol{L}_2(\boldsymbol{V}^{(1)}), \tag{8.33}$$

where $\boldsymbol{L}_2(\boldsymbol{V}^{(1)})$ is an arbitrary function of $\boldsymbol{V}^{(1)}$.

Accordingly, a general semi-parametric observed-data RAL estimator for $\boldsymbol{\beta}$ is the solution to an estimating equation of the form

$$\sum_{i=1}^{N} \left[\frac{R_i \boldsymbol{m}(\boldsymbol{V}_i, \boldsymbol{\beta})}{\varpi(\boldsymbol{V}_i^{(1)})} + \left\{ \frac{R_i - \varpi(\boldsymbol{V}_i^{(1)})}{\varpi(\boldsymbol{V}_i^{(1)})} \right\} \boldsymbol{L}_2(\boldsymbol{V}_i^{(1)}) \right] = \mathbf{0}, \tag{8.34}$$

where $\boldsymbol{m}(\boldsymbol{V}, \boldsymbol{\beta})$ is an arbitrary full-data estimating function and $\boldsymbol{L}_2(\boldsymbol{V}^{(1)})$ is an arbitrary function of $\boldsymbol{V}^{(1)}$). This result is the basis for the classes of estimators solving (8.8) and (8.16) in Examples 1 and 2, respectively, in Section 8.5.

8.7.4 Augmentation space Λ_2 with monotone missingness

A monotone pattern of missingness is common in practice. Writing the full data as $\boldsymbol{V} = (\boldsymbol{V}^{(1)'}, \ldots, \boldsymbol{V}^{(K)'})'$, under monotone missingness, if $\boldsymbol{V}^{(j)}$ is observed, then

$(\boldsymbol{V}^{(1)}, \ldots, \boldsymbol{V}^{(j-1)})$ are also observed for $j = 2, \ldots, K$. Thus, the vector of missing data indicators $\boldsymbol{R}$ takes on possible values $\boldsymbol{r}$ in $\{(1, 0, \ldots, 0), (1, 1, 0, \ldots, 0), \ldots, (1, \ldots, 1)\}$. Such a pattern of missingness often is seen in longitudinal studies in which the intent is to collect data at K pre-specified time points. Here, subjects may drop out during the course of the study, in which case data are observed for subjects who drop out up to the time of drop out, with data for all subsequent time points missing.

Under monotone missingness, it is convenient to represent the level of missingness using the scalar variable $D = \sum_{j=1}^{K} R_j$, so that if $D = j$, then $(\boldsymbol{V}^{(1)}, \ldots, \boldsymbol{V}^{(j)})$ is observed, and the complete-case indicator $I(\boldsymbol{R} = \boldsymbol{1}) = I(D = K)$. Define also the many-to-one function $\boldsymbol{G}_D(\boldsymbol{V}) = (\boldsymbol{V}^{(1)\prime}, \ldots, \boldsymbol{V}^{(D)\prime})'$.

When data are MAR, we have defined generically missingness probabilities as $P(\boldsymbol{R} = \boldsymbol{r}|\boldsymbol{V}) = \varpi\{\boldsymbol{r}, \boldsymbol{G_r}(\boldsymbol{V})\}$. Under monotone missingness, it is more convenient to define missingness probabilities using the discrete hazard function

$$\lambda_d(\boldsymbol{V}^{(1)}, \ldots, \boldsymbol{V}^{(d)}) = P(D = d | D \geq d, \boldsymbol{V}), \;\; d = 1, \ldots, K-1.$$

Defining

$$H_d(\boldsymbol{V}^{(1)}, \ldots, \boldsymbol{V}^{(d)}) = P(D > d|\boldsymbol{V}) = \prod_{j=1}^{d} \{1 - \lambda_j(\boldsymbol{V}^{(1)}, \ldots, \boldsymbol{V}^{(j)})\},$$

the probability of observing a complete case is given by

$$P(\boldsymbol{R} = \boldsymbol{1}|\boldsymbol{V}) = P(D > K-1|\boldsymbol{V}) = H_{K-1}(\boldsymbol{V}^{(1)}, \ldots, \boldsymbol{V}^{(K-1)}).$$

Using this notation, it is shown in Theorem 9.2 of Tsiatis (2006) that, under monotone missingness, a typical element of Λ_2 is

$$\sum_{j=1}^{K-1} \left[\frac{I(D = j) - \lambda_j(\boldsymbol{V}^{(1)}, \ldots, \boldsymbol{V}^{(j)}) I(D \geq j)}{H_j(\boldsymbol{V}^{(1)}, \ldots, \boldsymbol{V}^{(j)})} \right] \boldsymbol{L}_j(\boldsymbol{V}^{(1)}, \ldots, \boldsymbol{V}^{(j)}), \tag{8.35}$$

where $\boldsymbol{L}_j(\boldsymbol{V}^{(1)}, \ldots, \boldsymbol{V}^{(j)})$ denotes an arbitrary function of $(\boldsymbol{V}^{(1)}, \ldots, \boldsymbol{V}^{(j)})$, for $j = 1, \ldots, K-1$. Consequently, general semi-parametric observed-data estimators for $\boldsymbol{\beta}$ are given as solutions to the estimating equations of the form

$$\sum_{i=1}^{N} \left(\frac{I(D_i = K)\boldsymbol{m}(\boldsymbol{V}_i, \boldsymbol{\beta})}{H_{K-1}(\boldsymbol{V}_i^{(1)}, \ldots, \boldsymbol{V}_i^{(K-1)})} \right.$$
$$\left. + \sum_{j=1}^{K-1} \left[\frac{I(D_i = j) - \lambda_j(\boldsymbol{V}_i^{(1)}, \ldots, \boldsymbol{V}_i^{(j)}) I(D_i \geq j)}{H_j(\boldsymbol{V}_i^{(1)}, \ldots, \boldsymbol{V}_i^{(j)})} \right] \boldsymbol{L}_j(\boldsymbol{V}_i^{(1)}, \ldots, \boldsymbol{V}_i^{(j)}) \right) = \boldsymbol{0}$$

for an arbitrary full-data estimating function $\boldsymbol{m}(\boldsymbol{V}, \boldsymbol{\beta})$ and arbitrary functions $\boldsymbol{L}_j(\boldsymbol{V}^{(1)}, \ldots, \boldsymbol{V}^{(j)})$, $j = 1, \ldots, K-1$.

8.8 Optimal Observed-Data Estimating Function Associated with Full-Data Estimating Function

As demonstrated in the last section, identifying RAL observed-data estimators for $\boldsymbol{\beta}$ requires specification of a full-data estimating function $\boldsymbol{m}(\boldsymbol{V}, \boldsymbol{\beta})$ and functions $\boldsymbol{L}_{2\boldsymbol{r}}\{\boldsymbol{G_r}(\boldsymbol{V})\}$ for

$r \neq 1$. Ideally, we would like to characterize the semi-parametric efficient observed-data estimator for $\boldsymbol{\beta}$, which corresponds to the efficient influence function and achieves the semi-parametric efficiency bound, and deduce associated estimating equations that may be solved in practice. This entails identifying the optimal choices for $\boldsymbol{m}(\boldsymbol{V}, \boldsymbol{\beta})$ and $\boldsymbol{L}_{2r}\{\boldsymbol{G}_r(\boldsymbol{V})\}$. One would hope that the full-data estimating function $\boldsymbol{m}(\boldsymbol{V}, \boldsymbol{\beta})$ leading to the optimal full-data estimator would also be that leading to the optimal observed-data estimator. Unfortunately, this is not necessarily the case. Deriving the optimal choice of $\boldsymbol{m}(\boldsymbol{V}, \boldsymbol{\beta})$ is very challenging in all but the simplest problems and leads to estimators that are generally infeasible to implement in practice. See Section 8.10 and Chapter 11 of Tsiatis (2006) for further discussion.

Accordingly, even though the optimal estimator may be practically infeasible, the foregoing development using semi-parametric theory is useful for guiding construction of estimators that achieve high efficiency and may be implemented in practice. We now consider how to deduce such estimators based on a fixed, specified full-data estimating function $\boldsymbol{m}(\boldsymbol{V}, \boldsymbol{\beta})$.

Equation (8.32) provides the general form of an m-estimator for $\boldsymbol{\beta}$ using the estimating function

$$\frac{I(\boldsymbol{R}=1)\boldsymbol{m}(\boldsymbol{V},\boldsymbol{\beta})}{\varpi(1,\boldsymbol{V})} + \boldsymbol{L}_2\{\boldsymbol{R},\boldsymbol{G}_{\boldsymbol{R}}(\boldsymbol{V})\}, \tag{8.36}$$

where $\boldsymbol{L}_2\{\boldsymbol{R},\boldsymbol{G}_{\boldsymbol{R}}(\boldsymbol{V})\}$ is an element of Λ_2. Consider the class of observed-data estimating function of the form (8.36) for a fixed full-data estimating function $\boldsymbol{m}(\boldsymbol{V}, \boldsymbol{\beta})$. Given this fixed choice, a key issue is the corresponding choice of $\boldsymbol{L}_2\{\boldsymbol{R},\boldsymbol{G}_{\boldsymbol{R}}(\boldsymbol{V})\}$ in Λ_2 leading to the optimal estimator within this class.

From (8.6), the corresponding influence function for this m-estimator is

$$\begin{aligned}
\boldsymbol{\varphi}\{\boldsymbol{R},\boldsymbol{G}_{\boldsymbol{R}}(\boldsymbol{V}),\boldsymbol{\beta}\} &= \left(E\left[\frac{I(\boldsymbol{R}=1)\partial/\partial\boldsymbol{\beta}'\{\boldsymbol{m}(\boldsymbol{V},\boldsymbol{\beta})\}}{\varpi(1,\boldsymbol{V})}\right]\right)^{-1} \\
&\quad \times \left[\frac{I(\boldsymbol{R}=1)\boldsymbol{m}(\boldsymbol{V},\boldsymbol{\beta})}{\varpi(1,\boldsymbol{V})} + \boldsymbol{L}_2\{\boldsymbol{R},\boldsymbol{G}_{\boldsymbol{R}}(\boldsymbol{V})\}\right] \\
&= \left[E\left\{\frac{\partial \boldsymbol{m}(\boldsymbol{V},\boldsymbol{\beta})}{\partial\boldsymbol{\beta}'}\right\}\right]^{-1}\left[\frac{I(\boldsymbol{R}=1)\boldsymbol{m}(\boldsymbol{V},\boldsymbol{\beta})}{\varpi(1,\boldsymbol{V})} + \boldsymbol{L}_2\{\boldsymbol{R},\boldsymbol{G}_{\boldsymbol{R}}(\boldsymbol{V})\}\right].
\end{aligned} \tag{8.37}$$

The covariance matrix of the corresponding m-estimator is the covariance matrix of its influence function (8.37). Consequently, the optimal estimator within this class is found by choosing $\boldsymbol{L}_2 \in \Lambda_2$ to be the element $\boldsymbol{L}_2^{opt}$, say, such that the covariance matrix of

$$\left[\frac{I(\boldsymbol{R}=1)\boldsymbol{m}(\boldsymbol{V},\boldsymbol{\beta})}{\varpi(1,\boldsymbol{V})} + \boldsymbol{L}_2^{opt}\{\boldsymbol{R},\boldsymbol{G}_{\boldsymbol{R}}(\boldsymbol{V})\}\right]$$

is smaller than that for any other $\boldsymbol{L}_2 \in \Lambda_2$ in the sense of non-negative definiteness. Because Λ_2 is a closed linear subspace of the observed-data Hilbert space $\mathcal{H}$, by the properties of projections reviewed in Section 8.4, the element of Λ_2 that achieves this is the projection of

$$\frac{I(\boldsymbol{R}=1)\boldsymbol{m}(\boldsymbol{V},\boldsymbol{\beta})}{\varpi(1,\boldsymbol{V})}$$

onto Λ_2. That is, the optimal choice for $\boldsymbol{L}_2\{\boldsymbol{R},\boldsymbol{G}_{\boldsymbol{R}}(\boldsymbol{V})\}$ is

$$\boldsymbol{L}_2^{opt}\{\boldsymbol{R},\boldsymbol{G}_{\boldsymbol{R}}(\boldsymbol{V})\} = -\boldsymbol{\Pi}\left\{\left.\frac{I(\boldsymbol{R}=1)\boldsymbol{m}(\boldsymbol{V},\boldsymbol{\beta})}{\varpi(1,\boldsymbol{V})}\right|\Lambda_2\right\} \tag{8.38}$$

and is unique. In general, the projection onto Λ_2 cannot be obtained in a closed form.

However, when there are two levels of missingness or monotone missingness, closed form solutions do exist, which we now present.

Two levels of missingness. Consider the situation in Section 8.7.3. When $\boldsymbol{V} = (\boldsymbol{V}^{(1)'}, \boldsymbol{V}^{(2)'})'$, $\boldsymbol{V}^{(1)}$ is available for all individuals, and $\boldsymbol{V}^{(2)}$ may be missing, we showed that a typical element of Λ_2 may be written as in (8.33), that is, as

$$\left\{\frac{R - \varpi(\boldsymbol{V}^{(1)})}{\varpi(\boldsymbol{V}^{(1)})}\right\} \boldsymbol{L}_2(\boldsymbol{V}^{(1)}),$$

where $\boldsymbol{L}_2(\boldsymbol{V}^{(1)})$ is an arbitrary function of $\boldsymbol{V}^{(1)}$. The projection

$$\boldsymbol{\Pi}\left\{\left.\frac{R\boldsymbol{m}(\boldsymbol{V},\boldsymbol{\beta})}{\varpi(\boldsymbol{V}^{(1)})}\right| \Lambda_2\right\}$$

is the unique element of this form in Λ_2 such that the associated residual

$$\frac{R\boldsymbol{m}(\boldsymbol{V},\boldsymbol{\beta})}{\varpi(\boldsymbol{V}^{(1)})} - \left\{\frac{R - \varpi(\boldsymbol{V}^{(1)})}{\varpi(\boldsymbol{V}^{(1)})}\right\} \boldsymbol{L}_2^0(\boldsymbol{V}^{(1)}),$$

say, is orthogonal to every element in Λ_2. This requires that

$$E\left(\left[\frac{R\boldsymbol{m}(\boldsymbol{V},\boldsymbol{\beta})}{\varpi(\boldsymbol{V}^{(1)})} - \left\{\frac{R - \varpi(\boldsymbol{V}^{(1)})}{\varpi(\boldsymbol{V}^{(1)})}\right\} \boldsymbol{L}_2^0(\boldsymbol{V}^{(1)})\right]' \times \left\{\frac{R - \varpi(\boldsymbol{V}^{(1)})}{\varpi(\boldsymbol{V}^{(1)})}\right\} \boldsymbol{L}_2(\boldsymbol{V}^{(1)})\right) = 0$$

for all functions $\boldsymbol{L}_2(\boldsymbol{V}^{(1)})$. Using the law of iterated conditional expectations, it may be shown (Tsiatis, 2006, Theorem 10.2) that this is equivalent to requiring that

$$E\left[\left\{\frac{1 - \varpi(\boldsymbol{V}^{(1)})}{\varpi(\boldsymbol{V}^{(1)})}\right\}\left\{\boldsymbol{m}(\boldsymbol{V},\boldsymbol{\beta}) - \boldsymbol{L}_2^0(\boldsymbol{V}^{(1)})\right\}' \boldsymbol{L}_2(\boldsymbol{V}^{(1)})\right] = E\left(\left\{\frac{1 - \varpi(\boldsymbol{V}^{(1)})}{\varpi(\boldsymbol{V}^{(1)})}\right\}\left[E\{\boldsymbol{m}(\boldsymbol{V},\boldsymbol{\beta})|\boldsymbol{V}^{(1)}\} - \boldsymbol{L}_2^0(\boldsymbol{V}^{(1)})\right]' \boldsymbol{L}_2(\boldsymbol{V}^{(1)})\right) \tag{8.39}$$

be equal to zero for all $\boldsymbol{L}_2(\boldsymbol{V}^{(1)})$. Because $\varpi(\boldsymbol{V}^{(1)}) > 0$ for all $\boldsymbol{V}^{(1)}$ a.e. by assumption, the first term inside the expectation in (8.39) is bounded away from 0 and ∞. It follows that (8.39) is equal to zero for all $\boldsymbol{L}_2(\boldsymbol{V}^{(1)})$ if and only if $\boldsymbol{L}_2^0(\boldsymbol{V}^{(1)}) = E\{\boldsymbol{m}(\boldsymbol{V},\boldsymbol{\beta})|\boldsymbol{V}^{(1)}\}$.

Thus, with two levels of missingness, the optimal estimating function within the class of estimating functions (8.36) is found by choosing

$$\boldsymbol{L}_2\{\boldsymbol{R}, \boldsymbol{G}_{\boldsymbol{R}}(\boldsymbol{V}) = -\left\{\frac{R - \varpi(\boldsymbol{V}^{(1)})}{\varpi(\boldsymbol{V}^{(1)})}\right\} E\{\boldsymbol{m}(\boldsymbol{V},\boldsymbol{\beta})|\boldsymbol{V}^{(1)}\}.$$

Although interesting, this result is of little practical use, as the form of $E\{\boldsymbol{m}(\boldsymbol{V},\boldsymbol{\beta})|\boldsymbol{V}^{(1)}\}$ is unknown. However, it does suggest an adaptive approach in which the data are used to estimate this conditional expectation, as follows.

As in Example 1 of Section 8.5, one may posit a parametric model for the conditional expectation $E\{\boldsymbol{m}(\boldsymbol{V}, \boldsymbol{\beta})|\boldsymbol{V}^{(1)}\}$. Consider the model $\boldsymbol{\phi}(\boldsymbol{V}^{(1)}, \boldsymbol{\beta}, \boldsymbol{\psi})$, which depends on both $\boldsymbol{\beta}$ and an additional finite-dimensional parameter $\boldsymbol{\psi}$. Because of the MAR assumption, $E\{\boldsymbol{m}(\boldsymbol{V}, \boldsymbol{\beta})|\boldsymbol{V}^{(1)}\} = E\{\boldsymbol{m}(\boldsymbol{V}, \boldsymbol{\beta})|\boldsymbol{V}^{(1)}, R = 1\}$, and, the complete (full) data $\{i : R_i = 1\}$ may be used to fit this model. If $\boldsymbol{\beta}$ were known, $\boldsymbol{\psi}$ could be estimated by solving

$$\sum_{i=1}^{N} R_i \left\{ \frac{\partial \boldsymbol{\phi}(\boldsymbol{V}_i^{(1)}, \boldsymbol{\beta}, \boldsymbol{\psi})}{\partial \boldsymbol{\psi}} \right\} \{\boldsymbol{m}(\boldsymbol{V}_i, \boldsymbol{\beta}) - \boldsymbol{\phi}(\boldsymbol{V}_i^{(1)}, \boldsymbol{\beta}, \boldsymbol{\psi})\} = \boldsymbol{0}. \tag{8.40}$$

This suggests obtaining an adaptive estimator for $\boldsymbol{\beta}$ by simultaneously estimating $\boldsymbol{\beta}$ and $\boldsymbol{\psi}$ by solving the stacked estimating equation (8.40) jointly with the equation

$$\sum_{i=1}^{N} \left[\frac{R_i \boldsymbol{m}(\boldsymbol{V}_i, \boldsymbol{\beta})}{\varpi(\boldsymbol{V}_i^{(1)})} - \left\{ \frac{R_i - \varpi(\boldsymbol{V}_i^{(1)})}{\varpi(\boldsymbol{V}_i^{(1)})} \right\} \boldsymbol{\phi}(\boldsymbol{V}_i^{(1)}, \boldsymbol{\beta}, \boldsymbol{\psi}) \right] = \boldsymbol{0}. \tag{8.41}$$

Because (8.40) and (8.41) together determine an m-estimator for $\boldsymbol{\beta}$ and $\boldsymbol{\psi}$, the asymptotic covariance matrix of the resulting estimator may be derived using the standard sandwich estimator (8.7) for m-estimators.

Here, if the model $\boldsymbol{\phi}(\boldsymbol{V}_i^{(1)}, \boldsymbol{\beta}, \boldsymbol{\psi})$ for $E\{\boldsymbol{m}(\boldsymbol{V}, \boldsymbol{\beta})|\boldsymbol{V}^{(1)}\}$ is correctly specified, then the resulting estimator for $\boldsymbol{\beta}$ is the efficient estimator within the class (8.36) based on the fixed full-data estimating function $\boldsymbol{m}(\boldsymbol{V}, \boldsymbol{\beta})$. If, on the other hand, the model for $E\{\boldsymbol{m}(\boldsymbol{V}, \boldsymbol{\beta})|\boldsymbol{V}^{(1)}\}$ is misspecified, the resulting estimator $\widehat{\boldsymbol{\psi}}$ converges to some $\boldsymbol{\psi}^*$. In this case, the resulting estimator for $\boldsymbol{\beta}$ is consistent and asymptotically normal and asymptotically equivalent to the estimator solving

$$\sum_{i=1}^{N} \left[\frac{R_i \boldsymbol{m}(\boldsymbol{V}_i, \boldsymbol{\beta})}{\varpi(\boldsymbol{V}_i^{(1)})} - \left\{ \frac{R_i - \varpi(\boldsymbol{V}_i^{(1)})}{\varpi(\boldsymbol{V}_i^{(1)})} \right\} \boldsymbol{\phi}(\boldsymbol{V}_i^{(1)}, \boldsymbol{\beta}, \boldsymbol{\psi}^*) \right] = \boldsymbol{0}$$

as long as the model for missingness $P(R = 1|\boldsymbol{V}) = \varpi(V^{(1)})$ is correctly specified. However, it is relatively less efficient than that obtained using the correct model.

Monotone missingness. Consider the situation of Section 8.7.4. As shown in that section, with monotone missingness, the class of estimating functions given by (8.36) is

$$\frac{I(D = K)\boldsymbol{m}(\boldsymbol{V}, \boldsymbol{\beta})}{H_{K-1}(\boldsymbol{V}^{(1)}, \ldots, \boldsymbol{V}^{(K-1)})} + \sum_{j=1}^{K-1} \left[\frac{I(D = j) - \lambda_j(\boldsymbol{V}^{(1)}, \ldots, \boldsymbol{V}^{(j)}) I(D \geq j)}{H_j(\boldsymbol{V}^{(1)}, \ldots, \boldsymbol{V}^{(j)})} \right] \boldsymbol{L}_j(\boldsymbol{V}^{(1)}, \ldots, \boldsymbol{V}^{(j)}),$$

where the first term is the IPWCC estimating function, and the second term represents a typical element in Λ_2. Then, by Theorem 10.4 of Tsiatis (2006), the projection of

$$\frac{I(D = K)\boldsymbol{m}(\boldsymbol{V}, \boldsymbol{\beta})}{H_{K-1}(\boldsymbol{V}^{(1)}, \ldots, \boldsymbol{V}^{(K-1)})}$$

onto Λ_2 is obtained by taking

$$\boldsymbol{L}_j(\boldsymbol{V}^{(1)}, \ldots, \boldsymbol{V}^{(j)}) = -E\{\boldsymbol{m}(\boldsymbol{V}, \boldsymbol{\beta})|\boldsymbol{V}^{(1)}, \ldots, \boldsymbol{V}^{(j)}\}.$$

Thus, as in the case of two levels of missingness, the optimal choices of $\boldsymbol{L}_j(\boldsymbol{V}^{(1)}, \ldots, \boldsymbol{V}^{(j)})$,

$j = 1, \ldots, K-1$, within the class (8.36) are conditional expectations whose forms are unknown. Again, it is possible to consider an adaptive approach in which models are posited for these conditional expectations in terms of a parameter $\boldsymbol{\psi}$, which is then estimated from the data. One way of doing this was suggested by Bang and Robins (2005) and Tsiatis, Davidian, and Cao (2011). Specifically, consider models

$$\boldsymbol{\phi}_j(\boldsymbol{V}^{(1)}, \ldots, \boldsymbol{V}^{(j)}, \boldsymbol{\beta}, \boldsymbol{\psi})$$

for $E\{\boldsymbol{m}(\boldsymbol{V}, \boldsymbol{\beta}) | \boldsymbol{V}^{(1)}, \ldots, \boldsymbol{V}^{(j)}\}$, $j = 1, \ldots, K-1$, where $\boldsymbol{\psi}$ represents the vector of all parameters across all j. This allows separate parameterizations for each j; that is, $E\{\boldsymbol{m}(\boldsymbol{V}, \boldsymbol{\beta}) | \boldsymbol{V}^{(1)}, \ldots, \boldsymbol{V}^{(j)}\}$ modeled as $\boldsymbol{\phi}_j(\boldsymbol{V}^{(1)}, \ldots, \boldsymbol{V}^{(j)}, \boldsymbol{\beta}, \boldsymbol{\psi}_j)$, in which case $\boldsymbol{\psi} = (\boldsymbol{\psi}_1', \ldots, \boldsymbol{\psi}_{K-1}')'$; or models where parameters may be shared across j. Estimation of $\boldsymbol{\psi}$ takes advantage of the fact that, if these models were correct, then

$$\begin{aligned} \boldsymbol{\phi}_j(\boldsymbol{V}^{(1)}, \ldots, \boldsymbol{V}^{(j)}, \boldsymbol{\beta}, \boldsymbol{\psi}) &= E\{\boldsymbol{m}(\boldsymbol{V}, \boldsymbol{\beta}) | \boldsymbol{V}^{(1)}, \ldots, \boldsymbol{V}^{(j)}\} \\ &= E\{\boldsymbol{\phi}_{j+1}(\boldsymbol{V}^{(1)}, \ldots, \boldsymbol{V}^{(j+1)}, \boldsymbol{\beta}, \boldsymbol{\psi}) | \boldsymbol{V}^{(1)}, \ldots, \boldsymbol{V}^{(j)}, \boldsymbol{\beta}, \boldsymbol{\psi}), D > j\}. \end{aligned} \tag{8.42}$$

Equation (8.42) holds by the law of iterated conditional expectations and the MAR assumption.

Thus, to estimate $\boldsymbol{\beta}$ and $\boldsymbol{\psi}$, consider the stacked estimating functions

$$\begin{aligned} &\frac{I(D = K)\boldsymbol{m}(\boldsymbol{V}, \boldsymbol{\beta})}{H_{K-1}(\boldsymbol{V}^{(1)}, \ldots, \boldsymbol{V}^{(K-1)})} \\ &\quad + \sum_{j=1}^{K-1} \left[\frac{I(D = j) - \lambda_j(\boldsymbol{V}^{(1)}, \ldots, \boldsymbol{V}^{(j)}) I(D \geq j)}{H_j(\boldsymbol{V}^{(1)}, \ldots, \boldsymbol{V}^{(j)})} \right] \boldsymbol{\phi}_j(\boldsymbol{V}^{(1)}, \ldots, \boldsymbol{V}^{(j)}, \boldsymbol{\beta}, \boldsymbol{\psi}) \end{aligned} \tag{8.43}$$

and

$$\begin{aligned} &\sum_{j=1}^{K-1} I(D > j) \left[\frac{\partial \boldsymbol{\phi}_j(\boldsymbol{V}^{(1)}, \ldots, \boldsymbol{V}^{(j)})}{\partial \boldsymbol{\psi}'} \right]' \\ &\qquad \times \left\{ \boldsymbol{\phi}_{j+1}(\boldsymbol{V}^{(1)}, \ldots, \boldsymbol{V}^{(j+1)}, \boldsymbol{\beta}, \boldsymbol{\psi}) - \boldsymbol{h}_j(\boldsymbol{V}^{(1)}, \ldots, \boldsymbol{V}^{(j)}, \boldsymbol{\beta}, \boldsymbol{\psi}) \right\}, \end{aligned} \tag{8.44}$$

where $\boldsymbol{\phi}_K(\boldsymbol{V}^{(1)}, \ldots, \boldsymbol{V}^{(K)}, \boldsymbol{\beta}, \boldsymbol{\psi})$ is defined to be $\boldsymbol{m}(\boldsymbol{V}, \boldsymbol{\beta})$. Denoting the stacked estimating function given jointly by (8.43) and (8.44) as $\widetilde{\boldsymbol{m}}(D, \boldsymbol{V}^{(1)}, \ldots, \boldsymbol{V}^{(D)}, \boldsymbol{\beta}, \boldsymbol{\psi})$, $\boldsymbol{\beta}$ and $\boldsymbol{\psi}$ can be estimated by solving jointly in $\boldsymbol{\beta}$ and $\boldsymbol{\psi}$

$$\sum_{i=1}^{N} \widetilde{\boldsymbol{m}}(D_i, \boldsymbol{V}_i^{(1)}, \ldots, \boldsymbol{V}_i^{(D)}, \boldsymbol{\beta}, \boldsymbol{\psi}) = \boldsymbol{0},$$

and the corresponding asymptotic covariance matrix can be estimated using the sandwich estimator (8.7).

As long as the discrete hazard functions $\lambda_j(\boldsymbol{V}^{(1)}, \ldots, \boldsymbol{V}^{(j)})$, $j = 1, \ldots, K-1$, are correctly specified, the estimator for $\boldsymbol{\beta}$ so obtained is consistent and asymptotically normal, even if the models $\boldsymbol{\phi}_j(\boldsymbol{V}^{(1)}, \ldots, \boldsymbol{V}^{(j)}, \boldsymbol{\beta}, \boldsymbol{\psi})$ are misspecified, and will be optimal within the class (8.36) if these are correct.

Remark. The case of two levels of missingness is a special case of monotone missingness;

thus, the foregoing developments apply to this particular missingness pattern. When the pattern of missingness is non-monotone yet MAR, finding the projection onto Λ_2 is more difficult, with no closed form solution. In Section 10.5 of Tsiatis (2006) an iterative method for deriving this projection is discussed.

8.9 Estimating the Missing Data Process

Under the MAR mechanism, we have defined the missingness probabilities $P(\boldsymbol{R} = \boldsymbol{r}|\boldsymbol{V}) = \varpi\{\boldsymbol{r}, \boldsymbol{G_r}(\boldsymbol{V})\}$. So far in our development of the semi-parametric theory for missing data, we have taken the form of these missingness probabilities to be known. As noted previously, this would be the case if missingness were dictated by design, as illustrated in Example 2 of Section 8.5. However, for most missing data problems, missingness is by happenstance. This requires that models be developed for the missingness probabilities. We now consider methods for modeling and fitting these probabilities and elucidate the consequences for the properties of the AIPWCC estimators for $\boldsymbol{\beta}$.

Suppose we posit a model $\varpi\{\boldsymbol{r}, \boldsymbol{G_r}(\boldsymbol{V}), \boldsymbol{\gamma}\}$, say, for $P(\boldsymbol{R} = \boldsymbol{r}|\boldsymbol{V})$ as a function of a finite number of parameters $\boldsymbol{\gamma}$. Then an obvious approach to estimating $\boldsymbol{\gamma}$ based on the observed data $\{\boldsymbol{R}_i, \boldsymbol{G}_{\boldsymbol{R}_i}(\boldsymbol{V}_i)\}$, $i = 1 \ldots, N$, is by maximizing the likelihood

$$\prod_{i=1}^{N} \varpi\{\boldsymbol{R}_i, \boldsymbol{G}_{\boldsymbol{R}_i}(\boldsymbol{V}_i), \boldsymbol{\gamma}\} \text{ or } \prod_{i=1}^{N} \Big(\prod_{\boldsymbol{r}} [\varpi\{\boldsymbol{r}, \boldsymbol{G_r}(\boldsymbol{V}_i), \boldsymbol{\gamma}\}]^{I(\boldsymbol{R}_i = \boldsymbol{r})} \Big), \tag{8.45}$$

in $\boldsymbol{\gamma}$. Denote the resulting maximum likelihood estimator (MLE) as $\widehat{\boldsymbol{\gamma}}$. Then the general AIPWCC estimator for $\boldsymbol{\beta}$ would be obtained by substituting $\varpi\{\boldsymbol{r}, \boldsymbol{G_r}(\boldsymbol{V}_i), \widehat{\boldsymbol{\gamma}}\}$ for $\varpi\{\boldsymbol{r}, \boldsymbol{G_r}(\boldsymbol{V}_i)\}$ in equation (8.32).

An interesting feature is that the resulting AIPWCC estimator that uses $\varpi\{\boldsymbol{r}, \boldsymbol{G_r}(\boldsymbol{V}_i), \widehat{\boldsymbol{\gamma}}\}$ is more efficient (i.e., has smaller asymptotic covariance matrix in the sense of non-negative definiteness) then the same AIPWCC estimator had $\varpi\{\boldsymbol{r}, \boldsymbol{G_r}(\boldsymbol{V}_i)\}$ been known. This is a consequence of the fact that the influence function of the AIPWCC estimator with estimated missingness probabilities is the residual after projecting the corresponding influence function with known missingness probabilities onto the linear space spanned by the score vector induced by the missingness model in $\boldsymbol{\gamma}$. Details of this result are given in Theorem 9.1 of Tsiatis (2006).

Developing models for the missingness probabilities is not necessarily an easy task. In fact, when the missingness pattern is non-monotone, there is no natural way of building coherent models for the missingness probabilities. Robins and Gill (1997) have suggested developing non-monotone missingness models using what they refer to as *randomized monotone missingness* models. These models do not arise naturally and are difficult to fit. See Chapter 17 for a discussion. However, when there are two levels of missingness or monotone missingness, then there are natural ways to model these patterns, as we now describe.

Two levels of missingness. Here again, $\boldsymbol{V} = (\boldsymbol{V}^{(1)'}, \boldsymbol{V}^{(2)'})'$, $\boldsymbol{V}^{(1)}$ is always observed and $\boldsymbol{V}^{(2)}$ is observed when $R = 1$. Consider models for $P(R = 1|\boldsymbol{V})$ of the form $\varpi(\boldsymbol{V}^{(1)}, \boldsymbol{\gamma})$.

The MLE for γ is obtained by maximizing in γ

$$\prod_{i=1}^{N}\{\varpi(\boldsymbol{V}_i^{(1)},\gamma)\}^{R_i}\{1-\varpi(\boldsymbol{V}_i^{(1)},\gamma)\}^{1-R_i}.$$

Because the missingness indicator R is a binary, a natural model is

$$\varpi(\boldsymbol{V}^{(1)},\gamma)=\frac{\exp\{\gamma^T\boldsymbol{\zeta}(\boldsymbol{V}^{(1)})\}}{1+\exp\{\gamma^T\boldsymbol{\zeta}(\boldsymbol{V}^{(1)})\}}, \tag{8.46}$$

a logistic regression model with $\boldsymbol{\zeta}(\boldsymbol{V}^{(1)})$ is a vector of functions of $\boldsymbol{V}^{(1)}$. This formulation includes usual linear logistic regression models as well as those with predictor depending on polynomials, interactions, or even splines in the elements of $\boldsymbol{V}^{(1)}$. Here, the MLE for γ solves the score equation

$$\sum_{i=1}^{N}\boldsymbol{\zeta}(\boldsymbol{V}_i^{(1)})\left\{R_i-\frac{\exp\{\gamma^T\boldsymbol{\zeta}(\boldsymbol{V}_i^{(1)})\}}{1+\exp\{\gamma^T\boldsymbol{\zeta}(\boldsymbol{V}_i^{(1)})\}}\right\}=\mathbf{0}. \tag{8.47}$$

Consequently, with two levels of missingness, where the missingness probabilities must be modeled, the suggested strategy for deriving an estimator for $\boldsymbol{\beta}$ is to

1. Choose a full-data estimating function $\boldsymbol{m}(\boldsymbol{V},\boldsymbol{\beta})$.
2. Posit a model $\boldsymbol{\phi}(\boldsymbol{V}^{(1)},\boldsymbol{\beta},\boldsymbol{\psi})$ for $E\{\boldsymbol{m}(\boldsymbol{V},\boldsymbol{\beta})|\boldsymbol{V}^{(1)}\}$.
3. Posit a model $\varpi(\boldsymbol{V}^{(1)},\gamma)$ for $P(R=1|\boldsymbol{V})$; for example, a logistic model (8.46).
4. Simultaneously estimate the parameters $\boldsymbol{\beta}$, $\boldsymbol{\psi}$, and γ by solving the set of stacked estimating equations consisting of (8.40), (8.47) and

$$\sum_{i=1}^{N}\left[\frac{R_i\boldsymbol{m}(\boldsymbol{V}_i,\boldsymbol{\beta})}{\varpi(\boldsymbol{V}_i^{(1)},\gamma)}-\left\{\frac{R_i-\varpi(\boldsymbol{V}_i^{(1)},\gamma)}{\varpi(\boldsymbol{V}_i^{(1)},\gamma)}\right\}\boldsymbol{\phi}(\boldsymbol{V}_i^{(1)},\boldsymbol{\beta},\boldsymbol{\psi})\right]=\mathbf{0}.$$

Because the solution to this stacked set of estimating equations is an m-estimator for the parameters $\boldsymbol{\beta}$, $\boldsymbol{\psi}$, and γ, the corresponding asymptotic covariance matrix may be derived using the sandwich variance estimator (8.7).

The estimator resulting from the algorithm has the double robustness property discussed previously. That is, the estimator for $\boldsymbol{\beta}$ is a consistent, asymptotically normal estimator for $\boldsymbol{\beta}$ if either the model $\boldsymbol{\phi}(\boldsymbol{V}^{(1)},\boldsymbol{\beta},\boldsymbol{\psi})$ for a $E\{\boldsymbol{m}(\boldsymbol{V},\boldsymbol{\beta})|\boldsymbol{V}^{(1)}\}$ or the model for $\varpi(\boldsymbol{V}^{(1)},\gamma)$ for $P(R=1|\boldsymbol{V})$ is correctly specified, even if the other model is not. Moreover, the sandwich estimator for the asymptotic covariance matrix yields a consistent estimator for the asymptotic covariance matrix of the estimator $\widehat{\boldsymbol{\beta}}$ if either model is correct. A more detailed discussion of double robustness is given in Chapter 9.

Monotone missingness. Recall that here $\boldsymbol{V}$ comprises the ordered variables $(\boldsymbol{V}^{(1)'},\ldots,\boldsymbol{V}^{(K)'})'$, and the level of missingness is given by D taking integer values 1 to K so that if $D=j$, then the subset of $\boldsymbol{V}$ $(\boldsymbol{V}^{(1)'},\ldots,\boldsymbol{V}^{(j)'})'$ is observed. As we argued in Section 8.7.4, the missingness probabilities are conveniently defined through the discrete hazard functions

$$\lambda_j(\boldsymbol{V}^{(1)},\ldots,\boldsymbol{V}^{(j)}) = P(D=j|D\geq j,\boldsymbol{V}),\ j=1,\ldots,K-1.$$

Accordingly, it is natural to consider models for the discrete hazard functions in terms of parameters $\boldsymbol{\gamma}$, which we denote as $\lambda_j(\boldsymbol{V}^{(1)},\ldots,\boldsymbol{V}^{(j},\boldsymbol{\gamma})$. The likelihood under these models may be expressed as

$$\prod_{j=1}^{K-1}\prod_{i:D_i\geq j}\left\{\lambda_j(\boldsymbol{V}_i^{(1)},\ldots,\boldsymbol{V}_i^{(j)},\boldsymbol{\gamma})\right\}^{I(D_i=j)}\left\{1-\lambda_j(\boldsymbol{V}_i^{(1)},\ldots,\boldsymbol{V}_i^{(j)},\boldsymbol{\gamma})\right\}^{I(D_i>j)}.$$

Conditional on $D\geq j$, D either equals j or is greater than j. Consequently, a discrete hazard models the probability of a binary decision, and it is natural to use logistic regression models. Similar to the models for two levels of missingness above and writing $\boldsymbol{\gamma}=(\boldsymbol{\gamma}_1',\ldots,\boldsymbol{\gamma}_{K-1}')'$, we may consider models of the form

$$\lambda_j(\boldsymbol{V}^{(1)},\ldots,\boldsymbol{V}^{(j)},\boldsymbol{\gamma}) = \frac{\exp\{\boldsymbol{\gamma}_j'\boldsymbol{\zeta}_j(\boldsymbol{V}^{(1)},\ldots,\boldsymbol{V}^{(j)})\}}{1+\exp\{\boldsymbol{\gamma}_j'\boldsymbol{\zeta}_j(\boldsymbol{V}^{(1)},\ldots,\boldsymbol{V}^{(j)})\}}, \tag{8.48}$$

where, for each j, $\boldsymbol{\zeta}_j(\boldsymbol{V}^{(1)},\ldots,\boldsymbol{V}^{(j)})$ is a vector of functions of $(\boldsymbol{V}^{(1)},\ldots,\boldsymbol{V}^{(j)})$.

The corresponding likelihood is

$$\prod_{j=1}^{K-1}\prod_{i:D_i\geq j}\frac{\exp\{\boldsymbol{\gamma}_j'\boldsymbol{\zeta}_j(\boldsymbol{V}_i^{(1)},\ldots,\boldsymbol{V}_i^{(j)})\}I(D_i=j)}{1+\exp\{\boldsymbol{\gamma}_j'\boldsymbol{\zeta}_j(\boldsymbol{V}_i^{(1)},\ldots,\boldsymbol{V}_i^{(j)})\}},$$

and the MLE is the solution to the score equation

$$\sum_{j=1}^{K-1}\sum_{i:D_i\geq j}\boldsymbol{\zeta}_j(\boldsymbol{V}_i^{(1)},\ldots,\boldsymbol{V}_i^{(j)}) \tag{8.49}$$
$$\times\left\{I(D_i=j)-\frac{\exp\{\boldsymbol{\gamma}_j'\boldsymbol{\zeta}_j(\boldsymbol{V}_i^{(1)},\ldots,\boldsymbol{V}_i^{(j)})\}I(D_i=j)}{1+\exp\{\boldsymbol{\gamma}_j'\boldsymbol{\zeta}_j(\boldsymbol{V}_i^{(1)},\ldots,\boldsymbol{V}_i^{(j)})\}}\right\}=\mathbf{0}.$$

As in the case of two levels of missingness, the following algorithm may be employed to derive an AIPWCC estimator for $\boldsymbol{\beta}$.

1. Choose a full-data estimating function $\boldsymbol{m}(\boldsymbol{V},\boldsymbol{\beta})$.
2. Posit models for the conditional expectation of $\boldsymbol{m}(\boldsymbol{V},\boldsymbol{\beta})$ given $\boldsymbol{V}^{(1)},\ldots,\boldsymbol{V}^{(j)}$ in terms of additional parameters $\boldsymbol{\psi}$ for each j as suggested in equation (8.42).
3. Develop models for the discrete hazard functions, such as the logistic regression models in (8.48), in terms of additional parameters $\boldsymbol{\gamma}$.
4. Simultaneously estimate the parameters $\boldsymbol{\beta}$, $\boldsymbol{\psi}$, and $\boldsymbol{\gamma}$ by solving the stacked estimating equations

$$\sum_{i=1}^{N}\left(\frac{I(D_i=K)\boldsymbol{m}(\boldsymbol{V}_i,\boldsymbol{\beta})}{H_{K-1}(\boldsymbol{V}_i^{(1)},\ldots,\boldsymbol{V}_i^{(K-1)},\boldsymbol{\gamma})}\right.$$
$$+\sum_{j=1}^{K-1}\left[\frac{I(D_i=j)-\lambda_j(\boldsymbol{V}_i^{(1)},\ldots,\boldsymbol{V}_i^{(j)},\boldsymbol{\gamma})I(D_i\geq j)}{H_j(\boldsymbol{V}_i^{(1)},\ldots,\boldsymbol{V}_i^{(j)},\boldsymbol{\gamma})}\right]$$
$$\left.\times\boldsymbol{\phi}_j(\boldsymbol{V}_i^{(1)},\ldots,\boldsymbol{V}_i^{(j)},\boldsymbol{\beta},\boldsymbol{\psi})\right)=\mathbf{0},$$

where $H_j(\boldsymbol{V}^{(1)}, \ldots, \boldsymbol{V}^{(j)}, \gamma) = \prod_{\ell=1}^{j}[1 - \lambda_\ell(\boldsymbol{V}^{(1)}, \ldots, \boldsymbol{V}^{(j\ell}, \gamma)]$;

$$\sum_{i=1}^{N} \left(\sum_{j=1}^{K-1} I(D_i > j) \left[\frac{\partial \boldsymbol{\phi}_j(\boldsymbol{V}_i^{(1)}, \ldots, \boldsymbol{V}_i^{(j)})}{\partial \boldsymbol{\psi}'} \right]' \right.$$
$$\left. \times \left\{ \boldsymbol{\phi}_{j+1}(\boldsymbol{V}_i^{(1)}, \ldots, \boldsymbol{V}_i^{(j+1)}, \boldsymbol{\beta}, \boldsymbol{\psi}) - \boldsymbol{\phi}_j(\boldsymbol{V}_i^{(1)}, \ldots, \boldsymbol{V}_i^{(j)}, \boldsymbol{\beta}, \boldsymbol{\psi}) \right\} \right) = \mathbf{0};$$

and (8.50).

The resulting estimator for $\boldsymbol{\beta}$ is doubly robust and has asymptotic covariance matrix that can estimated using the sandwich formula applied to the stacked m-estimating equations.

8.10 Summary and Concluding Remarks

In this chapter, we have reviewed how estimators for parameters in semi-parametric models when some of the data are missing at random may be derived using the geometric theory of semi-parametrics. This is based on characterizing the class of observed-data influence functions for such missing data problems. As we noted at the outset, these developments demonstrate that the resulting AIPWCC estimators are not ad hoc but rather are based on a formal semi-parametric statistical model.

AIPWCC estimators under the assumption of MAR involve three elements:

1. A model for the missing data mechanism.
2. A full-data influence function or full-data estimating function.
3. The augmentation term. In our account here, we refer to the space containing all possible such terms as Λ_2. A typical element of Λ_2 is given by (8.31) and for the special cases of two levels of missingness and monotone missingness by (8.33) and (8.35), respectively.

Remarks regarding 1. If missingness is by design, then the probability distribution for the missing data mechanism is known. Otherwise, such distributions must be modeled, which is accomplished through introduction of additional parameters γ. In Section 8.9, we described methods that may be used to develop such models in the case where the missingness mechanism is MAR and includes two levels of missingness or monotone missingness. We also discussed briefly the difficulty involved in developing models in the case of non-monotone missingness. In fact, if this exercise is challenging, the analyst may question whether or not the assumption of MAR is tenable. Although likelihood methods for estimating parameters in parametric models do not require the analyst to posit a model for the missing mechanism, even under non-monotone missingness, these methods are valid only under the assumption of MAR. Because the analyst need not consider a missingness model, there is some risk that the fact that these methods are predicated on the MAR assumption may be overlooked. Having to model these probabilities may force one to think more carefully about the plausibility of the MAR assumption.

Remarks regarding 2. To implement the AIPWCC methodology, a full-data influence function or full-data estimating function must be chosen. The choice of this influence function will affect the efficiency of the resulting AIPWCC estimator. For non-parametric full-data models, as is the case when considering the mean of a random variable with no assumptions on its underlying density, there is only one such influence function, and there is no choice to be made. In this case, one is led to the estimator proposed in Example 1 of Section 8.5. However, as discussed in Section 8.8, for other full-data semi-parametric models, the class of full-data influence functions has more than one element, including a theoretically optimal choice. Unfortunately, as discussed in Section 8.8, the optimal full-data influence function for construction of AIPWCC estimators with missing data is not necessarily the same as that when there are no missing data. Moreover, the optimal choice is very difficult to derive and often involves the need to solve a complicated integral equation. Iterative methods are described in Chapter 11 of Tsiatis (2006), but the bottom line is that these would be infeasible to implement in most practical situations. Our experience has shown that attempting to derive such optimal full-data influence functions and corresponding estimating functions to achieve the optimal AIPWCC estimator may not be worth the effort. We recommend that, in practice, one use the same full-data estimating function that would be used if there were no missing data. There is no theoretical basis for this choice, but it may be advantageous from the perspective of implementation in available statistical software.

Remarks regarding 3. The augmentation term may be used to gain efficiency regardless of the choice of the full-data estimating function used. The theoretically optimal choice for the augmentation term was derived in Section 8.8. However, this optimal choice depends on unknown probability distributions. This leads to our advocacy for an adaptive approach, where models for the optimal choice are posited in terms of additional parameters ψ, which are then estimated from the data. Specific recommendations for doing this with two levels of missingness and monotone missingness were given in Section 8.8. Such adaptive methods also lead to estimators with the property of double robustness. That is, the resulting AIPWCC estimator is a consistent and asymptotically normal estimator for the parameter of interest in the full-data semi-parametric model if either one of the models for the optimal augmentation term based on ψ or the missingness model based on γ are correctly specified even if the other is not. Chapter 9 discusses doubly robust estimators in detail.

The developments in this chapter are predicated on the assumption that the missing data mechanism is MAR. As noted above, if missingness is by design, then it is MAR of necessity. However, if data are missing by happenstance, then the MAR assumption may or may not hold. The difficulty is that one cannot assess whether or not the MAR assumption is plausible based on the observed data because of identifiability issues. Consequently, such an assumption must be made using subject matter judgment on a case-by-case basis. When it is believed that missingness may not be at random (NMAR), then one approach is to develop models that describe how the probability of missingness depends on parts of the data that are not observed. However, the results will depend strongly on these assumptions, and, because of the inherent non-identifiability, the model adopted cannot be distinguished from another class of MAR models (e.g., Molenberghs, Verbeke, and Kenward, 2009). A more reasonable approach is to conduct sensitivity analyses in which one varies the deviation from MAR and examines how much this changes the results. This approach is discussed in detail in Chapter 17 of this book.

References

Bang, H. and Robins, J.M. (2005). Doubly robust estimation in missing data and causal inference models. *Biometrics* **61**, 962–972.

Bickel, P.J., Klaassen, C.A.J., Ritov, Y., and Wellner, J.A. (1993). *Efficient and Adaptive Estimation for Semi-parametric Models.* Baltimore: The Johns Hopkins University Press.

Casella, G. and Berger, R.L. (2002). *Statistical Inference, Second Edition.* New York: Duxbury Press.

Davidian, M., Tsiatis, A.A., and Leon, S. (2005). Semi-parametric estimation of treatment effect in a pretest-posttest study with missing data (with discussion). *Statistical Science* **20**, 261–301.

Horvitz, D.G. and Thompson, D.J. (1952) A generalization of sampling without replacement from a finite universe. *Journal of the American Statistical Association* **47**, 663–685.

Liang, K.Y. and Zeger, S.L. (1986) Longitudinal data analysis using generalized linear models. *Biometrika* **73**, 13–22.

Luenberger. D.G. (1969) *Optimization by Vector Space Methods.* New York: Wiley.

Molenberghs, G., Verbeke, G., and Kenward, M.G. (2009). Sensitivity analysis for incomplete data. In *Longitudinal Data Analysis*, G. Fitzmaurice, M. Davidian, G. Molenberghs, and G. Verbeke (eds.) Handbooks of Modern Statistical Methods. New York: Chapman & Hall/CRC., Chapter 22.

Newey, W.K. (1990). Semi-parametric efficiency bounds. *Journal of Applied Economics* **5**, 99–135.

Robins, J.M. and Gill, R.D. (1997). Non-response models for the analysis of non-monotone ignorable missing data. *Statistics in Medicine* **16**, 39–56.

Robins, J.M., Rotnitzky, A., and Zhao, L.P. (1994). Estimation of regression coefficients when some of the regressors are not always observed. *Journal of the American Statistical Association* **89,** 846–866.

Robins, J.M., Rotnitzky, A., and Zhao, L.P. (1995). Analysis of semi-parametric regression models for repeated outcomes in the presence of missing data. *Journal of the American Statistical Association* **90**, 106–121.

Stefanski, L.A. and Boos, D.D. (2002). The calculus of M-estimation. *The American Statistician* **56**, 29–38.

Tsiatis, A.A. (2006). *Semi-parametric Theory and Missing Data.* New York: Springer.

Tsiatis, A.A., Davidian, M., and Cao, W. (2011). Improved doubly robust estimation when data are monotonely coarsened, with application to longitudinal studies with dropout. *Biometrics*, **67**, 536–545.

9

Double-Robust Methods

Andrea Rotnitzky

Di Tella University, Buenos Aires, Argentina

Stijn Vansteelandt

Ghent University, Belgium

CONTENTS

9.1 Introduction

We review double-robust methods for the analysis of incomplete datasets, with an emphasis on studies with monotone missingness in the outcome. Double-robust methods demand modeling the missingness process, as well as the incomplete data, but they produce valid inferences so long as one of the models is correctly specified, but not necessarily both. We describe advantages and limitations of double-robust methods. We review state-of-the-art strategies to fit the models for the missingness process and for the incomplete data with the aim of obtaining a double-robust performance with finite-sample stability and good asymptotic efficiency. We end with a discussion of open problems.

Most statistical analyses are centered around one or multiple models of scientific interest: models indexed by parameters that provide direct insight into the scientific question at stake. While being a simplification of a complex reality, such models are useful and of interest because they make the scientific problem tractable and the answers to the scientific question informative and insightful. When the data are incomplete, additional assumptions are needed which, while not being of scientific interest, are nevertheless essential to make up for the lost information and the resulting loss of identifiability. Some of these assumptions characterize the degree to which *unobserved* prognostic factors are associated with missingness, e.g., missing at random (Rubin, 1976) or specific missing not at random assumptions (Scharfstein, Rotnitzky, and Robins, 1999). They are inescapable: even if infinitely many data were available, information about the veracity of these assumptions would still be absent because these assumptions relate to the unobserved data. Other assumptions characterize the degree to which *observed* prognostic factors are associated with either the incomplete data or missingness. For instance, below we review imputation estimators which rely on a model for the dependence of the incomplete data on observed prognostic factors, and inverse probability weighted estimators which instead rely on models for the missing data mechanism, i.e., for the dependence of the probability of missingness on observed prognostic factors. Such modeling assumptions would be avoidable if infinitely many data were available, i.e., if the distribution of the observables were known, but in reality they are inescapable due to the curse of dimensionality. Since these assumptions do not relate to the scientific question of interest, it is desirable to avoid them to the extent possible, so as to come as closely as possible to the analysis one would perform in the absence of missing data. The search for procedures that relax these assumptions has been a key activity in recent research on incomplete data analysis, and has in particular lead to the development of so-called double-robust procedures.

Double-robust procedures weaken reliance on modeling assumptions about the degree to which the missing data are selective, by offering the opportunity to avoid committing to one specific modeling strategy: modeling the dependence of the incomplete data on observed prognostic factors, or modeling the missing data mechanism. Double-robust procedures require that the two aforementioned dimension-reducing models be postulated but they produce valid inferences so long as one of the models is correctly specified, but not necessarily both. Thus, they offer the data analyst two independent chances for drawing valid inference.

Research on double-robust procedures in the missing data literature proliferated and became increasingly popular after Scharfstein, Rotnitzky, and Robins (1999) noted that an estimator of the outcome mean of an infinite population originally developed and identified as locally efficient in Robins, Rotnitzky, and Zhao (1994), was double-robust. This estimator turned out to be the extension to i.i.d. sampling from an infinite population of the so-called model assisted regression estimator of a finite population mean studied extensively in the survey sampling literature (Cassel, Sarndal, and Wretman, 1976; Sarndal, 1980, 1982; Isaki and Fuller, 1982; Wright, 1983 and Sarndal, Swensson, and Wretman, 1992). Kott (1994) and Firth and Bennet (1998) noted that the model assisted regression estimator was *design consistent*, a property which is essentially tantamount to double-robustness for estimation of finite population parameters. Since the Scharfstein et al. (1999) chapter, many estimators with the double-robust property have been proposed, not just for population mean parameters but also for a variety of complete-data target parameters, several of which we review in this article. Notwithstanding this, double-robust procedures have also become the subject of recent debate.

Kang and Schafer (2007) demonstrated that data-generating mechanisms can be constructed where both dimension-reducing models are (mildly) misspecified, yet under which double-robust procedures have a disastrous performance relative to standard imputation procedures. The virtue of double-robust procedures has been further put to question in view of competing proposals that use flexible models for either the dependence of either the incomplete data or missingness on observed prognostic factors (e.g., spline or boosted regression) in traditional imputation estimators or inverse probability weighted estimators (see, e.g., Little and An, 2004; Demirtas, 2005; McCaffrey, Ridgeway, and Morral, 2004).

The aforementioned critiques and alternatives have spurred the development of double-robust procedures with drastically improved efficiency and finite-sample performance (relative to the original double-robust proposals). The goal of this chapter is to give an overview of these procedures, primarily focusing on the problem of incomplete outcomes in cross-sectional and longitudinal studies. In addition, we will provide insight into the benefits and limitations of double-robust procedures versus non-double-robust procedures and present a discussion of open problems that are the focus of active and vigorous research, such as model selection and the development of double-robust procedures for complex data structures.

9.2 Data Configuration, Goal of Inference, and Identifying Assumptions

Double-robust methods can be best understood in the simple setting, which we shall assume until Section 9.5, of a cross-sectional study with outcome that is missing in some units and with covariates that are always observed. In particular, suppose that a random sample of N units is obtained. The intended, i.e., complete, data $\mathbf{Z}$ to be recorded in each unit is comprised of an outcome Y and a vector $\mathbf{W}$ which contains covariates that enter a regression for the outcome Y and/or are variables collected for secondary analyses. For instance, $\mathbf{Z}$ may encode the answers to a questionnaire conducted on a random sample of adolescents from a target population with the goal of investigating the prevalence of various risky behaviors. For the specific investigation of practice of unprotected sex, the outcome Y is then the indicator of such practice and $\mathbf{W}$ includes socioeconomic and demographic variables.

The scientific interest lies in a parameter β indexing a model $\mathcal{M}_{full}$ for the distribution of $\mathbf{Z}$, which we will refer to as the full data model. For instance, our interest may lie in the mean outcome

$$\beta = E(Y), \tag{9.1}$$

(e.g., the prevalence of unprotected sex). If so, then to avoid the possibility of incorrect inference due to model misspecification, we may choose to make no assumptions about the complete data and thus work under the non-parametric model $\mathcal{M}_{full}$. Alternatively, our interest may lie in investigating how the prevalence varies with socioeconomic level. In that case, we might choose to assume a logistic regression model $\mathcal{M}_{full}$ given as

$$E(Y|\mathbf{X}) = \text{expit}\left\{\left[1, \mathbf{X}^T\right]\beta\right\}, \tag{9.2}$$

where $\text{expit}(x) = \{1 + \exp(-x)\}^{-1}$ and $\mathbf{X}$ is the subset of $\mathbf{W}$ comprised of socioeconomic variables. This model is semi-parametric since it parameterizes the distribution of Y given $\mathbf{X}$, but leaves the distribution of $\mathbf{X}$ and the distribution of the remaining components of $\mathbf{W}$ given Y and $\mathbf{X}$ unspecified.

Suppose now that Y is not recorded in everyone (e.g., because some kids refuse to provide information about their sexual behavior). Let R equal 1 if Y is observed and be 0 otherwise. Then the observed data are N i.i.d. copies of the vector $\mathbf{O}$ where $\mathbf{O} = (R, \mathbf{Z})$ if $R = 1$ and $\mathbf{O} = (R, \mathbf{W})$ if $R = 0$. Now, the assumptions encoded in the full data model $\mathcal{M}_{full}$ will generally not suffice to identify the target parameter β. Identification will require that assumptions be made on the missingness process, which encodes how selective the missing data are. The popular missing at random assumption (MAR) (Rubin, 1976) is one such assumption which, in the simple setup considered here, reduces to the statement that

$$R \text{ and } Y \text{ are conditionally independent given } \mathbf{W}.$$

In the context of the preceding example, this assumption would hold if the variables in $\mathbf{W}$ are the only predictors of unsafe risk practice that are associated with refusal to respond. When the full data model $\mathcal{M}_{full}$ is non-parametric, investigators entertaining the MAR assumption cannot count on obtaining evidence against it since the observed data will always fit the MAR model perfectly. This happens because the observed data model implied by $\mathcal{M}_{full}$ and the MAR assumption is itself non-parametric (Gill, Robins, and van der Laan, 1997).

9.3 Dimension Reducing Strategies

Even with the MAR assumption in place, further dimension reducing assumptions are generally required to proceed with inference about β. For instance, suppose that interest lies in $\beta = E(Y)$ under the non-parametric model $\mathcal{M}_{full}$. MAR implies that

$$E(Y|\mathbf{X}) = E\{E(Y|\mathbf{W},R=1)|\mathbf{X}\}. \tag{9.3}$$

Thus, the maximum likelihood estimator of β under the induced non-parametric model $\mathcal{M}_{sub}$ for the observed data should be $\widehat{\beta}_{NPML} = E_N\{E_N(Y|\mathbf{W},R=1)\}$, where $E_N(A) = N^{-1}\sum_{i=1}^N A_i$ denotes the empirical mean operator and $E_N(Y|\mathbf{W}=\mathbf{w},R=1)$ denotes the empirical mean of Y in the subsample of responders with $\mathbf{W}=\mathbf{w}$. However, $\widehat{\beta}_{NPML}$ is not a feasible estimator because $E_N(Y|\mathbf{W}=\mathbf{W}_i,R=1)$ does not typically exist for subjects i with missing Y_i when the dimension of $\mathbf{W}$ is large and/or $\mathbf{W}$ has continuous components as in such scenario the subsample of responders with $\mathbf{W}=\mathbf{W}_i$ is empty.

Two types of dimension reducing assumptions are available. The first type models the degree to which observed prognostic factors $\mathbf{W}$ are associated with the incomplete data. The second type models the dependence of missingness on $\mathbf{W}$. Regression imputation estimators, discussed below, rely on assumptions of the first type; inverse probability weighted estimators, also discussed below, rely on assumptions of the second type. Unlike MAR, these two types of assumptions impose restrictions on the distribution of observables and consequently, are testable.

9.3.1 Regression imputation

Regression imputation (RI) estimators $\widehat{\beta}_{RI}(\widehat{m})$ are computed like the default complete-data estimators but with Y for each individual replaced by a prediction $\widehat{Y} = \widehat{m}(\mathbf{W})$, using a consistent estimator $\widehat{m}(\mathbf{W})$ of

$$m_0(\mathbf{W}) = E(Y|\mathbf{W},R=1) \tag{9.4}$$

under some model for (9.4). Thus, the RI estimator of a population mean is $E_N\{\widehat{m}(\mathbf{W})\}$, whereas the RI estimator of the coefficient of a complete data regression model (such as (9.2)) is obtained by regressing $\widehat{Y}$ on $\mathbf{X}$ using the default ordinary or weighted least squares procedure. Let $\mathcal{M}_{sub}$ be the observed data model induced by model $\mathcal{M}_{full}$, the MAR assumption and the imputation model, i.e., the model for (9.4). Then provided the imputation model is small enough (i.e., sufficiently smooth), an estimator $\widehat{m}(\cdot)$ of $m_0(\cdot)$ can be constructed (e.g., a parametric regression estimator or a non-parametric smoother), which converges at a sufficiently fast rate so that $\widehat{\beta}_{RI}(\widehat{m})$ is a Consistent and Asymptotically Normal (CAN) estimator for β under model $\mathcal{M}_{sub}$. That is, $\sqrt{N}\left\{\widehat{\beta}_{RI}(\widehat{m}) - \beta\right\}$ converges in law to a mean zero normal distribution whenever model $\mathcal{M}_{sub}$ holds. The essence for the justification of this lies in the identity (9.3).

RI estimators based on efficient procedures are efficient among estimators of β that are CAN under $\mathcal{M}_{sub}$. Specifically, suppose that the default complete data estimator on which the RI estimator is based is the maximum likelihood estimator (MLE) of β in model $\mathcal{M}_{full}$, e.g. the MLE of β under model (9.2), and the imputation $\widehat{m}(\mathbf{W})$ is the MLE of $m_0(\mathbf{W})$ under the assumed model for it. Then the resulting RI estimator of β is the MLE of β under model $\mathcal{M}_{sub}$ and consequently, asymptotically efficient in the class of estimators of β that are CAN under $\mathcal{M}_{sub}$ (Bang and Robins, 2005).

RI estimators have pros and cons. On the one hand, when the imputation model is correct, RI estimators typically have small bias relative to their standard error in finite samples so long as the complete data procedures on which they are based do. Also, RI estimators of a sample mean have the advantage of falling in the parameter space so long as the range of $\widehat{m}$ does, a property not satisfied by some of the estimators discussed in the next section.

On the other hand, when the distributions of covariates $\mathbf{W}$ in respondents versus non-respondents are concentrated on different regions, then even a hard to detect misspecification of the model for $m_0(\mathbf{W})$ can yield a RI estimator with large bias because the imputation of a missing Y with $\widehat{m}(\mathbf{W})$ relies heavily on extrapolation of a misspecified regression model. In the extreme case in which the distributions of covariates $\mathbf{W}$ in respondents versus non-respondents are concentrated on non-overlapping regions, RI estimation will rely entirely on model extrapolation. Corresponding bootstrap or normal theory standard error estimators do not express the uncertainty due to model extrapolation. They may therefore effectively hide the fact that the observed data alone carry no genuine information about the target parameter. A further complication arises for estimation of the coefficients of regression models for the outcomes on a subset $\mathbf{X}$ of $\mathbf{W}$. Formulating the imputation model may be tricky in such settings because off-the-shelf model choices may conflict with the scientific model for the expected outcome given $\mathbf{X}$. For instance, under the logistic regression model (9.2), a logistic regression model for Y given $\mathbf{W}$ will generally be misspecified because $E\left[\text{expit}\left\{\gamma^T\left[1,\mathbf{W}^T\right]^T\right\}|\mathbf{X}\right]$ is not usually of the form $\text{expit}\left\{\beta^T\left[1,\mathbf{X}^T\right]^T\right\}$.

9.3.2 Inverse probability weighting

In contrast to the RI estimators of the previous section, inverse probability weighted (IPW) estimators rely on a consistent estimator $\widehat{\pi}(\mathbf{W})$ of

$$\pi_0(\mathbf{W}) = \Pr(R = 1|\mathbf{W}) \tag{9.5}$$

under some model for (9.5). With missing outcomes, two distinct types of IPW estimators exist. The first type, denoted $\widehat{\beta}_{IPW}(\widehat{\pi})$, are computed like the default complete data estimators but with Y replaced by $\widetilde{Y} = R\widehat{\pi}(\mathbf{W})^{-1}Y$ for all individuals in the sample. The second type, denoted $\widehat{\beta}_{B,IPW}(\widehat{\pi})$, are computed like the default complete data estimator discarding the non-respondents and weighting each respondent contribution by $\widehat{\pi}(\mathbf{W})^{-1}$. Thus, for estimation of $\beta = E(Y)$, $\widehat{\beta}_{IPW}(\widehat{\pi}) = E_N\left\{R\widehat{\pi}(\mathbf{W})^{-1}Y\right\}$, whereas for estimation of the coefficient β of a complete data regression model, $\widehat{\beta}_{IPW}(\widehat{\pi})$ is obtained by regressing $\widetilde{Y}$ on $\mathbf{X}$. On the other hand, $\widehat{\beta}_{B,IPW}(\widehat{\pi})$ solves

$$E_N\left\{R\widehat{\pi}(\mathbf{W})^{-1}d(\mathbf{Z};\beta)\right\}=0, \tag{9.6}$$

where $d(\mathbf{Z};\beta)$ is the default complete data score or estimating function for β. For instance for estimation of a population mean, $d(\mathbf{Z};\beta)=Y-\beta$ yielding

$$\widehat{\beta}_{B,IPW}(\widehat{\pi})=E_N\left\{R\widehat{\pi}(\mathbf{W})^{-1}Y\right\}/E_N\left\{R\widehat{\pi}(\mathbf{W})^{-1}\right\},$$

whereas for estimation of the parameter β in model (9.2), $d(\mathbf{Z};\beta) = \left[1,\mathbf{X}^T\right]^T\left\{Y-\text{expit}\left\{\beta^T\left[1,\mathbf{X}^T\right]^T\right\}\right\}$, yielding the weighted maximum likelihood estimator based on respondents' data with weights $\widehat{\pi}(\mathbf{W})^{-1}$. Let $\mathcal{G}_{sub}$ be the observed data model induced by model $\mathcal{M}_{full}$, the MAR assumption and the model for the response probabilities (9.5). Then provided the missingness model for $\pi_0(\mathbf{W})$ is small enough (i.e., sufficiently smooth) and $\pi_0(\mathbf{W})$ is bounded away from 0 with probability 1, an estimator $\widehat{\pi}(\cdot)$ of $\pi_0(\cdot)$ can be constructed (e.g., a parametric regression estimator or a non-parametric smoother), which converges at a sufficiently fast rate so that $\widehat{\beta}_{IPW}(\widehat{\pi})$ and $\widehat{\beta}_{B,IPW}(\widehat{\pi})$ are CAN estimators for β under model $\mathcal{G}_{sub}$. The essence for the justification of this lies in the identity:

$$E(Y|\mathbf{X})=E\left\{R\pi_0(\mathbf{W})^{-1}Y|\mathbf{X}\right\}. \tag{9.7}$$

IPW estimators are not efficient in large samples under model $\mathcal{G}_{sub}$. In fact, their asymptotic efficiency is sometimes substantially outperformed by that of specific so-called inverse probability weighted *augmented* (IPWA) estimators (Robins et al., 1994). IPWA estimators require as input a function $m(\mathbf{W})$ that is either a priori specified or estimated from the available data. Given $m(\cdot)$, the IPWA estimator $\widehat{\beta}(\widehat{\pi},m)$ is computed by solving

$$E_N\left[R\widehat{\pi}(\mathbf{W})^{-1}d(\mathbf{Z};\beta)-\left\{R\widehat{\pi}(\mathbf{W})^{-1}-1\right\}d^0(\mathbf{W};\beta)\right]=0, \tag{9.8}$$

where $d^0(\mathbf{W};\beta)$ is computed like $d(\mathbf{Z};\beta)$ but with Y replaced by $m(\mathbf{W})$. For estimation of a population mean, $\widehat{\beta}(\widehat{\pi},m)$ thus reduces to

$$\widehat{\beta}(\widehat{\pi},m)=\widehat{\beta}_{IPW}(\widehat{\pi})-E_N\left[\left\{R\widehat{\pi}(\mathbf{W})^{-1}-1\right\}m(\mathbf{W})\right]. \tag{9.9}$$

Like IPW estimators, IPWA estimators are CAN for β under model $\mathcal{G}_{sub}$ (provided $\pi_0(\mathbf{W})$ is bounded away from 0). This can be seen from the earlier results for IPW estimators and the fact that the augmentation term in (9.8) (i.e., the term involving $d^0(\mathbf{W};\beta)$) has mean zero when $\widehat{\pi}$ is substituted by π_0.

Interestingly, the class of IPWA estimators obtained as m varies over all functions of $\mathbf{W}$ and $d(\mathbf{Z};\beta)$ varies over all possible complete data unbiased estimating functions is, up to asymptotic equivalence, the class of all CAN estimators for β under model $\mathcal{G}_{sub}$ (Robins et al., 1994). Specifically, for any estimator $\widetilde{\beta}$ of β that is CAN when model $\mathcal{G}_{sub}$ holds, there exist functions d and m such that $\sqrt{N}\left\{\widehat{\beta}(\widehat{\pi},m)-\widetilde{\beta}\right\}=o_p(1)$; i.e., both estimators are asymptotically equivalent. Furthermore, if $\widehat{\pi}(\mathbf{w})$ is the MLE of $\pi_0(\mathbf{w})$ under $\mathcal{G}_{sub}$, then among estimators $\widehat{\beta}(\widehat{\pi},m)$ that use a given full data estimating function $d(\mathbf{Z};\beta)$, the estimator $\widehat{\beta}(\widehat{\pi}, m_0)$ is the one with the smallest asymptotic variance under $\mathcal{G}_{sub}$ (Robins et al., 1994). While such IPWA estimator cannot be computed because m_0 is unknown, an IPWA estimator with the same asymptotic variance under model $\mathcal{G}_{sub}$ is obtained when m_0 is substituted by a (sufficiently fast converging) consistent estimator $\widehat{m}$ of m_0 (Robins et al., 1994). When the consistency of $\widehat{m}$ is dependent on model $\mathcal{M}_{sub}$ being correct, the estimator $\widehat{\beta}(\widehat{\pi},\widehat{m})$ is said to be locally efficient in the class of estimators $\widehat{\beta}(\widehat{\pi}, m)$ that use the same function d, at the sub-model $\mathcal{G}_{sub} \cap \mathcal{M}_{sub}$. This local efficiency property asserts that whenever both $\mathcal{G}_{sub}$ and $\mathcal{M}_{sub}$ hold, this estimator $\widehat{\beta}(\widehat{\pi},\widehat{m})$ has asymptotic variance no greater than that of any estimator $\widehat{\beta}(\widehat{\pi}, m)$ that use the same function d, and thus in particular than the estimators $\widehat{\beta}_{IPW}(\widehat{\pi})$ and $\widehat{\beta}_{B,IPW}(\widehat{\pi})$ that use the same estimating function d. This follows because $\widehat{\beta}_{IPW}(\widehat{\pi})$ and $\widehat{\beta}_{B,IPW}(\widehat{\pi})$ are asymptotically equivalent to IPWA estimators $\widehat{\beta}(\widehat{\pi}, m)$ that use the same estimating function d for specific functions m (Robins et al., 1994).

IPW estimators are attractive because they can easily be implemented with available software and easily generalize to the estimation of arbitrary full data target parameters of models for complex, not necessarily cross-sectional data configurations. Another appeal, also shared by IPWA estimators, is that they are not vulnerable to the extrapolation problem of RI estimators. Moreover, for estimation of regression parameters, their consistency relies on the validity of a model for the missingness probabilities which, unlike the imputation model on which RI estimators rely, never conflicts with the complete data regression model. However, an important drawback is that, even when the missingness model is correctly specified, both IPW and IPWA estimators can have poor finite sample behavior — in particular non-negligible bias. This is because a slight underestimate $\widehat{\pi}(\mathbf{w})$ of a $\pi_0(\mathbf{w})$ close to 0 may result in a respondent i with $\mathbf{W}_i = \mathbf{w}$ that carries an unduly large weight. This poor behavior is less severe for the IPW estimator $\widehat{\beta}_{B,IPW}(\widehat{\pi})$ because it involves weighting residuals rather than outcomes. For estimation of $\beta = E(Y)$ this is reflected in that $\widehat{\beta}_{B,IPW}(\widehat{\pi})$, but not $\widehat{\beta}_{IPW}(\widehat{\pi})$, is sample bounded (Tan, 2010) (hence the subscript 'B'): being a weighted sample average of Y among respondents, it takes values between the minimum and maximum observed values of Y.

9.4 Double-Robust Estimation with Missing-at-Random Data

As we have seen, the locally efficient IPWA estimator $\widehat{\beta}(\widehat{\pi}, \widehat{m})$ of the population mean β is CAN under model $\mathcal{G}_{sub}$. Remarkably, it remains CAN for β even if $\mathcal{G}_{sub}$ is wrong, provided $\mathcal{M}_{sub}$ is correct and $\widehat{m}$ converges to m_0 under $\mathcal{M}_{sub}$. For estimation of a population mean this can be seen by rewriting

$$\widehat{\beta}(\widehat{\pi},\widehat{m}) = \widehat{\beta}_{RI}(\widehat{m}) + E_N\left[R\widehat{\pi}(\mathbf{W})^{-1}\{Y - \widehat{m}(\mathbf{W})\}\right]. \tag{9.10}$$

Then $\widehat{\beta}(\widehat{\pi},\widehat{m})$ is CAN under $\mathcal{M}_{sub}$ as the term added to $\widehat{\beta}_{RI}(\widehat{m})$ in the preceding display (9.10) converges to zero (at the usual 1 over root N rate) provided $\widehat{m}(\cdot)$ converges sufficiently fast to $m_0(\cdot)$, since $E\left[R\pi(\mathbf{W})^{-1}\{Y - m_0(\mathbf{W})\}\right] = 0$ for any $\pi(\mathbf{W})$. For estimation of regression parameters β, consistency of $\widehat{\beta}(\widehat{\pi}, \widehat{m})$ under model $\mathcal{M}_{sub}$ is seen after re-arranging the terms in the left-hand side of (9.8), yielding the equation

$$E_N\{d^0(\mathbf{W};\beta)\} + E_N\left[R\widehat{\pi}(\mathbf{W})^{-1}\{d(\mathbf{Z};\beta) - d^0(\mathbf{W};\beta)\}\right] = 0. \tag{9.11}$$

For instance, for estimation of β in the logistic regression model (9.2), the last display (9.11) is

$$E_N\left[\left[1,\mathbf{X}^T\right]^T\left\{\widehat{m}\left(\mathbf{W}\right)-\text{expit}\left\{\beta^T\left[1,\mathbf{X}^T\right]^T\right\}\right\}\right] \tag{9.12}$$
$$+E_N\left[\left[1,\mathbf{X}^T\right]^T R\widehat{\pi}\left(\mathbf{W}\right)^{-1}\left\{Y-\ \widehat{m}\left(\mathbf{W}\right)\right\}\right]=0.$$

The first empirical mean agrees with the score for β in the logistic regression model except that observed and missing outcomes Y_i are replaced with $\widehat{m}\left(\mathbf{W}_i\right)$; when evaluated at the population regression parameter β, it henceforth converges to zero when $\mathcal{M}_{sub}$ is correct. As before, also the second empirical mean converges to zero (at the usual 1 over root N rate) under model $\mathcal{M}_{sub}$.

It follows from the above that $\widehat{\beta}\left(\widehat{\pi},\widehat{m}\right)$ is CAN for β so long as one of the models $\mathcal{M}_{sub}$ or $\mathcal{G}_{sub}$ is correctly specified, but not necessarily both. Unlike RI or IPWA estimators, this estimator thus offers the opportunity to avoid committing to one specific model $\mathcal{M}_{sub}$ or $\mathcal{G}_{sub}$: it yields valid large sample inference so long as one of the models is right, regardless of which one. It is therefore said to be double-robust CAN for β in the union model $\mathcal{M}_{sub}\cup\mathcal{G}_{sub}$ (i.e., CAN for β when at least one of the models $\mathcal{M}_{sub}$ or $\mathcal{G}_{sub}$ is correctly specified).

The double-robust property of the locally efficient IPWA estimator of a population mean was first noticed in Scharfstein et al. (1999). These authors studied the asymptotic properties of the IPWA estimator when $m_0\left(\cdot\right)$ and $\pi_0\left(\cdot\right)$ were estimated under parametric models, the former by ordinary or weighted least squares of Y on $\mathbf{W}$ among respondents and the latter by maximum likelihood. Such IPWA estimator is sometimes referred to as the standard double-robust estimator (Cao, Tsiatis, and Davidian, 2009).

Robins, Rotnitzky, and van der Laan (2000) and Robins and Rotnitzky (2001) showed that the double-robust property of the locally efficient IPWA estimator is not fortuitous. It is indeed a consequence of the particular factorization of the likelihood that takes place under MAR data into one factor that depends solely on the missingness probabilities and a second factor that does not depend on them.

Despite being locally efficient and double-robust, like other IPWA estimators, $\widehat{\beta}\left(\widehat{\pi},\widehat{m}\right)$ may have several deficiencies. These drawbacks, which can be lessened by using clever choices of $\widehat{\pi}$ and/or $\widehat{m}$ (as we will see in subsequent sections) include:

a. Poor finite-sample behavior when $\pi_0\left(\mathbf{W}\right)$ is close to zero for some sample units.

b. Poor large sample efficiency under model $\mathcal{G}_{sub}$ when model $\mathcal{M}_{sub}$ is misspecified. That is, if $\mathcal{G}_{sub}$ holds but $\mathcal{M}_{sub}$ does not, $\widehat{\beta}\left(\widehat{\pi},\widehat{m}\right)$ is no longer guaranteed to have asymptotic variance smaller than that of any estimator $\widehat{\beta}\left(\widehat{\pi},m\right)$ that uses the same estimating function d; in particular, $\widehat{\beta}\left(\widehat{\pi},\widehat{m}\right)$ may even have worse asymptotic efficiency than $\widehat{\beta}_{B,IPW}\left(\widehat{\pi}\right)$ and/or $\widehat{\beta}_{IPW}\left(\widehat{\pi}\right)$.

c. The estimator $\widehat{\beta}\left(\widehat{\pi},\widehat{m}\right)$ of a population mean may fall outside the parameter space.

d. The calculation of the estimator $\widehat{\beta}\left(\widehat{\pi},\widehat{m}\right)$ of a regression parameter may involve solving estimating equations that may not have a unique solution.

In the next sections we give an account of existing proposals for computing $\widehat{\pi}$ and/or $\widehat{m}$ so as to prevent or ameliorate these potential deficiencies. We do so following a historical perspective and thus introducing them, to the extent possible, in the same order as they appeared in the literature. Throughout, we omit discussion of double-robust estimators of the variance of the proposed estimators of β, i.e., variance estimators that are consistent when either $\mathcal{G}_{sub}$ or $\mathcal{M}_{sub}$ holds. The availability of such variance estimators is nonetheless important since confidence intervals and hypothesis tests based on a double-robust estimator of β, but a non-double-robust estimator of its variance, are not themselves guaranteed to be valid when either $\mathcal{G}_{sub}$ or $\mathcal{M}_{sub}$ holds, but not both. Estimators of β discussed below whose construction relies on parametric models for m_0 and π_0 ultimately solve, together with the parameters of these models, a set of M-estimating equations. An analytic expression for their asymptotic variance regardless of the validity of the models can then be obtained from standard Taylor expansions techniques for deriving the limiting distribution of M-estimators. An estimator of the asymptotic variance consistent under either $\mathcal{G}_{sub}$ or $\mathcal{M}_{sub}$ can then be constructed using standard sandwich type estimators (Stefanski and Boos, 2002). However, derivation of the asymptotic variance is in some cases involved. An alternative approach is to recognize that under regularity conditions, the estimators of β discussed below, even those that rely on models for m_0 and/or π_0 that make (just) enough smoothness assumptions, are regular and asymptotically linear. Then we can obtain consistent estimators of their asymptotic variance by the non-parametric bootstrap (Gill, 1989).

9.4.1 Double-robust regression imputation estimators

A number of double-robust estimators take the form of a RI estimator. These estimators are attractive because of their simplicity and the fact that they remedy deficiencies (c) and (d) and, to some extent, (a).

For estimation of a population mean, the construction of a double-robust RI estimator is based on regression imputations $\widehat{m}(\mathbf{w})$ computed in a manner that ensures that

$$E_N\left[R\widehat{\pi}(\mathbf{W})^{-1}\{Y-\widehat{m}(\mathbf{W})\}\right]=0. \tag{9.13}$$

By expression (9.10), it then follows that the resulting RI estimator $\widehat{\beta}_{RI}(\widehat{m})$ is equal to $\widehat{\beta}(\widehat{\pi},\widehat{m})$ and consequently double-robust. A similar strategy can be followed to construct double-robust RI estimators of regression parameters. For instance, double-robust RI estimators of the parameter β in the logistic regression model (9.2) are obtained upon using imputations $\widehat{m}(\mathbf{w})$ that satisfy

$$E_N\left[\left[1,\mathbf{X}^T\right]^T R\widehat{\pi}(\mathbf{W})^{-1}\{Y-\widehat{m}(\mathbf{W})\}\right]=0 \tag{9.14}$$

so that the RI-estimator of β obtained by regressing the imputed outcome $\widehat{Y}=\widehat{m}(\mathbf{W})$ on the covariate $\mathbf{X}$ is equal to the IPWA estimator $\widehat{\beta}(\widehat{\pi},\widehat{m})$ solving (9.12). In the following paragraphs, we discuss various proposals that achieve this.

To our knowledge, the first double-robust RI estimator of a population mean was proposed in Scharfstein et al. (1999). These authors adopted a model $\mathcal{G}_{sub}$ which encodes a parametric missingness model

$$\pi_0(\mathbf{W})=\pi(\mathbf{W};\alpha_0) \tag{9.15}$$

(for instance, $\pi(\mathbf{W};\alpha_0)=\text{expit}\left([1,\mathbf{W}^T]\alpha_0\right)$) under which they obtained the MLE $\widehat{\pi}_{ML}(\mathbf{w})$ of $\pi_0(\mathbf{w})$ for each $\mathbf{w}$. They supplemented this with a model $\mathcal{M}_{sub}$ that encodes a generalized linear model with canonical link Ψ:

$$m_0(\mathbf{W})=\Psi^{-1}\left\{\tau_0^T v(\mathbf{W})\right\} \tag{9.16}$$

for some given vector $v(\mathbf{W})$ of functions of $\mathbf{W}$. To ensure that (9.13) holds, they then used the default maximum likelihood procedure to fit the extended model

$$m_{ext}(\mathbf{W};\tau,\delta) = \Psi^{-1}\left\{\tau^T v(\mathbf{W}) + \delta\widehat{\pi}_{ML}(\mathbf{W})^{-1}\right\}, \tag{9.17}$$

which is correctly specified with $\delta = 0$ whenever model $\mathcal{M}_{sub}$ holds. They finally calculated the RI estimator with $\widehat{m}(\mathbf{w})$ set to the fitted value under the extended model, evaluated at $\mathbf{W} = \mathbf{w}$. Justification for the double-robustness of the resulting estimator (in model $\mathcal{M}_{sub} \cup \mathcal{G}_{sub}$) comes from the fact that the score equation for δ under the extended model implies restriction (9.13). This double-robust RI procedure was later extended in Bang and Robins (2005) to the estimation of the mean of an outcome at the end of follow-up in longitudinal studies with drop-outs. It can also be easily extended to the estimation of parameters of regression models. For instance, the RI estimator of β in model (9.2) that bases the imputations $\widehat{m}(\mathbf{W})$ on an extended model which uses $\delta^T\widehat{\pi}_{ML}(\mathbf{W})^{-1}\left[1, \mathbf{X}^T\right]^T$ instead of $\delta\widehat{\pi}_{ML}(\mathbf{W})^{-1}$, is double-robust.

A second double-robust RI estimator (Kang and Schafer, 2007) is the RI estimator in which $\widehat{m}(\mathbf{w})$ is the weighted least squares fit of model (9.16) based on respondents' data, with weights equal to $\widehat{\pi}_{ML}(\mathbf{W})^{-1}$ and with the model including an intercept. Justification for the double-robustness of the resulting estimator (in model $\mathcal{M}_{sub} \cup \mathcal{G}_{sub}$) comes from the fact that the score equation for the intercept in model $\mathcal{M}_{sub}$ implies restriction (9.13). Also this estimator extends easily to the estimation of regression parameters. For instance, the RI estimator of β in model (9.2) that bases the imputations $\widehat{m}(\mathbf{W})$ on the weighted least squares fit of model (9.16) to respondents' data, with weights equal to $\widehat{\pi}_{ML}(\mathbf{W})^{-1}$ and with the model including an intercept and the covariate vector $\mathbf{X}$, is also double-robust.

A third double-robust RI estimator is the Targeted Maximum Likelihood Estimator (TMLE) proposed first by van der Laan and Rubin (2006) and later refined in Gruber and van der Laan (2010). This estimator chooses $\widehat{m}$ to be the maximum likelihood fitted value from an extended one-parameter model for m_0:

$$m_{TML}(\mathbf{W};\delta) = \Psi^{-1}\left[\Psi\left\{\widehat{m}_{init}(\mathbf{W})\right\} + \delta\widehat{\pi}_{ML}(\mathbf{W})^{-1}\right], \tag{9.18}$$

which includes a single covariate $\widehat{\pi}_{ML}(\mathbf{W})^{-1}$ and the offset $\Psi\left\{\widehat{m}_{init}(\mathbf{W})\right\}$ for some preliminary arbitrary, not necessarily ML, estimator $\widehat{m}_{init}(\mathbf{W})$ of $m_0(\mathbf{W})$. It is thus closely related to the double-robust RI estimator of Scharfstein et al. (1999), but differs in that the extended model for m_0 is a one-parameter model. This estimator is double-robust in model $\mathcal{M}_{sub} \cup \mathcal{G}_{sub}$ provided for each $\mathbf{w}$, $\widehat{m}_{init}(\mathbf{w})$ is a consistent estimator of $m_0(\mathbf{w})$ under model $\mathcal{M}_{sub}$ (which converges at a sufficiently fast rate). Double-robust TML estimators of regression parameters can likewise be easily obtained. For instance, the RI estimator of β in model (9.2) that bases the imputations $\widehat{m}(\mathbf{W})$ on the extended model (9.18) which additionally includes the covariate vector $\widehat{\pi}_{ML}(\mathbf{W})^{-1}\mathbf{X}$, is double-robust.

While all three double-robust RI estimators have the same asymptotic distribution when both models $\mathcal{G}_{sub}$ and $\mathcal{M}_{sub}$ hold, their performance can be drastically different when one of these models is misspecified. In particular, simulation studies reported by Kang and Schafer (2007) revealed that when the identity link was used (i.e., $\Psi^{-1}(x) = x$), the first double-robust RI estimator of a population mean can have disastrous performance under mild departures from both models $\mathcal{G}_{sub}$ and $\mathcal{M}_{sub}$. Its mean squared error can be substantially larger than that of the second double-robust RI estimator, which in turn has worse performance than the non-double-robust RI estimator that uses an ordinary least squares estimator $\widehat{m}$ of m_0. This is not surprising. RI estimators rely on model extrapolation when, as in the simulation conducted by Kang and Schafer, the distributions of covariates in respondents and nonrespondents are concentrated on different regions. Incorrect model extrapolation can be exacerbated for the double-robust RI estimators when the model for the missingness probabilities is incorrect since then $\widehat{\pi}(\mathbf{W})^{-1}$ entering equation (9.13) may overweight regions of the covariate space that are far away from the regions where the covariates of the non-respondents lie. The problem is even worse for the first and third double-robust estimators because $\widehat{\pi}(\mathbf{W})^{-1}$ is an added covariate in the model for m_0 and thus causes respondents with high weights to exert undue influence on the fit of the model. To reduce the instability of the TML estimator of a population mean, Gruber and van der Laan (2010) estimate it by first transforming the outcome data as $Y^{\dagger} = (Y - a) / (b - a)$, where a and b are the minimum and the maximum values of Y in the sample. Next, they calculate the TML estimator $\widehat{\beta}^{\dagger}_{RI}(\widehat{m}_{TML})$ of $\beta^{\dagger} = E(Y^{\dagger})$ using the logistic link (i.e., $\Psi(x) = \text{expit}(x)$) and back-transform the result to $a + (b - a)\widehat{\beta}^{\dagger}_{RI}(\widehat{m}_{TML})$. The boundedness of the function expit(x) between 0 and 1 ensures that the resulting estimator of β falls in the interval (a, b), i.e., it is sample bounded. This strategy can be applied with any other estimator, e.g., the first double-robust RI estimator and the standard IPWA estimator, to ensure the sample boundedness of the resulting estimator of the population mean.

Unlike the first two double-robust RI estimators, the TMLE does not require parametric restrictions on $\mathcal{M}_{sub}$. It allows for $\mathcal{M}_{sub}$ being a model that imposes just smoothness restrictions on $m_0(\cdot)$. This can be accomplished because TML estimates $m_0(\mathbf{w})$ greedily, i.e., first an initial estimator $\widehat{m}_{init}(\mathbf{w})$ is obtained using for instance non-parametric smoothing techniques or ensemble learners, e.g., super-learning (van der Laan, Polley, and Hubbard, 2008); subsequently $m_0(\mathbf{w})$ is re-estimated under an extended model that regards $\widehat{m}_{init}(\mathbf{w})$ as an offset. Of course, the initial estimator $\widehat{m}_{init}(\mathbf{w})$ could also be used to construct a double-robust IPWA estimator $\widehat{\beta}(\widehat{\pi}, \widehat{m}_{init})$. Yet, the appeal of TMLE is that it delivers a RI estimator and thus overcomes drawbacks (c) and (d).

9.4.2 Double-robust sample bounded IPW estimators

A number of double-robust estimators take the form of a sample bounded IPW estimator. Like double-robust RI estimators, these are attractive because of their simplicity and the fact that they remedy deficiencies (c) and (d) and, to some extent, (a).

To achieve double-robustness, the estimator $\widehat{\pi}$ is computed in a manner that ensures that

$$E_N\left[\left\{R\widehat{\pi}(\mathbf{W})^{-1} - 1\right\} d^0\left(\mathbf{W}; \widehat{\beta}_{RI}(\widehat{m})\right)\right] = 0. \tag{9.19}$$

For such $\widehat{\pi}$, the bounded IPW-estimator $\widehat{\beta}_{B,IPW}(\widehat{\pi})$ obtained by solving (9.6), also solves the estimating equation

$$E_N\left[R\widehat{\pi}(\mathbf{W})^{-1} d(\mathbf{Z}; \beta) - \left\{R\widehat{\pi}(\mathbf{W})^{-1} - 1\right\} d^0\left(\mathbf{W}; \widehat{\beta}_{RI}(\widehat{m})\right)\right] = 0. \tag{9.20}$$

While this resembles the form of the IPWA estimating equation (9.8), a more subtle reasoning is needed to demonstrate the double-robustness of $\widehat{\beta}_{B,IPW}(\widehat{\pi})$ because of the augmentation being evaluated at a preliminary double-robust estimator $\widehat{\beta}_{RI}(\widehat{m})$. Consistency of $\widehat{\beta}_{B,IPW}(\widehat{\pi})$ for β under model $\mathcal{G}_{sub}$ follows because when $\widehat{\pi}$ converges to π_0 at a sufficiently fast rate, the augmentation term in (9.20) converges to 0 (at the usual 1 over root N rate) so long as $\widehat{\beta}_{RI}(\widehat{m})$ converges in probability to some value, regardless of whether convergence is towards the true β. Consistency of $\widehat{\beta}_{B,IPW}(\widehat{\pi})$ under $\mathcal{M}_{sub}$ can be seen after re-writing

(9.20) as

$$E_N\left[d^0\left(\mathbf{W};\widehat{\beta}_{RI}(\widehat{m})\right)\right.$$
$$\left.+R\widehat{\pi}(\mathbf{W})^{-1}\left\{d(\mathbf{Z};\beta)-d^0\left(\mathbf{W};\widehat{\beta}_{RI}(\widehat{m})\right)\right\}\right]=0. \tag{9.21}$$

The first term $E_N\left[d^0\left(\mathbf{W};\widehat{\beta}_{RI}(\widehat{m})\right)\right]$ is identically equal to 0 because, by definition, $\widehat{\beta}_{RI}(\widehat{m})$ is the value of β that solves the equation $E_N\left[d^0(\mathbf{W};\beta)\right]=0$. The second term converges to the expectation of an unbiased estimating function of β. For instance, for estimation of a population mean, equation (9.21) reduces to

$$E_N\left[R\widehat{\pi}(\mathbf{W})^{-1}\{Y-\widehat{m}(\mathbf{W})\}\right]-E_N\left\{R\widehat{\pi}(\mathbf{W})^{-1}\right\}\left\{\beta-\widehat{\beta}_{RI}(\widehat{m})\right\}=0,$$

from where the consistency of its solution $\widehat{\beta}_{B,IPW}(\widehat{\pi})$ when $\widehat{m}$ converges to m_0 is deduced: under $\mathcal{M}_{sub}$, the first term in the left-hand side converges to 0 and $\widehat{\beta}_{RI}(\widehat{m})$ in the second term converges to the true value of β.

For estimation of a population mean, Robins et al. (2007) derived a strategy to compute $\widehat{\pi}$ in a manner that ensures (9.19), so as to obtain sample bounded IPW estimators that are double-robust. Their estimator chooses $\widehat{\pi}$ to be the fitted value from an extended one-parameter model for π_0:

$$\pi_{ext}(\mathbf{W};\varphi)=\text{expit}\left[\text{logit}\{\widehat{\pi}_{ML}(\mathbf{W})\}+\varphi\left\{\widehat{m}(\mathbf{W})-\widehat{\beta}_{RI}(\widehat{m})\right\}\right], \tag{9.22}$$

which includes a single covariate $\widehat{m}(\mathbf{W})-\widehat{\beta}_{RI}(\widehat{m})$ and the offset logit$\{\widehat{\pi}_{ML}(\mathbf{W})\}$ where $\widehat{\pi}_{ML}(\mathbf{w})$ is the MLE of $\pi_0(\mathbf{w})$ under a parametric model encoded in $\mathcal{G}_{sub}$. To ensure that $\widehat{\pi}(\mathbf{w})$ satisfies (9.19), the estimator $\widehat{\varphi}$ is computed as the solution to

$$E_N\left[\left\{\widehat{m}(\mathbf{W})-\widehat{\beta}_{RI}(\widehat{m})\right\}\left\{R\pi_{ext}(\mathbf{W};\varphi)^{-1}-1\right\}\right]=0 \tag{9.23}$$

rather than by maximum likelihood. Because the left-hand side of equation (9.23) is a strictly decreasing continuous function of φ unbounded from above and below, it has a unique solution. By the unbiasedness of the estimating equation (9.23), this solution converges to 0. Consequently $\widehat{\pi}(\mathbf{w})=\pi_{ext}(\mathbf{w};\widehat{\varphi})$ converges to $\pi_0(\mathbf{w})$ if $\mathcal{G}_{sub}$ holds for then, in large samples, the extended model is correctly specified with $\varphi=0$. Robins et al. (2007) described their procedure for models $\mathcal{M}_{sub}$ and $\mathcal{G}_{sub}$ that imposed parametric forms on $m_0(\cdot)$ and $\pi_0(\cdot)$. Yet replacing $\widehat{\pi}_{ML}(\cdot)$ and $\widehat{m}(\cdot)$ with estimators obtained as the result of data adaptive machine learning procedures, the resulting estimator $\widehat{\beta}_{B,IPW}(\widehat{\pi})$ is double-robust in union models with non-parametric components $\mathcal{M}_{sub}$ and $\mathcal{G}_{sub}$ restricted only by smoothness conditions.

Double-robust sample bounded IPW estimators of regression parameters can likewise be easily obtained. For instance, the sample bounded IPW estimator of β in model (9.2) with $\widehat{\pi}(\mathbf{W})$ computed under an extended model like (9.22) but with the covariate vector $\left[1,\mathbf{X}^T\right]^T\left\{\widehat{m}(\mathbf{W})-\text{expit}\left\{\left[1,\mathbf{X}^T\right]\widehat{\beta}_{RI}(\widehat{m})\right\}\right\}$ instead of $\widehat{m}(\mathbf{W})-\widehat{\beta}_{RI}(\widehat{m})$ and with φ estimated as the solution of

$$E_N\left[\left[\begin{array}{c}1\\ \mathbf{X}\end{array}\right]\left\{\widehat{m}(\mathbf{W})-\text{expit}\left\{\left[1,\mathbf{X}^T\right]\widehat{\beta}_{RI}(\widehat{m})\right\}\right\}\left\{\frac{R}{\pi_{ext}(\mathbf{W};\varphi)}-1\right\}\right]=0,$$

is double-robust.

Double-robust sample bounded IPW estimators of an infinite population mean are closely related to the so-called calibration estimators of a finite population mean in the survey sampling literature (Sarndal, Swensson and Wretman, 1989, Deville and Sarndal, 1992). Given an auxiliary variable whose finite population mean is known, an estimator of the finite population mean of an outcome is said to be a calibration estimator if, when computed on the auxiliary variable instead of on the outcome, it yields the known finite population mean of the auxiliary. The review article of Lumley, Shaw, and Dai (2011) draws connections between double-robust and calibration estimators. In a recent article (Tan, 2013) adapts the sample bounded IPW improved efficient estimator reviewed below in Section 9.4.5 to obtain a calibration estimator of a finite population mean under Poisson rejective sampling or high-entropy sampling in the absence of non-response. The estimator is "design-efficient," i.e., it is asymptotically as efficient as the optimal estimator in the class of the so-called regression estimators (Fuller and Isaki, 1981). IPW sample bounded estimators are also closely related to estimators that were specifically proposed in the survey sampling literature for handling non-response. For instance, Kott and Liao (2012) discuss estimators that, under simple random sampling designs, agree with $\widehat{\beta}_{B,IPW}(\widehat{\pi})$. These authors use estimated response probabilities $\widehat{\pi}(\mathbf{W}) = \pi(\mathbf{W};\widehat{\alpha})$ for a certain parametric model $\pi(\mathbf{W};\alpha)$ for $\pi_0(\mathbf{W})$ and an estimator $\widehat{\alpha}$ which solves

$$E_N\left[\left\{R\pi(\mathbf{W};\alpha)^{-1} - 1\right\}\left[1, \mathbf{W}^T\right]^T\right] = 0.$$

They consider the inverse linear model $\pi(\mathbf{W};\alpha) = \left\{1 + \left[1, \mathbf{W}^T\right]^T \alpha\right\}^{-1}$, the exponential model $\pi(\mathbf{W};\alpha) = \exp\left\{-\left[1, \mathbf{W}^T\right]^T \alpha\right\}$ and the logistic model $\pi(\mathbf{W};\alpha) = \text{expit}\left\{\left[1, \mathbf{W}^T\right]^T \alpha\right\}$.Using such $\widehat{\alpha}$ returns estimators that are *sample calibrated*, i.e., such that $E_N\left[\left\{R\widehat{\pi}(\mathbf{W})^{-1}\mathbf{W}\right\}\right] = E_N(\mathbf{W})$ and which also calibrate the sample size, i.e., such that $E_N\left[\left\{R\widehat{\pi}(\mathbf{W})^{-1}\right\}\right] = 1$. In addition, the estimator is double-robust under a model $\mathcal{M}_{sub} \cup \mathcal{G}_{sub}$ in which $\mathcal{M}_{sub}$ encodes a linear imputation model, i.e., $m_0(\mathbf{W}) = \left[1, \mathbf{W}^T\right]^T \tau_0$ (as (9.19) holds for $\widehat{\pi}(\mathbf{w})$ in such case).

9.4.3 Double-robust estimators that are never less efficient than IPW estimators

The previous estimators resolve deficiencies (c), (d) and, to some extent, deficiency (a) (either by being an RI estimator, or by being sample bounded). Yet, all of them still suffer from drawback (b). For models $\mathcal{M}_{sub}$ and $\mathcal{G}_{sub}$ that encode parametric specifications (9.15) and (9.16), Tan (2006a) proposed a double-robust estimator of a population mean (in $\mathcal{M}_{sub} \cup \mathcal{G}_{sub}$) which partially remedies deficiency (b) at a very modest extra computational effort over the standard IPWA estimators. Specifically, under $\mathcal{G}_{sub}$ and regardless of whether or not $\mathcal{M}_{sub}$ holds, the proposed estimator is asymptotically at least as efficient as any IPWA estimator of the form $\widehat{\beta}(\widehat{\pi}_{ML}, \nu_1 + \nu_2 m_{\widehat{\tau}})$ for any fixed constant vector $\nu = (\nu_2, \nu_2)^T$, where $m_\tau(\mathbf{w}) = \Psi^{-1}\left\{\tau^T v(\mathbf{w})\right\}$ and $\widehat{\tau}$ is any given consistent estimator of τ_0 under model (9.16). It follows that when model $\mathcal{G}_{sub}$ is correctly specified and regardless of whether or not $\mathcal{M}_{sub}$ holds, the estimator is at least as efficient as the locally efficient IPWA estimator $\widehat{\beta}(\widehat{\pi}_{ML}, m_{\widehat{\tau}})$ (which corresponds with $\nu_1 = 0, \nu_2 = 1$), as the IPW estimator $\widehat{\beta}_{IPW}(\widehat{\pi}_{ML})$ (which corresponds with $\nu_1 = \nu_2 = 0$) and as the bounded IPW estimator $\widehat{\beta}_{B,IPW}(\widehat{\pi}_{ML})$ (which corresponds with $\nu_1 = \beta, \nu_2 = 0$).

To achieve the aforementioned efficiency property, the vector $\widehat{\nu} = (\widehat{\nu}_1, \widehat{\nu}_2)^T$ is cleverly constructed so that it converges to the minimizer ν_{opt} of the asymptotic variance of $\widehat{\beta}\left(\widehat{\pi}_{ML}, \nu_1 + \nu_2 m_{\widehat{\tau}}\right)$ under model $\mathcal{G}_{sub}$. For estimation of the population mean, this variance can be written as $V\left(\nu, \tau^{\dagger}, \rho_{\nu,\tau^{\dagger}}\right)$, where $\tau^{\dagger}$ is the probability limit of $\widehat{\tau}$, and where for any (ν, τ, ρ), $V(\nu, \tau, \rho)$ is given by

$$\operatorname{var}\left[\frac{RY}{\pi_0(\mathbf{W})} - \left\{\frac{R}{\pi_0(\mathbf{W})} - 1\right\}\{\nu_1 + \nu_2 m_{\tau}(\mathbf{W}) + l(\mathbf{W})\rho\}\right] \tag{9.24}$$

with $l(\mathbf{W}) = l(\alpha_0; \mathbf{W})$, $l(\alpha; \mathbf{W}) = \{1 - \pi_0(\mathbf{W})\}^{-1} \partial\pi(\mathbf{W};\alpha)/\partial\alpha^T$, and $\rho_{\nu,\tau}$ a constant conformable vector minimizing the asymptotic variance $V(\nu, \tau, \rho)$ for given ν and τ over all ρ (Robins et al., 1994). Viewing (9.24) as the variance of the residual of a regression of $R\pi_0(\mathbf{W})^{-1}Y$ on $\left\{R\pi_0(\mathbf{W})^{-1} - 1\right\}[1, m_{\tau}(\mathbf{W}), l(\mathbf{W})]$, it follows by standard least squares theory that $V\left(\nu, \tau^{\dagger}, \rho_{\nu,\tau^{\dagger}}\right)$ is minimized at $\left[\nu^T, \rho^T_{\nu,\tau^{\dagger}}\right]^T$ equal to the population least squares coefficient $\left[\nu^T_{opt}, \rho^T_{\nu_{opt},\tau^{\dagger}}\right]^T = C^{-1}B$ where

$$C = E\left[\left\{R\pi_0(\mathbf{W})^{-1} - 1\right\}^2 [1, m_{\tau^{\dagger}}(\mathbf{W}), l(\mathbf{W})]^{\otimes 2}\right]^{-1}, \tag{9.25}$$

$$B = E\left[R\pi_0(\mathbf{W})^{-1}Y\left\{\pi_0(\mathbf{W})^{-1} - 1\right\}[1, m_{\tau^{\dagger}}(\mathbf{W}), l(\mathbf{W})]^T\right] \tag{9.26}$$

and for any V, $V^{\otimes 2} = V^TV$. Suppose that C and B are estimated with $\breve{C}$ and $\widehat{B}$ obtained by replacing population means with sample means, $\pi_0(\mathbf{W})$ and $l(\mathbf{W})$ with their ML estimators and $m_{\tau^{\dagger}}$ with $m_{\widehat{\tau}}$. When $\mathcal{G}_{sub}$ holds the resulting estimator $\left[\breve{\nu}^T, \breve{\rho}^T_{\breve{\nu}}\right]^T = \widehat{C}^{-1}\widehat{B}$ converges to $\left[\nu^T_{opt}, \rho^T_{\nu_{opt},\tau^{\dagger}}\right]^T$ and, in spite of $\breve{\nu}$ being random, the estimator $\widehat{\beta}\left(\widehat{\pi}_{ML}, \breve{\nu}_1 + \breve{\nu}_2 m_{\widehat{\tau}}\right)$ has the same limiting distribution as $\widehat{\beta}\left(\widehat{\pi}_{ML}, \nu_{opt,1} + \nu_{opt,2} m_{\widehat{\tau}}\right)$, where $\nu_{opt} = \left(\nu_{opt,1}, \nu_{opt,2}\right)^T$ and $\breve{\nu} = \left(\breve{\nu}_1, \breve{\nu}_2\right)^T$, so it has the aforementioned efficiency properties (Robins, Rotnitzky, and Zhao, 1995; van der Laan, and Robins, 2003; Tsiatis, 2006). However, it is not double-robust because $\breve{\nu}$ may not converge to the vector $[0, 1]^T$ when model $\mathcal{G}_{sub}$ is misspecified, even when $\mathcal{M}_{sub}$ is correct, in which case the estimator does not converge to β when $\mathcal{G}_{sub}$ does not hold regardless of the validity of $\mathcal{M}_{sub}$.

Remarkably, this loss of double-robustness can be resolved. Specifically, Tan (2006b) proposes an estimator $\widehat{\nu}$ of ν_{opt}, which is consistent under $\mathcal{G}_{sub}$ and converges to the vector $[0, 1]^T$ under $\mathcal{M}_{sub}$. The resulting estimator $\widehat{\beta}\left(\widehat{\pi}_{ML}, \widehat{\nu}_1 + \widehat{\nu}_2 m_{\widehat{\tau}}\right)$ is thus double-robust and has the aforementioned efficiency properties. This estimator $\widehat{\nu}$ of ν_{opt} uses the same estimator $\widehat{B}$ of B as described before, but a different estimator of C. The key idea underlying the estimation of C is that it can be re-expressed under model $\mathcal{G}_{sub}$ as

$$C = E\left[R\pi_0(\mathbf{W})^{-1}\left\{\pi_0(\mathbf{W})^{-1} - 1\right\}[1, m_{\tau^{\dagger}}(\mathbf{W}), l(\mathbf{W})]^{\otimes 2}\right]^{-1}$$

and can thus be consistently estimated under $\mathcal{G}_{sub}$ with $\widehat{C}$ obtained by replacing in the right-hand side of the last display population means with sample means, $\pi_0(\mathbf{W})$ and $l(\mathbf{W})$ with their ML estimators and $m_{\tau^{\dagger}}$ with $m_{\widehat{\tau}}$. When model $\mathcal{M}_{sub}$ holds, but $\mathcal{G}_{sub}$ is misspecified, then $\widehat{\nu}$ converges to the vector $[0, 1]^T$ because $\widehat{B}$ and the second row of $\widehat{C}$ then have the same probability limit. Still focusing on outcomes missing at random, Tan (2011) extended this idea to the estimation of parameters of regression models and Wang, Rotnitzky, and Lin (2010) to estimation of non-parametric regression mean functions.

9.4.4 Double-robust estimators that are efficient over a parametric class of IPWA estimators

When $\mathcal{G}_{sub}$ and $\mathcal{M}_{sub}$ are as in the preceding section, with some extra computational effort, it is possible to construct another double-robust estimator $\widehat{\beta}(\widehat{\pi}_{ML}, \widetilde{\nu}_1 + \widetilde{\nu}_2 m_{\widetilde{\tau}})$ of the population mean (in $\mathcal{G}_{sub} \cup \mathcal{M}_{sub}$), for cleverly constructed estimators $\widetilde{\nu}$ and $\widetilde{\tau}$, that has even more advantageous efficiency properties under $\mathcal{G}_{sub}$. In particular, it is not only at least as efficient asymptotically as the estimator $\widehat{\beta}(\widehat{\pi}_{ML}, \widehat{\nu}_1 + \widehat{\nu}_2 m_{\widehat{\tau}})$ of the previous section under model $\mathcal{G}_{sub}$, but also at least as efficient asymptotically as any IPWA estimator of the form $\widehat{\beta}(\widehat{\pi}_{ML}, \nu_1 + \nu_2 m_\tau)$ for arbitrary constants ν_1, ν_2 and τ. This is so regardless of whether model $\mathcal{M}_{sub}$ is correctly specified. To achieve the aforementioned efficiency property, the estimators $\widetilde{\nu}$ and $\widetilde{\tau}$ are cleverly constructed so that they converge to the minimizer (ν^*, τ^*) of the asymptotic variance of $\widehat{\beta}(\widehat{\pi}_{ML}, \nu_1 + \nu_2 m_\tau)$ under model $\mathcal{G}_{sub}$ over all (ν, τ). In the previous section, this minimization was done w.r.t. ν only, corresponding to a fixed data dependent choice $\widehat{\tau}$ of τ; here, the minimization is done w.r.t. both ν and τ.

To derive $\widetilde{\nu}$ and $\widetilde{\tau}$, a key observation is that the asymptotic variance $V(\nu, \tau, \rho_{\nu,\tau})$ of $\widehat{\beta}(\widehat{\pi}_{ML}, \nu_1 + \nu_2 m_\tau)$ under $\mathcal{G}_{sub}$ (where $V(\nu, \tau, \rho_{\nu,\tau})$ is defined as (9.24) in the preceding section) can be written as $\text{var}(Y) + \mathcal{E}(\nu, \tau, \rho_{\nu,\tau})$, where $\mathcal{E}(\nu, \tau, \rho)$ equals (Cao et al., 2009)

$$E\left[R\pi_0(\mathbf{W})^{-1}\left\{\pi_0(\mathbf{W})^{-1} - 1\right\}\{Y - \nu_1 - \nu_2 m_\tau(\mathbf{W}) - l(\mathbf{W})\rho\}^2\right].$$

Thus the minimizer of $V(\nu, \tau, \rho_{\nu,\tau})$ and the minimizer of $\mathcal{E}(\nu, \tau, \rho_{\nu,\tau})$ over (ν, τ) agree. Since $\rho_{\nu,\tau}$ minimizes $V(\nu, \tau, \rho_{\nu,\tau})$ and, consequently $\mathcal{E}(\nu, \tau, \rho)$, over ρ we can now consistently estimate (ν^*, τ^*) under model $\mathcal{G}_{sub}$ with $(\widetilde{\nu}, \widetilde{\tau})$ constructed as the first part of the minimizer $(\widetilde{\nu}, \widetilde{\tau}, \widetilde{\rho})$ of $\widehat{\mathcal{E}}(\nu, \tau, \rho)$, which is obtained by replacing the population mean in the expression for $\mathcal{E}(\nu, \tau, \rho)$ with a sample mean, and $\pi_0(\mathbf{W})$ and $l(\mathbf{W})$ with their ML estimators. The estimator $\widehat{\beta}(\widehat{\pi}_{ML}, \widetilde{\nu}_1 + \widetilde{\nu}_2 m_{\widetilde{\tau}})$ is CAN for β and, in spite of $(\widetilde{\nu}, \widetilde{\tau})$ being random, it has asymptotic variance equal to $V(\nu^*, \tau^*, \rho_{\nu^*,\tau^*})$ under $\mathcal{G}_{sub}$. It thus has the desired efficiency benefits under $\mathcal{G}_{sub}$: it is at least as efficient asymptotically as $\widehat{\beta}(\widehat{\pi}_{ML}, \widehat{\nu}_1 + \widehat{\nu}_2 m_{\widehat{\tau}})$ whose asymptotic variance is $V\left(\nu_{opt}, \tau^\dagger, \rho_{\nu_{opt},\tau^\dagger}\right)$, and as any $\widehat{\beta}(\widehat{\pi}_{ML}, \nu_1 + \nu_2 m_\tau)$, whose asymptotic variance is $V(\nu, \tau, \rho_{\nu,\tau})$. Remarkably, $\widehat{\beta}(\widehat{\pi}_{ML}, \widetilde{\nu}_1 + \widetilde{\nu}_2 m_{\widetilde{\tau}})$ is double-robust (in $\mathcal{G}_{sub} \cup \mathcal{M}_{sub}$). To assess this, suppose now that model $\mathcal{G}_{sub}$ is incorrect but $\mathcal{M}_{sub}$ is correct. Then the probability limit of $\widehat{\mathcal{E}}(\nu, \tau, \rho)$ is

$$E\left[R\pi^\dagger(\mathbf{W})^{-1}\left\{\pi^\dagger(\mathbf{W})^{-1} - 1\right\}\{Y - \nu_1 - \nu_2 m_\tau(\mathbf{W}) - l(\mathbf{W})\rho\}^2\right],$$

which is minimized at $(\nu_0, \tau_0, 0)$ where ν_0 is the vector $[0, 1]^T$ since $m_{\tau_0}(\mathbf{W})$ is the mean of Y given $\mathbf{W}$ in the respondents. It thus follows that $\widetilde{\nu}_1 + \widetilde{\nu}_2 m_{\widetilde{\tau}}$ converges to $m_{\tau_0} = m_0$ and, henceforth, that $\widehat{\beta}(\widehat{\pi}_{ML}, \widetilde{\nu}_1 + \widetilde{\nu}_2 m_{\widetilde{\tau}})$ converges to β.

A slightly simplified version of $\widehat{\beta}(\widehat{\pi}_{ML}, \widetilde{\nu}_1 + \widetilde{\nu}_2 m_{\widetilde{\tau}})$ was first derived in Rubin and van der Laan (2008). Their estimator was specifically designed to achieve the smallest asymptotic variance of any $\widehat{\beta}(\pi^*, m_\tau)$ when model $\mathcal{M}_{sub}$ is wrong in settings where the missingness probabilities are known. It thus uses specified missingness probabilities π^* instead of $\widehat{\pi}_{ML}$ and computes $\check{\tau}$ like $\widetilde{\tau}$, but forcing ν_1 and ρ to be 0 and ν_2 to be 1. In his discussion of Rubin and van der Laan's paper, Tan (2008) derived the estimator $\widehat{\beta}(\widehat{\pi}_{ML}, \check{\nu}_2 m_{\check{\tau}})$, where $\check{\nu}_2$ and $\check{\tau}$ are computed like $\widetilde{\nu}_2$ and $\widetilde{\tau}$ while forcing ν_1 to 0. Tan pointed out that $\widehat{\beta}(\widehat{\pi}_{ML}, \check{\nu}_2 m_{\check{\tau}})$ has superior efficiency under $\mathcal{G}_{sub}$ than $\widehat{\beta}(\widehat{\pi}_{ML}, \nu_2 m_\tau)$ for any τ and ν_2. He also remarked that it is double-robust, a point not noted in Rubin and van der Laan (2008). Cao et al.

(2009) independently derived $\widehat{\beta}(\widehat{\pi}_{ML}, \widetilde{\nu}_1 + \widetilde{\nu}_2 m_{\widetilde{\tau}})$. In a subsequent article, Tsiatis, Davidian, and Cao (2011) extended this approach to estimation of the mean of an outcome from longitudinal studies with dropout. Their methodology can also be applied to the double-robust estimation of a regression parameter vector β, e.g., the parameter of model (9.2). However, because the methodology relies on estimation of the minimizer (ν^*, τ^*) of a scalar function $V(\nu, \tau, \rho_{\nu,\tau})$, it is not available to produce an estimator of the entire vector β with efficiency, under $\mathcal{G}_{sub}$, never inferior than that of any $\widehat{\beta}(\widehat{\pi}_{ML}, \nu_1 + \nu_2 m_\tau)$. Rather, the methodology can produce a double-robust locally efficient estimator of β that, under $\mathcal{G}_{sub}$, estimates one, a-priori specified, linear combination $\sigma^T\beta$ at least as efficiently as any $\sigma^T\widehat{\beta}(\widehat{\pi}_{ML}, \nu_1 + \nu_2 m_\tau)$ does.

Even for the estimation of a population mean, $\widehat{\beta}(\widehat{\pi}_{ML}, \widetilde{\nu}_1 + \widetilde{\nu}_2 m_{\widetilde{\tau}})$ may still suffer from the poor finite sample performance of standard IPWA estimators. In an effort to reduce the possibility of unduly large weights, Cao et al. (2009) proposed a clever extension of the original missingness model so that in the extended model the MLE of $\pi_0(\mathbf{W})$ satisfies the constraint $E_N\{R/\widehat{\pi}_{ML}(\mathbf{W})\} = 1$. This modification, however, does not ensure that the estimator falls in the parameter space for β. Two subsequently discovered double-robust estimators have almost identical efficiency benefits as $\widehat{\beta}(\widehat{\pi}_{ML}, \widetilde{\nu}_1 + \widetilde{\nu}_2 m_{\widetilde{\tau}})$ and are guaranteed to fall in the parameter space of β. The first, due to Tan (2010a), takes the form of an sample bounded IPW estimator; the second, due to Rotnitzky et al. (2012), takes the form of a RI estimator. We review these estimators in the next subsections.

9.4.5 Double-robust sample bounded IPW estimators with enhanced efficiency under the missingness model

Tan (2010a) derived a double-robust estimator of a population mean which, like the estimator $\widehat{\beta}(\widehat{\pi}_{ML}, \widetilde{\nu}_1 + \widetilde{\nu}_2 m_{\widetilde{\tau}})$ of the preceding section, is at least as efficient under $\mathcal{G}_{sub}$ as $\widehat{\beta}(\widehat{\pi}_{ML}, \nu_1 + \nu_2 m_\tau)$ for any ν_1, ν_2 and τ but which, unlike $\widehat{\beta}(\widehat{\pi}_{ML}, \widetilde{\nu}_1 + \widetilde{\nu}_2 m_{\widetilde{\tau}})$, is sample bounded IPW (here and throughout this section $\mathcal{G}_{sub}$ encodes the missingness model (9.15)). This estimator thus resolves drawback (c) and to some extent, also (a) and (b).

Like previous sample bounded IPW double-robust estimators, the construction is based on an extended model for the missing probabilities. In particular, given an estimator $\widehat{\tau}$ which is consistent for τ under model $\mathcal{M}_{sub}$. Tan (2010a) considers the following model

$$\begin{aligned}\pi_{Tan}(\mathbf{W}; \alpha, \lambda) &= \pi(\mathbf{W};\alpha) + \{1 - \pi(\mathbf{W};\alpha)\} \times \\ &\quad \times \{\lambda_0 + \lambda_1 m_{\widehat{\tau}}(\mathbf{W}) + \lambda_2^T l(\alpha; \mathbf{W})\},\end{aligned} \tag{9.27}$$

which is correctly specified with $\lambda = 0$ when the missingness model (9.15) is correct. In a first stage, he estimates α and λ as follows: first the MLE $\widehat{\alpha}_{ML}$ of α is computed while fixing λ at 0; next, an estimator $\widetilde{\lambda}$ of λ is computed by maximum likelihood pretending that α is known and equal to $\widehat{\alpha}_{ML}$. The rationale for this is that through the specific way of estimating the missingness probabilities as $\widetilde{\pi}_{Tan}(\mathbf{W}) = \pi_{Tan}\left(\mathbf{W}; \widehat{\alpha}_{ML}, \widetilde{\lambda}\right)$ one obtains a sample bounded IPW estimator $\widehat{\beta}_{B,IPW}(\widetilde{\pi}_{Tan})$, which has asymptotic variance, under $\mathcal{G}_{sub}$, given by the minimum of $V(\nu, \tau^\dagger, \rho_{\nu,\tau^\dagger})$ over all ν where $\tau^\dagger$ is the probability limit of $\widehat{\tau}$. In particular, one can choose $\widehat{\tau}$ equal to the estimator $\widetilde{\tau}$ given in the preceding section, and thus obtain an estimator $\widehat{\beta}_{B,IPW}(\widetilde{\pi}_{Tan})$ which has asymptotic variance, under $\mathcal{G}_{sub}$, equal to the minimum $V(\nu^*, \tau^*, \rho_{\nu^*,\tau^*})$ of $V(\nu, \tau, \rho_{\nu,\tau})$ over all ν and τ, i.e., at least as efficient as any $\widehat{\beta}(\widehat{\pi}_{ML}, \nu_1 + \nu_2 m_\tau)$. However, because $\widehat{\beta}_{B,IPW}(\widetilde{\pi}_{Tan})$ is not double-robust in $\mathcal{G}_{sub} \cup \mathcal{M}_{sub}$, in a second stage, Tan (2010a) re-estimates the parameters λ_0 and λ_1 with the values $\left(\widehat{\lambda}_0, \widehat{\lambda}_1\right)$ that maximize

$$E_N \left[\frac{R \log \left\{ \pi_{Tan} \left(\mathbf{W}; \widehat{\alpha}_{ML}, \lambda_0, \lambda_1, \widetilde{\lambda}_2 \right) \right\}}{1 - \pi \left(\mathbf{W}; \widehat{\alpha}_{ML} \right)} - \lambda_0 - \lambda_1 m_{\widetilde{\tau}} \left(\mathbf{W} \right) \right]. \tag{9.28}$$

Tan (2010a) shows that under fairly weak conditions, not only the maximization can be done effectively by using a globally convergent optimization algorithm (such as the R package trust) because this objective function is strictly concave with a unique global maximum, but also the maximum is attained at values $\left(\widehat{\lambda}_0, \widehat{\lambda}_1\right)$ that yield $\widehat{\pi}_{Tan}(\mathbf{W}) = \pi_{Tan}\left(\mathbf{W}; \widehat{\alpha}_{ML}, \widehat{\lambda}_0, \widehat{\lambda}_1, \widetilde{\lambda}_2\right)$ strictly positive for the respondents. He also shows that by re-estimating the missingness probabilities with $\widehat{\pi}_{Tan}(\mathbf{W})$, one obtains a new sample bounded IPW estimator $\widehat{\beta}_{B,IPW}(\widehat{\pi}_{Tan})$ with the same asymptotic distribution as the previous sample bounded IPW estimator $\widehat{\beta}_{B,IPW}(\widetilde{\pi}_{Tan})$ under $\mathcal{G}_{sub}$. Remarkably, the new estimator $\widehat{\beta}_{B,IPW}(\widehat{\pi}_{Tan})$ is double-robust in $\mathcal{G}_{sub} \cup \mathcal{M}_{sub}$ because it is indeed equal to $\widehat{\beta}(\widehat{\pi}_{Tan}, m_{\widetilde{\tau}})$. This can be seen because $\left(\widehat{\lambda}_0, \widehat{\lambda}_1\right)$ solves the estimating equations that set the gradient of the objective function (9.28) to zero:

$$E_N \left[\left[1, m_{\widetilde{\tau}}(\mathbf{W})\right]^T \left\{ \frac{R}{\pi_{Tan}\left(\mathbf{W}; \widehat{\alpha}_{ML}, \lambda_0, \lambda_1, \widetilde{\lambda}_2\right)} - 1 \right\} \right] = 0.$$

It follows that the equations (9.19) hold for $\widehat{\pi}(\mathbf{W}) = \widehat{\pi}_{Tan}(\mathbf{W})$, and henceforth that the sample bounded IPW estimator $\widehat{\beta}_{B,IPW}(\widehat{\pi}_{Tan})$ is equal to the double-robust estimator $\widehat{\beta}(\widehat{\pi}_{Tan}, m_{\widetilde{\tau}})$.

9.4.6 Double-robust regression imputation estimators with enhanced efficiency under the missingness model

For models $\mathcal{G}_{sub}$ and $\mathcal{M}_{sub}$ that encode the parametric missingness model (9.15) and the imputation model (9.16) with an intercept, Rotnitzky et al. (2012) derived a double-robust RI estimator (in model $\mathcal{G}_{sub} \cup \mathcal{M}_{sub}$) which, under $\mathcal{G}_{sub}$, is at least as efficient asymptotically as $\widehat{\beta}(\widehat{\pi}_{ML}, m_\tau)$ for any τ and as $\widehat{\beta}_{B,IPW}(\widehat{\pi}_{ML})$. We here focus on estimation of a population mean; results for estimation of parameters of regression models (e.g., model (9.2)) with missing outcomes as well as of models for repeated outcomes with missing at random dropouts are given in Rotnitzky et al. (2012).

Let $\widehat{\tau}_{WLS}$ be the weighted least squares estimator of imputation model (9.16) based on respondents' data, with weights equal to $\widehat{\pi}_{ML}(\mathbf{W})^{-1}$, and let $\widehat{\beta}_{RI}(m_{\widehat{\tau}_{WLS}})$ be the corresponding double-robust RI estimator of Section 9.4.1. Further, let $\widetilde{\tau}$ be the estimator of τ considered in Section 9.4.4 except that it is computed upon fixing ν_1 at 0 and ν_2 at 1. Rotnitzky et al. (2012) proposed the RI estimator $\widehat{\beta}_{RI}(m_{\widetilde{\tau}_{WLS}})$ where $\widetilde{\tau}_{WLS}$ is the weighted least squares estimator of model (9.16) based on respondents' data, with weights equal to $\widehat{\pi}_{ext,ML}(\mathbf{W})^{-1}$, where $\widehat{\pi}_{ext,ML}(\mathbf{W})$ is the maximum likelihood fit under the extended model for the missingness probabilities

$$\text{expit}\left\{\text{logit}\left\{\pi\left(\mathbf{W};\alpha\right)\right\} + \delta_1 A_1 + \delta_2 A_2\right\}, \tag{9.29}$$

with covariates

$$\begin{aligned} A_1 &= \widehat{\pi}_{ML}(\mathbf{W})^{-1}\left\{m_{\widetilde{\tau}_{WLS}}(\mathbf{W}) - \widehat{\beta}_{RI}\left(m_{\widetilde{\tau}_{WLS}}\right)\right\}, \\ A_2 &= \widehat{\pi}_{ML}(\mathbf{W})^{-1}\left\{m_{\widetilde{\tau}}(\mathbf{W}) - \widehat{\beta}_{RI}\left(m_{\widetilde{\tau}_{WLS}}\right)\right\}. \end{aligned}$$

The estimator $\widehat{\beta}_{RI}\left(m_{\widetilde{\tau}_{WLS}}\right)$ is double-robust in model $\mathcal{G}_{sub} \cup \mathcal{M}_{sub}$. Specifically, because the extended model (9.29) is correctly specified with $\delta_1 = \delta_2 = 0$ when the missingness model (9.15) is correct then the estimator $\widehat{\beta}_{RI}\left(m_{\widetilde{\tau}_{WLS}}\right)$,being equal to the IPWA estimator $\widehat{\beta}\left(\widehat{\pi}_{ext,ML}, m_{\widetilde{\tau}_{WLS}}\right)$, converges to β under $\mathcal{G}_{sub}$. Furthermore, because $\widetilde{\tau}_{WLS}$ is a weighted least squares estimator among respondents, it converges to the true value of τ when model $\mathcal{M}_{sub}$ is correct. Thus, $\widehat{\beta}_{RI}\left(m_{\widetilde{\tau}_{WLS}}\right)$, being a regression imputation estimator, converges to β under $\mathcal{M}_{sub}$.

Remarkably, through the specific way of estimating the missingness probabilities, $\widehat{\beta}_{RI}\left(m_{\widetilde{\tau}_{WLS}}\right)$ has the aforementioned efficiency properties. Specifically, under $\mathcal{G}_{sub}$, its asymptotic variance is the minimum of the variance of

$$\begin{aligned} &\frac{R(Y-\beta)}{\pi_0(\mathbf{W})} - \left\{\frac{R}{\pi_0(\mathbf{W})} - 1\right\} \times \\ &\qquad \times \left\{\kappa_1\left\{m_{\tau^*}(\mathbf{W}) - \beta\right\} + \kappa_2\left\{m_{\tau^{\dagger\dagger}}(\mathbf{W}) - \beta\right\} + l(\mathbf{W})\rho\right\} \end{aligned} \tag{9.30}$$

over all κ_1, κ_2 and ρ, where τ^* and $\tau^{\dagger\dagger}$ are the probability limits of $\widetilde{\tau}$ and $\widetilde{\tau}_{WLS}$, respectively (Rotnitzky et al., 2012). The asymptotic variance of $\widehat{\beta}_{RI}\left(m_{\widetilde{\tau}_{WLS}}\right)$ is thus smaller than the asymptotic variance of $\widehat{\beta}\left(\widehat{\pi}_{ML}, m_{\widetilde{\tau}}\right)$, since the latter is given by the minimum of the variance of display (9.30) over all ρ, with $\kappa_1 = 1$ and $\kappa_2 = 0$. This, in turn, implies that $\widehat{\beta}_{RI}\left(m_{\widetilde{\tau}_{WLS}}\right)$ is at least as efficient asymptotically as $\widehat{\beta}\left(\widehat{\pi}_{ML}, m_{\tau}\right)$ for any τ, since $\widehat{\beta}\left(\widehat{\pi}_{ML}, m_{\widetilde{\tau}}\right)$ has asymptotic variance no greater than that of $\widehat{\beta}\left(\widehat{\pi}_{ML}, m_{\tau}\right)$. Also, $\widehat{\beta}_{RI}\left(m_{\widetilde{\tau}_{WLS}}\right)$ is at least as efficient asymptotically as $\widehat{\beta}_{B,IPW}\left(\widehat{\pi}_{ML}\right)$ because the latter has asymptotic variance equal to the minimum of the variance in display (9.30) over all ρ, with $\kappa_1 = 0$ and $\kappa_2 = 0$.

9.5 Double-Robust Estimation in Longitudinal Studies with Attrition

In this section, we review the generalization of RI, IPWA and double-robust procedures to the estimation of mean and regression parameters from planned longitudinal studies that suffer from attrition. Typically in such studies, the intended data on each of N randomly selected units are $\mathbf{Z} = (\mathbf{W}, \mathbf{Z}_1, \ldots, \mathbf{Z}_n)$, where $\mathbf{W}$ is measured at baseline, and $\mathbf{Z}_j$ is a vector to be measured at study cycle j, $j \geq 1$. The scientific interest lies in a parameter β indexing a model $\mathcal{M}_{full}$ for the distribution of a scalar or vector valued function $Y = g(\mathbf{Z})$, perhaps conditional on a subset $\mathbf{X}$ of $\mathbf{W}$. For instance, in a clinical trial $\mathbf{W}$ may be baseline data (including the treatment arm indicator T) and $\mathbf{Z}_j$ the data to be recorded at study cycle j on each subject. The trial endpoint Y may be the number of cycles in which adverse events occurred, i.e., $Y = \sum_{j=1}^{n} Y_j$ where Y_j, a component of $\mathbf{Z}_j$, is the indicator of an adverse event at cycle j. With the goal of quantifying treatment effects and effect modification by a subset $\mathbf{B}$ of $\mathbf{W}$, we might choose to assume a log-linear model $\mathcal{M}_{full}$,

$$E(Y|\mathbf{X}) = \exp\left(\left[1, \mathbf{X}^T\right]\beta\right), \tag{9.31}$$

where$\mathbf{X} = (T, \mathbf{B}, T\mathbf{B})^T$. To facilitate the discussion we will assume that, as in this example, the endpoint Y of interest is scalar.

When the study suffers from attrition, the vector $\mathbf{Z}_j$ is observed if and only if the individual is still on study at cycle j, in which case we let the missingness indicator R_j equal 1 ($R_j = 0$ otherwise). In this setup, β is identified under the missing at random assumption (MAR) (Rubin, 1976), which states that the decision to leave the study at time j is conditionally independent of the missing data $(\mathbf{Z}_j, \ldots, \mathbf{Z}_n)$ among those with covariate history $\overline{\mathbf{Z}}_{j-1} = (\mathbf{W}, \mathbf{Z}_1, \ldots, \mathbf{Z}_{j-1})$ who were still on study at cycle $j-1$; that is, for all j,

$$R_j \text{ and } (\mathbf{Z}_j, \ldots, \mathbf{Z}_n) \text{ are conditionally independent given } R_{j-1} = 1, \overline{\mathbf{Z}}_{j-1}.$$

Like with cross-sectional data, RI estimators are computed as the default complete data estimators but replacing Y with an imputation $\widehat{m}_1(\mathbf{W})$ in all study subjects. However, unlike estimation with cross-sectional data, $\widehat{m}_1(\mathbf{W})$ cannot be chosen as an estimator of the mean of Y given $\mathbf{W}$ among subjects with Y observed, i.e., those with $R_n = 1$, because MAR does not imply that $E(Y) = E[E(Y|R_n = 1, \mathbf{W})]$. Instead, to arrive at $\widehat{m}_1(\mathbf{W})$, we must postulate for each $j = 1, \ldots, n$, a model for the expected outcome $m(\overline{\mathbf{Z}}_{j-1}) \equiv E(Y|\overline{\mathbf{Z}}_{j-1})$ given the observed data history up to time $j-1$, where $\overline{\mathbf{Z}}_0 = \mathbf{W}$. Then, starting from $j = n$, we must recursively compute estimators $\widehat{m}_j(\overline{\mathbf{Z}}_{j-1})$ of $E(Y|\overline{\mathbf{Z}}_{j-1})$ for $j = n, \ldots, 1$, using the default complete data estimator applied to the subsample with $R_j = 1$ in which Y is replaced with the estimate $\widehat{m}_{j+1}(\overline{\mathbf{Z}}_j)$ from the previous step in the recursion (except when $j = n$ in which case no replacement is needed). This algorithm yields consistent estimators $\widehat{m}_j(\overline{\mathbf{z}}_{j-1})$ of $E(Y|\overline{\mathbf{Z}}_{j-1} = \overline{\mathbf{z}}_{j-1})$ under the assumed models because under MAR,

$$E(Y|\overline{\mathbf{Z}}_{j-1}) = E\{E(Y|\overline{\mathbf{Z}}_j)|R_j = 1, \overline{\mathbf{Z}}_{j-1}\}, j = 1, \ldots, n. \tag{9.32}$$

The resulting RI estimator of β, which we denote with $\widehat{\beta}_{RI}(\widehat{\mathbf{m}})$ (as a reminder that it ultimately depends on the collection of fitted outcome models $\widehat{\mathbf{m}} = \{\widehat{m}_j; j = 1, \ldots, n\}$) is thus CAN under the observed data model $\mathcal{M}_{sub}$ induced by $\mathcal{M}_{full}$ and the assumed models for $E(Y|\overline{\mathbf{Z}}_j)$ (for $j = 0, \ldots, n-1$) provided each assumed model is sufficiently smooth or parametric.

IPW and IPWA (Robins et al., 1995) estimators rely on estimators $\widehat{\lambda}_j(\overline{\mathbf{Z}}_{j-1})$ of the discrete conditional hazards $\lambda_{0,j}(\overline{\mathbf{Z}}_{j-1}) = \Pr(R_j = 1|R_{j-1} = 1, \overline{\mathbf{Z}}_{j-1})$ of continuing on study at cycle $j = 1, \ldots, n$. An IPW bounded estimator $\widehat{\beta}_{B,IPW}(\widehat{\pi})$ is computed by solving

$$E_N\{R_n \widehat{\pi}_n^{-1}(\overline{\mathbf{Z}}_{n-1}) d(\mathbf{Z};\beta)\} = 0,$$

where for any λ_j (for $j = 1, \ldots, n$), $\pi_j(\overline{\mathbf{Z}}_{j-1}) = \lambda_1(\mathbf{Z}_0) \times \lambda_2(\overline{\mathbf{Z}}_1) \times \cdots \times \lambda_j(\overline{\mathbf{Z}}_{j-1})$, $\widehat{\pi}_j(\overline{\mathbf{Z}}_{j-1}) = \widehat{\lambda}_1(\mathbf{Z}_0) \times \widehat{\lambda}_2(\overline{\mathbf{Z}}_1) \times \cdots \times \widehat{\lambda}_j(\overline{\mathbf{Z}}_{j-1})$ and $d(\mathbf{Z};\beta)$ is a default complete data estimating function for β; for instance in the log-linear model (9.31), $d(\mathbf{Z};\beta) = [1, \mathbf{X}^T]\{Y - \exp([1, \mathbf{X}^T]\beta)\}$. An IPWA estimator $\widehat{\beta}(\widehat{\pi}, \mathbf{m})$ depends on a set of functions $\mathbf{m} = \{m_j(\overline{\mathbf{Z}}_{j-1}); j = 1, \ldots, n\}$ either specified a-priori or estimated from the available data, and is computed by solving

$$\begin{aligned} 0 = {} & E_N\left[\frac{R_n}{\widehat{\pi}_n(\overline{\mathbf{Z}}_{n-1})} d(\mathbf{Z};\beta)\right] \\ & - \sum_{j=1}^{n} E_N\left[\left\{\frac{R_j}{\widehat{\pi}_j(\overline{\mathbf{Z}}_{j-1})} - \frac{R_{j-1}}{\widehat{\pi}_{j-1}(\overline{\mathbf{Z}}_{j-2})}\right\} d^{j-1}(\overline{\mathbf{Z}}_{j-1};\beta)\right], \end{aligned} \tag{9.33}$$

where $d^{j-1}(\overline{\mathbf{Z}}_{j-1};\beta)$ is the same as $d(\mathbf{Z};\beta)$ but with Y replaced by $m_j(\overline{\mathbf{Z}}_{j-1})$ and, by convention, $R_0 = \widehat{\pi}_0\left(\overline{\mathbf{Z}}_{-1}\right) = 1$. Both $\widehat{\beta}_{B,IPW}\left(\widehat{\pi}\right)$ and $\widehat{\beta}\left(\widehat{\pi},\mathbf{m}\right)$ are CAN for β under the observed data model $\mathcal{G}_{sub}$ induced by $\mathcal{M}_{full}$ and the assumed models for $\lambda_{0,j}\left(\overline{\mathbf{Z}}_{j-1}\right)$ $(j = 1,\ldots,n)$, provided each assumed model is sufficiently smooth or parametric and each $\lambda_{0,j}\left(\overline{\mathbf{Z}}_{j-1}\right)$ is bounded away from 0 with probability 1. Justification for this is a consequence of the identities $E\left\{R_j\pi_j\left(\overline{\mathbf{Z}}_{j-1}\right)^{-1}|\mathbf{Z}\right\} = 1$ for all j, which hold under MAR. Notice that these identities imply that under $\mathcal{G}_{sub}$, the summation in the left-hand side of (9.33) converges to 0 for any β, whereas the first empirical mean converges to $E\left\{d\left(\mathbf{Z};\beta\right)\right\}$.

The IPWA estimator $\widehat{\beta}\left(\widehat{\pi},\widehat{\mathbf{m}}\right)$, where $\widehat{\mathbf{m}}$ is the collection of functions defined above, has properties similar to the corresponding IPWA estimator previously described for cross-sectional data structures. First, if $\mathcal{G}_{sub}$ encodes a parametric model for each $\lambda_{0,j}\left(\overline{\mathbf{Z}}_{j-1}\right)$ and $\widehat{\lambda}_j$ is an efficient estimator of λ_j under the model, then $\widehat{\beta}\left(\widehat{\pi},\widehat{\mathbf{m}}\right)$ is locally efficient in the class of IPWA estimators $\left\{\widehat{\beta}\left(\widehat{\pi},\mathbf{m}\right) : \mathbf{m} \text{ arbitrary}\right\}$ that use the same function d, at the sub-model $\mathcal{G}_{sub}\cap\mathcal{M}_{sub}$. That is, of all estimators of the form $\widehat{\beta}\left(\widehat{\pi},\mathbf{m}\right)$ for an arbitrary collection $\mathbf{m}$ which use the same function d, the estimator $\widehat{\beta}\left(\widehat{\pi},\widehat{\mathbf{m}}\right)$ has the smallest asymptotic variance when in fact, model $\mathcal{M}_{sub}$ is correct.

Second, the estimator $\widehat{\beta}\left(\widehat{\pi},\widehat{\mathbf{m}}\right)$ is double-robust in model $\mathcal{G}_{sub}\cup\mathcal{M}_{sub}$, i.e., it is CAN for β if model $\mathcal{G}_{sub}$ or model $\mathcal{M}_{sub}$ holds, regardless of which. Consistency of $\widehat{\beta}\left(\widehat{\pi},\widehat{\mathbf{m}}\right)$ under model $\mathcal{M}_{sub}$ is seen after re-arranging the terms in the left-hand side of (9.33), yielding the equation

$$\begin{aligned} 0 &= E_N\left[d^0\left(\overline{\mathbf{Z}}_0;\beta\right)\right] \\ &\quad +\sum_{j=1}^{n} E_N\left[\frac{R_j}{\widehat{\pi}_j\left(\overline{\mathbf{Z}}_{j-1}\right)}\left\{d^j\left(\overline{\mathbf{Z}}_j;\beta\right) - d^{j-1}\left(\overline{\mathbf{Z}}_{j-1};\beta\right)\right\}\right], \end{aligned} \tag{9.34}$$

where $d^n\left(\overline{\mathbf{Z}}_n;\beta\right) = d\left(\mathbf{Z};\beta\right)$. Under $\mathcal{M}_{sub}$ and regularity conditions, the term $E_N\left[d^0\left(\overline{\mathbf{Z}}_0;\beta\right)\right]$ converges to $E\left[d\left(\overline{\mathbf{Z}};\beta\right)\right]$, whereas each of the remaining empirical means in the display converges to 0. For instance, for estimation of β in the log-linear model (9.31), the last display (9.34) is

$$\begin{aligned} &E_N\left[\left[1,\mathbf{X}^T\right]\left\{\widehat{m}_1\left(\overline{\mathbf{Z}}_0\right) - \exp\left(\left[1,\mathbf{X}^T\right]\beta\right)\right\}\right] \\ &+\sum_{j=1}^{n} E_N\left[\left[1,\mathbf{X}^T\right]R_j\widehat{\pi}_j\left(\overline{\mathbf{Z}}_{j-1}\right)^{-1}\left\{\widehat{m}_{j+1}\left(\overline{\mathbf{Z}}_j\right) - \widehat{m}_j\left(\overline{\mathbf{Z}}_{j-1}\right)\right\}\right] = 0, \end{aligned} \tag{9.35}$$

where $\widehat{m}_{n+1}\left(\overline{\mathbf{Z}}_n\right) = Y$. When $\mathcal{M}_{sub}$ holds, $\widehat{m}_l\left(\overline{\mathbf{Z}}_{l-1}\right)$ converges to $E\left(Y|\overline{\mathbf{Z}}_l\right)$ for each $l = 0,\ldots,n$, so by (9.32), if each $\widehat{m}_l\left(\overline{\mathbf{Z}}_{l-1}\right)$ converges sufficiently fast, each empirical mean in the summation of the second term in the right-hand side converges to 0 regardless of the probability limit of $\widehat{\pi}_j\left(\overline{\mathbf{Z}}_{j-1}\right)$, whereas the first empirical mean converges to $E\left[\left[1,\mathbf{X}^T\right]\left\{E\left(Y|\overline{\mathbf{Z}}_0\right) - \exp\left(\left[1,\mathbf{X}^T\right]\beta\right)\right\}\right]$ which, in turn, is equal to $E\left[d\left(\overline{\mathbf{Z}};\beta\right)\right]$.

Double-robust RI estimators and double-robust sample bounded IPW estimators can be obtained along the same principles previously laid out for the cross-sectional data case. In particular, double-robust RI estimators can be obtained by using estimates $\widehat{m}_j, j = 1,\ldots,n$, that yield the summation from $j = 1$ to n in (9.35) equal to 0. This can be achieved via specific extensions of the distinct proposals of Section 9.4.1, which have been described by Bang and Robins (2005), Stitelman, De Gruttola, and van der Laan (2012) and Rotnitzky et al. (2012). For instance, for estimation of β in model (9.31), $\widehat{\beta}\left(\widehat{\pi},\widehat{\mathbf{m}}\right)$ is a double-robust

RI estimator if the estimates $\widehat{m}_j$ satisfy

$$\sum_{j=1}^{n} E_N \left[\left[1, \mathbf{X}^T\right] R_j \widehat{\pi}_j \left(\overline{\mathbf{Z}}_{j-1}\right)^{-1} \left\{ \widehat{m}_{j+1} \left(\overline{\mathbf{Z}}_j\right) - \widehat{m}_j \left(\overline{\mathbf{Z}}_{j-1}\right) \right\} \right] = 0; \tag{9.36}$$

this equation agrees with equation (9.14) when $n = 1$. Double-robust sample bounded IPW estimators can in principle be obtained by using estimates $\widehat{\lambda}_j$ (for $j = 1, \ldots, n$) that yield the summation from $j = 1$ to n in (9.33) equal to 0, while replacing the unknown β with $\widehat{\beta}_{RI}(\widehat{\mathbf{m}})$. To our knowledge, these have not been yet been described in the literature.

Focusing on models $\mathcal{M}_{sub}$ and $\mathcal{G}_{sub}$ that encode parametric specifications for $E\left(Y|\overline{\mathbf{Z}}_j\right)$ and $\lambda_{0,j}\left(\overline{\mathbf{Z}}_{j-1}\right)$ for all j, Tsiatis et al. (2011) derived IPWA double-robust estimators and Rotnitzky *et al.* (2012) derived double-robust RI estimators with the following asymptotic efficiency benefits. They are locally efficient in the class $\widehat{\beta}\left(\widehat{\pi}_{ML}, \mathbf{m}\right)$ at $\mathcal{G}_{sub} \cap \mathcal{M}_{sub}$, where $\widehat{\pi}_{ML}$ is a shortcut for the MLEs of $\lambda_{0,j}$ for each j. Furthermore, when model $\mathcal{G}_{sub}$ holds and regardless of the validity of model $\mathcal{M}_{sub}$, they have asymptotic variance Σ that satisfies (i) $\Sigma - \Sigma_{B,IPW}$ is semi-positive definite, where $\Sigma_{B,IPW}$ is the asymptotic variance of $\widehat{\beta}_{B,IPW}\left(\widehat{\pi}_{ML}\right)$, and, (ii) for a given conformable constant vector κ and all $\tau = \left(\tau_1, \ldots, \tau_n\right)$, $\kappa^T \Sigma \kappa - \kappa^T \Sigma_\tau \kappa$ is semi-positive definitive, where Σ_τ is the asymptotic variance of $\widehat{\beta}\left(\widehat{\pi}_{ML}, \mathbf{m}_\tau\right)$ where $\mathbf{m}_\tau = \{m_{1,\tau}, \ldots m_{n,\tau}\}$. Property (ii) implies that for estimating the linear combination $\kappa^T \beta$, when $\mathcal{G}_{sub}$ holds, the estimators of Tsiatis et al. (2011) and Rotnitzky et al. (2012) are never less efficient, asymptotically, than the IPWA estimators $\kappa^T \widehat{\beta}\left(\widehat{\pi}_{ML}, \mathbf{m}_\tau\right)$ for any τ.

9.6 Discussion

Missing data harm statistical analyses in many ways: not only through the loss of information and power they incur, but also by demanding assumptions that one would not otherwise posit in the absence of missing data. Some of these assumptions are unavoidable as they relate to the incomplete data; others relate to the observed data and are avoidable to a certain degree, depending on the sparsity in these data. Double-robust procedures weaken reliance on such assumptions by offering the opportunity to avoid committing to one specific modeling strategy: they deliver a consistent estimator of the scientific parameter of interest so long as either a model for the missingness process or the incomplete data is correctly specified. As such, they share the virtues of imputation estimators, which rely on a model for the incomplete data, as well as IPW estimators, which rely on a model for the missingness process. In particular, by relying on a model for the missingness process, double-robust estimators are not as susceptible to extrapolation bias as imputation estimators; by relying on a model for the incomplete data, double-robust estimators have efficiency benefits relative to IPW estimators.

The choice of estimators for the imputation and missingness process models entering the construction of a double-robust estimator can have a dramatic impact on its finite sample performance. It can also affect its asymptotic efficiency when either the model for the missingness probability or the imputation model is incorrect. In this chapter, we have reviewed recently proposed strategies specifically targeted to deliver double-robust estimators with finite-sample performance and/or large sample efficiency benefits when the model for the missingness probability is correct but the imputation model is wrong. These strategies are based on parametric models for the imputation and missingness process or on non-parametric models restricted only by smoothness conditions. Of these, some are designed to deliver regression imputation estimators (see Section 9.4.1) or sample-bounded IPW estimators (see Section 9.4.2). Such double-robust estimators have the advantage that they generally guarantee results that fall within the parameter space and have better finite-sample stability. For scalar scientific parameters, some of the considered strategies are designed to guarantee that the double-robust estimator is more efficient (under the missingness model) than each member of a given class of IPW augmented estimators, even when the model for the incomplete data is misspecified (see Sections 9.4.3 and 9.4.4). Of these, some deliver regression imputation estimators (see Section 9.4.6) and others deliver sample-bounded IPW estimators (see Section 9.4.5).

In spite of these advances, which have lead to drastic improvements in the performance of double-robust procedures, important open questions remain. For instance, these improved double-robust procedures guarantee efficiency benefits only when the missingness model is correct. It remains to be seen whether it is possible to quantify a lowest bound for the asymptotic variance of the double-robust estimators that have the improved efficiency properties of Section 9.4.4, under data-generating laws under which the imputation model is correct. If such bound can be quantified, it moreover remains to be seen whether double-robust estimators can be constructed that, while retaining the improved efficiency properties of Section 9.4.4 when the missingness model is correct, also have asymptotic variance close to the lowest bound when the imputation model is correct. Further open problems will be discussed in the following sections.

9.6.1 Double-robustness in other data structures

Double-robust methods have been developed for a variety of missing data configurations, including cross-sectional studies with ignorable and non-ignorable missingness in the outcome, follow-up studies with ignorable drop-out or with non-ignorable intermittent missingness. Most of the focus has been on the estimation of population means or the parameters indexing conditional mean models or proportional hazards models (Hyun, Lee, and Sun, 2012), but attention has also been devoted to non-parametric regression (Wang et al., 2010), to estimation of the area under the receiver-operating characteristic curve (Rotnitzky, Faraggi, and Schisterman, 2006) and, more generally, the mean of a K-sample U-statistic (Schisterman and Rotnitzky, 2001); in the causal inference literature, double-robust methods have been developed for a much wider variety of scientific parameters, such as average effects of time invariant and time varying treatments (Bang and Robins, 2005; Murphy, van der Laan, and Robins, 1997), quantile-based treatment effects (Zhang et al., 2012), optimal treatment regimes (Orellana, Rotnitzky, and Robins, 2010), statistical interaction parameters (Vansteelandt et al., 2008), controlled direct effects (Goetgeluk et al., 2008) and natural direct and indirect effects in mediation analysis (Tchetgen Tchetgen and Spitser, 2012) and causal parameters in instrumental variables analysis (Tan, 2006b, 2010b; Okui et al., 2012).

In cross-sectional studies with missing outcome, the double-robust methods here considered under MAR have been generalized to a broader class of exponential tilt models which relate the distribution of Y in the respondents and non-respondents populations, i.e.,

$$f(Y|R=0,\mathbf{W}) = f(Y|R=1,\mathbf{W})\exp(\gamma Y)/c(\mathbf{W}) \tag{9.37}$$

with $c(\mathbf{W}) = E\{\exp(\gamma Y)|R=1,\mathbf{W}\}$ the normalizing constant (Robins and Rotnitzky, 2001). Since (only) $\gamma = 0$ corresponds to MAR, increasing values of γ (in absolute value) encode increasing deviations from MAR. This class of exponential tilt models has been further generalized to follow-up studies with non-monotone missing data and double-robust methods have been developed under these models (Vansteelandt, Rotnitzky, and Robins, 2007). The same authors are currently finishing a manuscript that develops double-robust estimation with monotone missing not at random data structures.

In spite of these many proposals, the development of double-robust methods for data configurations with non-monotone missing covariates has stayed much more limited. This is on the one hand so because of the difficulty of formulating missing data assumptions for the analysis of non-monotone missing data that are sufficient for identification of the parameter of interest, yet impose no testable restrictions on the observed data distribution. While MAR is such an assumption, it is difficult to interpret for non-monotone missing data (Robins and Gill, 1997). On the other hand, even if one is willing to accept the MAR assumption, it turns out to be a nearly impossible task to postulate a model for the missingness probabilities that is sufficiently general so as not to a priori impose additional conditional independence restrictions than those embodied in the MAR assumption (Robins and Gill, 1997); this hinders the development of IPW-estimators and thus in particular double-robust estimators. In follow-up studies with intermittent missingness and missing covariates, Chen and Zhou (2011) make progress under an assumption that imposes that missingness at each time does not depend on any data recorded at that time or in the future, conditional on the observed past. This assumption is stronger than MAR and likely implausible. Furthermore, the imputation model on which their double-robust estimator is based, presumes that covariates at each time have no residual association with the outcome at that time, given the observed past; this is likely violated, making the property of double-robustness somewhat illusory. In cross-sectional studies with missing covariates, Tchetgen Tchetgen and Rotnitzky (2011) make progress for the estimation of parameters indexing the conditional odds ratio between a binary outcome and exposure under the assumption that the probability of observing complete data has no residual dependence on the incomplete covariates, after adjustment for the outcome and completely observed covariates; also this assumption implies MAR and can be partially testable.

9.6.2 Model selection

Throughout this chapter, we have for the most part assumed that working models $\mathcal{M}_{sub}$ and $\mathcal{G}_{sub}$ for the outcome and missingness process were a priori given. In practice, this is rarely the case. The selection of these models is usually data-driven, with forward or backward elimination in (semi-)parametric models for the missingness probabilities and for the outcome being the default strategies. Just like the choice of estimators, the choice of selection strategy can have a major impact on the resulting inference. Selection of these working models is therefore an area of growing research.

Suppose first that a correctly specified parametric model $\mathcal{G}_{sub}$ for the missingness probabilities is a priori given, as would be the case in studies with missingness by design. Suppose the outcome model is unknown and model search over a set of covariates, some of which may be associated with the outcome, is carried out using standard model selection strategies. In such case, the efficiency and thus the distribution of the estimator depends on the covariates that are being selected, and thus in particular depends on the model selection strategy. In fact, standard selection strategies are not ideal because they are not targeted towards the quality of the double-robust estimator. In particular, suppose that the considered set of working models for the outcome to which the selection process is confined, does not include the true outcome distribution. Then these strategies do not guarantee to deliver an estimator with minimal variance in the considered class; the same is true when more flexible non-parametric smoothing techniques or ensemble learners are adopted, as in TML estimation. If the search is limited to parametric models for the outcome, both these limitations can be somewhat mitigated by avoiding the use of stepwise selection procedures altogether; instead, one may choose a large parametric model for the outcome and then estimate the parameters indexing it as those that minimize the variance of the double-robust estimator along the lines described in Sections 9.4.4, 9.4.5 and 9.4.6; however, this strategy may not work well in small samples.

In the more likely case that no model $\mathcal{G}_{sub}$ for the missingness probabilities is a priori given, the selection of working models for the outcome and missingness probabilities becomes especially tricky. Stepwise elimination procedures applied to missingness models have a tendency to prioritize covariates that are strongly predictive of missingness. This can make the inverse probability weights highly variable, which in turn may drastically reduce precision (Brookhart and van der Laan, 2006) and increase finite-sample bias if some of the covariates selected into the missingness model are not associated with the outcome. Some proposals are now emerging in the literature that acknowledge these difficulties and attempt to resolve them. Brookhart and van der Laan (2006) start with specified large outcome and missingness models and then select the working models as those that minimize the mean squared error of the considered (double-robust) estimator, with the mean squared error estimated via cross-validation. Vansteelandt, Bekaert, and Claeskens (2012) base the mean squared error on asymptotic approximations and supplement the resulting procedure with plots of the obtained estimates of the target parameter along with confidence intervals in function of different considered working models. This enables a better appreciation of the extent to which covariate adjustment reduces bias and/or increases imprecision and of the sensitivity of the results to the choice of model. van der Laan and Gruber (2010) develop Collaborative Targeted Maximum Likelihood Estimation procedures (C-TMLE; see also Gruber and van der Laan, 2010), which combine TMLE with selection of a missingness model in a way that targets the quality of the double-robust estimator. This involves constructing TMLEs for a nested sequence of working models for the missingness probabilities, of increasing complexity, and retaining the model that minimizes a penalized log-likelihood criterion for the targeted MLE fit. An example of such loss function is the empirical sum of the squared residuals at the fitted missingness working model plus a penalty term equal to the mean squared error of the double-robust estimator, as estimated using V-fold cross-validation. By adding this penalty, missingness models that cause excessive variability in the inverse probability weights, are less likely to be selected. Suppose for instance that some of the covariates are solely predictive of missingness, but not the outcome. Then, in large samples, these will not be selected by the C-TMLE procedure so that the C-TMLE will have the same asymptotic efficiency as if it were a priori known that there is no need to adjust for these covariates. However, a drawback of all foregoing proposals is that the calculation of mean squared error is based on a potentially poor initial estimate of the target parameter under a large model, and that it remains so far unclear how to conduct inference for the resulting non-regular estimators that they deliver. In particular, the non-parametric bootstrap is no longer justified for such estimators, as well as for double-robust estimators obtained following other model selection procedures.

Acknowledgment

The first author acknowledges support from NIH grant 2 R37 AI032475 16-A and the second author from IAP research network grant P07/06 from the Belgian government (Belgian Science Policy).

References

Bang, H. and Robins, J.M. (2005). Double-robust estimation in missing data and causal inference models. *Biometrics* **61**, 962–973.

Brookhart, A.M. and van der Laan, M.J. (2006). A semiparametric model selection criterion with applications to the marginal structural model. *Computational Statistics and Data Analysis* **50**, 475–498.

Cao, W.H., Tsiatis, A.A., and Davidian, M. (2009). Improving efficiency and robustness of the double-robust estimator for a population mean with incomplete data. *Biometrika* **96**, 723–734.

Cassel, C.M., Sarndal, C.E., and Wretman, J.H. (1976). Some results on generalized difference estimation and generalized regression estimation for finite populations. *Biometrika* **63**, 615–620.

Chen, B.J. and Zhou, X.H. (2011). Double robust estimates for binary longitudinal data analysis with missing response and missing covariates. *Biometrics* **67**, 830–842.

Demirtas, H. (2005). Multiple imputation under Bayesianly smoothed pattern-mixture models for non-ignorable drop-out. *Statistics in Medicine* **24**, 2345–2363.

Deville, J.-C. and Särndal, C.-E. (1992). Calibration estimators in survey sampling. *Journal of the American Statistical Association* **87**, 376–382.

Firth, D. and Bennett, K.E. (1998). Robust models in probability sampling. *Journal of the Royal Statistical Society, Series B* **60**, 3–21.

Gill, R.D. (1989). Non- and semi-parametric maximum likelihood estimators and the von Mises method. *Scandinavian Journal of Statistics* **16**, 97–128.

Gill, R.D, van der Laan, M.J., and Robins, J.M. (1997). Coarsening at random: Characterizations, conjectures and counterexamples. *Proceedings of the First Seattle Symposium on Survival Analysis*, pp. 255–294.

Goetgeluk, S., Vansteelandt, S., and Goetghebeur, E. (2008). Estimation of controlled direct effects. *Journal of the Royal Statistical Society, Series B* **70**, 1049–1066.

Gruber, S. and van der Laan, M.J. (2010). A targeted maximum likelihood estimator of a causal effect on a bounded continuous outcome. *The International Journal of Biostatistics* **6**, Article 5.

Gruber, S. and van der Laan, M.J. (2010). An application of collaborative targeted maximum likelihood estimation in causal inference and genomics. *The International Journal of Biostatistics*, Article **6**, Article 6.

Hyun, S., Lee J., and Sun, Y. (2012). Proportional hazards model for competing risks data with missing cause of failure. *Journal of Statistical Planning and Inference* **142**, 1767–1779.

Isaki, C.T. and Fuller, W.A. (1982). Survey design under the regression superpopulation model. *Journal of the American Statistical Association* **77**, 89–96.

Kang, J.D.Y. and Schafer, J.L. (2007). Demystifying double-robustness: A comparison of alternative strategies for estimating a population mean from incomplete data. *Statistical Science* **22**, 523–539.

Kott, P.S. (1994). A note on handling nonresponse in surveys. *Journal of the American Statistical Association* **89**, 693–696.

Kott, P.S. and Liao, D. (2012). Providing double protection for unit nonresponse with a nonlinear calibration-weighting routine. *Survey Research Methods* **6**, 105–111.

Little, R. and An, H.G. (2004). Robust likelihood-based analysis of multivariate data with missing values. *Statistica Sinica* **14**, 949–968.

Lumley, T., Shaw, P.A., and Dai, J.Y. (2011). Connections between survey calibration estimators and semiparametric models for incomplete data. *International Statistical Review* **79**, 200–220.

McCaffrey, D.F., Ridgeway, G. and Morral, A.R. (2004). Propensity score estimation with boosted regression for evaluating causal effects in observational studies. *Psychological Methods* **9**, 403–425.

Murphy, S.A., van der Laan, M.J., Robins, J.M., and CPPRG (2001). Marginal mean models for dynamic regimes. *Journal of the American Statistical Association* **96**, 1410–1423.

Okui, R., Small, D.S., Tan, Z., and Robins, J.M. (2012). Doubly robust instrumental variable regression. *Statistica Sinica* **22**, 173–205.

Orellana, L., Rotnitzky, A., and Robins, J.M. (2010). Dynamic regime marginal structural mean models for estimation of optimal dynamic treatment regimes, Part I: Main content. *International Journal of Biostatistics* **6**, Article 8.

Robins, J.M. and Gill, R. (1997). Non-response models for the analysis of non-monotone ignorable missing data. *Statistics in Medicine* **16**, 39–56.

Robins, J.M., Rotnitzky, A., and Zhao, L.P. (1994). Estimation of regression coefficients when some regressors are not always observed. *Journal of the American Statistical Association* **89**, 846–866.

Robins, J.M., Rotnitzky, A., and Zhao, L.P. (1995). Analysis of semiparametric regression models for repeated outcomes in the presence of missing data. *Journal of the American Statistical Association* **90**, 106–121.

Robins, J.M., Rotnitzky, A. and van der Laan, M. (2000). Comment on "On Profile Likelihood" by S.A. Murphy and A.W. van der Vaart. *Journal of the American Statistical Association* **95**, 431–435.

Robins, J.M. and Rotnitzky, A. (2001). Comment on the Bickel and Kwon article, "Inference for semiparametric models: Some questions and an answer." *Statistica Sinica* **11**, 920–936.

Robins, J.M., Sued, M., Lei-Gomez, Q., and Rotnitzky, A. (2007). Comment: Performance of double-robust estimators when "'inverse probability" weights are highly variable. *Statistical Science* **22**, 544–559.

Rotnitzky, A., Faraggi, D., and Schisterman, E. (2006). Double-robust estimation of the area under the receiver-operating characteristic curve in the presence of verification bias. *Journal of the American Statistical Association*, **101** 1276–1288.

Rotnitzky, A., Lei, Q.H., Sued, M., and Robins, J.M. (2012). Improved double-robust estimation in missing data and causal inference models. *Biometrika* **99**, 439–456.

Rubin, D.B. (1976). Inference and missing data. *Biometrika* **63**, 581–592.

Rubin, D.B. and van der Laan, M.J. (2008). Empirical efficiency maximization: Improved locally efficient covariate adjustment in randomized experiments and survival analysis. *International Journal of Biostatistics* **4**, 1–40.

Sarndal, C.E. (1980). On π inverse weighting versus best linear unbiased weighting in probability sampling. *Biometrika* **67**, 639–650.

Sarndal, C.E. (1982). Implications of survey design for generalized regression estimation of linear functions. *Journal of the American Statistical Association* **7**, 155–170.

Sarndal, C.E., Swensson, B., and Wretman, J. (1989). The weighted residual technique for estimating the variance of the general regression estimator. *Biometrika* 76, 527–537.

Sarndal, C.E., Swensson, B., and Wretman, J. (1992). *Model Assisted Survey Sampling.* New York: Springer.

Scharfstein, D.O., Rotnitzky, A., and Robins, J.M. (1999). Adjusting for non-ignorable drop-out using semiparametric non-response models: Rejoinder. *Journal of the American Statistical Association* **94**, 1135–1146.

Schisterman, E. and Rotnitzky, A. (2001). Estimation of the mean of a K-sample U-statistic with missing outcomes and auxiliaries. *Biometrika* **88**, 713–725.

Stefanski, L.A. and Boos, D.D. (2002). The calculus of M-estimation. *The American Statistician* **56**, 29–38.

Stitelman, O.M., De Gruttola V., and van der Laan, M.J. (2012). A general implementation of TMLE for longitudinal data applied to causal inference in survival analysis. *International Journal of Biostatistics* **8**, Article 1.

Tan, Z. (2006a). A distributional approach for causal inference using propensity scores. *Journal of the American Statistical Association* **101**, 1619–1637.

Tan, Z. (2006b). Regression and weighting methods for causal inference using instrumental variables. *Journal of the American Statistical Association* **101**, 1607–1618.

Tan, Z. (2008). Comment: Improved local efficiency and double robustness. *International Journal of Biostatistics* **4**, article 10.

Tan, Z. (2010a). Bounded, efficient, and double-robust estimation with inverse weighting. *Biometrika* **97**, 661–682.

Tan, Z. (2010b). Marginal and nested structural models using instrumental variables. *Journal of the American Statistical Association* **105**, 157–169.

Tan, Z. (2011). Efficient restricted estimators for conditional mean models with missing data. *Biometrika* **98**, 663–684.

Tan, Z. (2013). Simple design-efficient calibration estimators for rejective and high-entropy sampling. *Biometrika* **100**, 399–415.

Tchetgen Tchetgen, E.J. and Rotnitzky, A. (2011). On protected estimation of an odds ratio model with missing binary exposure and confounders. *Biometrika* **98**, 749–754.

Tchetgen Tchetgen, E.J. and Spitser, I. (2012). Semiparametric theory for causal mediation analysis: Efficiency bounds, multiple robustness, and sensitivity analysis. *Annals of Statistics* **40**, 1816–1845.

Tsiatis, A.A. (2006). *Semiparametric Theory and Missing Data.* New York: Springer.

Tsiatis, A.A., Davidian, M., and Cao, W. (2011). Improved double-robust estimation when data are monotonely coarsened, with application to longitudinal studies with dropout. *Biometrics* **67**, 536–545.

van der Laan, M.J. and Gruber, S. (2010). Collaborative double-robust targeted maximum likelihood estimation. *International Journal of Biostatistics* **6**, Article 17.

van der Laan, M.J., Polley, E.C., and Hubbard, A.E. (2008). Super learner. *Statistical Applications of Genetics and Molecular Biology* **6**, Article 25.

van der Laan, M.J. and Robins, J.M. (2003). *Unified Methods for Censored Longitudinal Data and Causality.* New York: Springer.

van der Laan, M.J. and Rubin, D.B. (2006). Targeted maximum likelihood learning. *International Journal of Biostatistics* **2**, Article 1.

Vansteelandt, S., Bekaert, M., and Claeskens, G. (2012). On model selection and model misspecification in causal inference. *Statistical Methods in Medical Research* **21**, 7–30.

Vansteelandt, S., Rotnitzky, A., and Robins, J.M. (2007). Estimation of regression models for the mean of repeated outcomes under non-ignorable non-monotone non-response. *Biometrika* **94**, 841–860.

Vansteelandt, S., Vanderweele, T.J., Tchetgen Tchetgen, E.J., and Robins, J.M. (2008). Multiply robust inference for statistical interactions. *Journal of the American Statistical Association* **103**, 1693–1704.

Wang, L., Rotnitzky, A., and Lin, X. (2010). Nonparametric regression with missing outcomes using weighted kernel estimating equations. *Journal of the American Statistical Association* **105**, 1135–1146.

Wright, R.L. (1983). Finite population sampling with multivariate auxiliary information. *Journal of the American Statistical Association* **78**, 879–884.

Zhang, Z.W., Chen, Z., Troendle, J.F., and Zhang, J. (2012). Causal inference on quantiles with an obstetric application. *Biometrics* **68**, 697–706.

10

Pseudo-Likelihood Methods for Incomplete Data

Geert Molenberghs

Universiteit Hasselt & KU Leuven, Belgium

Michael G. Kenward

London School of Hygiene and Tropical Medicine, London, UK

CONTENTS

10.1 Introduction

In preceding chapters, various inferential approaches based on incomplete data have been discussed. In Chapters 3 and 4, a general framework is sketched, with emphasis on likelihood, also called direct likelihood. Bayesian methods are considered in Chapter 5. Earlier chapters in Part III on semi-parametric methods use generalized estimating equations (GEE, Liang and Zeger 1986) as a vehicle for the implementation of inverse probability weighting (IPW) methods and for their doubly robust (DR) extensions. An alternative semi-parametric method is based on pseudo-likelihood (PL, le Cessie and van Houwelingen 1991, Geys, Molenberghs, and Lipsitz 1998, Geys, Molenberghs, and Ryan 1999, Aerts et al. 2002). The area is also known as composite likelihood. While GEE methods replace score equations with alternative functions, in PL, the likelihood itself is replaced by a more tractable expression. Also, in contrast to GEE, PL methods can easily account for association (Yi, Zeng, and Cook 2011, He and Yi 2011). Broadly speaking, one can distinguish between marginal and conditional PL. In the former, the likelihood for an n_i-dimensional response vector is replaced by the product of all pairs (pairwise likelihood), or all triples, or all p-tuples of outcomes. In many practically relevant settings, it has reasonably good computational and statistical efficiency. In the latter, conditional, version, the likelihood contributions take the form of conditional densities of a subset of outcomes within a sequence or cluster, given another subset of outcomes. The choice of which route to take, marginal or conditional, depends chiefly on the inferential goals. For example, if three-way association is of interest, pairwise likelihood is not an appropriate choice. In addition, the choice of pseudo-likelihood function also depends to some extent on computational considerations.

While PL is closely related to full likelihood, it is not equivalent and so the validity under MAR that come with likelihood-based analyses does not carry over, excepting that in some special cases it may do, because Rubin's (1976) conditions for ignorability are sufficient, but not always necessary. Yi, Zeng, and Cook (2011) address this issue through a pairwise (pseudo-)likelihood method for incomplete longitudinal binary clustered data which does not require the missing data process to be modeled, and in this way circumvent the need for the MAR assumption. By contrast Molenberghs *et al.* (2011) and Birhanu et al. (2013) exploit IPW to recover validity under MAR. They propose a suite of corrections to PL in its standard form, also to ensure its validity under MAR. These corrections hold for PL in general and follow both single and double robustness concepts, supplemented in some cases with a predictive model for the unobserved outcomes given the observed.

This chapter is organized as follows. In Section 10.2, the necessary concepts and notation are introduced. Sufficient background on PL is provided in Section 10.3. PL methodology for incomplete data is the subject of Section 10.4 and the methodology is exemplified and illustrated through two data analyses in Section 10.5.

10.2 Notation and Concepts

Let the random variable Y_{ij} denote the outcome or response for the ith study subject on the jth occasion ($i = 1, \ldots, N; j = 1, \ldots, n_i$). Independence across subjects is assumed. We group the outcomes into a vector $\boldsymbol{Y}_i = (y_{i1}, \ldots, y_{in_i})'$, which is typically divided into its observed ($\boldsymbol{Y}_i^o$) and missing ($\boldsymbol{Y}_i^m$) components. We further define a vector of missingness indicators $\boldsymbol{R}_i = (R_{i1}, R_{i2}, \ldots, R_{in_i})'$, with $R_{ij} = 1$ if y_{ij} is observed and 0 otherwise. In the specific case of dropout in longitudinal studies, the vector $\boldsymbol{R}_i$ can be replaced by a categorical variable for dropout pattern $D_i = 1 + \sum_{j=1}^{n_i} R_{ij}$, denoting the time at which subject i drops out.

In analyzing incomplete data, one would need to consider, in general, the full data density $f(\boldsymbol{y}_i, \boldsymbol{r}_i|\boldsymbol{\theta}, \boldsymbol{\psi})$, where the parameters $\boldsymbol{\theta}$ and $\boldsymbol{\psi}$ describe the measurement and missing data processes, respectively. When appropriate, the vector $\boldsymbol{\theta}$ will be divided into the mean regression parameters $\boldsymbol{\beta}$ and the association parameters $\boldsymbol{\alpha}$, i.e., $\boldsymbol{\theta} = (\boldsymbol{\beta}', \boldsymbol{\alpha}')'$. Covariates are assumed to be measured and grouped into $\boldsymbol{x}_i$, although, for notational simplicity, this is sometimes dropped from notation in later sections. This full data density function can be factored as follows (Rubin 1976, Little and Rubin 2002):

$$f(\boldsymbol{y}_i, \boldsymbol{r}_i|\boldsymbol{x}_i, \boldsymbol{\theta}, \boldsymbol{\psi}) = f(\boldsymbol{y}_i|\boldsymbol{x}_i, \boldsymbol{\theta}) f(\boldsymbol{r}_i|\boldsymbol{x}_i, \boldsymbol{y}_i, \boldsymbol{\psi}). \tag{10.1}$$

This is the selection model factorization, in which the first factor is the marginal density of the measurement process, while the second is the density of the missingness process, conditional on the outcomes.

10.3 Pseudo-Likelihood

In this section, we review general concepts of PL. The principal idea behind this is to replace a numerically challenging joint density (and hence likelihood) by a simpler function assembled from suitable factors. For example, when a joint density contains a computationally intractable normalizing constant, one might calculate a suitable product of conditional densities which does not involve such a complicated function. A bivariate distribution $f(y_1, y_2)$, for instance, can be replaced by the product of both conditionals $f(y_1|y_2)f(y_2|y_1)$, even though this is not the correct factorization, in the sense of recovering the original joint distribution. While the method can achieve important computational economies by changing the method of estimation, it does not affect the interpretation of the model parameters, which can be chosen in the same way as with full likelihood and retain their meaning. Because the above product does not lead to a likelihood function, appropriate modifications are needed to guarantee correct inferences. Although the example cited above is of a conditional nature, marginal forms can be used as well. For example, an intractable three-way density $f(y_1, y_2, y_3)$ can be replaced by the product $f(y_1, y_2)f(y_1, y_3)f(y_2, y_3)$. To introduce PL in a more formal way, we use the general definition given by Arnold and Strauss (1991).

10.3.1 Definition and properties

Suppose that we have an outcome sequence of length n, such as n repeated measurements on a subject, or a multivariate outcome vector with n components. Let S be the set of all $2^n - 1$ vectors of length n consisting solely of zeros and ones, with each vector having at least one non-zero entry. Denote by $\boldsymbol{y}_i^{(s)}$ the sub-vector of $\boldsymbol{y}_i$ that corresponds to the components of s that are non-zero. The associated joint density is $f_s(\boldsymbol{y}_i^{(s)}; \boldsymbol{\theta})$. To define a PL function, one chooses a set $\delta = \{\delta_s | s \in S\}$ of real numbers, with at least one non-zero component. The log of the pseudo-likelihood is then

$$p\ell = \sum_{i=1}^{N} \sum_{s \in S} \delta_s \ln f_s(\boldsymbol{y}_i^{(s)}; \boldsymbol{\theta}). \tag{10.2}$$

Adequate regularity conditions have to be invoked to ensure that (10.2) can be maximized by solving the PL (score) equations, the latter obtained by differentiating the logarithmic PL and equating its derivative to zero. These regularity conditions are spelled out in Arnold and Strauss (1991), Geys, Molenberghs, and Ryan (1999), and Molenberghs et al. (2011), for example. In particular, when the components in (10.2) result from a combination of marginal and conditional distributions of the original distribution, then a valid PL function results. In particular, the conventional log-likelihood function is obtained by setting $\delta_s = 1$ if s is the vector consisting solely of ones, and 0 otherwise. More details can be found in Varin (2008), Lindsay (1988), and Joe and Lee (2008). Note that Joe and Lee (2008) use weighting for reasons of efficiency in pairwise likelihood, in a similar spirit to that of Geys, Molenberghs, and Lipsitz (1998), but that is distinct from its use here, where we focus on bias correction when data are incomplete. Another important reference is Cox and Reid (2004).

Let θ_0 be the true parameter. Under the required regularity conditions, maximizing (10.2) produces a consistent and asymptotically normal estimator $\widetilde{\boldsymbol{\theta}}_0$ so that $\sqrt{N}(\tilde{\boldsymbol{\theta}}_N - \boldsymbol{\theta}_0)$ converges in distribution to

$$N_p[\mathbf{0}, I_0(\boldsymbol{\theta}_0)^{-1} I_1(\boldsymbol{\theta}_0) I_0(\boldsymbol{\theta}_0)^{-1}], \tag{10.3}$$

with $I_0(\boldsymbol{\theta})$ defined by

$$I_{0,k\ell}(\boldsymbol{\theta}) = -\sum_{s \in S} \delta_s E_{\boldsymbol{\theta}} \left(\frac{\partial^2 \ln f_s(\boldsymbol{y}^{(s)}; \boldsymbol{\theta})}{\partial \theta_k \partial \theta_\ell} \right) \tag{10.4}$$

and $I_1(\boldsymbol{\theta})$ by

$$I_{1,k\ell}(\boldsymbol{\theta}) = \sum_{s,t \in S} \delta_s \delta_t E_{\boldsymbol{\theta}} \left(\frac{\partial \ln f_s(\boldsymbol{y}^{(s)}; \boldsymbol{\theta})}{\partial \theta_k} \frac{\partial \ln f_t(\boldsymbol{y}^{(t)}; \boldsymbol{\theta})}{\partial \theta_\ell} \right). \tag{10.5}$$

10.3.2 Pairwise pseudo-likelihood

Marginal models for non-Gaussian data can be impracticable for full maximum likelihood inference, especially with large within-unit replication (n_i). le Cessie and van Houwelingen (1991) and Geys, Molenberghs, and Lipsitz (1998) replace the true contribution of a vector of correlated binary data to the full likelihood, written as $f(y_{i1}, \ldots, y_{in_i})$, by the product of all pairwise contributions $f(y_{ij}, y_{ik})$ $(1 \le j < k \le n_i)$, and so derive a PL function. Note that the term *composite likelihood* is also used in this context. Renard, Molenberghs, and Geys (2004) refer to this particular instance of PL as *pairwise likelihood.* Grouping the outcomes for subject i into a vector $\boldsymbol{Y}_i$, the contribution of the ith subject to the log PL becomes

$$p\ell_i = \sum_{j<k} \ln f(y_{ij}, y_{ik}), \tag{10.6}$$

if it contains more than one observation. Otherwise $p\ell_i = f(y_{i1})$. Extension to three-way and higher-order pseudo-likelihood is straightforward. All of these are special cases of (10.2).

10.3.3 Full conditional pseudo-likelihood

Some models, such as log-linear models (Molenberghs and Verbeke 2005, Ch. 12), lend themselves more easily to conditioning than to marginalization. Given that

$$f(y_{ij} | y_{ik}, k \ne j) = \frac{f(y_{i1}, \ldots, y_{in_i})}{f(y_{i1}, \ldots, y_{i,j-1}, y_{i,j+1}, \ldots, y_{in_i})} = \frac{f_{\mathbf{1}}(\boldsymbol{y}_i^{(\mathbf{1})})}{f_{s_j}(\boldsymbol{y}_i^{(s_j)})},$$

a full conditional likelihood contribution becomes:

$$p\ell_i = n_i \cdot \ln f_{\mathbf{1}}(\boldsymbol{y}_i^{(\mathbf{1})}) - \sum_{j=1}^{n_i} \ln f_{s_j}(\boldsymbol{y}_i^{(s_j)}).$$

Here, $\mathbf{1}$ is a vector of ones and s_j is a vector of ones, with a single 0 in the jth entry. Alternative versions of conditional PL are also possible. For example, one could consider all pairs, conditioning upon the remaining $n_i - 2$ outcomes. This setting has been considered by Geys, Molenberghs, and Ryan (1999) and by Molenberghs et al. (2011). The latter authors study the incomplete-data aspects of conditional PL.

10.4 Pseudo-Likelihood for Incomplete Data

In the case of incomplete data, pseudo-likelihood is valid when the missing data mechanism is MCAR, but this validity does not extend to MAR mechanisms in general. There are however a limited number of specific situations, such as full exchangeability, as shown in Molenberghs et al. (2011) for which it does lead to valid inferences under MAR. Parzen et al. (2006) apply PL methods to settings where the outcome vector is grouped with covariates and missing-data indicators. In this chapter, we extend the ideas in Molenberghs et al. (2011), using inverse probability weighting and double robustness concepts (Scharfstein, Rotnitzky, and Robins 1999, Van der Laan and Robins 2003, Bang and Robins 2005), Rotnitzky 2009). While Molenberghs et al. (2011) considered specific case studies involving Gaussian and exchangeable binary outcomes, Birhanu et al. (2013) addressed more general types of correlation structure.

In Molenberghs et al. (2011) and Birhanu et al. (2013), general forms of estimating equations for incomplete data, as well as specific forms for the case of pseudo-likelihood, were presented and their validity established, for marginal and conditional PL. We present their general expressions, and develop them for two specific cases: a marginal model for binary data and a multivariate normal linear model.

10.4.1 Estimating equations for pairwise likelihood

Suppose that $\boldsymbol{U}$ denotes the estimating equations resulting from pairwise pseudo-likelihood, with $E(\boldsymbol{U}) = \boldsymbol{0}$.

Let us first consider the "naive" cases.

$$\boldsymbol{U}_{\text{nCC}} = \sum_{i=1}^{N} \widetilde{R}_i \sum_{j<k} \boldsymbol{U}_i(y_{ij}, y_{ik}), \tag{10.7}$$

$$\boldsymbol{U}_{\text{nCP}} = \sum_{i=1}^{N} \sum_{j<k<d_i} \boldsymbol{U}_i(y_{ij}, y_{ik}), \tag{10.8}$$

$$\boldsymbol{U}_{\text{nAC}} = \sum_{i=1}^{N} \left[\sum_{j<k<d_i} \boldsymbol{U}_i(y_{ij}, y_{ik}) + \sum_{j=1}^{d_i-1} (n_i - d_i + 1)\boldsymbol{U}_i(y_{ij}) \right], \tag{10.9}$$

where $\widetilde{R}_i = 1$ if subject i is fully observed and 0 otherwise.

In the above, "naive" refers to the fact that these estimating equations would generally lead to biased estimators under MAR. In the first case, the pairwise pseudo-likelihood (score) contributions in (10.7) are from all pairs of outcomes for subjects with fully observed $\boldsymbol{Y}_i$; hence, the notation "CC" for complete cases. The contributions in (10.8), on the other hand, come from all pairs ("CP" denoting complete pairs) of outcomes for subjects having at least 2 outcomes, including of course, the completers. Note that, while the contributions in (10.7) come only from completers, in (10.8), these are supplemented with contributions from dropouts having incomplete $\boldsymbol{Y}_i$ but with at least two outcomes observed. Finally, for the available case ("AC") version in (10.9), the contributions in (10.8) are further supplemented with contributions from each observed outcome, i.e., from the so-called widows.

In what follows, robust versions will be introduced. By singly robust, we mean robust against misspecification of the likelihood function as a pseudo-likelihood function, even when missingness is MAR, provided inverse probability weights are correctly specified. The doubly robust versions are valid when the conditional distribution of the unobserved outcomes given the observed ones is correctly specified or the weights are correctly specified, but not necessarily both. For a detailed account of doubly robust methods, we refer to Chapter 9 of this volume.

Singly robust versions of (10.7) to (10.9) are based on IPW methods, in which each subject's contribution is weighted by the inverse probability, either of being fully observed (IPWCC), or of being observed up to a certain time (IPWAC). Suppose that the probabilities of subject i being completely observed, and of being observed up to and including occasion j, are respectively expressed as:

$$\pi_i = \prod_{\ell=2}^{n_i}(1-p_{i\ell}) \qquad \text{and} \qquad \pi_{ij} = \prod_{\ell=2}^{j}(1-p_{i\ell}),$$

where $p_{i\ell} = P(D_i = \ell | D_i \geq \ell, \boldsymbol{y}_{i\bar{\ell}}, \boldsymbol{x}_{i\bar{\ell}})$ are the component probabilities of dropping out at occasion ℓ, given the subject is still in the study, the covariate history $\boldsymbol{x}_{i\bar{\ell}}$ and the outcome history $\boldsymbol{y}_{i\bar{\ell}}$. Then, singly robust versions of (10.7) to (10.9) would take the form:

$$\boldsymbol{U}_{\text{IPWCC}} = \sum_{i=1}^{N} \frac{\widetilde{R}_i}{\pi_i} \sum_{j<k} \boldsymbol{U}_i(y_{ij}, y_{ik}), \tag{10.10}$$

$$\boldsymbol{U}_{\text{IPWCP}} = \sum_{i=1}^{N} \sum_{j<k<d_i} \frac{R_{ijk}}{\pi_{ijk}} \boldsymbol{U}_i(y_{ij}, y_{ik}), \tag{10.11}$$

$$\boldsymbol{U}_{\text{IPWAC}} = \sum_{i=1}^{N} \sum_{j<k} \left[\frac{R_{ij}}{\pi_{ij}} \boldsymbol{U}_i(y_{ij}) + \frac{R_{ik}}{\pi_{ik}} \boldsymbol{U}_i(y_{ik}|y_{ij}) \right], \tag{10.12}$$

where R_{ijk} and π_{ijk} are the indicator and probability, respectively, for observing both y_{ij} and y_{ik}. Further, $\boldsymbol{U}_i(y_{ik}|y_{ij})$ is the conditional score for the outcome at occasion k, given the outcome at occasion j. Single robustness is established in Molenberghs et al. (2011). Note that for the case of monotone missingness or dropout, whenever $j < k$,

$$R_{ijk} \equiv R_{ik} \qquad \text{and} \qquad \pi_{ijk} \equiv \pi_{ik} = \prod_{\ell=2}^{k}(1-p_{i\ell}),$$

in which case (10.11) can be re-expressed as:

$$\boldsymbol{U}_{\text{IPWCP}} = \sum_{i=1}^{N} \sum_{j<k<d_i} \frac{R_{ik}}{\pi_{ik}} \boldsymbol{U}_i(y_{ij}, y_{ik}).$$

Finally, doubly robust versions incorporate, into the singly robust expressions, predictive models for the unobserved outcomes given the observed ones, yielding

$$\begin{aligned}
\boldsymbol{U}_{\text{IPWCC, dr}} &= \sum_{i=1}^{N}\sum_{j<k}\left[\frac{\widetilde{R}_i}{\pi_i}\boldsymbol{U}_i(y_{ij},y_{ik})\right.\\
&\quad\left.+\left(1-\frac{\widetilde{R}_i}{\pi_i}\right)E_{\boldsymbol{Y}^m|\boldsymbol{y}^o}\boldsymbol{U}_i(y_{ij},y_{ik})\right], \qquad (10.13)\\
\boldsymbol{U}_{\text{IPWCP, dr}} &= \sum_{i=1}^{N}\sum_{j<k<n_i}\left[\frac{R_{ijk}}{\pi_{ijk}}\boldsymbol{U}_i(y_{ij},y_{ik})\right.\\
&\quad\left.+\left(1-\frac{R_{ijk}}{\pi_{ijk}}\right)E_{\boldsymbol{Y}^m|\boldsymbol{y}^o}\boldsymbol{U}_i(y_{ij},y_{ik})\right], \qquad (10.14)\\
\boldsymbol{U}_{\text{IPWAC, dr}} &= \sum_{i=1}^{N}\sum_{j<k}\left[\frac{R_{ij}}{\pi_{ij}}\boldsymbol{U}_i(y_{ij})+\frac{R_{ik}}{\pi_{ik}}\boldsymbol{U}_i(y_{ik}|y_{ij})\right.\\
&\quad+\left(1-\frac{R_{ij}}{\pi_{ij}}\right)E_{\boldsymbol{Y}^m|\boldsymbol{y}^o}\boldsymbol{U}_i(y_{ij})\\
&\quad\left.+\left(1-\frac{R_{ik}}{\pi_{ik}}\right)E_{\boldsymbol{Y}^m|\boldsymbol{y}^o}\boldsymbol{U}_i(y_{ik}|y_{ij})\right]. \qquad (10.15)
\end{aligned}$$

Double robustness is shown in Molenberghs et al. (2011).

It can be shown that all three doubly robust versions coincide, that is,

$$\begin{aligned}
\boldsymbol{U}_{\text{IPWCC, dr}} &= \boldsymbol{U}_{\text{IPWCP, dr}}=\boldsymbol{U}_{\text{IPWAC, dr}}=\\
&= \sum_{i=1}^{N}\left\{\sum_{j<k<d_i}\boldsymbol{U}_i(y_{ij},y_{ik})+\sum_{j=1}^{d_i-1}(n_i-d_i+1)\boldsymbol{U}_i(y_{ij})\right.\\
&\quad\left.+\sum_{j<d_i\le k}E\left[\boldsymbol{U}_i(y_{ik}|y_{ij})\right]+\sum_{d_i\le j<k}E\left[\boldsymbol{U}_i(y_{ij},y_{ik})\right]\right\}. \qquad (10.16)
\end{aligned}$$

Several observations can be made. First, an implication of (10.16) is that there is no need to specify the missing-data model, because the weights are no longer present in the common expression (10.16). A predictive model, however, is still necessary for the unobserved outcomes. The last two terms of (10.16) relate to the predictive models, all of which involve two types of contributions: (a) $E\left[\boldsymbol{U}_i(y_{ik}|y_{ij})\right]$ for pairs with the first component y_{ij} observed and the second one y_{ik} missing, and (b) $E\left[\boldsymbol{U}_i(y_{ij},y_{ik})\right]$ for pairs with both components y_{ij} and y_{ik} missing. Second, the equivalence of the three doubly robust versions holds for pseudo-likelihood in general, not just for pairwise (pseudo-)likelihood as presented here. Third, it can be shown that the expectations in (10.15) vanish under exchangeability, rendering (10.16) essentially equivalent to (10.9), thereby making the naive available case version not only valid, but actually doubly robust in this special setting (Molenberghs et al. 2011). An intuitive argument for this result is that under exchangeability the expectations of an unobserved measurement given the history can be replaced, consistently, by the expectation given the *observed* portion of the history, which then vanishes.

10.4.2 Precision estimation

For the singly and doubly robust versions (10.10) to (10.15), when one posits a parametric model for dropout, then the uncertainty induced by the estimation of the ψ parameters of the latter needs to be accommodated. Suppose we write $\boldsymbol{U} = \sum_{i=1}^{N} \boldsymbol{V}_i(\boldsymbol{\theta})$, and let the $\boldsymbol{\psi}$ parameters be estimated from score or estimating equations $\boldsymbol{W} = \sum_{i=1}^{N} \boldsymbol{W}_i(\boldsymbol{\psi})$. The entire score for subject i is thus $\boldsymbol{S}_i = (\boldsymbol{V}_i', \boldsymbol{W}_i')'$. The asymptotic variance-covariance matrix can be estimated consistently by $\widehat{I}_0^{-1}\widehat{I}_1\widehat{I}_0^{-1}$, with

$$I_0 = \sum_{i=1}^{N} \begin{pmatrix} \dfrac{\partial \boldsymbol{V}_i}{\partial \boldsymbol{\theta}} & \dfrac{\partial \boldsymbol{V}_i}{\partial \boldsymbol{\psi}} \\ \mathbf{0} & \dfrac{\partial \boldsymbol{W}_i}{\partial \boldsymbol{\psi}} \end{pmatrix} \quad \text{and} \quad I_1 = \sum_{i=1}^{N} \boldsymbol{S}_i(\widehat{\boldsymbol{\theta}}, \widehat{\boldsymbol{\psi}})\boldsymbol{S}_i'(\widehat{\boldsymbol{\theta}}, \widehat{\boldsymbol{\psi}}). \tag{10.17}$$

See also Bang and Robins (2005) and Rotnitzky (2009).

Note, however, that for the doubly robust case, by virtue of (10.16), an explicit model for dropout is not needed, which would in turn imply that these modifications to I_0 and I_1 may actually be unnecessary. Hence,

$$I_0 = \sum_{i=1}^{N} \frac{\partial \boldsymbol{V}_i}{\partial \boldsymbol{\theta}} \quad \text{and} \quad I_1 = \sum_{i=1}^{N} \boldsymbol{S}_i(\widehat{\boldsymbol{\theta}})\boldsymbol{S}_i'(\widehat{\boldsymbol{\theta}}). \tag{10.18}$$

This is also the case for the naive versions, in which no weighting is used in the estimating equations.

10.4.3 Marginal pseudo-likelihood for binary data

Assume that we have a model for multivariate and hence also for bivariate binary data. For instance, Bahadur (1961) proposed a marginal model that accounts for the association via marginal correlations. Using the notation $\nu_{ij} = P(Y_{ij} = 1), \nu_{ijk} = P(Y_{ij} = 1, Y_{ik} = 1)$, and $\nu_{ik|j} = P(Y_{ik} = 1|y_{ij} = \ell)(\ell = 0, 1)$, pairwise Bahadur probabilities take the form

$$\nu_{ijk} = \nu_{ij}\nu_{ik}\left[1 + \rho_{ijk}\frac{1-\nu_{ij}}{\sqrt{\nu_{ij}(1-\nu_{ij})}}\frac{1-\nu_{ik}}{\sqrt{\nu_{ik}(1-\nu_{ik})}}\right]. \tag{10.19}$$

Here, the ρ parameters are pairwise and higher-order correlations. The expressions are implicit and fitting the model is challenging from a computational perspective. The multivariate Bahadur model can be written as $f(\boldsymbol{y}_i) = f_1(\boldsymbol{y}_i)c(\boldsymbol{y}_i)$, with

$$f_1(\boldsymbol{y}_i) = \prod_{j=1}^{n_i} \nu_{ij}^{y_{ij}}(1-\nu_{ij})^{1-y_{ij}}, \tag{10.20}$$

$$c(\boldsymbol{y}_i) = 1 + \sum_{j_1<j_2} \rho_{ij_1j_2}e_{ij_1}e_{ij_2} + \sum_{j_1<j_2<j_3} \rho_{ij_1j_2j_3}e_{ij_1}e_{ij_2}e_{ij_3} + \cdots + \rho_{ij_1j_2\ldots j_{n_i}}e_{ij_1}e_{ij_2}\cdots e_{ij_{n_i}}, \tag{10.21}$$

where $e_{ij} = \dfrac{y_{ij}-\nu_{ij}}{\sqrt{\nu_{ij}(1-\nu_{ij})}}$.

Even though the model admits a convenient and concise closed form, its fitting is less than trivial, owing to strong and intractable constraints on the parameter space, be it in fully general or second-order form (where the third- and higher-order correlations are set equal to zero). This makes pseudo-likelihood attractive. A generic contribution to the pairwise log-likelihood takes the form:

$$\begin{aligned} p\ell_{ijk} &= y_{ij}y_{ik}\ln\nu_{ijk} + y_{ij}(1-y_{ik})\ln(\nu_{ij}-\nu_{ijk}) \\ &\quad +(1-y_{ij})y_{ik}\ln(\nu_{ik}-\nu_{ijk}) \\ &\quad + (1-y_{ij})(1-y_{ik})\ln(1-\nu_{ij}-\nu_{ik}+\nu_{ijk}). \end{aligned}$$

As before, let $\boldsymbol{\theta} = (\boldsymbol{\beta}', \boldsymbol{\alpha}')'$, where $\nu_{ij} = \nu_{ij}(\boldsymbol{\beta})$ and the association parameters are functions of $\boldsymbol{\alpha}$. Hence, $\nu_{ijk} = \nu_{ijk}(\boldsymbol{\beta}, \boldsymbol{\alpha})$. Pairwise and conditional contributions to the score take the form:

$$\begin{aligned} \boldsymbol{U}_{ijk} &= \frac{y_{ij}y_{ik}}{\nu_{ijk}}\frac{\partial}{\partial\boldsymbol{\theta}}\nu_{ijk} + \frac{y_{ij}(1-y_{ik})}{\nu_{ij}-\nu_{ijk}}\frac{\partial}{\partial\boldsymbol{\theta}}(\nu_{ij}-\nu_{ijk}) \\ &+ \frac{(1-y_{ij})y_{ik}}{\nu_{ik}-\nu_{ijk}}\frac{\partial}{\partial\boldsymbol{\theta}}(\nu_{ik}-\nu_{ijk}) \\ &+ \frac{(1-y_{ij})(1-y_{ik})}{1-\nu_{ij}-\nu_{ik}+\nu_{ijk}}\frac{\partial}{\partial\boldsymbol{\theta}}(1-\nu_{ij}-\nu_{ik}+\nu_{ijk}) \end{aligned} \tag{10.22}$$

and

$$\begin{aligned} \boldsymbol{U}_{ik|j} = & \frac{y_{ij}y_{ik}\nu_{ij}}{\nu_{ijk}}\frac{\partial}{\partial\boldsymbol{\theta}}\left(\frac{\nu_{ijk}}{\nu_{ij}}\right) + \frac{y_{ij}(1-y_{ik})\nu_{ij}}{\nu_{ij}-\nu_{ijk}}\frac{\partial}{\partial\boldsymbol{\theta}}\left(\frac{\nu_{ij}-\nu_{ijk}}{\nu_{ij}}\right) \\ &+ \frac{(1-y_{ij})y_{ik}(1-\nu_{ij})}{\nu_{ik}-\nu_{ijk}}\frac{\partial}{\partial\boldsymbol{\theta}}\left(\frac{\nu_{ik}-\nu_{ijk}}{1-\nu_{ij}}\right) \\ &+ \frac{(1-y_{ij})(1-y_{ik})(1-\nu_{ij})}{1-\nu_{ij}-\nu_{ik}+\nu_{ijk}}\frac{\partial}{\partial\boldsymbol{\theta}}\left(\frac{1-\nu_{ij}-\nu_{ik}+\nu_{ijk}}{1-\nu_{ij}}\right). \end{aligned} \tag{10.23}$$

In addition, we need expectations of these over the conditional distribution of the unobserved outcomes given the observed ones. Evidently, because (10.22) and (10.23) are linear in the triplet y_{ij}, y_{ik} and $y_{ij}y_{ik}$, it suffices to calculate the expectations over these. Their corresponding probabilities are

$$\nu_{ij|\overline{d}} = \frac{\nu_{i\overline{d}j}}{\nu_{i\overline{d}}} \qquad \text{and} \qquad \nu_{ijk|\overline{d}} = \frac{\nu_{i\overline{d}jk}}{\nu_{i\overline{d}}}, \tag{10.24}$$

where $\overline{d}$ refers to the set of indices $(1, 2, \ldots, d-1)$, corresponding to the observed portion of $\boldsymbol{y}$.

Combining (10.22) and (10.23) with (10.24) leads to:

$$\begin{aligned}
E(\boldsymbol{U}_{ijk}) &= \frac{\nu_{i\bar{d}jk}}{\nu_{i\bar{d}}\nu_{ijk}}\frac{\partial}{\partial\boldsymbol{\theta}}\nu_{ijk} + \frac{\nu_{i\bar{d}j}-\nu_{i\bar{d}jk}}{\nu_{i\bar{d}}(\nu_{ij}-\nu_{ijk})}\frac{\partial}{\partial\boldsymbol{\theta}}(\nu_{ij}-\nu_{ijk}) \\
&+ \frac{\nu_{i\bar{d}k}-\nu_{i\bar{d}jk}}{\nu_{i\bar{d}}(\nu_{ik}-\nu_{ijk})}\frac{\partial}{\partial\boldsymbol{\theta}}(\nu_{ik}-\nu_{ijk}) \\
&+ \frac{\nu_{i\bar{d}}-\nu_{i\bar{d}j}-\nu_{i\bar{d}k}+\nu_{i\bar{d}jk}}{\nu_{i\bar{d}}(1-\nu_{ij}-\nu_{ik}+\nu_{ijk})}\frac{\partial}{\partial\boldsymbol{\theta}}(1-\nu_{ij}-\nu_{ik}+\nu_{ijk})
\end{aligned} \tag{10.25}$$

and

$$\begin{aligned}
E(\boldsymbol{U}_{ik|j}) &= \frac{y_{ij}\nu_{i\bar{d}k}\nu_{ij}}{\nu_{i\bar{d}}\nu_{ijk}}\frac{\partial}{\partial\boldsymbol{\theta}}\left(\frac{\nu_{ijk}}{\nu_{ij}}\right) + \frac{y_{ij}(\nu_{i\bar{d}}-\nu_{i\bar{d}k})\nu_{ij}}{\nu_{i\bar{d}}(\nu_{ij}-\nu_{ijk})}\frac{\partial}{\partial\boldsymbol{\theta}}\left(\frac{\nu_{ij}-\nu_{ijk}}{\nu_{ij}}\right) \\
&+ \frac{(1-y_{ij})\nu_{i\bar{d}k}(1-\nu_{ij})}{\nu_{i\bar{d}}(\nu_{ik}-\nu_{ijk})}\frac{\partial}{\partial\boldsymbol{\theta}}\left(\frac{\nu_{ik}-\nu_{ijk}}{1-\nu_{ij}}\right) \\
&+ \frac{(1-y_{ij})(\nu_{i\bar{d}}-\nu_{i\bar{d}k})(1-\nu_{ij})}{\nu_{i\bar{d}}(1-\nu_{ij}-\nu_{ik}+\nu_{ijk})}\frac{\partial}{\partial\boldsymbol{\theta}}\left(\frac{1-\nu_{ij}-\nu_{ik}+\nu_{ijk}}{1-\nu_{ij}}\right).
\end{aligned} \tag{10.26}$$

All probabilities involving $\bar{d}$ are potentially high-dimensional, and follow from the multivariate Bahadur model. We have seen, however, that several alternative routes are possible. For example, here, one could simply resort to the singly robust version. Alternatively, the expectations could be replaced by simple models, for example the logistic: $E_{\boldsymbol{Y}_i^m|\boldsymbol{y}_i^o}(y_{ij})$ could be written as a standard logistic model, in which the existing covariates are supplemented by $\boldsymbol{y}_{i\bar{d}}$, whereas for $E_{\boldsymbol{Y}_i^m|\boldsymbol{y}_i^o}(y_{ij}y_{ik})$ the pairwise model under consideration can be used, again supplementing the covariate information with $\boldsymbol{y}_{i\bar{d}}$. Another approach would be to evaluate expressions (10.25) and (10.26) using the classical definition of an expectation for discrete distributions. That is,

$$\begin{aligned}
E(\boldsymbol{U}_{ijk}) &\equiv E\left[\boldsymbol{U}_i(y_{ij},y_{ik})\right] \\
&= \sum_{y_{ij}=0}^{1}\sum_{y_{ik}=0}^{1}\boldsymbol{U}_i(y_{ij},y_{ik})P(Y_{ij}=y_{ij},Y_{ik}=y_{ik}), \qquad (10.27) \\
E(\boldsymbol{U}_{ik|j}) &\equiv E\left[\boldsymbol{U}_i(y_{ik}|y_{ij})\right] \\
&= \sum_{y_{ik}=0}^{1}\boldsymbol{U}_i(y_{ik}|y_{ij})P(Y_{ik}=y_{ik}|Y_{ij}=y_{ij}), \qquad (10.28)
\end{aligned}$$

where $\boldsymbol{U}_i(y_{ij},y_{ik})$ and $\boldsymbol{U}_i(y_{ik}|y_{ij})$ are as defined in (10.22) and (10.23), while $P(Y_{ij}=y_{ij},Y_{ik}=y_{ik})$ and $P(Y_{ik}=y_{ik}|Y_{ij}=y_{ij})$ are the pairwise and conditional probabilities for the Bahadur model, respectively. We note as well that expressions (10.22)–(10.28) require derivatives with respect to the univariate and pairwise probabilities. For most pairwise models, such as the Bahadur model, these are reasonably straightforward to evaluate, and have been derived by various authors. See Molenberghs and Verbeke (2005) for details.

10.4.4 The multivariate normal model

While the multivariate normal model parameters are generally easy to estimate with full maximum likelihood, there is potential value in developing PL for this setting because closed-form expressions for the estimating equations can be derived. There are also practical settings, such as high-dimensional problems, and those for which there is doubt about the correct specification of the covariance structure, for which PL has practical advantages over full likelihood.

Assume $\boldsymbol{Y}_i \sim N(\boldsymbol{\mu}, \Sigma)$. Denote the elements of $\boldsymbol{\mu}$ by μ_j and the entries in Σ by σ_{jk}. Conditional means and variances are denoted by $\mu_{k|j}$ and $\sigma_{kk|j}$, respectively. Then first, suppressing the index i from notation, and writing down the expressions for observed values, we find:

$$\begin{aligned}
&\boldsymbol{U}(y_k|y_j)\\
&\quad= \frac{\partial(\mu_{k|j}, \sigma_{kk|j})}{\partial(\mu_j, \mu_k, \sigma_{jj}, \sigma_{jk}, \sigma_{kk})} \cdot \frac{\partial \ln \phi(y_k|y_j; \mu_{k|j}, \sigma_{kk|j})}{\partial(\mu_{k|j}, \sigma_{kk|j})}\\
&\quad= \begin{pmatrix} -\frac{\sigma_{jk}}{\sigma_{jj}} & 0 \\ 1 & 0 \\ -\frac{\sigma_{jk}}{\sigma_{jj}^2}(y_j - \mu_j) & \frac{\sigma_{jk}^2}{\sigma_{jj}^2} \\ \frac{y_j - \mu_j}{\sigma_{jj}} & -\frac{2\sigma_{jk}}{\sigma_{jj}} \\ 0 & 1 \end{pmatrix} \begin{pmatrix} \frac{y_k - \mu_{k|j}}{\sigma_{kk|j}} \\ -\frac{1}{2\sigma_{kk|j}} + \frac{1}{2}\frac{(y_k - \mu_{k|j})^2}{\sigma_{kk|j}^2} \end{pmatrix},
\end{aligned} \tag{10.29}$$

where $\phi(\cdot)$ is the normal density with mean and variance given by:

$$\mu_{k|j} = \mu_k + \frac{\sigma_{jk}}{\sigma_{jj}}(y_j - \mu_j) \quad \text{and} \quad \sigma_{kk|j} = \frac{\sigma_{jj}\sigma_{kk} - \sigma_{jk}^2}{\sigma_{jj}}.$$

The only stochastic elements in (10.29) are the conditional residual and its square. We need to take their expectation conditional upon the observed outcomes, producing for the second factor in (10.29):

$$\begin{pmatrix} \frac{\sigma_{jj}\Sigma_{k\overline{d}}\Sigma_{\overline{d}\,\overline{d}}^{-1}(\boldsymbol{Y}_{\overline{d}} - \boldsymbol{\mu}_{\overline{d}}) - \sigma_{jk}(y_j - \mu_j)}{\sigma_{jj}\sigma_{kk} - \sigma_{jk}^2} \\ \frac{\sigma_{jj}(\sigma_{jk}^2 - \sigma_{jj}\Sigma_{k\overline{d}}\Sigma_{\overline{d}\,\overline{d}}^{-1}\Sigma_{\overline{d}k}) + \left[\sigma_{jj}\Sigma k\overline{d}\,\Sigma_{\overline{d}\,\overline{d}}^{-1}(\boldsymbol{Y}_{\overline{d}} - \boldsymbol{\mu}_{\overline{d}}) - \sigma_{jk}(y_j - \mu_j)\right]^2}{2(\sigma_{jj}\sigma_{kk} - \sigma_{jk}^2)^2} \end{pmatrix}. \tag{10.30}$$

Here, $\overline{d}$ refers to the set of indices $(1, 2, \ldots, d-1)$, corresponding to the observed portion of $\boldsymbol{Y}$. Turning to the other expectation, we find:

$$\begin{aligned}
\boldsymbol{U}(y_j, y_k) &= \frac{\partial \ln \phi(y_j, y_k; \mu_j, \mu_k, \sigma_{jj}, \sigma_{jk}, \sigma_{kk})}{\partial(\mu_j, \mu_k, \sigma_{jj}, \sigma_{jk}, \sigma_{kk})}\\
&= \begin{pmatrix} \Sigma^{-1}(\boldsymbol{y} - \boldsymbol{\mu}) \\ h_{jj} + Q_{jj} \\ h_{jk} + Q_{jk} \\ h_{kk} + Q_{kk} \end{pmatrix},
\end{aligned} \tag{10.31}$$

where

$$h_{jj} = -\frac{1}{2}\frac{\sigma_{kk}}{\varphi}, \qquad h_{jk} = \frac{\sigma_{jk}}{\varphi}, \qquad h_{kk} = -\frac{1}{2}\frac{\sigma_{jj}}{\varphi},$$

$$\varphi = \sigma_{jj}\sigma_{kk} - \sigma_{jk}^2,$$

$$Q_\sigma = \tfrac{1}{2}(\boldsymbol{y} - \boldsymbol{\mu})'\Sigma^{-1}S_\sigma\Sigma^{-1}(\boldsymbol{y} - \boldsymbol{\mu}),$$

$$S_{jj} = \begin{pmatrix} 1 & 0 \\ 0 & 0 \end{pmatrix}, \qquad S_{jk} = \begin{pmatrix} 0 & 1 \\ 1 & 0 \end{pmatrix}, \qquad S_{kk} = \begin{pmatrix} 0 & 0 \\ 0 & 1 \end{pmatrix}.$$

Here, S_σ is generic notation for either one of the three pairs (j,j), (j,k), and (k,k).

To calculate the expectation of (10.31), we need:

$$E(\boldsymbol{Y}|\boldsymbol{y}_{\overline{d}}) = \boldsymbol{\mu}^c_{jk} = \boldsymbol{\mu} + \Sigma_{jk,\overline{d}}\Sigma^{-1}_{\overline{d},\overline{d}}(\boldsymbol{y}_{\overline{d}} - \boldsymbol{\mu}_{\overline{d}}), \tag{10.32}$$

$$\text{var}(\boldsymbol{Y}|\boldsymbol{y}_{i\overline{d}}) = \Sigma_{jk,jk} - \Sigma_{jk,\overline{d}}\Sigma^{-1}_{\overline{d},\overline{d}}\Sigma_{\overline{d},jk}. \tag{10.33}$$

It now follows that

$$E\left[\boldsymbol{U}(y_j,y_k)|\boldsymbol{y}_{\overline{d}}\right] = \begin{pmatrix} \Sigma^{-1}_{jk,jk}\Sigma_{jk,\overline{d}}\Sigma^{-1}_{\overline{d},\overline{d}}(\boldsymbol{y}_{\overline{d}} - \boldsymbol{\mu}_{\overline{d}}) \\ h_{jj} + E[Q_{jj}|\boldsymbol{y}_{\overline{d}}] \\ h_{jk} + E[Q_{jk}|\boldsymbol{y}_{\overline{d}}] \\ h_{kk} + E[Q_{kk}|\boldsymbol{y}_{\overline{d}}] \end{pmatrix}, \tag{10.34}$$

where some straightforward algebra produces:

$$E[Q_\sigma|\boldsymbol{y}_{\overline{d}}] = \frac{1}{2}\text{tr}\left\{\Sigma^{-1}_{jk,jk}S_\sigma\Sigma^{-1}_{jk,jk}\left[\Sigma_{jk,jk} + \Sigma_{jk,\overline{d}}\Sigma^{-1}_{\overline{d},\overline{d}} \times \right.\right.$$
$$\left.\left. \times \left((\boldsymbol{y}_{\overline{d}} - \boldsymbol{\mu}_{\overline{d}})(\boldsymbol{y}_{\overline{d}} - \boldsymbol{\mu}_{\overline{d}})' - \Sigma_{\overline{d},\overline{d}}\right)\Sigma^{-1}_{\overline{d},\overline{d}}\Sigma_{\overline{d},jk}\right]\right\}. \tag{10.35}$$

In the special case of two measurements, the first of which always observed, $\overline{d} = 1$ in (10.30), i.e., it refers to the first measurement. Hence, both expectations in (10.30) reduce to 0, implying in turn that then $E_{y^m|y^o}\boldsymbol{U}(y_2|y_1) = E_{y_2|y_1}\boldsymbol{U}(y_2|y_1) = \boldsymbol{0}$, as it should because in this simple case pseudo-likelihood coincides with full likelihood.

For each of the estimators, the sandwich estimator can be computed. For the case of IPWCC and its doubly robust version, Molenberghs et al. (2011) provide generic expressions.

10.5 Case Studies

10.5.1 A clinical trial in onychomycosis

These data come from a randomized, double-blind, parallel group study for the comparison of two oral treatments (coded as A and B) for toenail dermatophyte onychomycosis (TDO), described in full detail by De Backer et al. (1996). See also Verbeke and Molenberghs (2000) and Molenberghs and Verbeke (2005). The aim of the trial was to compare the efficacy and safety of 12 weeks of continuous therapy with treatment A, or with treatment B. We consider data from 146 patients in arm A and 148 in arm B, for whom the big toenail was the target nail. Patients were assessed at 0, 1, 2, 3, 6, 9, and 12 months. The response is the unaffected nail length, measured from the nail bed to the infected part of the nail, which is always at the free end of the nail, expressed in *mm*.

The design and data type of this study are sufficiently simple to allow for full likelihood, providing a basis for comparison with the proposed pseudo-likelihood methods. We used several forms of pairwise marginal likelihood, in particular with the multivariate normal versions as in Section 10.4.4.

For the unaffected nail length Y_{ij}, measured at occasion j for patient i, we specify a linear

TABLE 10.1
Toenail Data. (Unaffected nail length outcome). Parameter estimates (purely model-based standard errors; empirically corrected standard errors) for full likelihood, and naive, singly robust, and doubly robust pairwise likelihood.

Effect	Par.	$U_{\text{full.lik.}}$	U_{nCC}	U_{nCP}	U_{nAC}
Int.A	β_0	2.52(0.25;0.23)	2.77(0.09;0.27)	2.70(0.08;0.25)	2.56(0.08;0.23)
Int.B	β_1	2.77(0.24;0.25)	2.82(0.08;0.27)	2.81(0.08;0.25)	2.77(0.07;0.25)
Sl.A	β_2	0.56(0.02;0.05)	0.55(0.01;0.05)	0.56(0.01;0.05)	0.57(0.01;0.05)
Sl.B	β_3	0.61(0.02;0.04)	0.60(0.01;0.04)	0.61(0.01;0.04)	0.61(0.01;0.04)
R.I.v.	τ^2	6.49(0.63;0.63)	6.71(0.23;0.73)	6.67(0.21;0.68)	6.41(0.20;0.65)
Res.v.	σ^2	6.94(0.25;0.47)	7.31(0.15;0.52)	7.13(0.14;0.48)	7.05(0.14;0.47)
Effect	**Par.**	$U_{\text{wt.lik.}}$	U_{IPWCC}	U_{IPWCP}	U_{IPWAC}
Int.A	β_0	1.85(0.09;0.30)	2.71(0.07;0.27)	2.77(0.08;0.27)	2.59(0.07;0.24)
Int.B	β_1	2.65(0.09;0.52)	2.78(0.07;0.27)	2.82(0.08;0.27)	2.77(0.07;0.25)
Sl.A	β_2	0.68(0.01;0.07)	0.54(0.01;0.05)	0.53(0.01;0.04)	0.55(0.01;0.05)
Sl.B	β_3	0.73(0.01;0.10)	0.60(0.01;0.04)	0.59(0.01;0.04)	0.60(0.01;0.04)
R.I.v.	τ^2	6.21(0.24;1.03)	6.66(0.20;0.72)	6.72(0.21;0.75)	6.44(0.19;0.67)
Res.v.	σ^2	5.05(0.09;0.60)	7.29(0.13;0.51)	7.59(0.14;0.56)	7.35(0.13;0.51)
Effect	**Par.**			$U_{\text{IPWCC,dr}} = U_{\text{IPWCP,dr}} = U_{\text{IPWAC,dr}}$	
Int.A	β_0			2.52(0.07;0.23)	
Int.B	β_1			2.77(0.07;0.25)	
Sl.A	β_2			0.56(0.01;0.05)	
Sl.B	β_3			0.61(0.01;0.04)	
R.I.v.	τ^2			6.23(0.20;0.64)	
Res.v.	σ^2			7.09(0.14;0.48)	

mixed-effects model:

$$\begin{aligned} Y_{ij}|b_i &\sim N[b_i + \beta_0 \cdot I(T_i = 0) + \beta_1 \cdot I(T_i = 1) + \beta_2 t_j \cdot I(T_i = 0) \\ &\quad +\beta_3 t_j \cdot I(T_i = 1), \sigma^2], \\ b_i &\sim N(0, \tau^2), \end{aligned} \tag{10.36}$$

where $T_i = 0$ if patient i received standard treatment and 1 for experimental therapy ($i = 1, \ldots, 298$). Further, t_j is the time at which the jth measurement is taken ($j = 1, \ldots, 7$). Finally, $I(\cdot)$ is an indicator function. Parameter estimates and standard errors, obtained through maximum likelihood and pairwise likelihood, are presented in Table 10.1.

Observe that all point estimates are relatively close to each other, except for some deviation in the weighted likelihood analysis, a weighted version of conventional likelihood analysis. Note that, with likelihood, there is little rationale for using weights, here leading to a worse fit.

The model-based standard errors are appropriate only in the conventional full likelihood case, where they are reasonably close to the empirically corrected ones. They are not appropriate for the weighted analyses, as they are based on the incorrect assumption that the weights represent replication at the subject (or pair) level. Furthermore, naive standard errors in the pseudo-likelihood case are based on the entirely incorrect assumption that every pair results from independent replication whereas, for example, in a completely observed sequence every measurement is used in six different pairs. It is important therefore to use the empirically corrected standard errors for inferential purposes.

TABLE 10.2
Analgesic Trial. Absolute and relative frequencies of the five GSA categories for each of the four follow-up times.

GSA	Month 3		Month 6		Month 9		Month 12	
1	55	14.3%	38	12.6%	40	17.6%	30	13.5%
2	112	29.1%	84	27.8%	67	29.5%	66	29.6%
3	151	39.2%	115	38.1%	76	33.5%	97	43.5%
4	52	13.5%	51	16.9%	33	14.5%	27	12.1%
5	15	3.9%	14	4.6%	11	4.9%	3	1.4%
Tot	385		302		227		223	

It is clear that using complete cases only resulted in a small loss of efficiency in the naive and IPW cases, whereas the available-case approach makes optimal use of the data. Turning to the doubly robust versions, not only was it confirmed that all three coincide, they were also very close to full likelihood, both in terms of point estimates and precision.

In a relatively large dataset with continuous outcomes, like this one, treating the weights in the weighted analysis as either fixed or random leads to the same standard errors. In the next example, however, this is not the case. The weights were based on a logistic model for dropout in the toenail study:

$$\begin{aligned}
&\text{logit}[P(D_i = j|D_i \geq j, T_i, t_j, Y_{i,j-1})] \\
&= \ -3.17(0.24) - 0.28(0.24)T_i + 0.072(0.036)t_j \\
&\quad -0.035(0.036)Y_{i,j-1}. \qquad (10.37)
\end{aligned}$$

Note that, while the effect of the previous measurement was not significant, the weighted analyses were different from the unweighted ones. In this sense, it is an advantage that the doubly robust versions removes the need for using weights, as long as the expectations are included. This is not always the case, because it is a consequence of the pairwise marginal nature of the likelihood contributions.

10.5.2 Analgesic trial

These data come from a single-arm clinical trial in 395 patients who are given analgesic treatment for pain caused by chronic non-malignant disease. Treatment was to be administered for 12 months and assessed by means of a five-point "Global Satisfaction Assessment" (GSA) scale: (1) very good; (2) good; (3) indifferent; (4) bad; (5) very bad. Some analyses have been done on a dichotomized version, GSABIN, which is 1 if GSA$\leq$ 3 and 0 otherwise. Apart from the outcome of interest, a number of covariates are available, such as age, sex, weight, duration of pain in years prior to the start of the study, type of pain, physical functioning, psychiatric condition, respiratory problems, etc.

GSA was rated by each person four times during the trial: at months 3, 6, 9, and 12. An overview of the frequencies per follow up time is given in Table 10.2. Inspecting Table 10.2 reveals that the total per column is variable. This is due to missingness. At three months, 10 subjects lack a measure, with these numbers being 93, 168, and 172 at subsequent times. Although subjects with intermittent values exist, this study focuses only on subjects with monotone missingness or dropout.

TABLE 10.3
Analgesic Trial. Overview of missingness patterns and the frequencies with which they occur. "O" indicates observed and "M" indicates missing.

	Measurement Occasion					
	Month 3	Month 6	Month 9	Month 12	N	%
Completers	O	O	O	O	163	41.2
Dropouts	O	O	O	M	51	12.91
	O	O	M	M	51	12.91
	O	M	M	M	63	15.95
Non-monotone Missingness	O	O	M	O	30	7.59
	O	M	O	O	7	1.77
	O	M	O	M	2	0.51
	O	M	M	O	18	4.56
	M	O	O	O	2	0.51
	M	O	O	M	1	0.25
	M	O	M	O	1	0.25
	M	O	M	M	3	0.76

An overview of the extent of missingness is shown in Table 10.3. Note that only around 40% of the subjects have complete data. Both dropout and intermittent pattern of missingness occur, the former amounting to roughly 40%, with close to 20% for the latter. This example underscores that a satisfactory longitudinal analysis will oftentimes have to address the missing data problem.

For all ensuing analyses on the analgesic trial data, we consider only completers and dropouts, i.e., a subset of 328 patients from the original dataset. We first build a logistic regression for the dropout indicator, in terms of the previous outcome and pain control assessment at baseline, i.e.,

$$\text{logit } P(D_i = j | D_i \geq j, x_i, y_{i,j-1}) = \psi_0 + \psi_x x_i + \psi_{prev} y_{i,j-1}. \tag{10.38}$$

The highly significant p-value ($p < .0001$) for the parameter ψ_{prev} corresponding to the previous outcome provides evidence against MCAR in favor of MAR. Weights are then calculated based on predicted probabilities from this logistic model.

Preliminary analyses have indicated that, among a set of potential covariates, the linear and square effects of time t_{ij}, as well as the effect of baseline pain control assessment (PCA_0, denoted x_i) are of importance. The marginal regression model for the dichotomized GSA score, GSABIN, denoted as Y, is thus specified as

$$\text{logit } P(Y_{ij} = 1 | t_{ij}, x_i) = \beta_0 + \beta_1 t_{ij} + \beta_2 t_{ij}^2 + \beta_3 x_i. \tag{10.39}$$

For the correlation across the within-subject outcomes, we posit a Toeplitz type correlation structure.

The resulting parameter estimates, along with corresponding standard errors, for model specification (10.39), with a Toeplitz correlation structure, using full likelihood and estimating equations (10.7) to (10.12) and (10.16) are presented in Table 10.4. There are two panels for the IPW cases: the first panel provides the results for IPW when the weights obtained from the dropout model are considered as fixed, while the second panel shows the corresponding results considering that the weights are estimated, in which case, the variability in the estimated weights is incorporated in the computation of the standard errors.

The high degree of similarity in the results in these two panels indicates that the additional variability induced by estimation of the weight model does not seem to impact largely on either the estimates or their standard errors.

Fairly comparable results are also observed for the parameter estimates under full likelihood, naive AC and the doubly robust cases. Moreover, substantial efficiency over full likelihood seems to be gained under the naive AC and doubly robust approaches. Whereas these observations are not surprising for the doubly robust case, precisely because of their property, the relatively good performance of the naive AC case seems counterintuitive. However, under exchangeability, as shown before, the naive AC can be seen as a doubly robust estimator, given that then the expectation in (10.15) can be removed because observed and unobserved components from a subject's history are interchangeable. To this effect, we assessed the plausibility of the Toeplitz correlation structure of the analgesic trial data, using full likelihood, and determined that the three correlation parameters $\rho^{(k)}$, $k = 1, 2, 3$, were not significantly different ($p = 0.8091$), which implies that the underlying correlation structure might very well be exchangeable. This explains the excellent behavior of the naive AC estimator.

There are two versions for the doubly robust estimators. (A) refers to predictive distributions based on logistic regression, whereas with (B) the proper conditional is used for these predictive distributions. The difference in the estimates from these two approaches is quite noticeable, especially for the correlation parameters.

Next, we consider the CC and CP versions, both naive and singly robust (IPW). For the CC approaches, while the estimates for the parameters β_1, β_2 and β_3 are reasonably close to those under full likelihood, some disparity can be seen in the intercept β_0 and in the correlation parameters, the latter is particularly apparent for the IPW cases. In addition, the standard errors under the naive CC approach are generally larger than those for full likelihood. For IPWCC, in contrast, smaller standard errors are observed, a result that could be attributed to the single robustness of IPWCC. For the CP cases, in either the naive or the singly robust situations, the CP results seem to fall in between the CC and the AC results, implying somewhat of a compromise between the latter two. This can be inferred from the incremental nature of the contributions in expressions (10.7), (10.8), (10.9) and (10.10), (10.11), (10.12).

For the singly robust versions, throughout, all IPW versions yield correlation parameter estimates that are quite different to those obtained under full likelihood. This might be a result of the misspecified correlation structure, as mentioned earlier. There is apparent protection in the sense that the regression model parameters are generally reasonable, but the association parameters are not as well-protected.

10.6 Concluding Remarks

Based upon work by Molenberghs et al. (2011) and Birhanu et al. (2013), we have described the use of PL methodology when data are incomplete. PL shares with GEE its semi-parametric basis, so its use under MAR generally requires corrections, of a singly robust or doubly robust type. Particular emphasis has been placed on marginal PL and, in particular, its pairwise form. Molenberghs et al. (2011) also give an example with conditional likelihood.

TABLE 10.4
Analgesic Trial. Parameter estimates (empirically corrected standard errors) for naive, singly and doubly robust pairwise likelihood and for full likelihood. (A) refers to predictive distributions based on logistic regression, whereas with (B) the proper conditional is used.

Effect	Par.	U_{nCC}	U_{nCP}	U_{nAC}	$U_{\text{full.lik.}}$
Inter.	β_0	3.131 (0.703)	2.962 (0.562)	2.691 (0.370)	2.636 (0.523)
Time	β_1	-0.913 (0.504)	-0.908 (0.407)	-0.825 (0.304)	-0.763 (0.379)
Time^2	β_2	0.170 (0.098)	0.177 (0.081)	0.183 (0.066)	0.167 (0.078)
PCA$_0$	β_3	-0.130 (0.136)	-0.125 (0.119)	-0.195 (0.069)	-0.187 (0.103)
corr$_1$	$\rho^{(1)}$	0.217 (0.069)	0.244 (0.056)	0.210 (0.056)	0.192 (0.474)
corr$_2$	$\rho^{(2)}$	0.199 (0.075)	0.234 (0.068)	0.178 (0.068)	0.160 (0.068)
corr$_3$	$\rho^{(3)}$	0.224 (0.102)	0.232 (0.103)	0.116 (0.096)	0.123 (0.102)

Considering Weights as Fixed

Effect	Par.	U_{IPWCC}	U_{IPWCP}	U_{IPWAC}
Inter.	β_0	3.090 (0.297)	2.717 (0.519)	2.763 (0.381)
Time	β_1	-0.997 (0.200)	-0.774 (0.368)	-0.690 (0.253)
Time^2	β_2	0.193 (0.039)	0.154 (0.072)	0.131 (0.047)
PCA$_0$	β_3	-0.195 (0.061)	-0.141 (0108)	-0.155 (0.074)
corr$_1$	$\rho^{(1)}$	0.263 (0.028)	0.275 (0.048)	0.286 (0.031)
corr$_2$	$\rho^{(2)}$	0.257 (0.031)	0.255 (0.065)	0.264 (0.033)
corr$_3$	$\rho^{(3)}$	0.295 (0.041)	0.267 (0.106)	0.291 (0.042)

Incorporating Variability from Estimated Weights

Effect	Par.	U_{IPWCC}	U_{IPWCP}	U_{IPWAC}
Inter.	β_0	3.079 (0.299)	2.714 (0.521)	2.767 (0.385)
Time	β_1	-0.999 (0.200)	-0.775 (0.368)	-0.701 (0.253)
Time^2	β_2	0.194 (0.039)	0.154 (0.072)	0.134 (0.047)
PCA$_0$	β_3	-0.193 (0.061)	-0.141 (0.108)	-0.155 (0.074)
corr$_1$	$\rho^{(1)}$	0.258 (0.028)	0.275 (0.049)	0.284 (0.031)
corr$_2$	$\rho^{(2)}$	0.252 (0.031)	0.256 (0.065)	0.260 (0.033)
corr$_3$	$\rho^{(3)}$	0.284 (0.041)	0.266 (0.107)	0.276 (0.042)

Effect	Par.	$U_{\text{IPW, dr (A)}}$	$U_{\text{IPW, dr (B)}}$
Inter.	β_0	2.637 (0.272)	2.644 (0.301)
Time	β_1	-0.763 (0.182)	-0.761 (0.193)
Time^2	β_2	0.167 (0.029)	0.169 (0.033)
PCA$_0$	β_3	-0.187 (0.017)	-0.188 (0.046)
corr$_1$	$\rho^{(1)}$	0.192 (0.002)	0.194 (0.053)
corr$_2$	$\rho^{(2)}$	0.160 (0.003)	0.161 (0.061)
corr$_3$	$\rho^{(3)}$	0.123 (0.006)	0.126 (0.092)

Having shown that, under MAR, naive complete-case and available-case estimating equations are biased, we have formulated several alternative versions that overcome this problem, including both singly and doubly robust forms. The second of these requires evaluation of conditional expectations of the unobserved outcomes given the observed ones, which in turn may require joint distributions of a higher order than those used in the singly robust version. While at first sight this seems to undermine the appeal of pseudo-likelihood, the role of such joint distributions is solely to construct expectations, and with considerably less computational burden. Sometimes, this might still be impractical, but then the model-based expectation can be replaced by a simpler, but sufficiently rich, model in line with Bang and Robins (2005) and Meng (1994).

While in general doubly robust versions require the specification of both a weight and a predictive model, considerable simplification applies to the important special case of marginal pairwise (or, more generally, n-way) likelihood, also known as composite likelihood. In this case, the doubly robust versions merely require the formulation of a predictive model. In many models these are relatively easy to compute or approximate, as was illustrated for the normal and binary cases. This is a strong advantage of the combined use of doubly-robust and composite likelihood ideas.

For the estimation of precision we have indicated how a conventional sandwich-type estimator can be used. Should the derivation of explicit forms be deemed cumbersome, one could resort to sampling-based methods such as stochastic EM, multiple imputation, the bootstrap, and MCMC machinery.

Acknowledgments

The authors gratefully acknowledge support from IAP research Network P7/06 of the Belgian Government (Belgian Science Policy).

References

Aerts, M., Geys, H., Molenberghs, G., and Ryan, L.M. (2002). *Topics in Modelling of Clustered Data.* London: CRC/Chapman & Hall.

Arnold, B.C. and Strauss, D. (1991). Pseudolikelihood estimation: Some examples. *Sankhya B* **53**, 233–243.

Bahadur, R.R. (1961) A representation of the joint distribution of responses of n dichotomous items. In: *Studies in Item Analysis and Prediction*, H. Solomon (Ed.), Stanford Mathematical Studies in the Social Sciences VI. Stanford, California: Stanford University Press.

Bang, H. and Robins, J.M. (2005). Doubly robust estimation in missing data and causal inference models. *Biometrics* **61**, 962–972.

Birhanu, T., Sotto, C., Molenberghs, G., Kenward, M.G., and Verbeke, G. (2013). Doubly robust composite likelihood for hierarchical categorical data. *Submitted for publication.*

Cox, D. and Reid, N. (2004). A note on pseudolikelihood constructed from marginal densities. *Biometrika* **91**, 729–737.

De Backer, M., De Keyser, P., De Vroey, C., and Lesaffre, E. (1996). A 12-week treatment for dermatophyte toe onychomycosis: Terbinafine 250mg/day vs. itraconazole 200mg/day–a double-blind comparative trial. *British Journal of Dermatology* **134**, 16–17.

Geys, H., Molenberghs, G., and Lipsitz, S.R. (1998). A note on the comparison of pseudo-likelihood and generalized estimating equations for marginal odds ratio models. *Journal of Statistical Computation and Simulation* **62**, 45–72.

Geys, H., Molenberghs, G. and Ryan, L. (1999). Pseudo-likelihood modelling of multivariate outcomes in developmental toxicology. *Journal of the American Statistical Association* **94**, 34–745.

He, W. and Yi, G.Y. (2011). A pairwise likelihood method for correlated binary data with/without missing observations under generalized partially linear single-index models. *Statistica Sinica* **21**, 207–229.

Joe, H. and Lee, Y. (2008). On weighting of bivariate margins in pairwise likelihood. *Journal of Multivariate Analysis* **100**, 670–685.

le Cessie, S. and van Houwelingen, J.C. (1991). A goodness-of-fit test for binary regression models, based on smoothing methods. *Biometrics* **47**, 1267–1282.

Liang, K.-Y. and Zeger, S.L. (1986). Longitudinal data analysis using generalized linear models. *Biometrika* **73**, 13–22.

Lindsay, B.G. (1988). Composite likelihood methods. *Contemporary Mathematics* **80**, 221–239.

Little, R.J.A. and Rubin, D.B. (2002). *Statistical Analysis with Missing Data.* New York: John Wiley & Sons.

Meng, X.-L. (1994). Multiple-imputation inferences with uncongenial sources of input. *Statistical Science* **9**, 538–558.

Molenberghs, G., Kenward, M.G., Verbeke, G., and Teshome Ayele, B. (2011). Pseudo-likelihood estimation for incomplete data. *Statistica Sinica,* **21**, 187–206.

Molenberghs, G. and Verbeke, G. (2005). *Models for Discrete Longitudinal Data.* New York: Springer.

Parzen, M., Lipsitz, S.R., Fitzmaurice, G.M., Ibrahim, J.G., and Troxel, A. (2006). Pseudo-likelihood methods for longitudinal binary data with non-ignorable missing responses and covariates. *Statistics in Medicine* **25**, 2784–2796.

Renard, D., Molenberghs, G., and Geys, H. (2004) A pairwise likelihood approach to estimation in multilevel probit models. *Computational Statistics and Data Analysis,* **44**, 649–667.

Rotnitzky, A. (2009). Inverse probability weighted methods. In: *Longitudinal Data Analysis* (G. Fitzmaurice, M. Davidian, G. Verbeke, and G. Molenberghs eds.), 453–476. CRC/Chapman & Hall, Boca Raton.

Rubin, D.B. (1976). Inference and missing data. *Biometrika* **63**, 581–592.

Scharfstein, D.O., Rotnitzky, A., and Robins, J.M. (1999). Adjusting for nonignorable drop-out using semi-parametric nonresponse models. *Journal of the American Statistical Association* **94**, 1096–1120 (with Rejoinder, 1135–146).

Van der Laan, M.J. and Robins, J.M. (2003). *Unified Methods for Censored Longitudinal Data and Causality.* New York: Springer.

Varin, C. (2008). On composite marginal likelihoods. *Advances in Statistical Analysis* **92**, 1–28.

Verbeke, G. and Molenberghs, G. (2000). *Linear Mixed Models for Longitudinal Data.* New York: Springer.

Yi, G.Y., Zeng, L., and Cook, R.J. (2011). A robust pairwise likelihood method for incomplete longitudinal binary data arising in clusters. *Canadian Journal of Statistics* **39**, 34–51.

Part IV

Multiple Imputation

11

Introduction

Michael G. Kenward

London School of Hygiene and Tropical Medicine

CONTENTS

11.1 Introduction

Multiple Imputation (MI) was developed by Donald Rubin, and first described by him in a 1977 manuscript prepared for the United States Social Survey Administration. This is reproduced as an Appendix in his 1987 book *Multiple Imputation for Nonresponse in Surveys* (Rubin, 1987). The origins of MI clearly lie in the analysis of survey data, and the dominant (but not exclusive) paradigm for developing and justifying the approach, as set out in both the original manuscript and the subsequent book, is that of sample surveys. In the decades that followed, the use of MI has extended to many other application areas, most notably in biostatistics. As such it is now commonly framed in a likelihood-based frequentist paradigm.

When first introduced to MI it is tempting to see it as mainly a computational technique: most introductions begin by setting out the steps taken when carrying out an MI analysis, breaking it down into what Rubin originally called the 'tasks.' While obviously useful for the user, it is important at the same time to separate out the conceptual and computational aspects of the approach. To be an effective user, it is important to have a good grasp of the underlying rationale, and so understand when MI can be expected to work well and when not. The justification for the approach is a subtle interplay of Bayesian and frequentist arguments, although ultimately its performance must be judged through the frequentist paradigm. It is certainly true that alternative methods of analysis can usually be formulated for any given problem for which MI might be considered. The main strengths of MI lie in ability to retain the analysis, in particular the estimates and variances, that would have been used had there been no missing data. These are combined using very simple formulae, sometimes called Rubin's rules, to produce appropriate inferences for the incomplete data. In non-trivial settings, the alternatives to MI often have one-off, or bespoke, aspects that require considerable extra work, and which put them beyond the capability of the less specialized user. The relative simplicity of MI relies, among other things, on the clear separation of the 'substantive' model, the original analysis model for the complete data, and the 'imputation' model. The latter is determined by the conditional distribution of the missing data given the observed. The relationship between these two models is self-evidently key. An important class of MI analyses arise when the two models are said to be 'congenial'

(Meng, 1994). In such situations quite general asymptotic results can be obtained about the behaviour of the MI estimators and accompanying variances (Wang and Robins 1998, Nielsen 2003). Roughly, congeniality means that the two models can both be derived from a common overarching one, but in detail it is more subtle than this. There are important uses of MI, however, where uncongenial models are required and the justification for the approach then rests more on empirical results and arguments by analogy. Congeniality is another key concept that the user of MI needs to appreciate. The introduction and overview in Chapter 12 provides details of both the main computational steps that make up an MI analysis, and a careful and thorough discussion of these important conceptual issues.

The simplicity of Rubin's rules does rest on the requirement that imputations are drawn from an appropriate model for the missing data using the Bayesian paradigm. It is this part of the procedure that typically presents the most challenges for the user. In many problems there will be missing data in several variables of different type: continuous, binary, ordinal, and so on. In principle, these need to be modelled jointly and, if the original data are structured, for example hierarchically, this also needs to be accommodated in the model. This can be a complex problem in non-trivial settings. Unsurprisingly much recent research in MI has been devoted to methods for drawing suitable imputations. At the risk of oversimplification, these methods can be grouped into two main types. Each has its advantages and disadvantages. In the first, the so-called "Fully Conditional Specification," no attempt is made to construct the required joint model. Rather, a univariate imputation model is formulated for each partially observed variable with some, or all, of the the other variables, both complete and incomplete, included as covariates. A Gibbs type sampling procedure is then used to draw imputations from the joint distribution of the partially observed variables. The univariate models are relatively simple to construct, and the sampler merely requires repeated fitting of these models. So, computationally, this approach has many advantages. It does appear to work well in practice in a wide variety of settings, but it should be noted that there is no rigorous justification beyond those very restrictive settings where the conditional and joint distributions are known to be compatible, namely those derived from the multivariate Gaussian distribution or saturated log-linear model. The main limitation of this approach at the moment is in structured data problems. Although modifications have been derived for simple clustered problems, there do not exist versions for more complex multi-level structures. In Chapter 13 a thorough account is provided of this approach. The alternative uses a well defined multilevel model for the joint distribution of the various types of outcome, which is constructed on a latent multivariate Gaussian structure for the non-continuous outcomes. This has the advantages of distributional rigour, and great flexibility in extending the scope of missing data problems to situations such as measurement error and data linkage, but is much more demanding computationally. For practical use it is very dependent on efficient implementation. This approach is the subject of Chapter 14.

As seen below in Part V of this handbook, sensitivity analysis has an important role in the handling of missing data. Rubin identified the potential of MI for sensitivity analysis at an early stage of its development (Rubin, 1987, Ch. 6). The separation of imputation and substantive models allows the user to vary assumptions about the missing data mechanism through the changes to the imputation model, and so explore the implications of these alternatives for the conclusions. This can be done in various ways, and several of these are explored in Chapter 19.

References

Meng X-L. (1994). Multiple-imputation inferences with uncongenial sources of input. *Statistical Science* **9**, 538–558.

Nielsen, S.F. (2003). Proper and improper multiple imputation. *International Journal of Statistics* **28**, 593–627.

Rubin, D.B. (1987). *Multiple Imputation for Nonresponse in Surveys.* New Jersey: John Wiley & Sons.

Wang, N. and Robins, J.M. (1998). Large-sample theory for parametric multiple imputation procedures. *Biometrika* **85**, 935–948.

12

Multiple Imputation: A Perspective and Historical Overview

John B. Carlin

Murdoch Children's Research Institute & The University of Melbourne, Melbourne, Australia

CONTENTS

12.1 What Is Multiple Imputation?

12.1.1 Introduction

This chapter provides an introduction to the method of multiple imputation, encompassing an overview of its historical origins, theoretical foundations and application in practice. In presenting such a broad overview there are inevitable gaps and a lack of depth on many topics, but subsequent chapters provide more detail in many areas. The first section provides a simple description of the method, accompanied by a detailed illustrative example, and a brief historical account of its origins in large-scale sample survey methodology. The following section then gives an explanation of the formal basis of the method in Bayesian statistical inference, as developed by Rubin in the 1970s, and a review of some of the key theoretical concepts. The application of multiple imputation to practical data analysis has developed within the framework of "ignorable" models (see Chapter 3), and Section 12.3 reviews the methods and software tools that appeared from the mid-1990s to implement the approach within this framework. These tools have sparked much more widespread uptake of the method in practice. The final section reviews the range of problems to which multiple imputation has been applied, beginning with several large computational efforts in the late 1980s and early 1990s and concluding with a brief overview of the current range of applications and outstanding methodological research problems.

12.1.2 A brief overview

The method of multiple imputation (MI) can be described simply in algorithmic terms. To facilitate discussion of the basic concepts assume for now that the focus of interest is a rectangular dataset of n 'units' by p 'variables,' which is represented as a matrix Y of n exchangeable rows $\boldsymbol{Y}_i'$ each of length p. The data analyst intends to conduct an analysis using all of the variables considered; this would often be a regression analysis relating one of the variables to the others, but for purposes of describing the method of multiple imputation initially there is no need to distinguish between "response" or "outcome" and "predictors" or "covariates." Likewise it may on occasion be useful to distinguish any variables that are not subject to missing values, but for the moment it is assumed that any of the values in the Y array may be missing, while the desired analysis requires all values to be available. In this scenario, when almost any standard (packaged) statistical procedure is invoked it will proceed by including in the analysis only those units that contain no missing values: this is the so-called "complete-case" analysis or method of "listwise deletion." For analyses such as multiple regression involving numerous covariates all of which may be subject to missing data, the complete-case analysis may use only a small subset of the data rows, raising concerns about both potential biases and loss of precision in inferences for the target parameters.

Multiple imputation is a procedure that enables the user to carry on with their standard analysis *almost* as if the dataset contained no missing values. The method proceeds as follows:

1. Create a number (m) of copies of the incomplete dataset, and use an appropriate procedure to *impute* (fill in) the missing values in each of these copies. Since we do not know the true values that are missing it seems reasonable that the imputed values used in each copy should in general be different from each other. The choice of m is discussed in Section 12.2.5.

2. For each completed (imputed) copy of the dataset, carry out the standard analysis that would have been performed in the absence of missing values, and store the parameter estimates of interest, along with their estimated standard errors (or variance-covariance matrix in the case of a multivariate parameter of interest). Here and for most of this chapter, the focus will be on a single (univariate) parameter of interest, which we denote β. The estimate of β obtained from the $k^{th} (k = 1, \ldots, m)$ completed dataset is denoted $\widehat{\beta}^{(k)}$ and its (estimated) variance $V^{(k)}$. (Note that this step implicitly assumes that the inference desired from the analysis can be effectively reduced to a parameter estimate and standard error, an assumption that will be addressed below.)

3. Use a pair of formulas widely known as "Rubin's rules," firstly to create a combined

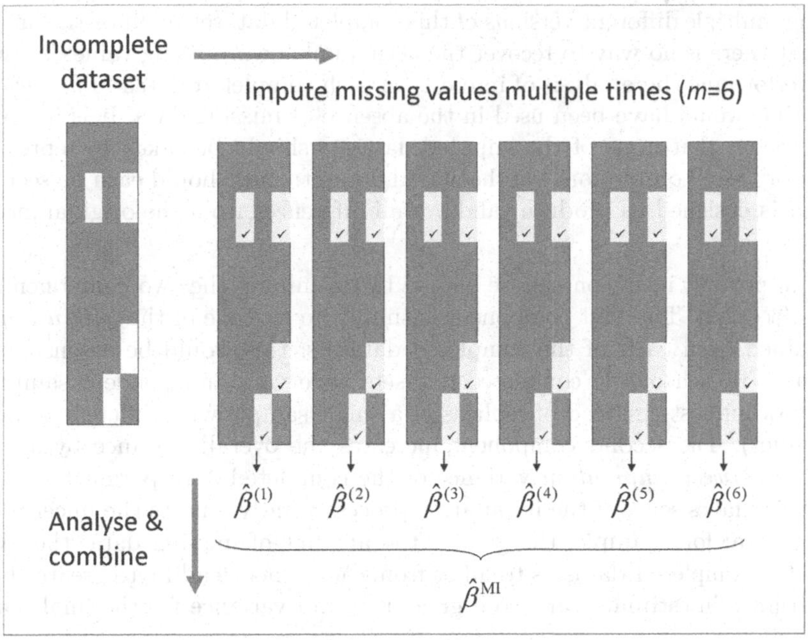

FIGURE 12.1
Schematic illustration of the method of multiple imputation.

estimate of the parameter, as the average of the m separate estimates:

$$\widehat{\beta}^{\text{MI}} = \frac{1}{m} \sum_{1}^{m} \widehat{\beta}^{(k)}, \tag{12.1}$$

and then to obtain a standard error for this estimate, as the square root of the following combined variance estimate:

$$V^{\text{MI}} = \bar{V} + \left(1 + \frac{1}{m}\right) B, \tag{12.2}$$

where $\bar{V} = \sum_{1}^{m} V^{(k)}/m$, and $B = \sum_{1}^{m} (\widehat{\beta}^{(k)} - \widehat{\beta}^{\text{MI}})^2/(m-1)$, which is an estimate of the *between-imputation* variance of the parameter of interest. Section 12.2 reviews the basic theory underlying these formulas, but in general multiple-imputation inference proceeds in the usual way from these results, by forming test statistics and confidence intervals under the assumption that $(\widehat{\beta}^{\text{MI}} - \beta)/\sqrt{V^{\text{MI}}}$ follows either a standard normal or t distribution.

The process of statistical inference using multiple imputation is illustrated in Figure 12.1. The next section examines the theoretical basis of the method in some detail, and in particular the derivation of the variance calculation as an approximate Bayesian inference. However, at this stage it is worth noting some attractive heuristic and practical properties of the approach, which generally make it more appealing in practice than a fully-fledged Bayesian modelling approach. Firstly, the process of filling in or imputing the missing values has the intuitive appeal of "restoring" the full dataset that was desired in the beginning, while obtaining multiple different versions of this completed dataset reminds the analyst appropriately that there is no way to recover the actual unknown missing values. Second, the core work of performing the analysis of interest (in each completed dataset) follows exactly the approach that would have been used in the absence of missing data. It is important to emphasize, however, that none of the imputed datasets should be taken to represent true substitutes for or true "completions" of the actual dataset; they should each be seen as part of a whole that is designed to produce valid overall inferences from the original incomplete data.

Finally, some important intuition can be gained by examining the two components of the variance formula (12.2). The first component is simply an average of the *within-imputation* variances obtained from each of the completed datasets: this would be a sensible overall variance estimate if each of the completed datasets were valid independent samples from the population of interest (instead of replicas of a single sample with different imputations for missing values). The second component increases the overall variance by an amount that reflects the *between-imputation* variance of the completed-data parameter estimates. Heuristically this makes sense if the imputation process validly reflects the uncertainty due to the missing data: for example, the greater the amount of missing data, the more one should expect the completed datasets to differ from each other, leading to greater between-imputation variance in estimates and so a greater overall variance for the final parameter estimate.

These heuristic characteristics of the MI procedure will be discussed in greater depth and with greater formality later in the chapter. For now, a key additional observation is that one might not expect the final results to be valid unless the process of imputation is performed appropriately; this notion has been formalized in the concept of "proper" imputation, which is discussed in Section 12.2.2.

12.1.3 Illustrative example

In this section the method of multiple imputation is illustrated with an example involving three variables, providing a simple but non-trivial application. Figure 12.2 displays a scatter plot of the (log-transformed) level of a serum antibody level versus age (in months) for 93 acutely ill children who had one of two possible illnesses (labelled M and E), indicated by distinct markers in the plot. (Note that these data are based on a real dataset but the details and context are not important for the purposes of this illustration.)

A question of central interest in analysing these data was whether the antibody levels in the children tended to be higher in one disease group than the other. Both background knowledge and simple inspection of the scatter plot suggest that age may be an important confounding factor in making this comparison, because children with disease M tend to be older than children with disease E, while antibody levels also increase with age. Therefore a natural approach to analysing the data for the question of interest is to use a regression adjustment for age while estimating the mean difference between the groups, i.e. to use the method of analysis of covariance (ANCOVA). This analysis is illustrated by the two fitted lines in the scatter plot, the vertical distance between which is the estimated adjusted mean difference between the groups. The estimated mean difference, with and without adjustment for age, is displayed in Table 12.1, and the results show clear evidence of the confounding effect in question, in that the crude mean difference is much larger (and much more statistically significant) than the adjusted estimate. From this analysis it would be

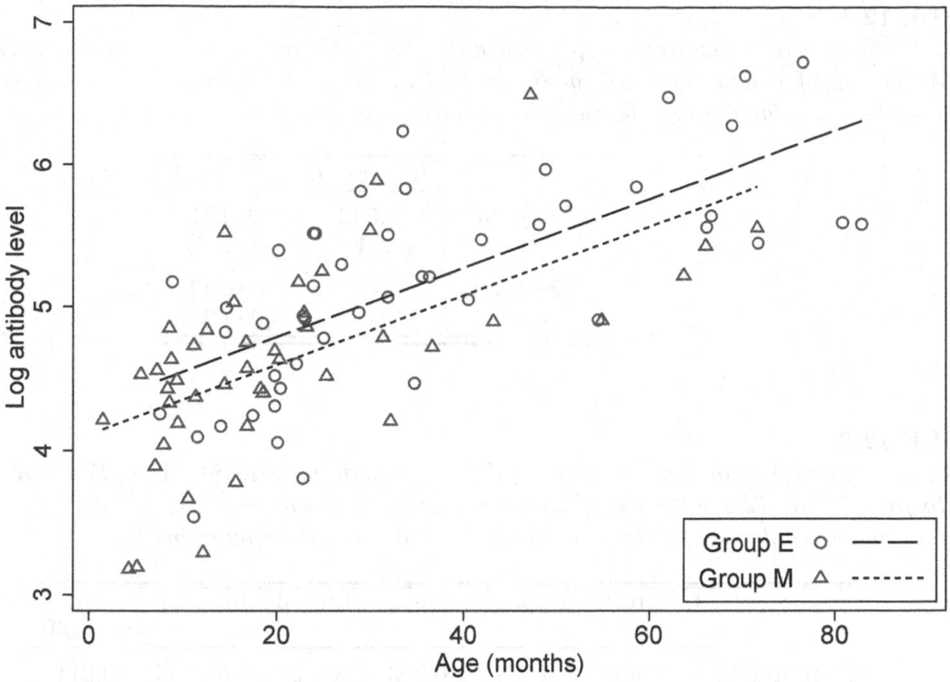

FIGURE 12.2
Scatter plot of (log) antibody level versus age, for two disease groups, with fitted regression lines from ANCOVA model. The vertical distance between the lines is the ANCOVA estimate of group mean difference (Table 12.1).

concluded that there is little evidence of a difference in antibody levels between the diseases, once age is taken into account.

Missing data are now introduced into this example. Figure 12.3 displays a rather extreme hypothetical scenario in which 50% of the age values have been lost to the analyst (in a manner designed for illustrative purposes, discussed further below). In this context, it is possible to perform several analyses: we could decide to use all the outcome values and forget about adjusting for age, which as already discussed leads to a biased conclusion if age is a confounding factor for the "true" effect of disease, or we could perform the adjusted analysis using just the so-called complete cases, the 50% of the sample for whom the age value is present. An inspection of the distribution of the outcome values for which age is missing suggests, however, that the latter approach may also lead to a biased conclusion, because age tends to be missing more often for children with higher values of the outcome. This is confirmed in the complete-case analysis shown in Table 12.2 and illustrated with the ANCOVA fit superimposed on Figure 12.3.

It is intuitively appealing to think that if the missing values of age could be reliably imputed, then an unbiased estimate of the group difference adjusted for age, based on the entire sample, might be recovered. Figure 12.4 shows four datasets in which the missing age values have been imputed, using the regression method described below in Section 12.3.1, with estimated regression lines from the ANCOVA model superimposed. It can be seen that the imputation process generally reproduces a scatter that looks similar to the original observed

TABLE 12.1
Estimates (standard errors) of group mean difference with and without adjustment (ANCOVA) for age, for data displayed in Figure 12.2 (n=93). These are the "complete-data" estimates to which later results based on incomplete data are compared.

	Unadjusted	ANCOVA
Group diff.	-0.542	-0.194
(s.e.)	(0.145)	(0.117)
Age effect	-	0.0242
(s.e)	-	(0.0029)

TABLE 12.2
Estimates (standard errors) of group mean difference with adjustment (ANCOVA) for age, in data from Figure 12.2 with 50% of age values missing, as shown in Figure 12.3 (n = 47). (Imp. 1: completed dataset for imputation 1; m: – number of imputations.)

	Complete cases	Imp. 1	Imp. 2	Imp. 3	Imp. 43	MI (m=100)
Group diff.	-0.051	-0.179	-0.303	-0.328	-0.275	-0.211
(s.e.)	(0.167)	(0.121)	(0.117)	(0.123)	(0.113)	(0.147)
Age effect	0.0318	0.0308	0.0303	0.0285	0.0250	0.0293
(s.e.)	(0.0051)	(0.0039)	(0.0039)	(0.0042)	(0.0029)	(0.0045)

data, but there is considerable variation between the completed datasets. (It may also be noted that one of the sets of imputations produced an imputed age value that is negative, reinforcing the point that the imputed values cannot be treated as true substitutes for the true values. As an aside, this also raises the question of whether it is reasonable that an imputation model should produce an imputed value that is outside the range of the variable being imputed; the answer to this question is not as clear-cut as might be expected.)

Notably, in each of the imputed datasets the estimated group difference is greater than it was in the complete-case analysis, and is thus closer to the original ANCOVA result (Table 12.2). Applying the method of multiple imputation means repeating the imputation procedure several times and obtaining a fitted line based on the average of the coefficient estimates over the single imputations, using Rubin's rules. We display the results obtained from this analysis in Figure 12.5 and also in Table 12.2. For this particular pattern of missing data, the MI procedure does remarkably well in obtaining a final result that is close to the analysis in the original complete data. It should also be noted that the standard errors of the MI estimates are considerably larger than those from the complete data but at the same time substantially smaller than those from the (biased) complete-case analysis. This provides prima facie evidence that the MI procedure has recovered information for the analysis.

The missing data in this example were created for illustrative purposes by a probabilistic mechanism that ensured the data were missing at random (MAR) given the disease groups and the outcome. See Chapters 1–3 for background on this important assumption. As discussed below (Section 12.2.3), MAR is a key assumption in ensuring the validity of

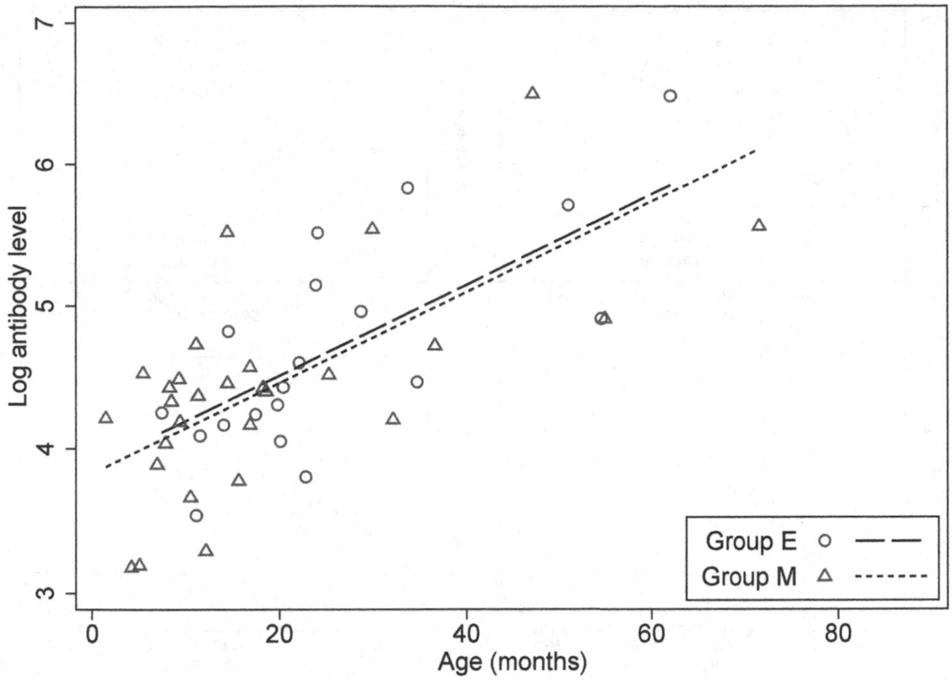

FIGURE 12.3
Scatter plot of (log) antibody level versus age for two disease groups, in sub-sample of "complete cases," after losing 50% of age values, with fitted regression lines from ANCOVA model based on the complete cases.

standard approaches to multiple imputation, so this example was deliberately constructed in order to illustrate a successful application of the method. The illustration is also artificial in comparing analyses of simulated incomplete data with the analysis of the original complete dataset. A full theoretical investigation of the repeated-sampling properties of the method would evaluate the ability of MI to improve estimation of population parameters in the context of sampling variability in both the complete data and the missing data process. However, the example does show that MI has the potential to help, in the sense of producing results closer to what we would have obtained in the complete data were it not for the missing data. Whether the apparently convincing results obtained for a single hypothetical pattern of missing data are typical of what might be expected in this illustrative example was investigated by reproducing missing data under exactly the same probability model as was used for the example shown. The results showed that on average over repeated 'draws' of the missing data process the MI results were not quite as successful as under the initial 'draw' in producing results close to those from the original data: the mean estimated group difference under MI over 1000 simulations of 50% missing age values was -0.136 with average standard error 0.143). However, the MI results remained both closer to the complete-data point estimate and more precise than the complete-case results (in which the mean estimate and standard error were -0.095, 0.164, respectively).

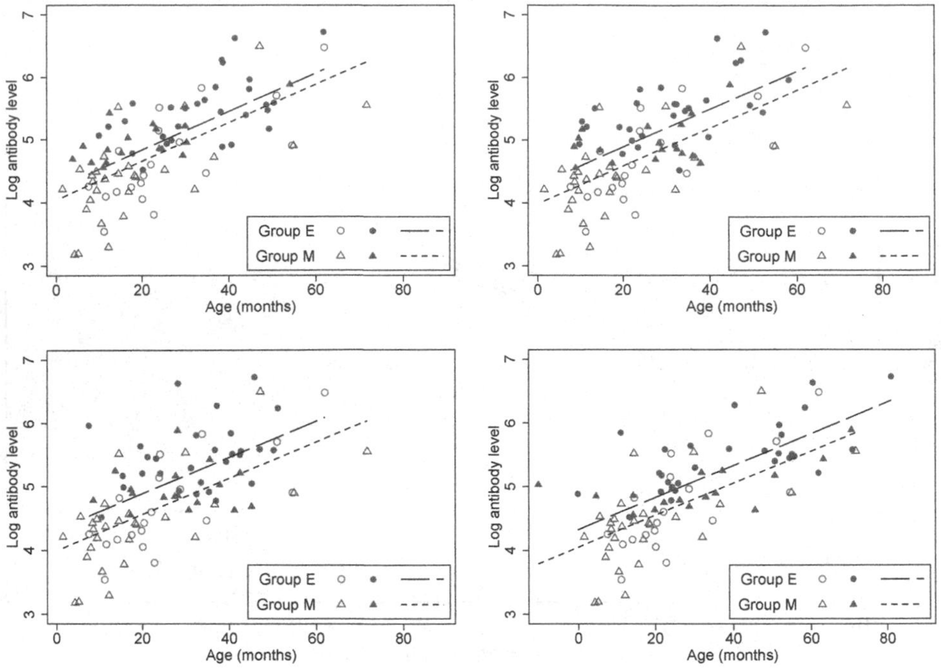

FIGURE 12.4
Scatter plot of (log) antibody level versus age for two disease groups after imputation of missing age values (points shown with solid symbols), with fitted regression lines from ANCOVA model based on each imputed dataset.

12.1.4 Early history

The method of multiple imputation was invented by Donald Rubin in the 1970s. The first recorded account is an unpublished report prepared for the U.S. Social Security Administration (reprinted in the *American Statistician*, 2004), which was followed by a more comprehensive theoretical framework published in the Proceedings of the American Statistical Association (Rubin 1978). Rubin had become involved in survey sampling problems at the U.S. Bureau of the Census, which has a long history of pioneering work in survey methodology. As recounted by Scheuren (2005), official government surveys of the 1940s and 1950s were largely untroubled by non-response problems as "trust in government was very high," but this happy state of affairs did not last into the latter part of the 20th century, which saw increasing levels of non-response both at unit (whole person or occasion of survey) and item (variable) levels. With small amounts of missing data and an emerging tradition of making available the raw data from large national surveys for "public use," survey statisticians had developed methods of filling in or imputing missing values. A method called hot-deck imputation was particularly popular and continues to be used for some applications, although the term now seems quaintly historical (referring to imputing values from those found on a computer card that is selected by matching on key sociodemographic characteristics, either directly or with sampling, from a "deck" of cards representing fully responding individuals; Little and Rubin 2002).

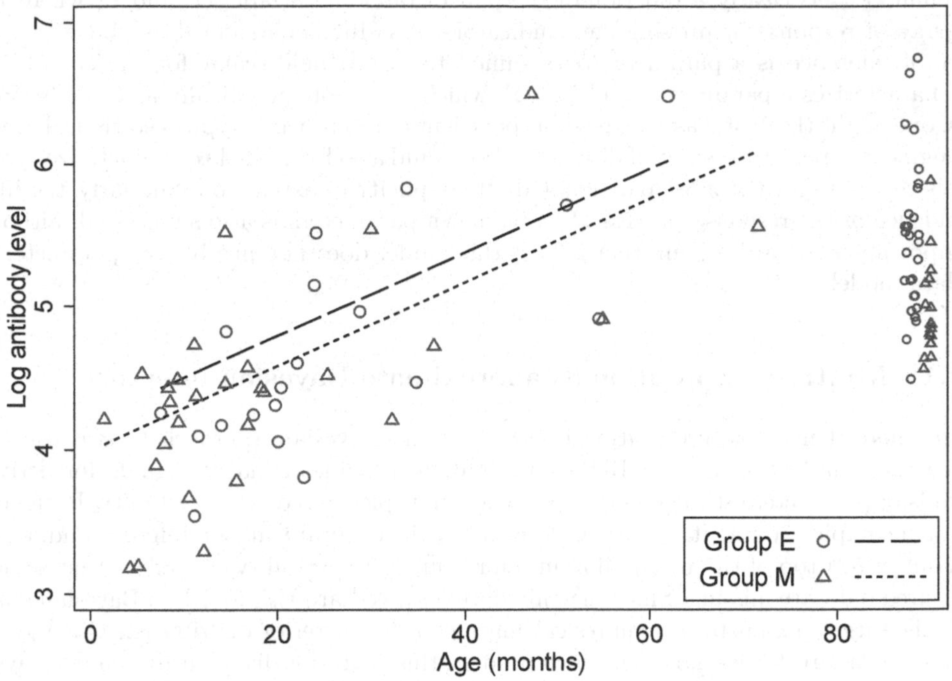

FIGURE 12.5
Scatter plot of (log) antibody level versus age for two disease groups, with fitted regression lines from multiple-imputation-based ANCOVA model. Data points for which age was missing, and so was multiply imputed for the analysis, are shown ("jittered") at the right-hand side of the scatter plot.

With the increasing use of hot-deck imputation, especially for problems with a substantial amount of missing data, statisticians became increasingly concerned about how to provide valid estimates of variance for sample survey estimates in the presence of non-response. Rubin developed multiple imputation as an attractive method for addressing this problem. Section 12.4 provides a historical overview of the practical application of multiple imputation; meanwhile the underlying general theory and some technical issues are reviewed.

12.2 Foundations and Basic Theory

To introduce the basic theory underlying multiple imputation, further notation is needed. A widely used convention is to let Y_{obs} represent the (irregular) array of data values that are observed, so Y_{obs} can be thought of as the concatenation of the individual rows of observed values $\boldsymbol{Y}_i^{o\prime}$, in the notation introduced in Chapter 1. The complementary array of missing values is denoted Y_{mis}. When these arrays are thought of as realized data, these definitions are clear (apart from the fact that all values of Y_{mis} are undefined!), but when they are considered to be random variables for modeling purposes some care may be needed. In particular, Y_{obs} is clearly a function of the pair of random variables Y and R, where R is the array of response or missing data indicators. It is further assumed as above that the target of inference is a parameter β (assumed to be a single scalar for the time being) that characterizes a parametric model for Y, which we denote generically as $P(Y|\beta)$. To fix ideas, one might think of β as a regression coefficient of interest in a typical statistical model; although other parameters not of direct interest would also be needed to characterize typical models these are for the most part ignored for simplicity of exposition. Similarly, the likely dependence of the response process R on unknown parameters is also suppressed, although an important standard assumption is that this model does not involve the parameters of the data model.

12.2.1 Multiple imputation as approximate Bayesian inference

The method of multiple imputation is based on the Bayesian approach to statistical inference, essentially because the Bayesian paradigm provides a natural guide for deriving methods of parameter estimation that allow for multiple sources of uncertainty. In the context of incomplete data, it is clearly desirable to incorporate in our inference additional uncertainty (on top of the usual allowance for sampling variability) to reflect that some of the desired data are missing. Since the only data observed are Y_{obs} and R, a Bayesian analysis for the target parameter β requires calculation of the posterior distribution $P(\beta|Y_{obs}, R)$. In principle, a direct Bayesian analysis to produce this posterior distribution could always be performed, but in practice this may require tailored programming to perform the necessary computations and will often be out of the reach of the statistician working on a practical data analysis. However, a simple standard application of conditional probability shows that the posterior distribution may be expressed in a way that separates out the missing data problem from the "analysis model" of interest:

$$P(\beta|Y_{obs}, R) = \int P(\beta|Y_{obs}, Y_{mis})P(Y_{mis}|Y_{obs}, R)dY_{mis}. \qquad (12.3)$$

This integral should be thought of as a sum in the case of discrete Y_{mis}. The attraction of this representation is that the first term in the integral represents the posterior distribution for β given a complete dataset (and R, which is redundant there), and this calculation is often relatively simple to perform: it is the Bayesian version of the standard complete-data analysis. The key insight of multiple imputation is to see that if a way can be found to generate or impute a sample of m values $Y_{mis}^{(k)}$ from the predictive distribution for the missing data, i.e. from the distribution $P(Y_{mis}|Y_{obs}, R)$, then the integral (12.3) may be approximated by an average over the complete-data posterior distributions evaluated at each of the imputed values of Y_{mis}, as:

$$P(\beta|Y_{obs}, R) \approx \frac{1}{m}\sum_{1}^{m} P(\beta|Y_{obs}, Y_{mis}^{(k)}). \qquad (12.4)$$

Note that the formulas above have retained a dependence on R in the predictive distribution for Y_{mis}. This emphasizes that in full generality the predictive distribution used for imputation may need to take account not only of the values Y_{obs} themselves but also of where they appear in the dataset and why those values were observed while others were not. However, as will be seen, in practical applications of MI the assumption of *ignorability* (of the missing data process) or 'missing at random' (MAR) is generally invoked, in which case the issue is 'assumed away' and imputation is performed without explicitly modelling the process that led to the missing data.

In practice, it is generally only feasible, and is quite sufficient, to work with just the first two moments of the posterior distribution, the mean and variance of β, which may be obtained by applying the rules of iterated expectations (see, e.g., Gelman et al. 2004, Little and Rubin 2002):

$$\mathrm{E}(\beta|Y_{obs}, R) = \mathrm{E}[\mathrm{E}(\beta|Y_{obs}, Y_{mis})|Y_{obs}, R], \tag{12.5}$$

and

$$\begin{aligned}\mathrm{Var}(\beta|Y_{obs}, R) &= \mathrm{E}[\mathrm{Var}(\beta|Y_{obs}, Y_{mis})|Y_{obs}, R] \\ &\quad +\mathrm{Var}[\mathrm{E}(\beta|Y_{obs}, Y_{mis})|Y_{obs}, R].\end{aligned} \tag{12.6}$$

These quantities may be estimated from the imputed data by approximating the integrals in (12.5) and (12.6) by sums over the sample of drawn values of $Y_{mis}^{(k)}$, recalling that the inner expectation (or variance) calculation in each of these formulas corresponds to obtaining a standard complete-data statistic from each completed dataset while the outer expectation corresponds to averaging over the posterior distribution of the missing data Y_{mis}. Substituting the complete-data estimate of the posterior mean of β in (12.5) and averaging the result over the imputed values $Y_{mis}^{(k)}$ gives:

$$\mathrm{E}(\beta|Y_{obs}, R) \approx \frac{1}{m}\sum_{1}^{m}\widehat{\beta}^{(k)}, \tag{12.7}$$

with the estimate denoted $\widehat{\beta}^{\mathrm{MI}}$, as already defined in (12.1). The variance calculation contains two terms, with the second one obtained as the standard unbiased estimate, i.e., the "sample variance" of $\widehat{\beta}^{(k)}$ across the m completed datasets:

$$\mathrm{Var}(\beta|Y_{obs}, R) \approx \frac{1}{m}\sum_{1}^{m}V^{(k)} + \frac{1}{m-1}\sum_{1}^{m}(\widehat{\beta}^{(k)} - \widehat{\beta}^{\mathrm{MI}})^2. \tag{12.8}$$

The right-hand side of (12.8) can be expressed as $\bar{V} + B$, where $\bar{V} = \sum_1^m V^K/m$, and $B = \sum_1^m(\widehat{\beta}^{(k)} - \widehat{\beta}^{\mathrm{MI}})^2/(m-1)$, which is an estimate of the *between-imputation* variance of the parameter of interest. These quantities are valid approximations to the first two moments of the posterior distribution for large m. For small values of m the variance formula needs to include an additional term, proportional to $1/m$, in order to reflect uncertainty in the value of $\widehat{\beta}^{\mathrm{MI}}$ as an estimate of the true posterior mean (Schafer 1997). This leads directly to expression (12.2) as the recommended approximation for the posterior variance of β.

Given these approximate posterior moments, approximate inferences may be created in the usual way. In particular, if the posterior distribution were normal, one would use $\widehat{\beta}^{\mathrm{MI}} \pm z_{(1-\alpha)}\sqrt{V^{\mathrm{MI}}}$ (with $z_{(1-\alpha)}$ representing the standard normal quantile) as a $100(1-\alpha)\%$ credible interval for the unknown parameter value. With small m the uncertainty in the variance parameters may be allowed for by using a t reference distribution, for which the choice of degrees of freedom is discussed below in Section 12.2.4.

It is more complicated to carry through these arguments formally in the presence of additional (nuisance) parameters in the model of interest. However, a clear avenue to this simple Bayesian view of multiple imputation remains as long as one can define a statistic $\widehat{\beta}$ that provides a valid (large-sample) estimate of the posterior mean of β, irrespective of nuisance parameters, in the (hypothetical) complete data, i.e., in the full Y array that was intended to be observed. More specifically, the posterior distribution must be normal with mean $\widehat{\beta}$ in large samples, and there must exist a valid large-sample estimate V of the posterior variance of β; with these assumptions the results above flow through exactly as before. The complementary frequentist properties will also be key assumptions for interpreting multiple imputation from a sampling theory point of view, in which context the desired property of the estimate in question is defined as one of consistency for estimating β in repeated samples; see next subsection. It is important to remember that $\widehat{\beta}$ represents a function that can only be calculated from a fully observed (or completed) Y array.

12.2.2 Proper imputation and repeated sampling properties

The previous subsection referred to the need to generate "a sample of m values $Y_{mis}^{(k)}$ from the predictive distribution for the missing data, i.e., from the distribution $P(Y_{mis}|Y_{obs}, R)$." From the Bayesian point of view this is all that is required for the subsequent theory to flow, and if working entirely within a Bayesian paradigm then the task of multiple imputation becomes a computational one of developing methods for sampling from the posterior distribution of the missing data. Section 12.3 reviews the computational challenges of developing algorithms for such sampling, but first the concept of *proper* imputation and its connection to the frequentist or sampling-theory validity of multiple imputation is discussed.

As pointed out repeatedly by Rubin (e.g., Rubin 1996; see also Gelman 2011), the Bayesian paradigm is incomplete for real-world statistical practice because all Bayesian analysis is predicated on the assumption that the proposed models (whether sampling models or prior distributions) are correct. In practice, a critical perspective with respect to modeling assumptions is central to applied statistics; model checking is an essential feature of sound statistical analysis. Another implication is that even Bayesians should avoid using statistical methods that can be expected to perform poorly when considered within a framework of repeated sampling. To address this issue in the context of multiple imputation, Rubin coined the term "proper" imputation (Rubin 1987).

For repeated-sampling validity, MI-based point estimates should be consistent for the target parameter(s) and interval estimates should achieve at least the nominal coverage. Although Rubin's formal definitions of the properties of an imputation analysis that characterize "properness" do not seem particularly useful in themselves, their essence is to specify the properties of the complete-data estimator and the imputation-based inference procedure (including the within- and between-imputation variance estimates) that guarantee repeated-sampling validity. A key point is that he then showed that if the imputation method accounts appropriately for all sources of variability, which is guaranteed if it is performed by full Bayesian inference, i.e., sampling from a properly defined posterior distribution under a "correct" model, then it will be proper. Rubin has given several examples of approaches to imputation that are not proper in this sense and illustrates how these fail to produce reliable inferences. For example, the classical hot-deck method of imputation is not proper, but can be modified to be so, in the form of the "approximate Bayesian bootstrap" (Rubin and Schenker 1986). The essential logic here is that if one simply draws imputed values from the pool of observed values, then no allowance has been made for the sampling variability inherent in the original values; the bootstrap overcomes this by resampling the original values in order to create the hot deck for imputation draws.

A source of potential confusion is that Rubin's original presentation of multiple imputation (Rubin 1987) and most of his subsequent writing (e.g., Rubin 1996) framed the method and its theory within the context of finite population (survey sampling) inference. So instead of focusing on a parametric sampling model, as traditionally considered within mathematical statistics, Rubin's work addressed the task of estimating explicitly definable finite-population quantities (e.g., the mean income in a defined population). In this context, repeated-sampling validity is referred to as "randomization validity" by Rubin. However, the common thread across the different settings of finite-population sampling theory and standard frequentist mathematical statistics is that repeated-sampling validity is tied to the existence of a complete-data statistic $\widehat{\beta}$ that is normally distributed in large samples. If the imputation procedure is proper, then the MI-based version of this statistic inherits the same asymptotic distributional properties, and its variance is conservatively estimated (i.e., possibly over- but not under-estimated) by the Rubin formula, i.e., expression (12.2).

12.2.3 The role of the missing data model

In Section 12.2.1 it was explained that multiple imputation requires obtaining draws from the distribution $P(Y_{mis}|Y_{obs}, R)$, which is a conditional distribution for Y_{mis} that depends on the response mechanism R. In this general "nonignorable" setting, the value we impute for a particular element of Y_{mis} could depend not only on the values of other variables observed for that unit, and possibly other units (i.e., elements of Y_{obs}), but also on which variables are missing and which are observed. The key assumption that enables multiple imputation to be implemented in general-use computer packages is that of "ignorability," which in simple terms implies that R can be omitted from the imputation model, i.e., that $P(Y_{mis}|Y_{obs}, R) = P(Y_{mis}|Y_{obs})$. This assumption is equivalent for most practical purposes to the concept of "missing at random" (MAR), as defined in Chapter 1, although there are a number of subtleties underlying these concepts (Lu and Copas 2004, Seaman et al. 2013).

Under ignorability, the expression for the predictive distribution of the missing values, $P(Y_{mis}|Y_{obs})$, shows that imputation can in principle proceed by fitting a joint model to the full data array Y, and then computing the necessary conditional distributions for sampling values of Y_{mis}. Of course, the usual approach to such model fitting requires a parametric specification: we specify a model parameterized by θ (say) for the complete data Y, and obtain the posterior distribution for θ given the observed component of Y (assuming ignorability). We then draw values from this posterior distribution and for each drawn value obtain an imputation of Y_{mis} from the conditional distribution $P(Y_{mis}|Y_{obs}, \theta)$. Formally, this amounts to drawing values from the following distribution:

$$\int P(Y_{mis}|Y_{obs}, \theta)P(\theta|Y_{obs})d\theta. \tag{12.9}$$

In many settings, where the model for Y specifies independent and identically distributed rows, drawing from $P(Y_{mis}|Y_{obs}, \theta)$ is straightforward, since it amounts to a regression prediction at the unit level, with fixed parameters. However, the calculation of, or sampling from $P(\theta|Y_{obs})$ is often not straightforward at all. Specific approaches to performing these calculations are discussed in Section 12.3.

12.2.4 Further technical aspects of multiple imputation inference

Leaving aside the specifics of creating proper imputations for the moment, this section briefly reviews some additional definitions and technical issues that arise in performing multiple imputation inference.

As discussed above, because of the finite number of imputations used in practice, it is appropriate to use a t distribution for creating inferences from the point estimate $\widehat{\beta}^{\text{MI}}$ and its estimated variance V^{MI}. The appropriate degrees-of-freedom for this t-distribution were derived by Rubin, using a standard Satterthwaite-type approximation to the posterior distribution of the parameter of interest β given the information contained in the m imputed datasets (Rubin 1987) as:

$$\nu = (m-1)\left[1 + \frac{\bar{V}}{(1+m^{-1})B}\right]^2 \tag{12.10}$$

A refinement to this approximation, appropriate to problems in which the completed datasets have small sample size and complete-data estimates would attract degrees-of-freedom adjustments because of imprecise estimates of variance parameters, was derived by Barnard and Rubin (1999). Although historically important, these formulas are less relevant in modern practice, in which the value of m is usually in a range (20 or more) where the t and normal distributions are not materially different; see Section 12.2.5.

A key component of the degrees-of-freedom calculation is the ratio

$$r_m = \frac{(1+m^{-1})B}{\bar{V}}, \tag{12.11}$$

which is sometimes referred to as the "relative variance increase" (RVI) due to missing data, since $\bar{V}$ represents the variance of the parameter estimate in the absence of any missing data (i.e., when there is no variation between the $\widehat{\beta}^{(k)}$s, in which case $B = 0$).

A closely related quantity, termed the "fraction of missing information" (FMI), may be derived by comparing the "information" in the approximate posterior (t) density, defined as the negative of the second derivative of the log-posterior density, with that in the hypothetical complete data posterior density (i.e., $\bar{V}^{-1}$), giving (Rubin 1987):

$$\gamma_m = \frac{r_m + 2/(\nu+3)}{r_m + 1}. \tag{12.12}$$

Each of these quantities provides an indication of the extent to which missing data has affected the parameter estimate of interest, and they are related as their names suggest, with $\gamma_m \to r_m/(r_m+1)$ as $m \to \infty$. An important aspect of their interpretation is that the effect of missing data is a combination of the actual amount of missing data *and* the degree to which information in the incomplete cases contributes to the estimate of interest via the imputation modeling.

These concepts may be illustrated with the example from Section 12.1.3. In the MI analysis displayed in Table 12.2, the values of r_m and γ_m for the parameter estimating the group difference were 0.55 and 0.37, respectively. So, although half the observations were incomplete, the fraction of missing information was substantially less than 0.5, reflecting the fact that the incomplete cases contained considerable information on the parameter of interest (the adjusted mean difference between groups), which was recovered by the imputation model.

In contrast, it is instructive to perform a similar hypothetical analysis in which, instead of having missing age values, the indicator for group membership is missing (in the same 50% of the sample). In this case the complete-case analysis is of course the same as before, but applying multiple imputation using a logistic regression model for imputation of disease group membership (see Section 12.3.1), we obtain values of r_m and γ_m of 0.88 and 0.48, respectively. This indicates that the variance of the group difference is around twice what it might have been if there had been no missing data, which is exactly what would be expected if only the complete cases were analysed, since they comprise half of the sample. So the calculations suggest that multiple imputation has not recovered any useful information on the parameter of interest in this case. These calculations only relate to variance estimates of course, and it is possible that an MI analysis may correct bias with no gain in precision. In the present example, this is not the case, because the estimated group difference from the MI analysis with 50% missing group identifiers was -0.052, effectively the same as the complete-case estimate. This example illustrates that multiple imputation can by no means always be expected to improve inference, despite the appealing illusion of recovering incomplete cases for analysis; see White, Royston, and Wood (2011) and Lee et al. (2012) for further discussion.

Most of this chapter considers MI inference for single parameters only, but of course in many problems it is important to consider multi-parameter estimands. In essence much of the theory carries over to the multi-parameter case with standard extensions of the formulas to matrix/vector form. Clearly, the MI estimate of a $p \times 1$ vector $\boldsymbol{\beta}$ will, under the appropriate assumptions, still be the mean of the m completed-data estimates, as in expression (12.1). Similarly, we would expect a matrix extension of (12.2) also to be appropriate, with $\bar{V} = \sum_1^m V^{(k)}/m$ now representing the matrix average of m variance-covariance matrices $V^{(k)}$, and $B = \sum_1^m (\widehat{\boldsymbol{\beta}}^{(k)} - \widehat{\boldsymbol{\beta}}^{\text{MI}})(\widehat{\boldsymbol{\beta}}^{(k)} - \widehat{\boldsymbol{\beta}}^{\text{MI}})'/(m-1)$ an estimate of the between-imputation variance-covariance matrix of $\widehat{\boldsymbol{\beta}}$. However, when m is small and p is even moderately large, this estimate of B is unstable (and may not even be positive definite) because of the large number of parameters in the matrix. A number of proposals have been made to address this problem, in particular assuming that the within- and between-imputation variances are proportional to each other, across all the components of $\boldsymbol{\beta}$ (Li, Raghunathan, and Rubin 1987; Reiter 2005). However, with recent trends towards using much larger values of m (see next section), the need for these approximations has reduced.

12.2.5 How many imputations?

A major practical issue in using multiple imputation is choosing how many imputations (m) to run. The early literature suggested that small values of m, generally in the range 3–5, were adequate because there are sharply diminishing gains in efficiency as m increases. The expected variance of the MI estimate, reflected in (12.2) reduces only very slightly as m increases beyond values in the range 3–5, even for quite large fractions of missing information. Schafer (1997, p. 107) explained this in terms of the Monte Carlo error associated with the imputation process being "a relatively small portion of the overall inferential uncertainty" (Schafer 1997).

However, recent publications have pointed out that small values of m lead to results that are still visibly affected by Monte Carlo error: point estimates and confidence intervals may vary quite widely if the MI analysis is repeated with a different set of imputations using the same m (Royston 2004, White, Royston, and Wood 2011, and van Buuren 2012). Given that modern computational technology allows larger values of m to be used with minimal cost (except perhaps with extremely large datasets), it seems reasonable to choose m in such a way that the final results are reproducible, so that "a repeat analysis of the same data would produce essentially the same results" (White, Royston, and Wood 2011). This was shown by White, Royston, and Wood (2011) to lead to a rule of thumb that m should be no less than $100 \times \text{FMI}$, or, given that FMI itself is unknown and generally not estimated precisely, that m be set to a value at least as great as the percentage of incomplete cases. In the example of Section 12.1.3, this would suggest setting m to be at least 50. Our choice of $m = 100$ in that example is likely to be an upper bound on the number of imputations worth doing in any practical setting.

12.3 Imputation Methods

Although many different approaches to performing imputation have been reported in the literature, the most important of these can be described within three major categories. These are (i) univariate (generally regression-based) imputation methods for single variables and for monotone missingness patterns, (ii) data-augmentation imputation using a joint probability model for Y, and (iii) the method of fully conditional univariate regression specification (also known as chained equations or sequential regression imputation). This section provides a brief summary of each of these approaches and concludes with a general discussion of the principles that should be considered when building an imputation model.

12.3.1 Regression-based imputation

When only one variable is subject to missing data, imputation modelling can be built around the familiar concept of prediction from a regression model. In the framework summarized above by expression (12.9), if we introduce a p-dimensional vector of (fully observed) covariates, a standard regression model would be specified in terms of a p-dimensional parameter vector $\boldsymbol{\eta}$, consisting of the regression coefficients and any scale parameters that may be needed (such as the error variance in a normal linear regression specification). Specifically, we now use the notation $\boldsymbol{Y}$ to refer to the data vector of values Y_i $(i = 1 \ldots n)$ of the single variable that is subject to missing data, and $\boldsymbol{Z}_i$ to refer to all variables that are considered important for explaining variation in Y_i (which for simplicity of notation we take to include a constant term representing a model intercept). Then if a normal linear regression model is appropriate, imputation can be based on predictions from the standard model:

$$Y_i \sim \mathrm{N}(\boldsymbol{Z}_i'\boldsymbol{\eta}, \sigma^2). \tag{12.13}$$

To describe the imputation process explicitly, we partition the complete data random vector $\boldsymbol{Y}$ into sub-vectors $\boldsymbol{Y}^0$ (of length n_0) and $\boldsymbol{Y}^1$ (of length $n_1 = n - n_0$) representing the observed and missing components of $\boldsymbol{Y}$, respectively, where n_0 and n_1 denote the number of observed and missing values, respectively, in $\boldsymbol{Y}$. (Note that $\boldsymbol{Y}^0$ and $\boldsymbol{Y}^1$ provide alternative notation to the generic Y_{obs} and Y_{obs}, respectively; the superscript notation is more useful in the following.) Similarly, the covariate matrix Z is partitioned into Z^0 and Z^1, representing the rows corresponding to observed and missing Y values, respectively. Then multiple imputation may be performed as follows, using the formulation adopted by one major software package (StataCorp 2009):

1. Fit the regression model (12.13) to the observed data $\boldsymbol{Y}^0, Z^0$—which conventionally means obtaining point estimates $\widehat{\boldsymbol{\eta}}$ and $\widehat{\sigma}^2$, but in this context should be thought of as obtaining values that characterize the (joint) posterior distribution of $\boldsymbol{\eta}$ and σ^2 under a conventional "non-informative" prior distribution (Gelman *et al* 2004).

2. Draw values $\boldsymbol{\eta}_*$ and σ_*^2 from their joint posterior distribution. This is done in two steps, firstly sampling the variance parameter from an inverse gamma (inverse chi-squared) distribution:

$$\sigma_*^2 \sim \widehat{\sigma}^2 \frac{(n_0 - p)}{\chi^2_{n_0 - p}},$$

   and then drawing the regression parameters from a (multivariate) normal distribution conditional on the drawn value of σ_*^2:

$$\boldsymbol{\eta}_* \sim \mathrm{N}\left(\widehat{\boldsymbol{\eta}}, \sigma_*^2 (Z_0' Z_0)^{-1}\right).$$

3. Obtain a set of imputed values for $\boldsymbol{Y}^1$ by drawing from $\mathrm{N}(Z^1 \boldsymbol{\eta}_*, \sigma_*^2 I_{n_1 \times n_1})$.
4. Repeat steps 2 and 3 m times, to obtain m sets of imputed values for $\boldsymbol{Y}^1$.

Steps 2 and 3 of the process described are just those required to obtain simulated values from the posterior predictive distribution of the missing values (e.g., Gelman *et al* 2004, Chapter 14). Most readers will recognize step 3 as a standard normal linear regression prediction, but this would often be done by substituting the point estimates $\widehat{\boldsymbol{\eta}}$ and $\widehat{\sigma}^2$, rather than using values drawn from (approximate) posterior distributions. It is the latter step (i.e., step 2 above) that is key to making this imputation procedure proper in Rubin's sense.

Many variations of the basic idea of regression-based imputation have been proposed. A natural set of extensions arises when Y is not continuously scaled. For example if Y is dichotomous, then a logistic regression model would be more appropriate than the normal linear specification, and extensions to multinomial and ordinal logistic regressions are natural for categorical variables. For each of these, the key steps 1–4 are essentially the same, except that it is more difficult to derive posterior distributions for the required regression parameters, so asymptotic normal approximations are invariably used.

A different class of extensions, to semi-parametric and non-parametric methods, has been developed to deal with imputing variables that have continuous or semi-continuous distributions for which a normal linear model does not fit, even after transformation. Many of these may be regarded as variants of the hot-deck method (Andridge and Little 2010). The most widely used semi-parametric approach is probably the method of "predictive mean matching." In this approach step 3 above is replaced by imputing each missing value by the observed Y value that is obtained by identifying the unit with observed Y for which the predicted (expected) value (i.e., mean) given by $Z^0 \boldsymbol{\eta}_*$ matches most closely to the predicted value for the missing observation, obtained from $Z^1 \boldsymbol{\eta}_*$ with the same draw of $\boldsymbol{\eta}_*$ (Heitjan and Little 1991, Schenker and Taylor 1996).

The methods of imputation based on univariate regression approaches are important, not only in their own right, but also because they form the basis of many more complex methods that have been developed to handle problems in which more than one variable is subject to missing data. There are two main directions for such extensions, the more general of which is discussed below in Section 12.3.3. The other important extension is to problems where there is a *monotone* pattern of missingness, which means that the variables that are subject to missing data can be ordered as $Y^{(1)}, \ldots, Y^{(k)}$ in such a way that if $Y^{(1)}$ is missing (for a particular unit) then $Y^{(2)}$ is also missing (and may be missing for other units), as are other variables further down the order. The last variable in the ordering has the largest number of missing values. With such a pattern of missing values, imputation may be performed by first using a regression-based imputation for $Y^{(1)}$ conditional on all fully observed covariates, then using a similar approach for $Y^{(2)}$ conditional on the imputed $Y^{(1)}$ values and the fully observed covariates, and so on down the list. If all of the univariate imputations are proper then the overall process will also be proper. This is a powerful approach but problems in which multivariate missingness has a monotone structure seem to be relatively uncommon in practice, motivating the need for more general approaches.

12.3.2 Imputation under a joint model

Many missing data problems involve complex non-monotone patterns of missingness. In general, in such settings the specification and estimation of a parametric model for all of the variables that are subject to incomplete data, for the purpose of generating imputations, is very challenging. Indeed, if it were readily possible to build and fit such models, then many inference problems with missing data could be solved directly, by performing inferences for the parameters of the joint model for Y_{obs} and Y_{mis}. A major advance in the practical application of multiple imputation was Schafer's development of computational algorithms for imputation under joint probability models, in particular the multivariate normal distribution (Schafer 1997).

The general algorithm developed by Schafer was an application of "data augmentation," which is a specific version of the Monte Carlo Markov chain approach to Bayesian computation, and was first discussed in detail by Tanner and Wong (1987). Referring back to the general parametric formulation of the imputation problem given in the text surrounding (12.9), the data augmentation (DA) algorithm is an iterative sampling scheme that proceeds as follows:

1. Begin with an estimate of the (multidimensional) parameter θ that is required to characterize the joint model for Y_{obs} and Y_{mis}, say $\theta^{(t)}$ (where $t = 0$ at the first iteration).

2. Draw a value of the missing data from the conditional distribution $Y_{mis}^{(t)} \sim P(Y_{mis}|Y_{obs}, \theta^{(t)})$ (the I-step, for "Imputation").

3. Draw a new value of the parameter from its complete-data posterior distribution, "plugging in" the newly drawn value of Y_{mis}: $\theta^{(t+1)} \sim P(\theta|Y_{obs}, Y_{mis}^{(t)})$ (the P-step, for "Posterior").

4. Repeat the I- and P-steps until an appropriate criterion for convergence is satisfied.

As already remarked, the I-step is not difficult to implement in a number of standard parametric models. The P-step is generally much more difficult, but Schafer showed how the algorithm could be implemented for a number of specific models, the most important of these being the multivariate normal. In this model, θ consists of a vector of means and a variance-covariance matrix for all of the variables included in the imputation model. Schafer highlighted the close connection between the DA algorithm and the widely known EM algorithm (Dempster, Laird, and Rubin 1997), which proceeds in a similar series of alternating steps. The main distinction is that instead of drawing values of the parameter from its posterior distribution (P-step), in EM the posterior distribution or likelihood function is maximized (the M-step). In Schafer's implementation of the DA algorithm for the multivariate normal distribution, the link with the EM algorithm is central, with the latter being used to identify an estimate of θ with which to initiate the iterative simulations.

The DA algorithm is only feasible for a limited number of joint probability distributions, with the multivariate normal being the only one that has achieved extensive application in practice (see Section 12.4). The underlying assumption, that all variables subject to missing values should follow a joint normal distribution, is patently wrong for almost all real applications, but the algorithm itself is very stable, since it is essentially based only on first and second moments of the data. Schafer and others have also produced good evidence that the end-results are not necessarily badly affected by failure of the normality assumption. In particular, the algorithm appears to perform quite well for dichotomous variables, unless they are severely skewed (Schafer 1997, Bernaards, Belin, and Schafer 2007, Lee and Carlin 2010). In this context, it is important to remember that the final results of an analysis based on multiple imputation may be reasonable even if some individual imputed values do not appear to be so (Schafer 1997). (Recall for example the negative imputed age value in Figure 12.4.)

A specific issue that arises in using multivariate-normal imputation with categorical variables is that imputed values will vary continuously, and a natural question arises as to whether and how these should be rounded to values that reflect the original categorization. It has been argued that for some purposes rounding is not necessary and bias may be minimized by using unrounded imputed values. For many purposes, however, rounding seems desirable and best methods for rounding are a subject of current research (Galati et al. 2012, Lee and Carlin 2012).

Beyond the unstructured multivariate normal model reviewed here, considerable work has been done on extensions that allow for multilevel structure in the data. Recognizing and modeling multilevel structure in the imputation process may be important, especially for unbalanced data: see Chapter 14.

12.3.3 Imputation using fully conditional specification

Many incomplete data problems involve numerous variables of a variety of types subject to missing data in non-monotone patterns. It is intuitively appealing to apply the univariate regression approach to each of the variables that has missing values, except that the non-monotone pattern means that the variables that are desired as predictors in each of the univariate regression imputation models will often themselves have missing values for units that require an imputation prediction. An approach to circumvent this difficulty was proposed independently by van Buuren (1999) and Raghunathan et al. (2001), who developed implementations in the packages S-PLUS and SAS, respectively, followed by Royston (2004) in Stata. This approach is described in more detail in the next chapter, but briefly it involves repeatedly cycling through all of the variables that require imputation and performing a univariate regression imputation for each variable sequentially after having filled in any missing values for variables that are required in the current imputation model. Initially, missing covariate values are filled by a simple convenient approach such as random sampling from observed values on other units, but once an initial set of imputations has been created for all variables, these imputed values are then used where needed in the imputation models for subsequent cycles.

This approach was originally called "multiple imputation using chained equations" (MICE), but other terms have also been used, such as "sequential regression multiple imputation" and "fully conditional specification," which emphasizes that the method relies entirely on specifying univariate regression models for each variable, conditioning on all other variables in the dataset (in contrast to the joint modeling approach, which requires an explicit joint probability model for all variables that are subject to missingness). The appeal of the method is its flexibility in allowing an appropriate univariate regression specification for each variable, which not only allows appropriate scaling and modeling of univariate error distributions, but also allows the univariate models to incorporate non-linear terms and interaction effects. Against these advantages there is the lack of a complete theoretical basis for the method, and in particular the potential danger that successive univariate model specifications may be incompatible with each other. These issues are still the subject of research although little evidence of substantial difficulty arising from these theoretical problems has yet emerged and the method has become widely used in practice.

12.3.4 General principles for building imputation models

Left unstated in the overview of the theory of multiple imputation given in Section 12.2 was that the modeling process used to perform imputation must be compatible with the modeling that will be used in the final analysis. Formal mathematical developments tend to imply the existence of a "true" model, but in real applications the data analyst may have many variables available for model building and needs to make choices, at best arriving at "defensible" rather than "true" models. The important principle is that the modeling choices made in performing imputation should be compatible with those that will be made in the ultimate analysis of interest. This concept has been studied formally under the label of "congeniality" (between imputation and analysis models; Meng 1994), but as with many aspects of multiple imputation the practical implications of these theoretical investigations remain somewhat elusive.

The generally accepted view, however, is that imputation models should be as large as possible, in the sense of including as many variables and as much potential richness of dependency between variables as possible. Intuitively, there is a clear danger that an overly simple imputation model will fail to represent key relationships between variables, which may lead to downward biases in estimates of association in final analyses. The best-known illustration of this danger is when imputation modeling only includes variables that will be used as predictors or covariates in the final analysis, and excludes the outcome(s) of interest (Sterne et al. 2009). On the other hand, Rubin and others have emphasized that imputation modeling should not be approached in the same way as a prediction modeling problem: "Our actual objective is valid statistical inference not optimal point prediction under some loss function [...] Judging the quality of missing data procedures by their ability to recreate the individual missing values (according to hit-rate, mean square error, etc.) does not lead to choosing procedures that result in valid inference..." (Rubin 1996). Furthermore, imputation models that are too complex (e.g., include too many variables) may lead to unstable algorithms and produce unreliable imputations.

12.4 Multiple Imputation in Practice

As discussed above, the method of multiple imputation arose in the context of statisticians seeking to find ways of making available for public use large survey datasets that were subject to incompleteness due to non-response, in such a way that users could perform valid inferences that allowed appropriately for the missing values. This rationale for multiple imputation has been widely promulgated by Rubin and his colleagues and former students, although only a very limited number of such public-use imputed datasets appear to have been published. With the publication and free dissemination of Schafer's software for imputation under the multivariate normal model in the late 1990s, followed very shortly by the various versions of the chained equations approach, multiple imputation became feasible for applied statisticians and data analysts to apply within the context of more or less routine data analysis. The resulting development of applications in the first decade of the 2000s has been rapid. This section provides a brief overview of these historical developments.

12.4.1 Early applications: Expert imputation of large-scale survey data and other tailored applications

Around the time of the publication of Rubin's seminal 1987 book, a number of applications of multiple imputation appeared in the statistical literature, tailored to specific problems. The most well known of these was a substantial project undertaken by Rubin and colleagues that sought to enable analysis to be performed on U.S. census data across two decennial waves (1970, 1980) of the census, between which substantial changes had occurred in the classification of industries and occupations (Rubin and Schenker 1991, Treiman et al. 1998, Clogg et al. 1991). Fewer than a third of the 1970 occupation categories had an exact match in the 1980 codes. For the analysis of changes over time, the 1980 classification of these cases was treated as missing in the 1970 data and these missing values were multiply imputed, using a model that was developed from a subsample of the 1970 census on which 'double-coding' (using both 1970 and 1980 categories) had been performed. Logistic regression models were used for imputation on a combined total of 1.6 million cases for which the 1970 codes were missing. This was an unusual missing-data problem in two ways: firstly, that the double-coded sample was randomly selected (so the missing data mechanism was clearly ignorable), and secondly, that the double-coded sample contained fewer than 10% of the desired cases, so the fraction of missing information was high. Solving the problem required a number of technical issues to be addressed, including the fitting of logistic regression models with many predictors to often sparse data. The methods used to address this problem (Clogg et al. 1991) have recently been extended and made more widely available with modern imputation software (White, Daniel, and Royston 2010).

It is difficult to determine the extent to which the U.S. Census Bureau, or other national statistical agencies, have adopted Rubin's intended use of multiple imputation for the publication of public-use datasets. However, one national agency that has continued to publish multiply imputed public-use data is the U.S. National Centre for Health Statistics (NCHS). A large-scale project in the early 1990s was undertaken for the National Health and Nutrition Examination Survey (NHANES), which is a major national survey conducted at regular intervals by the NCHS for assessing the health and nutritional status of the U.S. population. Multiple imputation was performed for more than 60 variables with missingness proportions of 30% or more on a sample of over 30,000 individuals (Schafer 1996). This has been followed by more recent work on the imputation of missing income data in the National Health Interview Survey (Schenker et al. 2006). For both of these projects, publicly available imputed datasets ($m = 5$) are available (at the time of writing) on the NCHS website.

Another major imputation modeling effort was undertaken in the late 1980s to address substantial missing data problems in the Fatal Accident Reporting System, a national database maintained by the U.S. National Highway Traffic Safety Administration (NHTSA; Heitjan and Little 1991). This work was again intended to produce a public-use dataset, and was interesting as the first major application of imputation using the method of prediction mean matching. This was applied in a two-stage algorithm to impute blood alcohol content, which is zero in a substantial proportion of cases and has a severely skewed distribution when not zero. Other early applications of multiple imputation included an analysis in which the rounding or "coarsening" of reported ages in children led to various biases that could be addressed by imputation of the true ages of the children (Heitjan and Rubin 1990) and an analysis that aimed to estimate the distribution of time from HIV seroconversion to the onset of AIDS, in which imputation was a convenient approach to estimating variability for longer follow-up times, for which many of the observation times were censored (so effectively missing; Taylor et al. 1990).

The early applications of multiple imputation that have been cited above were published in statistics journals with a primary focus on the methodology (with the possible exception of the last-mentioned study, published in a journal that is intended for clinicians as well as statisticians), and it is not clear how much these developments penetrated the statistical analyses performed by social scientists and policy makers. There seems to be little evidence that public-use imputed datasets have been widely used in substantive empirical research practice.

12.4.2 Multiple imputation enters the toolkit for practical data analysis

From the late 1990s onwards, software tools for implementing multiple imputation started to become available in packaged form, and the method gradually entered the range of more or less standard tools for applied statisticians and data analysts to use in practice. This trend began with the web-publication of NORM, a stand-alone (Microsoft Windows) implementation of the multivariate-normal data augmentation algorithm by Joe Schafer, who also distributed implementations of the other joint modeling approaches described in his 1997 book (Schafer 1997). Further impetus followed shortly thereafter with the publication of MICE for S-PLUS (van Buuren, Boshuizen, and Knook 1999), IVEware for SAS (Raghunathan et al. 2001), and `ice` for Stata (Royston 2004), each of which implemented the chained equations approach. These initiatives led in the first decade of the 2000s to the major statistical packages incorporating various capabilities for imputation in their standard releases, beginning with experimental versions of procedures including multivariate normal imputation that were released by SAS (version 8.1) in 2000.

A parallel line of development recognized that multiple imputation would be difficult to use in standard analyses without software tools that could facilitate the manipulation of multiply imputed datasets. This was necessary not only to perform the multiple analyses and combination of results using Rubin's rules, required to perform inference, but also to allow flexible manipulation of the multiple datasets, for example to allow recoding and transformations of variables, as is commonly needed in practice. A prototype system for managing multiple imputation analysis in the package Stata was first published in 2003 (Carlin, Greenwood, and Coffey 2003), with a substantially improved version (Carlin, Galati, and Royston 2008), shortly after which an extensive suite of multiple imputation tools became available as an official component of Release 11 (2009) of Stata (StataCorp 2009). At the time of writing, the tools available in Stata, which now include all of the major approaches to imputation described in Section 12.3, are the most comprehensive of those accessible in standard software, although active development also continues in R (Su, Gelman, Hill, and Yajima 2011; van Buuren 2012). The availability of multiple imputation as a tool within standard software packages has led to a rapid increase in the number of applications. These reflect a substantial move away from Rubin's original concept of the 'expert imputer' to a 'self-service' approach where imputation is performed by the same analyst (or at least within the same research team) as the final analysis itself. Applications have generally involved rather smaller datasets than the tailored applications from the earlier period reviewed in the previous section. Reviews of the increasing usage of multiple imputation in major medical research journals have charted especially dramatic growth between 2005 and 2010 (Sterne et al. 2009; Mackinnon 2010). The author is less familiar with other fields of empirical research but there is anecdotal evidence of a similar proliferation of applications within the social and behavioural sciences.

When the imputer and analyst are the same person, there appears to be less danger of "uncongeniality" between the imputation and analysis models, although this assumes an appropriate level of expertise and understanding on the part of the user, to ensure that the imputation model is appropriately specified. There is at least one published example of incorrect conclusions arising from naive application of multiple imputation (Hippisley-Cox et al. 2007; Sterne et al. 2009). A recent case study noted that "the implementation of good imputation routines is a difficult and cumbersome task and an unsupervised imputation using standard imputation software can harm the results more than simple complete case analysis" (Drechsler 2011). The increased availability of MI software tools may in fact have led to some over-use of the method: it is important to consider carefully whether there is genuine scope for information to be recovered from incomplete cases before launching into an application (Lee and Carlin 2012, Graham 2012). Guidelines are now starting to appear to assist less sophisticated users, but these require further development (Sterne et al. 2009; White, Royston, and Wood 2011; van Buuren 2012). There has also been renewed emphasis on the use of multiple imputation as a tool for analysis of the sensitivity of results to the range of assumptions that might be made about the processes underlying missing data—in particular the MAR assumption (Heraud-Bosquet et al. 2012, Carpenter and Kenward 2013).

12.4.3 New applications and outstanding problems

Research on multiple imputation continues in numerous directions. An extensive technical review has highlighted the need for further work in relation to variance estimation (Reiter and Raghunathan 2007). The authors discuss concerns that the Rubin variance estimate (as given in expression (12.2)) may be upwardly biased in some circumstances (Robins and Wang 2000; Nielsen 2003) although conclude, as have others, that this conservative bias may not be a major problem in practice. They go on to review several variations of the method of multiple imputation beyond the classical setting of large-scale survey non-response, such as correcting for measurement error (Cole, Chu, and Greenland 2006) and the production and analysis of synthetic multiply-imputed datasets that are released for public use, in place of the original survey data, in order to protect confidentiality (Reiter 2005). Considerable research is also still needed to better understand the practical value and reliability of the available packaged imputation methods. At present, the multivariate normal and fully conditional (chained equations) methods are the only ones that can be—and are—widely applied, and these appear likely to dominate applications for the foreseeable future. Relatively few direct comparisons between these approaches have been published, perhaps because of the difficulty of the underlying theory and the challenge of designing simulation studies that allow convincing generalization (Lee and Carlin 2010). Users of the multivariate normal approach need to make difficult decisions about how to handle variables that clearly do not follow a normal distribution, even marginally. Current advice is to eliminate skewness in the margins as far as possible and to consider carefully the implications of rounding and truncation of imputed values to the original scale of the variable being imputed. A practical advantage of the method is that the computational algorithm is remarkably stable and apparently reasonable results can generally be obtained for quite large numbers of variables (100 or more). The fully conditional method has probably overtaken the multivariate normal as the most widely applied in practice, because of the flexibility of specification of each conditional imputation model; see next chapter. This flexibility comes at the cost, however, of potential instabilities and inconsistencies when handling large numbers of variables. Challenges remain in how to deal with the many varieties of skewed, multi-modal and truncated distributions that arise in practice, as well as with interaction effects and other forms of non-linearity. Further research is needed on the extent to which incompatible conditional distributions may arise and lead to invalid inferences.

Relatively little development of new approaches to generating imputed data has taken place recently. Some methods that have appeared in the statistical literature do not appear to have been translated into practice, perhaps because their underlying assumptions may be too restrictive, as with a method for tackling the important problem of imputing 'semi-continuous' variables, which take on a single discrete value with positive probability but are otherwise continuously distributed. (Javaras and Van Dyk 2003). An especially challenging problem, in view of the widespread dissemination and uptake of multiple imputation by relatively unsophisticated users, is that of providing tools for diagnosing potential problems in imputation models. Several authors have proposed approaches ranging from simple graphical checks (Abayomi, Gelman, and Levy 2008) to the use of posterior predictive model checking, which assesses the extent to which the imputation model is capable of replicating data structures that are similar to the observed data (He and Zaslavsky 2012). This work is challenging because of the fundamental fact that where data are incomplete we can never be sure of the extent to which the missing values should be consistent with any proposed model. In particular, it is important to remember that the packaged tools for imputation all assume ignorability of the missing data mechanism (Section 12.2.3), which is likely never to be strictly true, although it may be a reasonable working assumption for producing analyses that improve on complete-case or other simple methods. In this regard, further development and dissemination of tools for assessing the sensitivity of inferences to imputation model assumptions will be valuable.

In conclusion, the method of multiple imputation has reached the mainstream of statistical practice over the past decade and is now an important tool for the analysis of incomplete data. However, a sound understanding of its underlying assumptions and limitations is needed in order to avoid unproductive or even misleading applications.

References

Abayomi, K., Gelman, A., and Levy, M. (2008). Diagnostics for multivariate imputations. *Applied Statistics*, **57**, 273–291.

Andridge, R.R. and Little, R.J.A. (2010). A review of hot deck imputation for survey non-response. *International Statistical Review*, **78**, 40–64, 2010.

Barnard, J. and Rubin, D.B. (1999). Small-sample degrees of freedom with multiple imputation. *Biometrika*, **86**, 948–955.

Bernaards, C.A., Belin, T.R., and Schafer, J.L. (2007). Robustness of a multivariate normal approximation for imputation of incomplete binary data. *Statistics in Medicine*, **26**, 1368–1382.

Carlin, J.B., Galati, J.C., and Royston, P. (2008). A new framework for managing and analysing multiply imputed datasets in STATA. *The STATA Journal*, **8**, 49–67.

Carlin, J.B., Greenwood, P., Li, N., and Coffey, C. (2003). Tools for analysing multiple imputed datasets. *The STATA Journal*, **3**, 226–244.

Carpenter, J.R. and Kenward, M.G. (2013). *Multiple Imputation and Its Application.* Chichester: John Wiley & Sons.

Clogg, C.C., Rubin, D.B., Schenker, N., Schultz, B., and Weidman, L. (1991). Multiple imputation of industry and occupational codes in census public-use samples using Bayesian logistic regression. *Journal of the American Statistical Association*, **86**, 68–78.

Cole, S.R., Chu, H., and Greenland, S. (2006). Multiple-imputation for measurement-error correction. *International Journal of Epidemiology*, **35**, 1074–1081.

Dempster, A.P., Laird, N.M., and Rubin, D.B. (1977). Maximum likelihood from incomplete data via the EM algorithm. *Journal of the Royal Statistical Society, Series B*, **39**, 1–38.

Drechsler, J. (2011). Multiple imputation in practice — a case study using a complex German establishment survey. *AStA Advances in Statistical Analysis*, **95**, 1–26.

Galati, J.C., Seaton, K.A., Lee, K.J., Simpson, J.A., and Carlin, J.B. (2012). Rounding non-binary categorical variables following multivariate normal imputation: evaluation of simple methods and implications for practice. *Journal of Statistical Computation and Simulation*, **00**, 000–000.

Gelman, A. (2011). Induction and deduction in Bayesian data analysis. *Rationality, Markets and Morals*, **2**, 67-78.

Gelman, A., Carlin, J.B., Stern, H.S., and Rubin, D.B. (2004). *Bayesian Data Analysis* (2nd ed.). Boca Raton: Chapman & Hall/CRC.

Graham, J.W. (2012). *Missing Data: Analysis and Design.* New York: Springer.

He, Y. and Zaslavsky, A.M. (2012). Diagnosing imputation models by applying target analyses to posterior replicates of completed data. *Statistics in Medicine*, **31**, 1–18.

Heitjan, D.F. and Little, R.J.A. (1991). Multiple imputation for the fatal accident reporting system. *Applied Statistics*, **40**, 13–29.

Heitjan, D.F. and Rubin, D.B. (1990). Inference from coarse data via multiple imputation with application to age heaping. *Journal of the American Statistical Association*, **85**, 304–314.

Heraud-Bousquet, V., Larsen, C., Carpenter, J., Desenclos, J.-C., and Le Strat, Y. (2012). Practical considerations for sensitivity analysis after multiple imputation applied to epidemiological studies with incomplete data. *BMC Medical Research Methodology*, **12**, 73.

Hippisley-Cox, J., Coupland, C., Vinogradova, Y., Robson, J., May, M., and Brindle, P. (2007). Derivation and validation of QRISK: A new cardiovascular disease risk score for the united kingdom: prospective open cohort study. *British Medical Journal*, **335**, 136.

Javaras, K.N. and Van Dyk, D.A. (2003). Multiple imputation for incomplete data with semicontinuous variables. *Journal of the American Statistical Association*, **98**, 703–715.

Lee, K.J. and Carlin, J.B. (2010). Multiple imputation for missing data: Fully conditional specification versus multivariate normal imputation. *American Journal of Epidemiology*, **171**, 624–632.

Lee, K.J. and Carlin, J.B. (2012). Recovery of information from multiple imputation: A simulation study. *Emerging Themes in Epidemiology*, **9**, 3.

Lee, K.J., Galati, J.C., Simpson, J.A., and Carlin, J.B. (2012). Comparison of methods for imputing ordinal data using multivariate normal imputation: A case study of non-linear effects in a large cohort study. *Statistics in Medicine*, **31**, 4164–4174.

Li, K.H., Raghunathan, T.E., and Rubin, D.B. (1991). Large-sample significance levels from multiply-imputed data using moment-based statistics and an F reference distribution. *Journal of the American Statistical Association*, **86**, 1065–1073.

Little, R.J.A. and Rubin, D.B. (2002). *Statistical Analysis with Missing Data* (2nd ed.). New York: John Wiley & Sons.

Lu, G. and Copas, J.B. (2004). Missing at random, likelihood ignorability and model completeness. *Annals of Statistics*, **32**, 754–765.

Mackinnon, A.. (2010) The use and reporting of multiple imputation in medical research—A review. *Journal of Internal Medicine*, **268**, 586–593.

Meng, X.-L. (1994). Multiple-imputation inferences with uncongenial sources of input. *Statistical Science*, **9**, 538–573.

Nielsen, S.F. (2003). Proper and improper multiple imputation. *International Statistical Review*, **71**, 593–607.

Raghunathan, T.E., Lepkowski, J.M., Van Hoewyk, J., and Solenberger, P. (2001). A multivariate technique for multiply imputing missing values using a sequence of regression models. *Survey Methodology*, **27**, 85–95.

Reiter, J.P. (2005). Releasing multiply imputed, synthetic public use microdata: An illustration and empirical study. *Journal of the Royal Statistical Society, Series A*, **168**, 185–205.

Reiter, J.P. (2007). Small-sample degrees of freedom for multi-component significance tests with multiple imputation for missing data. *Biometrika*, **94**, 502–508.

Reiter, J.P. and Raghunathan, T.E. (2007). The multiple adaptations of multiple imputation. *Journal of the American Statistical Association*, **102**, 1462–1471.

Robins, J.M. and Wang, N. (2000). Inference for imputation estimators. *Biometrika*, **87**, 113–124.

Royston, P. (2004). Multiple imputation of missing values. *The STATA Journal*, **4**, 227–241.

Rubin, D.B. and Schenker, N. (1991). Multiple imputation in health-care databases: an overview and some applications. *Statistics in Medicine*, **10**, 585–598.

Rubin, D.B. (1978). Multiple imputations in sample surveys: A phenomenological Bayesian approach to nonresponse. *Proceedings of the Survey Research Methods Section, American Statistical Association*, pp. 20–28.

Rubin, D.B. (1987). *Multiple Imputation for Nonresponse in Surveys.* New York: John Wiley & Sons.

Rubin, D.B (1996). Multiple imputation after 18+ years. *Journal of the American Statistical Association*, **91**, 473–489.

Rubin, D.B. and Schenker, N. (1986). Multiple imputation for interval estimation from simple random samples with ignorable nonresponse. *Journal of the American Statistical Association*, **81**, 366–374.

Schafer, J.L. (1997). *Analysis of Incomplete Multivariate Data.* London: CRC / Chapman & Hall.

Schafer, J.L., Ezzati-Rice, T.M., Johnson, W., Khare, M., Little, R.J.A., and Rubin, D.B. (1996). The NHANES III multiple imputation project. *Proceedings of the Survey Research Methods Section of the American Statistical Association*, pp. 28–37.

Schenker, N., Raghunathan, T.E., Chiu, P.-L., Makuc, D.M., Zhang, G., and Cohen, A.J. (2006). Multiple imputation of missing income data in the national health interview survey. *Journal of the American Statistical Association*, **101**, 924–933.

Schenker, N. and Taylor, J.M.G. (1996). Partially parametric techniques for multiple imputation. *Computational Statistics and Data Analysis*, **22**, 425–446.

Scheuren, F. (2005). Multiple imputation: How it began and continues. *The American Statistician*, **59**, 315–319.

Seaman, S., Galati, J.C., Jackson, D., and Carlin, J.B. (2013). What is meant by "missing at random"? *Statistical Science*, **28**, 257–268.

StataCorp. (2009). *Multiple Imputation Reference Manual.* StataCorp LP, College Station, TX.

Sterne, J.A.C., White, I.R., Carlin, J.B., Spratt, M., Royston, P., Kenward, M.G., Wood, A.M., and Carpenter, J.R. (2009). Multiple imputation for missing data in epidemiological and clinical research: Potential and pitfalls. *British Medical Journal*, **338**, b2393.

Su, Y.-S., Gelman, A., Hill, J.L., and Yajima, M. (2011). Multiple imputation with diagnostics (mi) in R: Opening windows into the black box. *Journal of Statistical Software*, **45**, 1–31.

Tanner, M.A. and Wong, W.H. (1987). The calculation of posterior distributions by data augmentation. *Journal of the American Statistical Association*, **82**, 528–528.

Taylor, J.M.G., Muñoz, A., Bass, S.M., Saah, A.J., Cmiel, J.S., and Kingsley, L.A. (1990). Estimating the distribution of times from HIV seroconversion to AIDS using multiple imputation. *Statistics in Medicine*, **9**, 505–514.

Treiman, D.J., Bielby, W.T., and Cheng, M.-T. (1998). Evaluating a multiple-imputation method for recalibrating 1970 U.S. census detailed industry codes to the 1980 standard. *Sociological Methodology*, **18**, 309–345.

van Buuren, S. (2012). *Flexible Imputation of Missing Data.* Boca Raton: CRC / Chapman & Hall.

van Buuren, S., Boshuizen, H.C., and Knook, D.L. (1999). Multiple imputation of missing blood pressure covariates in survival analysis. *Statistics in Medicine*, **18**, 681–694.

White, I.R. and Carlin, J.B. (2010). Bias and efficiency of multiple imputation compared with complete-case analysis for missing covariate values. *Statistics in Medicine*, **29**, 2920–2931.

White, I.R., Daniel, R., and Royston, P. (2010). Avoiding bias due to perfect prediction in multiple imputation of incomplete categorical variables. *Computational Statistics and Data Analysis*, **54**, 2267–2275.

White, I.R., Royston, P., and Wood, A.M. (2011). Multiple imputation using chained equations: issues and guidance for practice. *Statistics in Medicine*, **30**, 377–399.

13

Fully Conditional Specification

Stef van Buuren
Netherlands Organisation for Applied Scientific Research TNO, Leiden & University of Utrecht, the Netherlands

CONTENTS

13.1 Introduction

13.1.1 Overview

The term *fully conditional specification* (FCS) refers to a class of imputation models for non-monotone multivariate missing data. Other names for this class of models include *sequential regression multivariate imputation* and *chained equations*. As non-monotone missing data frequently occur in practice, FCS covers a wide range of missing data problems.

The material presented here builds upon the foundations of multiple imputation as laid out in Chapter 11, and draws heavily on Chapters 4 and 5 of my book *Flexible Imputation of Missing Data* (van Buuren 2012). Additional background, computer code to apply FCS in practice, and an overview of current software can be found on www.multiple-imputation.com.

The present chapter focuses on fully conditional specification, an approach for multivariate multiple imputation that has become very popular with practitioners thanks to its ease of use and flexibility. Section 13.2 outlines a number of practical problems that appear when trying to impute multivariate missing data. Section 13.3 distinguishes various multivariate missing data patterns, and introduces four linkage measures that aid in setting up multivariate imputation models. Section 13.4 briefly reviews three general strategies to impute multivariate missing data. Section 13.5 describes the FCS approach, its assumptions, its history, the Multivariate Imputation by Chained Equations (MICE) algorithm, and discusses issues surrounding compatibility and performance. Section 13.6 provides a systematic account of seven choices that need to be made when applying FCS in practice. Section 13.7 highlights the role of diagnostics in imputation.

13.1.2 Notation

Let Y denote the $N \times p$ matrix containing the data values on p variables for all N units in the sample. The response indicator R is an $N \times p$ binary matrix. Elements in R are denoted by r_{ij} with $i = 1, \ldots, N$ and $j = 1, \ldots, p$. Element $r_{ij} = 1$ if the corresponding data value in Y is observed, and $r_{ij} = 0$ if it is missing. We assume that we know where the missing data are, so R is always completely observed. The observed data in Y are collectively denoted by Y^o. The missing data are collectively denoted as Y^m, and contain all Y-values that we do not see because they are missing. Notation Y_j denotes the j-th column in Y, and Y_{-j} indicates the complement of Y_j, that is, all columns in Y except Y_j. When taken together $Y = (Y^o, Y^m)$ contains the hypothetically complete data values. However, the values of the part Y^m are unknown to us, and the data are thus incomplete. Notation Y_j^o and Y_j^m stand for the observed and missing data in Y_j, respectively. Symbol $\dot{Y}_j$ stands for imputations of Y_j^m.

13.2 Practical Problems in Multivariate Imputation

There are various practical problems that may occur when one tries to impute multivariate missing data. Many imputation models for Y_j use the remaining columns Y_{-j} as predictors. The rationale is that conditioning on Y_{-j} preserves the relations among the Y_j in the imputed data. This section considers some potential difficulties in setting up imputation models of this type.

Suppose that we want to impute variable Y_j given other predictors in the data Y_{-j}. An obvious difficulty is that any of the predictors Y_{-j} may also contain missing data. In that case, it is not possible to calculate a linear predictor for cases that have missing data, and consequently such cases cannot be imputed.

A second difficulty is that circular dependence can occur, where Y_j^m depends on Y_h^m and Y_h^m depends on Y_j^m with $h \neq j$. In general, Y_j and Y_h are correlated, even given other variables. The limiting case occurs if Y_h^m is a function of Y_j^m for example, a transformation. When ignored, such circularities may lead to inconsistencies in the imputed data, or to solutions with absurd values.

Third, variables may have different measurement levels, e.g., binary, unordered categorical, ordered categorical, continuous, or censored data. Properly accounting for such features of the data is not possible using the application of theoretically convenient models, such as the multivariate normal, potentially leading to impossible values, e.g., negative counts. Distributions can take many forms. If the scientific interest focuses on extreme quantiles of the distribution, the imputation model should represent fairly the shape of the entire distribution.

Many datasets consist of hundreds, or even thousands, of variables. This creates problems in setting up imputation models. If the number of incomplete variables is large, problems with collinearity, unstable regression weights and empty cells occur. The general advice is to condition on as many variables as possible, but this may easily lead to imputation models that have more parameters than data points. A good selection of predictors and a judicious choice of constraints will often substantially improve imputations.

The ordering of rows and columns can be meaningful, e.g., as in longitudinal or spatial data. Data closer in space or time are typically more useful as predictors. With monotone missing data, imputation needs to progress from the most complete to least complete variable, so one may wish to regulate the sequence in which variables are imputed. Also, modeling could be done efficiently if it respects the known ordering in the data.

The relation between Y_j and predictors Y_{-j} can be complex, e.g., nonlinear, subject to censoring processes, or functionally dependent. For example, if the complete-data model requires a linear and a quadratic term, then both terms should be present in the imputation model. Also, the contribution of a given predictor may depend on the value of another one. If such relations exist in the observed data, it makes sense to preserve these in the imputed data. Taking care of complex relations is not automatic and requires careful treatment on behalf of the imputer.

Imputation of multivariate missing data can create impossible combinations, such as pregnant grand-fathers, or "quality of life" of the deceased. We would generally like to avoid combinations of data values that can never occur in reality (e.g., a diastolic blood pressure that exceeds a systolic blood pressure), but achieving this requires a careful analysis and setup of the imputation model.

In practice, it can also happen that the total of a set of variables is known (e.g., a budget total), but that some of the components are missing. If two components are missing, then imputation of one implies imputation of the other, since both values should add up to a known total. A more complex problem surfaces when two or more components are missing.

Finally, there are often different causes for the missing data in the same dataset. For example, the missing data could result from a failure to submit the questionnaire, from non-contact, from the fact that the respondent skipped the question, and so on. Depending on the subject matter, each of these multiple causes could require its own imputation model.

Other complexities may appear in real life. Properly addressing such issues is not only challenging, but also vital to creating high quality and believable imputations.

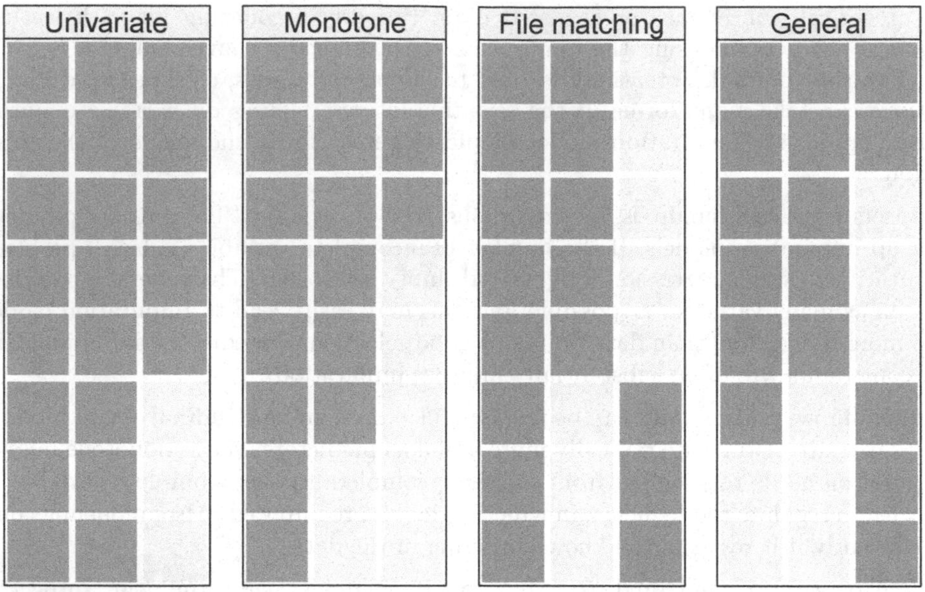

FIGURE 13.1
Some missing data patterns in multivariate data. Dark is observed, light is missing.

13.3 Missing Data Patterns

13.3.1 Overview

It is useful to study the missing data pattern for several reasons. For monotone missing data, we have convergence in one step, so there is no need to iterate. Also, the missing data pattern informs us which variables can contain information for imputation, and hence plays an important role in the setup of the imputation model.

Figure 13.1 illustrates four missing data patterns. The simplest type is the monotone pattern, which can result from drop-out in longitudinal studies. If a pattern is monotone, the variables can be sorted conveniently according to the percentage of missing data. Imputation can then proceed variable by variable from left to right with one pass through the data (Little and Rubin 2002).

The patterns displayed in Figure 13.1 are connected since it is possible to travel to all dark cells by horizontal or vertical moves, just like the moves of the rook in chess. Connected patterns are needed to estimate parameters. For example, in order to be able to estimate a correlation coefficient between two variables, they need to be connected, either directly by a set of cases that have scores on both, or indirectly through their relation with a third set of connected data. Unconnected patterns may arise in particular data collection designs, like data combination of different variables and samples, or potential outcomes.

More intricate missing data patterns can occur for data organized in the "long" format, where different visits of the same subject form different rows in the data. van Buuren and Groothuis-Oudshoorn (2011) contains examples for hierarchical data.

13.3.2 Ways to quantify the linkage pattern

The missing data pattern influences the amount of information that can be transferred between variables. Imputation can be more precise if other variables are present for those cases that are to be imputed. By contrast, predictors are potentially more powerful if they are present in rows that are very incomplete in other columns. This section introduces four measures of linkage of the missing data pattern. Note that degree of missingness is only one factor to consider, so the material presented is very much a partial guide to decisions faced by the imputer.

The *proportion of usable cases* (van Buuren, Boshuizen, and Knook 1999) for imputing variable Y_j from variable Y_k is defined as

$$I_{jk} = \frac{\sum_i^N (1 - r_{ij}) r_{ik}}{\sum_i^N (1 - r_{ij})}. \tag{13.1}$$

This quantity can be interpreted as the number of pairs (Y_j, Y_k) with Y_j missing and Y_k observed, divided by the total number of missing cases in Y_j. The proportion of usable cases I_{jk} equals 1 if variable Y_k is observed in all records where Y_j is missing. The statistic can be used to quickly select potential predictors Y_k for imputing Y_j based on the missing data pattern. High values of I_{jk} are preferred.

Conversely, we can also measure how well observed values in variable Y_j connect to missing data in other variables as

$$O_{jk} = \frac{\sum_i^N r_{ij}(1 - r_{ik})}{\sum_i^N r_{ij}}. \tag{13.2}$$

This quantity is the number of observed pairs (Y_j, Y_k) with Y_j observed and Y_k missing, divided by the total number of observed cases in Y_j. The quantity O_{jk} equals 1 if variable Y_j is observed in all records where Y_k is missing. The statistic can be used to evaluate whether Y_j is a potential predictors for imputing Y_k.

The statistics in (13.1) and (13.2) are specific for the variable pair (Y_j, Y_k). We can define overall measures of how variable Y_j connects to all other variables Y_{-j} by aggregating over the variable pairs.

The *influx coefficient* I_j is defined as

$$I_j = \frac{\sum_k^p \sum_i^N (1 - r_{ij}) r_{ik}}{\sum_k^p \sum_i^N r_{ik}}. \tag{13.3}$$

The coefficient is equal to the number of variable pairs (Y_j, Y_k) with Y_j missing and Y_k observed, divided by the total number of observed data cells. The value of I_j depends on the proportion of missing data of the variable. Influx of a completely observed variable is equal to 0, whereas for completely missing variables we have $I_j = 1$. For two variables with the same proportion of missing data, the variable with higher influx is better connected to the observed data, and might thus be easier to impute.

The *outflux coefficient* O_j is defined in an analogous way as

$$O_j = \frac{\sum_k^p \sum_i^N r_{ij}(1 - r_{ik})}{\sum_k^p \sum_i^N (1 - r_{ik})}. \tag{13.4}$$

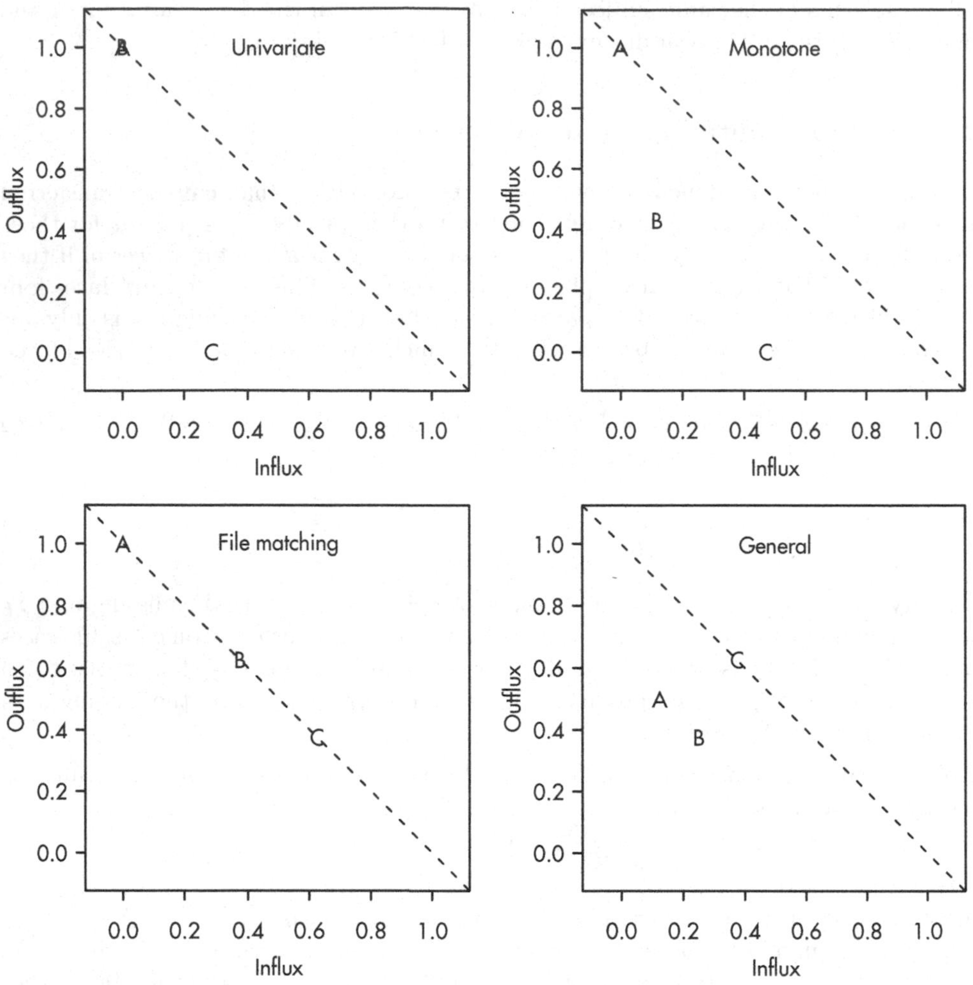

FIGURE 13.2
Fluxplot: Outflux versus influx in the four missing data patterns from Figure 13.1.

The quantity O_j is the number of variable pairs with Y_j observed and Y_k missing, divided by the total number of incomplete data cells. Outflux is an indicator of the potential usefulness of Y_j for imputing other variables. Outflux depends on the proportion of missing data of the variable. Outflux of a completely observed variable is equal to 1, whereas outflux of a completely missing variable is equal to 0. For two variables having the same proportion of missing data, the variable with higher outflux is better connected to the missing data, and thus potentially more useful for imputing other variables. Note the word "potentially," since the actual usefulness will also depend on the amount of association between the variables.

The influx of a variable quantifies how well its missing data connect to the observed data on other variables. The outflux of a variable quantifies how well its observed data connect to the missing data on other variables. Higher influx and outflux values are preferred. Figure 13.2 plots outflux against influx. In general, variables that are located higher up in the display are more complete and thus potentially more useful for imputation. In practice, variables closer to the sub-diagonal are better connected than those further away. The fluxplot can be used to spot variables that clutter the imputation model. Variables that are located in the lower regions (especially near the left-lower corner) *and* that are uninteresting for later analysis are better removed from the data prior to imputation.

Influx and outflux are summaries of the missing data pattern intended to aid in the construction of imputation models. Keeping everything else constant, variables with high influx and outflux are preferred. Realize that outflux indicates the potential (and not actual) contribution to impute other variables. A variable with high O_j may turn out to be useless for imputation if it is fully unrelated to the incomplete variables, e.g., an administrative person identifier. On the other hand, the usefulness of a highly predictive variable is severely limited by a low O_j.

13.4 Multivariate Imputation Models

13.4.1 Overview

Rubin (1987, pp. 160–166) distinguished three tasks for creating imputations: the modeling task, the imputation task, and the estimation task. The modeling task is to provide a specification for the joint distribution $P(Y) = P(Y^o, Y^m)$ of the hypothetically complete data. The issues that arise with incomplete data are essentially the same as for complete data, but in imputation the emphasis is on getting accurate predictions of the missing values. The imputation task is to specify the posterior predictive distribution $P(Y^m|Y^o)$ of the missing values given the observed data and given the assumed model $P(Y)$. The estimation task consists of calculating the posterior distribution of the parameters of this distribution given the observed data, so that random draws can be made from it.

In Rubin's framework, the imputations follow from the specification of the joint distribution $P(Y)$. van Buuren (2007) distinguished three strategies to specify the model used to impute multivariate missing data.

- *Monotone data imputation.* Given a monotone missing data pattern, imputations are created by drawing for a sequence of univariate conditional distributions $P(Y_j|Y_1, \ldots, Y_{j-1})$ for $j = 1, \ldots, p$.
- *Fully conditional specification (FCS).* For general patterns, the user specifies a conditional distribution $P(Y_j|Y_{-j})$ directly for each variable Y_j, and assumes this distribution to be the same for the observed and missing Y_j (ignorability assumption). Imputations are created by iterative drawing from these conditional distributions. The multivariate model $P(Y)$ is implicitly specified by the given sets of conditional models.
- *Joint modeling.* For general patterns, imputations are drawn from a multivariate model $P(Y)$ fitted to the data, usually per missing data pattern, from the derived conditional distributions.

Chapter 12 reviews methods for multiple imputation for monotone missing data, whereas Chapter 14 covers joint modeling in great detail. The present chapter concentrates on FCS. The remainder of this section briefly reviews monotone data imputation and joint modeling.

13.4.2 Imputation of monotone missing data

If the missing data pattern in Y is monotone, then the variables can be ordered as $Y_1, \ldots, Y_j, \ldots, Y_p$ according to their missingness. The joint distribution $P(Y) = (Y^o, Y^m)$ decomposes as (Rubin 1987, p. 174)

$$P(Y|\phi) = P(Y_1|\phi_1)P(Y_2|Y_1, \phi_2) \ldots P(Y_p|Y_1, \ldots, Y_{p-1}, \phi_p), \tag{13.5}$$

where $\phi_1, \ldots, \phi_j, \ldots, \phi_p$ are the parameters of the model to describes the distribution of Y. The ϕ_j parameters only serve to create imputations, and are generally not of any scientific interest or relevance. The decomposition requires that the missing data pattern is monotone. In addition, there is a second, more technical requirement: the parameters of the imputation models should be *distinct* (Rubin 1987, pp. 174–178).

Monotone data imputation is fast and provides great modeling flexibility. Depending on the data, Y_1 can be imputed by a logistic model, Y_2 by a linear model, Y_3 by a proportional odds model, and so on. In practice, a dataset may be near-monotone, and may become monotone if a small fraction of the missing data were imputed (Li, 1988; Rubin and Schafer, 1990; Schafer, 1997; and Rubin, 2003a). See Chapter 12 for more detail.

13.4.3 Imputation by joint modeling

Joint modeling starts from the assumption that the hypothetically complete data can be described by a multivariate distribution. Assuming ignorability, imputations are created as draws under the assumed model. Joint modeling describes the data Y by the multivariate distribution $P(Y|\theta)$, where θ is a vector of unknown parameters of the distribution. The model for $P(Y|\theta)$ can be any multivariate distribution, but the multivariate normal distribution, with $\theta = (\mu, \Sigma)$ for the mean μ and covariance Σ, is a convenient and popular choice.

Within the joint modeling framework, the parameters of scientific interest are functions of θ (Schafer 1997, Ch. 4). Observe that the θ parameters are conceptually different from the ϕ_j parameters used in Section 13.4.2. The θ parameters derive from the multivariate model specification, whereas the ϕ_j parameters are just unknown parameters of the imputation model, and have no scientific relevance.

When the assumptions hold, joint modeling is elegant and efficient. For example, under multivariate normality, the *sweep operator* and *reverse sweep operator* are highly efficient computational tools for converting outcomes into predictors and vice versa. See Little and Rubin (2002, pp. 148–156) and Schafer (1997, pp. 157–163) for details.

The major limitation of joint modeling is that the specified multivariate model may not be a good fit to the data. For example, if the data are skewed or if dependencies occur in the data, it could prove to be difficult to find an appropriate multivariate model. Schafer (1997, pp. 211–218) reported simulations that showed that imputations drawn under the multivariate normal model are generally robust to non-normal data. Joint modeling by a multivariate distribution can often be made more realistic through data transformations, or through the use of specific rounding techniques. Nevertheless, in many practical situations where imputations are desired (c.f. Section 13.2), there will be no reasonable multivariate model for the data.

13.5 Fully Conditional Specification (FCS)

13.5.1 Overview

In contrast to joint modeling, FCS specifies the multivariate distribution $P(Y|\theta)$ through a set of conditional densities $P(Y_j|Y_{-j}, \phi_j)$, where ϕ_j are unknown parameters of the imputation model. As in Section 13.4.2, the ϕ_j parameters are not of scientific interest, and only serve to model conditional relations used for imputation. The key assumption is that the conditional densities for Y_j^o and Y_j^m are the same (ignorability assumption). This conditional density is used to impute Y_j^m given the other information. Starting from simple random draws from the marginal distribution, imputations $\dot{Y}_1$ are drawn for Y_1^m given the information in the other columns. Then, Y_2 is imputed given the currently imputed data, and so on until all variables are imputed with one pass through the data. Then, Y_1^m is re-imputed during the second iteration using the imputation draw in iteration one, and so on. In practice, the iteration process can already be stopped after five or ten passes through the data. FCS is a generalization of univariate imputation for monotone data, and borrows the idea of Gibbs sampling from the joint modeling framework.

FCS bypasses task 1 of the procedure of Section 13.4.1, the specification of the joint distribution $P(Y|\theta)$. Instead, the user specifies the conditional distribution $P(Y_j|Y_{-j})$ directly for each variable Y_j. Imputations are created by iterative draws from these conditional distributions. The multivariate model $P(Y|\theta)$ is only implicitly specified by the specified set of conditional models.

The idea of conditionally specified models is quite old. Conditional probability distributions follow naturally from the theory of stochastic Markov chains (Barlett 1978, pp. 231–236). For spatial data analysis, Besag preferred the use of conditional probability models over joint probability models, since "the conditional probability approach has greater intuitive appeal to the practising statistician" (Besag 1974, p. 223). Buck (1960) proposed a procedure for calculating estimates for all missing entries by multiple regression. For example, to impute missing data in the first variable, Y_1 was regressed on $Y_2, \ldots, Y_p$, where the regression coefficients are computed using the complete cases. Buck's method does not iterate and requires a large sized sample of complete cases. Gleason and Staelin (1975) extended Buck's method to include multivariate regression, and noted that their ad-hoc method could also be derived more formally from the multivariate normal distribution. These authors also brought up the possibility of an iterated version of Buck's method, suggesting that missing entries from one iteration could be used to form an improved estimate of the correlation matrix for use in a subsequent iteration. Due to a lack of computational resources at that time, they were unable to evaluate the idea, but later work along these lines has been put forward by Finkbeiner (1979), Raymond and Roberts (1987), Jinn and Sedransk (1989), van Buuren and Rijckevorsel (1992), and Gold and Bentler (2000).

Multiple imputation is different from this literature because it draws imputations from a distribution instead of calculating optimal predictive values. Ideas similar to FCS have surfaced under a variety of names: stochastic relaxation (Kennickell 1991), variable-by-variable imputation (Brand 1999), switching regressions (van Buuren, Booshuizen, and Knook 1999), sequential regressions (Raghunathan et al. 2001), ordered pseudo-Gibbs sampler (Heckerman et al. 2001), partially incompatible MCMC (Rubin 2003b), iterated univariate imputation (Gelman 2004) and chained equations (van Buuren and Groothuis-Oudshoorn 2000).

FIGURE 13.3
MICE algorithm for imputation of multivariate missing data.

1. Specify an imputation model $P(Y_j^m|Y_j^o, Y_{-j})$ for variable Y_j with $j = 1, \ldots, p$.
2. For each j, fill in starting imputations $\dot{Y}_j^0$ by random draws from Y_j^o.
3. Repeat for $t = 1, \ldots, T$:
4. Repeat for $j = 1, \ldots, p$:
5. Define $\dot{Y}_{-j}^t = (\dot{Y}_1^t, \ldots, \dot{Y}_{j-1}^t, \dot{Y}_{j+1}^{t-1}, \ldots, \dot{Y}_p^{t-1})$ as the currently complete data except Y_j.
6. Draw $\dot{\phi}_j^t \sim P(\phi_j^t|Y_j^o, \dot{Y}_{-j}^t)$.
7. Draw imputations $\dot{Y}_j^t \sim P(Y_j^m|Y_j^o, \dot{Y}_{-j}^t, \dot{\phi}_j^t)$.
8. End repeat j.
9. End repeat t.

The main reasons for using FCS is increased flexibility and ease of use. Little (2013) explains the advantages of conditional modeling as follows:

> When modeling, it can be useful to factor a multivariate distribution into sequence of conditional distributions. Univariate regression is easier to understand, and a sequence of univariate conditional regressions is more easily elaborated, for example, by including interactions, polynomials, or splines, or modeling heteroscedasticity.

13.5.2 Chained equations: The MICE algorithm

There are several ways to implement imputation under conditionally specified models. The algorithm in Figure 13.3 describes one particular instance: the MICE algorithm (van Buuren and Groothuis-Oudshoorn 2000, 2011), which divides the multivariate data in columns. The algorithm starts with a random draw from the observed data, and imputes the incomplete data in a variable-by-variable fashion. One iteration consists of one cycle through all Y_j. The number of iterations can often be low, say 5 or 10. The MICE algorithm generates multiple imputations by executing the algorithm in Figure 13.3 in parallel m times.

13.5.3 Properties

The MICE algorithm is a Markov chain Monte Carlo (MCMC) method, where the state space is the collection of all imputed values. If the conditionals are compatible, the MICE algorithm is a Gibbs sampler, a Bayesian simulation technique that samples from the conditional distributions in order to obtain samples from the joint distribution (Gelfand and

Smith, 1990; Casella and George, 1992). In conventional applications of the Gibbs sampler the full conditional distributions are derived from the joint probability distribution (Gilks 1996). In MICE however, the conditional distributions are directly specified by the user, and so the joint distribution is only implicitly known, and may not even exist. While the latter is clearly undesirable from a theoretical point of view (since we do not know the joint distribution to which the algorithm converges), in practice it does not seem to hinder useful applications of the method (cf. Section 13.5.4).

In order to converge to a stationary distribution, a Markov chain needs to satisfy three important properties (Roberts 1996; Tierney 1996):

- *irreducible*, the chain must be able to reach all interesting parts of the state space;
- *aperiodic*, the chain should not oscillate between states;
- *recurrence*, all interesting parts can be reached infinitely often, at least from almost all starting points.

With the MICE algorithm, irreducibility is generally not a problem since the user has large control over the interesting parts of the state space. This flexibility is actually the main rationale for FCS instead of a joint model.

Periodicity is a potential problem, and can arise in the situation where imputation models are clearly inconsistent. A rather artificial example of oscillatory behavior occurs when Y_1 is imputed by $Y_2\beta + \epsilon_1$ and Y_2 is imputed by $-Y_1\beta + \epsilon_2$ for some constant β. The sampler will oscillate between two qualitatively different states, so this is a periodic procedure. The problem with periodic sampler is that the result will depend on the stopping point. In general, we would like the statistical inferences to be independent of the stopping point. A way to diagnose the *ping-pong* problem is to stop the chain at different points. The stopping point should not affect the statistical inferences. The addition of noise to create imputations is a safeguard against periodicity, and allows the sampler to "break out" more easily.

Non-recurrence may also be a potential difficulty, manifesting itself as explosive or non-stationary behavior. For example, if imputations are made through deterministic functions, the Markov chain may lock up. Such cases can sometimes be diagnosed from the trace lines of the sampler. See van Buuren and Groothuis-Oudshoorn (2011) for examples and remedies. As long as the parameters of imputation models are estimated from the data, non-recurrence is likely to be mild or absent.

13.5.4 Compatibility

Gibbs sampling is based on the idea that knowledge of the conditional distributions is sufficient to determine a joint distribution, if it exists. Two conditional densities $P(Y_1|Y_2)$ and $P(Y_2|Y_1)$ are said to be *compatible* if a joint distribution $P(Y_1, Y_2)$ exists that has $P(Y_1|Y_2)$ and $P(Y_2|Y_1)$ as its conditional densities. More precisely, the two conditional densities are compatible if and only if their density ratio $P(Y_1|Y_2)/P(Y_2|Y_1)$ factorizes into the product $u(Y_1)v(Y_2)$ for some integrable functions u and v (Besag 1974). So, the joint distribution either exists and is unique, or does not exist.

The MICE algorithm is ignorant of the non-existence of the joint distribution, and happily produces imputations whether the joint distribution exists or not. The question is whether the imputed data can be trusted when we cannot find a joint distribution $P(Y_1, Y_2)$ that has $P(Y_1|Y_2)$ and $P(Y_2|Y_1)$ as its conditionals.

For the trivariate case, the joint distribution $P(Y_1, Y_2, Y_3)$, if it exists, is uniquely specified by the following set of three conditionals: $P(Y_1|Y_2, Y_3)$, $P(Y_2|Y_3)$ and $P(Y_3|Y_1)$ (Gelman and Speed 1993). Imputation under FCS typically specifies general forms for $P(Y_1|Y_2, Y_3)$, $P(Y_2|Y_1, Y_3)$ and $P(Y_3|Y_1, Y_2)$, which is different, and estimates the free parameters for these conditionals from the data. Typically, the number of parameters in imputation is much larger than needed to uniquely determine $P(Y_1, Y_2, Y_3)$. While perhaps inefficient as a parameterization, it is not easy to see why that in itself would introduce bias or affect the accuracy of the imputations.

Not much is known about the consequences of incompatibility on the quality of imputations. Simulations with strongly incompatible models found no adverse effects on the estimates after multiple imputation (van Buuren et al. 2006). Somewhat surprisingly, several methods based on deliberately specified incompatible methods outperformed complete case analysis. Imputation using the partially compatible Gibbs sampler seems to be robust against incompatible conditionals in terms of bias and precision, thus suggesting that incompatibility may be a relatively minor problem in multivariate imputation. More work is needed to verify such claims in more general and more realistic settings though.

In cases where the multivariate density is of genuine scientific interest, incompatibility clearly represents an issue because the data cannot be represented by a formal model. For example, incompatible conditionals could produce a ridge (or spike) in an otherwise smooth density, and the location of the ridge may actually depend on the stopping point. If such is the case, then we should have a reason to favor a particular stopping point. Alternatively, we might try to reformulate the imputation model so that the stopping point effect disappears. In imputation the objective is to make correct statistical inferences by augmenting the data and preserving the relations and uncertainty in the data. In that case, having a joint distribution may be convenient theoretically, but the price may be lack of fit. (Gelman and Raghunathan 2001) noted:

> One may argue that having a joint distribution in the imputation is less important than incorporating information from other variables and unique features of the dataset (e.g., zero/nonzero features in income components, bounds, skip patterns, nonlinearity, interactions).

In practice, incompatibility issues could arise in MICE if deterministic functions of the data are imputed along with their originals. For example, the imputation model may contain interaction terms, data summaries or nonlinear functions of the data. Such terms may introduce feedback loops and impossible combinations into the system, which can invalidate the imputations. van Buuren and Groothuis-Oudshoorn (2011) provide ways to avoid this behavior and to eliminate feedback loops from the system. Section 13.6.5 describes the tools to do this. Apart from potential feedback problems, it appears that incompatibility is a relatively minor problem in practice, especially if the amount of missing data is modest.

Further theoretical work has been done by Arnold, Castillo, and Sarabia (2002). The field has recently become active. Several methods for identifying compatibility from actual data have been developed in the last few years (Tian et al. 2009; Ip and Wang 2009; Tan, Tian, and Ng 2010; Wang and Kuo 2010; Kuo and Wang 2011; Chen 2011). It is not yet known what the added value of such methods will be in the context of missing data.

13.5.5 Number of iterations

When m sampling streams are calculated in parallel, monitoring convergence is done by plotting one or more statistics of interest in each stream against iteration number t. Common statistics to be plotted are the mean and standard deviation of the synthetic data, as well as the correlation between different variables. The pattern should be free of trend, and the variance within a chain should approximate the variance between chains.

In practice, a low number of iterations appears to be enough. Brand (1999) and van Buuren, Boshuizen, and Knook (1999) set the number of iterations T quite low, usually somewhere between 5 to 20 iterations. This number is much lower than in other applications of MCMC methods, which often require thousands of iterations.

The explanation for the pleasant property is that the imputed data $\dot{Y}^m$ form the only memory of the MICE algorithm. Imputations are created in a stepwise optimal fashion that adds a proper amount of random noise to the predicted values (depending on the strength of the relations between the variables), which helps to reduce the autocorrelation between successive draws. Hence, convergence will be rapid, and in fact immediate if all variables are independent. Thus, the incorporation of noise into the multiply-imputed data has the pleasant side effect of speeding up convergence. Situations to watch out for include:

- the correlations between the Y_js are high;
- the missing data rates are high;
- constraints on parameters across different variables exist.

The first two conditions directly affects the amount of autocorrelation in the system. The latter condition becomes relevant for customized imputation models. A useful trick for reducing the amount of autocorrelation in highly correlated repeated measures Y_1 and Y_2 is to draw imputations $\dot{\delta}$ for the increment $Y_2 - Y_1$ rather than for Y_2. Imputations are then calculated as the sum of the previous value and the increment, $Y_1 + \dot{\delta}$.

Simulation work suggests that FCS can work well using no more than just five iterations, but many more iterations might be needed in problems with high correlations and high proportions of missing data (van Buuren 2007, 2012).

It is important to investigate convergence by inspecting traces lines of critical parameters the Gibbs samplers, as these might point to anomalies in the imputed data. van Buuren and Groothuis-Oudshoorn (2011) show several cases with problematic convergence of the MICE algorithm, and where it may even be entirely stuck because of circularities. Also, imputing large blocks of correlated data may produce degenerate solutions (van Buuren 2012, p. 208). Such cases can often be prevented by simplifying the prediction model.

In general, we should be careful about convergence in missing data problems with high correlations and high missing data rates. On the other hand, we really have to push the MICE algorithm to its limits to see adverse effect. Of course, it never hurts to do a couple of extra iterations, but in most applications good results can be obtained with a small number of iterations.

13.5.6 Performance

Each conditional density has to be specified separately, so FCS requires (sometimes considerable) modeling effort on the part of the user. Most software provides reasonable defaults for standard situations, so the actual effort required may be small.

A number of simulation studies provide evidence that FCS generally yields estimates that are unbiased and that possess appropriate coverage (Brand 1999; Raghunathan et al. 2001; Brand et al. 2003; Tang, Belin, and Unutzer 2005; van Buuren et al. 2006; Horton and Kleinman 2007; Yu, Burton, and Rivero-Arias 2007). FCS and joint modeling will often find similar estimates, especially for estimates that depend on the center of the distribution, like mean, median, regression estimates, and so on. Lee and Carlin (2010) contrasted the multivariate normal joint model to FCS using a simulation of a typical epidemiological set-up. They found that both the FCS and joint modeling provided substantial gains over complete-case analysis (CCA) when estimating the binary intervention effect. The joint model appeared to perform well even in the presence of binary and ordinal variables. The regression estimates pertaining to a skewed variable were biased when normality was assumed. Transforming to normality (in joint modeling or FCS) or using predictive mean matching (in FCS) could resolve this problem.

The studies by van der Palm, van der Ark, and Vermunt (2012) and Gebregziabher et al. (2012) compared various imputation methods for fully categorical data, based on both joint modeling and FCS. They reported some mild improvements of a more recent latent class model over a standard FCS model based on logistic regression. Both studies indicate that the number of latent classes needs to be large. Additional detail can be found in the original papers.

Although the substantive conclusions are generally robust to the precise form of the imputation model, the use of the multivariate normal model, whether rounded or not, is generally inappropriate for the imputation of categorical data (van Buuren 2007). The problem is that the imputation model is more restrictive than the complete-data model, an undesirable situation known as uncongeniality (Meng 1994; Schafer 2003). More particularly, the multivariate normal model assumes that categories are equidistant, that the relations between all pairs of variables is linear, that the residual variance is the same at every predicted value, and that no interactions between variables exist. Without appropriate assessment of the validity of these assumptions, imputation may actually introduce systematic biases into the data that we may not be aware of. For example, Lee et al. (2012) demonstrated that imputing ordinal variables as continuous can lead to bias in the estimation of the exposure outcome association in the presence of a non-linear relationship. It may seem a trivial remark, but continuous data are best imputed by methods designed for continuous data, and categorical data are best imputed by methods designed for categorical data.

13.6 Modeling in FCS

13.6.1 Overview

The specification of the imputation model is the most challenging step in multiple imputation. The imputation model should

- account for the process that created the missing data,
- preserve the relations in the data, and
- preserve the uncertainty about these relations.

The idea is that adherence to these principles will yield proper imputations, and thus result in valid statistical inferences (Rubin 1987, pp. 118–128). van Buuren and Groothuis-Oudshoorn (2011) list the following seven choices:

1. First, we should decide whether the MAR assumption is plausible. Chained equations can handle both MAR and NMAR, but multiple imputation under NMAR requires additional modeling assumptions that influence the generated imputations.

2. The second choice refers to the form of the imputation model. The form encompasses both the structural part and the assumed error distribution. In FCS the form needs to be specified for each incomplete column in the data. The choice will be steered by the scale of the variable to be imputed, and preferably incorporates knowledge about the relation among the variables.

3. A third choice concerns the set of variables to include as predictors in the imputation model. The general advice is to include as many relevant variables as possible, including their interactions. This may, however, lead to unwieldy model specifications.

4. The fourth choice is whether we should impute variables that are functions of other (incomplete) variables. Many datasets contain derived variables, sum scores, ratios and so on. It can be useful to incorporate the transformed variables into the multiple imputation algorithm.

5. The fifth choice concerns the order in which variables should be imputed. The visit sequence may affect the convergence of the algorithm and the synchronization between derived variables.

6. The sixth choice concerns the setup of the starting imputations and the number of iterations. The convergence of the MICE algorithm can be monitored by trace lines.

7. The seventh choice is m, the number of multiply imputed datasets. Setting m too low may result in large simulation error and statistical inefficiency, especially if the fraction of missing information is high.

The above points are by no means exhaustive. Much sensible advice on modeling in multiple imputation in an FCS context also can be found in Sterne et al. (2009), White, Royston, and Wood (2011), and Carpenter and Kenward (2013). The remainder of this section discusses points 1–5. Point 6 was already addressed in Section 13.5.5, while point 7 was discussed in Chapter 12.

13.6.2 MAR or NMAR?

The most important decision in setting up an imputation model is to determine whether the available data are enough to solve the missing data problem at hand. The MAR assumption is essentially the belief that the available data are sufficient to correct for the missing data. Unfortunately, the distinction between MAR and NMAR cannot, in general, be made from the data. In practice, 99% of analysts assume MAR, sometimes explicitly, but often more so implicitly. While MAR is often useful as a starting point, the actual causes of the missingness may be related to the quantities of scientific interest, even after accounting for the data. An incorrect MAR assumption may then produce biased estimates.

Collins, Schafer, and Kam (2001) investigated the role of “lurking” variables Z that are correlated with the variables of interest Y and with the missingness of Y. For linear regression, they found that if the missing data rate did not exceed 25% and if the correlation between the Z and Y was 0.4, omitting Z from the imputation model had a negligible effect. For more extreme situations (50% missing data and/or a correlation of 0.9) the effect depended strongly on the form of the missing data mechanism. When the probability of being missing was linear in Z, then omitting Z from the imputation model only affected the intercept, whereas the regression weights and variance estimates were unaffected. When more missing data were created in the extremes, the reverse occurred: omitting Z biased the regression coefficients and variance estimates, but the intercept was unbiased with the correct confidence interval. In summary, they found that all estimates under multiple imputation appeared robust against NMAR. Beyond a correlation of 0.4, or for a missing data rate over 25%, it is the form of the missing data mechanism that determines which parameters will be biased.

While these results are generally comforting, there are three main strategies that we might pursue if the response mechanism is non-ignorable:

- Expand the variables in the imputation model in the hope of making the missing data mechanism closer to MAR.
- Formulate an explicit nonresponse model in combination with a complete-data model, and estimate the parameters of interest.
- Formulate and fit a series of nonignorable imputation models, and perform sensitivity analysis on the critical parameters.

In the first strategy, the MAR assumption is the natural starting point. MAR could be made more plausible by finding additional variables that are strongly predictive of the missingness, and include these into the imputation model. The fact that these are included is more important than the precise form in which that is done (Jolani et al. 2013).

There is a large literature on the second option, often starting from the Heckman model (Heckman 1979). There has been some recent work to generate multiple imputations under this model, as well as generalizations thereof. The key idea is to extend the imputation model with a model for the missing data process, where the probability of being missing depends on the variable Y_j to be imputed. The FCS framework can be used to generate imputations under the combined model by drawing imputations for Y_j and R_j. See Jolani (2012) for details.

Finally, one might perform a concise simulation study as in Collins, Schafer, and Kam (2001) customized for the problem at hand with the goal of finding out how extreme the NMAR mechanism needs to be to influence the parameters of scientific interest. More generally, the use of sensitivity is advocated by the Panel on Handling Missing Data in Clinical Trials of the National Research Council (Little et al. 2010). Chapter 20 of the Handbook deals with sensitivity analysis using multiple imputation.

13.6.3 Model form

The MICE algorithm requires a specification of a univariate imputation method separately for each incomplete variable. It is important to select univariate imputation methods that have correct statistical coverage for the scientific parameters of interest, and that yield sensible imputed values. The measurement level of a variable largely determines the form of the univariate imputation model. There are special methods for continuous, dichotomous, ordered categories, unordered categories, count data, semi-continuous data, censored data, truncated data and rounded data. In addition, there are regression tree imputation methods aimed at preserving interactions. Chapter 3 of van Buuren (2012) contains an in-depth treatment of many univariate imputation methods. Predictive mean matching (Little 1988) is an all-round imputation method that works well in many cases. It is the default method in MICE for imputing continuous data (van Buuren and Groothuis-Oudshoorn 2011).

Model specification is straightforward when the data are cross-sectional or longitudinal, where in the longitudinal setting, different time points coded as different columns, i.e., as a *broad* matrix. Genuinely hierarchical data are typically coded as a *long* matrix, with time points or nested observations are coded as distinct rows. van Buuren and Groothuis-Oudshoorn (2011) discusses and compares three imputation methods for hierarchical data:

- Ignore any clustering structure in the data, and use standard imputation techniques tools for non-hierarchical data.
- Add the cluster allocation as a fixed factor, thus allowing for between class variation by a fixed effects model.
- Draw imputations under a linear mixed-effect model by a Markov Chain Monte Carlo algorithm.

Ignoring the multilevel structure when in fact it is present will bias the intra-class correlation downwards, adding a fixed factor will bias it upwards, while the linear mixed-effects model is about right. In general, smaller class sizes complicate the imputation problem. Overall, imputation under the linear mixed-effects model is superior to the two other methods, but it is not yet ideal as the coverage may fail to achieve the nominal level. Computational details can be found in van Buuren (2012, pp. 84–87). Moreover, Chapter 9 of that book contains two applications on longitudinal data, one using a broad matrix in a repeated measures problem, and the other using the linear mixed-effects imputation model on the long matrix.

13.6.4 Predictors

The general advice is to include as many variables in the imputation model as possible (Meng 1994, Collins, Schafer, and Kam 2001), but there are necessarily computational limitations that must be taken into account. Conditioning on all other data is often reasonable for small to medium datasets, containing up to, say, 20–30 variables, without derived variables, interactions effects and other complexities. Including as many predictors as possible tends to make the MAR assumption more plausible, thus reducing the need to make special adjustments for NMAR.

For datasets containing hundreds or thousands of variables, using all predictors may not be feasible (because of multicollinearity and computational problems). It is also not necessary. In practice, the increase in explained variance in linear regression is typically negligible after the best, say, 15 variables have been included. For imputation purposes, it is expedient to select a suitable subset of data that contains no more than 15 to 25 variables. Hardt, Herke, and Leonhart (2012) suggested that the number of complete rows in the imputation model should be at least three times the number of variables. van Buuren and Groothuis-Oudshoorn (1999) provide the following strategy for selecting predictor variables from a large database:

1. Include all variables that appear in the complete data model, i.e., the model that will be applied to the data after imputation, including the outcome (Little 1992; Moons et al. 2006). Failure to include the outcome will bias the complete data analysis, especially if the complete data model contains strong predictive relations. Note that this

step is somewhat counter-intuitive, as it may seem that imputation would artificially strengthen the relations of the complete data model, which would be clearly undesirable. If done properly however, this is not the case. On the contrary, not including the complete data model variables will tend to bias the results toward zero. Note that interactions of scientific interest also need to be included in the imputation model.

2. In addition, include the variables that are related to the nonresponse. Factors that are known to have influenced the occurrence of missing data (stratification, reasons for nonresponse) are to be included on substantive grounds. Other variables of interest are those for which the distributions differ between the response and nonresponse groups. These can be found by inspecting their correlation with the response indicator of the variable to be imputed. If the magnitude of this correlation exceeds a certain level, then the variable should be included.

3. In addition, include variables that explain a considerable amount of variance. Such predictors help reduce the uncertainty of the imputations. They are basically identified by their correlation with the target variable. Only include predictors with a relatively high outflux coefficient (cf. Section 13.3).

4. Remove from the variables selected in steps 2 and 3 those variables that have too many missing values within the subgroup of incomplete cases. A simple indicator is the percentage of observed cases within this subgroup, the percentage of usable cases (cf. Section 13.3).

Most predictors used for imputation are incomplete themselves. In principle, one could apply the above modeling steps for each incomplete predictor in turn, but this may lead to a cascade of auxiliary imputation problems. In doing so, one runs the risk that every variable needs to be included after all.

In practice, there is often a small set of key variables, for which imputations are needed, which suggests that steps 1 through 4 are to be performed for key variables only. This was the approach taken in van Buuren, Boshuizen, and Knook (1999), but it may miss important predictors. A safer and more efficient, though more laborious, strategy is to perform the modeling steps also for the predictors of predictors of key variables. This is done in Groothuis-Oudshoorn, van Buuren, and van Rijckevorsel (1999). At the terminal node, one can apply a simple method, like sampling from the marginal, that does not need any predictors for itself.

By default, most computer programs impute a variable Y_j from all other variables Y_{-j} in the data. Some programs, however, have the ability to specify the set of predictors to be used per incomplete variable (Su et al. 2011; van Buuren and Groothuis-Oudshoorn 2011; Royston and White 2011). These facilities are highly useful for refining the imputation model and for customizing the imputations to the data. Ridge regression (Hoerl and Kennard 1970; Tibshirani 1996) provides an alternative way to control the estimation process, making the algorithm more robust at the expense of bias.

Although it might seem somewhat laborious, the quality of the imputed values can be enhanced considerably by a judicious specification of the set of predictors that enter the imputation model. It is generally worthwhile to set apart some time to set up the imputation model, often in combination with the use of suitable diagnostics (c.f. Section 13.7).

13.6.5 Derived variables

Derived variables (transformations, recodes, interaction terms, and so on) pose special challenges for the imputation model. There are three general strategies to impute derived variables:

1. Leave derived variables out of the imputation, and calculate them afterwards from the multiply-imputed data.
2. Calculate derived variables before imputation, and impute them as usual.
3. Update derived variables within the imputation algorithm as soon as one of the original variables is imputed.

Method 1 is easy, but the generated imputations do not account for the relationship between the derived variables and other variables in the data, potentially resulting is biased estimates in the complete-data analysis. Method 2 repairs this deficit, but at the expense of creating inconsistencies between the imputations of the originals and of the derived versions. Method 3 can address both problems, but some care is needed in setting up the predictor matrix. This section looks briefly at various types of derived variables.

In practice, there is often extra knowledge about the data that is not modeled explicitly. For example, consider the weight/height ratio, defined as weight divided by height (kg/m). If any one of the triplet height, weight or weight/height ratio is missing, then the missing value can be calculated with certainty by a simple deterministic rule. Unless we specify otherwise, the default imputation model is, however, unaware of the relation between the three variables, and will produce imputations that are inconsistent with the rule. Inconsistent imputations are undesirable since they yield combinations of data values that are impossible had the data been observed.

The easiest way to deal with the problem is to leave any derived data outside the imputation process (Method 1). For example, we may impute any missing height and weight data, and append the weight/height ratio to the imputed data afterward. The disadvantage in this post-imputation method is that the derived variable is not available for imputation, potentially resulting in incorrect statistical inferences.

Another possibility is to calculate the weight/height ratio before imputation, and impute it as "just another variable," an approach known as JAV (Method 2). Although JAV may yield valid statistical inferences in particular cases (for linear regression weights under MCAR, see von Hippel 2009), it invariably produces impossible combinations of imputed values, and may be biased under MAR.

A solution for this problem is *passive imputation*, a method that calculates the derived variable on the fly once one of its components is imputed (Method 3). Passive imputation maintains the consistency among different transformations of the same data (thus solving the problem of JAV) and makes the derived variable available for imputation (thus solving the problem of the post-imputation method).

Care is needed in setting up the predictor matrix when using passive imputation. In particular, we may not use the derived variable as a predictor for its components, so feedback loops between the derived variables and their originals should be broken. In the above example, we would thus need to remove the weight/height ratio from the imputation models for height and weight. Failing to do so may result in absurd imputations and problematic convergence.

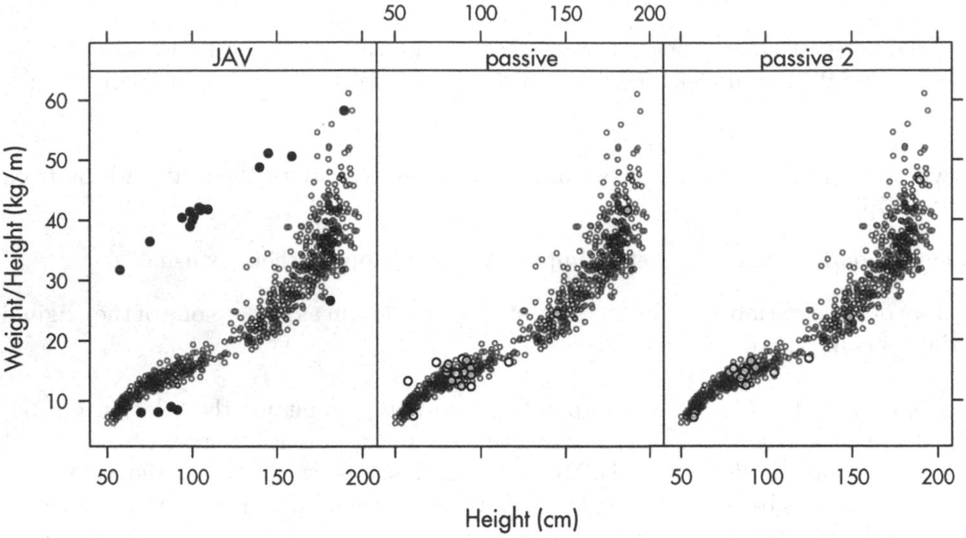

FIGURE 13.4
Three different imputation models to impute weight/height ratio. The relation between the weight/height ratio and height is not respected under "just another variable" (JAV). Both passive methods yield imputations that are close to the observed data. "Passive 2" does not allow for models in which weight/height ratio and BMI are simultaneous predictors.

Figure 13.4 compares JAV to passive imputation on real data. The leftmost panel in Figure 13.4 shows the results of JAV. The imputations are far off any of the observed data, since JAV ignores the fact that the weight/height ratio is a function of height and weight. The middle panel shows that passive imputation represents an improvement over JAV. The values are generally similar to the real data and adhere to the derived rules. The rightmost panel shows that somewhat improved imputations can be obtained by preventing that the body mass index (BMI) and weight/height ratio (which have an exact nonlinear relationship) are simultaneous predictors.

The *sum score* is another type of derived variable. The sum score undefined if one of the original variables is missing. Sum scores of imputed variables are useful within the MICE algorithm to economize on the number of predictors. van Buuren (2010) reports a simulation on sub scale scores from imputed questionnaire items that shows that plain multiple imputation using sum scores improves upon dedicated imputation methods. See Sections 7.3 and 9.2 in van Buuren (2012) for applications on real data.

Interaction terms are also derived variables. The standard MICE algorithm only accommodates main effects. Sometimes the *interaction* between variables is of scientific interest. For example, in a longitudinal study we could be interested in assessing whether the rate of change differs between two treatment groups, in other words, the treatment-by-group interaction. The standard algorithm does not take interactions into account, so the interactions of interest should be added to the imputation model. Interactions can be added using passive imputation. An alternative is to impute the data in separate groups.

In some cases it makes sense to restrict the imputations, possibly conditional on other data. For example, if we impute "male," we can skip questions particular to females, e.g. about pregnancy. Such *conditional imputation* could reset the imputed data in the pregnancy block to missing, thus imputing only part depending on gender. Of course, appropriate care is needed when the pregnancy variables are used later as predictors to restrict to females. Such alterations to the imputations can be implemented easily within a FCS framework by *post-processing imputations* within the iterative algorithm.

Compositional data are another form of derived data, and often occur in household and business surveys. Sometimes we know that a set of variables should add up to a given total. If one of the additive terms is missing, we can directly calculate its value with certainty by deducting the known terms from the total. However, if two additive terms are missing, imputing one of these terms uses the available one degree of freedom, and hence implicitly determines the other term. Imputation of compositional data has only recently received attention (Tempelman 2007, Hron, Templ, and Flzmoser 2010; de Waal, Pannekoek, and Scholtus 2011), and can be implemented conveniently with an FCS framework. See Section 5.4.5 in van Buuren (2012) for an illustration of the main idea.

Nonlinear relations are often modeled using a linear model by adding quadratic or cubic terms of the explanatory variables. Creating imputed values that adhere to a *quadratic relation* poses some challenges. Current imputation methodology either preserves the quadratic relation in the data and biases the estimates of interest, or provides unbiased estimates but does not preserve the quadratic relation (von Hippel 2009; White et al. 2011). An alternative approach is to define a *polynomial combination* Z as $Z = Y\beta_1 + Y^2\beta_2$ for some β_1 and β_2. The idea is to impute Z instead of Y and Y^2, followed by a decomposition of the imputed data Z into components Y and Y^2. Section 5.4.6 in van Buuren (2012) provided an algorithm that does these calculations. Simulations indicate that the quadratic method worked well in a variety of situations (Vink and van Buuren 2013).

In all cases, feedback between different versions of the same variable should be prevented. Failing to do so may lock up the MICE algorithm or produce erratic imputations.

13.6.6 Visit sequence

The MICE algorithm as described in Section 13.5.2 imputes incomplete variables in the data from left to right. Theoretically, the visit sequence of the MICE algorithm is irrelevant as long as each column is visited often enough, though some schemes are more efficient than others. In practice, there are small order effects of the MICE algorithm, where the parameter estimates depend on the sequence of the variables. To date, there is little evidence that this matters in practice, even for clearly incompatible imputation models (van Buuren et al. 2006). For monotone missing data, convergence is immediate if variables are ordered according to their missing data rate. Rather than reordering the data itself, it is more convenient to change the visit sequence of the algorithm.

It may also be useful to visit a given variable more than once within the same iteration. For example, weight/height ratio can be recalculated immediately after the missing data in weight and after the missing data in height are imputed. This ensures that the weight/height ratio remains properly synchronized with both weight and height at all times.

13.7 Diagnostics

An important and unique advantage of multiple imputation over other statistical techniques is that we can easily infer the plausibility of the statistical (imputation) model. This is straightforward because the imputation model produces data, and we are very well equipped to look at data.

One of the best tools for assessing the plausibility of imputations is to study the discrepancy between the observed and imputed data. The idea is that high quality imputed data will have distributions similar to the observed data. Except under MCAR, the distributions do not need to be identical, as strong MAR mechanisms may induce systematic differences between the two distributions. However, any dramatic differences between the imputed and observed data (such as seeing a body mass index of 300 in the imputed data) should certainly alert us to the possibility that something is wrong with the imputation model. It is reassuring when the synthetic data could have been real values had they not been missing.

Suppose we compare density estimates of the observed and imputed data. Some type of discrepancies that are of interest are

- the points have different means;
- the points have different spreads;
- the points have different scales;
- the points have different relations;
- the points do not overlap and they defy common sense.

Such differences between the densities may suggest a problem that needs to be further checked. Other useful graphic representations include the box plot, the stripplot, the histogram, and the scattergram of variables, stratified according to whether the data are real or imputed.

Figure 13.5 shows kernel density estimates of imputed and observed data. In this case, the distributions match up well. Other imputation diagnostics have been suggested by Gelman et al. (2005), Raghunathan and Bondarenko (2007), Abayomi, Gelman, and Levy (2008), and Su et al. (2011).

Compared to diagnostic methods for conventional statistical models, imputation comes with the advantage that we can directly compare the observed and imputed values. Unfortunately, diagnostics are currently underused. One reason is that not all software properly supports diagnostics. Another reason is that the imputer may put too much trust into the appropriateness of the defaults of the software for the data at hand. Absurd imputations are however easy to spot by simple methods, and should be repaired before attempting complete-data analysis.

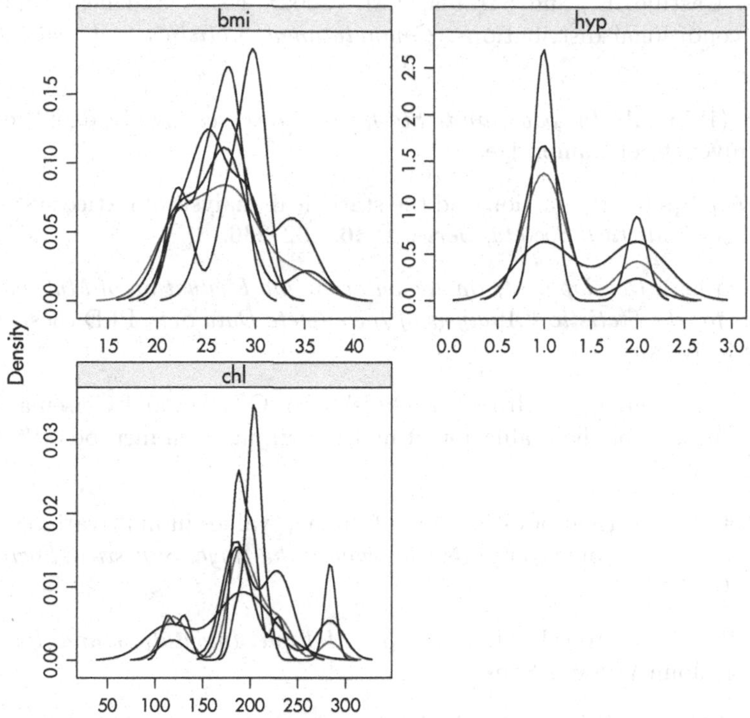

FIGURE 13.5
Kernel density estimates for the marginal distributions of the observed data (thick line) and the $m = 5$ densities per variable calculated from the imputed data (thin lines).

13.8 Conclusion

FCS has rapidly been adopted by applied researchers in many branches of science. FCS remains close to the data and is easy to apply. The relevant software is now widespread, and available in all major statistical packages. Appendix A of van Buuren (2012) is an overview of software for FCS.

The technology has now evolved into the standard way of creating multiple imputations. Of course, there are still open issues, and more experience is needed with practical application of FCS. Nevertheless, FCS is an open and modular technology that will continue to attract the attention of researchers who want to solve their missing data problems.

References

Abayomi, K., Gelman, A., and Levy, M. (2008). Diagnostics for multivariate imputations. *Applied Statistics* **57**, 273–291.

Arnold, B.C., Castillo, E., and Sarabia, J.M. (2002). Exact and near compatibility of discrete conditional distributions. *Computational Statistics and Data Analysis* **40**, 231–252.

Bartlett, M.S. (1978). *An Introduction to Stochastic Processes.* Cambridge: Press Syndicate of the University of Cambridge.

Besag, J. (1974). Spatial interaction and the statistical analysis of lattice systems. *Journal of the Royal Statistical Society, Series B* **36**, 192–236.

Brand, J.P.L. (1999). *Development, Implementation and Evaluation of Multiple Imputation Strategies for the Statistical Analysis of Incomplete Data Sets.* PhD thesis. Rotterdam: Erasmus University.

Brand, J.P.L., van Buuren, S., Groothuis-Oudshoord, C.G.M., and Gelsema, E.S. (2003). A toolkit in SAS for the evaluation of multiple imputation methods. *Statistica Neerlandica* **57**, 36–45.

Buck, S.F. (1960). A method of estimation of missing values in multivariate data suitable for use with an electronic computer. *Journal of the Royal Statistical Society, Series B* **22**, 302–306.

Carpenter, J.R. and Kenward, M.G. (2013). *Multiple Imputation and Its Application.* Chichester: John Wiley & Sons.

Casella, G. and George, E.I. (1992). Explaining the Gibbs sampler. *The American Statistician* **46**, 167–174.

Chen, H.Y. (2011). Compatibility of conditionally specified models. *Statistics and Probability Letters* **80**, 670–677.

Collins, L.M., Schafer, J.L., and Kam, C.M. (2001). A comparison of inclusive and restrictive strategies in modern missing data procedures. *Psychological Methods* **6**, 330–351.

de Waal, T., Pannekoek, J., and Scholtus, S. (2011). *Handbook of Statistical Data Editing and Imputation.* Hoboken, NJ: John Wiley & Sons.

Finkbeiner, C. (1979). Estimation for the multiple factor model when data are missing. *Psychometrika* **44**, 409–420.

Gebregziabher, M., Zhao, Y., Axon, N., Gilbert, G.E., Echols, C., and Egede, L.E. (2012). Lessons learned in dealing with missing race data: An empirical investigation. *Journal of Biometrics & Biostatistics*, **3**, Article 138.

Gelfand, A.E. and Smith, A.F.M. (1990). Sampling-based approaches to calculating marginal densities. *Journal of the American Statistical Association* **85**, 398–409.

Gelman, A. (2004). Parameterization and Bayesian modeling. *Journal of the American Statistical Association* **99**, 537–545.

Gelman, A. and Raghunathan, T.E. (2001). Discussion of Arnold et al. "Conditionally Specified Distributions." *Statistical Science* **16**, 249–274.

Gelman, A. and Speed, T.P. (1993). Characterizing a joint probability distribution by conditionals. *Journal of the Royal Statistical Society, Series B* **55**, 185–188.

Gelman, A., Van Mechelen, I., Meulders, M., Verbeke, G., and Heitjan, D.F. (2005). Multiple imputation for model checking: Completed-data plots with missing and latent data. *Biometrics* **61**, 74–85.

Gilks, W.R. (1996). Full conditional distributions. In: W.R. Gilks, S. Richardson, and J. Spiegelhalter, D. (Eds.). *Markov Chain Monte Carlo in Practice*, pp. 75–88. London: CRC / Chapman & Hall

Gleason, T.C. and Staelin, R. (1975). A proposal for handling missing data. *Psychometrika* **40**, 229–252.

Gold, M.S. and Bentler, P.M. (2000). Treatments of missing data: A Monte Carlo comparison of RBHDI, iterative stochastic regression imputation, and expectation-maximization. *Structural Equation Modeling* **7**, 319–355.

Groothuis-Oudshoorn, C.G.M., van Buuren, S. and van Rijckevorsel, J.L.A. (1999). *Flexible Multiple Imputation by Chained Equations of the AVO-95 Survey*, volume (PG/VGZ/00.045). Leiden: TNO Prevention and Health.

Hardt, J., Herke, M., and Leonhart, R. (2012). Auxiliary variables in multiple imputation in regression with missing X: A warning against including too many in small sample research. *BMC Medical Research Methodology* **12**, 184.

Heckerman, D., Chickering, D.M., Meek, C., Rounthwaite, R., and Kadie, C. (2001). Dependency networks for inference, collaborative filtering, and data visualisation. *Journal of Machine Learning Research* **1**, 49–75.

Heckman, J.J. (1979). Sample selection bias as a specification error. *Econometrica* **47**, 153–161.

Hoerl, A.,E. and Kennard, R.W. (1970). Ridge regression: Biased estimation for nonorthogonal problems. *Technometrics* **12**, 55–67.

Horton, N.J. and Kleinman, K.P. (2007). Much ado about nothing: A comparison of missing data methods and software to fit incomplete data regression models. *The American Statistician* **61**, 79–90.

Hron, K., Templ, M., and Filzmoser, P. (2010). Imputation of missing values for compositional data using classical and robust methods. *Computational Statistics and Data Analysis* **54**, 3095–3107

Ip, E.H. and Wang, Y.J. (2009). Canonical representation of conditionally specified multivariate discrete distributions. *Journal of Multivariate Analysis* **100**, 1282–1290.

Jinn, J.-H. and Sedransk, J. (1989). Effect on secondary data analysis of common imputation methods. *Sociological Methodology* **19**, 213–241.

Jolani, S. (2012). *Dual Imputation Strategies for Analyzing Incomplete Data.* PhD thesis. Utrecht: University of Utrecht.

Jolani, S., van Buuren, S., and Frank, L.E. (2013). Combining the complete-data and nonresponse models for drawing imputations under MAR. *Journal of Statistical Computation and Simulation* **83**, 868–879.

Kennickell, A.B. (1991). Imputation of the 1989 survey of consumer finances: Stochastic relaxation and multiple imputation. In: *Proceedings of the Section on Survey Research Methods. Joint Statistical Meeting 1991*, pp. 1–10. Alexandria, VA: American Statistical Association.

Kuo, K.-L. and Wang, Y.J. (2011). A simple algorithm for checking compatibility among discrete conditional distributions. *Computational Statistics and Data Analysis* **55**, 2457–2462.

Lee, K.J. and Carlin, J.B. (2010). Multiple imputation for missing data: Fully conditional specification versus multivariate normal imputation. *American Journal of Epidemiology* **171**, 624–632.

Lee, K.J., Galati, J.C., Simpson, J.A., and Carlin, J.B. (2012). Comparison of methods for imputing ordinal data using multivariate normal imputation: A case study of non-linear effects in a large cohort study. *Statistics in Medicine* **31**, 4164–4174.

Li, K.-H. (1988). Imputation using Markov chains. *Journal of Statistical Computation and Simulation* **30**, 57–79.

Little, R.J. (2013). In praise of simplicity not mathematistry! Ten simple powerful ideas for the statistical scientist. *Journal of the American Statistical Association* **108**, 359–369.

Little, R.J.A. (1988). Missing-data adjustments in large surveys (with discussion). *Journal of Business Economics and Statistics* **6**, 287–301.

Little, R.J.A. (1992). Regression with missing Xs: A review. *Journal of the American Statistical Association* **87**, 1227–1237.

Little, R.J.A. and Rubin, D.B. (2002). *Statistical Analysis with Missing Data* (2nd ed.). New York: John Wiley & Sons.

Little, R.J.A., D'Agostino, R., Dickersin, K., Emerson, S.S., Farrar, J.T., Frangakis, C., Hogan, J.W., Molenberghs, G., Murphy, S.A., Neaton, J.D., Rotnitzky, A., Scharfstein, D., Shih, W., Siegel, J.P., and Stern, H. National Research Council (2010). *The Prevention and Treatment of Missing Data in Clinical Trials. Panel on Handling Missing Data in Clinical Trials.* Committee on National Statistics, Division of Behavioral and Social Sciences and Education. Washington, D.C.: The National Academies Press.

Meng, X.-L. (1994). Multiple-imputation inferences with uncongenial sources of input. *Statistical Science* **9**, 538–573.

Moons, K.G.M., Donders, A.R.T., Stijnen, T., and Harrell, F.E. (2006). Using the outcome for imputation of missing predictor values was preferred. *Journal of Clinical Epidemiology* **59**, 1092–1101.

Raghunathan, T.E. and Bondarenko, I. (2007). Diagnostics for multiple imputations. *Unpublished manuscript.*

Raghunathan, T.E., Lepkowski, J.M., van Hoewyk, J., and Solenberger, P.W. (2001). A multivariate technique for multiply imputing missing values using a sequence of regression models. *Survey Methodology* **27**, 85–95.

Raymond, M.R. and Roberts, D.M. (1987). A comparison of methods for treating incomplete data in selection research. *Educational and Psychological Measurement* **47**, 13–26.

Roberts, G.O. (1996). Markov chain concepts related to sampling algorithms. In: W.R. Gilks, S. Richardson, and J. Spiegelhalter, D. (Eds.). *Markov Chain Monte Carlo in Practice*, pp. 45–57. London: CRC / Chapman & Hall.

Royston, P., and White, I.R. (2011). Multiple imputation by chained equations (MICE): Implementation in STATA. *Journal of Statistical Software* **45**, 1–20.

Rubin, D.B. (1987). *Multiple Imputation for Nonresponse in Surveys.* New York: John Wiley & Sons.

Rubin, D.B. (2003a). Discussion on multiple imputation. *International Statistical Review* **71**, 619–623.

Rubin, D.B. (2003b). Nested multiple imputation of NMES via partially incompatible MCMC. *Statistica Neerlandica* **57**, 3–18.

Rubin, D.,B. and Schafer, J.L. (1990). Efficiently creating multiple imputations for incomplete multivariate normal data. In: *ASA 1990 Proceedings of the Statistical Computing Section*, pp. 83–88. Alexandria, VA: American Statistical Association.

Schafer, J.L. (1997). *Analysis of Incomplete Multivariate Data.* London: CRC / Chapman & Hall.

Schafer, J.L. (2003). Multiple imputation in multivariate problems when the imputation and analysis models differ. *Statistica Neerlandica* **57**, 19–35.

Sterne, J.A.C., White, I.R., Carlin, J.B., Spratt, M., Royston, P., Kenward, M.G., Wood, A.M., and Carpenter, J.R. (2009). Multiple imputation for missing data in epidemiological and clinical research: potential and pitfalls. *British Medical Journal* **338**, b2393.

Su, Y.-S., Gelman, A., Hill, J.L., and Yajima, M. (2011). Multiple imputation with diagnostics (mi) in R: Opening windows into the black box. *Journal of Statistical Software* **45**, 1–31.

Tan, M.T., Tian, G.-L., and Ng, K.W. (2010). *Bayesian Missing Data Problem: EM, Data Augmentation and Noniterative Computation.* Boca Raton: Chapman & Hall/CRC.

Tang, L., Song, J., Belin, T.R., and Unutzer, J. (2005). A comparison of imputation methods in a longitudinal randomized clinical trial. *Statistics in Medicine* **24**, 2111–2128.

Tempelman, D.C.G. (2007). *Imputation of Restricted Data.* PhD thesis. Groningen: University of Groningen.

Tian, G.-L., Tan, M.T., Ng, K.W., and Tang, M.-L. (2009). A unified method for checking compatibility and uniqueness for finite discrete conditional distributions. *Communications in Statistics, Theory and Methods* **38**, 115–129.

Tibshirani, R. (1996). Regression shrinkage and selection via the lasso. *Journal of the Royal Statistical Society, Series B* **58**, 267–288.

Tierney, L. (1996). Introduction to general state-space Markov chain theory. In: W.R. Gilks, S. Richardson, and J. Spiegelhalter, D. (Eds.). *Markov Chain Monte Carlo in Practice*, pp. 59–74. London: CRC / Chapman & Hall.

van Buuren, S. (2007). Multiple imputation of discrete and continuous data by fully conditional specification. *Statistical Methods in Medical Research* **16**, 219–242.

van Buuren, S. (2010). Item imputation without specifying scale structure. *Methodology* **6**, 31–36.

van Buuren, S. (2012). *Flexible Imputation of Missing Data.* Boca Raton: Chapman & Hall/CRC.

van Buuren, S., Boshuizen, H.C., and Knook, D.L. (1999). Multiple imputation of missing blood pressure covariates in survival analysis. *Statistics in Medicine* **18**, 681–694.

van Buuren, S., Brand, J.P.L., Groothuis-Oudshoorn, C.G.M., and Rubin, D.B. (2006). Fully conditional specification in multivariate imputation. *Journal of Statistical Computation and Simulation* **76**, 1049–1064.

van Buuren, S. and Groothuis-Oudshoorn, C.G.M. (1999). *Flexible Multivariate Imputation by MICE,* volume PG/VGZ/99.054. Leiden: TNO Prevention and Health.

van Buuren, S. and Groothuis-Oudshoorn, C.G.M. (2000). *Multivariate Imputation by Chained Equations: MICE V1.0 User Manual,* volume PG/VGZ/00.038. Leiden: TNO Prevention and Health.

van Buuren, S. and Groothuis-Oudshoorn, K. (2011). MICE: Multivariate imputation by chained equations in R. *Journal of Statistical Software* **45**, 1–67.

van Buuren, S. and van Rijckevorsel, J.L.A. (1992). Imputation of missing categorical data by maximizing internal consistency. *Psychometrika* **57**, 567–580.

van der Palm, D.W., van der Ark, L.A., and Vermunt, J.K. (2012). A comparison of incomplete-data methods for categorical data. *Statistical Methods in Medical Research* **00**, 000–000.

Vink, G. and van Buuren, S. (2013). Multiple imputation of squared terms. *Sociological Methods and Research* **42**, 598–607.

von Hippel, P.T. (2009). How to impute interactions, squares, and other transformed variables. *Sociological Methodology* **39**, 265–291.

Wang, Y.J. and Kuo, K.-L. (2010). Compatibility of discrete conditional distributions with structural zeros. *Journal of Multivariate Analysis* **101**, 191–199.

White, I.R., Horton, N.J., Carpenter, J.R., and Pocock, S.J. (2011). Strategy for intention to treat analysis in randomised trials with missing outcome data. *British Medical Journal* **342**, d40.

White, I.R., Royston, P., and Wood, A.M. (2011). Multiple imputation using chained equations: Issues and guidance for practice. *Statistics in Medicine* **30**, 377–399.

Yu, L.-M., Burton, A., and Rivero-Arias, O. (2007). Evaluation of software for multiple imputation of semi-continuous data. *Statistical Methods in Medical Research* **16**, 243–258.

14

Multilevel Multiple Imputation

Harvey Goldstein

University of Bristol, Bristol, UK

James R. Carpenter

London School of Hygiene and Tropical Medicine and MRC Clinical Trials Unit at UCL, London, UK

CONTENTS

14.1 Introduction

Multiple Imputation (MI), introduced in Chapter 11, is now firmly established as a broadly applicable, practical method for the analysis of partially observed data. While it is probably true that in many missing datasettings there are alternative approaches that can be taken to MI—and these may be more efficient or have a stronger justification in a strictly statistical sense—MI has the advantage of generality and practicality (Carpenter and Kenward, 2013, p. 73).

In this chapter, we describe and illustrate a general joint-modelling approach to MI with mixed response types and multilevel structure. We will see that this has great flexibility and wide applicability. It naturally handles partially observed variables at different levels of the hierarchy, and allows wide flexibility in modelling covariance structures. We show how it extends to incorporate missing values in non-linear effects and interactions, in a way that is consistent with the substantive model. We further show how it extends to handle coarsened data, and record linkage. We present two examples to illustrate the flexibility of the approach. The first is educational, where we explore the non-linear relationship between class size and educational achievement among children in their first year of school. The second is individual patient data meta-analysis and prognostic modelling, where besides handling sporadic missing data within studies, we can incorporate variables measured on different scales in different studies, and impute variables that were not collected in certain studies.

Our joint modelling approach builds on use of the multivariate normal model for imputation, which is described in detail in Schafer (1997). For single level data structures, imputation using joint models can be well approximated by using a linked series of conditional regressions, i.e., the "Full Conditional Specification" (FCS) approach (as set out in Chapter 13). Specifically, Hughes et al. (2014b) show that, for appropriate choice of priors, FCS and joint modelling are strictly equivalent for unstructured multivariate normal models and saturated log-linear models. In other cross-sectional settings FCS is likely to provide a good approximation to joint modelling, but the equivalence is not exact. However, once we consider multilevel structures, particularly with unbalanced data and partially observed variables at several levels of the hierarchy, then FCS loses its simplicity and computational attraction (Carpenter and Kenward, 2013, p. 220–222).

We therefore focus on the joint modelling approach to MI in this chapter, assuming the missingness mechanism is ignorable (see Chapter 3). The use of MI for exploring the robustness of inferences to departures from ignorability is discussed in Chapter 19.

We begin, in Section 14.2, with a straightforward multivariate normal 2-level model, in order to illustrate our basic approach. We show how multiple imputation can be carried out for missing values in responses and covariates measured at either level 1 or level 2. The methodology is then extended to generalised linear models where the response or predictor variables may be continuous or categorical, as is typically the case in practice. This avoids having to make strong assumptions about the appropriateness of normal approximations in such cases. Traditionally with multiple imputation, problems occur when interaction terms are present with missing values in one of the constituent variables. Recent work has shown how these may be incorporated within a fully Bayesian model and we set out the methodology to do this in Section 14.3. Having described a general framework for handling missing data, we show how apparently disparate settings that can usefully be formulated as missing data problems and how our methodology provides an adequate solution. Specifically, in Section 14.4 we consider "coarsened data," which is intermediate between fully missing and fully known: one example being where some respondents to a survey can supply an accurate value and others only an interval estimate or none at all. Another example that we deal with in Section 14.5 arises in record linkage where, for example, data values are to be transferred to records in a survey data file from an administrative database and where we can attach *a priori* probabilities to a subset of possible matches from the administrative database. We present two illustrative applications in Section 14.6, and summarize the chapter in Section 14.7.

14.2 Multilevel Multiple Imputation with Mixed Response Types

In this section, we describe multilevel multiple imputation for continuous data and a two level hierarchy, with missing data at the lower level. We then show how this approach can be extended to handle a mix of response types, which may be partially observed at both levels of the hierarchy.

Consider a generic substantive multilevel model for a two-level dataset:

$$
\begin{aligned}
Y_{ij} &= \boldsymbol{X}_{ij}^{(1)}\boldsymbol{\alpha}_1 + \boldsymbol{X}_j^{(2)}\boldsymbol{\alpha}_2 + \boldsymbol{Z}_{ij}\boldsymbol{u}_j + e_{ij}, \\
\boldsymbol{u}_j &\sim N(0, \boldsymbol{\Omega}_u) \\
e_{ij} &\sim N(0, \sigma_e^2),
\end{aligned} \tag{14.1}
$$

We use a standard subscript convention for multilevel models (Goldstein, 2011) with $j = 1, \ldots, J$ indexing level 2 units and $i = 1, \ldots, n$ indexing level 1 units. For ease of exposition, we assume the same number of level 1 units for each level 2 unit, but this is not necessary in practice. Here $\boldsymbol{X}_{ij}^{(1)}$ is a $(1 \times p_1)$ vector of level 1 covariates (including the constant) with fixed coefficient vector $\boldsymbol{\alpha}_1$, $\boldsymbol{X}_j^{(2)}$ is a $(1 \times p_2)$ vector of level 2 covariates (constant within level 1 units) with fixed coefficient vector $\boldsymbol{\alpha}_2$ and $\boldsymbol{Z}_{ij}$ is a $(1 \times q)$ vector of covariates with random coefficients at level 2, which for ease of exposition, we assume is a subset of $\boldsymbol{X}_{ij}^{(1)}$. As usual we assume the $(q \times 1)$ vector $\boldsymbol{u}_j$ follows a multivariate normal distribution, independent of the level 1 residuals e_{ij}. Typically, interest focuses on inference for $\boldsymbol{\alpha}_1, \boldsymbol{\alpha}_2, \boldsymbol{\Omega}_u$ and σ_e^2.

Suppose there may be missing observations in any of Y_{ij}, $\boldsymbol{X}_{ij}^{(1)}$, $\boldsymbol{X}_j^{(2)}$. We assume the missingness mechanism is ignorable (cf. Chapter 3). We assume for now that these variables are jointly multivariate normal. To form an imputation model congenial with the substantive model (14.1), we take this model—in which the response has a normal distribution conditional on the covariates—together with a multivariate normal distribution for the covariates, to form a joint multilevel multivariate normal model for the response and covariates from the substantive model.

Our strategy is to fit this model to the observed data using Markov Chain Monte Carlo (MCMC), treating missing values as parameters and updating them appropriately as the MCMC algorithm proceeds. Once we judge that the MCMC sampler has converged (with the help of the usual MCMC sampler diagnostics), we retain the current draws of the missing values, together with the observed values, as our first imputed dataset. We then update the sampler further (an update consists of a new draw of all the parameters and missing values), and then retain the current draws of the missing values, together with the observed values, as the second imputed dataset, and so on. In order for Rubin's MI combination rules to apply (see Chapter 12), successive imputed datasets should be—at least approximately—independent draws from the distribution of the missing given the observed data. To ensure this, the MCMC sampler needs to be updated a sufficient number of times between retaining values for successive imputed datasets. In practice we have found 100–1000 updates is sufficient, with values towards the upper end of this range required as the proportion of missing values and/or the complexity of the imputation model increases. Utilizing MCMC sampling in this way provides a convenient method for providing approximately independent draws of the missing data from the Bayesian predictive distribution.

We now describe this in more detail. First, to express the multivariate multilevel normal imputation model algebraically, we note that for level 1 unit i, nested within level 2 unit j, there are p_1 level 1 responses consisting of the level 1 response Y_{ij} and the $(p_1 - 1)$ level 1 covariates obtained from $\boldsymbol{X}_{ij}^{(1)}$ by omitting the constant. We stack these and denote them by the $(p_1 \times 1)$ column vector $\boldsymbol{Y}_{ij}^{(1)}$, $i = 1, \ldots, n$. We also have p_2 level 2 responses consisting of the p_2 covariates $\boldsymbol{X}_j^{(2)}$, which we denote by the $(p_2 \times 1)$ vector $\boldsymbol{Y}_j^{(2)}$. We can then write the joint multivariate multilevel normal imputation model as

$$\begin{aligned}
\boldsymbol{Y}_{ij}^{(1)} &= \boldsymbol{\beta}^{(1)} + \boldsymbol{u}_j^{(1)} + \boldsymbol{e}_{1ij}, \\
\boldsymbol{Y}_j^{(2)} &= \boldsymbol{\beta}^{(2)} + \boldsymbol{u}_j^{(2)}, \\
\boldsymbol{e}_{1ij} &\sim N(0, \boldsymbol{\Omega}_1), \\
\boldsymbol{u}_j = (\boldsymbol{u}_j^{(1)^T}, \boldsymbol{u}_j^{(2)^T})^T &\sim N(0, \boldsymbol{\Omega}_2).
\end{aligned} \tag{14.2}$$

Here $\boldsymbol{\beta}^{(1)}$ is a $(p_1 \times 1)$ column vector of means for the level 1 units and $\boldsymbol{\beta}^{(2)}$ is a $(p_2 \times 1)$ column vector of the means for the level 2 units.

Carpenter and Kenward (2013, p. 214) draw on Goldstein et al. (2009) to give details of an MCMC sampler for fitting (14.2). As usual in such MCMC algorithms, within each update cycle missing values are drawn from the distribution of the missing given observed data at the current parameter values. To complete the cycle, they are then conditioned on when updating the parameters and random effects in (14.2).

To fit (14.2), we choose some initial values and then update the MCMC sampler a number of times (typically at least 1000 times) so that it converges to the posterior distribution (a process often referred to as "burning in" the MCMC sampler). Subsequent draws from the MCMC sampler are then correlated draws from the posterior distribution of the missing values and parameters given the observed data. After the burn-in, we can therefore choose a particular update and retain the current draw of the missing data from the posterior together with the observed data to form an imputed dataset (which has no missing values). Notice that we are not interested in the parameters of (14.2) themselves, so these are not retained. As we noted above, in order to apply Rubin's rules, we need the draws of the missing data in the imputed datasets to be stochastically independent given the observed data. To achieve this, we update the sampler a number of times (typically at least 100 times) between imputed datasets. Once we have a set of imputed datasets then as—usual with MI—the substantive model is fitted to each imputed dataset in turn and the resulting parameter estimates combined for final inference using to Rubin's rules (Rubin, 1987).

While the discussion has focussed on a two level setting, as with multilevel modelling more generally, the extension of this above approach to more than two levels involves no additional conceptual issues.

14.2.1 Congeniality

For reliable results from MI, it is important that the imputation model is at least approximately congenial with the analyst's substantive model. A congenial imputation model needs to contain all the variables in the substantive model in a way that is consistent with the interrelationships among the variables in the substantive model. For example, if the substantive model is a linear regression of Y on continuous X and Z, with X partially observed, a congenial imputation model could be a linear regression of X on Y and Z. If we included an auxiliary variable W in the imputation model, it would no longer be strictly congenial, but may give better inferences. Essentially, strict congeniality is needed for Rubin's MI combination rules for estimating the variance to be valid, but mild uncongeniality that arises, for example, from including auxiliary variables is practically inconsequential. However, uncongeniality arising because the conditional relationships implied by the imputation model differ from those of the substantive model should be avoided. We elaborate on this in Section 14.4. For a thorough discussion of these points see Carpenter and Kenward (2013) pp. 55–73 and Hughes et al. (2014a).

The attraction of our formulation (14.2) is that, if we do not add auxiliary variables in (14.2), it ensures congeniality, and as such imputes missing variables at both level 1 and level 2 appropriately. In the multivariate normal case, if we include auxiliary variables in (14.2), these are additional responses so marginalising over them gives relationships compatible with the substantive model (14.1).

14.2.2 Conditioning on fully observed variables in imputation

In many settings it will be convenient to modify the multivariate multilevel normal imputation model (14.2) by transferring some or all of the fully observed variables within $\boldsymbol{Y}_{ij}^{(1)}, \boldsymbol{Y}_{j}^{(2)}$ to the right-hand side, and conditioning on them.

For example, if the level 1 units are repeated observations in time on the level 2 units, and particularly if the observation times are irregular, it is much more natural to condition on time as a covariate in (14.2). However, as discussed in Section 14.2.1, if we have auxiliary variables which are not in the substantive model but which we wish to include in the imputation model, to preserve relationships after marginalization it is generally better to include these as additional responses in (14.2).

14.2.3 Choice of auxiliary variables

Auxiliary variables should always be predictive of the missing values. If they are also predictive of the chance of the data being missing, so that including them increases the plausibility of the ignorability assumption, they may also reduce bias in the estimation of the parameters of the substantive model (Carpenter and Kenward, 2013, p. 72). The ability to flexibly and appropriately exploit the information in auxiliary data, sets MI apart from other approaches, and generally ensures results are more precise than those obtained from restricting to the subset of complete records, even when the assumptions for the latter approach to be valid hold.

14.3 Imputing Mixed Response Types

Fully observed categorical variables from the substantive model can be handled using the approach above if they are included as covariates in the imputation model. More generally, though, they will have missing values. We therefore extend the above approach using a latent normal model for the categorical variables, as detailed below. The latent normal variables that are derived from the categorical variables are then included in the imputation model alongside the continuous variables using the multivariate multilevel normal model described above. We outline this approach where the substantive model is a regression of a continuous variable on (I) a binary variable; (II) an ordinal variable and (III) an unordered categorical variable.

Case I: Binary variable

Consider a special case of model (14.1), where we only have a single binary covariate X_{ij}, and we have dropped the superscripts for simplicity:

$$\begin{aligned} Y_{ij} &= \alpha_0 + \alpha_1 X_{ij} + u_j + e_{ij} \\ u_j &\sim N(0, \sigma_u^2) \\ e_{ij} &\sim N(0, \sigma_e^2). \end{aligned} \tag{14.3}$$

To incorporate binary X_{ij}, we modify imputation model (14.2), modelling X_{ij} by treating it as though derived from a latent normally distributed random variable Z_{ij} with $X_{ij} = 1$ corresponding to $Z_{ij} > 0$. The imputation model (14.2) then becomes:

$$\begin{aligned} Y_{ij} &= \beta_{01} + u_{1j} + e_{1ij}, \\ Z_{ij} &= \beta_{02} + u_{2j} + e_{2ij}, \text{ where} \\ \Pr(X_{ij} = 1) &= \Pr(Z_{ij} > 0), \text{ and} \\ \begin{pmatrix} u_{1j} \\ u_{2j} \end{pmatrix} \sim N(0, \mathbf{\Omega}_u), & \begin{pmatrix} e_{1ij} \\ e_{2ij} \end{pmatrix} \sim N\left\{0, \mathbf{\Omega}_e = \begin{pmatrix} \sigma_1^2 & \\ \sigma_{12} & 1 \end{pmatrix}\right\}. \end{aligned} \tag{14.4}$$

Here, the variance of the latent normal variable at level 1 is constrained to 1 so the model is identifiable. It follows that

$$\Pr(X_{ij} = 1) = \Pr(Z_{ij} > 0) = \Pr\{e_{2ij} > -(\beta_{02} + u_{2j})\} = \Phi(\beta_{02} + u_{2j}),$$

where Φ is the cumulative distribution function of the standard normal. In other words, the latent normal structure in (14.4) corresponds to a probit link in the imputation model, as opposed to the more usual logit link.

Using this latent normal approach makes the MCMC procedure considerably more straightforward, while the difference between the predicted probabilities, and hence imputed values from the logit and probit models, is of no practical importance in our experience.

Given the MCMC algorithm for fitting the multilevel multivariate normal imputation model (14.2), the extension to a latent normal structure is natural:

1. Initialize any missing X_{ij} randomly as either 1 or 0.
2. Initialize the latent Z_{ij} with $Z_{ij} = 1$ if $X_{ij} = 1$ and $Z_{ij} = -1$ if $X_{ij} = 0$.
3. Conditional on the Z_{ij}, use the MCMC algorithm for the multivariate normal to update the parameters of (14.4) exactly as before, and impute any missing values of the continuous variable Y_{ij} from the appropriate conditional distribution.
4. To update X_{ij}:
   (a) *If X_{ij} is missing:* update Z_{ij} for each i, j, by drawing from the conditional normal distribution of $Z_{ij} \mid Y_{ij}$ given the current parameter values. Set $X_{ij} = 1$ if $Z_{ij} > 0$ and $X_{ij} = 0$ otherwise.
   (b) *If X_{ij} is observed:* first draw a proposal, $\tilde{Z}_{ij}$, from the conditional normal distribution of $Z_{ij}|Y_{ij}$ given the current parameter values. Accept the proposal if $\text{sign}(\tilde{Z}_{ij}) = \text{sign}(X_{ij} - 0.5)$. If $\tilde{Z}_{ij}$ is rejected, draw again until it is accepted; if it is accepted, set $Z_{ij} = \tilde{Z}_{ij}$.
5. Iterate steps 3–4 until convergence.

As mentioned above, the MCMC algorithm in step 3 needs to respect the constraint that the variance of the latent normal variable $\sigma_2^2 = 1$. This is most easily accomplished by updating the elements of the level 1 covariance matrix in the imputation model one at a time. The details are given by Browne (2006) and Carpenter and Kenward (2013) p. 97. If desired, the rejection step 4b can be replaced by an appropriate draw from a truncated normal, as described by Carpenter and Kenward, (2013) p. 96.

Case II: Ordinal variable
For ordinal data we use a proportional probit model. Suppose that the covariate X_{ij} in the substantive model (14.3) is ordinal, taking values $m = 1, 2, \ldots, M$. The imputation model again uses a latent normal variable, Z_{ij}, whose variance is constrained to be 1. However, to distinguish the various categories, in equation (14.4) we additionally constrain the mean of Z_{ij} to be zero, and then replace β_{02} with $(M-1)$ "cut point" parameters, $\gamma_1, \ldots, \gamma_{(M-1)}$, such that

$$\Pr(X_{ij} = 1) = \Pr(Z_{ij} < \gamma_1 + u_{2j}) = \Phi(\gamma_1 + u_{2j}),$$

for $m = 2, \ldots, M-1$,

$$\begin{aligned}\Pr(X_{ij} = m) &= \Pr(\gamma_{m-1} + u_{2j} < Z_{ij} < \gamma_m + u_{2j}) \\ &= \Phi(\gamma_m + u_{2j}) - \Phi(\gamma_{m-1} + u_{2j}),\end{aligned}$$

and

$$\Pr(X_{ij} = M) = \Pr(Z_{ij} > \gamma_M + u_{2j}) = 1 - \Phi(\gamma_M + u_{2j}),$$

where Φ is the standard cumulative normal density function. With a continuous and ordinal variable in (14.4), the MCMC algorithm to fit the model now has to update the latent $\{Z_{ij}\}$ as well as $\boldsymbol{\gamma} = (\gamma_1, \ldots, \gamma_M)$. To do this we write

$$\begin{aligned}&f\left(\{Z_{ij}\}, \boldsymbol{\gamma} \mid \beta_{01}, \boldsymbol{u}, \boldsymbol{\Omega}_u, \boldsymbol{\Omega}_e, \{Y_{ij}\}, \{X_{ij}\}\right) \\ &\quad = f\left(\{Z_{ij}\} \mid \boldsymbol{\gamma}, \beta_{01}, \boldsymbol{u}, \boldsymbol{\Omega}_u, \boldsymbol{\Omega}_e, \{Y_{ij}\}, \{X_{ij}\}\right) \times \\ &\qquad f\left(\boldsymbol{\gamma} \mid \beta_{01}, \boldsymbol{u}, \boldsymbol{\Omega}_u, \boldsymbol{\Omega}_e, \{Y_{ij}\}, \{X_{ij}\}\right),\end{aligned}$$

where $\boldsymbol{\gamma}$ is updated using a Metropolis step, and the $\{Z_{ij}\}$ by rejection sampling, as before. The details are given by Carpenter and Kenward (2013, p. 102).

Case III: Unordered categorical variable
We now consider the case where the covariate in the substantive model, X_{ij}, is an unordered categorical variable with M levels, and for simplicity we assume this is defined at level 1 and have dropped the superscripts. This would typically be included in the substantive model through $(M-1)$ dummy variables. To illustrate how our imputation model extends to this case, we suppose the categorical variable has 3 levels (the extension to M levels involves no additional concepts). Our substantive model can now be written

$$\begin{aligned}Y_{ij} &= \alpha_0 + \alpha_1 X_{2ij} + \alpha_2 X_{3ij} + u_j + e_{ij}, \\ u_j &\sim N(0, \sigma_u^2), \\ e_{ij} &\sim N(0, \sigma_e^2),\end{aligned} \tag{14.5}$$

where $X_{2ij} = 1$ if $X_{ij} = 2$ and 0 otherwise, and $X_{3ij} = 1$ if $X_{ij} = 3$ and 0 otherwise. Then we introduce $(3-1) = 2$ latent normal variables into the imputation model (14.4), which becomes:

$$
\begin{aligned}
Y_{ij} &= \beta_{01} + u_{1j} + e_{1ij}, \\
Z_{1ij} &= \beta_{02} + u_{2j} + e_{2ij}, \\
Z_{2ij} &= \beta_{03} + u_{3j} + e_{3ij},
\end{aligned}
$$

$$
\begin{pmatrix} u_{1j} \\ u_{2j} \\ u_{3j} \end{pmatrix} \sim N(0, \mathbf{\Omega}_u), \quad \begin{pmatrix} e_{1ij} \\ e_{2ij} \\ e_{3ij} \end{pmatrix} \sim N\left\{0, \mathbf{\Omega}_e = \begin{pmatrix} \sigma_1^2 & & \\ \sigma_{12} & 1 & \\ \sigma_{13} & 0 & 1 \end{pmatrix}\right\}, \text{ where}
$$

$$
\begin{aligned}
\Pr(X_{ij} = 1) &= \Pr(Z_{1ij} > Z_{2ij} > 0), \\
\Pr(X_{ij} = 2) &= \Pr(Z_{2ij} > Z_{1ij} > 0), \\
\Pr(X_{ij} = 3) &= \Pr(Z_{1ij} < 0 \text{ and } Z_{2ij} < 0).
\end{aligned} \tag{14.6}
$$

Notice that in the level 1 covariance matrix the variance of the latent normal variables is constrained to 1. In our analyses, we typically also constrain their covariance to zero, although as discussed in Carpenter and Kenward (2013, p. 115), the derivation from the maximum indicant model gives a covariance of 0.5. Fitting model (14.6) by MCMC is a natural extension of the algorithm for the binary case above, where

(i) in step 4a (when X_{ij} is missing) we draw (Z_{1ij}, Z_{2ij}) from the conditional normal distribution and then derive the corresponding value for X_{ij}, and

(ii) if X_{ij} is observed, we draw proposals from the conditional normal distribution, accepting the first proposal that is compatible with the category value taken by X_{ij}.

If there are three or more categorical variables, then we have the choice of the including each of them separately in the latent normal formulation, or of including them into one variable. Suppose we have three variables, with K, L, and M categories, respectively. The former approach corresponds approximately to imputation that is consistent with a partial association log-linear model. This will be adequate for most settings. The latter approach creates a KLM level categorical variable and includes this; this approximately corresponds to saturating the log-linear model (Carpenter and Kenward, 2013, p. 121).

14.3.1 Some comments

We have described a multivariate response multilevel normal model, which can be used to impute continuous (approximately joint normal) data in a multilevel dataset. In particular, model (14.2) gives a valid framework for imputing missing values at all levels of the hierarchy. This approach extends, via the latent normal formulation, to binary, ordered and unordered categorical variables. We have shown how this works for missing values of discrete variables at level 1 in the hierarchy, but we can bring these ideas directly into model (14.2) to handle missing values at any level of the hierarchy. We also note that if we start from a congenial imputation model, and include auxiliary variables (of whatever type) as additional responses, then marginalising over these variables gives the congenial imputation model. This suggests that in applications it may be preferable to include fully observed auxiliary variables as additional responses, rather than covariates.

We can use the latent normal approach more widely for variables from other discrete distributions such as the Poisson. Skew continuous distributions may be handled via a Box-Cox transformation (see Goldstein, 2011, Ch. 16, for details).

In totality, these extensions enable us to impute missing data for a very wide range of variable types at any level of a data hierarchy. They can be viewed as an approximation to the general location model, in which the categorical data are modelled using the latent normal structure, resulting in a computationally tractable joint likelihood (see also Carpenter and Kenward, 2013, p. 121). There is potential for further extension, with random covariance matrices at level 1 (Carpenter and Kenward, 2013, p. 226; Yucel, 2011).

We next describe a further extension to the imputation procedure that enables imputation consistent with a substantive model that includes interactions and general functions of partially observed covariates.

14.4 Interactions and General Functions of Covariates

Existing approaches to missing data, including that described in the previous section cannot properly handle interactions among predictors, including polynomials. This is because they all assume conditional linear relationships between the variables. Data analysts have adopted different approaches. In one approach the interaction or polynomial terms are ignored when missing values are imputed and then computed by multiplying the relevant values post-imputation (cf. Section 13.6.5). Another approach is simply to treat each interaction or polynomial term as "Just Another Variable" (JAV) and include it in the imputation model as such, disregarding its deterministic relationship to other variables. A third approach (so called predictive mean matching imputation) imputes from a donor pool identified by the imputation model, thus attempting to preserve non-linear relationships by imputing missing values from comparable records with no missing data. Von Hippel (2009) criticizes the first and third approaches, and advocates JAV as the most satisfactory method. However, Seaman, Bartlett, and White (2012) show that under the missing at random (MAR) assumption JAV may perform poorly when the substantive model is a linear regression, and that it is generally unsatisfactory when the substantive model is logistic regression.

Our approach is to define the (joint) imputation model as the product of the substantive model (typically of a response conditional on covariates) and a marginal model for the covariates. We can use this approach to create imputed datasets in the usual way. However, as we make explicit below, under this approach the substantive model is explicitly used in building—effectively embedded in—the imputation model. It follows that fitting the imputation model gives estimates of the parameters in the substantive model. This means that for the substantive model used to formulate the imputation model, the two-stage process of multiple imputation, which separates imputation and fitting the substantive model, is not necessary. Nevertheless, the imputation approach will often still be preferable because:

- we can include auxiliary variables in the imputation model, to increase the plausibility of MAR and improve the prediction of missing values;
- imputation and analysis can be done separately;
- we can impute for the most general model we are entertaining, and then use the imputed data to perform model selection, and
- imputation provides a natural route for sensitivity analysis, as described in Chapter 19.

As discussed in Sections 14.2.1–14.2.3, it may often be preferable to include auxiliary variables as additional responses in a multivariate response extension of the conditional model of Y on X, rather than as additional covariates in the conditional model for Y. This is because in the former case marginalising the multivariate response imputation model over the auxiliary variables will give the substantive model.

We now describe how this approach works for multivariate normal data. As throughout this chapter, we assume the missingness mechanism is ignorable.

Consider the substantive model

$$Y_i = \alpha_0 + \alpha_1 X_i + \alpha_2 X_i^2 + e_i, \quad e_i \overset{\text{i.i.d.}}{\sim} N(0, \sigma^2_{Y|X}), \tag{14.7}$$

i.e., a non-linear regression model for Y on X, with X partially observed, and that the marginal distribution of X is normal. Then, the joint distribution of Y and X does not take a known form, but can be written as

$$f(Y, X) = f(Y \mid X) f(X). \tag{14.8}$$

From (14.7) the likelihood for record i in the sample can be written as

$$\begin{aligned} \mathcal{L}(Y, X) &= f(Y \mid X; \theta_{Y\mid X}) f(X; \theta_X) \\ &= \left[\frac{1}{\sqrt{2\pi\sigma^2_{Y\mid X}}} \exp\left\{ -\frac{(Y_i - (\alpha_0 + \alpha_1 X_i + \alpha_2 X_i^2))^2}{2\sigma^2_{Y\mid X}} \right\} \right] \times \\ & \quad \left[\frac{1}{\sqrt{2\pi\sigma^2_X}} \exp\left\{ -\frac{(X - \mu_X)^2}{2\sigma^2_X} \right\} \right], \end{aligned} \tag{14.9}$$

where $(\theta_{Y\mid X}, \theta_X)$ are parameter vectors for the regression of Y on X and the marginal distribution of X, respectively. Notice that if α_2 were zero, then the joint distribution of (Y, X) would be bivariate normal and we could adopt the approach set out in Section 14.2. The advantage of formulation (14.8) comes into play when the substantive model $f(Y \mid X)$ involves a non-linear function of the covariates, possibly including interactions. Such relationships are not captured in imputation model (14.2).

We now sketch the steps for an MCMC algorithm to fit (14.8) and create imputed datasets:

1. Choose initial values for all the parameters and all the missing observations.
2. Conditional on X, use an appropriate MCMC algorithm to update the parameters $\theta_{Y\mid X}$; then draw (impute) missing values of Y.
3. For each missing value of the covariate X in turn:
   (a) Draw a proposal $X^\star$ from a suitable symmetric proposal distribution, say $N(X_{\text{current}}, c\sigma^2_X)$, for a suitable choice of constant c, typically 0.5.
   (b) Use a Metropolis-Hastings step, accepting $X^\star$ with probability
   $$\min\left\{1, \frac{\mathcal{L}(X^\star)}{\mathcal{L}(X)}\right\},$$
   where $\mathcal{L}$ is given by $\mathcal{L}(X) = f(Y \mid X, \theta_{Y\mid X}) f(X \mid \theta_X)$ which in our example (14.7) takes the form (14.9).
4. Update the parameters θ_X given current values of X.

As with the MCMC algorithms for imputation described above, steps 2–4 are repeated a number of times to 'burn in' the sampler; then an imputed dataset is recorded; then the sampler is updated a smaller number of times before a second imputed dataset is recorded, and so on. Our substantive model is then fitted to each imputed dataset and the results combined using Rubin's rules in the usual way; see also Chapter 12. Alternatively, as discussed above, if the substantive model is exactly $f(Y \mid X)$ in (14.8), we may obtain posterior summaries of the parameters $\theta_{Y \mid X}$ direct from the MCMC sampler.

Once we have seen the structure for imputing consistent with a non-linear model such as (14.7) using formulation (14.8) explicitly in the MCMC algorithm above, the extensions follow naturally. Model (14.7) can be a multilevel model, whose linear predictor contains non-linear relationships involving continuous and discrete covariates. The marginal model for these covariates can then be a itself a multilevel multivariate model, of the sort described in Section 14.2, with a mix of responses at different levels of the hierarchy. The practical utility of this approach has been demonstrated by Goldstein, Carpenter, and Browne (2014). They illustrate using simulation that, in line with theory, this approach produces unbiased efficient estimators.

We next show how the approach we have outlined in Sections 14.2–14.3 extends to handle partially known variables in Section 14.4 and record linkage in Section 14.5, before considering examples in Section 14.6.

14.5 Coarsened Data

Measurement error is an example of data coarsening, formalised by Heitjan and Rubin (1991), where what is observed is intermediate between fully missing and fully known. Multiple imputation is an attractive approach in this setting, and has been considered by for example Ghosh-Dastidar and Schafer (2003), Yucel and Zaslavsky (2005), Raghunathan (2006), and Cole et al. (2006). Reiter (2008) gives a version of Rubin's combination rules for the situation where the analyst wishes not to combine the substantive dataset and the validation dataset in the substantive analysis. We set out various cases below, and then describe how MI applies to them.

A common setting is to have a response where some data values are known accurately but others are only known to lie within a certain range, within which the values are distributed according to a particular probability distribution. Truncation is a simple example of this. Another example arises with retrospectively collected data where the time since an event is measured, and some individuals can only provide an interval estimate. Another example, from the individual patient data meta-analysis we consider below, arises when a key prognostic variable is recorded on a continuous scale in some studies, but only on a binary or ordinal scale in others. In this setting appropriate MI for the underlying values of the coarsened data permits the underlying values to be used in regression adjustment, which may be particularly important if the variable is a key confounder.

We base our discussion on an idealized setting; more general settings can be readily developed from this. Suppose we wish to regress Y_1 on a possibly non-linear function $g(Y_2)$, but for a proportion of individuals Y_2 is missing. However, we observe Z, a coarsened version of Y_2, for all individuals. We assume that the distribution of $Y_1 \mid g(Y_2)$ is normal. We assume the coarsening process is ignorable for the analysis. We distinguish the following cases:

1. *Case I:* Z is discrete and defines the interval in which Y_2 lies; this includes truncation as a special case;
2. *Case II:* either the distribution of $Y_2 \mid Z$ is known, or the distribution of $Z \mid Y_2$ is known, or we have observed some values of Y_2 (as well as Z), and
3. *Case III:* no gold standard: all values of Y_2 are missing.

Case I:
The joint distribution of (Y_1, Y_2) is

$$f[Y_1 \mid g(Y_2)]f(Y_2).$$

We need to assume a marginal distribution $f(Y_2)$ to proceed. Typically, some values of Y_2 will be observed and these, together with contextual knowledge, can guide this choice. Once this is done, we can use the MCMC approach described in Section 14.3 above to fit the model (consistent with possibly non-linear function g). As the MCMC algorithm to fit the model updates, a proposal for missing Y_{i2} is only accepted if it lies in the range of values consistent with the observed Z_{i2}. In this way the MCMC algorithm gives imputed datasets with no missing values of Y_{i2}. The substantive model can then be fitted to the imputed datasets in the usual way and the results combined for final inference using Rubin's rules. If g is linear, we can use the bivariate normal imputation model

$$\begin{aligned} Y_{i1} &= \beta_{01} + e_{i1} \\ Y_{i2} &= \beta_{02} + e_{i2} \\ \begin{pmatrix} e_{i1} \\ e_{i2} \end{pmatrix} &\sim N(0, \boldsymbol{\Omega}_e). \end{aligned}$$

As the MCMC algorithm to fit the model updates, a proposal for missing Y_{i2} is only accepted if it lies in the range of values consistent with the observed Z_{i2}.

Case II:
In this case we need to consider the joint distribution of (Y_1, Y_2, Z). If $f(Y_2 \mid Z)$ is known, we can write this as $f(Y_1, Y_2, Z) = f[Y_1 \mid g(Y_2)]f(Y_2 \mid Z)f(Z)$. This assumes that, given $g(Y_2)$, Y_1 is independent of Z. This is reasonable, as Z is a coarsened version of Y_2. If, as we have assumed, Z is fully observed then we do not need to specify a marginal distribution for Z. We can apply the approach described in Section 14.3 directly to impute the missing Y_2, drawing on information in Y_1 and Z. If the parameters of $f(Y_2 \mid Z)$ are unknown, these can be estimated within the MCMC framework provided we have some individuals with both Y_2 and Z observed. If the form of $f(Y_2 \mid Z)$ is unknown, we can choose a model for it, consistent with the data and our contextual knowledge.

Alternatively, it may be that $f(Z \mid Y_2)$ is known, or easier to model than $f(Y_2 \mid Z)$. Then we write $f(Y_1, Y_2, Z) = f[Y_1 \mid g(Y_2)]f(Z \mid Y_2)f(Y_2)$, where we again assume that, given $g(Y_2)$, Y_1 is independent of Z. In this case we need to choose an appropriate marginal distribution for partially observed Y_2. Then we can apply the approach described in Section 14.3 directly to impute the missing Y_2, drawing on information in Y_1 and Z. If the parameters of $f(Z \mid Y_2)$ are unknown, these can be estimated as part of the MCMC process alongside parameters of the marginal distribution of Y_2. Or, if the form of $f(Z \mid Y_2)$ is unknown, provided we have some individuals with both Y_2 and Z observed we can model it, consistent with the data and our contextual knowledge. We then apply the MCMC approach described in Section 14.3 to fit the imputation model and impute the underlying Y_2 values.

Before leaving Case II, we note that there may be situations where our primary dataset does not contain Y_2, but that a secondary validation dataset contains both Y_2 and Z. In this setting we may either:

(i) Take Z from the principal dataset and Y_2 and Z from the validation dataset. Impute Y_2 using only Z, creating K imputed vectors for Y_2. Use these to fit the substantive model in each of the K principal datasets and combine the results for inference using Rubin's rules, or

(ii) Put the principal and validation datasets together, impute missing values of all variables (Y_1 is missing in the validation portion, Y_2 in the principal portion), fit the substantive model to the imputed datasets and combine the results for inference using Rubin's rules.

Under option (i), the imputation model is uncongenial with the substantive model, as it does not contain all the variables and structure in the substantive model. In this case, inference from Rubin's rules may be unreliable. By contrast, under option (ii), provided we fit the substantive model to the combined, imputed, principal and validation data, Rubin's rules should be applicable. If we use option (ii), but only wish to fit the substantive model to the principal dataset (excluding the validation dataset), then we need to use a modification of Rubin's rules proposed by Reiter (2008).

Case III:
In this case, we can only recover an estimate of the underlying, un-coarsened observations by hypothesising a model relating the underlying and the observed, coarsened, values. We now discuss a simple version of such a model and its implication for MI. We suppose that both Y_2 and Y_3 are measured with error about a true underlying value. In order to proceed we need to choose a model for this measurement error process; we consider a simple linear model for illustration. Our joint substantive and measurement error model is:

$$\begin{aligned} Y_{i1} &= \beta_0 + \beta_1(\gamma_0 + u_i) + e_{i1}, \\ Y_{ij} &= \gamma_0 + u_i + e_{ij}, \quad \text{for } j = 2, 3, \end{aligned}$$

where $u_i \sim N(0, \sigma_u^2)$, independent of $(e_{i2}, e_{i3})^T \sim N(0, \mathbf{\Omega}_e)$, typically independent of $e_{i1} \sim N(0, \sigma^2)$.

For subject i, the substantive model is a linear regression of Y_{i1} on the estimated underlying value of the covariate for subject i, which is $(\gamma_0 + u_i)$. Of course, γ_0 may be omitted from the model for Y_{i1} and absorbed into the intercept. This model is again a special case of the multilevel multiple imputation model presented in Section 14.2 above. As above, since the substantive model is explicitly specified within the imputation model, imputation is not necessary here. However, in practice we would typically wish to include more variables in the imputation model than the substantive model. In this case, multiple imputation would be a natural way to proceed.

14.6 Record Linkage

Consider the case of linking two data files, a primary file of interest (FOI) and a secondary, typically larger, linking data file (LDF). We assume that for modelling purposes we wish to carry variables from 'linked' records from the LDF to the FOI. We outline how multiple imputation can be used in this setting, where it has the potential to improve on standard approaches to probabilistic record linkage.

Assume we have a set of matching variables (MV). For many records in the FOI these will identify a unique match in the LDF. For the remainder, standard probabilistic record linkage uses the MV to determine the probability that record i in the FOI matches record j in the LDF, for $j = 1, \ldots, M(i)$. Here $M(i)$ is the number of LDF records entertained as possible links for record i in the FOI. For each individual i in the FOI with pattern g of the MV, a probabilistic matching procedure (see, for example, Jaro, 1995) typically calculates (often using data from other sources)

$$m_{ij} = \Pr(g \mid \text{FOI record } i \text{ matches LDF record } j)$$

and

$$\bar{m}_{ij} = \Pr(g \mid \text{ FOI record } i \text{ does not match LDF record } j),$$

for $j = 1, \ldots, M(i)$. It then calculates

$$R_{ij} = \frac{m_{ij}}{\bar{m}_{ij}},$$

and derives a weight $W_{ij} = \log_2 R_{ij}$. The traditional method sets a threshold for linkage; for each record i where any W_{ij} exceeds the threshold, the LDF entry with the highest W_{ij} is linked.

In most cases, the MV will be present in both the FOI and the LDF; however, for confidentiality reasons it can also happen that the MV are withheld from the analyst, who may only have the weights. In both cases we may improve on this procedure, provided (as will typically be the case) we have additional variables in the LDF that are not in the MV set. The idea is to normalise the weights W_{ij} to obtain a prior, and then use the distribution of the LDF variables given the FOI variables (estimated from the true matches) to provide a likelihood, which in turn allows us to compute posterior weights. A multiple imputation approach then allows us to propagate the uncertainty.

For individuals i with definite matches, let Y_i^{LDF} denote the variables in the LDF (excluding the MV in the linkage file), and Y_i^{FOI} denote the variables in the FOI (including the matching variables). We proceed as follows:

1. For records with definite linkage, use the multivariate response model introduced in Section 14.2, fitted by MCMC, to estimate $f(Y_i^{\text{LDF}} \mid Y_i^{\text{FOI}})$;

2. For each record i that is not definitively linked, derive the weights as above and normalize them, setting:

$$p_{ij} = \frac{W_{ij}}{\sum_{j=1}^{M(i)} W_{ij}},$$

   and

3. For each record i that is not definitively linked, calculate

$$\gamma_{ij} = f(Y_j^{\text{LDF}} \mid Y_i^{\text{FOI}}) p_{ij},$$

   and normalise, setting

$$\pi_{ij} = \frac{\gamma_{ij}}{\sum_{j=1}^{M(i)} \gamma_{ij}}.$$

We can then use the π_{ij} to generate imputed dataset $k = 1, \ldots, K$, as follows. For each record i that is not definitively linked, sample a record from the set of $M(i)$ possible links with probability π_{ij}.

Taking this approach, we obtain K imputed datasets which represent the uncertainty in the matching. Alternatively, we may threshold the π_{ij} and directly impute unlinked data for records i for which all π_{ij} lie below the threshold.

When adopting this approach, we need to be careful not to use the information in the MVs twice; once in the prior and once in the likelihood. Since by definition we exclude the MV in the LDF from Y_i^{LDF}, this is reasonable if

$$f(Y_i^{\text{LDF}}, MV_i \,|\, Y_i^{\text{FOI}}) f(Y_i^{\text{LDF}} \,|\, Y_i^{\text{FOI}}) f(MV_i^{\text{LDF}} \,|\, Y_i^{\text{FOI}}).$$

In other words, the MV and the other variables in the LDF are approximately conditionally independent given Y_i^{FOI}. If this is not the case, it will often be preferable to discard the prior p_{ij} and simply include the matching variables in Y_i^{LDF}.

Goldstein, Harron, and Wade (2012) describe this approach in detail, and demonstrate in a simulation study how (i) the approach outlined above can improve on probabilistic matching alone and (ii) that, direct imputation of unmatched records will often outperform both probabilistic matching and the approach described above. Linkage again illustrates attraction of MI, which provides a practical method for analysis that appropriately propagates the uncertainty and unifies the approach to linkage and missing data.

14.7 Applications

Above, we have described how multiple imputation provides a flexible, framework for missing data in multilevel analyses, where the substantive model may include a non-linear function of the covariates, where data may be coarsened, and where record linkage may be uncertain. We now illustrate some of these issues with the analysis of data.

14.7.1 Modelling class size data with missing values

We use data from a study set up to explore the effect of class size on learning among children entering primary school in England. For details of the study see Blatchford et al. (2002). Literacy and numeracy skills were measured when children started their reception class and at the end of their reception year, and the child's class size is the average size of the class the child belonged to over the reception year (in many schools about one half of the reception class starts in September, and the remaining younger children in the same school year join the class after Christmas). The literacy and numeracy test scores (i) before entry into the reception class and (ii) at the end of the reception year have been normalized to create the variables used in the analyses below.

We model the relationship between our chosen outcome, mathematics score at the end of the reception year (the first year of school), against pre-reception mathematics x_1, pre-reception literacy x_2, term of entry to school (0 if first term, 1 otherwise, x_3), eligibility for free school meals (1 if yes, 0 otherwise, x_4), gender (girl=1, boy=0, x_5), age in months (centred at 51 months, x_6) class size (centred at 25 children, x_7) together with the interaction of pre-reception mathematics and literacy scores ($x_1 x_2$), the square of pre-reception mathematics

(x_1^2), the square of pre-reception literacy (x_2^2), the square of class size (centred at 25) (x_7^2), and a regression spline term, defined as follows:

$$zx_7^2, \text{ where } z = \begin{cases} 0 \text{ if } x_7 < 0 \\ 1 \text{ if } x_7 \geq 0 \end{cases}.$$

This adds a smoothly joining quadratic term at a class size of 25 to the overall quadratic term for class size.

As we noted above, each child will generally experience a different average class size over the year due to other children entering at different times of year so that class size becomes a level 1 variable.

Let j index school and i index children within schools. The substantive model is

$$\begin{aligned} Y_{ij} = \beta_0 + \sum_{k=1}^{7} \beta_k x_{kij} + \beta_8 x_{1ij} x_{2ij} + \beta_9 x_{2ij}^2 + \beta_1 0 x_{1ij}^2 + \beta_{11} x_{7ij}^2 + \\ \beta_{12} z_{ij} x_{7ij}^2 + u_j + e_{ij}, \\ u_j \sim N(0, \sigma_u^2) \text{ independent of } e_{ij} \sim N(0, \sigma_e^2). \end{aligned} \tag{14.10}$$

In the illustrative dataset we have extracted from the class size study for the analysis here, only class size is missing.

We impute the missing class size data, consistent with the multilevel structure and the non-linear relationship in (14.10), using the approach we described in Section 14.3. Auxiliary variables can be included as additional responses, or—if they are fully observed—as additional covariates. As we remarked in Section 14.3, in the case where there are no auxiliary variables, and we impute with precisely those covariates which are in our final substantive model, then the imputation process gives posterior distributions for the parameters of interest in the substantive model. Thus we do not need to go through the full procedure of generating imputed datasets, fitting the substantive model to each, and combining the results using Rubin's rules. In this illustrative analysis of the class size data we are in exactly this situation, since we include no auxiliary variables, and impute with the variables shown in (14.10).

The approach set out in Section 14.3 requires us to choose a marginal distribution for the partially observed covariate(s) (here class size) in our substantive/imputation model (14.10). We chose a normal distribution for class size, even though it is slightly positively skewed. Both the complete records analysis and the full Bayesian model are fitted using MCMC, with a burn in of 500 updates and 2500 updates used to estimate the posterior mean and standard error of the parameters. Uniform independent default priors were used for the fixed parameters, inverse gamma (0.001, 0.001) priors for the variances and minimally informative inverse Wishart priors for the covariance matrices (Browne and Draper, 2000). Table 14.1 shows the parameter estimates. The main differences between the multiple imputation estimates and those obtained by restricting analysis to the subset of complete records lie in the smaller standard errors associated with the former, up to 25% in some cases, reflecting the additional information brought into the analysis from the partially observed records. Nevertheless, we also see a much smaller coefficient for free school meals, the coefficient for the interaction between pre-reception mathematics and literacy scores changes sign and becomes formally significant at the 5% level, and the coefficient of the quadratic term for pre-reception mathematics changes sign and becomes not significant.

Figure 14.1 plots the estimated relationship between (adjusted) maths score and class size from both the complete records analysis and the analysis including all the observed data.

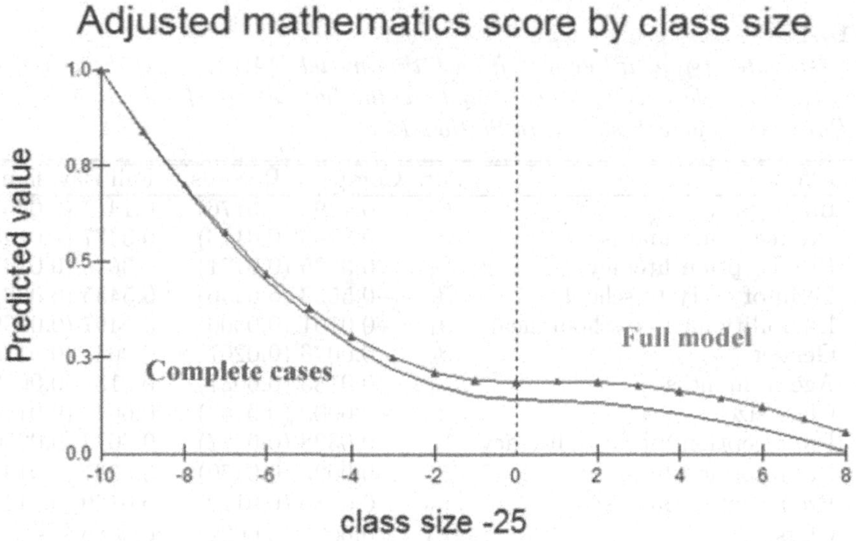

FIGURE 14.1
Adjusted relationship between mathematics score at the end of the reception year and class size, estimated using model (14.10). *All other covariates have been set at their 'reference' level, and the effect of class size age 15 set to 1.0.*

To calculate this relationship, all categorical covariate values are set to their reference value (zero), normalised pre-reception literacy and numeracy are also set at zero, and the effect at a class size of 15 set to 1.0. Both analyses show a similar relationship, though including all the partially observed records suggests that the effect of class size is slightly reduced at the right-hand end. The between child residual variance for the full model is 0.6^2 so that the reduction in mathematics score between a class size of 15 and class sizes of between 23 and 30 is approximately equal to the residual standard deviation between children; a non-trivial effect.

In summary, we have used the approach set out in Section 14.3 to impute consistent with the highly non-linear relationship with class size in this multilevel dataset. We are therefore able to make full use of the observed data to confirm this practically important relationship, which standard imputation procedures cannot do. A simulation study illustrating this approach, together with an analysis of predictors of educational attainment in the UK 1958 birth cohort study are reported by Goldstein et al. (2014). A version of this imputation algorithm using full conditional specification (suitable when there is no multilevel structure) is described by Bartlett et al. (2014).

14.7.2 Individual patient data meta-analysis

Our second example illustrates handling coarsened data, as discussed in Section 14.4. The data come from an individual patient data meta analysis of survival of patients with heart failure, known as MAGGIC (MAGGIC, 2012). Heart failure (HF) is a leading cause of cardiovascular morbidity and mortality and arises as a consequence of many cardiovascular conditions, including coronary artery disease, valve disease and hypertension. HF has been traditionally viewed as a failure of heart pump function, and the left ventricular ejection

TABLE 14.1
Parameter estimates (standard errors) from fitting model (14.10) *to (i) the subset of complete records (3791 children, 233 schools) and (ii) the full data (8041 children, 347 schools) using the Bayesian approach set out in Section 14.3.*

Effect	Par.	Complete Records	Full Bayesian
Intercept	β_0	0.1495 (0.0176)	0.1417 (0.0357)
Pre-reception maths	β_1	0.3505 (0.0176)	0.3477 (0.0145)
Pre-reception literacy	β_2	0.3725 (0.0171)	0.3640 (0.0147)
Term of entry to school	β_3	−0.5533 (0.0446)	−0.5435 (0.0381)
Eligibility for free school meals	β_4	−0.0791 (0.0306)	−0.0497 (0.0265)
Gender	β_5	−0.0076 (0.0207)	0.0041 (0.0170)
Age in months	β_6	0.0133 (0.0037)	0.0130 (0.0030)
Class size	β_7	−0.0009 (0.0193)	0.0046 (0.0156)
Pre-reception maths × literacy	β_8	0.0328 (0.0297)	−0.0672 (0.0250)
Pre-reception literacy^2	β_9	0.0025 (0.0170)	0.0261 (0.0141)
Pre-reception maths^2	β_{10}	−0.0380 (0.0172)	0.0129 (0.0140)
Class size^2	β_{11}	0.0085 (0.0029)	0.0086 (0.0024)
Spline term	β_{12}	−0.0105 (0.0055)	−0.0112 (0.0046)
Between-class variance	σ_u^2	0.2172 (0.0234)	0.2063 (0.0197)
Between-child variance	σ_e^2	0.3694 (0.0088)	0.3618 (0.0073)

fraction (LVEF) measurement (taken by echocardiogram) has been widely used to define systolic function, assess prognosis and select patients for therapeutic interventions. Twenty nine studies with 43,000 patient records formed the basis of the published analysis, which developed a prognostic model for survival in this patient population.

We present an illustrative analysis using individual patient data from three of the studies. Our substantive model is a linear regression of LVEF on age, gender, history of ischaemic heart disease, history of hypertension, ischaemic aetiology of heart failure, and a measure of cardiothoracic ratio (ratio of heart to chest diameter). This is categorised using the New York Heart Association (NYHA) ordinal scale of 1, 2, 3, and 4 (MAGGIC, 2012).

Ischaemic aetiology is entirely missing from this study, and in addition all the variables except age, gender and study have sporadically missing data. In addition, LVEF is measured as a percentage on a continuous, essentially normal, scale. However, often only a categorized value is available. Here, studies 1 and 2 only have an estimate of LVEF categorised as greater or less than 50% (and even this is missing for some patients).

We include study identification as a covariate with two dummy variables in both the imputation model and the substantive model. If we were modelling a large number of studies we may wish to incorporate these as level 2 random effects. In the imputation model, the three fully observed covariates (study, age and gender) are included as covariates, and the remaining variables are treated as responses with appropriate latent normal sampling for the categorical values, as described in Section 14.2. For the LVEF data we have both normal and binary responses (and some where both are missing). Where we have a normal response, then that is the one that is used. Where we have a binary response, we are in the first case considered in Section 14.4 on coarsened data. As described there, we sample from the appropriate normal distribution conditioning on the remaining responses, but now the sampled value is only accepted if it has a value below or above the cut point of 50% corresponding to what is observed. In other words, if we observe that a value is above 50%

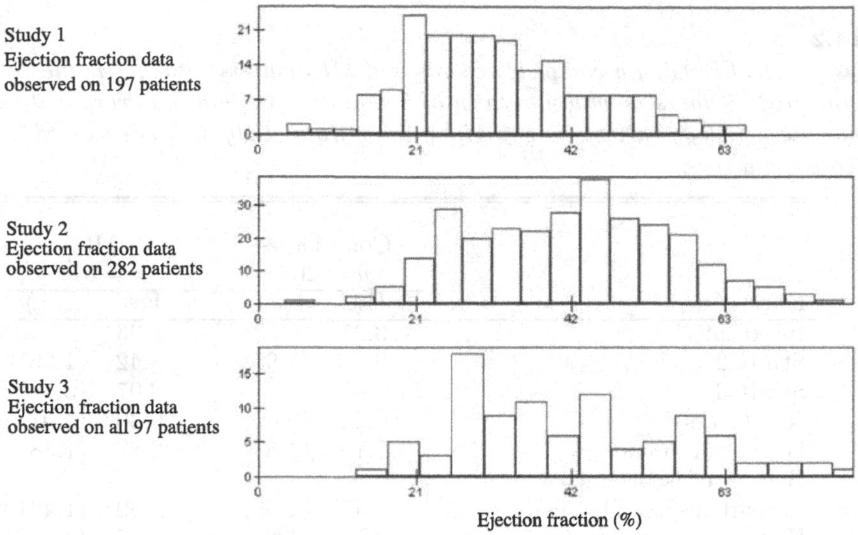

FIGURE 14.2
Results from one of the imputed datasets: Observed and imputed ejection fraction values for studies 1 and 2, compared with observed values from study 3.

this is equivalent to assigning a prior probability of 1 to being above 50% and zero for being below zero.

Figure 2 shows histograms for imputed continuous LVEF values for studies 1 and 2 from one of the imputed datasets. For reference, the LVEF values from study 3, which were fully observed, are also included. Study 1 appears to have a lower distribution of LVEF values, indicating patients with more pronounced heart failure.

Table 2 shows the results of fitting the substantive linear regression model of LVEF on the predictors, using the complete records and after multiple imputation. As study 3 has no data on aetiology it is excluded from the complete records analysis. In the imputation analysis information on this study is recovered and all the standard errors have been reduced, with the gender effect now being formally significant at the 5% level. This analysis illustrates the flexibility MI for handling coarsened data, which frequently arises in individual patient data meta-analysis, since the studies are not conducted with an a-priori agreed protocol, but instead the data is assembled after the studies have been completed.

14.8 Conclusions and Software

This chapter has extended the standard multiple imputation algorithm in several directions. We have described how a mix of continuous and discrete data can be imputed in a principled manner consistent with multilevel, or hierarchical, data. This permits valid imputation of missing data at all levels of the hierarchy. We have shown how this extends to enable multiple imputation consistent with non-linear effects and interactions. Next, we described how this can be applied to handle coarsened data, in a range of settings covering many situations that

TABLE 14.2
Meta analysis of LVEF using a complete records and MI analysis. Study 1 is the reference study. We imputed 10 datasets with a burn in of 500 and an imputation every 250 updates. As the ischaemic aetiology variable is entirely missing from study 3, it is excluded from the complete records analysis.

	Completers $n = 208$		MI $n = 576$	
Effect	Est.	(s.e.)	Est.	(s.e.)
Intercept	3.27		8.93	
Study 2	9.28	(1.95)	8.42	(1.65)
Study 3	—	—	8.07	(3.52)
Age (years)	0.34	(0.09)	0.30	(0.07)
Gender (1=male)	−2.10	(1.55)	−2.87	(1.38)
History of ischaemic heart disease (1=yes)	4.45	(1.68)	2.82	(1.49)
History of hypertension (1=yes)	0.75	(1.65)	0.63	(1.41)
Ischaemic aetiology of heart failure (1=no)	6.16	(1.54)	5.31	(1.40)
NYHA	−0.93	(0.84)	−1.05	(0.78)
Residual variance	165.8	(13.6)	179.9	(11.6)

are likely to arise in applications. We then outlined a further extension to record linkage. Together these illustrate how multiple imputation provides a unified approach to a range of apparently disparate problems. We have illustrated the practicality and utility of these developments with educational data collected to investigate the effects of class size, and an individual patient data meta-analysis of patients with heart failure.

Software implementing the approach developed here, known as REALCOM, is described in Carpenter, Goldstein and Kenward (2011) and freely available from http://www.bristol.ac.uk/cmm/software/realcom/, with a Stata interface from www.missingdata.org.uk. This approach has also been incorporated in the software Stat-JR http://www.bristol.ac.uk/cmm/.

Acknowledgments

We are grateful to Peter Blatchford for permission to use data from the class size study, and to the MAGGIC investigators for permission to use data from the MAGGIC individual patient data meta-analysis.

References

Bartlett, J.W., Seaman, S.R., White, I.R., and Carpenter, J.R. (2014). Multiple imputation of covariates by fully conditional specification: Accommodating the substantive model. *Statistical Methods in Medical Research*, in press. doi:10.1177/0962280214521348.

Blatchford, P., Goldstein, H., Martin, C., and Browne, W. (2002). A study of class size effects in English school reception year classes. *British Educational Research Journal* **28**, 169–185.

Browne, W.J. (2006). MCMC algorithms for constrained variance matrices. *Computational Statistics and Data Analysis* **50**, 1655–1677.

Browne, W.J. and Draper, D. (2000). Implementation and performance issues in the Bayesian and likelihood fitting of multilevel models. *Computational Statistics* **15** 391–420.

Carpenter, J.R., Goldstein, H., and Kenward, M.G. (2011). REALCOM-IMPUTE software for multilevel multiple imputation with mixed response types. *Journal of Statistical Software* **45**, 1–14.

Carpenter, J.R. and Kenward, M.G. (2013). *Multiple Imputation and Its Application.* Chichester: John Wiley & Sons.

Cole, S.R., Chu, H. and Greenland, S. (2006). Multiple imputation for measurement error correction. *International Journal of Epidemiology* **35**, 1074–1081.

Goldstein, H. (2011). *Multilevel Statistical Models* (4th Ed.) Chichester: John Wiley & Sons.

Goldstein, H., Carpenter, J.R., and Browne, W. (2014). Fitting multilevel multivariate models with missing data in responses and covariates that may include interactions and nonlinear terms. *Journal of the Royal Statistical Society, Series A* **177**, 553–564.

Goldstein, H., Carpenter, J.R., Kenward, M.G., and Levin, K. (2009). Multilevel models with multivariate mixed response types. *Statistical Modelling* **9**, 173–197.

Goldstein, H., Harron, K., and Wade, A. (2012). The analysis of record-linked data using multiple imputation with data value priors. *Statistics in Medicine* **31**, 3481–93.

Gosh-Dastida, B. and Schafer, J.L. (2003). Multiple edit/imputation for multivariate continuous data. *Journal of the American Statistical Association*, **98** 807–817.

Heitjan, D.F. and Rubin, D.B. (1991). Ignorability and coarse data. *The Annals of Statistics* **19**, 2244–2253.

Hughes, R.A, Sterne, J.A.C., and Tilling, K. (2014a). Comparison of imputation variance estimators. *Statistical Methods in Medical Research*, in press. doi: 10.1177/0962280214526216.

Hughes, R.A., White, I.R., Carpenter, J.R., Tilling, K., and Sterne, J.A.C. (2014b). Joint modelling rationale for chained equations imputation *BMC Medical Research Methodology* **14**, Article 28. doi: 10.1186/1471-2288-14-28.

Jaro, M. (1995). Probabilistic linkage of large public health data files. *Statistics in Medicine* **14**, 491–498.

Meta-analysis Global Group in Chronic Heart Failure (MAGGIC) (2012). The survival of patients with heart failure with preserved or reduced left ventricular ejection fraction: An individual patient data meta-analysis. *European Heart Journal* **33**, 1750–1757.

Raghunathan, T. (2006). Combining information from multiple surveys for assessing health disparities. *Allgemeines Statistisches Archiv* **95**, 933–946.

Reiter, J.P. (2008). Multiple imputation when records used to imputation are not used or disseminated for analysis. *Biometrika 95*, 933–946.

Rubin, D.B. (1987). *Multiple Imputation for Non-Response in Surveys.* Chichester: John Wiley & Sons.

Seaman, S.R., Bartlett, J.W., and White, I.R. (2012). Multiple imputation of missing covariates with non-linear effects and interactions: Evaluation of statistical methods. *BMC Methodology* **12**, Article 46.

Schafer, J.L. (1997). *Analysis of Incomplete Multivariate Data.* London: CRC/Chapman and Hall.

Von Hippel, P.T. (2009). How to impute interactions, squares and other transformed variables. *Sociological Methodology* **39**, 265–291.

Yucel, R.M. (2011). Random covariances and mixed-effect models for imputing multivariate multilevel continuous data. *Statistical Modelling* **11**, 351–370.

Yucel, R.M. and Zaslavsky, A.M. (2005). Imputation of binary treatment variables with measurement error in administrative data. *Journal of the American Statistical Association* **100**, 1123–1132.

Part V

Sensitivity Analysis

15

Sensitivity Analysis: Introduction and Overview

Geert Molenberghs

Universiteit Hasselt & KU Leuven, Belgium

Geert Verbeke

KU Leuven & Universiteit Hasselt, Belgium

Michael G. Kenward

London School of Hygiene and Tropical Medicine, London, UK

CONTENTS

15.1 Sensitivity

It has been seen in Chapters 1 and 3 how, given an MAR missing data mechanism, a valid analysis that ignores the missing value mechanism can be obtained within a likelihood or Bayesian framework, provided that mild regularity conditions hold. For simple non-likelihood-based frequentist inferences, the more stringent MCAR assumption is required. An important class of modified frequentist methods are also valid under MAR however; these include inverse probability weighted and doubly robust semi-parametric approaches (Robins, Rotnitzky, and Zhao 1994, Bang and Robins 2005) (see Chapters 8 and 9). Further, semi-parametric and other frequentist approaches can be combined with multiple imputation to yield inferences that hold under MAR (see Part IV of this volume). An increasing number of software implementations of these methods are now appearing in standard software.

One implication of this is that there exists a wide range of potential analyses that can be used under the MAR assumption. This does not however remove the issues that surround this key assumption. In practice, one can never fully rule out NMAR, under which explicit or implicit modeling of the missing data mechanism is required to provide valid inferences. But the data alone do not provide all the information necessary for this (Jansen et al. 2006). Different NMAR models can provide identical fits to the observed data, but have quite different implications for the unobserved data, and so lead to different conclusions (Molenberghs et al. 2008). It follows that the conclusions drawn can depend critically on assumptions that cannot be assessed from the data at hand: without additional information one can only distinguish between such models using their fit to the observed data, making entire classes of models indistinguishable from each other. In other words, goodness-of-fit tools alone can never provide the adequate means of choosing between such models and so, in turn, of choosing between the statistical plausibility of the different potential conclusions. We refer to this as *sensitivity* to the so-called predictive distribution, i.e., the conditional distribution of the unobserved data, given what is observed. An important consequence is that one cannot definitively distinguish between MAR and NMAR models (Molenberghs et al. 2008).

15.2 Sensitivity Analysis

The issues described above, which surround the necessary ambiguity in modeling with missing data, have led to a broad consensus that sensitivity analysis is an important component of such modeling procedures. We informally define a sensitivity analysis as one in which several statistical models are considered simultaneously and/or where a statistical model is further investigated using specialized tools, such as diagnostic measures. These alternative models may be incorporated in a variety of ways, for example in terms of different hypothesized data generating mechanisms. This admittedly loose and general definition leaves room for a wide variety of potential approaches. The simplest procedure is to fit a selected number of (NMAR) models, all of which are deemed plausible. Alternatively, a preferred (primary) analysis can be supplemented with a range of modified versions. The degree to which conclusions (inferences) are stable across such ranges provides an indication of the confidence that can be placed in them. Or, in a so-called tipping point analysis, a boundary can be identified in a key quantitative assumption beyond which the conclusions change in some substantively significant way. Modifications to a basic model can be conceived in various ways. An obvious strategy is to consider various dependencies of the missing data process on the outcomes and/or on covariates. Also, one can choose to supplement a selection-model-based analysis with one or several in the pattern-mixture modeling framework. Yet another strategy is to alter the distributional assumptions on which the model rests. It is impossible, and probably not even desirable, to attempt to provide a definitive overview of all possible sensitivity analyzes for missing datasettings. Research in this area is very active and relatively recent. Already the alternative proposals are wide-ranging, and it is very likely that others will appear in the not too distant future, some perhaps very different to the tools we have now.

15.3 Sensitivity Analysis for Parametric Models

Early references to sensitivity analyses in a fully parametric setting include Nordheim (1984), Little (1994), Rubin (1994), Laird (1994), Vach and Blettner (1995), Fitzmaurice, Molenberghs, and Lipsitz (1995), Molenberghs et al. (1999). Many of these are *ad hoc*, but nonetheless useful, approaches. Whereas such informal sensitivity analyses are an indispensable step in the analysis of incomplete data, it is worthwhile to have more formal frameworks within which to develop such analyses. A fairly clear distinction exists between methods that use fully parameterized models, broadly within the selection, pattern-mixture, and shared-parameter frameworks introduced in Chapter 4, and those that use semi-parametric or restricted moment approaches, as developed in Chapter 8. In both approaches the dominant theme for sensitivity analysis is the representation of departures from MAR. In the semi-parametric setting, distributional assumptions about the observed data are minimized as far as is practical, which has some implications for the implicit representation of MAR, but both approaches share the requirement for careful construction of NMAR model features that cannot be justified from the data alone. In some examples, e.g., Scharfstein, Rotnitzky, and Robins (1999) and Daniels and Hogan (2008), there is an absolute separation between substantive model parameters that are wholly estimable from the observed data, and carefully defined sensitivity parameters about which the data tell us nothing. In other examples, e.g., Kenward (1998), this distinction is less clear-cut.

There are two main approaches to the construction of fully parametric NMAR models, and these mirror the two ways in which departures from MAR can be expressed. In the first, the departures are expressed through the missing data mechanism itself. Parameters can be included in a model for the mechanism that link the probability of missingness to the values taken by potentially unobserved variables. Selection models provide the natural framework for this. Sensitivity is explored through the variation of these parameters that govern the degree of deviation from MAR. Such methods are described in Chapter 16. By contrast, NMAR can be expressed indirectly, or implicitly, through the statistical behavior of the potentially unobserved data. Under MAR, the conditional behavior of these data given the always observed data is the same whether missing or not. By introducing differences in these conditional distributions between the missing and observed portions, a NMAR mechanism is created. The natural setting for this is the pattern-mixture model (PMM). In addition, some authors have considered PMM as a useful counterpoint to selection models to either answer the same scientific question, such as marginal treatment effect or time evolution, based on these two rather different modeling strategies, or to gain additional insight by supplementing the selection model results with those from a pattern-mixture approach. Again, PMM-based sensitivity analyses are explored in Chapter 16, and under a Bayesian paradigm in Chapter 18. Although not strictly required, multiple imputation (MI) has proved to be a convenient tool for constructing sensitivity analysis based on pattern-mixture models, as described by Thijs et al. (2002) and Kenward, Molenberghs, and Thijs (2003) for example. For this reason we also meet such methods in the MI context in Chapter 19.

A complementary approach to this explicit modelling of potential NMAR mechanisms, is to quantify the impact of one or a few observations on the substantive conclusions in addition to those related to the missing data mechanism, e.g., Verbeke et al. (2001), Copas and Li (1997), Troxel, Harrington, and Lipsitz (1998). Such methods are considered in Section 16.5.

A third parametric framework met in Chapter 4 is that of the shared-parameter model (SPM). By construction, these typically correspond to NMAR mechanisms because of the presence of latent, unobserved, variables that link the outcomes to the missing value mechanism. They have a long history of use within the broader framework of structural equation models (e.g., Muthén, Kaplan, and Hollis 1987), for which they represent a natural component, and are particularly convenient for handling non-monotone missingness. Nevertheless, these models are based on very strong parametric assumptions, such as normality of the shared random effects, which exacerbates the potential sensitivities of NMAR models to underlying assumptions. An overview is given by Tsonaka, Verbeke, and Lesaffre (2009), who consider shared parameter models without any parametric assumptions for the shared parameters. A theoretical assessment of the sensitivity with respect to these parametric assumptions is presented in Rizopoulos, Verbeke, and Molenberghs (2007). Finally Beunckens et al. (2007) proposed a so-called *latent-class mixture model*, bringing together features of the SeM, PMM, and SPM frameworks. SEM and latent-class mixture models are discussed in Chapter 16.

15.4 Sensitivity Analysis in a Semi-Parametric Setting

Within the semi-parametric paradigm, weighted generalized estimating equations (WGEE), proposed by Robins, Rotnitzky, and Zhao (1994) and Robins and Rotnitzky (1995) play a central role. Rather than jointly modeling the outcome and missingness processes, the centerpiece is inverse probability weighting of a subject's contribution, where the weights are specified in terms of factors influencing missingness, such as covariates and observed outcomes. These concepts are developed in Robins, Rotnitzky and Scharfstein (1998) and Scharfstein, Rotnitzky, and Robins (1999). Robins, Rotnitzky, and Scharfstein (2000) and Rotnitzky et al. (2001) use this modeling framework to conduct sensitivity analysis. They allow for the dropout mechanism to depend on potentially unobserved outcomes through the specification of a non-identifiable sensitivity parameter. An important special case for such a sensitivity parameter, $\boldsymbol{\tau}$ say, is $\boldsymbol{\tau} = \mathbf{0}$, which the authors term explainable censoring; it is essentially a sequential version of MAR. Conditional upon $\boldsymbol{\tau}$, key parameters, such as treatment effect, are identifiable. By varying $\boldsymbol{\tau}$, sensitivity can be assessed. As such, there is similarity between this approach and the interval of ignorance concept (Section 16.5). There is a connection with pattern-mixture models too, in the sense that, for subjects with the same observed history until a given time $t-1$, the distribution for those who drop at t for a given cause is related to the distribution of subjects who remain on study at time t. Sensitivity analysis in a semi-parametric context is discussed in Chapter 17.

As argued at the beginning of this chapter, in the absence of at least some of the data that one planned to collect, it is even more vital than otherwise to use as much substantive knowledge as possible. The use of expert knowledge is discussed in Chapter 20.

References

Bang, H. and Robins, J.M. (2006). Double-robust estimation in missing data and causal inference models. *Biometrics* **61**, 962–973.

Beunckens, C., Molenberghs, G., Verbeke, G., and Mallinckrodt, C. (2007). A latent-class mixture model for incomplete longitudinal Gaussian data. *Biometrics* **63**, 96–105.

Copas, J.B. and Li, H.G. (1997). Inference from non-random samples (with discussion). *Journal of the Royal Statistical Society, Series B* **59**, 55–96.

Daniels, M. J. and Hogan, J. W. (2008) *Missing Data in Longitudinal Studies: Strategies for Bayesian Modeling and Sensitivity Analysis.* Chapman & Hall/CRC.

Fitzmaurice, G.M., Molenberghs, G., and Lipsitz, S.R. (1995). Regression models for longitudinal binary responses with informative dropouts. *Journal of the Royal Statistical Society, Series B* **57**, 691–704.

Jansen, I., Hens, N., Molenberghs, G., Aerts, M., Verbeke, G., and Kenward, M.G. (2006). The nature of sensitivity in missing not at random models. *Computational Statistics and Data Analysis* **50**, 830–858.

Kenward, M.G. (1998). Selection models for repeated measurements with nonrandom dropout: An illustration of sensitivity. *Statistics in Medicine* **17**, 2723–2732.

Kenward, M.G., Molenberghs, G., and Thijs, H. (2003). Pattern-mixture models with proper time dependence. *Biometrika* **90**, 53–71.

Laird, N.M. (1994). Discussion of Diggle, P.J. and Kenward, M.G.: "Informative dropout in longitudinal data analysis." *Applied Statistics* **43**, 84.

Little, R.J.A. (1994). Discussion of Diggle, P.J. and Kenward, M.G.: "Informative dropout in longitudinal data analysis." *Applied Statistics* **43**, 78.

Molenberghs, G., Beunckens, C., Sotto, C., and Kenward, M.G. (2008). Every missing not at random model has got a missing at random counterpart with equal fit. *Journal of the Royal Statistical Society, Series B* **70**, 371–388.

Molenberghs, G., Goetghebeur, E.J.T., Lipsitz, S.R., Kenward, M.G. (1999). Non-random missingness in categorical data: Strengths and limitations. *The American Statistician* **53**, 110–118.

Muthén, B., Kaplan, D., and Hollis, M. (1987). On structural equation modeling with data that are not missing completely at random. *Psychometrika* **52**, 431–462.

Nordheim, E.V. (1984). Inference from nonrandomly missing categorical data: An example from a genetic study on Turner's syndrome. *Journal of the American Statistical Association* **79**, 772–780.

Rizopoulos D., Verbeke G., and Molenberghs G. (2007). Shared parameter models under random-effects misspecification. *Biometrika* **94**, 000–000.

Robins, J.M. and Rotnitzky, A. (1995). Semiparametric efficiency in multivariate regression models with missing data. *Journal of the American Statistical Association* **90**, 122–129.

Robins, J.M., Rotnitzky, A., and Scharfstein, D.O. (1998). Semiparametric regression for repeated outcomes with non-ignorable non-response. *Journal of the American Statistical Association* **93**, 1321–1339.

Robins, J.M., Rotnitzky, A., and Scharfstein, D.O. (2000). Sensitivity analysis for selection bias and unmeasured confounding in missing data and causal inference models. In: *Statistical Models in Epidemiology, the Environment, and Clinical Trials*, M.E. Halloran and D.A. Berry, eds. New York: Springer, pp. 1–94.

Robins, J.M., Rotnitzky, A., and Zhao, L.P. (1994). Estimation of regression coefficients when some regressors are not always observed. *Journal of the American Statistical Association* **89**, 846–866.

Rotnitzky, A., Scharfstein, D., Su, T.L., and Robins, J.M. (2001). Methods for conducting sensitivity analysis of trials with potentially nonignorable competing causes of censoring. *Biometrics* **57**, 103–113.

Rubin, D.B. (1994). Discussion of Diggle, P.J. and Kenward, M.G.: "Informative dropout in longitudinal data analysis." *Applied Statistics* **43**, 80–82.

Scharfstein, D.O., Rotnitzky, A., and Robins, J.M. (1999). Adjusting for nonignorable drop-out using semiparametric nonresponse models (with discussion). *Journal of the American Statistical Association* **94**, 1096–1146.

Thijs, H., Molenberghs, G., Michiels, B., Verbeke, G., and Curran, D. (2002). Strategies to fit pattern-mixture models. *Biostatistics* **3**, 245–265.

Troxel, A.B., Harrington, D.P., and Lipsitz, S.R. (1998). Analysis of longitudinal data with non-ignorable non-monotone missing values. *Applied Statistics* **47**, 425–438.

Tsonaka R., Verbeke G., and Lesaffre E. (2009). A semi-parametric shared parameter model to handle non-monotone non-ignorable missingness. *Biometrics*, **65**, 81–87.

Vach, W. and Blettner, M. (1995). Logistic regression with incompletely observed categorical covariates: investigating the sensitivity against violation of the missing at random assumption. *Statistics in Medicine* **12**, 1315–1330.

Verbeke, G., Molenberghs, G., Thijs, H., Lesaffre, E., and Kenward, M.G. (2001). Sensitivity analysis for non-random dropout: A local influence approach. *Biometrics* **57**, 7–14.

16

A Likelihood-Based Perspective on Sensitivity Analysis

Geert Verbeke
KU Leuven & Hasselt University, Belgium

Geert Molenberghs
Hasselt University & KU Leuven, Belgium

Michael G. Kenward
London School of Hygiene and Tropical Medicine, London, UK

CONTENTS

16.1 Introduction

While likelihood, Bayesian, and semi-parametric methods, under the assumption of MAR, have been embraced as primary analyses for incomplete data (Little et al. 2010), all models make assumptions about the so-called predictive distribution, i.e., the distribution governing the missing data, given the observed ones. Such assumptions are by their very nature unverifiable from the data, while they may have an impact on the inferences drawn. It is therefore imperative to explore how sensitive the conclusions drawn are to the unverifiable assumptions. In this chapter, the focus is on likelihood-based methods. In subsequent chapters, sensitivity analysis in the context of semi-parametric methods (Chapter 17), Bayesian approaches (Chapter 18), and multiple imputation (Chapter 19) is discussed. Many sensitivity analyses will take the form of exploring how deviations from MAR towards NMAR change the conclusions.

We broadly define a sensitivity analysis as one in which several statistical models are considered simultaneously and/or where a statistical model is further scrutinized using specialized tools, such as diagnostic measures. This qualitative definition encompasses a wide variety of useful approaches. The simplest procedure is to fit a selected number of (NMAR) models which are all deemed plausible; alternatively, a preferred (primary) analysis can be supplemented with a number of modifications. The degree to which conclusions (inferences) are stable across such ranges provides an indication of the confidence that can be placed in them. Modifications to a basic model can be constructed in different ways. One obvious strategy is to consider various dependencies of the missing data process on the outcomes and/or on covariates. One can choose to supplement an analysis within the selection modeling framework, say, with one or several in the pattern-mixture modeling framework (Chapter 4). Also, the distributional assumptions of the models can be altered.

Early references pointing to the aforementioned sensitivities and responses to them include Nordheim (1984), Little (1994), Rubin (1994), Laird (1994), Vach and Blettner (1995), Fitzmaurice, Molenberghs, and Lipsitz (1995), Molenberghs et al. (1999), Kenward (1998), and Kenward and Molenberghs (1999). Many of these are to be considered potentially useful but *ad hoc* approaches. Whereas such informal sensitivity analyses are an indispensable step in the analysis of incomplete longitudinal data, it is desirable to have more formal frameworks within which to develop such analyses. Such frameworks can be found in Thijs, Molenberghs, and Verbeke (2000), Verbeke et al. (2001), Molenberghs et al. (2001), Molenberghs, Kenward, and Goetghebeur (2001), Kenward, Goetghebeur, and Molenberghs (2001), Van Steen et al. (2001), and Jansen et al. (2006). Furthermore, a number of authors have aimed at quantifying the impact of one or a few observations on the substantive and missing data mechanism related conclusions (Verbeke et al. 2001, Copas and Li,1997, Troxel, Harrington, and Lipsitz 1998).

In Section 16.2, four motivating datasets are introduced, while in Section 16.3, key notation and concepts are briefly reviewed. Specific issues that occur when modeling incomplete data are discussed in Section 16.4. Section 16.5 is devoted to a triple of likelihood-based sensitivity analysis tools: interval of ignorance, global influence, and local influence. A sensitivity analysis approach in the shared-parameter model setting is the topic of Section 16.6. A flexible latent-class mixture modeling framework is discussed in Section 16.7. Some further methods are briefly reviewed in Section 16.8.

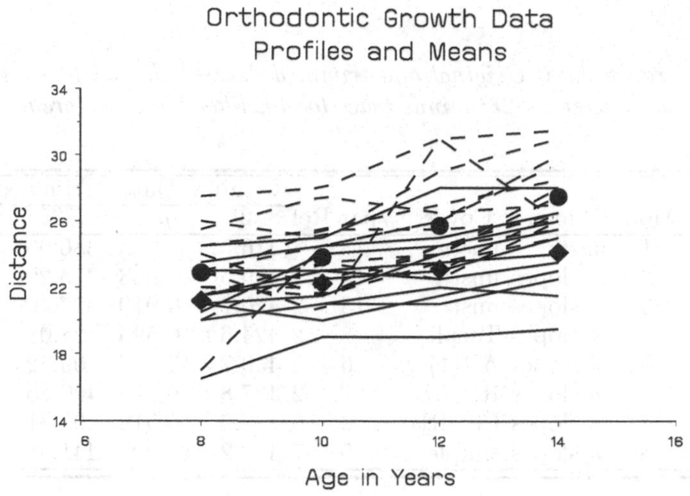

FIGURE 16.1
The Orthodontic Growth Data. Observed profiles and group by age means. Solid lines and diamonds are for girls, dashed lines and bullets are for boys.

16.2 Motivating Examples

We will introduce five examples that will be used throughout the remainder of the chapter. For the first and the second, the orthodontic growth data and the Slovenian Public Opinion Survey, initial analyses are provided here, too.

16.2.1 The orthodontic growth data

These data, introduced by Pothoff and Roy (1964), contain growth measurements for 11 girls and 16 boys. For each subject, the distance from the center of the pituitary to the pteryqomaxillary fissure was recorded at ages 8, 10, 12, and 14. The data were used by Jennrich and Schluchter (1986) to illustrate estimation methods for unbalanced data, where imbalance is now to be interpreted in the sense of an unequal number of boys and girls. Individual profiles and sex group by age means are plotted in Figure 16.1.

Little and Rubin (2002) deleted 9 of the $[(11+16)\times 4]$ observations, thereby producing 9 subjects with incomplete data, in particular, with a missing measurement at age 10. Their missingness generating mechanism was such that subjects with a low value at age 8 were more likely to have a missing value at age 10. We first focus on the analysis of the original complete dataset.

Jennrich and Schluchter (1986), Little and Rubin (2002), and Verbeke and Molenberghs (2000) each fitted the same eight models, which can be expressed within the general linear mixed models family (Verbeke and Molenberghs 2000):

$$\boldsymbol{Y}_i = X_i\boldsymbol{\beta} + Z_i\boldsymbol{b}_i + \boldsymbol{\varepsilon}_i, \tag{16.1}$$

where $\boldsymbol{b}_i \sim N(\boldsymbol{0}, G)$, and $\boldsymbol{\varepsilon}_i \sim N(\boldsymbol{0}, V_i)$, and $\boldsymbol{b}_i$ and $\boldsymbol{\varepsilon}_i$ are statistically independent. Here, $\boldsymbol{Y}_i$ is the (4×1) response vector, X_i is a $(4\times p)$ design matrix for the fixed effects, $\boldsymbol{\beta}$

TABLE 16.1
The orthodontic growth data. Original and trimmed dataset. Model fit summary. ('#par': number of model parameters; -2ℓ: minus twice log-likelihood; Ref: reference model for likelihood ratio test).

					Original Data		Trimmed Data	
Mod.	Mean	Covar.	#par	Ref	-2ℓ	p	-2ℓ	p
1	unstr.	unstr.	18		416.5		386.96	
2	$\neq$ slopes	unstr.	14	1	419.5	0.563	393.29	0.176
3	$=$ slopes	unstr.	13	2	426.2	0.010	397.40	0.043
4	$\neq$ slopes	Toepl.	8	2	424.6	0.523	398.03	0.577
5	$\neq$ slopes	AR(1)	6	2	440.7	0.007	409.52	0.034
6	$\neq$ slopes	RI+RS	8	2	427.8	0.215	400.45	0.306
7	$\neq$ slopes	CS (RI)	6	6	428.6	0.510	401.31	0.502
8	$\neq$ slopes	simple	5	7	478.2	<0.001	441.58	<0.001

is a vector of unknown fixed regression coefficients, Z_i is a $(4 \times q)$ design matrix for the random effects, $\boldsymbol{b}_i$ is a $(q \times 1)$ vector of normally distributed random parameters, with covariance matrix G, and $\boldsymbol{\varepsilon}_i$ is a normally distributed (4×1) random error vector, with covariance matrix Σ. Estimation and inference is traditionally obtained from likelihood principles based on the marginal distribution $\boldsymbol{Y}_i \sim N(X_i\boldsymbol{\beta}, Z_iGZ_i' + \Sigma_i)$. In our example, every subject contributes exactly four measurements at exactly the same time points. It is therefore possible to drop the subscript i from the error covariance matrix Σ_i unless, for example, sex is thought to influence the residual covariance structure. The random error $\boldsymbol{\varepsilon}_i$ encompasses both within-subject variability and serial correlation. The mean $X_i\boldsymbol{\beta}$ will be a function of age, sex, and/or the interaction between both.

Table 16.1 summarizes model fitting and comparison for the eight models originally considered by Jennrich and Schluchter (1986). Model 1 starts from an unstructured time by sex mean structure (8 parameters) and an unstructured covariance matrix (10 parameters). Successive model simplifications show that the mean profiles are best captured by non-parallel straight lines, and that the variance-covariance structure can be simplified to compound symmetry. This makes Model 7 the preferred one. Fits of the same eight models to the trimmed, incomplete, version of the dataset, as presented by Little and Rubin (2002), using direct-likelihood methods, are summarized in the same table. Also now, Model 7 is selected. These analyses are discussed at length in Verbeke and Molenberghs (1997, 2000).

16.2.2 The Slovenian Public Opinion Survey

In 1991, Slovenians voted for independence from Yugoslavia in a plebiscite. In anticipation of this result, the Slovenian government collected data on its possible outcome by inserting questions in the so-called Slovenian Public Opinion (SPO) Survey. The survey, administered to 2074 voting-age Slovenians, was conducted a month prior to the plebiscite using a three-stage sampling design (Barnett 2002). Rubin, Stern, and Vehovar (1995) studied the three fundamental questions added to the SPO and, in comparison to the plebiscite's outcome, drew conclusions about the missing data process. We will analyze the data as an ordinary contingency table with incomplete margins, ignoring sample survey features, but in line with analyses conducted by Rubin, Stern, and Vehovar (1995).

The three questions added were: (1) Are you in favor of Slovenian independence? (2) Are

TABLE 16.2
Theoretical distribution of the probability mass over full and observed cells, respectively, for a bivariate binary outcome with missingness in none, one, or both responses.

(a) Complete cells

$\pi_{11,11}$	$\pi_{11,12}$		$\pi_{10,11}$	$\pi_{10,12}$		$\pi_{01,11}$	$\pi_{01,12}$		$\pi_{00,11}$	$\pi_{00,12}$
$\pi_{11,21}$	$\pi_{11,22}$		$\pi_{10,21}$	$\pi_{10,22}$		$\pi_{01,21}$	$\pi_{01,22}$		$\pi_{00,21}$	$\pi_{00,22}$

(b) Observed cells

$\pi_{11,11}$	$\pi_{11,12}$		$\pi_{10,1+}$		$\pi_{01,+1}$	$\pi_{01,+2}$		$\pi_{00,++}$
$\pi_{11,21}$	$\pi_{11,22}$		$\pi_{10,2+}$					

you in favor of Slovenia's secession from Yugoslavia? (3) Will you attend the plebiscite? In spite of their apparent equivalence, questions (1) and (2) are different because independence had been possible in a transition from the existing federal structure to a loose, confederal, form as well, and therefore the secession question is added. Question (3) is highly relevant because the political decision was taken that not attending was treated as an effective NO to question (1). Thus, the primary estimand is the proportion θ of people that will be considered as voting YES, and in the context of the three questions, this can be defined as the fraction of people answering YES on both the independence and attendance questions (1) and (3), respectively, regardless of their response to question (2). A cross-classification of the attendance versus independence questions is presented in Table 16.3.

Clearly, the data are incomplete, hampering the straightforward estimation of θ. Missingness adds a source of uncertainty and it is useful to distinguish between two types of *statistical uncertainty*. The first one, *statistical imprecision*, stems from finite sampling. The Slovenian Public Opinion Survey included not all Slovenians but only 2074 respondents. However, even if all would have been included, there would have been residual uncertainty because some fail to report at least one answer. This second source of uncertainty, stemming from incompleteness, will be called *statistical ignorance*. Statistical imprecision is classically quantified by means of estimators (standard error and variance, confidence region,...) and properties of estimators (consistency, asymptotic distribution, efficiency,...). In order to quantify statistical ignorance, it is useful to distinguish between full and observed data.

We focus on two binary questions, such as the independence and attendance questions in the Slovenian Public Opinion Survey. The 16 theoretical full cell probabilities are as in Table 16.2a, producing 15 full data degrees of freedom. The generic expression for the cell probabilities is $\pi_{r_1 r_2, j_1 j_2} = P(R_1 = r_1, R_2 = r_2, J_1 = j_1, J_2 = j_2)$, where R_1 and R_2 denote the missingness indicators for the independence and attendance questions, respectively, while J_1 and J_2 denote the actual responses for the independence and attendance questions, respectively. Further, $r_1 = 0$ (1) if the answer to the independence question is missing (observed) and $j_1 = 1$ (2) if the answer to the independence question is yes (no). The indices r_2 and j_2 are similarly defined for the attendance question.

Similarly, the 9 observed cells are shown in Table 16.2b. Presenting cell counts instead of probabilities gives Table 16.3, which also shows the cell indexing system used in the rest of the paper.

TABLE 16.3
Observed cells for the Slovenian Public Opinion Survey, collapsed over the secession question. A simplified cell indexing system is shown in relation to the original cell indexing system. The columns refer to "independence" while the rows refer to "attendance."

Cell 1 $Z_{11,11} = 1439$	Cell 2 $Z_{11,12} = 78$		Cell 5 $Z_{10,1+} = 159$
Cell 3 $Z_{11,21} = 16$	Cell 4 $Z_{11,22} = 16$		Cell 6 $Z_{10,2+} = 32$
Cell 7 $Z_{01,+1} = 144$	Cell 8 $Z_{01,+2} = 54$		Cell 9 $Z_{00,++} = 136$

The full cell probabilities in Table 16.2a are four-way joint probabilities of the two missingness indicators and the two binary responses. Moreover, the parameter of interest, θ, i.e., the proportion of people answering YES to *both* the independence and attendance questions, is the marginal probability that the two responses are both equal to 1 (marginalized over the missingness indicators), and can be expressed as:

$$\theta = \sum_{r_1=0}^{1}\sum_{r_2=0}^{1} \pi_{r_1 r_2,11} = \sum_{r_1=0}^{1}\sum_{r_2=0}^{1} P(R_1 = r_1, R_2 = r_2, J_1 = 1, J_2 = 1).$$

It is clear from the above expression that when the data are complete, evaluation of θ would simply entail summing the upper-left cells from each the four (2×2) tables in Table 16.2a. However, when the data are incomplete, as in Table 16.2b, the relation is not straightforward as the collapsed cells have to be split in order to obtain values for the cells pertinent to the estimation of θ.

At various points in this chapter, we conduct a number of sensitivity analyses on this set of data. The data were used by Molenberghs, Kenward, and Goetghebeur (2001) to illustrate their proposed sensitivity analysis tool, the interval of ignorance. Molenberghs et al. (2007) used the data to exemplify results about the relationship between MAR and NMAR models. An overview of various analyses can be found in Molenberghs and Kenward (2007) and Beunckens et al. (2009). These authors used the models proposed by Baker, Rosenberger, and DerSimonian (1992) for the setting of two-way contingency tables, subject to missingness in either none, one, or both responses. Baker, Rosenberger, and DerSimonian (1992) conducted several analyses of the both responses. Molenberghs, Kenward, and Goetghebeur (2001) conducted several analyses of the both responses. Baker, Rosenberger, and DerSimonian (1992) conducted several analyses of the data. Their main emphasis was on determining the proportion θ of the population that would attend the plebiscite and vote for independence. Their estimates are reproduced in Table 16.4. Some will be discussed here, others we return to later in the chapter.

The pessimistic (optimistic) bounds, or non-parametric bounds, are obtained by setting all incomplete data that can be considered a yes (no), as yes (no). The complete case estimate for θ is based on the subjects answering all three questions and the available case estimate is based on the subjects answering the two questions of interest here. It is noteworthy that both of these estimates are out of bounds. This is not a mistake: it should be recalled that the political decision was made to treat a NO on the attendance question as an effective NO vote in the plebiscite. Disregarding incomplete cases ignores this aspect and thus discards available information, thereby causing the estimate to exceed the bounds. Note that the

TABLE 16.4
The Slovenian Public Opinion Survey. Some estimates of the proportion θ attending the plebiscite and voting for independence, as presented in Rubin, Stern, and Vehovar (1995) and Molenberghs, Kenward, and Goetghebeur (2001), as well as sensitivity-analysis results.

Estimation Method	Voting in Favor of Independence: $\widehat{\theta}$
Standard Analyses	
Non-parametric bounds	[0.694;0.905]
Complete cases	0.928
Available cases	0.929
MAR (2 questions)	0.892
MAR (3 questions)	0.883
NMAR	0.782
Sensitivity Analyses	
BRD1–BRD9	[0.741;0.892]
Well-fitting BRD6–BRD9	[0.741;0.867]
BRD1(MAR)–BRD9(MAR)	[0.892;0.892]
II: Model 10	[0.762;0.893]
II: Model 11	[0.766;0.883]
II: Model 12	[0.694;0.905]
Plebiscite	0.885

bounds apply to only those estimators making use of all available data, not to estimators based on subsets. Rubin, Stern, and Vehovar (1995) considered two MAR models, also reported in Table 16.4, the first one solely based on the two questions of direct interest, the second one using all three. Finally, they considered a single NMAR model, based on the assumption that missingness on a question depends on the answer to that question but not on the other questions.

The data will be analyzed further in Sections 16.4 and 16.5, where also the sensitivity analyses, reported in Table 16.4, will be discussed.

16.2.3 The rat data

The data come from a randomized experiment, designed to study the effect of the inhibition of testosterone production in rats (Department of Orthodontics of the University of Leuven) in Belgium; Verdonck et al. 1998). A total of 50 male Wistar rats have been randomized to either control or one of two treatment groups (low or high dose of the drug Decapeptyl; an inhibitor for the testosterone production). The treatment started at the age of 45 days, and measurements were taken every 10 days, with the first observation taken at the age of 50 days. Our response is a characterization of the height of the skull, taken under anesthesia. Many rats do not survive anesthesia implying that for only 22 (44%) rats all 7 designed measurements could have been taken. The investigators' impression is that dropout is independent of the measurements.

The individual profiles are shown in Figure 16.2. To linearize, we use the logarithmic transformation $t = \ln(1 + (\text{age} - 45)/10)$ for the time scale. Let y_{ij} denote the jth measurement for the ith rat, taken at $t = t_{ij}$, $j = 1, \ldots, n_i$, $i = 1, \ldots, N$. A simple statistical model,

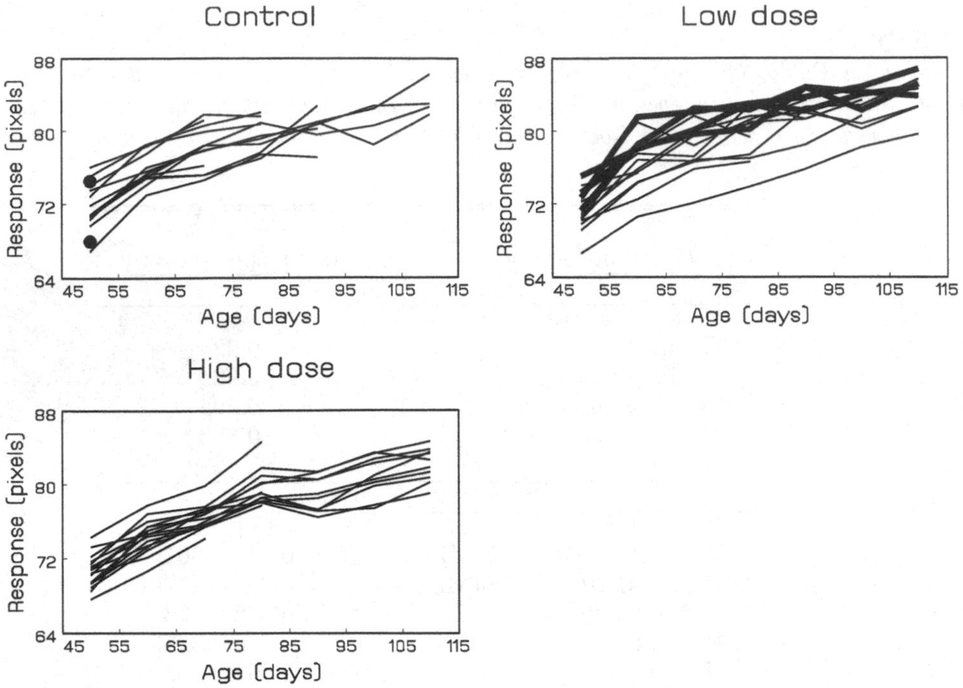

FIGURE 16.2
Rat data. Individual growth curves for the three treatment groups separately. Influential subjects are highlighted.

as considered by Verbeke et al. (2001), then assumes that y_{ij} satisfies a model of the form (16.10) with common average intercept β_0 for all three groups, average slopes β_1, β_2 and β_3 for the three treatment groups, respectively, and assuming a so-called compound symmetry covariance structure, i.e., with common variance $\sigma^2 + \tau^2$ and common covariance τ^2.

16.2.4 A clinical trial in onychomycosis

The data were obtained from a randomized, double-blind, parallel group, multi-center study for the comparison of two oral treatments (in the sequel coded as A and B) for toenail dermatophyte onychomycosis (TDO), described in full detail by De Backer et al. (1996). TDO is a common toenail infection, difficult to treat, affecting more than 2 out of 100 persons (Roberts 1992). Anti-fungal compounds, classically used for treatment of TDO, need to be taken until the whole nail has grown out healthy. The development of new such compounds, however, has reduced the treatment duration to 3 months. The aim of the present study was to compare the efficacy and safety of 12 weeks of continuous therapy with treatment A or with treatment B.

In total, 2×189 patients, distributed over 36 centers, were randomized. Subjects were followed during 12 weeks (3 months) of treatment and followed further, up to a total of 48 weeks (12 months). Measurements were taken at baseline, every month during treatment, and every 3 months afterwards, resulting in a maximum of 7 measurements per subject. At the first occasion, the treating physician indicates one of the affected toenails as the target nail, the nail which will be followed over time. We will restrict our analyses to only those

TABLE 16.5
Toenail data. Number of available repeated measurements per subject, for each treatment arm.

	Group A		Group B	
# Obs.	N	%	N	%
1	4	2.74%	1	0.68%
2	2	1.37%	1	0.68%
3	4	2.74%	3	2.03%
4	2	1.37%	4	2.70%
5	2	1.37%	8	5.41%
6	25	17.12%	14	9.46%
7	107	73.29%	117	79.05%
Total:	146	100%	148	100%

patients for which the target nail was one of the two big toenails. This reduces our sample under consideration to 146 and 148 subjects, in group A and group B, respectively.

One of the responses of interest was the unaffected nail length, measured from the nail bed to the infected part of the nail, which is always at the free end of the nail, expressed in *mm*. This outcome has been studied extensively in Verbeke and Molenberghs (2000). Owing to a variety of reasons, the outcome has been measured at all 7 scheduled time points for only 224 (76%) out of the 294 participants. Table 16.5 summarizes the number of repeated measurements available per subject, per treatment group. We observe that the occurrence of missingness is similar in both treatment groups for longer, but not for shorter sequences.

16.2.5 A depression trial

The data arise a randomized, double-blind psychiatric clinical trial in patients with depression, conducted in the United States. The primary objective of this trial was to compare the efficacy of an experimental anti-depressant with a non-experimental one. In these retrospective analyses, data from 170 patients are considered. The Hamilton Depression Rating Scale ($HAMD_{17}$) is used to measure the depression status of the patients. For each patient, a baseline assessment is available, as well as 5 post-baseline visits going from visit 4 to 8. The data are analyzed in Section 16.7.2.

16.3 Notation and Concepts

Let the random variable Y_{ij} denote the response of interest, for the ith study subject, designed to be measured at occasions t_{ij}, $i = 1, \ldots, N$, $j = 1, \ldots, n_i$. Independence across subjects is assumed. The outcomes can be grouped into a vector $\boldsymbol{Y}_i = (Y_{i1}, \ldots, Y_{in_i})'$. Define a vector of missingness indicators $\boldsymbol{R}_i = (R_{i1}, \ldots, R_{in_i})'$ with $R_{ij} = 1$ if Y_{ij} is observed and 0 otherwise. In the specific case of dropout, $\boldsymbol{R}_i$ can usefully be replaced by the dropout indicator $D_i = \sum_{j=1}^{n_i} R_{ij}$. Note that the concept of dropout refers to time-ordered variables, such as in longitudinal studies. For a complete sequence, $\boldsymbol{R}_i = \boldsymbol{1}$ and/or

$D_i = n_i$. It is customary to split the vector $\boldsymbol{Y}_i$ into observed ($\boldsymbol{Y}_i^o$) and missing ($\boldsymbol{Y}_i^m$) components, respectively.

We will make use of the missing data mechanisms (MCAR, MAR, NMAR), modeling frameworks (selection—SEM, pattern-mixture—PMM, and shared-parameter—SPM models), and of ignorability, as defined in Chapters 3 and 4 of this volume.

16.4 What Is Different When Data Are Incomplete?

Thanks to the ignorability result of the previous section, likelihood and Bayesian inferences are viable candidates for the status of primary analysis in clinical trials and a variety of other settings (Molenberghs et al. 2004, Molenberghs and Kenward 2007). Nevertheless, a number of issues arising when analyzing such incomplete data, under MAR as well as under NMAR exist and are enlisted in Section 16.4.1. Ways of addressing these are reviewed in Section 16.4.3.

16.4.1 Problems with model selection and assessment with incomplete data

In spite of the appeal of ignorability for likelihood-based analysis of incomplete data under MAR, Molenberghs, Verbeke, and Beunckens (2007) have brought forward generic issues arising when fitting models to incomplete data: (i) the classical relationship between observed and expected features is convoluted since one observes the data only partially while the model describes all data; (ii) the independence of mean and variance parameters in a (multivariate) normal is lost, implying increased sensitivity, even under MAR; (iii) also the well-known agreement between the frequentist ordinary least squares (OLS) approach and maximum likelihood estimation methods for normal models is lost, as soon as the missing data mechanism is not of the MCAR type, with related results holding in the non-normal case; (iv) in a likelihood-based context, deviances and related information criteria cannot be used in the same vein as with complete data since they provide no information about a model's prediction of the unobserved data; and, in particular, (v) several models may saturate the observed-data degrees of freedom, while providing a different fit to the complete data, i.e., they only coincide in as far as they describe the observed data; as a consequence, different inferences may result from different saturated models, where "saturation" is to be understood in terms of the observed but not the full data.

Based on these considerations, it follows that model assessment should always proceed in two steps. In the first step, the fit of a model to the *observed* data should be carefully assessed, while in the second step the sensitivity of the conclusions to the *unobserved data given the observed data* should be addressed. Gelman et al. (2005) proposed an approach to this effect, the essence of which will be described in Section 16.4.3.

In what follows, we will address some of the issues brought forward. We first present a model family needed for the analysis of the Slovenian Public Opinion survey.

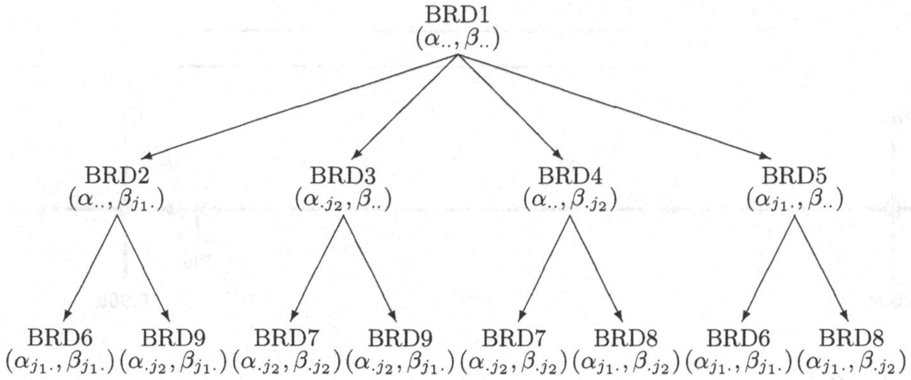

FIGURE 16.3
Graphical representation of the BRD model nesting structure.

16.4.2 The BRD family of models

To conduct further analyses of the Slovenian Public Opinion Survey, we first introduce the log-linear model family of Baker, Rosenberger, and DerSimonian (1992), based upon a four-way classification of both outcomes, together with their missingness indicators. In line with the notation in Table 16.2a, we denote the cell counts by $Z_{r_1r_2,j_1j_2}$, where $r_1, r_2 = 0, 1$ indicates, respectively, whether the measurement is missing or observed at occasions 1, 2, and $j_1, j_2 = 1, 2$ indicates the response categories for both outcomes. The models are written as:

$$\begin{array}{ll} E(Y_{11,j_1j_2}) = Y_{++,++}\pi_{11,j_1j_2} = m_{j_1j_2} & E(Y_{01,j_1j_2}) = m_{j_1j_2}\alpha_{j_1j_2} \\ E(Y_{10,j_1j_2}) = m_{j_1j_2}\beta_{j_1j_2} & E(Y_{00,j_1j_2}) = m_{j_1j_2}\alpha_{j_1j_2}\beta_{j_1j_2}\gamma. \end{array}$$

The α (β) parameters describe missingness in the independence (attendance) question as the proportion of subjects with a missing response on the independence (attendance) question relative to the proportion of subjects with both responses present, given a particular response combination (j_1, j_2). The γ parameter, on the other hand, captures the interaction between the two missingness indicators via the (conditional) odds ratio for a given response combination. The subscripts are missing from γ since Baker, Rosenberger, and DerSimonian (1992) have shown that this quantity needs to be independent of j_1 and j_2 in any identifiable model.

Baker, Rosenberger, and DerSimonian (1992) considered nine models, based on setting $\alpha_{j_1j_2}$ and $\beta_{j_1j_2}$ constant in one or more indices, and can be enumerated using the "BRD" abbreviation as displayed in Figure 16.3, together with their nesting structure. Interpretation is straightforward; for example, in BRD1, both missingness indicators do not depend on the responses, thereby characterizing an MCAR mechanism. In view, however, of the nesting structure among the different types of mechanisms described in Chapter 1, although BRD1 is more precisely classified as MCAR, it can also be considered a special case of MAR, or even NMAR. For instance, in BRD4, missingness in the first variable is constant, while missingness in the second variable depends on its value. Moreover, BRD6–BRD9 saturate the observed data degrees of freedom, while the lower numbered ones leave room for evaluating the goodness-of-fit of the model to the observed data.

Molenberghs, Kenward, and Goetghebeur (2001) and Molenberghs et al. (2007) fitted the

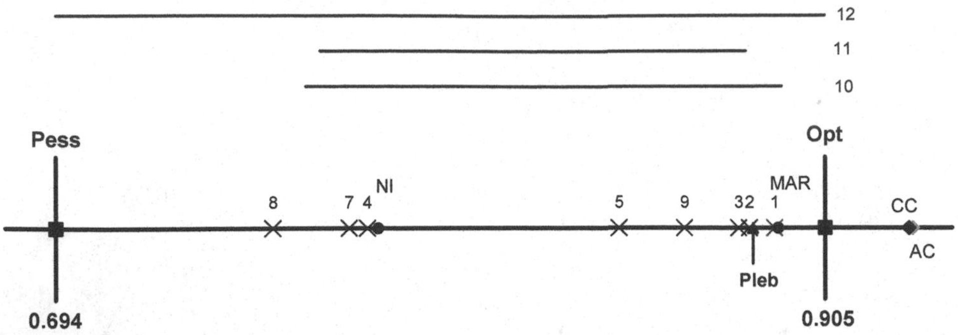

FIGURE 16.4
The Slovenian Public Opinion Survey. Relative position for the estimates of "proportion of YES votes," based on the models considered in Rubin, Stern, and Vehovar (1995) and on the BRD Models. The vertical lines indicate the nonparametric pessimistic-optimistic Bounds. (Pess(■): pessimistic boundary; Opt(■): optimistic boundary; MAR(•): Rubin et al.'s MAR model; NI(•): Rubin et al. 's NMAR model; AC(♦): available cases; CC(♦): complete cases; Pleb(▲): plebiscite outcome. Crosses (×) and the numbers above them refer to the BRD models. Intervals of ignorance (Models 10–12) are represented by horizontal bars.)

TABLE 16.6
The Slovenian Public Opinion Survey. Analysis restricted to the independence and attendance questions. Summaries on each of the Models BRD1–BRD9 are presented, with obvious column labels. The column labeled '$\widehat{\theta}_{MAR}$' refers to the model corresponding to the given one, with the same fit to the observed data, but with missing data mechanism of the MAR type.

Model	Structure	d.f.	loglik	$\widehat{\theta}$	C.I.	$\widehat{\theta}_{\text{MAR}}$
BRD1	(α, β)	6	−2495.29	0.892	[0.878;0.906]	0.8920
BRD2	(α, β_{j_1})	7	−2467.43	0.884	[0.869;0.900]	0.8915
BRD3	(α_{j_2}, β)	7	−2463.10	0.881	[0.866;0.897]	0.8915
BRD4	(α, β_{j_2})	7	−2467.43	0.765	[0.674;0.856]	0.8915
BRD5	(α_{j_1}, β)	7	−2463.10	0.844	[0.806;0.882]	0.8915
BRD6	$(\alpha_{j_1}, \beta_{j_1})$	8	−2431.06	0.819	[0.788;0.849]	0.8919
BRD7	$(\alpha_{j_2}, \beta_{j_2})$	8	−2431.06	0.764	[0.697;0.832]	0.8919
BRD8	$(\alpha_{j_1}, \beta_{j_2})$	8	−2431.06	0.741	[0.657;0.826]	0.8919
BRD9	$(\alpha_{j_2}, \beta_{j_1})$	8	−2431.06	0.867	[0.851;0.884]	0.8919

BRD models; Table 16.6 summarizes the results for the various models, with overviews presented in Table 16.4. BRD1 produces $\widehat{\theta} = 0.892$, exactly the same as the first MAR estimate obtained by Rubin, Stern, and Vehovar (1995). This does not come as a surprise, because BRD1, though MCAR, belongs to the MAR family, as does Rubin, Stern, and Vehovar's (1995) model; both use information from the two main questions. A graphical representation of the original analyses and the BRD models combined is given in Figure 16.4.

(a)		**model**	
		complete	incomplete
raw	complete		
data	incomplete		

(b)		**model**	
		complete	incomplete
raw	complete	*	
data	incomplete		

(c)		**model**	
		complete	incomplete
raw	complete		
data	incomplete	*	

(d)		**model**	
		complete	incomplete
raw	complete		
data	incomplete		*

FIGURE 16.5
Model assessment when data are incomplete. (a) Two dimensions in model (assessment) exercise when data are incomplete. (b) Ideal situation. (c) Dangerous situation, bound to happen in practice. (d) Comparison of data and model at coarsened, observable level.

16.4.3 Model selection and assessment with incomplete data

The five issues of the previous section originate from the need, when fitting models to incomplete data, to manage two aspects rather than a single one, as schematically represented in Figure 16.5: the contrast between data and model is supplemented with a second contrast between their complete and incomplete versions.

Ideally, we should like to consider Figure 16.5(b), where the comparison is entirely made at the complete level. Because the complete data are, by definition, beyond reach, it is tempting but dangerous to settle for Figure 16.5(c). This would happen when we would conclude that Model 1 fit poorly to the orthodontic growth data, as elucidated in Figure 16.6. Such a conclusion would ignore that the model fit is at the complete-data level, accounting for 16 boys and 11 girls at the age of 10, whereas the data only represent the residual 11 boys and 7 girls at the age of 10. Thus, a fair model assessment should be confined to the situations laid out in Figure 16.5(b) and (d). We will start out with the simpler (d) and then return to (b).

Assessing whether Model 1 fits the incomplete version of the growth dataset well can be achieved by comparing the observed means at the age of 10 to its prediction by the model. This implies we have to confine model fit to those children actually observed at the age of 10.

We will study the two important tasks in turn: (i) model fit to the *observed* data and (ii) sensitivity analysis. The first of these is discussed next; the latter throughout the remainder of this chapter.

As stated before, model fit to the observed data can be checked by means of either what

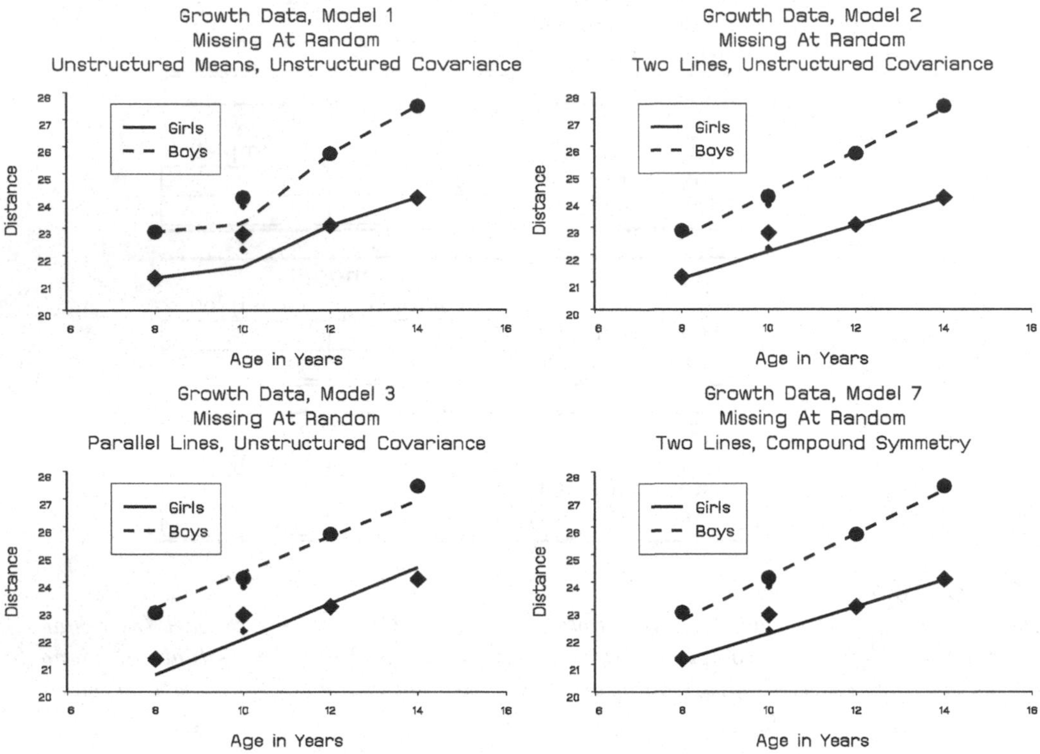

FIGURE 16.6
The orthodontic growth data. Profiles for the growth dataset, from a selected set of models. MAR analysis. (The small symbols at age 10 are the observed group means for the complete dataset.)

we will label Scenario I, as laid out in Figure 16.5(b), or by means of Scenario II of Figure 16.5(d).

Under Scenario I, we conclude BRD6–9 or their MAR counterpart fit perfectly. There is nothing wrong with such a conclusion, as long as we realize *there is more than one model* with this very same property, while at the same time they lead to different substantive conclusions.

Turning to the orthodontic growth data, considering the fit of Model 1 to the data has some interesting ramifications. When the OLS fit is considered, only valid under MCAR, one would conclude that there is a perfect fit to the observed means, also at the age of 10. The fit using ML would apparently show a discrepancy, because the observed mean refers to a reduced sample size while the fitted mean is based on the entire design.

These considerations suggest that caution is required when we consider the fit of a model to an incomplete set of data; moreover, extension and/or modification of the classical model

TABLE 16.7
The Slovenian Public Opinion Survey. Analysis restricted to the independence and attendance questions. The fit of models BRD1, BRD2, BRD7, and BRD9, and their MAR counterparts, to the observed data are shown.

Fit of BRD7, BRD7(MAR), BRD9, and BRD9(MAR) to incomplete data

1439	78	159	144	54	136
16	16	32			

Fit of BRD1 and BRD1(MAR) to incomplete data

1381.6	101.7	182.9	179.7	18.3	136.0
24.2	41.4	8.1			

Fit of BRD2 and BRD2(MAR) to incomplete data

1402.2	108.9	159.0	181.2	16.8	136.0
15.6	22.3	32.0			

assessment paradigms may be needed. In particular, it is of interest to consider assessment under Scenario II.

Gelman et al. (2005) proposed a Scenario II method. The essence of their approach is as follows. First, a model, saturated or non-saturated, is fitted to the observed data. Under the fitted model, and assuming ignorable missingness, datasets simulated from the fitted model should "look similar" to the actual data. Therefore, multiple sets of data are sampled from the fitted model, and compared to the dataset at hand. Because what one actually observes consists of, not only the actually observed outcome data, but also realizations of the missingness process, comparison with the simulated data would also require simulation from, hence full specification of, the missingness process. This added complexity is avoided by augmenting the observed outcomes with imputations drawn from the fitted model, conditional on the observed responses, and by comparing the so-obtained completed dataset with the multiple versions of simulated complete datasets. Such a comparison will usually be based on relevant summary characteristics such as time-specific averages or standard deviations. As suggested by Gelman et al. (2005), this so-called data-augmentation step could be done multiple times, along multiple-imputation ideas from Rubin (1987). However, in cases with a limited amount of missing observations, the between-imputation variability will be far less important than the variability observed between multiple simulated datasets. This is in contrast to other contexts to which the technique of Gelman et al. (2005) has been applied, e.g., situations where latent unobservable variables are treated as "missing."

16.4.4 Model assessment for the orthodontic growth data

We first apply the method to the orthodontic growth data. The first model considered assumes a saturated mean structure (as in Model 1), with a compound-symmetric covariance structure. Twenty datasets are simulated from the fitted model, and time-specific sample averages are compared to the averages obtained from augmenting the observed data based on the fitted model. The results are shown in Figure 16.7(a). The sample average at age 10, for the girls, is relatively low compared to what would be expected under the fitted model. Because the mean structure is saturated, this may indicate lack of fit of the covariance struc-

TABLE 16.8
The Slovenian Public Opinion Survey. Analysis restricted to the independence and attendance questions. The fit of models BRD1, BRD2, BRD7, and BRD9, and their MAR counterparts, to the hypothetical complete data is shown.

Fit of BRD1 and BRD1(MAR) to complete data							
1381.6	101.7	170.4	12.5	176.6	13.0	121.3	9.0
24.2	41.4	3.0	5.1	3.1	5.3	2.1	3.6
Fit of BRD2 to complete data							
1402.2	108.9	147.5	11.5	179.2	13.9	105.0	8.2
15.6	22.3	13.2	18.8	2.0	2.9	9.4	13.4
Fit of BRD2(MAR) to complete data							
1402.2	108.9	147.7	11.3	177.9	12.5	121.2	9.3
15.6	22.3	13.3	18.7	3.3	4.3	2.3	3.2
Fit of BRD7 to complete data							
1439	78	3.2	155.8	142.4	44.8	0.4	112.5
16	16	0.0	32.0	1.6	9.2	0.0	23.1
Fit of BRD9 to complete data							
1439	78	150.8	8.2	142.4	44.8	66.8	21.0
16	16	16.0	16.0	1.6	9.2	7.1	41.1
Fit of BRD7(MAR) and BRD9(MAR) to complete data							
1439	78	148.1	10.9	141.5	38.4	121.3	9.0
16	18	11.8	20.2	2.5	15.6	2.1	3.6

ture. We therefore extend the model by allowing for gender-specific covariance structures. The results under this new model are presented in Figure 16.7(b). The observed data are now less extreme compared to what is expected under the fitted model. Formal comparison of the two models, based on a likelihood ratio test, indeed rejects the first model in favor of the second one ($p = 0.0003$), with much more between-subject variability for the girls than for the boys, while the opposite is true for the within-subject variability.

16.4.5 Model assessment for the Slovenian Public Opinion Survey

Molenberghs et al. (2007) showed that, strictly speaking, the correctness of an alternative, NMAR, model can only be verified in as far as it fits the *observed* data. Thus, evidence for or against NMAR can only be provided within a particular, predefined parametric family, whose plausibility cannot be verified in empirical terms alone. This implies that an overall (omnibus) assessment of MAR *versus* NMAR is not possible, because every NMAR model can be doubled up with a uniquely defined MAR counterpart, producing exactly the same fit as the original NMAR model, in the sense that it leads to entirely the same predictions to the observed data (e.g., fitted counts in an incomplete contingency table) as the original NMAR model, and depending on exactly the same parameter vector. "Unique" means that, for a given NMAR model, there is one and only one MAR model with the specified features

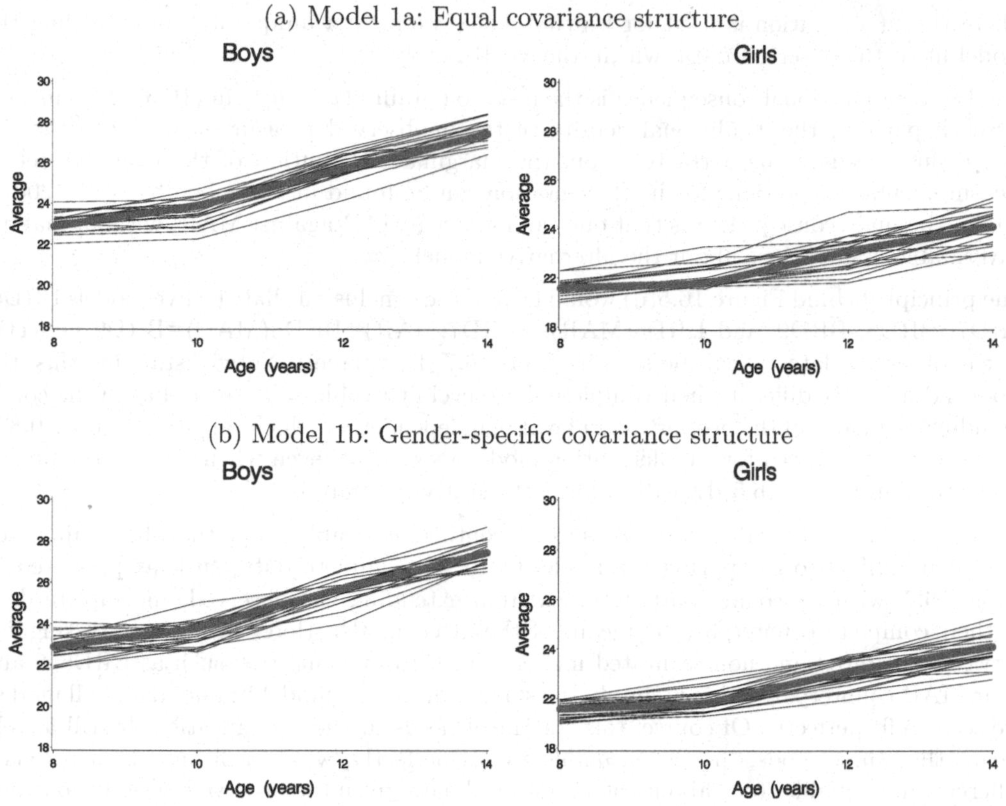

FIGURE 16.7
The orthodontic growth data. Sample averages for the augmented data (bold line type), compared to sample averages from 20 simulated datasets, based on the method of Gelman et al. (2005). Both models assume a saturated mean structure and compound symmetric covariance. Model 1a assumes the same covariance structure for boys and girls, while Model 1b allows gender-specific covariances.

corresponding to it. However, an entire class of NMAR models will share such an MAR counterpart, we have a many-to-one map (cf. related ideas in Tsiatis 1975). We construct the companion as follows: (1) an NMAR model is fitted to the data; (2) the fitted model is reformulated in a pattern-mixture model form; (3) the density or distribution of the unobserved measurements given the observed ones, and given a particular response pattern, is replaced by its MAR counterpart; and (4) it is established that such an MAR counterpart uniquely exists.

First, an NMAR model is fitted to the observed set of data. Second, the likelihood contribution for subject i, expressed in pattern-mixture form, is:

$$f(\boldsymbol{y_i}^o|\boldsymbol{r}_i, \widehat{\boldsymbol{\theta}}, \widehat{\boldsymbol{\psi}})f(\boldsymbol{r}_i|\widehat{\boldsymbol{\theta}}, \widehat{\boldsymbol{\psi}})f(\boldsymbol{y_i}^m|\boldsymbol{y_i}^o, \boldsymbol{r}_i, \widehat{\boldsymbol{\theta}}, \widehat{\boldsymbol{\psi}}). \tag{16.2}$$

Third, based on Molenberghs et al. (1998b), we use the fact that MAR within the PMM framework means: $f(\boldsymbol{y}_i^m|\boldsymbol{y}_i^o, \boldsymbol{r}_i, \boldsymbol{\theta}) = f(\boldsymbol{y}_i^m|\boldsymbol{y}_i^o, \boldsymbol{\theta})$. Hence, clearly $f(\boldsymbol{y_i}^m|\boldsymbol{y_i}^o, \boldsymbol{r}_i, \widehat{\boldsymbol{\theta}}, \widehat{\boldsymbol{\psi}})$ needs to be replaced by

$$h(\boldsymbol{y_i}^m|\boldsymbol{y_i}^o, \boldsymbol{r}_i) = h(\boldsymbol{y_i}^m|\boldsymbol{y_i}^o) = f(\boldsymbol{y_i}^m|\boldsymbol{y_i}^o, \widehat{\boldsymbol{\theta}}, \widehat{\boldsymbol{\psi}}), \tag{16.3}$$

where the $h(\cdot)$ notation is used for brevity. Thus, we have a unique way of extending the model fit to the observed data, within the MAR family.

The key computational consequence is the need to obtain $h(\boldsymbol{y}_i{}^m|\boldsymbol{y}_i{}^o)$ in (16.3). This means, for each pattern, the conditional density of the unobserved measurements given the observed ones needs to be extracted from the marginal distribution of the complete set of measurements. Suggestions for implementation can be found in Molenberghs et al. (2007). Again, the main consequence is that one cannot test NMAR against MAR, without making unverifiable assumptions about the alternative model.

The principle behind Figure 16.5(d) would lead to the conclusion that the five models BRD6, BRD7, BRD8, BRD9, and BRD6(MAR)≡ BRD7(MAR)≡BRD8(MAR)≡BRD9 perfectly fit the observed data, as can be seen in Table 16.7 (first panel). Notwithstanding this, the models drastically differ in their complete-data level fit (Table 16.8) as well as in the corresponding estimates of the proportion in favor of independence, which ranges over [0.74; 0.89]. This points to the need for supplementing model assessment, even when done in the preferable situation Figure 16.5(d), with a form of sensitivity analysis.

We now turn to the SPO data. In such a contingency table case, the above approach can be simplified to comparing the model fit to the complete data, such as presented in Table 16.7, with its counterpart obtained from extending the observed, incomplete data to their complete counterpart by means of the fitted model. Here, we have to distinguish between saturated and non-saturated models. For saturated models, such as BRD6–9 and their MAR counterparts, this is simply the same table as the model fit and again, all models are seen to fit perfectly. Of course, this statement needs further qualification. It still merely means that these models fit the *incomplete* data perfectly, while each one of them tells a different, unverifiable story about the unobserved data given the observed ones. In contrast, for the non-saturated models, such as BRD1–5 and their MAR counterparts, a so-completed table is different from the fitted one. To illustrate this, the completed tables are presented in Table 16.9, for the same set of models as in Tables 16.7 and 16.8.

A number of noteworthy observations can be made. First, BRD1≡BRD1(MAR) exhibit the poorest fit (i.e., the largest discrepancies between this completed table and the model fit as presented in Table 16.8), with an intermediate quality fit for a model with 7 degrees of freedom, such as BRD2, and a perfect fit for BRD7, BRD9, and their MAR counterparts. Second, compare the data completed using BRD1 (Table 16.9) to the fit of BRD1 (Table 16.8): the data for the group of completers is evidently equal to the original data (Table 16.7) because here no completion takes place; the complete data for the subjects without observations is entirely equal to the model fit (Table 16.8), as here there are no data from which to start; the complete data for the two partially classified tables takes a position in between and hence is not exactly equal to the model fit. Third, note that the above statement needs amending for BRD2 and BRD2(MAR). Now, the first subtable of partially classified subjects exhibits an exact match between completed data and model fit, while this is not true for the second subtable. The reason is that BRD2 allows missingness on the second question to depend on the first one, leading to saturation of the first subtable, whereas missingness on the first question is independent of one's opinion on either question.

While the method is elegant and gives us a handle regarding the quality of the model fit to the incomplete data while contemplating the completed data and the full model fit, the method is unable to distinguish between the saturated models BRD6–9 and the MAR counterpart, as any method would. This phenomenon points to the need for sensitivity analysis, a topic to which the remainder of the chapter is devoted.

TABLE 16.9

The Slovenian Public Opinion Survey. Analysis restricted to the independence and attendance questions. Completed versions of the observed data, using the fit of the modelsBRD1, BRD2, BRD7, and BRD9, and their MAR counterparts.

Completed data using BRD1≡BRD1(MAR) fit							
1439	78	148.1	10.9	141.5	38.4	121.3	9.0
16	16	11.9	20.1	2.5	15.6	2.1	3.6
Completed data using BRD2 fit							
1439	78	147.5	11.5	142.4	44.7	105.0	8.2
16	16	13.2	18.8	1.6	9.3	9.4	13.4
Completed data using BRD2(MAR) fit							
1439	78	147.7	11.3	141.4	40.2	121.2	9.3
16	16	13.3	18.7	2.6	13.8	2.3	3.2
Completed data using BRD7 fit							
1439	78	3.2	155.8	142.4	44.8	0.4	112.5
16	16	0.0	32.0	1.6	9.2	0.0	23.1
Completed data using BRD9 fit							
1439	78	150.8	8.2	142.4	44.8	66.8	21.0
16	16	16.0	16.0	1.6	9.2	7.1	41.1
Completed data using BRD7(MAR)≡BRD9(MAR) fit							
1439	78	148.1	10.9	141.5	38.4	121.3	9.0
16	18	11.8	20.2	2.5	15.6	2.1	3.6

16.5 Interval of Ignorance, Global Influence, and Local Influence

We will use the working definition that a sensitivity analysis is one in which several statistical models are considered simultaneously and/or where a statistical model is further scrutinized using specialized tools, such as diagnostic measures. This informal definition encompasses a wide variety of useful approaches. The simplest procedure is to fit a selected number of (non-random) models which are all deemed plausible or in which a preferred (primary) analysis is supplemented with a number of variations. The extent to which conclusions (inferences) are stable across such ranges provides an indication about the degree to which they are robust to inherently untestable assumptions about the missingness mechanism. Variations to a basic model can be constructed in different ways. The most obvious strategy, cast within the selection model paradigm, is to consider various dependencies of the missing data process on the outcomes and/or on covariates. Alternatively, the distributional assumptions of the models can be changed. This route will be followed in Section 16.5.1. Thijs et al. (2002) consider sensitivity analysis within the context of pattern-mixture models.

Related to this, we can assess how an NMAR model, or a collection of NMAR models, differ from the set of models with equal fit to the observed data but that are of a MAR nature. This path is followed in Section 16.4.5.

Additionally, a sensitivity analysis can also be performed on the level of individual observations instead of on the level of the models. In that case, interest is directed towards finding those individuals who drive the conclusions towards one or more NMAR models. Therefore, the influence of every individual separately will be explored. Two techniques exist, i.e., global influence (Section 16.5.2) and local influence (Section 16.5.3, Cook 1986). The global influence methodology, also known as the case-deletion method Cook and Weisberg (1982), is introduced by Cook (1979, 1986) in linear regression, and by Molenberghs et al. (2003) and Thijs, Molenberghs, and Verbeke (2000) in linear mixed models. Verbeke et al. (2001) and Thijs, Molenberghs, and Verbeke (2000) already used local influence on the Diggle and Kenward (1994) model, which is based on a selection model, integrating a linear mixed model for continuous outcomes with logistic regression for dropout. Later, Van Steen et al. (2001) adapted these ideas to the model of Molenberghs, Kenward, and Lesaffre (1997), for monotone repeated ordinal data.

16.5.1 Interval of ignorance

A sample from Table 16.2 produces empirical proportions representing the π's with error. This imprecision disappears asymptotically. What remains is ignorance regarding the redistribution of all but the first four π's over the missing values. This leaves ignorance regarding any probability in which at least one of the first or second indices is equal to 0, and hence regarding any derived parameter of scientific interest. For such a parameter, θ say, a region of possible values which is consistent with Table 16.2 is called a region of ignorance. Evidently, such a region will depend, not only on the data and the way it is incomplete, but also on the model for which it is constructed. Analogously, an observed incomplete table leaves ignorance regarding the would-be observed full table, which leaves imprecision regarding the true full probabilities. The region of estimators for θ consistent with the observed data provides an estimated region of ignorance. For a single parameter, the region becomes the *interval of ignorance.* Various ways of constructing regions of ignorance are conceivable. Practically, one selects the largest possible set of identifiable parameters. The remaining ones are then termed sensitivity parameters. For every value chosen for the latter, the former can be estimated by means of, for example, maximum likelihood. Repeating this for all values of the sensitivity parameters or, practically speaking, a sufficiently refined grid, one effectively obtains a region or, in the univariate case, an interval of estimates. The $(1-\alpha)100\%$ *region of uncertainty* is a larger region, encompassing the region of ignorance, in the spirit of a confidence region, designed to capture the combined effects of imprecision and ignorance. Practically, for every point in the region of ignorance, a confidence region is constructed, the union of which then produces the interval of uncertainty. Details regarding construction and asymptotic properties can be found in Molenberghs, Kenward, and Goetghebeur (2001), Kenward, Goetghebeur, and Molenberghs (2001), and Vansteelandt et al. (2006).

The estimated intervals of ignorance and intervals of uncertainty are shown in Table 16.10, while a graphical representation of the YES votes is given in Figure 16.4. Model 10 is defined as $(\alpha_{j_2}, \beta_{j_1 j_2})$ with

$$\beta_{j_1 j_2} = \beta_0 + \beta_{j_1} + \beta_{j_2}, \tag{16.4}$$

while Model 11 assumes $(\alpha_{j_1 j_2}, \beta_{j_1})$ and uses

$$\alpha_{j_1 j_2} = \alpha_0 + \alpha_{j_1} + \alpha_{j_2}, \tag{16.5}$$

Finally, Model 12 is defined as $(\alpha_{j_1 j_2}, \beta_{j_1 j_2})$, a combination of both (16.4) and (16.5). Model 10 shows an interval of ignorance which is very close to $[0.741, 0.892]$, the range produced

TABLE 16.10
The Slovenian Public Opinion Survey. Intervals of ignorance and intervals of uncertainty for the proportion θ (confidence interval) attending the plebiscite, following from fitting overspecified Models 10, 11, and 12.

			$\widehat{\theta}$	
Model	d.f.	loglik	II	IU
10	9	−2431.06	[0.762;0.893]	[0.744;0.907]
11	9	−2431.06	[0.766;0.883]	[0.715;0.920]
12	10	−2431.06	[0.694;0.905]	

by the models BRD1–BRD9, while Model 11 is somewhat sharper and just fails to cover the plebiscite value. However, the corresponding intervals of uncertainty contain the true value.

Interestingly, Model 12 virtually coincides with the non-parametric range even though it does not saturate the complete data degrees of freedom. To do so, not 2 but in fact 7 sensitivity parameters would have to be included. Thus, it appears that a relatively simple sensitivity analysis is adequate to increase insight in the information provided by the incomplete data about the proportion of valid YES votes, in the study under consideration here.

16.5.2 Global influence

One sensitivity-analysis tool is global influence, which starts from case deletion and is based on the difference in log-likelihood between the model fitted to the entire dataset $\ell(\boldsymbol{\phi}) = \sum_{i=1}^{N} \ell_i(\boldsymbol{\phi})$, with $\ell_i(\boldsymbol{\phi})$ the contribution of the ith individual, on the one hand, and the dataset minus one subject or a subject doubled, $\ell_{(\pm i)}(\boldsymbol{\phi})$, on the other hand. Here, $\boldsymbol{\phi}$ parameterizes the particular BRD model. Cook's distances (CD) are based on measuring the discrepancy, induced by deletion or doubling, in either the likelihood or the parameter vector:

$$CD_{1i} = 2\left[\widehat{\ell}(\boldsymbol{\phi}) - \widehat{\ell}_{(\pm i)}(\boldsymbol{\phi})\right] \quad \text{or} \quad CD_{2i} = 2(\widehat{\boldsymbol{\phi}} - \widehat{\boldsymbol{\phi}}_{(\pm i)})' \, \ddot{L}^{-1} \, (\widehat{\boldsymbol{\phi}} - \widehat{\boldsymbol{\phi}}_{(\pm i)}),$$

with $\ddot{L}$ the matrix of second derivatives of $\ell(\boldsymbol{\phi})$, with respect to $\boldsymbol{\phi}$, evaluated at $\widehat{\boldsymbol{\phi}}$. Our focus is on the latter version, because interest is on changes in the parameter estimates rather than on the likelihood. Both measures can be constructed for the entire parameter vector or for sub-vectors thereof; this includes, of course, a single parameter. Performing a global influence analysis on data with categorical outcomes is less time consuming than on data with continuous outcomes, since the data can then be organized into cells, as in Table 16.3.

Figure 16.8 shows a selection of the results for the global influence analysis on the SPO survey data. Inasmuch as Cook's distance measure CD_{2i} was approximately zero for all cells, indicating no substantial influence when adding or removing a single case from a particular cell, for all other models, only the results for BRD4, BRD7, and BRD8 are presented. Observe that, for BRD4, adding a single observation to cell 3 has a large influence on the parameters, as well as deletion from either cells 3 or 5. Cell 3 represents subjects with a NO on the attendance question and a YES on the independence question. An addition or removal of one such respondent can largely affect the parameters of BRD4. Similarly, exclusion of a

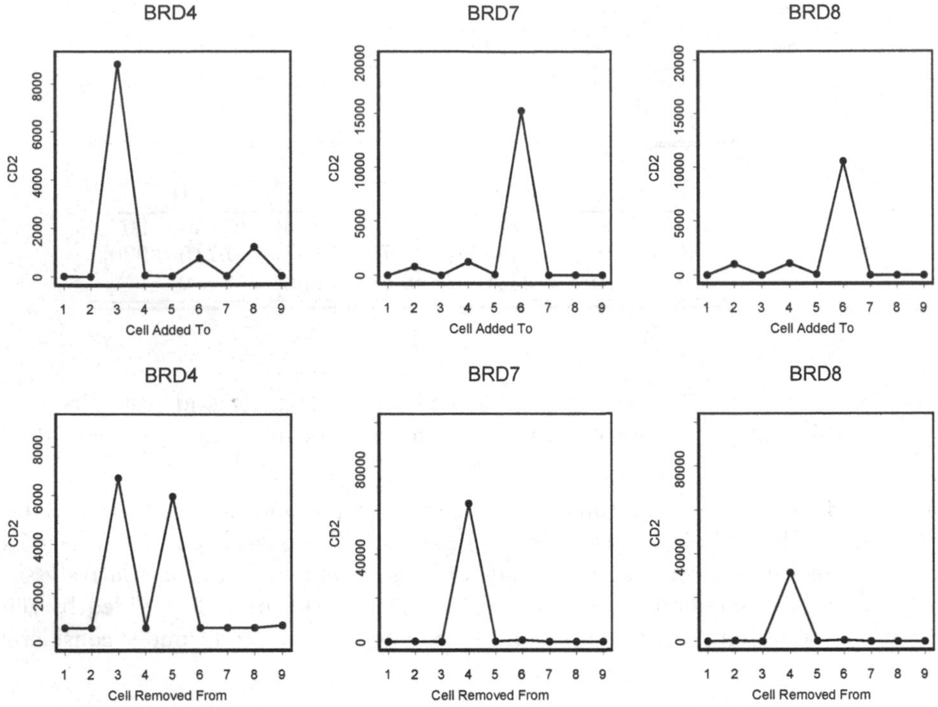

FIGURE 16.8
The Slovenian Public Opinion Survey. Global influence analysis for BRD4, BRD7 and BRD8. Cook's distance measure, CD_{2i}, is evaluated when an observation is added to a specific cell (first row) and when an observation is deleted from a specific cell (second row).

single respondent with a NO on the independence question but a missing response on the attendance question (cell 5), also influences BRD4's model parameters, though to a lesser extent. For models BRD7 and BRD8, an additional observation in cell 6 or a deletion from cell 4 leads to significant influence on these models' parameters. Thus, adding a subject with a YES for independence and a missing attendance response, or excluding a respondent with YES on both questions, yields changes in the model parameters of BRD7 and BRD8. These findings hint on the influential nature of subjects with a YES on the independence question, which is likely related with this group's sparseness.

16.5.3 Local influence

A drawback of global influence is that the specific cause of the influence is hard to retrieve because, by deleting or adding a subject, all types of influence stemming from it are lumped together. Local influence (Cook 1986, Verbeke et al. 2001) allegedly is more suitable for this purpose, in particular also because the method often leads to closed forms.

Denote the log-likelihood corresponding to a particular BRD model by $\ell(\boldsymbol{\phi}|\boldsymbol{\omega}) = \sum_{i=1}^{N} \ell_i(\boldsymbol{\phi}|\omega_i)$, where $\boldsymbol{\phi}=(\boldsymbol{\theta},\boldsymbol{\psi})$ is the s-dimensional vector, grouping, respectively, the parameters of the measurement and dropout models. Further, $\boldsymbol{\omega} = (\omega_1, \omega_2, \ldots, \omega_N)'$, belonging to an open subset Ω of $\mathbb{R}^N$, is a vector defining infinitesimal perturbations around the model studied. Obviously, $\boldsymbol{\omega}_o = (0, 0, \ldots, 0)'$ corresponds to the original model.

Let $\widehat{\boldsymbol{\phi}}$ be the maximum likelihood estimator for $\boldsymbol{\phi}$, obtained by maximizing $\ell(\boldsymbol{\phi}|\boldsymbol{\omega}_o)$, and let $\widehat{\boldsymbol{\phi}}_{\boldsymbol{\omega}}$ denote the maximum likelihood estimator for $\boldsymbol{\phi}$ under $\ell(\boldsymbol{\phi}|\boldsymbol{\omega})$. The relative change in likelihood is a measure for influence; Cook (1986) captured this through the likelihood displacement: $LD(\boldsymbol{\omega}) = 2[\ell(\widehat{\boldsymbol{\phi}}|\boldsymbol{\omega}_o) - \ell(\widehat{\boldsymbol{\phi}}_{\boldsymbol{\omega}}|\boldsymbol{\omega}_o)]$. A graph of $LD(\boldsymbol{\omega})$ versus $\boldsymbol{\omega}$, i.e., the geometric surface formed by values of the $N+1$ dimensional vector $\zeta(\boldsymbol{\omega}) = (\boldsymbol{\omega}', LD(\boldsymbol{\omega}))'$, depicts the influence of perturbations. Because this so-called *influence graph* Lesaffre and Verbeke (1998) can only be depicted when $N = 2$, Cook (1986) proposed to consider local influence, i.e., the normal curvatures $C_{\boldsymbol{h}}$ of $\zeta(\boldsymbol{\omega})$ in $\boldsymbol{\omega}_o$, in the direction of some N-dimensional vector $\boldsymbol{h}$ of unit length. A general expression is

$$C_{\boldsymbol{h}} = 2\left|\boldsymbol{h}'\,\boldsymbol{\Delta}'\,(\ddot{L})^{-1}\,\boldsymbol{\Delta}\,\boldsymbol{h}\right|, \tag{16.6}$$

with

$$\boldsymbol{\Delta}_i = \left.\frac{\partial^2 \ell_i(\boldsymbol{\phi}|\omega_i)}{\partial\omega_i \partial\boldsymbol{\phi}}\right|_{\boldsymbol{\phi}=\widehat{\boldsymbol{\phi}},\omega_i=0},$$

$\boldsymbol{\Delta}$ the $(s \times N)$ matrix with $\boldsymbol{\Delta}_i$ as its ith column, and $\ddot{L}$ the $(s \times s)$ matrix of second-order derivatives of $\ell(\boldsymbol{\phi}|\boldsymbol{\omega}_o)$ with respect to $\boldsymbol{\phi}$, evaluated at $\boldsymbol{\phi} = \widehat{\boldsymbol{\phi}}$. A sensible choice for $\boldsymbol{h}_i$ is the vector with a one in the ith position and zero elsewhere, corresponding to the perturbation of the ith subject only. Another important direction is the direction $\boldsymbol{h}_{\max}$ of maximal normal curvature $C_{\max}$. It shows how to perturb the model to obtain the largest local changes in the likelihood displacement. Details can be found in Verbeke and Molenberghs (2000).

The above development is geared towards studying the influence on the likelihood function. Other choices are possible, too. In our contingency-table setting, it is instructive to study influence in predicted cell counts, $Y_{r_1r_2,j_1j_2}$. In such cases, when $Z(\boldsymbol{\phi})$ denotes a particular function of the model parameters, the expression for $C_{\boldsymbol{h}}$ can be further generalized as:

$$C_{\boldsymbol{h}} = 2\left|h'\,\boldsymbol{\Delta}'\,(\ddot{\boldsymbol{L}})^{-1}\,\ddot{\boldsymbol{Z}}\,(\ddot{\boldsymbol{L}})^{-1}\,\boldsymbol{\Delta}\,h\right|, \tag{16.7}$$

with $\|h\| = 1$ and $\boldsymbol{\Delta}$, $\ddot{\boldsymbol{L}}$, and $\ddot{\boldsymbol{Z}}$ defined as:

$$\boldsymbol{\Delta}_{ij} = \left.\frac{\partial^2 \ell(\boldsymbol{\phi}|\boldsymbol{\omega})}{\partial\phi_i \partial\omega_j}\right|_{\boldsymbol{\phi}=\widehat{\boldsymbol{\phi}},\boldsymbol{\omega}=\boldsymbol{\omega}_o}, \quad \ddot{\boldsymbol{L}}_{i\ell} = \left.\frac{\partial^2 \ell(\boldsymbol{\phi}|\boldsymbol{\omega}_o)}{\partial\phi_i \partial\phi_\ell}\right|_{\boldsymbol{\phi}=\widehat{\boldsymbol{\phi}}}, \quad \ddot{\boldsymbol{Z}}_{il} = \left.\frac{\partial^2 Z(\boldsymbol{\phi})}{\partial\phi_i \partial\phi_\ell}\right|_{\boldsymbol{\phi}=\widehat{\boldsymbol{\phi}}},$$

with $i, \ell = 1, \ldots, p$ and $j = 1, \ldots, q$.

Here, the local influence measure is not calibrated, because (16.6) takes the form of a squared second derivative, owing to the double occurrence of $\boldsymbol{\Delta}$, "divided by" another second derivative, $\ddot{L}$. Changing units, therefore, changes scale, unlike in the mixed-models application of Lesaffre and Verbeke (1998) and Verbeke and Molenberghs (2000), where the influence measures approximately sum to twice the sample size. As a consequence, when applying local influence as presented here, interpretation ought to be relative rather than absolute. The important issue remains then as to "how large is large?" It is extremely hard to provide firm guidelines, but it may be wise, as one possible rule of thumb, to scrutinize the subjects with the largest 5% of influence values. The issue has been studied in detail in Jansen et al. (2006).

Local influence is useful when assessing which (groups of) observations are most influential in driving the conclusions about the nature of the missing data mechanism in the direction of the more elaborate NMAR model. As Jansen et al. (2006) indicate, such a phenomenon should not be seen as evidence, let alone proof, that some observations are genuinely influenced by a complex NMAR mechanism rather than, for example, by a simpler MAR mechanism. Indeed, this would conflict with the MAR-counterpart results. Rather, such influence graphs are instructive when assessing which observations have the power to drive the conclusions towards a more complex mechanism. Often, the issue is that other outlying features, such as unusual values, unusual slopes in longitudinal observations, etc. are responsible for the apparent conclusion about the missing data mechanism. Like the MAR-counterpart result, this type of sensitivity analysis issues a cautionary warning against excessive confidence regarding the nature of the missing data mechanism. A limiting feature, unsurprising in view of the foregoing discussion, is the absence of a "yardstick," or a threshold, demarcating influential subjects; arguably, influence graphs will blow a whistle over subjects that need further scrutiny.

We first consider perturbations of a given BRD model in the direction of a more elaborate one. For example, BRD4 includes the parameter $\beta_{.j_2}, (j_2 = 1, 2)$, whereas BRD1 only includes $\beta_{..}$. For the influence analysis, ω_i is not a parameter but an infinitesimal perturbation of the simpler model towards the more complex one, confined to a single subject. For example, for the perturbation of BRD1 in the direction of BRD4, one considers $\beta_{..}$ and $\beta_{..} + \omega_i$. The vector of all ω_i's defines the direction in which such a perturbation is considered. The BRD log-likelihood is:

$$\begin{aligned}\ell(\phi|\omega) &= \sum_{j_1,j_2} Y_{11,j_1j_2} \ln \pi_{11,j_1j_2} + \sum_{j_1} Y_{10,j_1+} \ln \pi_{10,j_1+} + \\ &\quad \sum_{j_2} Y_{01,+j_2} \ln \pi_{01,+j_2} + Y_{00,++} \ln \pi_{00,++},\end{aligned} \tag{16.8}$$

where $\pi_{r_1r_2,j_1j_2} = p_{j_1j_2}\, q_{r_1r_2|j_1j_2}$, with $p_{11} = p_1, p_{12} = p_2, p_{21} = p_3, p_{22} = 1 - p_1 - p_2 - p_3$, and

$$q_{r_1r_2|j_1j_2} = \frac{\exp\{\alpha_{j_1j_2}(1-r_1) + \beta_{j_1j_2}(1-r_2) + \gamma(1-r_1)(1-r_2)\}}{1 + \exp(\alpha_{j_1j_2}) + \exp(\beta_{j_1j_2}) + \exp(\alpha_{j_1j_2} + \beta_{j_1j_2} + \gamma)}. \tag{16.9}$$

For BRD4, with $(\alpha_{..}, \beta_{.j_2})$, expression (16.9) yields:

$$\begin{aligned} q_{r_1r_2|j_11} &= \frac{\exp\{\alpha_{..}(1-r_1) + \beta_{..}(1-r_2) + \gamma(1-r_1)(1-r_2)\}}{1 + \exp(\alpha_{..}) + \exp(\beta_{..}) + \exp(\alpha_{..} + \beta_{..} + \gamma)}, \\ q_{r_1r_2|j2} &= \frac{\exp\{\alpha_{..}(1-r_1) + (\beta_{..} + \omega_i)(1-r_2) + \gamma(1-r_1)(1-r_2)\}}{1 + \exp(\alpha_{..}) + \exp(\beta_{..} + \omega_i) + \exp(\alpha_{..} + \beta_{..} + \omega_i + \gamma)}. \end{aligned}$$

The perturbation ω_i defines a difference between the dropout probabilities above, while under the simpler (null) model BRD1, the two expressions reduce to a single dropout probability. Local influence measures now follow from the general logic described above.

For the SPO data, we focus on the model pairs BRD1 *vs.* BRD4, BRD3 *vs.* BRD7, and BRD4 *vs.* BRD7. These pairs are precisely the ones with high influence on the likelihood displacement.

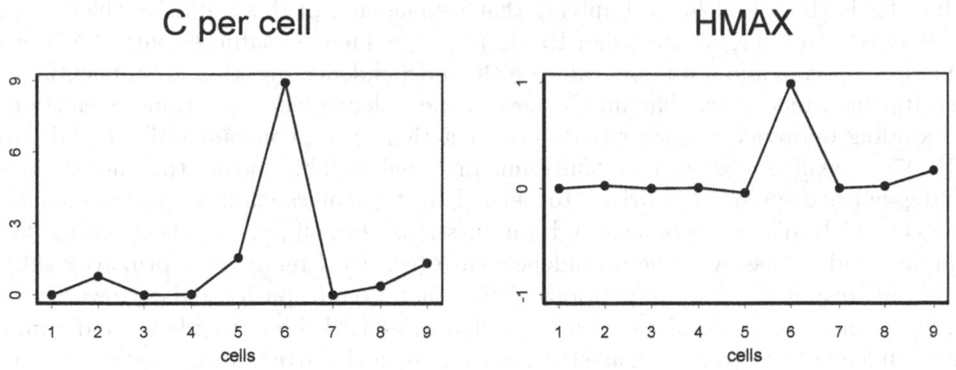

(a) *BRD1 vs BRD4*

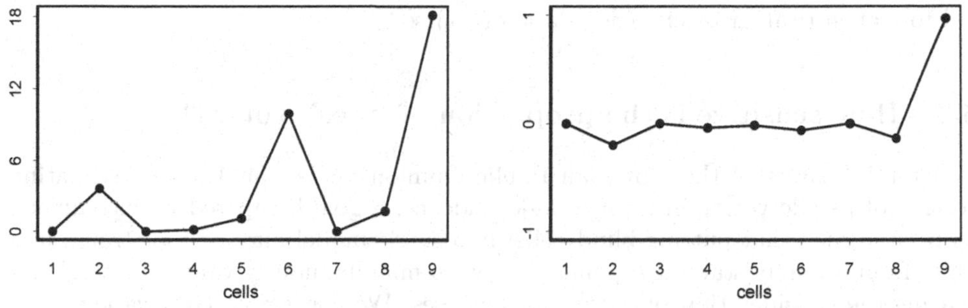

(b) *BRD3 vs. BRD7*

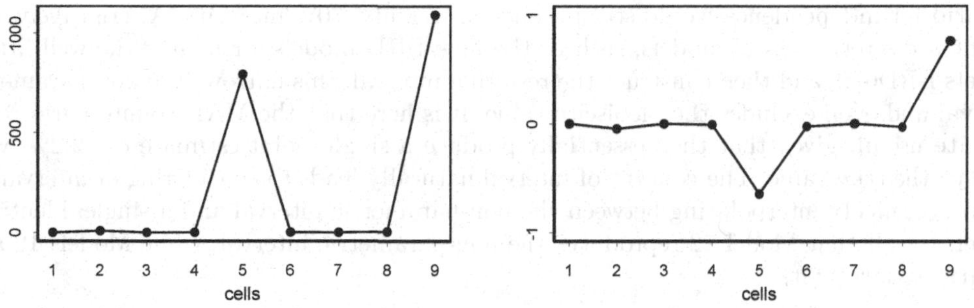

(c) *BRD4 vs. BRD7*

FIGURE 16.9
The Slovenian Public Opinion Survey. Local influence analysis on parameters for model pairs (a) BRD1 vs. BRD4, (b) BRD3 vs. BRD7, and (c) BRD4 vs. BRD7. The first column shows the local influence measure C_i *at the* i^{th} *observed cell; the second column shows* $\boldsymbol{h}_{max}$ *for the* i^{th} *observed cell.*

Figure 16.9 shows the influence measures C_i, plotted against the ith observed cell and $\boldsymbol{h}_{\max}$ against the ith observed cell. For the comparison of BRD1 *vs.* BRD4, a peak is observed at cell 6, for both C_i and $\boldsymbol{h}_{\max}$, implying that respondents in this cell drive the data more towards BRD4 $(\alpha_{..}, \beta_{.j_2})$ rather than BRD1 $(\alpha_{..}, \beta_{..})$. That is, subjects with a NO on the attendance question and a missing value on the independence question are influential when perturbing the model such that missingness in the independence question depends on the corresponding unobserved answer (BRD4) rather than being constant (BRD1). For BRD3 *vs.* BRD7, a peak is observed at "fully missing' 'cell 9, This means that missingness in the independence question is driven to depend on the corresponding unobserved answer by subjects with missing responses on both questions, and slightly by those with a NO on attendance and a missing value on independence (cell 6). Finally, it is primarily subjects with missing responses on both questions (cell 9) that seem to push the data towards BRD7 $(\alpha_{.j_2}, \beta_{.j_2})$. These subjects, along with those that have a YES on attendance and a missing value on independence (cell 5), make the missingness in the attendance question depend on the response of the independence question.

Jansen et al. (2006) also reported local influence analyses in terms of cell counts, which is useful in addition to the one reported above in terms of individual observations. It is one typical for categorical, cross-classified data, of course.

16.5.4 How sensitive is the proportion of "Yes" voters?

The inferential target of the Slovenian Public Opinion Survey analyses is estimating the proportion of people voting in favor of independence, a goal hampered by incompleteness. It has been argued that putting blind belief in a single model may be too strong; it is not possible, from a purely statistical point of view, to unambiguously validate a single model, motivating the consideration of sensitivity analyses. We contemplated a variety of these, going beyond conventional sensitivity analysis applications, which are often confined to a single sensitivity assessment tool. Table 16.4 summarizes various methods. The simplest analysis considers the non-parametric bounds and deduces that even the least supportive scenario for independence would still produce a, roughly, 70% majority. Alternatively, one can fit a discrete class of models, such as the nine BRD models, or merely the well-fitting models BRD6–9, and then construct the resulting interval; this narrows the non-parametric interval and even excludes the plebiscite value. It is here that the MAR counterparts come in quite useful, given that they essentially produce a single point estimate of 89.2%, very close to the true value. The concept of interval naturally leads to considering of intervals of ignorance, nicely interpolating between the non-parametric interval and a single, identified model. Recall that Model 12 reproduces the non-parametric interval, while Models 10 and 11 further narrow it.

Turning to substantive considerations, the estimates for θ (Table 16.6), can be split into two:

Optimistic: MAR bodyguards, BRD1, BRD3, (BRD5), (BRD6), and BRD9.

Pessimistic: BRD4, BRD7, and BRD8.

The parenthetical ones are slightly less pronounced than the others. The assumptions regarding the missingness mechanism, underpinning all twelve models, can be read from the second column in Table 16.6; they are also spelled out in Table 16.11. It is striking that the three fully identified, pessimistic estimates allow missingness in the independence question to depend on independence, whereas the other six do not: these three models support the

TABLE 16.11

Slovenian Public Opinion Survey. Meaning of the missingness mechanism in the nine identified Models BRD1–BRD9 and its three overspecified extensions Models 10–12.

Missingness in $\longrightarrow$	Attendance		Independence	
Depends on $\longrightarrow$	attendance	independence	attendance	independence
BRD1	—	—	—	—
BRD2	—	—	√	—
BRD3	—	√	—	—
BRD4	—	—	—	√
BRD5	√	—	—	—
BRD6	√	—	√	—
BRD7	—	√	—	√
BRD8	√	—	—	√
BRD9	—	√	√	—
Model 10	—	√	√	√
Model 11	√	√	√	—
Model 12	√	√	√	√

thesis that there is a large group of people, in favor of independence, that would not partake in the plebiscite, as can be seen from Table 16.6. We will study these in more detail, using BRD7, BRD9, and their MAR counterparts. The pessimistic fit of BRD7 asserts that the proportion of people in favor of independence, yet that would not attend the plebiscite, amounts to $(78 + 155.8 + 44.8 + 112.5)/2074$, i.e., 18.9%. The optimistic BRD9 predicts this fraction to be as low as 7.3%, while for the counterparts it goes down further to 6.6%. We infer with reasonable confidence that, in the actual plebiscite, people expressed their opinion, regardless of real or perceived pressure, and that an overwhelming majority favored independence, supporting the optimistic scenario.

There are more nuances in the overspecified Models 10–12. While Models 10 and 12 also include the "pessimistic relationship" of missingness in independence on the independence answer, the interval is not pessimistic. Rather, the three intervals encompass both very pessimistic and very optimistic scenarios. This is because, unlike the fully identified models, there is no need to sacrifice one type of dependence to maintain identifiability. Precisely, while none of BRD4, BRD7, and BRD8 allow missingness in the independence question to depend on the attendance response, the three II's do allow for such a dependence, a feature shared with BRD2, BRD6, and BRD9. It ought not to go unnoticed that, among the optimistic ones, BRD5 and BRD6 are somewhat less pronounced; these models allow missingness in the attendance question to depend on the respondent's attendance position. The effect of this is similar, but less sharp, than in the "independence on independence" scenario sketched above. It is therefore no longer a surprise that BRD8, where these two effects play together, produces the most pessimistic estimate.

The question remains why the impact of changing the assumptions is rather spectacular, in the sense that the most pessimistic scenarios are very close to the pessimistic bound, whereas the most optimistic scenarios are virtually at the optimistic bound. It is instructive to return to the global and local influence analyses. First and foremost, we learn that there is relatively little influence all together, *except in BRD4, BRD7, and BRD8*, i.e., the entire pessimistic group. This is clear from the discussion in Sections 16.5.2 and 16.5.3, and from Figures 16.8 and 16.9. From global influence we learn that there is strong impact of the (relatively small) no-on-independence cells, in the pessimistic models: changes in these counts can dramatically alter the way in which the incomplete cells are split over the hypothetical complete cells, rendering $\widehat{\theta}$ unstable. Local influence focuses on a different aspect: which observations/cells drive the conclusions away from a given null model? For BRD1 versus BRD4, this is Cell 6, the not-in-favor-of-independence respondents without declared attendance status. Thus, a small but influential count is simultaneously responsible for a move towards a pessimistic scenario and the extent of pessimism.

Sensitivity analyses combined and substantive considerations demonstrate that the optimistic scenarios, whether from the MAR counterparts or the optimistic group of BRD models, are plausible descriptions of the mechanisms operating during the plebiscite exercise. The influence analyses show that the pessimistic ones are rather different from their optimistic counterparts and constitute plausible scenarios only owing to the presence of one or a few influential cells.

16.5.5 Local influence for Gaussian data

Local influence methodology presented in Section 16.5.3, though applied to cross-classified data, is entirely generic. In fact, the earliest applications in a missing-data context were for Gaussian repeated measures (Thijs et al. 2002, Verbeke et al. 2001).

We consider the Diggle and Kenward selection model in some more detail. They combine a linear mixed model (Laird and Ware 1982) for the measurement process with a logistic regression model for the dropout process. The measurement model assumes that the vector $\boldsymbol{y_i}$ of repeated measurements for the ith subject satisfies the linear regression model

$$\boldsymbol{y_i} \sim N(X_i\boldsymbol{\beta}, V_i), \quad i = 1, \ldots, N \tag{16.10}$$

in which $\boldsymbol{\beta}$ is a vector of population-averaged regression coefficients called fixed effects, and where $V_i = Z_iGZ_i' + \Sigma_i$ (Verbeke and Molenberghs 2000) for positive definite matrices G and Σ_i. The parameters in $\boldsymbol{\beta}$, G, and Σ_i are assembled into $\boldsymbol{\theta}$.

Since no data would be observed otherwise, we assume that the first measurement y_{i1} is obtained for every subject in the study. The model for the dropout process is based on a logistic regression for the probability of dropout at occasion j (let D_i be the occasion at which dropout occurs), given the subject was still in the study up to occasion j. We denote this probability by $g(\boldsymbol{h}_{ij}, y_{ij})$ in which $\boldsymbol{h}_{ij}$ is a sub-vector of the history $\widetilde{\boldsymbol{h}}_{ij}$, containing all responses observed up to but not including occasion j, as well as covariates. We assume

$$\text{logit}[g(\boldsymbol{h}_{ij}, y_{ij})] = \text{logit}\left[\text{pr}(D_i = j|D_i \geq j, \boldsymbol{y}_i)\right] = \boldsymbol{h}_{ij}\boldsymbol{\psi} + \omega y_{ij} \tag{16.11}$$

$(i = 1, \ldots, N)$. In our case $\boldsymbol{h}_{ij}$ will contain the previous measurement $y_{i,j-1}$.

When ω equals zero and the model assumptions made are correct, the dropout model is random, and all parameters can be estimated using standard software since the measurement model and dropout model parameters can then be fitted separately. If $\omega \neq 0$, the dropout process is assumed to be non-random. Earlier, we pointed to the sensitivity of such an approach and a dropout model may be found to be non-random solely since one or a few influential subjects have driven the analysis.

We focus on (16.10) and (16.11). Details can be found in Verbeke et al. (2001). We need expressions for Δ and $\ddot{L}$. Straightforward derivation shows that the columns $\boldsymbol{\Delta}_i$ of Δ are given by

$$\left.\frac{\partial^2 \ell_{i\omega}}{\partial\boldsymbol{\theta}\partial\omega_i}\right|_{\omega_i=0} = 0, \quad (16.12)$$

$$\left.\frac{\partial^2 \ell_{i\omega}}{\partial\boldsymbol{\psi}\partial\omega_i}\right|_{\omega_i=0} = -\sum_{j=2}^{n_i} \boldsymbol{h}_{ij} y_{ij} g(\boldsymbol{h}_{ij})[1-g(\boldsymbol{h}_{ij})], \quad (16.13)$$

for complete sequences (no dropout) and by

$$\left.\frac{\partial^2 \ell_{i\omega}}{\partial\boldsymbol{\theta}\partial\omega_i}\right|_{\omega_i=0} = [1-g(\boldsymbol{h}_{id})]\frac{\partial\lambda(y_{id}|\boldsymbol{h}_{id})}{\partial\boldsymbol{\theta}}, \quad (16.14)$$

$$\begin{aligned}\left.\frac{\partial^2 \ell_{i\omega}}{\partial\boldsymbol{\psi}\partial\omega_i}\right|_{\omega_i=0} &= -\sum_{j=2}^{d-1} \boldsymbol{h}_{ij} y_{ij} g(\boldsymbol{h}_{ij})[1-g(\boldsymbol{h}_{ij})] \\ &\quad -\boldsymbol{h}_{id}\lambda(y_{id}|\boldsymbol{h}_{id})g(\boldsymbol{h}_{id})[1-g(\boldsymbol{h}_{id})],\end{aligned} \quad (16.15)$$

for incomplete sequences. All above expressions are evaluated at $\widehat{\gamma}$, and $g(\boldsymbol{h}_{ij}) = g(\boldsymbol{h}_{ij}, y_{ij})|_{\omega_i=0}$, is the MAR version of the dropout model. In (16.14), we make use of the conditional mean

$$\lambda(y_{id}|\boldsymbol{h}_{id}) = \lambda(y_{id}) + V_{i,21}V_{i,11}^{-1}[\boldsymbol{h}_{id} - \lambda(\boldsymbol{h}_{id})]. \quad (16.16)$$

The variance matrices follow from partitioning the responses as $(y_{i1},\ldots,y_{i,d-1}|y_{id})'$. The derivatives of (16.16) w.r.t. the measurement model parameters are

$$\frac{\partial\lambda(y_{id}|\boldsymbol{h}_{id})}{\partial\boldsymbol{\beta}} = \boldsymbol{x}_{id} - V_{i,21}V_{i,11}^{-1}X_{i,(d-1)},$$

$$\frac{\partial\lambda(y_{id}|\boldsymbol{h}_{id})}{\partial\alpha} = \left[\frac{\partial V_{i,21}}{\partial\boldsymbol{\alpha}} - V_{i,21}V_{i,11}^{-1}\frac{\partial V_{i,11}}{\partial\boldsymbol{\alpha}}\right]V_{i,11}^{-1}[\boldsymbol{h}_{id} - \lambda(\boldsymbol{h}_{id})]$$

where $\boldsymbol{x}_{id}'$ is the dth row of X_i, and where $X_{i,(d-1)}$ indicates the first $(d-1)$ rows X_i. Further, $\boldsymbol{\alpha}$ indicates the subvector of covariance parameters within the vector $\boldsymbol{\theta}$.

In practice, the parameter $\boldsymbol{\theta}$ in the measurement model is often of primary interest. Since $\ddot{L}$ is block-diagonal with blocks $\ddot{L}(\boldsymbol{\theta})$ and $\ddot{L}(\boldsymbol{\psi})$, we have that for any unit vector $\boldsymbol{h}$, $C_{\boldsymbol{h}}$ equals $C_{\boldsymbol{h}}(\boldsymbol{\theta}) + C_{\boldsymbol{h}}(\boldsymbol{\psi})$, with

$$C_{\boldsymbol{h}}(\boldsymbol{\theta}) = -2\boldsymbol{h}'\left[\left.\frac{\partial^2 \ell_{i\omega}}{\partial\boldsymbol{\theta}\partial\omega_i}\right|_{\omega_i=0}\right]'\ddot{L}^{-1}(\boldsymbol{\theta})\left[\left.\frac{\partial^2 \ell_{i\omega}}{\partial\boldsymbol{\theta}\partial\omega_i}\right|_{\omega_i=0}\right]\boldsymbol{h} \quad (16.17)$$

$$C_{\boldsymbol{h}}(\boldsymbol{\psi}) = -2\boldsymbol{h}'\left[\left.\frac{\partial^2 \ell_{i\omega}}{\partial\boldsymbol{\psi}\partial\omega_i}\right|_{\omega_i=0}\right]'\ddot{L}^{-1}(\boldsymbol{\psi})\left[\left.\frac{\partial^2 \ell_{i\omega}}{\partial\boldsymbol{\psi}\partial\omega_i}\right|_{\omega_i=0}\right]\boldsymbol{h}, \quad (16.18)$$

evaluated at $\boldsymbol{\gamma} = \widehat{\boldsymbol{\gamma}}$. It now immediately follows from (16.12) and (16.14) that *direct* influence on $\boldsymbol{\theta}$ only arises from those measurement occasions at which dropout occurs. In particular, from (16.14) it is clear that the corresponding contribution is large only if (1) the dropout probability was small but the subject disappeared nevertheless and (2) the conditional mean "strongly depends" on the parameter of interest. This implies that complete sequences cannot be influential in the strict sense ($C_i(\boldsymbol{\theta}) = 0$) and that incomplete sequences only contribute, in a direct fashion, at the actual dropout time. However, we make an important distinction between direct and indirect influence. It was shown that complete sequences can have an impact by changing the conditional expectation of the unobserved measurements given the observed ones *and given the dropout mechanism.* Thus, a complete observation which has a strong impact on the *dropout model parameters*, can still drastically change the measurement model parameters and functions thereof.

Expressions (16.17)–(16.18) can be simplified further in specific cases. For example, Verbeke et al. (2001) considered the compound-symmetric situation.

Next, the rat data will be analyzed and local influence applied to the results. An application of the same methodology to a clinical trial in depression is reported in Shen et al. (2006).

16.5.6 Analysis and sensitivity analysis of the rat data

The rat data, which are introduced in Section 16.2.3, are analyzed using model (16.10) with the following specific version of dropout model (16.11):

$$\text{logit}\left[\text{pr}(D_i = j | D_i \geq j, \boldsymbol{y}_i)\right] = \psi_0 + \psi_1 y_{i,j-1} + \psi_2 y_{ij}. \qquad (16.19)$$

Parameter estimates are shown in Table 16.12. More details about these estimates and the performance of a local influence analysis can be found in Verbeke et al. (2001). This section will focus on specific details of this local influence analysis.

Figure 16.10 displays overall C_i and influences for subvectors θ, β, α, and ψ. In addition, the direction $\boldsymbol{h}_{\max}$ corresponding to maximal local influence is given. Apart from the last one of these graphs, the scales are not unitless and therefore it would be hard to use a common one for all of the panels. This implies that the main emphasis should be on *relative* magnitudes.

The largest C_i are observed for rats #10, #16, #35, and #41, and virtually the same picture holds for $C_i(\boldsymbol{\psi})$. They are highlighted in Figure 16.2. All four belong to the low dose group. Arguably, their relatively large influence is caused by an interplay of three facts. First, the profiles are relatively high, and hence y_{ij} and h_{ij} are large. Second, since all four profiles are complete, there is a maximal number of large terms. Third, the computed v_{ij} are relatively large.

Turning attention to $C_i(\boldsymbol{\alpha})$ reveals peaks for rats #5 and #23. Both belong to the control group and drop out after a single measurement occasion. They are highlighted in the first panel of Figure 16.2. To explain this, observe that the relative magnitude of $C_i(\boldsymbol{\alpha})$ is determined by $1-g(h_{id})$ and $h_{id}-\lambda(h_{id})$. The first term is large when the probability of dropout is small. Now, when dropout occurs early in the sequence, the measurements are still relatively low, implying that the dropout probability is rather small (cf. Table 16.12). This feature is built into the model by writing the dropout probability in terms of the raw measurements with time-independent coefficients rather than, for example, in terms of residuals. Further, the residual $h_{id}-\lambda(h_{id})$ is large since these two rats are somewhat distant from the group by time mean. A practical implication of this is that the time-constant nature of the dropout model may be unlikely to hold. Therefore, a time-varying version was considered, where the logit of the dropout model takes form $\psi_0 + \psi_1 y_{i,j-1} + \nu_0 t_{ij} + \nu_1 t_{ij} y_{i,j-1}$. There is overwhelming evidence in favor of such a more elaborate MAR model (likelihood ratio statistic of 167.4 on 2 degrees of freedom). Thus, local influence can be used to call into question the posited MAR (and NMAR) models, and to guide further selection of more elaborate, perhaps MAR, models.

Because all deviations are rather moderate, we further explore our approach by considering a second analysis where all responses for rats #10, #16, #35, and #41 have been increased with 20 units. To check whether these findings are recovered by the local influence approach,

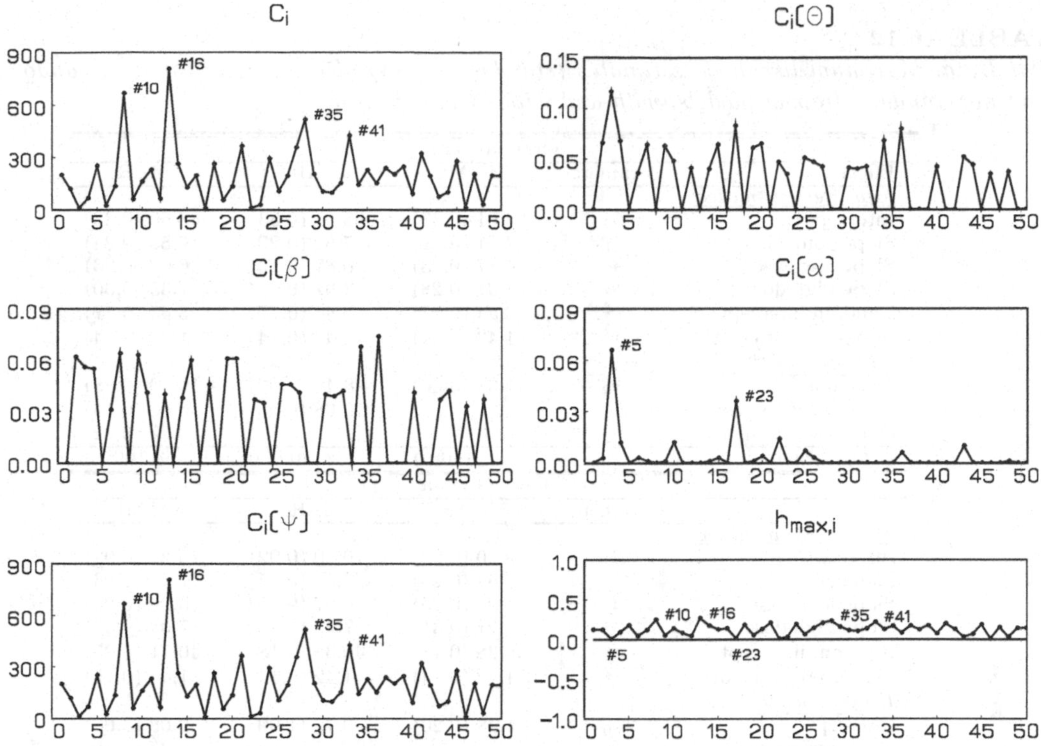

FIGURE 16.10
Rat data. Index plots of C_i, $C_i(\boldsymbol{\theta})$, $C_i(\boldsymbol{\beta})$, $C_i(\boldsymbol{\alpha})$, $C_i(\boldsymbol{\psi})$, and of the components of the direction $\boldsymbol{h}_{\max}$ of maximal curvature.

we examine Figure 16.11. In line with the changes in parameter estimates, $C_i(\boldsymbol{\beta})$ shows no peaks in these observations but peaks in $C_i(\boldsymbol{\alpha})$ and $C_i(\boldsymbol{\psi})$ indicate a relatively strong influence from the four extreme profiles.

It will be clear from the above that subjects may turn out to be influential, for reasons different from the nature of the dropout model. Indeed, increasing the profile by 20 units primarily changes the level of the random intercept and ultimately changes the form of the random-effects distribution. Nevertheless, this feature shows in our local influence analysis, where the perturbation is put into the dropout model and not, for example, in the measurement model. This feature was studied in Jansen et al. (2006) and summarized next.

16.5.7 Local influence methods and their behavior

A number of concerns have been raised, not only about sensitivity, but also about the tools used to assess sensitivity themselves. For example, Verbeke et al. (2001) noted, based on a case study, that the local influence tool, is able to pick up anomalous features of study subjects that are not necessarily related to the missingness mechanism. In particular, they found that subjects with an unusually high profile, or a somewhat atypical serial correlation behavior, are detected with the local influence tool. At first sight, this is a little disconcerting, because the ω_i parameter is placed in the dropout model and not in the measurement model,

TABLE 16.12
Rat Data. Maximum likelihood estimates (standard errors) of completely random, random and non-random dropout models, with and without modification.

Effect	Parameter	MCAR	MAR	NMAR
		Original Data		
Measurement model:				
Intercept	β_0	68.61 (0.33)	68.61 (0.33)	68.60 (0.33)
Slope control	β_1	7.51 (0.22)	7.51 (0.22)	7.53 (0.24)
Slope low dose	β_2	6.87 (0.23)	6.87 (0.23)	6.89 (0.23)
Slope high dose	β_3	7.31 (0.28)	7.31 (0.28)	7.35 (0.30)
Random intercept	τ^2	3.44 (0.77)	3.44 (0.77)	3.43 (0.77)
Measurement error	σ^2	1.43 (0.14)	1.43 (0.14)	1.43 (0.14)
Dropout model:				
Intercept	ψ_0	−1.98 (0.20)	−8.48 (4.00)	−10.30 (6.88)
Prev. measurement	ψ_1		0.08 (0.05)	0.03 (0.16)
Curr. measurement	ψ_2			0.07 (0.22)
-2 loglikelihood		1100.4	1097.6	1097.5
		Modified Data		
Effect	Parameter	MCAR	MAR	NMAR
Measurement model:				
Intercept	β_0	70.20 (0.92)	70.20 (0.92)	70.25 (0.92)
Slope control	β_1	7.52 (0.25)	7.52 (0.25)	7.42 (0.26)
Slope low dose	β_2	6.97 (0.25)	6.97 (0.25)	6.90 (0.25)
Slope high dose	β_3	7.21 (0.31)	7.21 (0.31)	7.04 (0.33)
Random intercept	τ^2	40.38 (0.18)	40.38 (0.18)	40.71 (8.25)
Measurement error	σ^2	1.42 (0.14)	1.42 (0.14)	1.44 (0.15)
Dropout model:				
Intercept	ψ_0	−1.98 (0.20)	-0.79 (1.99)	2.08 (3.08)
Prev. measurement	ψ_1		−0.015 (0.03)	0.23 (0.15)
Curr. measurement	ψ_2			−0.28 (0.17)
-2 loglikelihood		1218.0	1217.7	1214.8

necessitating further investigation regarding which effects are easy or difficult to detect with these local influence methods.

Jansen et al. (2006) studied the behavior of sensitivity assessment tools further, and reached the following conclusions.

The effect of sample size: Unlike in Lesaffre and Verbeke (1998), who applied local influence to the standard linear mixed model, the proposed influence measure for missing data is *not* calibrated, in the sense that the sum of the influence measures across the dataset does not approximately sum to $-2N$, like in Lesaffre and Verbeke (1998). This complicates drawing a line between subjects that are influential and those that are not. A solution, proposed by Jansen et al. (2006) is to construct the distribution of the influence measure under null circumstances, which can then be used to aid the delineation of influential subjects.

The effect of anomalies in the missingness mechanism: These authors showed that, when the true missing-data mechanism is NMAR rather than MAR, the behavior of the local influence measure, including the just mentioned empirical distributions, cannot be trusted. This in itself is not a surprise, given that local influence is designed to detect one or a few anomalies in set of data that otherwise follows the model.

The effect of anomalies in the measurement model: It has been reported by these authors and in the initial papers that anomalies in the measurement model may show

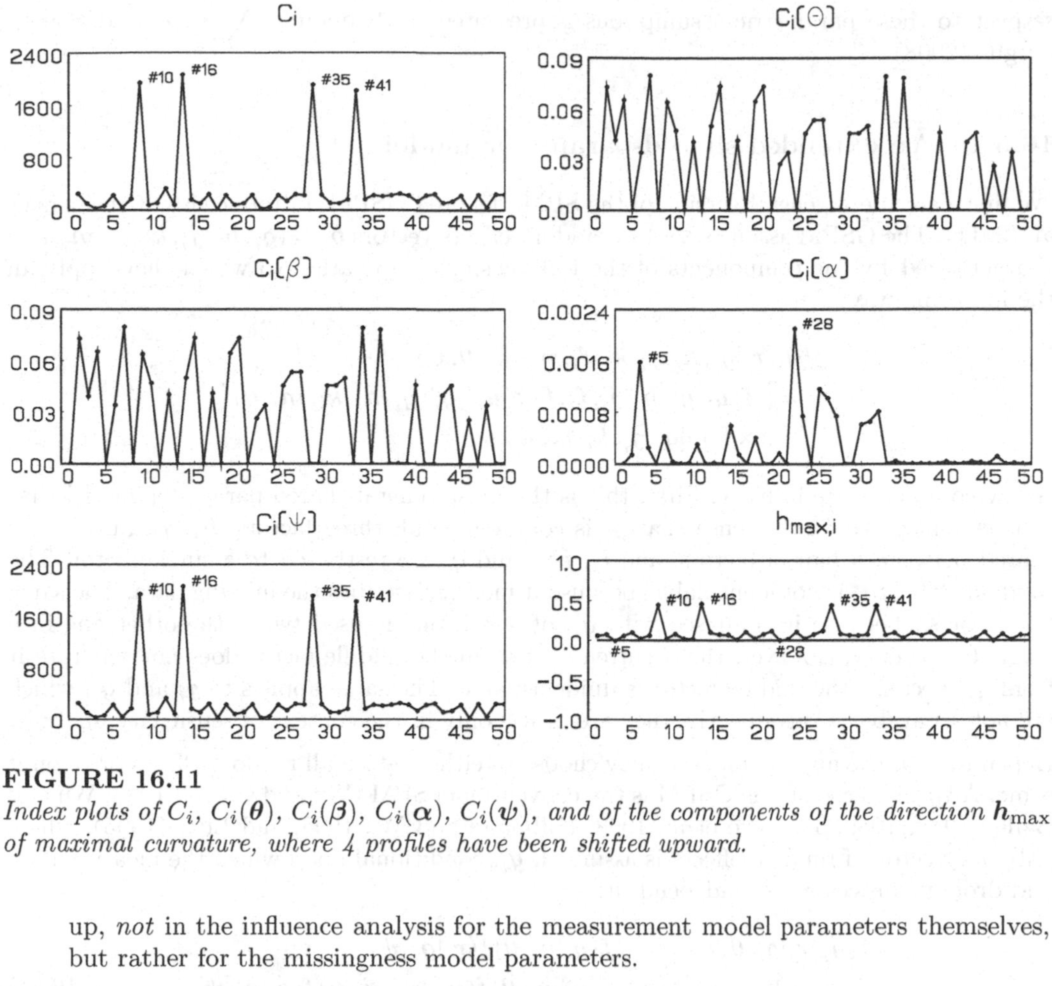

FIGURE 16.11
Index plots of C_i, $C_i(\boldsymbol{\theta})$, $C_i(\boldsymbol{\beta})$, $C_i(\boldsymbol{\alpha})$, $C_i(\boldsymbol{\psi})$, and of the components of the direction $\boldsymbol{h}_{\max}$ of maximal curvature, where 4 profiles have been shifted upward.

up, *not* in the influence analysis for the measurement model parameters themselves, but rather for the missingness model parameters.

Thus, the methodology, while useful, should be used with caution. The occurrence of influential subjects should be seen as a trigger for further investigation, but there are not automated rules as to what a particular deviation precisely means. Jansen et al. (2006) also discuss the non-standard nature of the likelihood ratio test when used to compare posited MAR and NMAR models.

16.6 A Sensitivity Analysis for Shared-Parameter Models

In this section, and based on the SPM work of Creemers et al. (2010, 2011), we present a sensitivity analysis paradigm in the shared-parameter modeling framework (SPM), which is both intuitively appealing and relatively easy to fit. Creemers et al. (2011) introduced a general shared-parameter framework, thereby allowing for the missing data mechanism to be MAR; Creemers et al. (2010) developed sensitivity tools within the framework, which we will briefly present in the rest of this section. We also refer to the overview in Tsonaka, Verbeke, and Lesaffre (2008), who consider shared parameter models without any parametric assumptions for the shared parameters. A theoretical assessment of the sensitivity with respect to these parametric assumptions is presented in Rizopoulos, Verbeke, and Molenberghs (2008).

16.6.1 An extended shared-parameter model

We first describe a general version of the SPM, denoted GSPM, introduced by Creemers et al. (2011). The GSPM assumes a set of random-effects vectors $\boldsymbol{b}_i = (\boldsymbol{g}_i, \boldsymbol{h}_i, \boldsymbol{j}_i, \boldsymbol{k}_i, \boldsymbol{\ell}_i, \boldsymbol{m}_i, \boldsymbol{q}_i)$, characterized by the components of the full-density factorization to which they apply, in the following way:

$$\begin{aligned} & f(\boldsymbol{y}_i, \boldsymbol{r}_i|\boldsymbol{g}_i, \boldsymbol{h}_i, \boldsymbol{j}_i, \boldsymbol{k}_i, \boldsymbol{\ell}_i, \boldsymbol{m}_i, \boldsymbol{q}_i, \boldsymbol{\theta}, \boldsymbol{\psi}) \\ = \; & f(\boldsymbol{y}_i^o|\boldsymbol{g}_i, \boldsymbol{h}_i, \boldsymbol{j}_i, \boldsymbol{\ell}_i, \boldsymbol{\theta}) f(\boldsymbol{y}_i^m|\boldsymbol{y}_i^o, \boldsymbol{g}_i, \boldsymbol{h}_i, \boldsymbol{k}_i, \boldsymbol{m}_i, \boldsymbol{\theta}) \\ & \times f(\boldsymbol{r}_i|\boldsymbol{g}_i, \boldsymbol{j}_i, \boldsymbol{k}_i, \boldsymbol{q}_i, \boldsymbol{\psi}). \end{aligned} \tag{16.20}$$

A few comments are in place. First, this is the most general shared-parameter model that can be considered in the sense that $\boldsymbol{g}_i$ is common to all three factors, $\boldsymbol{h}_i$, $\boldsymbol{j}_i$, and $\boldsymbol{k}_i$ are shared between a pair of factors, and $\boldsymbol{\ell}_i$, $\boldsymbol{m}_i$, and $\boldsymbol{q}_i$ are restricted to a single factor. The random effect $\boldsymbol{m}_i$ is not identifiable, because it merely describes the missing data. The same holds for $\boldsymbol{k}_i$ because it is aliased with $\boldsymbol{q}_i$, of which one is used twice, the other one in a single factor only. However, the occurrence of $\boldsymbol{k}_i$ in the middle factor does not separate it from $\boldsymbol{q}_i$, because the middle factor is unidentifiable. The same applies to $\boldsymbol{j}_i$ and $\boldsymbol{g}_i$, which are not separable. Consequently, they are of use only in the context of sensitivity analysis.

Depending on the application, one may choose to either retain all random effects or to omit some. A special case of this GSPM is the conventional SPM (Wu and Carroll 1988, Wu and Bailey 1988, 1989, TenHave et al. 1998, Follmann and Wu 1995, and Little 1995), where only one vector of random effects is assumed, $\boldsymbol{g}_i$, conditional upon which the measurement and dropout processes are independent:

$$\begin{aligned} f(\boldsymbol{y}_i, \boldsymbol{r}_i|\boldsymbol{g}_i, \boldsymbol{\theta}, \boldsymbol{\psi}) &= f(\boldsymbol{y}_i|\boldsymbol{g}_i, \boldsymbol{\theta}) f(\boldsymbol{r}_i|\boldsymbol{g}_i, \boldsymbol{\psi}) \\ &= f(\boldsymbol{y}_i^o|\boldsymbol{g}_i, \boldsymbol{\theta}) f(\boldsymbol{y}_i^m|\boldsymbol{y}_i^o, \boldsymbol{g}_i, \boldsymbol{\theta}) f(\boldsymbol{r}_i|\boldsymbol{g}_i, \boldsymbol{\psi}). \end{aligned} \tag{16.21}$$

It is useful to gauge the implications of such simplifications, preferably also in terms of the missing data mechanism operating. In particular, Creemers et al. (2011) established conditions under which MAR operates on the one hand, and missingness does not depend on future, unobserved measurements in a longitudinal context on the other hand. Second, in full generality, Model (16.20) may come across as somewhat contrived. It is not the objective to postulate GSPM (16.20) as a model of use in every possible application of SPM, but rather as the most general SPM from which substantively appropriate models follow as sub-classes. Related to this, it appears that general GSPM, represented by (16.20), assumes two different distributions for the outcome vector, i.e., divorcing the observed from the missing components. This is not entirely the case because $\boldsymbol{g}_i$ and $\boldsymbol{h}_i$ still tie both factors together. The impact of $\boldsymbol{j}_i$, $\boldsymbol{k}_i$, $\boldsymbol{\ell}_i$, and $\boldsymbol{m}_i$ is to modify one's latent process in terms of missingness. In other words, the most general model assumes that observed and missing components are governed in part by common processes and partly by separate processes. Third, in principle, one could expand GSPM (16.20) with the densities of the random effects. This is generally not necessary for our purposes, though. Fourth, the assumption of independent random-effects vectors is not restrictive, because association is captured through the sets common to at least two factors. Fifth, conventional SPM (16.21) follows by removing all random effects but $\boldsymbol{g}_i$. Sixth, in agreement with Creemers et al. (2010, 2011) and as mentioned above, also an MAR sub-class of the GSPM can be defined.

The missing-data taxonomy can now be presented symmetrically in all three frameworks. MCAR holds when the measurement and missingness processes are independent (conditional on possible covariates). In the aforementioned modeling frameworks, this means:

$$\begin{array}{rcrcl} \text{SeM} & : & f(\boldsymbol{r}_i|\boldsymbol{y}_i,\boldsymbol{\psi}) & = & f(\boldsymbol{r}_i|\boldsymbol{\psi}), \\ \text{PMM} & : & f(\boldsymbol{y}_i|\boldsymbol{r}_i,\boldsymbol{\theta}) & = & f(\boldsymbol{y}_i|\boldsymbol{\theta}), \\ \text{SPM} & : & f(\boldsymbol{y}_i,\boldsymbol{r}_i|\boldsymbol{b}_i,\boldsymbol{\theta},\boldsymbol{\psi}) & = & f(\boldsymbol{y}_i|\boldsymbol{\theta})f(\boldsymbol{r}_i|\boldsymbol{\psi}), \\ \text{GSPM} & : & \multicolumn{3}{l}{f(\boldsymbol{y}_i^o|\boldsymbol{g}_i,\boldsymbol{h}_i,\boldsymbol{j}_i,\boldsymbol{\ell}_i,\boldsymbol{\theta})f(\boldsymbol{y}_i^m|\boldsymbol{y}_i^o,\boldsymbol{g}_i,\boldsymbol{h}_i,\boldsymbol{k}_i,\boldsymbol{m}_i,\boldsymbol{\theta})\times} \\ & & \multicolumn{3}{r}{\times f(\boldsymbol{r}_i|\boldsymbol{g}_i,\boldsymbol{j}_i,\boldsymbol{k}_i,\boldsymbol{q}_i,\boldsymbol{\psi})} \\ & & \multicolumn{3}{l}{= f(\boldsymbol{y}_i^o|\boldsymbol{h}_i,\boldsymbol{\ell}_i,\boldsymbol{\theta})f(\boldsymbol{y}_i^m|\boldsymbol{y}_i^o,\boldsymbol{h}_i,\boldsymbol{m}_i,\boldsymbol{\theta})f(\boldsymbol{r}_i|\boldsymbol{q}_i,\boldsymbol{\psi}).} \end{array}$$

The mechanism is MAR when missingness depends on the observed outcomes and observed covariates, but not further on the unobserved outcomes. In the various frameworks, MAR applies if and only if

$$\begin{array}{rcl} \text{SeM} & : & f(\boldsymbol{r}_i|\boldsymbol{y}_i,\boldsymbol{\psi}) = f(\boldsymbol{r}_i|\boldsymbol{y}_i^o,\boldsymbol{\psi}), \\ \text{PMM} & : & f(\boldsymbol{y}_i^m|\boldsymbol{y}_i^o,\boldsymbol{r}_i,\boldsymbol{\theta}) = f(\boldsymbol{y}_i^m|\boldsymbol{y}_i^o,\boldsymbol{\theta}), \\ \text{SPM} & : & \int f(\boldsymbol{y}_i^o,\boldsymbol{r}_i,\boldsymbol{b}_i)\left\{f(\boldsymbol{y}_i^m|\boldsymbol{y}_i^o,\boldsymbol{b}_i) - f(\boldsymbol{y}_i^m|\boldsymbol{y}_i^o)\right\}d\boldsymbol{b}_i = 0, \\ \text{GSPM} & : & \frac{\int f(\boldsymbol{y}_i^o|\boldsymbol{g}_i,\boldsymbol{h}_i,\boldsymbol{j}_i)f(\boldsymbol{y}_i^m|\boldsymbol{y}_i^o,\boldsymbol{g}_i,\boldsymbol{h}_i,\boldsymbol{k}_i)f(\boldsymbol{r}_i|\boldsymbol{g}_i,\boldsymbol{j}_i,\boldsymbol{k}_i)f(\boldsymbol{b}_i)\,d\boldsymbol{b}_i}{\int f(\boldsymbol{y}_i^o|\boldsymbol{g}_i,\boldsymbol{j}_i)f(\boldsymbol{r}_i|\boldsymbol{g}_i,\boldsymbol{j}_i)f(\boldsymbol{b}_i)\,d\boldsymbol{b}_i} \\ & & = f(\boldsymbol{y}_i^o|\boldsymbol{g}_i,\boldsymbol{h}_i)f(\boldsymbol{y}_i^m|\boldsymbol{y}_i^o,\boldsymbol{g}_i,\boldsymbol{h}_i)f(\boldsymbol{b}_i)\,d\boldsymbol{b}_i f(\boldsymbol{y}_i^o). \end{array} \tag{16.22}$$

The PMM model in (16.22) states that the unobserved outcome $\boldsymbol{y}_i^m$ given the observed outcome $\boldsymbol{y}_i^o$ does not depend on the missingness. GSPM (16.22) was obtained by Creemers et al. (2011). A proof of the result can be found in their paper. These authors define the following MAR sub-class:

$$f(\boldsymbol{y}_i^o|\boldsymbol{j}_i,\boldsymbol{\ell}_i)f(\boldsymbol{y}_i^m|\boldsymbol{y}_i^o,\boldsymbol{m}_i)f(\boldsymbol{r}_i|\boldsymbol{j}_i,\boldsymbol{q}_i), \tag{16.23}$$

where the random effects $\boldsymbol{g}_i, \boldsymbol{h}_i$ and $\boldsymbol{k}_i$ vanish. Although this subclass does not contain all MAR models, it has intuitive appeal. Examples of models that satisfy GSPM (16.22) without belonging to (16.23) are described in Creemers et al. (2011).

The above considerations can also be applied to the conventional SPM framework. The corresponding MAR sub-class containing only MAR models in this case becomes

$$f(\boldsymbol{y}_i^m|\boldsymbol{y}_i^o,\boldsymbol{b}_i) = f(\boldsymbol{y}_i^m|\boldsymbol{y}_i^o). \tag{16.24}$$

As argued in Section 16.4.5, every NMAR model can be doubled up with an MAR counterpart. This also applies to the (G)SPM framework. Here, the MAR counterpart is found by replacing $f(\boldsymbol{y}_i^m|\boldsymbol{y}_i^o,\boldsymbol{g}_i,\boldsymbol{h}_i,\boldsymbol{k}_i,\boldsymbol{m}_i,\boldsymbol{\theta})$ in (16.20) with (Creemers et al. 2011)

$$h(\boldsymbol{y}_i^m|\boldsymbol{y}_i^o,\boldsymbol{m}_i) = \int_{\boldsymbol{g}_i}\int_{\boldsymbol{h}_i}\int_{\boldsymbol{k}_i} f(\boldsymbol{y}_i^m|\boldsymbol{y}_i^o,\boldsymbol{g}_i,\boldsymbol{h}_i,\boldsymbol{k}_i,\boldsymbol{m}_i)d\boldsymbol{g}_i d\boldsymbol{h}_i d\boldsymbol{k}_i. \tag{16.25}$$

It is clear that this marginalization only changes the predictions for the unobserved data and thus the choice of $h(\cdot)$ does not alter the fit of the model to the observed portion.

16.6.2 A collection of SPM models for the onychomycosis data

In this section, we will analyze the longitudinal profile of unaffected nail length using various sub-models of the Model (16.20), obtained by removing portions of the random-effects

TABLE 16.13
Overview: Sub-models of the general shared-parameter model considered.

GSPM_k	Model	MAR Counterpart
GSPM_1	$f(\boldsymbol{y}_i^o \mid \boldsymbol{g}_i)f(\boldsymbol{y}_i^m \mid \boldsymbol{y}_i^o, \boldsymbol{g}_i)f(\boldsymbol{r}_i \mid \boldsymbol{g}_i)$	$f(\boldsymbol{y}_i^o)f(\boldsymbol{y}_i^m \mid \boldsymbol{y}_i^o)f(\boldsymbol{r}_i)$
GSPM_2	$f(\boldsymbol{y}_i^o \mid \boldsymbol{g}_i, \boldsymbol{h}_i)f(\boldsymbol{y}_i^m \mid \boldsymbol{y}_i^o, \boldsymbol{g}_i, \boldsymbol{h}_i)f(\boldsymbol{r}_i \mid \boldsymbol{g}_i)$	$f(\boldsymbol{y}_i^o)f(\boldsymbol{y}_i^m \mid \boldsymbol{y}_i^o)f(\boldsymbol{r}_i)$
GSPM_3	$f(\boldsymbol{y}_i^o \mid \boldsymbol{g}_i)f(\boldsymbol{y}_i^m \mid \boldsymbol{y}_i^o, \boldsymbol{g}_i, \boldsymbol{k}_i)f(\boldsymbol{r}_i \mid \boldsymbol{g}_i, \boldsymbol{k}_i)$	$f(\boldsymbol{y}_i^o)f(\boldsymbol{y}_i^m \mid \boldsymbol{y}_i^o)f(\boldsymbol{r}_i)$
GSPM_4	$f(\boldsymbol{y}_i^o \mid \boldsymbol{g}_i, \boldsymbol{j}_i)f(\boldsymbol{y}_i^m \mid \boldsymbol{y}_i^o, \boldsymbol{g}_i,)f(\boldsymbol{r}_i \mid \boldsymbol{g}_i, \boldsymbol{j}_i)$	$f(\boldsymbol{y}_i^o \mid \boldsymbol{j}_i)f(\boldsymbol{y}_i^m \mid \boldsymbol{y}_i^o)f(\boldsymbol{r}_i \mid \boldsymbol{j}_i)$
GSPM_5	$f(\boldsymbol{y}_i^o \mid \boldsymbol{h}_i)f(\boldsymbol{y}_i^m \mid \boldsymbol{y}_i^o, \boldsymbol{h}_i, \boldsymbol{k}_i)f(\boldsymbol{r}_i \mid \boldsymbol{k}_i)$	$f(\boldsymbol{y}_i^o)f(\boldsymbol{y}_i^m \mid \boldsymbol{y}_i^o)f(\boldsymbol{r}_i)$
GSPM_6	$f(\boldsymbol{y}_i^o \mid \boldsymbol{j}_i)f(\boldsymbol{y}_i^m \mid \boldsymbol{y}_i^o, \boldsymbol{k}_i)f(\boldsymbol{r}_i \mid \boldsymbol{j}_i, \boldsymbol{k}_i)$	$f(\boldsymbol{y}_i^o \mid \boldsymbol{j}_i)f(\boldsymbol{y}_i^m \mid \boldsymbol{y}_i^o)f(\boldsymbol{r}_i \mid \boldsymbol{j}_i)$
GSPM_7	$f(\boldsymbol{y}_i^o \mid \boldsymbol{h}_i, \boldsymbol{j}_i)f(\boldsymbol{y}_i^m \mid \boldsymbol{y}_i^o, \boldsymbol{h}_i)f(\boldsymbol{r}_i \mid \boldsymbol{j}_i)$	$f(\boldsymbol{y}_i^o \mid \boldsymbol{j}_i)f(\boldsymbol{y}_i^m \mid \boldsymbol{y}_i^o)f(\boldsymbol{r}_i \mid \boldsymbol{j}_i)$
GSPM_8	$f(\boldsymbol{y}_i^o \mid \boldsymbol{h}_i)f(\boldsymbol{y}_i^m \mid \boldsymbol{y}_i^o, \boldsymbol{h}_i)f(\boldsymbol{r}_i)$	$f(\boldsymbol{y}_i^o)f(\boldsymbol{y}_i^m \mid \boldsymbol{y}_i^o)f(\boldsymbol{r}_i)$
GSPM_9	$f(\boldsymbol{y}_i^o \mid \boldsymbol{j}_i)f(\boldsymbol{y}_i^m \mid \boldsymbol{y}_i^o)f(\boldsymbol{r}_i \mid \boldsymbol{j}_i)$	$f(\boldsymbol{y}_i^o \mid \boldsymbol{j}_i)f(\boldsymbol{y}_i^m \mid \boldsymbol{y}_i^o)f(\boldsymbol{r}_i \mid \boldsymbol{j}_i)$
GSPM_{10}	$f(\boldsymbol{y}_i^o)f(\boldsymbol{y}_i^m \mid \boldsymbol{y}_i^o, \boldsymbol{k}_i)f(\boldsymbol{r}_i \mid \boldsymbol{k}_i)$	$f(\boldsymbol{y}_i^o)f(\boldsymbol{y}_i^m \mid \boldsymbol{y}_i^o)f(\boldsymbol{r}_i)$

structure. Table 16.13 presents ten different forms, together with their MAR counterparts. In the face of such a collection, two views can be taken. First, a particular model can be selected for the sake of its interpretation. In the same way, one or a few models can be excluded for this purpose. For example, scenario 10 assumes that the observed measurements are independent, which may be deemed implausible for wide classes of repeated measures designs. Second, models can be considered jointly, for the sake of sensitivity analysis. We will predominantly take the latter view. For illustration, linear mixed models will be considered.

Under all scenarios, the mean structure includes effects of treatment (T_i), time (t_j), their interaction, as well as a random effect. Further, we assume the variance-covariance matrix $\boldsymbol{\Sigma}_i$ to be of the form $\sigma^2 \boldsymbol{I}_7$ with $\boldsymbol{I}_7$ the 7×7 identity matrix. A complete overview for mean and variance-covariance structures in the ten scenarios is presented in Table 16.14. Even with these assumptions, some parameters will not be fully identified. In cases where the user wants to generalize $\boldsymbol{\Sigma}_i$, there is no problem in doing so. Of course, one then has to check carefully whether the resulting model is still identifiable.

Furthermore, a model for the missingness mechanisms needs to be formulated. We allow the sequence $\boldsymbol{r}_i$ to take one of two forms: either a length-7 vector of ones, for a completely observed subject, or a sequence of k ones followed by a sole zero, for someone dropping out at time k, $k = 1, \ldots, 6$. $k \geq 1$, because the initial measurement has been observed for all participants. For the missingness mechanism, we assume a logistic regression with the same parametric structure as for the mean structure. The various forms are presented in Table 16.15. γ_g, γ_k, and γ_j are scale factors for the shared random effects in the missingness model, to avoid forced equality of the variance in the measurement and dropout model. As a result, $\gamma_g g_i \sim N(0, \gamma_g^2 d_g^2)$, $\gamma_k k_i \sim N(0, \gamma_k^2 d_k^2)$, and $\gamma_j j_i \sim N(0, \gamma_j^2 d_j^2)$.

Scenario 1 coincides with the conventional SPM (16.21) and is studied in detail in Creemers et al. (2011). The models were fitted using the `SAS` procedure `NLMIXED`. Parameter estimates (standard errors) for all scenarios are displayed in Table 16.16. Although not much difference can be observed in the estimates for the mean structure parameters, the variance-covariance parameter estimates change considerably over the different scenarios. The total variance, roughly 13.5, is spread out over the residual variance and the random-effects variances. The scale parameters are all estimated to be negative, although non-significantly so. Note that the scale parameter γ_k is never identifiable, because the random effect k_i is the link between the missingness process and the missing observations given the observed ones, which can never be observed. The same problem holds when both random effects g_i and j_i are included in the model; then the scale parameter γ_j is not identifiable. For these analyses, they were

TABLE 16.14

Toenail data. Mean structures for the observed responses given the random effects, mean structures for the missing responses given the observed responses, and the random effects and the variance-covariance matrix for the random effects for the various scenarios. T_i is the treatment indicator for subject i and t_j is the time at which the jth measurement is taken.

Scen.	$E[Y_{ij}^o \mid \boldsymbol{b}_i]$	$E[Y_{ij}^m \mid \boldsymbol{y}_i^o, \boldsymbol{b}_i]$	$\boldsymbol{D}$
1	$\beta_0 + g_i + \beta_1 T_i + \beta_2 t_j + \beta_3 T_i t_j$	$\beta_0 + g_i + \beta_1 T_i + \beta_2 t_j + \beta_3 T_i t_j$	d_g^2
2	$\beta_0 + g_i + h_i + \beta_1 T_i + \beta_2 t_j + \beta_3 T_i t_j$	$\beta_0 + g_i + h_i + \beta_1 T_i + \beta_2 t_j + \beta_3 T_i t_j$	$\begin{pmatrix} d_g^2 & 0 \\ 0 & d_h^2 \end{pmatrix}$
3	$\beta_0 + g_i + \beta_1 T_i + \beta_2 t_j + \beta_3 T_i t_j$	$\beta_0 + g_i + k_i + \beta_1 T_i + \beta_2 t_j + \beta_3 T_i t_j$	$\begin{pmatrix} d_g^2 & 0 \\ 0 & d_k^2 \end{pmatrix}$
4	$\beta_0 + g_i + j_i + \beta_1 T_i + \beta_2 t_j + \beta_3 T_i t_j$	$\beta_0 + g_i + \beta_1 T_i + \beta_2 t_j + \beta_3 T_i t_j$	$\begin{pmatrix} d_g^2 & 0 \\ 0 & d_j^2 \end{pmatrix}$
5	$\beta_0 + h_i + \beta_1 T_i + \beta_2 t_j + \beta_3 T_i t_j$	$\beta_0 + h_i + k_i + \beta_1 T_i + \beta_2 t_j + \beta_3 T_i t_j$	$\begin{pmatrix} d_h^2 & 0 \\ 0 & d_k^2 \end{pmatrix}$
6	$\beta_0 + j_i + \beta_1 T_i + \beta_2 t_j + \beta_3 T_i t_j$	$\beta_0 + k_i + \beta_1 T_i + \beta_2 t_j + \beta_3 T_i t_j$	$\begin{pmatrix} d_j^2 & 0 \\ 0 & d_k^2 \end{pmatrix}$
7	$\beta_0 + h_i + j_i + \beta_1 T_i + \beta_2 t_j + \beta_3 T_i t_j$	$\beta_0 + h_i + \beta_1 T_i + \beta_2 t_j + \beta_3 T_i t_j$	$\begin{pmatrix} d_j^2 & 0 \\ 0 & d_h^2 \end{pmatrix}$
8	$\beta_0 + h_i + \beta_1 T_i + \beta_2 t_j + \beta_3 T_i t_j$	$\beta_0 + h_i + \beta_1 T_i + \beta_2 t_j + \beta_3 T_i t_j$	d_h^2
9	$\beta_0 + j_i + \beta_1 T_i + \beta_2 t_j + \beta_3 T_i t_j$	$\beta_0 + \beta_1 T_i + \beta_2 t_j + \beta_3 T_i t_j$	d_j^2
10	$\beta_0 + \beta_1 T_i + \beta_2 t_j + \beta_3 T_i t_j$	$\beta_0 + k_i + \beta_1 T_i + \beta_2 t_j + \beta_3 T_i t_j$	d_k^2

fixed to 1. However, this un-identifiability provides us with the opportunity to perform sensitivity analysis, as will be outlined in the next section. These parameters will then play the role of sensitivity parameters.

At first sight, negative estimates for the sensitivity parameters may appear to be counter-intuitive. However, they merely imply that, with an increase of unaffected nail length, patients are less inclined to remain on study. It ought to be kept in mind that the active period of the compound is the first three months only. Thereafter, maintenance visits take place. Those patients fully cured are less likely to remain faithful to the trial schedule.

Using empirical Bayes estimates, predictions for the incomplete profiles can be obtained. The MAR counterpart reduces all predictions to the same profile, while in most cases the NMAR prediction yields different evolutions per subjects. This is because the prediction of $\boldsymbol{Y}_i^m$ depends on $\boldsymbol{y}_i^o$ only, in case of the MAR model, whereas for NMAR also $\boldsymbol{g}_i$, $\boldsymbol{h}_i$, and $\boldsymbol{k}_i$ intervene. The fact that the various scenarios produce quite similar predictions enhances confidence in the conclusions reached about the evolution of toenail infection and the differences between both treatment arms. Furthermore, it is entirely possible to implement a more general variance-covariance structure. Again, one then has to check the identifiability of the resulting model.

For Scenarios 6, 9, and 10, NMAR and MAR predictions coincide. For Scenario 9 this is clearly due to the fact that the NMAR model is the same as its MAR counterpart (see Table 16.13). In Scenarios 6 and 10 this can be explained by the small (and highly non-significant) variance of the random effect k. While the fixed-effect estimates for Scenario 10 are somewhat different from the other scenarios, this is still minor in view of the precision estimates. Nevertheless, Scenario 10 is qualitatively different from the others, in the sense

TABLE 16.15
Toenail data. Models for the missingness mechanism in the different scenarios.

Scenario	logit $[P(R_{ij}=1\|R_{i,j-1}=0,b_i,T_i,t_j,\gamma)]$
1	$\gamma_0+\gamma_g g_i+\gamma_1 T_i+\gamma_2 t_j+\gamma_3 T_i t_j$
2	$\gamma_0+\gamma_g g_i+\gamma_1 T_i+\gamma_2 t_j+\gamma_3 T_i t_j$
3	$\gamma_0+\gamma_g g_i+\gamma_k k_i+\gamma_1 T_i+\gamma_2 t_j+\gamma_3 T_i t_j$
4	$\gamma_0+\gamma_g g_i+\gamma_j j_i+\gamma_1 T_i+\gamma_2 t_j+\gamma_3 T_i t_j$
5	$\gamma_0+\gamma_k k_i+\gamma_1 T_i+\gamma_2 t_j+\gamma_3 T_i t_j$
6	$\gamma_0+\gamma_k k_i+\gamma_j j_i+\gamma_1 T_i+\gamma_2 t_j+\gamma_3 T_i t_j$
7	$\gamma_0+\gamma_j j_i+\gamma_1 T_i+\gamma_2 t_j+\gamma_3 T_i t_j$
8	$\gamma_0+\gamma_1 T_i+\gamma_2 t_j+\gamma_3 T_i t_j$
9	$\gamma_0+\gamma_j j_i+\gamma_1 T_i+\gamma_2 t_j+\gamma_3 T_i t_j$
10	$\gamma_0+\gamma_k k_i+\gamma_1 T_i+\gamma_2 t_j+\gamma_3 T_i t_j$

that the random effects within the observed sequence are assumed independent; this is rather unrealistic.

16.6.3 A sensitivity analysis in the shared-parameter framework

In this section, we will propose a method for sensitivity analysis within the GSPM framework. This can, of course, take several forms. First, like in the previous section, a collection of sub-models can be considered and stability of inferences compared across them. Second, like will also be done here, sensitivity analysis can be placed within a particular model, making use of its non-identifiability. Technically, this means that a model actually corresponds to an infinite collection of models. For the sake of illustration, this will be done within Scenario 6.

As stated, some of the GSPM models in Section 16.6.2 were not fully identifiable. Consider, for example, Scenario 6, which only includes the random effects j and k. Because the second term in (16.20) $(f(\boldsymbol{y}_i^m|\boldsymbol{y}_i^o,\boldsymbol{g}_i,\boldsymbol{h}_i,\boldsymbol{k}_i,\boldsymbol{m}_i,\boldsymbol{\theta}))$ is never observed, the sensitivity parameter γ_k is not identifiable. Only by fixing this γ_k, the *sensitivity parameter*, can the model be fitted. The principle is to let γ_k take different values and compare the inferences across fits. To this end, a grid in the sensitivity parameter space will be defined, and γ_k will take all values in this grid. For each of the values, the model will be fitted and then, based on this model, y_i^m will be imputed multiple times. This will be done using conventional multiple imputation (Little and Rubin 2002, Molenberghs and Kenward 2007). Details can be found in Creemers et al. (2010). Re-fitting the model using these imputations and summarizing the different inferences resulting from different imputations into a single set of inferences, purports to sensitivity analysis. Creemers et al. (2010) outline in detail how this sensitivity analysis can be implemented. We apply it here to Scenario 6.

We will test the significance of the treatment effect, the time effect, and the treatment × time interaction. Next to this, we will also consider the variance parameters σ^2 and dg^2. Figure 16.12 shows results for the test of a treatment effect at the last time point, i.e., the seventh measurement, taken at month 12. The left plot contains the values for the test statistic, while the right plot presents the corresponding p-values. The points in both plots correspond to dataset-specific values. Per grid point, these 5 different values can be reduced to a single one using multiple imputation; they are indicated by the lines. The line with the overall p values is located above the 0.05 line, which indicates there is no significant

TABLE 16.16
Toenail data. Parameter estimates and standard errors for the SPM fits.

Effect	Par.	Scen. 1	Scen. 2	Scen. 3	Scen. 4	Scen. 5
				Mean Structure		
Measurement						
interc.	β_0	2.51(0.25)	2.51(0.25)	2.51(0.25)	2.51(0.25)	2.52(0.25)
treat	β_1	0.26(0.35)	0.26(0.35)	0.26(0.35)	0.26(0.35)	0.25(0.35)
time	β_2	0.56(0.02)	0.56(0.02)	0.56(0.03)	0.56(0.02)	0.56(0.02)
treat×time	β_3	0.05(0.03)	0.05(0.03)	0.05(0.03)	0.05(0.03)	0.05(0.03)
Dropout						
interc.	γ_0	-3.13(0.28)	-3.11(0.78)	-3.21(0.86)	-3.21(0.86)	-3.11(0.28)
treat	γ_1	-0.54(0.44)	-0.55(0.45)	-0.54(0.45)	-0.54(0.45)	-0.54(0.44)
time	γ_2	0.04(0.04)	0.04(0.06)	0.04(0.07)	0.04(0.07)	0.03(0.04)
treat×time	γ_3	0.04(0.06)	0.04(0.06)	0.04(0.06)	0.04(0.06)	0.04(0.06)
				Variance-Covariance Structure		
Measurement						
resid. var.	σ^2	6.94(0.25)	6.94(0.25)	6.94(0.25)	6.94(0.25)	6.94(0.25)
rand. int. var.	d_g^2	6.51(0.63)	1.26(8.55)	6.51(0.63)	6.37(1.53)	
	d_h^2		5.25(8.57)			6.49(0.63)
	d_k^2			0.17(1.71)		0.00(0.25)
	d_j^2				0.141(1.394)	
Dropout						
scale factor	γ_g	-0.08(0.06)	-0.41(2.89)	-0.08(0.07)	-0.10(0.28)	
	γ_k			fix:1		fix:1
	γ_j				fix:1	
rand. int. var.	$\gamma_g^2 d_g^2$	0.04(0.06)	0.21(1.56)	0.04(0.07)	0.07(0.35)	
Effect	**Par.**	**Scen. 6**	**Scen. 7**	**Scen. 8**	**Scen. 9**	**Scen. 10**
				Mean Structure		
Measurement						
interc.	β_0	2.510(0.25)	2.51(0.25)	2.52(0.25)	2.51(0.25)	2.55(0.18)
treat	β_1	0.26(0.35)	0.26(0.35)	0.25(0.35)	0.26(0.35)	0.22(0.25)
time	β_2	0.56(0.02)	0.56(0.02)	0.56(0.02)	0.56(0.02)	0.58(0.03)
treat×time	β_3	0.05(0.03)	0.05(0.03)	0.05(0.03)	0.05(0.03)	0.04(0.04)
Dropout						
interc.	γ_0	-3.21(0.86)	-3.21(0.78)	-3.11(0.28)	-3.13(0.28)	-3.11(0.28)
treat	γ_1	-0.54(0.45)	-0.55(0.45)	-0.54(0.44)	-0.54(0.45)	-0.54(0.44)
time	γ_2	0.04(0.07)	0.04(0.06)	0.03(0.04)	0.04(0.04)	0.03(0.04)
treat×time	γ_3	0.04(0.06)	0.04(0.06)	0.04(0.06)	0.04(0.06)	0.04(0.06)
				Variance-Covariance Structure		
Measurement						
resid. var.	σ^2	6.94(0.25)	6.94(0.25)	6.94(0.25)	6.94(0.25)	13.53(0.44)
rand. int. var.	d_g^2					
	d_h^2		5.25(1.86)	6.49(0.63)		
	d_k^2	0.16(1.7)				0.00(0.00)
	d_j^2	6.51(0.63)	0.26(8.51)		6.51(0.63)	
Dropout						
scale factor	γ_k	fix:1				fix:1
	γ_j	-0.08(0.07)	-0.41(2.88)		-0.08(0.06)	
rand. int. var.	$\gamma_j^2 d_j^2$	0.04(0.07)	0.21(1.55)		0.04(0.06)	

treatment difference in unaffected nail length at month 7. For large negative values of γ_k, some dataset-specific p values are below 0.05. This indicates that ranging over γ_k implies some fluctuation in the p-values, but it never becomes significant.

In Figures 16.13 and 16.14, parameter estimates and p-values to test their significance are presented. The variance parameters σ^2 and dg^2 have p-values of order, respectively, 10^{-5} and 10^{-4} and thus are highly significant. Mean structure parameters β_1 (treatment effect) and β_3 (interaction effect) are not significant. Again, from the plot with p-values for β_3, it turns out that varying γ_k controls the evidence for the effect of treatment, in as far as it is captured by the treatment-by-time interaction. It should be noted that, over the entire range

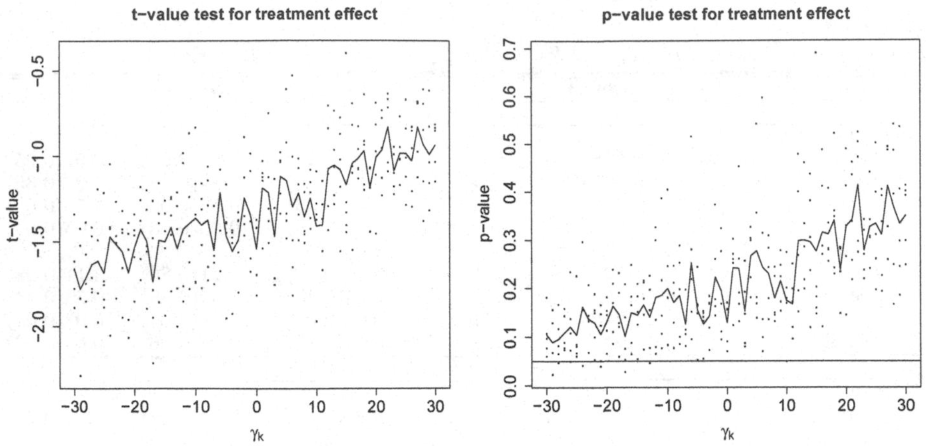

FIGURE 16.12
Toenail data. Results for sensitivity analysis applied under Scenario 6: t-test for difference between treatment effect at the last time point. (t-values (left) and p-values (right)). Points correspond to the dataset specific values (5 for every grid point), lines correspond to overall value.

of the sensitivity parameter, the p-values for β_3 remain consistent with no strong difference in evolution. Given that this is the main scientific parameter, this analysis strongly elevates the level of confidence with the conclusions reached. Further, a clear time effect (β_2) is observed (p values of the order 10^{-5}). Because parameter estimates for β_2 are increasing with increasing γ_k, the larger the sensitivity parameter, the bigger the time effect. Although the value of γ_k changes across models, the main result is qualitatively invariant. The inferences resulting from examining the p-values are only moderately sensitive to the choice of γ_k; hence they can interpreted with confidence, even though the confidence statements issue a somewhat more subtle message.

16.7 A Latent-Class Mixture Model for Incomplete Longitudinal Gaussian Data

Beunckens et al. (2008) proposed an extension of the SPM framework, capturing possible heterogeneity between the subjects not measured through covariates but rather through a latent variable. The authors call this model a *latent-class mixture model.* Next to one or more so-called shared parameters, $\boldsymbol{b}_i$, the model contains a latent variable, $\boldsymbol{Q}_i$, dividing the population into g subgroups. This latent variable is a vector of group indicators $\boldsymbol{Q}_i = (Q_{i1}, \ldots, Q_{ig})$, defined as $Q_{ik} = 1$, if subject i belongs to group k, and 0 otherwise. The measurement process as well as the dropout process depend on this latent variable, directly and through the subject-specific effects $\boldsymbol{b}_i$. The distribution of $\boldsymbol{Q}_i$ is multinomial and defined by $P(Q_{ik} = 1) = \pi_k$ $(k = 1, \ldots, g)$, where π_k denotes the group or component probability, also termed prior probabilities of the components. These are restricted through $\sum_{k=1}^{g} \pi_k = 1$.

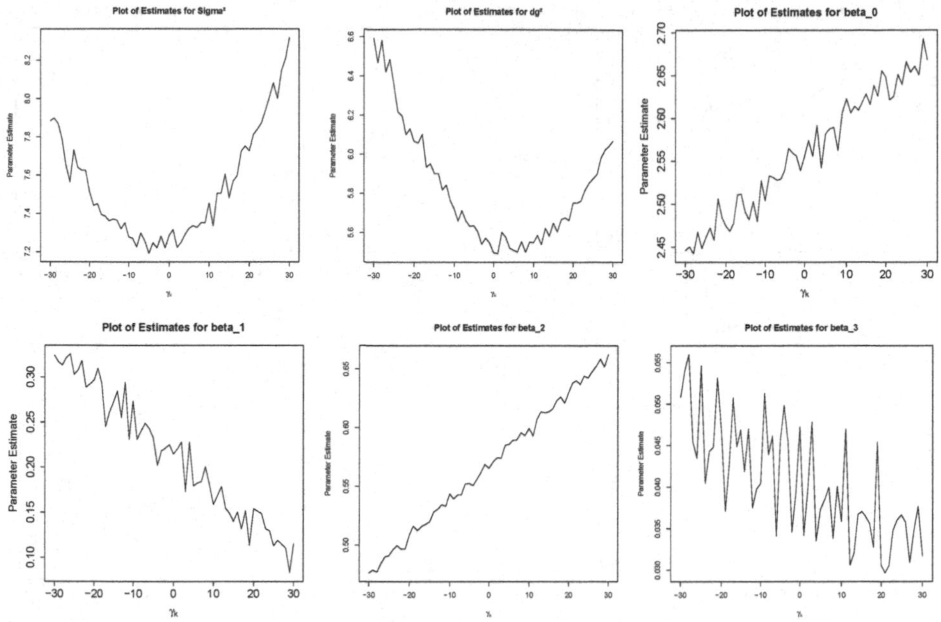

FIGURE 16.13
Toenail data. Results for sensitivity analysis applied to Scenario 6: parameter estimates for the mean and variance parameters.

The measurement process will be modeled by a heterogeneity linear mixed model (Verbeke and Lesaffre 1996, Verbeke and Molenberghs 2000): $\boldsymbol{Y}_i|q_{ik}=1,\boldsymbol{b}_i \sim N(\boldsymbol{X}_i\boldsymbol{\beta}_k+\boldsymbol{Z}_i\boldsymbol{b}_i,\boldsymbol{\Sigma}_i^{(k)})$, where $\boldsymbol{X}_i$ and $\boldsymbol{Z}_i$ are design matrices, $\boldsymbol{\beta}_k$ component-dependent fixed effects, $\boldsymbol{b}_i$ denote the shared parameters, following a mixture of g normal distributions with mean vectors $\boldsymbol{\mu}_k$ and covariance matrices $\boldsymbol{D}_k$, i.e., $\boldsymbol{b}_i|q_{ik}=1 \sim N(\boldsymbol{\mu}_k,\boldsymbol{D}_k)$ and thus $\boldsymbol{b}_i \sim \sum_{k=1}^{g}\pi_k N(\boldsymbol{\mu}_k,\boldsymbol{D}_k)$. The measurement error terms $\boldsymbol{\varepsilon}_i$ follow a normal distribution with mean zero and covariance matrix $\boldsymbol{\Sigma}_i^{(k)}$ and are independent of the shared parameters. The mean and the variance of $\boldsymbol{Y}_i$, assuming that the shared effects are "calibrated," i.e., $\sum_{k=1}^{g}\pi_k\boldsymbol{\mu}_k=\boldsymbol{0}$, take the form:

$$\begin{aligned} E(\boldsymbol{Y}_i) &= \boldsymbol{X}_i\sum_{k=1}^{g}\pi_k\boldsymbol{\beta}_k, \\ \text{Var}(\boldsymbol{Y}_i) &= \boldsymbol{Z}_i'\left[\sum_{k=1}^{g}\pi_k\boldsymbol{\mu}_k^2+\sum_{k=1}^{g}\pi_k\boldsymbol{D}_k\right]\boldsymbol{Z}_i+\sum_{k=1}^{g}\pi_k\boldsymbol{\Sigma}_i^{(k)}. \end{aligned}$$

Assuming that the first measurement Y_{i1} is obtained for every subject in the study, the model for the dropout process is based on a logistic regression for the probability of dropout at occasion j, given: (1) the subject was still in the study up to occasion j, (2) $\boldsymbol{b}_i$, and (3) that the subject belongs to the kth component. Denote this probability by $g_{ij}(\boldsymbol{w}_{ij},\boldsymbol{b}_i,q_{ik})$, in which $\boldsymbol{w}_{ij}$ is a vector containing all relevant covariates: $g_{ij}(\boldsymbol{w}_{ij},\boldsymbol{b}_i,q_{ik})=P(D_i=j|D_i\geq j,\boldsymbol{w}_{ij},\boldsymbol{b}_i,q_{ik}=1)$. We then assume that $g_{ij}(\boldsymbol{w}_{ij},\boldsymbol{b}_i,q_{ik})$ satisfies $\text{logit}[g_{ij}(\boldsymbol{w}_{ij},\boldsymbol{b}_i,q_{ik})]=\boldsymbol{w}_{ij}\gamma_k+\boldsymbol{\lambda}\boldsymbol{b}_i$. The joint likelihood of the measurement and dropout processes takes the form:

$$f(\boldsymbol{y}_i,d_i)=\sum_{k=1}^{g}\pi_k\int f(\boldsymbol{y}_i|q_{ik}=1,\boldsymbol{b}_i,\boldsymbol{X}_i,\boldsymbol{Z}_i)f(d_i|q_{ik}=1,\boldsymbol{b}_i,\boldsymbol{w}_i)f_k(\boldsymbol{b}_i)d\boldsymbol{b}_i,$$

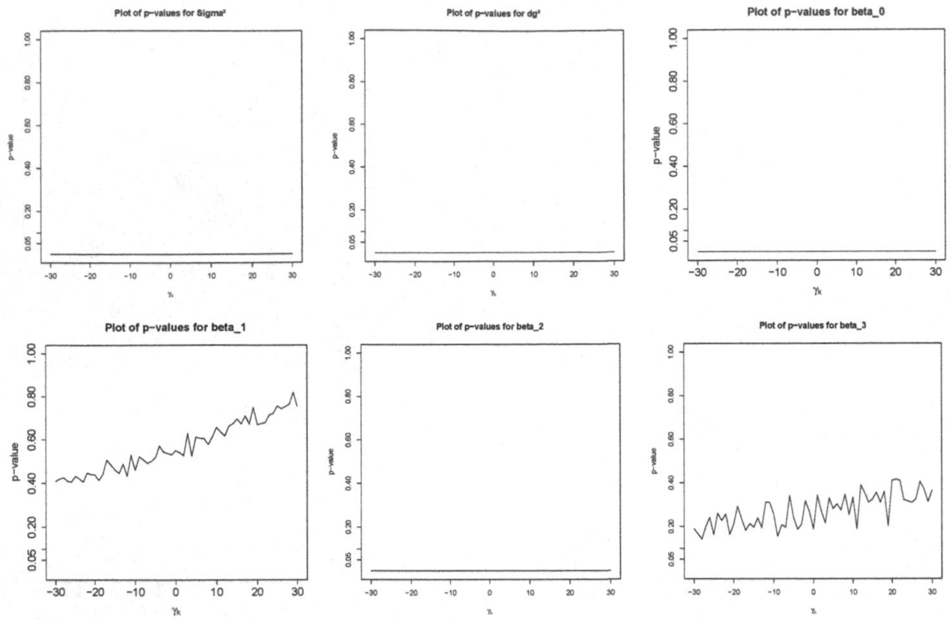

FIGURE 16.14
Toenail data. Results for sensitivity analysis applied to Scenario 6: p-values for significance for the mean and variance parameters.

with $f(\boldsymbol{y}_i|q_{ik} = 1, \boldsymbol{b}_i, \boldsymbol{X}_i, \boldsymbol{Z}_i)$ the density function of the normal distribution $N(\boldsymbol{X}_i\boldsymbol{\beta}_k + \boldsymbol{Z}_i\boldsymbol{b}_i, \boldsymbol{\Sigma}_i^{(k)})$, $f_k(\boldsymbol{b}_i)$ the density function of $N(\boldsymbol{\mu}_k, \boldsymbol{D}_k)$, and

$$f(d_i|q_{ik} = 1, \boldsymbol{b}_i, \boldsymbol{w}_i) = \begin{cases} g_{id_i}(\boldsymbol{w}_{id_i}, \boldsymbol{b}_i, q_{ik}) \times \prod_{j=2}^{d_i-1}[1 - g_{ij}(\boldsymbol{w}_{ij}, \boldsymbol{b}_i, q_{ik})] & \text{if incomplete,} \\ \prod_{j=2}^{n_i}[1 - g_{ij}(\boldsymbol{w}_{ij}, \boldsymbol{b}_i, q_{ik})] & \text{if complete.} \end{cases}$$

Note that the dropout model can depend, not only on the outcomes, but also on relevant covariates such as treatment allocation, time, gender, age, etc. Not all models that can be formulated in this way are identified, so restrictions might be needed. Beunckens et al. (2007) proposed and EM-based estimation strategy. They also conducted simulations to explore the behavior of the model and found it to be very satisfactory and stable from a numerical standpoint.

16.7.1 Classification

One can also classify the subjects into the different latent subgroups of the fitted model. Through the structure of the latent-class mixture model, the subdivision of the population in latent groups depends on the number of observed measurements. The *posterior probabilities*, used to this effect, are. We have that $P(Q_{ik} = 1) = \pi_k$, thus the component probabilities π_k, express how likely the ith subject is to belong to group k, without using information from outcomes and dropout pattern. For this reason, the component probabilities are often

TABLE 16.17
Depression trial. Information criteria AIC and BIC, for models with dropout model (16.26) or (16.27), and $g = 1, 2, 3$.

Model	Dropout Model	g	# Par	-2ℓ	AIC	BIC
1	$\gamma_{0,k} + \gamma_{1,k}\, t_j$	1	10	4676.07	4696.08	4727.44
2	$\gamma_{0,k} + \gamma_{1,k}\, t_j$	2	14	4662.37	**4690.37**	4734.27
3	$\gamma_{0,k} + \gamma_{1,k}\, t_j$	3	18	4662.03	4698.03	4754.48
4	$\gamma_{0,k} + \gamma_{1,k}\, t_j + \lambda\, b_i$	1	11	4669.12	4691.12	**4725.61**
5	$\gamma_{0,k} + \gamma_{1,k}\, t_j + \lambda\, b_i$	2	15	4662.02	4692.02	4739.06

called *prior* probabilities. The *posterior* probability for subject i to belong to the kth group is given by

$$\pi_{ik} = \left. \frac{\pi_k f_{ik}(\boldsymbol{y}_i^o, d_i|\boldsymbol{\theta}, \boldsymbol{\psi}, \boldsymbol{\alpha})}{\sum_{k=1}^{g} \pi_k f_{ik}(\boldsymbol{y}_i^o, d_i|\boldsymbol{\theta}, \boldsymbol{\psi}, \boldsymbol{\alpha})} \right|_{\widehat{\boldsymbol{\Omega}}}.$$

A subject is classified into the component with maximum posterior probability.

Clearly, we should be cautious with the resulting classification, since for a particular subject i, the vector of posterior probabilities is given by $\boldsymbol{\pi}_i = (\pi_{i1}, \ldots, \pi_{ig})$ with $\sum_{k=1}^{g} \pi_{ik} = 1$. For a good comfort level, one of these posterior probabilities for subject i would lie close to 1. However, a scenario would be that two or more posterior probabilities are almost equal, of which one is the maximum of all posterior probabilities for that particular subject. This makes classification nearly random and misclassification is likely to occur. Therefore, rather than merely considering the classification of subjects into the latent subgroups, it is instructive to inspect the posterior probabilities in full. Furthermore, we can vary the number of latent groups g and explore the sensitivity of the classification to the number of latent subgroups considered.

16.7.2 Analysis of the depression trial

Beunckens et al. (2007) modeled the $HAMD_{17}$ score using a model with fixed intercept, treatment indicator, the baseline $HAMD_{17}$ score, a linear and quadratic time variable, and the interaction between treatment and time. In a LCMM with two groups, the parameter values for these fixed effects are assumed to be equal for both latent subgroups. The measurement error terms are assumed to be independent and to follow a normal distribution with mean 0 and variance σ^2. A shared intercept, b_i, is included in the measurement model, which follows a mixture of g normal distributions with different means, $\mu_1, \ldots, \mu_g$ respectively, but with equal variance d^2.

Two versions of the dropout model are considered, without and with the shared random effect. The first one is

$$\text{logit}[g_{ij}(\boldsymbol{w}_{ij}, \boldsymbol{b}_i, q_{ik})] = \gamma_{0,k} + \gamma_{1,k}\, t_j \tag{16.26}$$

and the second one

$$\text{logit}[g_{ij}(\boldsymbol{w}_{ij}, \boldsymbol{b}_i, q_{ik})] = \gamma_{0,k} + \gamma_{1,k}\, t_j + \lambda\, b_i, \tag{16.27}$$

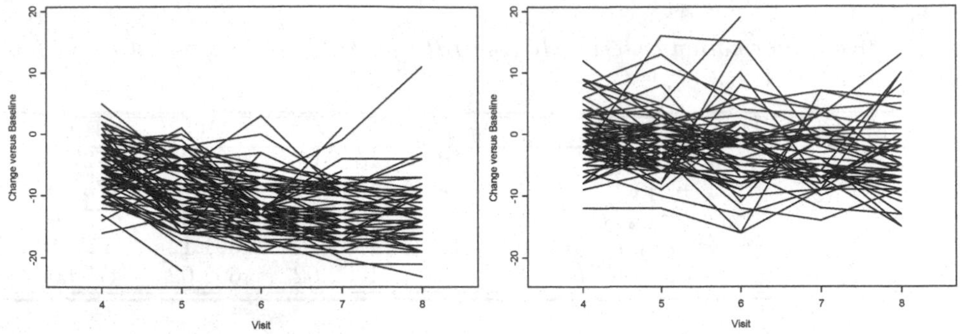

FIGURE 16.15
Depression trial. Classification of the subjects of the depression trial based on a latent-class mixture model. The left panel corresponds to patients classified into first group, he right panel to patients classified into second one.

TABLE 16.18
Depression trial. Parameter estimates (standard errors) and p-values for the latent-class mixture model applied to the depression trial.

Effect	Estimate	(s.e.)	p-value
Measurement Model			
Intercept : β_0	23.17	(3.75)	< 0.0001
Treatment : β_1	2.69	(1.49)	0.072
Time : β_2	-6.18	(1.18)	< 0.0001
Time $\times$ Treatment : β_3	-0.52	(0.24)	0.028
Baseline : β_4	-0.42	(0.07)	< 0.0001
Time $\times$ Time : β_5	0.41	(0.10)	< 0.0001
Measurement Error : σ	4.24	(0.13)	< 0.0001
Dropout Model			
Intercept Group 1 : $\gamma_{0,1}$	-8.58	(3.57)	0.009
Time Group 1 : $\gamma_{1,1}$	0.83	(0.44)	0.056
Intercept Group 2 : $\gamma_{0,2}$	-1.35	(1.28)	0.292
Time Group 1 : $\gamma_{1,2}$	-0.05	(0.20)	0.793
Shared Effects			
Mean Shared Intercept Group 1 : μ_1	-3.64	(0.43)	< 0.0001
Variance Shared Intercept : d	2.67	(0.50)	< 0.0001
Prior probability Group 1 : $\pi_1 = \pi$	0.48	(0.10)	< 0.0001
Loglikelihood		-2331.18	

where t_j is the jth visit.

An overview of the models considered is given in Table 16.17. Because assessing the number of components by a classical likelihood ratio test is not valid in the mixture model framework (McLachlan and Peel 2000), we informally assess the fit by Akaike's Information Criterion (AIC) and the Bayesian Information Criterion (BIC).

Table 16.17 shows that when assuming dropout model (16.26), AIC opts for the model with two latent subgroups (Model 2), whereas BIC gives preference to the shared-parameter

TABLE 16.19
Depression trial. Classification of subjects based on the magnitude of posterior probabilities π_{i1}.

π_{i1}	Classification	# Patients
$0.80 \rightarrow 1.00$	Clearly Group 1	61
$0.60 \rightarrow 0.80$	Group 1	8
$0.55 \rightarrow 0.60$	Doubtful, more likely Group 1	5
$0.45 \rightarrow 0.55$	Uncertain	8
$0.40 \rightarrow 0.45$	Doubtful, more likely Group 2	5
$0.20 \rightarrow 0.40$	Group 2	19
$0.00 \rightarrow 0.20$	Clearly Group 2	64

TABLE 16.20
Depression trial. Estimates, standard errors, and p-values for the treatment effect at visit 8, as well as the treatment-by-time interaction, for the latent-class mixture model (Model 4 in Table 16.17), the shared-parameter model (Model 2 in Table 16.17), the pattern-mixture model, and both selection models, assuming either MAR (which equals MCAR) or NMAR.

	Tr. Eff.(Endp.)		Tr.×Time	
Model	Est.(s.e.)	p	Est.(s.e.)	p
Latent-Class Mixture Model	−1.44(0.91)	0.114	−0.52(0.23)	0.028
Shared-Parameter Model	−1.69(0.93)	0.069	-0.50(0.24)	0.035
Pattern-Mixture Model	−2.01(1.20)	0.096	−0.55(0.31)	0.077
MCAR≡MAR Selection Model	−2.17(1.25)	0.082	−0.58(0.32)	0.068
NMAR Selection Model	−2.16(1.24)	0.081	−0.57(0.31)	0.068

model (Model 1). Further, in case of dropout model (16.27) however, both information criteria select the shared-parameter model (Model 4). Note that, since the dropout model in Model 1 does not depend on the shared intercept, the dropout model and the measurement model are independent, resulting in the MCAR assumption, whereas in Model 2, the dropout model is linked to the measurement model through the latent classes (NMAR). Overall, the AIC criterion prefers Model 2, the 2-component latent-class mixture model with no random effect in the dropout model, whereas BIC picks Model 4, the classical shared-parameter model. Because of these discrepancies, we take a more detailed look at the latent-class mixture model with two components, indicated by AIC, whereas we will consider the classical shared-parameter model in a sensitivity analysis in the next section.

Parameter estimates (standard errors) of the 2-component LCMM are shown in Table 16.18. The 170 patients split into two groups of 79 and 91 patients, respectively (Figure 16.15). The first group consists of patients with lower $HAMD_{17}$ scores, that continue to decrease over time (improvement). The second group contains patients with a higher change versus baseline compared to the patients from the first group. Their changes of $HAMD_{17}$ score fluctuate around 0, with more or less time-constant profiles. The differences between both groups are confirmed by formal tests, although the groups do neither differ in terms of average baseline $HAMD_{17}$, nor as a function of treatment. Beunckens et al. (2007) labeled the groups as acute versus chronic depression. The groups differ in terms of dropout group. The first latent group mainly contains patients who complete the study, 62 in total, the other

17 dropping out late. In the second group, 44 patients drop out, many of them at earlier visits. Further, patients in the acute group are younger. This is important insight, even though it may be hard to disentangle the causal relationship between age and chronicity.

To assess the stability of classification, we do so by following the guidelines of Table 16.19. Only 8 patients have an "uncertain" classification. For 152 patients, thee maximal posterior probability is above 0.60.

16.7.3 A sensitivity analysis for the depression trial

In this section, we apply LCMMs as a sensitivity analysis tool. In addition to the two-component latent-class mixture model of Section 16.7.2, a few other models are considered. All contain the same fixed effects as in the two-component LCMM, i.e., intercept, treatment, time, baseline, time^2, and treatment-by-time interaction.

The SPM, selected by the BIC in Section 16.7.2, includes a shared intercept, conditional upon which the measurement model follows a normal distribution, with the dropout process based on (16.27). Next, the Diggle-Kenward (DK) model combines a multivariate normal model for the measurement process with a logistic regression model for the dropout process. The logistic dropout model will take the form

$$\text{logit}\left[P(D_i = j \mid D_i \geq j, \boldsymbol{h}_{ij}, y_{ij}, \boldsymbol{\Omega})\right] = \psi_0 + \psi_1 y_{i,j-1} + \psi_2 y_{ij},$$

once under MAR ($\psi_2 = 0$), and once allowing for NMAR. Finally, a PMM, taking the same form as the DK models, except with separate intercepts and slopes for each of the dropout patterns, is considered. Note that classification into groups in the LCMM is data driven, whereas the PMM starts a priori from the dropout groups.

Since the main interest of the depression trial was in the treatment effect at the last visit, Table 16.20 shows results for this effect. The p-values resulting are very similar, between 0.07 and 0.11, yielding the same conclusion for the treatment effect at visit 8. Thus, the significance results here are not sensitive to the model used, and hence more trust can be put into the conclusion. This notwithstanding, note that using both the two-component latent-class mixture model and the classical SPM, the standard error is reduced by 0.3 units, compared to either selection model, or PMM, resulting in a more accurate confidence interval for the treatment effect at the last visit.

We also explore the sensitivity of the treatment-by-time interaction. The p-values are clearly moving around the significance level of 0.05. Whereas under the LCMM and the SPM the p-value is about 0.03, the p-value under both DK models and the PMM is around 0.07. While one should be cautious with over-interpretation of p-values, there are contexts, such as regulated clinical trials, where strict decision rules are implemented. In such a case and when in addition the treatment by time interaction is the primary effect, the LCMM and the shared-parameter model would lead to a claim of significance, whereas this would not be justified with neither the SEM nor the PMM.

16.8 Further Methods

In this section, we present a brief overview of some further approaches to sensitivity analysis classes deserving of attention.

Pattern-mixture models can be considered for their own sake, typically because they are appropriate, in given circumstances, to answer a particular scientific question. Furthermore, a range of authors have considered PMM as a useful contrast to selection models either (1) to answer the same scientific question, such as marginal treatment effect or time evolution, based on these two rather different modeling strategies, or (2) to gain additional insight by supplementing the selection model results with those from a pattern-mixture approach. Pattern-mixture models also have a special role in some multiple imputation-based sensitivity analyses.

Pattern-mixture models have the advantage that, strictly speaking, only the observed data need to be modeled. However, whenever aspect of the marginal distribution of the outcome is of interest (e.g., the marginal evolution over time), the unobserved measurements must be modeled jointly with the observed ones. Various methods have been proposed in the statistical literature (Molenberghs and Kenward 2007), all based on very specific assumptions about the conditional distribution of the missing outcomes, given the observed ones. This points to the so-called under-identification of the PMM: within a pattern, there is by definition no information on the unobserved outcomes given the observed ones. There are several ways forward (Molenberghs and Kenward, 2007). First, parametric extrapolation can be used, where the model fitted to the observed data is believed to also hold for the missing data in the same pattern, i.e., the parametric model is extrapolated. Second, so-called identifying restrictions (Little, 1993, 1994) can be used: inestimable parameters of the incomplete patterns are set equal to (functions of) the parameters describing the distribution of the completers. Although some authors perceive this under-identification as a drawback, it can be seen as bringing important advantages. First, assumptions need to be made very explicit. Second, PMM in this way aid our understanding of the precise nature of sensitivity in selection models. Third, and related to the previous point, PMM can serve important roles in a sensitivity analysis.

Examples of pattern-mixture applications can be found in Verbeke, Lesaffre, and Spiessens (2001) or Michiels et al. (2002) for continuous outcomes, and Michiels, Molenberghs, and Lipsitz (1999) for categorical outcomes. Further references include Cohen and Cohen (1983), Muthén, Kaplan, and Hollis (1987), Allison (1987), McArdle and Hamagani (1992), Little and Wang (1996), Little and Yau (1996), Hedeker and Gibbons (1997), Hogan and Laird (1997), Ekholm and Skinner (1998), Molenberghs, Michiels, and Kenward (1998), Verbeke and Molenberghs (2000), Thijs et al. (2002), and Molenberghs and Verbeke (2005). An example within the Bayesian framework is given by Rizopoulos, Verbeke, and Lesaffre (2007). Molenberghs, Michiels, Kenward, and Diggle (1998) and Kenward, Molenberghs, and Thijs (2003) studied the relationship between selection models and PMM within the context of missing data mechanisms. The earlier paper presents the PMM counterpart of MAR, and the later one states how pattern-mixture models can be constructed such that dropout does not depend on future points in time.

Chapter 18 of this volume is devoted to sensitivity analysis tools in a semi-parametric setting. Bayesian sensitivity analysis is studied in Chapter 19, while a multiple imputation point of view is taken in Chapter 20. The use of expert opinion is discussed in Chapter 21.

16.9 Concluding Remarks

First, this chapter has underscored the complexities arising when fitting models to incomplete data. Five generic issues, already arising under MAR, have been brought to the forefront: (i) the classical relationship between observed and expected features is convoluted because one observes the data only partially while the model describes all data; (ii) the independence of mean and variance parameters in a (multivariate) normal is lost, implying increased sensitivity, even under MAR; (iii) also the well-known agreement between the (frequentist) OLS and maximum likelihood estimation methods for normal models is lost, as soon as the missing data mechanism is not of the MCAR type, with related results holding in the non-normal case; (iv) in a likelihood-based context, deviances and related information criteria cannot be used in the same vein as with complete data because they provide no information about a model's prediction of the unobserved data and, in particular, (v) several models may saturate the observed-data degrees of freedom, while providing a different fit to the complete data, i.e., they only coincide in as far as they describe the observed data; as a consequence, different inferences may result from different saturated models.

Second, and based on these considerations, it has been argued that model assessment should always proceed in two steps. In the first step, the fit of a model to the *observed* data should be assessed carefully, while in the second step the sensitivity of the conclusions to the *unobserved data given the observed data* should be addressed. In the first step, one should ensure that the required assessment be done under one of two allowable scenarios, as represented by Figures 16.5(b) and (d), thereby carefully avoiding the scenario of Figure 16.5(c), where the model at the complete data level is compared to the incomplete data; apples and oranges as it were. These phenomena underscore the fact that fitting a model to incomplete data necessarily encompasses a part that cannot be assessed from the observed data. In particular, whether or not a dropout model is acceptable cannot be determined solely by mechanical model building exercises. Arbitrariness can be removed partly by careful consideration of the plausibility of a model. One should use as much context derived information as possible. Prior knowledge can give an idea of which models are more plausible. Covariate information can be explicitly included in the model to increase the range of plausible models which can be fit. Moreover, covariates can help explain the dependence between response mechanism and outcomes. Sensible non-response models should make use of the design and other information available. It is worth reiterating that *no* single analysis provides conclusions that are free of dependence in some way or other on untestable assumptions.

Third, and directly following on the previous point, a case has been made for supporting any analysis of incomplete data with carefully conceived and contextually relevant sensitivity analyses, a view that, incidentally, is consistent with the position of the International Conference on Harmonization guidelines (1999) and the NRC document on incomplete data in clinical trials (Little et al. 2010). A first family of sensitivity analyses is based on considering a collection of models, (1) through non-parametric bounds, (2) a family of identified models, or (3) intervals resulting from over-specified models. Next, we studied the interval of ignorance, and global and local influence. We paid attention to sensitivity analysis in a shared-parameter context, and indicated how very flexible models, such as the latent-class mixture model, can be used to assess the sensitivity in more conventional sub-models. Evidently, this collection is not exhaustive. In Section 16.8, additional likelihood-based methods are discussed. Also, Chapters 18–21 discuss sensitivity analyses in different paradigms.

Acknowledgments

The authors gratefully acknowledge support from IAP research Network P7/06 of the Belgian Government (Belgian Science Policy).

References

Allison, P.D. (1987). Estimation of linear models with incomplete data. *Sociology Methodology* **17**, 71–103.

Baker, S.G., Rosenberger, W.F., and DerSimonian, R. (1992). Closed-form estimates for missing counts in two-way contingency tables. *Statistics in Medicine* **11**, 643–657.

Barnett, V. (2002). *Sample Survey: Principles and Methods (3rd ed.)*. London: Arnold.

Beunckens, C., Molenberghs, G., Verbeke, G., and Mallinckrodt, C. (2008). A latent-class mixture model for incomplete longitudinal Gaussian data. *Biometrics* **64**, 96–105.

Beunckens, C., Sotto, C., Molenberghs, G., and Verbeke, G. (2009). A multi-faceted sensitivity analysis of the Slovenian Public Opinion Survey data. *Applied Statistics* **58**, 171–196.

Cohen, J. and Cohen, P. (1983). *Applied Multiple Regression/Correlation Analysis for the Behavioral Sciences* (2nd ed.). Hillsdale, NJ: Erlbaum.

Cook, R.D. (1979). Influential observations in linear regression. *Journal of the American Statistical Association* **74**, 169–174.

Cook, R.D. (1986). Assessment of local influence. *Journal of the Royal Statistical Society, Series B* **48**, 133–169.

Cook, R.D. and Weisberg, S. (1986). *Residuals and Influence in Regression*. London: Chapman & Hall.

Copas, J.B. and Li, H.G. (1997). Inference from non-random samples (with discussion). *Journal of the Royal Statistical Society, Series B* **59**, 55–96.

Creemers, A., Hens, N., Aerts, M., Molenberghs, G., Verbeke, G., and Kenward, M.G. (2010). A sensitivity analysis for shared-parameter models for incomplete longitudinal outcomes. *Biometrical Journal* **52**, 111–125.

Creemers, A., Hens, N., Aerts, M., Molenberghs, G., Verbeke, G., and Kenward, M.G. (2011). Generalized shared-parameter models and missingness at random. *Statistical Modeling* **11**, 279–311.

De Backer, M., De Keyser, P., De Vroey, C., and Lesaffre, E. (1996). A 12-week treatment for dermatophyte toe onychomycosis: Terbinafine 250mg/day vs. itraconazole 200mg/day: A double-blind comparative trial. *British Journal of Dermatology* **134**, 16–17.

Diggle, P.J. and Kenward, M.G. (1994). Informative drop-out in longitudinal data analysis (with discussion). *Applied Statistics* **43**, 49–93.

Ekholm, A. and Skinner, C. (1998). The muscatine children's obesity data reanalysed using pattern mixture models. *Applied Statistics* **47**, 251–263.

Fitzmaurice, G.M., Molenberghs, G., and Lipsitz, S.R. (1995). Regression models for longitudinal binary responses with informative dropouts. *Journal of the Royal Statistical Society, Series B* **57** 691–704.

Follmann, D. and Wu, M. (1995). An approximate generalized linear model with random effects for informative missing data. *Biometrics* **51**, 151-168.

Gelman, A., Van Mechelen, I., Verbeke, G., Heitjan, D.F., and Meulders, M. (2005). Multiple imputation for model checking: completed-data plots with missing and latent data. *Biometrics* **61**, 74–85.

Hedeker, D. and Gibbons, R.D. (1997). Application of random-effects pattern-mixture models for missing data in longitudinal studies. *Psychological Methods* **2**, 64–78.

Hogan, J.W. and Laird, N.M. (1997). Mixture models for the joint distribution of repeated measures and event times. *Statistics in Medicine* **16**, 239–258.

International Conference on Harmonisation E9 Expert Working Group (1999). Statistical principles for clinical trials: ICH Harmonised Tripartite Guideline. *Statistics in Medicine* **18**, 1905–1942.

Jansen, I., Hens, N., Molenberghs, G., Aerts, M., Verbeke, G., and Kenward, M.G. (2006). The nature of sensitivity in missing not at random models. *Computational Statistics and Data Analysis* **50**, 830–858.

Jennrich, R.I. and Schluchter, M.D. (1986). Unbalanced repeated measures models with structured covariance matrices. *Biometrics* **42**, 805–820.

Kenward, M.G. (1998). Selection models for repeated measurements with nonrandom dropout: an illustration of sensitivity. *Statistics in Medicine* **17**, 2723–2732.

Kenward, M.G, Goetghebeur, E.J.T., and Molenberghs, G. (2001). Sensitivity analysis of incomplete categorical data. *Statistical Modelling* **1**, 31–48.

Kenward, M.G. and Molenberghs, G. (1998). Likelihood based frequentist inference when data are missing at random. *Statistical Science* **12**, 236–247.

Kenward, M.G. and Molenberghs, G. (1999). Parametric models for incomplete continuous and categorical longitudinal studies data. *Statistical Methods in Medical Research* **8**, 51–83.

Kenward, M.G., Molenberghs, G., and Thijs, H. (2003). Pattern-mixture models with proper time dependence. *Biometrika* **90**, 53–71.

Laird, N.M. (1994). Discussion of Diggle, P.J. and Kenward, M.G.: Informative dropout in longitudinal data analysis. *Applied Statistics* **43**, 84.

Laird, N.M. and Ware, J.H. (1982). Random effects models for longitudinal data. *Biometrics* **38**, 963–974.

Lesaffre, E. and Verbeke, G. (1998). Local influence in linear mixed models. *Biometrics* **54**, 570–582.

Little, R.J.A. (1993). Pattern-mixture models for multivariate incomplete data. *Journal of the American Statistical Association* **88**, 125–134.

Little, R.J.A. (1994). Discussion of Diggle, P.J. and Kenward, M.G. "Informative dropout in longitudinal data analysis." *Applied Statistics* **43**, 78.

Little, R.J.A. (1995). Modeling the drop-out mechanism in repeated measures studies. *Journal of the American Statistical Association* **90**, 1112–1121.

Little, R.J.A., D'Agostino, R., Dickersin, K., Emerson, S.S., Farrar, J.T., Frangakis, C., Hogan, J.W., Molenberghs, G., Murphy, S.A., Neaton, J.D., Rotnitzky, A., Scharfstein, D., Shih, W., Siegel, J.P., and Stern, H. National Research Council (2010). *The Prevention and Treatment of Missing Data in Clinical Trials. Panel on Handling Missing Data in Clinical Trials.* Committee on National Statistics, Division of Behavioral and Social Sciences and Education. Washington, D.C.: The National Academies Press.

Little, R.J.A. and Rubin, D.B. (2002). *Statistical Analysis with Missing Data* (2nd ed.). New York: John Wiley & Sons.

Little, R.J.A. and Wang, Y. (1996). Pattern-mixture models for multivariate incomplete data with covariates. *Biometrics* **52**, 98–111.

Little, R.J.A. and Yau, L. (1996). Intent-to-treat analysis for longitudinal studies with drop-outs. *Biometrics* **52**, 1324–1333.

McArdle, J.J. and Hamagami, F. (1992). Modeling incomplete longitudinal and cross-sectional data using latent growth structural models. *Experimental Aging Research* **18**, 145–166.

McLachlan, G.J. and Peel, D. (2000) *Finite Mixture Models.* New York: John Wiley & Sons.

Michiels, B., Molenberghs, G., Bijnens, L., and Vangeneugden, T. (2002). Selection models and pattern-mixture models to analyze longitudinal quality of life data subject to dropout. *Statistics in Medicine* **21**, 1023–1041.

Michiels, B., Molenberghs, G., and Lipsitz, S.R. (1999). Selection models and pattern-mixture models for incomplete categorical data with covariates. *Biometrics* **55**, 978–983.

Molenberghs, G., Goetghebeur, E.J.T., Lipsitz, S.R., Kenward, M.G. (1999). Non-random missingness in categorical data: strengths and limitations. *The American Statistician* **53**, 110–118.

Molenberghs, G. and Kenward, M.G. (2007). *Missing Data in Clinical Studies.* Chichester: John Wiley & Sons.

Molenberghs, G., Kenward, M.G., and Goetghebeur, E. (2001). Sensitivity analysis for incomplete contingency tables: The Slovenian plebiscite case. *Applied Statistics* **50**, 15–29.

Molenberghs, G., Kenward, M.G., and Lesaffre, E. (1997). The analysis of longitudinal ordinal data with non-random dropout. *Biometrika* **84**, 33–44.

Molenberghs, G., Michiels, B., and Kenward, M.G. (1998a). Pseudo-likelihood for combined selection and pattern-mixture models for missing data problems. *Biometrical Journal* **40**, 557–572.

Molenberghs, G., Michiels, B., Kenward, M.G., and Diggle, P.J. (1998b). Missing data mechanisms and pattern-mixture models. *Statistica Neerlandica* **52**, 153–161.

Molenberghs, G., Thijs, H., Jansen, I., Beunckens, C., Kenward, M.G., Mallinckrodt, C., and Carroll, R.J. (2004). Analyzing incomplete longitudinal clinical trial data. *Biostatistics* **5**, 445-464.

Molenberghs, G., Thijs, H., Kenward, M.G. and Verbeke, G. (2003). Sensitivity analysis of continuous incomplete longitudinal outcomes. *Statistica Neerlandica* **57**, 112–135.

Molenberghs, G. and Verbeke, G. (2005). *Models for Discrete Longitudinal Data.* New York: Springer.

Molenberghs, G., Verbeke, G., and Beunckens, C. (2007). Formal and informal model selection with incomplete data. *Statistical Science* **23**, 201–218.

Molenberghs, G., Verbeke, G., Thijs, H., Lesaffre, E., and Kenward, M.G. (2001). Mastitis in dairy cattle: influence analysis to assess sensitivity of the dropout process. *Computational Statistics and Data Analysis* **37**, 93–113.

Muthén, B., Kaplan, D., and Hollis, M. (1987). On structural equation modeling with data that are not missing completely at random. *Psychometrika* **52**, 431–462.

Nordheim, E.V. (1984). Inference from nonrandomly missing categorical data: an example from a genetic study on Turner's syndrome. *Journal of the American Statistical Association* **79**, 772–780.

Potthoff, R.F. and Roy, S.N. (1964). A generalized multivariate analysis of variance model useful especially for growth curve problems. *Biometrika* **51**, 313–326.

Rizopoulos D., Verbeke G., and Lesaffre E. (2007). Sensitivity analysis in pattern mixture models using the extrapolation method. IAP Statistics Network. Technical report #0665.

Rizopoulos, D., Verbeke, G., and Molenberghs, G. (2008). Shared parameter models under random-effects misspecification. *Biometrika*, **95**, 63–74.

Roberts, D.T. (1992). Prevalence of dermatophyte onychomycosis in the United Kingdom: Results of an omnibus survey. *British Journal of Dermatology* **126 Suppl. 39**, 23–27.

Rubin, D.B. (1987). *Multiple Imputation for Nonresponse in Surveys.* New York: John Wiley.

Rubin, D.B. (1994). Discussion of Diggle, P.J. and Kenward, M.G.: "Informative dropout in longitudinal data analysis." *Applied Statistics* **43**, 80–82.

Rubin, D.B., Stern H.S., and Vehovar V. (1995). Handling "don't know" survey responses: the case of the Slovenian plebiscite. *Journal of the American Statistical Association* **90**, 822–828.

Shen, S., Beunckens, C., Mallinckrodt, C., and Molenberghs, G. (2006). A local influence sensitivity analysis for incomplete longitudinal depression data. *Journal of Biopharmaceutical Statistics* **16**, 365–384.

TenHave, T.R., Kunselman, A.R., Pulkstenis, E.P., and Landis, J.R. (1998). Mixed effects logistic regression models for longitudinal binary response data with informative dropout. *Biometrics* **54**, 367–383.

Thijs, H., Molenberghs, G., Michiels, B., Verbeke, G., and Curran, D. (2002). Strategies to fit pattern-mixture models. *Biostatistics* **3**, 245–265.

Thijs, H., Molenberghs, G., and Verbeke, G. (2000). The milk protein trial: influence analysis of the dropout process. *Biometrical Journal* **42**, 617–646.

Troxel, A.B., Harrington, D.P., and Lipsitz, S.R. (1998). Analysis of longitudinal data with non-ignorable non-monotone missing values. *Applied Statistics* **47**, 425–438.

Tsiatis, A.A. (1975). A nonidentifiability aspect of the problem of competing risks. *Proceedings of the National Academy of Science* **72**, 20–22.

Tsonaka R., Verbeke G., and Lesaffre E. (2008). A semi-parametric shared parameter model to handle non-monotone non-ignorable missingness. *Biometrics* **65**, 81–87.

Vach, W. and Blettner, M. (1995). Logistic regression with incompletely observed categorical covariates: Investigating the sensitivity against violation of the missing at random assumption. *Statistics in Medicine* **12**, 1315–1330.

Vansteelandt, S., Goetghebeur, E., Kenward, M.G., and Molenberghs, G. (2006). Ignorance and uncertainty regions as inferential tools in a sensitivity analysis. *Statistica Sinica* **16**, 953–979.

Van Steen, K., Molenberghs, G., Verbeke, G., and Thijs, H. (2001). A local influence approach to sensitivity analysis of incomplete longitudinal ordinal data. *Statistical Modelling: An International Journal* **1**, 125–142.

Verbeke, G. and Lesaffre, E. (1996). A linear mixed-effects model with heterogeneity in the random-effects population. *Journal of the American Statistical Association* **91**, 217–222.

Verbeke, G., Lesaffre, E., and Spiessens, B. (2001). The practical use of different strategies to handle dropout in longitudinal studies. *Drug Information Journal* **35**, 419–434.

Verbeke, G. and Molenberghs, G. (1997). *Linear Mixed Models in Practice: A SAS-Oriented Approach.* Lecture Notes in Statistics 126. New York: Springer.

Verbeke, G. and Molenberghs, G. (2000). *Linear Mixed Models for Longitudinal Data.* New York: Springer.

Verbeke, G., Molenberghs, G., Thijs, H., Lesaffre, E., and Kenward, M.G. (2001). Sensitivity analysis for non-random dropout: A local influence approach. *Biometrics* **57**, 7–14.

Verdonck, A., De Ridder, L., Verbeke, G., Bourguignon, J.P., Carels, C., Kuhn, E.R., Darras, V., and de Zegher, F. (1998). Comparative effects of neonatal and prepubertal castration on craniofacial growth in rats. *Archives of Oral Biology* **43**, 861–871.

Wu, M.C. and Bailey, K.R. (1988). Analysing changes in the presence of informative right censoring caused by death and withdrawal. *Statistics in Medicine* **7**, 337–346.

Wu, M.C. and Bailey, K.R. (1989). Estimation and comparison of changes in the presence of informative right censoring: conditional linear model. *Biometrics* **45**, 939–955.

Wu, M.C. and Carroll, R.J. (1988). Estimation and comparison of changes in the presence of informative right censoring by modeling the censoring process. *Biometrics* **44**, 175–188.

17

Sensitivity Analysis: A Semi-Parametric Perspective

Stijn Vansteelandt

Ghent University, Belgium

CONTENTS

We will present a philosophy for sensitivity analysis that has been set forward in pioneering work by James Robins, Andrea Rotnitzky and Daniel Scharfstein and that will lead us to the use of semi-parametric methods. We will then lay out a principled strategy for conducting semi-parametric sensitivity analyses for missing data. This will be illustrated in the analysis of a randomized clinical trial, on the basis of which we will compare quality-of-life between different chemotherapy regimes in premenopausal women with breast cancer.

17.1 Why Semi-Parametric Sensitivity Analyses?

Statistical analyses come in different colors and flavors. Some investigators fancy parametric approaches because they prioritize estimates with low variance or because they value their approachability, as partly ensured by off-the-shelf statistical software packages. Others are more in favor of semi-parametric approaches, as they worry more about bias due to model misspecification and prioritize methods that give an accurate reflection of the uncertainty in the results, as expressed through confidence intervals. The variation in tastes and existing methods contributes to the art of statistics and makes the field especially exciting. In this chapter, the focus will be on semi-parametric methods for sensitivity analysis. Before expanding on a principled strategy, we will lay out the main rationale for adopting a semi-parametric approach, which—we believe—transcends the usual tension between bias and variance.

The usual concerns for bias due to model misspecification indeed become more conspicuous when the data are incomplete. The reason is that parametric missing data analyses employ statistical models to "predict" the incomplete data; even when these models fit the observed data well, it can be difficult to diagnose whether they adequately predict the incomplete data. This can happen because the observed data distributions may be very different between the different missingness patterns, even when the default missing at random assumption is satisfied. For instance, in the simple case where missingness is limited to the outcome data, a model for the conditional distribution of the outcome, given completely measured covariates, may fit the responders' data well; yet, when responders and nonresponders are very different in their covariate distributions, it may serve as a poor model to predict the outcome of nonresponders. In such circumstances, parametric missing data analyses may have a tendency to extrapolate beyond the range of the observed data (Tan, 2007; Vansteelandt, Carpenter and Kenward, 2010). This may result in bias and, as importantly, in standard errors and confidence statements that ignore extrapolation uncertainty and are therefore overly optimistic.

These concerns for bias become even more pronounced when the incomplete data analysis becomes part of a sensitivity analysis. On the one hand this is because of the added labor for conducting a sensitivity analysis, which requires repeating the incomplete data analysis corresponding to different missing data assumptions, and which may thereby make the analyst less reluctant to repeat detailed model checks in each sub-analysis. On the other hand, it can be essentially impossible to postulate parametric selection models that are correctly specified for each of the considered missing data assumptions (Robins, Rotnitzky and Scharfstein, 1999ab; Scharfstein, Rotnitzky and Robins, 1999), as the following example illustrates.

Example 1. Consider a study design which involves collecting a scalar covariate X and outcome Y for a random sample of individuals. Suppose that the covariates are completely observed, but that the outcome is missing for some individuals. One may base a parametric sensitivity analysis (see Chapter 16) on the selection model:

$$\begin{aligned} Y|X &\sim N(\beta_0 + \beta_1 X, \sigma^2) \\ P(R=0|X,Y) &= \text{expit}(\alpha_0 + \alpha_1 X + \gamma Y), \end{aligned} \tag{17.1}$$

where $\text{expit}(x) = \exp(x)/\{1+\exp(x)\}$ and where γ measures the degree to which the missingness is informative. For fixed γ, maximum likelihood estimators for β_0 and β_1 can be obtained, e.g. via direct likelihood methods. A parametric sensitivity analysis may now involve repeating such maximum likelihood analysis for different fixed values of the parameter γ, which we will refer to as a sensitivity parameter.

Suppose that the above selection model is correctly specified under missing at random (i.e., when $\gamma = 0$). In particular, suppose that the conditional distribution of Y, given X, amongst responders is normal with mean X and variance 1. Then using Bayes' rule, it can be inferred from this and the fact that

$$\frac{\text{odds}(R=0|X,Y=y_1)}{\text{odds}(R=0|X,Y=y_2)} = \exp\{\gamma(y_1 - y_2)\},$$

where $\text{odds}(R=0|X,Y) \equiv P(R=0|X,Y)/P(R=1|X,Y)$, that

$$f(Y|R=0,X) = f(Y|R=1,X)\exp\{\gamma(Y-X) - \gamma^2/2\}. \tag{17.2}$$

Indeed, it is immediate from Bayes' rule that $f(Y|R=0,X)$ must be of the form $f(Y|R=1,X)\exp(\gamma Y)\,c(X)$ for some normalizing constant $c(X)$, which can be obtained using the moment generating function of the normal distribution. Let furthermore $P(R = 0|X) = \text{expit}(X)$, then Figure 17.1 displays the resulting marginal distribution

$$\begin{aligned} f(Y = y|X = x) &= f(Y = y|R = 1, X = x)\left[1 - \text{expit}(x)\right. \\ &\quad \left. + \text{expit}(x)\exp\left\{\gamma(y - x) - \gamma^2/2\right\}\right] \end{aligned} \tag{17.3}$$

for different values of x (corresponding to the different panels) and for different values of γ (corresponding to the different densities within each panel). It illustrates that, while (by assumption) the selection model is correctly specified at $\gamma = 0$, it is grossly misspecified away from the missing at random assumption where the marginal distribution of Y, given X, deviates from the postulated normal distribution.

The fact that the conditional distribution of Y, given X, matches the postulated normal distribution only at $\gamma = 0$, should not be taken as evidence that the missing at random assumption is satisfied. Indeed, as explained in previous chapters, the observed data can never provide evidence about the validity of the missing at random assumption. All that we may conclude from this analysis is that the postulated normal distribution of Y, given X, is misspecified unless $\gamma = 0$. For instance, it follows from (17.3) after some algebra that at $\gamma = 1$, the conditional density of Y, given X, ought to be of the form

$$f(Y = y|X = x) = \frac{1}{\sqrt{2\pi}}\exp\left\{-0.5(y - x)^2\right\}\frac{1 + \exp(y - 0.5)}{1 + \exp(x)}, \tag{17.4}$$

which is not contained within model (17.1). If this represented the data-generating model and $\gamma = 1$, it would however still give rise to the same normal observed data distribution with mean X and variance 1.

The important conclusion from the above example is that, even in simple settings, it may essentially be impossible to specify a parametric selection model which is correctly specified at each of the considered missing data assumptions in a sensitivity analysis. While evidently only a single missing data assumption matches the true data-generating distribution, since we do not know which it is, we would ideally want the model to be correctly specified at each of the considered missing data assumptions; for instance, we would like it to imply the same distribution of the observed data at each of the different values of γ in Example 1. Without this being the case, we are essentially guaranteed, even prior to seeing the data, that the sensitivity analysis will report biased results and/or prioritize specific missing data assumptions that lead to better fitting models. For instance, the parametric selection model of Example 1 was only correctly specified at $\gamma = 0$ as it could only reproduce the normal observed data distribution at $\gamma = 0$; this value may thus—misleadingly—appear better supported by the data. In the following section, we will see that this concern can be alleviated through semi-parametric sensitivity analyses.

17.2 Non-Parametric Identification

The selection modeling approach has a tendency to conflate statistical modeling assumptions and missing data assumptions. For instance, the model in Example 1 turned out to be correctly specified under missing at random, but not away from it; however, it was essentially impossible to tell whether this was effectively due to γ equaling zero as it might equally have been the case that $\gamma = 1$ and that the true distribution of Y, given X, was

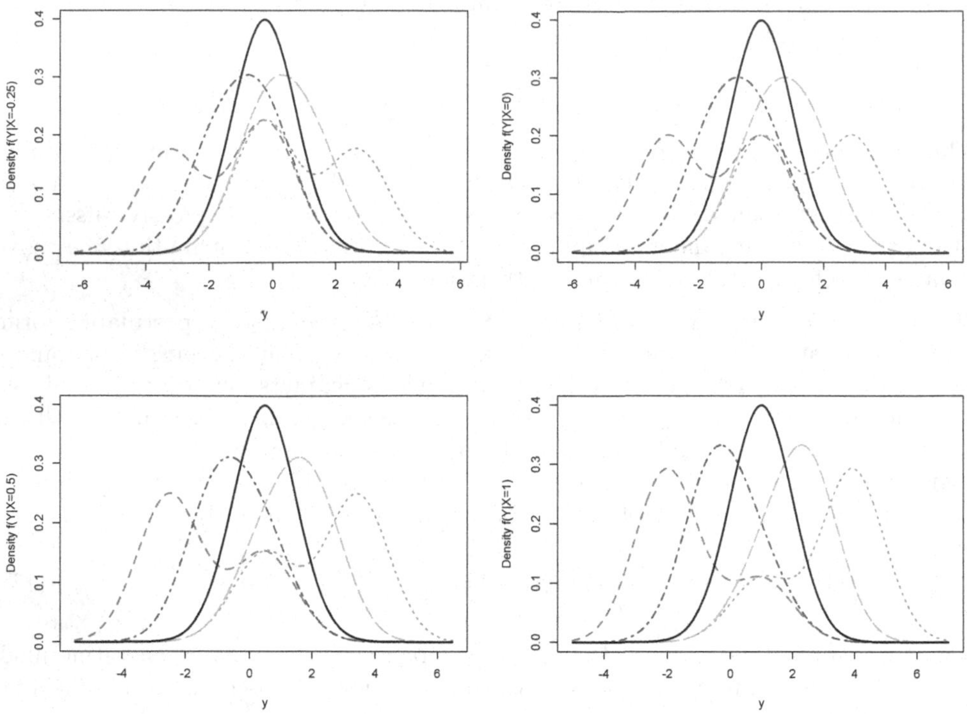

FIGURE 17.1
Density $f(Y|X=x)$ for $x=-0.25$ (upper left), $x=0$ (upper right), $x=0.5$ (bottom left) and $x=1$ (bottom right), and for $\gamma=-3,-1.5,0.,1.5$ and 3 (from left to right within each panel) in Example 1.

of the form (17.4) rather than normal. To be able to understand the results of a sensitivity analysis, it is therefore important to clearly disentangle assumptions that relate to the incomplete data from pure statistical modeling assumptions. This is most easily done within the class of so-called pattern-mixture models. For instance, reconsidering the missing outcome problem of Example 1, the pattern-mixture approach involves building separate models for $f(X), f(Y|R=1,X)$ and $f(Y|R=0,X)$, where the first two involve observed data only and are therefore identifiable without reliance on missing data assumptions; the data carry no information at all about the distribution $f(Y|R=0,X)$ so that any choice of model must come from pure missing data assumptions. In spite of this, we shall not adopt the pattern-mixture approach in this chapter because even parsimonious models for $f(X), f(Y|R=1,X)$ and $f(Y|R=0,X)$ may combine into rather complex models for $f(Y|X)$. This may not only make results difficult to interpret and report, but can also make interesting hypotheses difficult to test.

To clearly separate statistical modeling assumptions from missing data assumptions within the selection modeling approach, we will adopt the following two-stage process (Robins, Rotnitzky and Scharfstein, 1999). In the first stage, we will delineate the range of missing data assumptions that will be evaluated in the sensitivity analysis. In the second stage, we will detail the statistical modeling assumptions. The first stage involves a difficult and subtle process as it ideally requires striking a balance between making a "sufficient number" of missing data assumptions, yet "not too many." The following section explains this in more detail.

17.2.1 Why non-parametric identification?

In all statistical analyses, interest focuses on some target parameter, which can be viewed as a functional of the full data distribution. For instance in regression analysis, interest lies in the conditional mean $E(Y|X)$, which is a characteristic of the distribution $f(Y, X)$ of the full data. When the data are incomplete, we will demand that sufficient missing data assumptions are made to be able to identify the target parameter. This means that if we were given the true distribution of the observed data together with those missing data assumptions, then we should be able to calculate that parameter. The reader may be lead to think that failure to meet this requirement will automatically prevent the analyst from calculating the target parameter. However, this is not entirely true. The reason is that whenever insufficient assumptions are made to be able to identify that parameter, then statistical estimation procedures will extract information from other sources, such as from statistical modeling assumptions that we will impose in a later stage, or from prior distributions in a Bayesian analysis. This is undesirable as it makes the results sensitive to the correctness of the assumed statistical models, which are only partially verifiable when insufficient missing data assumptions are made. In Example 1, for instance, our ability to deduce that $\gamma = 0$ is heavily dependent on the assumed normal distribution of Y, given X, within the entire study population. Given the missing data, the observed data carry insufficient information to confirm that this distribution is normal and it would be very unlikely that subject-matter considerations can justify that particular choice. A different choice of model, like (17.4), would be indistinguishable on the basis of the observed data, but would imply a completely different conclusion that $\gamma = 1$.

We conclude that it is important for the user of missing data assumptions to understand whether these assumptions suffice to identify the target parameter. Without this being the case, the results may become very sensitive to the choice of statistical modeling assumptions or prior distributions, and it can be difficult for the user to appreciate precisely what aspects of the model and/or prior distributions become influential; see e.g., Diggle and Kenward (1994) and the discussions thereof.

We will furthermore try not to impose more missing data assumptions than required to be able to identify the parameter of interest. The reason is that when more missing data assumptions than strictly necessary are being made, they might also impose testable restrictions on the observed data and thus not be pure assumptions about the missing data. This may make the analysis more restrictive and, moreover, may make the missing data assumptions partially verifiable and thus refutable on the basis of the observed data. See for instance Vansteelandt (2009) who criticizes a study for presenting data analysis results obtained under missing data assumptions that turn out to be strongly rejected by the observed data.

In summary, we will strive to work with missing data assumptions that are sufficient to be able to identify the target parameter, yet impose no testable restrictions on the observed data (so long as one refrains from imposing other statistical modeling assumptions). We will say that such missing data assumptions enable non-parametric identification (Robins, 1997; Rotnitzky, Robins and Scharfstein, 1998; Robins, Rotnitzky and Scharfstein, 1999). Here, the term "identification" conveys that the assumptions are sufficient to identify the parameter of interest; the term "non-parametric" articulates that the assumptions are untestable in the sense that they impose no restrictions on the distribution of the observed data.

Example 2. Consider the statistical model defined by

$$P(R=0|Y,X) = \text{expit}\{h(X)+\gamma Y\}, \tag{17.5}$$

where $h(X)$ is an unknown function of X and γ is fixed and known. Then it follows from Bayes' rule that

$$\begin{aligned}\frac{P(R=0|Y,X)}{P(R=1|Y,X)} &= \exp\{h(X)\}\exp(\gamma Y)\\ &= \frac{f(Y|R=0,X)}{f(Y|R=1,X)}\frac{P(R=0|X)}{P(R=1|X)},\end{aligned}$$

from which

$$f(Y|R=0,X) = f(Y|R=1,X)\frac{\exp(\gamma Y)}{k(X)}, \tag{17.6}$$

where

$$k(X) \quad = \quad \frac{P(R=0|X)}{P(R=1|X)}\frac{1}{\exp\{h(X)\}} = E\{\exp(\gamma Y)|R=1,X\}; \tag{17.7}$$

the last equality follows because $f(Y|R=0,X)$ must normalize to 1. It thus follows that model (17.5) is equivalent with the pattern-mixture model defined by (17.6). Since the data carry no information about $f(Y|R=0,X)$, it can be inferred from this that knowledge of γ does not impose testable restrictions on the observed data. The statistical model defined by (17.5) is thus a non-parametric model for each fixed value of γ. Since furthermore

$$f(Y|X) = f(Y|R=1,X)\left[P(R=1|X)+P(R=0|X)\frac{\exp(\gamma Y)}{k(X)}\right]$$

and since for given γ, the terms $f(Y|R=1,X), P(R=1|X)$ and $k(X) = E\{\exp(\gamma Y)|R=1,X\}$ in this expression all describe features of the observed data alone and can thus be identified on the basis of the observed data, knowledge of γ is sufficient to identify $f(Y|X)$. We may conclude that the model given by (17.5) defines a non-parametric identified model for $f(Y|X)$.

17.2.2 The curse of dimensionality

Missing data assumptions that entail non-parametric identifiability are attractive in a sensitivity analysis: each such assumption is perfectly (and thus equally) compatible with the observed data; in that sense, each such assumption can be viewed as a pure missing data assumption. However, to obtain well behaved analysis results, additional modeling assumptions must typically be imposed: on the one hand, the number of variables may be so large, relative to the sample size, that some form of modeling is required to deal with the curse of dimensionality; on the other hand, some degree of modeling may be desired to end up with results that are easy to communicate.

Example 3. Reconsider Example 2. While we have seen that for each fixed γ in model (17.5), the joint distribution of the full data (Y,X)—and thus in particular the conditional mean $E(Y|X)$—can be non-parametrically identified on the basis of the observed data, we will choose to impose modeling assumptions in addition to model (17.5). Specifically, let us restrict the conditional mean of Y, given X, to be of the form

$$E(Y|X) = \beta_0 + \beta_1 X,$$

where β_0 and β_1 are unknown scalars. We will do so because we would ultimately like to obtain a parsimonious description of the relation between X and Y, which may be simplifying reality, but nonetheless captures the main signal and is easy to interpret. In addition, we will impose restrictions on the functional $h(X)$, which, from (17.7), can be identified as

$$h(X) = \log\left[\frac{P(R=0|X)}{P(R=1|X)}\frac{1}{E\{\exp(\gamma Y)|R=1,X\}}\right]. \tag{17.8}$$

We will do so because, unless X is discrete with few levels, unrealistically large sample sizes may be needed to estimate the components $P(R=1|X)$ and $E\{\exp(\gamma Y)|R=1,X\}$ in the above expression non-parametrically. In particular, we will postulate that

$$P(R=0|Y,X) = \text{expit}\left(\alpha_0 + \alpha_1 X + \gamma Y\right),$$

where α_0 and α_1 are unknown scalars. The model defined by these restrictions on $E(Y|X)$ and $P(R=0|Y,X)$ is no longer a non-parametrically identified model. Indeed, since for each fixed choice of γ, $E(Y|X)$ and $h(X)$ can be identified on the basis of the observed data, it follows in particular that the linearity restrictions $E(Y|X) = \beta_0 + \beta_1 X$ or $h(X) = \alpha_0 + \alpha_1 X$ may be refutable for some values of γ. This should not be taken as evidence that those values of γ are not supported by the observed data, but rather as evidence that the linearity restrictions are implausible at those values of γ.

For instance, suppose that the distribution of the observed data is as given in Example 1. Then it follows from (17.8) that

$$h(X) \quad = \quad X - \log\exp(\gamma X + \gamma^2/2) = (1-\gamma)X - \gamma^2/2,$$

which satisfies the postulated model restriction $h(X) = \alpha_0 + \alpha_1 X$ for all γ. It further follows that, as illustrated in Figure 17.2, the conditional mean of Y (as implied by the conditional distribution given in (17.3)) is linear in X when $\gamma = 0$, but not otherwise. We thus conclude that the postulated model does not fit the observed data equally well over the considered range of γ-values. It can however be seen from Figure 17.2 that the degree of model misspecification is relatively minor and (typically) not of the extent that fitting a linear model could be grossly misleading. In particular, approximating the true conditional mean of Y given X by a linear curve may be desirable for practical reporting. Note from Figure 17.1 that the degree of misspecification of the parametric model (17.1) is much more severe.

In summary, to understand missing data assumptions, it is important that they entail non-parametric identification when used in isolation (i.e., without further statistical modeling). Even though non-parametric identification may not be tenable in the ultimate analysis, demanding it prior to making modeling assumptions is the only guarantee that the analysis involves a sufficient number of missing data assumptions that are pure, in the sense of not being restrictions on the observed data. The final statistical analysis will typically impose additional modeling assumptions in view of the curse of dimensionality. This could be viewed as disadvantageous, but unavoidable, in a sensitivity analysis because it implies that the different models, corresponding to the different missing data assumptions, may not fit the observed data equally well to the extent that some of these models may be guaranteed to be misspecified. In this chapter, we will attempt to remedy this by working with semi-parametric models that impose sufficiently weak restrictions on the observed data so that any degree of model misspecification is not of practical concern (cfr. Figure 17.2).

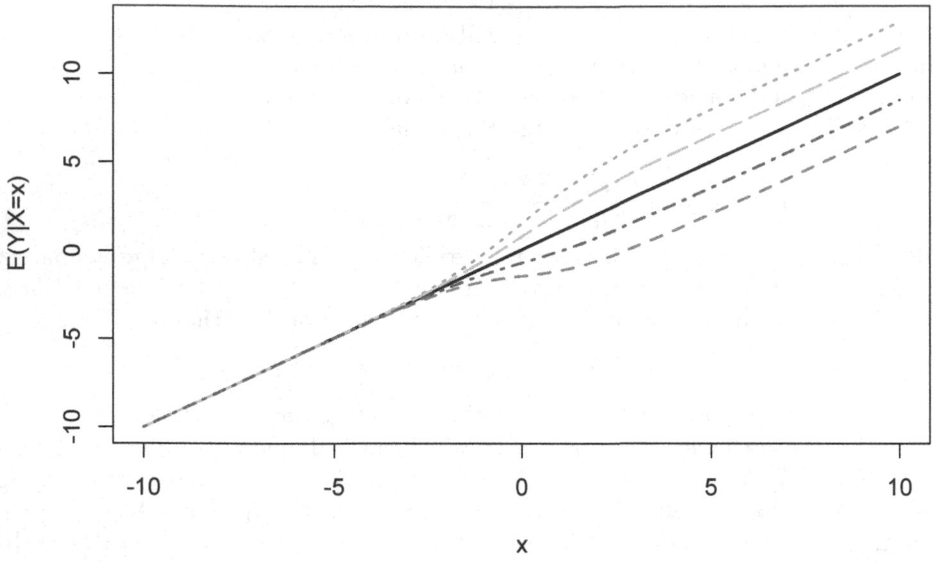

FIGURE 17.2
Mean $E(Y|X = x)$ for $\gamma = -3, -1.5, 0., 1.5$ and 3 (from bottom to top).

17.3 Case Study on Non-Monotone Missing Follow-Up Data

17.3.1 International Breast Cancer Study Group, Trial VI

As a leading application to illustrate semi-parametric sensitivity analysis methods, we will consider a re-analysis of data from clinical trial VI of the International Breast Cancer Study Group (IBCSG) (Hürny et al., 1992; Ibrahim et al., 2001). The goal of this study was to compare quality-of-life (QOL) between 4 different chemotherapy regimes in premenopausal women with breast cancer. Regimes A and B (C and D) involved a 6 (3) month episode of chemotherapy. Regimes B and D extended regimes A and C, respectively, with three 1 month episodes of chemotherapy, each 3 months apart.

Four hundred sixty three patients were asked to complete a QOL questionnaire at baseline and on 6 later occasions at 3-monthly intervals. The outcome is self-reported mood, from 0 (best) to 100 (worst). Patients did refuse, on occasion, to complete the questionnaire, but even so, they were invited to attend the next visit. This resulted in non-monotone missing data patterns, as visualized for 40 randomly selected patients in Figure 17.3. Outcomes were missing on at least 1 occasion for 80.5% of patients, on at least 2 occasions for 63.7% of patients, on at least 3 occasions for 49.2% of patients and on all occasions for 2.2% of patients.

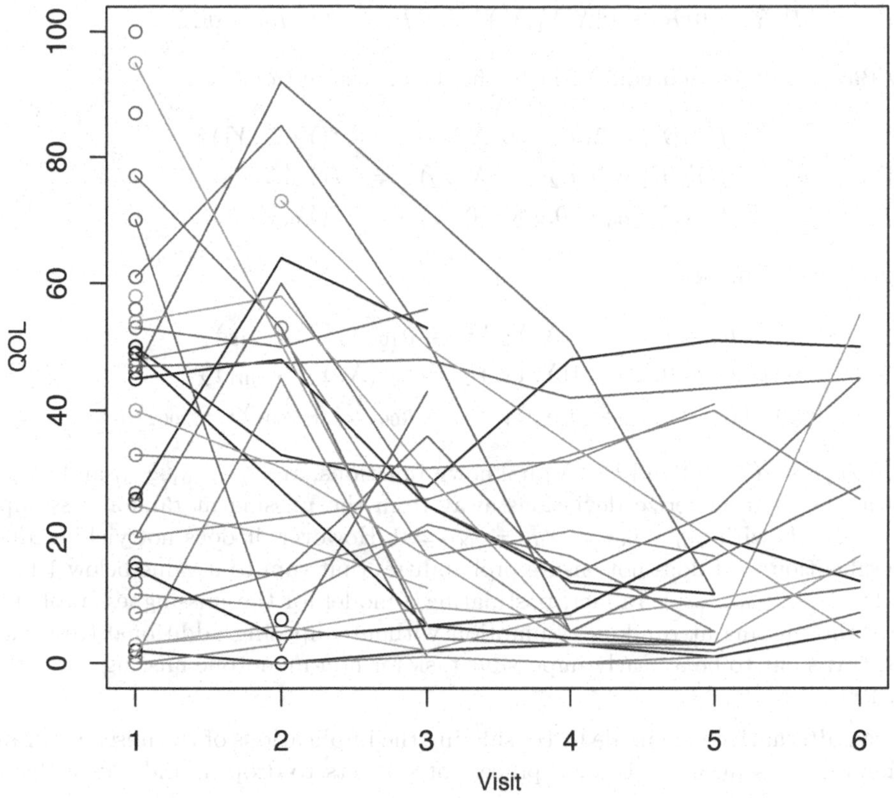

FIGURE 17.3
Profile plot of the quality-of-life score for 40 randomly sampled patients.

17.3.2 Missing data assumptions

It is reasonable to conjecture that a patient is more likely not to come to fill out the questionnaire when her mood is poor (Ibrahim et al., 2001). However, the degree to which is uncertain, so that we will ultimately attempt to perform a sensitivity analysis where we vary the dependence of missingness on the missing data over a plausible range. In this section, we will guide the reader through the difficult task of postulating a class of missing data assumptions that allow for varying degree of dependence of missingness on quality-of-life, thereby respecting the principles advocated in the previous sections. The following example illustrates some of the difficulties.

Example 4. Suppose that measurements are taken at 2 fixed time points. Let R_t be a missingness indicator at time $t, t = 1, 2$, which indicates 1 if the outcome Y_t at that time is observed, and 0 otherwise. Then the missing at random assumption (Rubin, 1976; Gill, van der Laan and Robins, 1997) (w.r.t. the full data X, Y_1, Y_2 with X a scalar baseline covariate) postulates that

$$\begin{aligned}
P(R_1=1,R_2=0|X,Y_1,Y_2) &= P(R_1=1,R_2=0|X,Y_1)\\
P(R_1=0,R_2=1|X,Y_1,Y_2) &= P(R_1=0,R_2=1|X,Y_2)\\
P(R_1=0,R_2=0|X,Y_1,Y_2) &= P(R_1=0,R_2=0|X),
\end{aligned}$$

which by Bayes' rule is itself equivalent to the set of assumptions

$$\begin{aligned}
f(Y_2|R_1=1,R_2=0,X,Y_1) &= f(Y_2|X,Y_1)\\
f(Y_1|R_1=0,R_2=1,X,Y_2) &= f(Y_1|X,Y_2)\\
f(Y_1,Y_2|R_1=0,R_2=0,X) &= f(Y_1,Y_2|X).
\end{aligned}$$

The following set of models

$$\begin{aligned}
\text{logit}P(R_1=1,R_2=0|X,Y_1,Y_2) &= h_{10}(X,Y_1)+\gamma_{10}Y_2\\
\text{logit}P(R_1=0,R_2=1|X,Y_1,Y_2) &= h_{01}(X,Y_2)+\gamma_{01}Y_2\\
\text{logit}P(R_1=0,R_2=0|X,Y_1,Y_2) &= h_{00}(X)+\gamma_{001}Y_1+\gamma_{002}Y_2,
\end{aligned}$$

with $h_{10}(X,Y_1), h_{01}(X,Y_2)$ and $h_{00}(X)$ unknown functions, and $\gamma_{10},\gamma_{01},\gamma_{001}$ and γ_{002} fixed and known, thus parameterize deviations away from the missing at random assumption, which corresponds with $\gamma_{10}=\gamma_{01}=\gamma_{001}=\gamma_{002}=0$. However, it does not yield a coherent model specification as it does not ensure probabilities that sum to a value below 1 for each choice of $(\gamma_{10},\gamma_{01},\gamma_{001},\gamma_{002})$. In fact, postulating a model for the missingness probabilities that obeys the missing at random assumption without imposing additional (testable) restrictions, turns out to be a nearly impossible task for non-monotone missing data (Robins and Gill, 1997).

Progress can alternatively be made by considering the implications of the missing at random assumption on the sequential decision process of subjects to drop in and out of the study at each time:

$$\begin{aligned}
P(R_1=0|X,Y_1,Y_2) &= P(R_1=0,R_2=1|X,Y_2)\\
&\quad +P(R_1=0,R_2=0|X)\\
&= P(R_1=0|X,Y_2)\\
P(R_2=0|R_1=0,X,Y_1,Y_2) &= \frac{P(R_1=0,R_2=0|X)}{P(R_1=0|X,Y_2)}\\
&= P(R_2=0|R_1=0,X,Y_2)\\
P(R_2=0|R_1=1,X,Y_1,Y_2) &= \frac{P(R_1=1,R_2=0|X,Y_1)}{P(R_1=1|X,Y_2)}.
\end{aligned}$$

The first two identities are comprehensible as conditional independence restrictions; however, the last identity illustrates that missing at random implies complex restrictions on the probability to drop out of the study at the second visit, which cannot be interpreted in terms of conditional independence relationships. With the additional (testable) assumption that $P(R_1=0|X,Y_2)=P(R_1=0|X)$, these restrictions simplify to:

$$P(R_1=0|X,Y_1,Y_2) = P(R_1=0|X) \tag{17.9}$$
$$P(R_2=0|R_1=0,X,Y_1,Y_2) = P(R_2=0|R_1=0,X) \tag{17.10}$$
$$P(R_2=0|R_1=1,X,Y_1,Y_2) = P(R_2=0|R_1=1,X,Y_1). \tag{17.11}$$

Deviations away from this more restrictive set of assumptions are easier to parameterize. For instance, consider the set of models

$$\text{logit}P(R_1 = 1|X, Y_1, Y_2) = h_1(X) + \gamma_{11}Y_1 + \gamma_{12}Y_2 \quad (17.12)$$

$$\text{logit}P(R_2 = 1|R_1 = 1, X, Y_1, Y_2) = h_{11}(X, Y_1) + \gamma_1 Y_2 \quad (17.13)$$

$$\text{logit}P(R_2 = 1|R_1 = 0, X, Y_1, Y_2) = h_{10}(X) + \gamma_{01}Y_1 + \gamma_{02}Y_2, \quad (17.14)$$

with $h_1(X), h_{11}(X, Y_1)$ and $h_{10}(X)$ unknown functions, and $\gamma_{11}, \gamma_{12}, \gamma_1, \gamma_{01}$ and γ_{02} fixed and known; here, the choice $\gamma_{11} = \gamma_{12} = \gamma_1 = \gamma_{01} = \gamma_{02} = 0$ implies assumptions (17.9)-(17.11). This set of models guarantees a coherent model specification for each choice of $(\gamma_{11}, \gamma_{12}, \gamma_1, \gamma_{01}, \gamma_{02})$. A drawback, however, is that this model imposes testable restrictions on the observed data in the sense that the observed data may carry evidence to refute certain tuples $(\gamma_{11}, \gamma_{12}, \gamma_1, \gamma_{01}, \gamma_{02})$. This can be intuitively understood by noting that the dependence of missingness on Y_2 is prespecified in the above models, even though the data contain partial information (from the stratum $R_1 = 0, R_2 = 1$) to identify this dependence. It can be shown that the model defined by (17.12)–(17.14) for given value of $(\gamma_{11}, \gamma_{12}, \gamma_1, \gamma_{01}, \gamma_{02})$ thus fails to yield a non-parametric identified model. This is of concern in view of the arguments given in Section 17.2.1. It is moreover of concern as it demands specification of more sensitivity parameters than effectively needed for identification; this can be a cumbersome task, especially when the number of time points is large and the number of sensitivity parameters thus excessive.

Example 4 illustrates that the missing at random assumption, in its most general form, is a complex assumption for non-monotone missing data; deviations away from it are difficult to parameterize (see Robins and Gill 1997, and Vansteelandt, Rotnitzky and Robins, 2007, for further detail). To overcome this difficulty, we will formulate an alternative class of missing data assumptions which entails non-parametric identification, like the missing at random assumption, and which furthermore lends itself naturally to a sensitivity analysis in the sense that deviations from it are easy to parameterize. To this end, we will view the observed data structure as a repetition of the structure of Example 2 at each time $t, t = 1, \ldots, T$. In particular, let R_t be a missingness indicator at time $t, t = 1, \ldots, T$, which indicates 1 if the outcome Y_t at that time is observed, and 0 otherwise. Let $\mathbf{W}_t$ be a set of extraneous covariates measured at time t and $\mathbf{W}_0$ be a vector of completely observed baseline covariates. Let $\mathbf{O}_t \equiv (R_t, R_t\mathbf{W}'_t, R_tY_t)'$ denote the observed data measured at time t, and $\overline{\mathbf{O}}_t \equiv (\mathbf{W}_0, \mathbf{O}'_1, \ldots, \mathbf{O}'_t)'$ denote the observed data measured up to time t. With the correspondences R_t for R, Y_t for Y and $\overline{\mathbf{O}}_{t-1}$ for X in Example 2, we then arrive at the following statistical model at each time $t = 1, \ldots, T$:

$$P(R_t = 0|Y_t, \overline{\mathbf{O}}_{t-1}) = \text{expit}\left\{h_t(\overline{\mathbf{O}}_{t-1}) + \gamma Y_t\right\}, \quad (17.15)$$

where $h_t(\overline{\mathbf{O}}_{t-1})$ is an unknown function of $\overline{\mathbf{O}}_{t-1}$ and γ is fixed and known. It can be seen along the same lines of the arguments in Example 2 that the model defined by these restrictions is a non-parametric identified model for $f(Y_t|\overline{\mathbf{O}}_{t-1})$: it imposes no restrictions on the observed data and is sufficient to identify the distribution $f(Y_t|\overline{\mathbf{O}}_{t-1})$ and thus in particular $E(Y_t|\mathbf{X})$ for arbitrary sub-vector $\mathbf{X}$ of $\mathbf{W}_0$.

The choice $\gamma = 0$ can be used as reference point in a sensitivity analysis: it encodes the assumption that at each time t, the decision not to attend the planned clinic visit at that time has no residual dependence on the quality-of-life at that time, given the history of the observed data at that time; that is:

$$P(R_t = 1|\overline{\mathbf{O}}_{t-1}, Y_t) = P(R_t = 1|\overline{\mathbf{O}}_{t-1}).$$

This assumption has been proposed by Lin, Scharfstein, and Rosenheck (2003) and was later termed "sequential explainability" by Vansteelandt, Rotnitzky and Robins (2007). While the assumption of sequential explainability may appear to resemble the missing at random assumption, it is not implied by it—and thus the result of a statistical analysis that is based on it must be viewed as a not missing at random analysis. This can be seen because the missing at random assumption imposes additional restrictions on the dependence of missingness at each time t on future outcomes at that time.

The assumption of sequential explainability could be realistic if the decision of individuals to attend a planned clinic visit were only based on information that was disclosed at previous clinic visits and recorded in the database. However, like the missing at random assumption, it is unlikely to hold in the IBSCG study because it assumes, for example, that if an individual chooses today to miss his next two visits and then to return, he/she will not reassess this decision based on evolving time-dependent covariates associated with the response. This is unlikely, as often the decision to miss a given study cycle is influenced by aspects of the patient's health and psychological status that evolved during earlier missed study cycles (Vansteelandt, Rotnitzky and Robins, 2007). In particular, the decision not to attend a planned clinic visit likely depends not only on the individual's mood as disclosed on previously attended clinic visits, but presumably also on the unrecorded mood close to the time of the planned visit. We will therefore allow for deviations from the assumption of sequential explainability by varying γ over a plausible range. In model (17.15), the magnitude of γ can be understood upon noting that

$$\exp(\gamma) = \frac{\text{odds}(R_t = 0|Y_t = y + 1, \overline{\mathbf{O}}_{t-1})}{\text{odds}(R_t = 0|Y_t = y, \overline{\mathbf{O}}_{t-1})} \tag{17.16}$$

for each y; thus $\exp(\gamma)$ expresses how much the odds of not attending the planned clinic visit at time t depends on the mood at that time, after adjusting for the observed past.

17.3.3 Estimation

Suppose that

$$E(Y_t|X) = \beta_0 + \beta_1 X + \beta_2 t + \beta_3 Xt, \tag{17.17}$$

where X denotes treatment group, which we consider to be dichotomous for now. Primary interest lies in β_3, which encodes the treatment effect. In this section, which can be skipped by the less technically interested reader, we will explain how semi-parametric estimators (Tsiatis, 2006) for the parameters indexing this model can be obtained for each fixed choice of γ in model (17.15). To allow for sufficient generality as well as for concise notation, we will give results for the more general model

$$E(Y_t|\mathbf{X}) = g_t(\mathbf{X}; \boldsymbol{\beta}),$$

for $t = 1, \ldots, T$, where $g_t(\mathbf{X}; \boldsymbol{\beta})$ is a known function, smooth in an unknown finite-dimensional parameter $\boldsymbol{\beta}$ and $\mathbf{X}$ is a sub-vector of $\mathbf{W}_0$; however, the reader is welcome to substitute $g_t(\mathbf{X}; \boldsymbol{\beta})$ with the specific choice $\beta_0 + \beta_1 X + \beta_2 t + \beta_3 Xt$ for better understanding.

It follows from the discussion in the previous section (and more formally from the results in Vansteelandt, Rotnitzky, and Robins (2007)) that, while $\boldsymbol{\beta}$ is identified for each fixed γ, additional modeling assumptions are generally needed in order to obtain well-behaved estimators in small to moderate sample sizes. Two options will be considered.

One possibility would be to assume a parametric model for the probability of missingness at each time $t = 1, \ldots, T$, e.g.,

$$P(R_t = 0|Y_t, \overline{\mathbf{O}}_{t-1}) \\ = \text{expit}\,(\alpha_0 + \alpha_1 t + \alpha_2 R_{t-1} + \alpha_3 R_{t-1} Y_{t-1} + \alpha_4 X + \gamma Y_t)\,. \qquad (17.18)$$

For conciseness and generality, we will present results for the more general missingness model

$$P(R_t = 0|Y_t, \overline{\mathbf{O}}_{t-1}) = \text{expit}\left\{h_t(\overline{\mathbf{O}}_{t-1}; \boldsymbol{\alpha}) + q_t(Y_t, \overline{\mathbf{O}}_{t-1}; \gamma)\right\},$$

for $t = 1, \ldots, T$. Here, $h_t(\overline{\mathbf{O}}_{t-1}; \boldsymbol{\alpha})$ is a known function, smooth in an unknown finite-dimensional parameter $\boldsymbol{\alpha}$, which parameterizes the dependence of missingness on the observed data. Further, $q_t(Y_t, \overline{\mathbf{O}}_{t-1}; \gamma)$ is a known function of Y_t and $\overline{\mathbf{O}}_{t-1}$, satisfying $q_t(0, \overline{\mathbf{O}}_{t-1}; \gamma) = 0$ and $q_t(Y_t, \overline{\mathbf{O}}_{t-1}; \mathbf{0}) = 0$ so that it parameterizes the dependence of missingness on the missing outcome itself (e.g., $q_t(Y_t, \overline{\mathbf{O}}_{t-1}; \gamma) = \gamma Y_t$); γ is assumed fixed and known.

An alternative possibility would be to assume a parametric model for the expected outcome in the non-responders at each time $t = 1, \ldots, T$, e.g.,

$$E(Y_t|R_t = 0, \overline{\mathbf{O}}_{t-1}) = \delta_1 + \delta_2 t + \delta_3 R_{t-1} + \delta_4 R_{t-1} Y_{t-1} + \delta_5 X.$$

For conciseness and generality, we will present results for the more general outcome model of the form

$$E(Y_t|R_t = 0, \overline{\mathbf{O}}_{t-1}) = m_t(\overline{\mathbf{O}}_{t-1}; \boldsymbol{\delta})$$

for $t = 1, \ldots, T$, where $m_t(\overline{\mathbf{O}}_{t-1}; \boldsymbol{\delta})$ is a known function, smooth in an unknown finite-dimensional parameter $\boldsymbol{\delta}$.

An estimator $\widehat{\boldsymbol{\alpha}}$ for the parameters indexing the missingness model can be obtained as the solution to an estimating equation of the form (Vansteelandt, Rotnitzky and Robins, 2007)

$$\mathbf{0} = \sum_{i=1}^{n} \sum_{t=1}^{T} \left(1 - R_{it}\left[1 + \exp\left\{h_t(\overline{\mathbf{O}}_{i,t-1}; \boldsymbol{\alpha}) + q_t(Y_{it}, \overline{\mathbf{O}}_{i,t-1}; \gamma)\right\}\right]\right) \psi_t^h(\overline{\mathbf{O}}_{i,t-1}),$$

where $\psi_t^h(\overline{\mathbf{O}}_{t-1})$ is an (essentially) arbitrary function of the observed past, which has the dimension of $\boldsymbol{\alpha}$. For instance, for fixed γ, $\boldsymbol{\alpha} = (\alpha_1, \alpha_2, \alpha_3, \alpha_4, \alpha_5)'$ in model (17.18) can be estimated as the solution to

$$\mathbf{0} = \sum_{i=1}^{n} \sum_{t=1}^{T} (1 \ t \ R_{i,t-1} \ R_{i,t-1} Y_{i,t-1} \ X_i)' \{1 - R_{it} \\ - R_{it} \exp\left(\alpha_0 + \alpha_1 t + \alpha_2 R_{i,t-1} + \alpha_3 R_{i,t-1} Y_{i,t-1} + \alpha_4 X_i + \gamma Y_{it}\right)\}.$$

The resulting estimator is consistent for $\boldsymbol{\alpha}$ (i.e., converges to the population value as the sample size approaches infinity) regardless of the choice of $\psi_t^h(\overline{\mathbf{O}}_{t-1})$. This is because R_{it} has (conditional) mean equal to the reciprocal of $1 + \exp\left\{h_t(\overline{\mathbf{O}}_{i,t-1}; \boldsymbol{\alpha}) + q_t(Y_{it}, \overline{\mathbf{O}}_{i,t-1}; \gamma)\right\}$, so that the above equation indeed has mean zero when evaluated at the population value of $\boldsymbol{\alpha}$.

For given estimator $\widehat{\boldsymbol{\alpha}}$ of $\boldsymbol{\alpha}$, the target parameter $\boldsymbol{\beta}$ may now be obtained as the solution $\widehat{\boldsymbol{\beta}}$ to an estimating equation of the form

$$\mathbf{0} = \sum_{i=1}^{n} \sum_{t=1}^{T} \mathbf{d}_t^h(\mathbf{X}_i) R_{it} \left\{Y_{it} - g_t(\mathbf{X}_i; \boldsymbol{\beta})\right\} \qquad (17.19) \\ \times \left[1 + \exp\left\{h_t(\overline{\mathbf{O}}_{i,t-1}; \widehat{\boldsymbol{\alpha}}) + q_t(Y_{it}, \overline{\mathbf{O}}_{i,t-1}; \gamma)\right\}\right] \\ + \left(1 - R_{it}\left[1 + \exp\left\{h_t(\overline{\mathbf{O}}_{i,t-1}; \widehat{\boldsymbol{\alpha}}) + q_t(Y_{it}, \overline{\mathbf{O}}_{i,t-1}; \gamma)\right\}\right]\right) \boldsymbol{\phi}_t^h(\overline{\mathbf{O}}_{i,t-1}),$$

for arbitrary functions $\mathbf{d}_t^h(\mathbf{X})$ and $\phi_t^h(\overline{\mathbf{O}}_{t-1})$ of the dimension of $\boldsymbol{\beta}$. Here, the terms on the first two lines involve taking residuals from the outcome model of interest, evaluating these for responders at each time t, but inversely weighting these to account for the selective nature of those individuals; an obvious choice for $\mathbf{d}_t^h(\mathbf{X})$ would be the vector of covariates in the model for $E(Y_t|\mathbf{X})$, i.e., $(1, X, t, Xt)'$ in model (17.17). The term on the third line adds a mean zero contribution, which is of the same form as the estimating functions previously used for estimating $\boldsymbol{\alpha}$. Depending on the choice of $\phi_t^h(\overline{\mathbf{O}}_{t-1})$, to which we will return later, this term may help to increase the precision of $\widehat{\boldsymbol{\beta}}$.

An estimator $\widehat{\boldsymbol{\delta}}$ of the parameters indexing the outcome regression model can be obtained as the solution to an estimating equation of the form

$$\mathbf{0} = \sum_{i=1}^{n}\sum_{t=1}^{T} R_{it} \exp\left\{q_t(Y_{it}, \overline{\mathbf{O}}_{i,t-1}; \gamma)\right\}\left\{Y_{it} - m_t(\overline{\mathbf{O}}_{i,t-1}; \boldsymbol{\delta})\right\}\psi_t^m(\overline{\mathbf{O}}_{i,t-1}),$$

for arbitrary index functions $\psi_t^m(\overline{\mathbf{O}}_{t-1})$ of the dimension of $\boldsymbol{\delta}$. For instance, for fixed γ, $\boldsymbol{\delta} = (\delta_1, \delta_2, \delta_3, \delta_4, \delta_5)'$ can be estimated as the solution to

$$\begin{aligned}\mathbf{0} = \sum_{i=1}^{n}\sum_{t=1}^{T} & R_{it} \exp(\gamma Y_{it})(Y_{it} - \delta_1 - \delta_2 t - \delta_3 R_{i,t-1} - \delta_4 R_{i,t-1}Y_{i,t-1} - \delta_5 X_i) \\ & \times (1 \; t \; R_{i,t-1} \; R_{i,t-1}Y_{i,t-1} \; X_i)'.\end{aligned}$$

The resulting estimator is consistent for $\boldsymbol{\beta}$ regardless of the choice of $\psi_t^m(\overline{\mathbf{O}}_{t-1})$. This can be understood by noting that, as in (17.2), the term $\exp\left\{q_t(Y_{it}, \overline{\mathbf{O}}_{i,t-1}; \gamma)\right\}$ operates as a tilt function which links the outcome distribution in responders to that in non-responders.

For given estimator $\widehat{\boldsymbol{\delta}}$ of $\boldsymbol{\delta}$, the target parameter $\boldsymbol{\beta}$ may now be estimated as the solution $\widehat{\boldsymbol{\beta}}$ to an estimating equation of the form

$$\begin{aligned}\sum_{i=1}^{n}\sum_{t=1}^{T} & \mathbf{d}_t^m(\mathbf{X}_i)\left\{R_{it}Y_{it} + (1 - R_{it})m_t(\overline{\mathbf{O}}_{i,t-1}; \widehat{\boldsymbol{\delta}}) - g_t(\mathbf{X}_i; \boldsymbol{\beta})\right\} \qquad (17.20)\\ & + R_{it}\exp\left\{q_t(Y_{it}, \overline{\mathbf{O}}_{i,t-1}; \gamma)\right\}\left\{Y_{it} - m_t(\overline{\mathbf{O}}_{i,t-1}; \widehat{\boldsymbol{\delta}})\right\}\phi_t^m(\overline{\mathbf{O}}_{i,t-1}) = \mathbf{0},\end{aligned}$$

for arbitrary functions $\mathbf{d}_t^m(\mathbf{X})$ and $\phi_t^m(\overline{\mathbf{O}}_{t-1})$ of the dimension of $\boldsymbol{\beta}$. Here, the first term involves a form of conditional mean imputation, which uses the observed outcome for non-responders and a predicted outcome for non-responders; an obvious choice for $\mathbf{d}_t^m(\mathbf{X})$ would again be the vector of covariates in the model for $E(Y_t|\mathbf{X})$, i.e., $(1, X, t, Xt)'$ in model (17.17). The second term adds a mean zero contribution, which is of the same form as the estimating functions previously used for estimating $\boldsymbol{\delta}$.

Of the above two strategies, we view the second as the less attractive one. The reason is that the model for $E(Y_t|R_t = 0, \overline{\mathbf{O}}_{t-1})$ imposes restrictions on the outcome mean at each time t and these restrictions might not be compatible with the restrictions imposed by the target model for $E(Y_t|\mathbf{X})$. In particular, there may be no joint distribution that satisfies the restrictions imposed by both the models for $E(Y_t|\mathbf{X})$ and $E(Y_t|R_t = 0, \overline{\mathbf{O}}_{t-1})$, in which case the model is essentially guaranteed to be misspecified. However, it is easily shown (Vansteelandt, Rotnitzky, and Robins, 2007) that the estimating equations (17.19) and (17.20) are mathematically identical when $\mathbf{d}_t^h(\mathbf{X}) = \mathbf{d}_t^m(\mathbf{X})$ and moreover

$$\phi_t^h(\overline{\mathbf{O}}_{t-1}) = m_t(\overline{\mathbf{O}}_{t-1}; \widehat{\boldsymbol{\delta}}) - g_t(\mathbf{X}; \boldsymbol{\beta}) \quad \text{and} \quad \phi_t^m(\overline{\mathbf{O}}_{t-1}) = \exp\left\{h_t(\overline{\mathbf{O}}_{t-1}; \widehat{\boldsymbol{\alpha}})\right\}.$$

These choices are recommended because they deliver a reasonably precise estimator, which is

TABLE 17.1
Data analysis results under sequential explainability.

Parameter	Estimate	95% CI
Time effect D	−3.7	−5.8 to −1.6
Time effect A-D	1.5	−1.3 to 4.3
Time effect B-D	2.0	−0.53 to 4.6
Time effect C-D	1.7	−1.2 to 4.7

not heavily dependent on parametric modeling assumptions. That they deliver a reasonably precise estimator can be intuitively understood from the form of equations (17.19) upon noting that not only subjects with complete information at time t contribute information. Indeed, also subjects with missing outcome at time t contribute the prediction $m_t(\overline{\mathbf{O}}_{t-1}; \widehat{\delta})$ of their missing outcome. That these choices deliver an estimator that is less dependent upon parametric modeling assumptions can be understood from the fact that the estimating equations (17.19) and (17.20) are mathematically identical at these choices of $\phi_t^h(\overline{\mathbf{O}}_{t-1})$ and $\phi_t^m(\overline{\mathbf{O}}_{t-1})$. The resulting estimator is henceforth consistent when the missingness models are correctly specified at each time t, even when the outcome regression models are misspecified (in view of equation (17.19)); it is moreover consistent when the outcome regression models are correctly specified at each time t, even when the missingness models are misspecified (in view of equation (17.20)). Such estimator is called doubly robust in view of its double "robustness" against model misspecification (Robins and Rotnitzky, 2001; Tsiatis, 2006); this property of the estimator is especially desirable in a sensitivity analysis in view of the discussion in Section 17.2.2.

17.3.4 Sensitivity analysis with a scalar sensitivity parameter

We will now apply the estimation strategy of the previous section in the analysis of clinical trial VI of the International Breast Cancer Study Group. Throughout, we will focus on model (17.17) with X substituted by a dummy indicator of treatment group. Table 1 shows the results of an analysis under sequential explainability (relative to the observed information on treatment, time, language, previous attendance, and previous outcome). It shows a significant improvement in quality-of-life over time under treatment D, but small and non-significant differences between treatments.

This analysis gives a natural reference point in sensitivity analysis. This is on the one hand so because the assumption of sequential explainability forms a breaking point between data-generating mechanisms where attendance is more likely when the quality-of-life is high, versus poor. Moreover, the assumption of sequential explainability can be made more plausible by collecting predictors of missingness, so that in some analyses it can be developed to be a reasonably plausible assumption. Here, however, the assumption of sequential explainability may not be very likely to hold. Indeed, even amongst patients with the same treatment and language, and with the same history of attendance and recorded quality-of-life outcomes, it is rather likely that non-attendance is more probable amongst those with poor quality-of-life.

In view of this, we will conduct a sensitivity analysis to allow for deviations from sequential explainability. For this illustration, the sensitivity analysis will assume that missingness at each time t is explained by the quality-of-time outcome just prior to time t, regardless of whether it was recorded or not. This will be done by postulating that

$$\text{logit}P(R_t = 0|\overline{\mathbf{O}}_{t-1}, Y_t) = h_t(\overline{\mathbf{O}}_{t-1}) + \gamma(1 - R_{t-1})Y_t.$$

Since the quality-of-time outcome just prior to time t is contained within the observed history $\overline{\mathbf{O}}_{t-1}$ when $R_{t-1} = 1$, this model only allows for a dependence on Y_t when the previous outcome was missing. The magnitude γ of this dependence is unknown and will be specified and varied within the sensitivity analysis. Specifically, under the above model, we repeated the analysis by varying $\gamma \in [-1/\text{SD}(Y), 1/\text{SD}(Y)]$, each time regarding γ as known. This choice of interval amounts to allowing the odds of missing a visit to be up to $\exp(1) = 2.72$ times higher/smaller per SD increase in QOL. The results of this analysis are displayed in Figure 17.4; similar sensitivity plots were proposed in for instance Rotnitzky, Robins, and Scharfstein (1998) and Scharfstein, Rotnitzky, and Robins (1999a).

Positive values of γ are a priori deemed to be more plausible as they express that the probability of missing a visit is more likely amongst those with a poor quality-of-life at the previous visit. With this in mind, Figure 17.4 shows that the improvement in quality-of-life over time under treatment D may be overestimated up to about 1 unit, but the evidence for a significant improvement remains in spite of the missing data uncertainty. The contrasts in time effect between the various treatments show little sensitivity to the missing data; regardless of the missing data mechanism (within the considered class), the evidence for a difference in effect between treatments stays absent.

17.3.5 Summarizing the sensitivity analysis results

Sensitivity plots are very informative in a sensitivity analysis, but they may be too detailed for the consumer of the study results who may have a difficult time to judge which values of the sensitivity parameter are plausible. These plots may moreover be difficult to construct when there is more than one sensitivity parameter, and this may in turn have the undesirable effect of promoting the use of overly simplistic missingness models. In the spirit that it is the statistician's duty to summarize the data, a preferable strategy for practical reporting may be to collect subject-matter knowledge on the plausible range of sensitivity parameters, and to subsequently use this information to summarize the sensitivity analysis results (see, e.g., Scharfstein, Daniels, and Robins, 2003). For instance, assuming that the odds of missing a visit can be up to 2.72 times higher/smaller per SD increase in QOL, it can be deduced from Figure 17.4 that the time effect under treatment D is estimated to lie between -2.7 and -4.5. Such intervals have been considered by many authors (see e.g., Horowitz and Manski, 2000; Manski, 1990, 2003; Kenward, Molenberghs, and Goetghebeur, 2001; Molenberghs, Kenward and Goetghebeur, 2001; Robins, 1989, p. 123; Romano and Shaikh, 2008; Vansteelandt et al., 2006). They are called (estimated) ignorance intervals by Kenward, Molenberghs and Goetghebeur (2001), Molenberghs, Kenward and Goetghebeur (2001) and Vansteelandt et al. (2006) and are termed identified sets by Romano and Shaikh (2008). Such intervals express uncertainty due to data incompleteness, but ignore finite-sample imprecision.

Uncertainty due to both data incompleteness and finite-sample variability can be expressed through so-called uncertainty intervals (Imbens and Manski, 2004; Vansteelandt et al., 2006). Just like confidence intervals express finite-sample uncertainty on a point estimate, these express finite-sample uncertainty on an ignorance interval. Various definitions of uncertainty intervals, or more generally uncertainty regions, have been proposed. For a given set Γ of sensitivity parameters γ, a 95% pointwise uncertainty region (Vansteelandt et al., 2006) for some parameter $\boldsymbol{\beta}$ is defined to be a region $\mathcal{U}$ which covers $\boldsymbol{\beta}$ with at least 95% chance uniformly over all $\gamma \in \Gamma$. Specifically, let $\boldsymbol{\beta}(\gamma)$ be the value of $\boldsymbol{\beta}$ as would be cal-

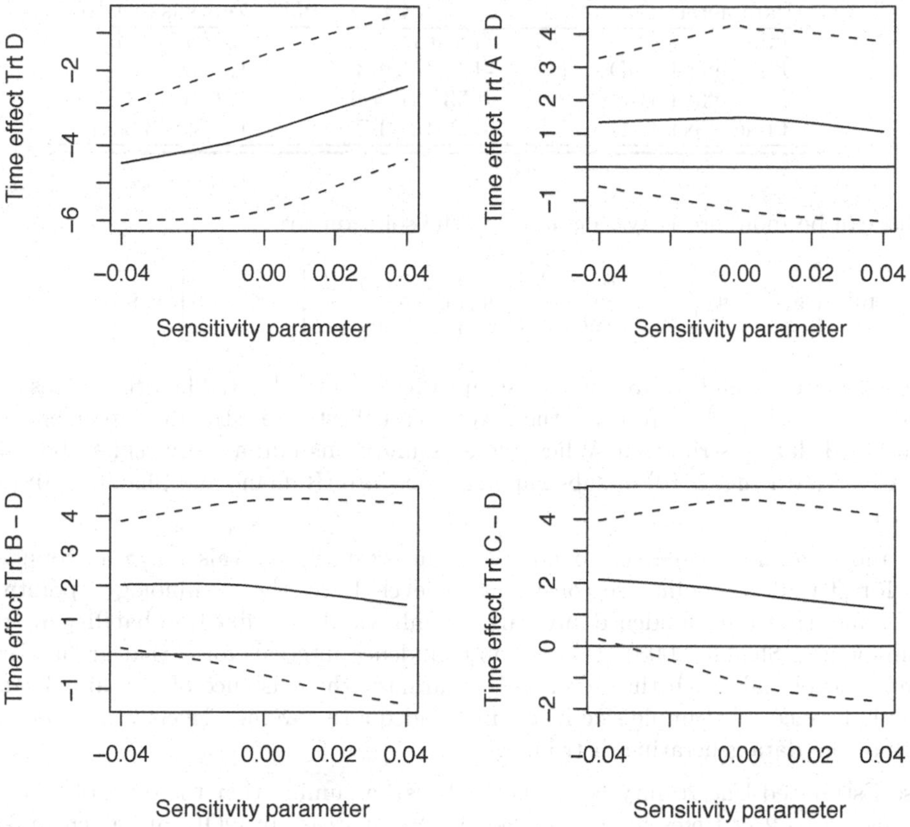

FIGURE 17.4
Sensitivity plot: parameter estimates (solid line) and 95% confidence intervals (dashed lines) in function of the sensitivity parameter γ.

culated on the basis of the population distribution of the observed data if the sensitivity parameter were to take the value γ. Then $\mathcal{U}$ must satisfy:

$$\inf_{\gamma \in \Gamma} P\{\beta(\gamma) \in \mathcal{U}\} \geq 0.95.$$

For scalar parameters β, Vansteelandt et al. (2006) estimate such uncertainty regions as

$$\left[\widehat{\beta}_l - c \times \text{se}(\widehat{\beta}_l), \widehat{\beta}_u + c \times \text{se}(\widehat{\beta}_u)\right],$$

where $\widehat{\beta}_l$ is the lower bound of the estimated ignorance interval (with standard error $\text{se}(\widehat{\beta}_l)$) and where $\widehat{\beta}_u$ is the upper bound of the estimated ignorance interval (with standard error $\text{se}(\widehat{\beta}_u)$); the constant c approximates the 95th percentile of the standard normal distribution,

TABLE 17.2
Sensitivity analysis results.

Parameter	Est.	95% CI	95% Pointwise EURO
Time effect D	-3.7	-5.8 to -1.6	-5.7 to -0.81
Time effect A-D	1.5	-1.3 to 4.3	-1.4 to 4.3
Time effect B-D	2.0	-0.53 to 4.6	-1.2 to 4.1
Time effect C-D	1.7	-1.2 to 4.7	-1.5 to 3.9

1.645, and can be more precisely calculated as the solution to

$$\min\left[\Phi(c)-\Phi\left\{-c-\frac{\widehat{\beta}_u-\widehat{\beta}_l}{\text{se}(\widehat{\beta}_u)}\right\},\Phi\left\{c+\frac{\widehat{\beta}_u-\widehat{\beta}_l}{\text{se}(\widehat{\beta}_l)}\right\}-\Phi(-c)\right]=95\%,$$

when the estimates $\widehat{\beta}_l$ and $\widehat{\beta}_u$ are jointly asymptotically normal and when the values of the sensitivity parameter γ which result in these worst-case/best-case estimates are independent of the observed data distribution. When these assumptions fail, a more refined bootstrap (Todem, Fine, and Peng, 2010) or subsampling procedure (Romano and Shaikh, 2008) may be adopted.

Romano and Shaikh (2008) refer to pointwise uncertainty intervals for $\boldsymbol{\beta}$ as confidence intervals for $\boldsymbol{\beta}$ that are pointwise consistent in level. Here, the terminology "pointwise" reflects the fact that the confidence intervals are only valid for a fixed probability distribution. Romano and Shaikh (2008) also develop confidence intervals for $\boldsymbol{\beta}$ that are uniformly consistent in level and which thus essentially guarantee the existence of a minimal sample size such that, once this sample size is attained, adequate coverage levels will be attained regardless of the data-generating distribution.

Pointwise Estimated Uncertainty RegiOns (EUROs), assuming that the odds of missing a visit can be up to 2.72 times higher/smaller per SD increase in QOL, are given in Table 17.2. They are not much broader than the 95% confidence intervals that assume sequential explainability. This is on the one hand because the estimates are not very sensitive to the missing data assumptions under the considered model, and on the other hand because the width of the estimated ignorance interval partly compensates for the uncertainty due to sampling variability. The reported 95% pointwise EUROs enable us to conclude that, in spite of the missing data, the monthly improvement in quality-of-life under treatment D amounts to reductions in quality-of-life between 0.81 and 5.7, with at least 95% chance. The attraction of these intervals lies in the fact that they can be used and interpreted similarly as the usual 95% confidence intervals, even though they express additional missing data uncertainty. In particular, from the fact that the 95% pointwise EURO for the time effect under treatment D excludes zero, one may deduce that there is a significant improvement in quality-of-life under treatment D at the 5% significance level; the Type I error rate corresponding to this hypothesis test is then at most 5% (Vansteelandt et al., 2006).

17.3.6 Sensitivity analysis with a vector of sensitivity parameters

Estimation of the treatment effect is likely most affected by incomplete data uncertainty when different missingness mechanisms operate on different arms. In view of this, we reanalyzed the data under model

$$\text{logit}P(R_t=0|\overline{\mathbf{O}}_{t-1},Y_t)=h_t(\overline{\mathbf{O}}_{t-1})+\gamma_1 Y_t I(T=D)+\gamma_2 Y_t I(T\neq D),$$

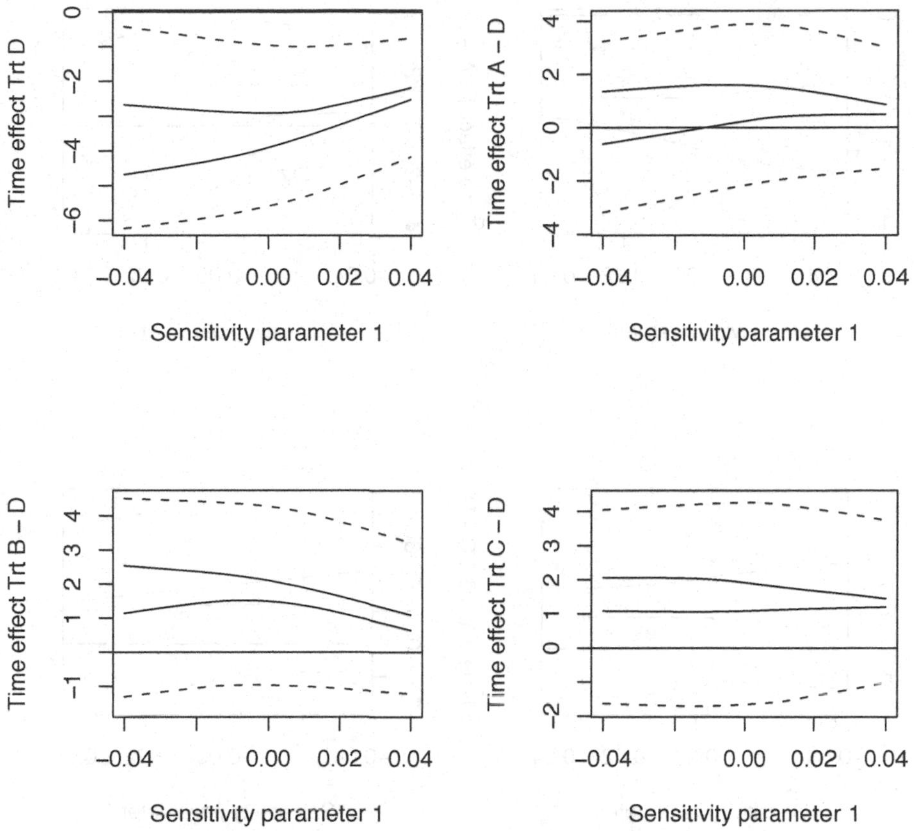

FIGURE 17.5
Sensitivity plot for sensitivity parameter γ_1 with estimated ignorance intervals (solid lines) and 95% pointwise uncertainty intervals (dashed lines).

which allows for missingness at each time t to depend differentially on the previous outcome in arm D relative to the other arms. In a sensitivity analysis, we varied γ_1, γ_2 independently in $[-1/25, 1/25]$.

Displaying the results is now more cumbersome because more than one parameter is driving the missingness mechanism. In Figures 17.5 and 17.6 we therefore visualize the effect of changes in γ_1 in terms of ignorance and 95% pointwise uncertainty intervals corresponding to each value of γ_1, while varying γ_2 over $[-1/25, 1/25]$. The figures again show relatively little sensitivity, and lead us to the same conclusions as before.

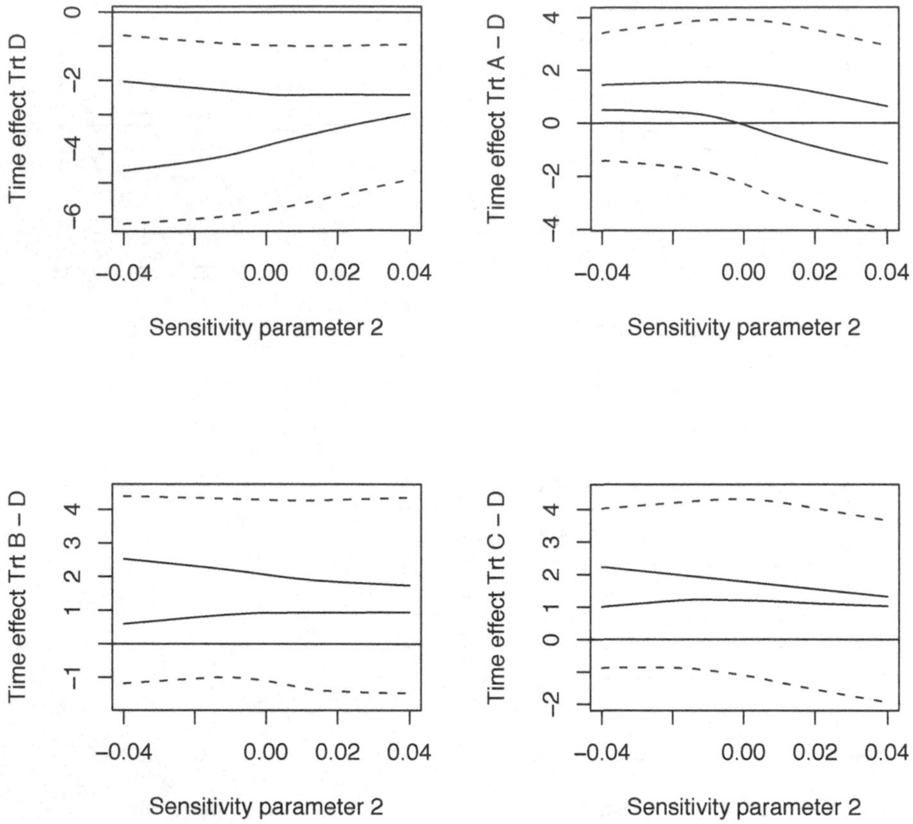

FIGURE 17.6
Sensitivity plot for sensitivity parameter γ_2 *with estimated ignorance intervals (solid lines) and 95% pointwise uncertainty intervals (dashed lines).*

17.4 Discussion

In this chapter, we have focused on sensitivity analysis strategies based on selection models for incomplete data. We have highlighted two principal complications for such analysis. The first complication is that it may be difficult to come up with a class of selection models that parameterize different degrees of informativeness of the incomplete data. We have illustrated this in the context of non-monotone missing follow-up data, where deviations away from the default missing at random assumption turned out difficult to parameterize (see also Robins and Gill, 1997). The second complication is that different selection models, corresponding to different degrees of informativeness of the incomplete data, may fit the observed data to varying degrees, to the extent that some selection models may be strongly rejected by the observed data. This need not express that those degrees of informativeness of the incomplete data are implausible, as it may instead point towards the implausibility of the postulated parametric modeling assumptions at those degrees of informativeness. This suggests that different parametric modeling assumptions may be required depending on the chosen missing data assumption. Because using different parametric models at each choice of missing data assumption is generally an impossible task, we have instead opted for semi-parametric procedures that refrain from imposing strong parametric assumptions.

We have illustrated such semi-parametric sensitivity analysis approach in a re-analysis of the International Breast Cancer Study Group, Trial VI, which involved non-monotone missing follow-up data. This analysis has shown that the construction of a class of missing data assumptions that allow for varying degrees of dependence of missingness on the missing data, can be challenging. However, this is not different for parametric sensitivity analyses (see Chapter 17) if one wishes to respect the principles outlined in Section 17.2.1 that missing data assumptions should be pure assumptions about the incomplete data, not the observed data. Related semi-parametric sensitivity analysis approaches have been developed for missing covariate problems (Egleston and Wong, 2009), for monotone missing follow-up data with drop-out at fixed time points (Rotnitzky, Robins, and Scharfstein, 1998; Birmingham, Rotnitzky, and Fitzmaurice, 2003), for monotone missing follow-up data with drop-out at random time points (Scharfstein, Robins, and Rotnitzky, 1999a), for informatively outcome-dependent follow-up data (Buzkova and Lumley, 2009), for time-to-event data subject to informative censoring (Scharfstein et al., 2001; Scharfstein and Robins, 2002), for informatively interval-censored time-to-event data (Shardell, Scharfstein, and Bozzette, 2007), for informatively censored quality-of-life adjusted failure time data (Rotnitzky, Bergesio, and Farall, 2009) and for time-to-event data subject to competing risks (Shardell, Scharfstein, and Bozzette, 2007). Rotnitzky et al. (2001) and Rotnitzky et al. (2009) extend these results to settings with multiple causes of missingness or censoring when the outcome is continuous or an event time, respectively. Scharfstein, Daniels, and Robins (2003) consider Bayesian analogs. When one faces different data structures than those in these references, the problem of delineating a suitable class of missing data assumptions must be reconsidered.

For practical reporting, we recommend that analysis results obtained under a reference missing data assumption (e.g., MAR or sequential explainability) are disclosed together with 95% pointwise uncertainty intervals corresponding to a set of sensitivity parameters, which contains all missing data mechanisms that are deemed plausible by subject-matter experts. Ideally, such set of sensitivity parameters is pre-specified prior to data collection, but can be widened as information on the missing data mechanism is accrued which contradicts the preset choices. The advantage of reporting pointwise uncertainty intervals as a summary of the data analysis results is that they can essentially be interpreted and used like the usual confidence intervals, but give a more honest expression of the actual uncertainty about the study results.

In some cases, subject-matter experts may prefer to express their prior beliefs on the missing data mechanism in terms of a prior distribution on the sensitivity parameter γ, rather than in terms of an interval of values that are deemed plausible. For instance, one might be willing to express one's prior beliefs about a scalar sensitivity parameter γ in terms of a normal distribution centered at $1/\{2\text{SD}(Y)\}$ and with standard deviation given by $1/\{4\text{SD}(Y)\}$, so that most of the probability mass corresponds to values between 0 and $1/\text{SD}(Y)$. More generally, such prior distributions can be elicited from subject-matter experts, e.g., using trial roulettes (Gore, 1987); see e.g., Scharfstein, Daniels, and Robins (2003) and Shardell et al. (2008a) for detailed sensitivity analyses that involve eliciting expert knowledge on the missingness mechanism. Given such prior distribution $f(\gamma)$, a 95% uncertainty region for the target parameter β may alternatively be defined as a region $\mathcal{U}$ which satisfies

$$\int P\{\beta(\gamma) \in \mathcal{U}\} f(\gamma) d\gamma \geq 95\%;$$

it may be termed a "95% posterior uncertainty region'," in analogy to the Bayesian terminology. How it is best estimated, requires further study.

In spite of these developments, specifying a plausible set of sensitivity parameters remains a formidable task in each sensitivity analysis. This is especially so because of the non-collapsibility (Greenland, Robins, and Pearl, 1999) of the odds ratio sensitivity parameters that index the considered selection models. This non-collapsibility implies that the meaning and magnitude of these sensitivity parameters may change upon adjusting for a measured extraneous covariate, even if that covariate is unrelated to the outcome. While intuitively, one would expect the addition of extraneous covariates to the selection model to attenuate the dependence of missingness on the incomplete outcome, an obstacle in choosing a plausible set of sensitivity parameters is that such attenuation is not guaranteed in view of this non-collapsibility (see, e.g., Scharfstein, Robins, and Rotnitzky, 1999b). A partial remedy to this problem would be to focus the interpretation on standardized sensitivity parameters. These can for instance be obtained by substituting conditional probabilities $P(R_t = 0|Y_t = y, \overline{\mathbf{O}}_{t-1})$ in the expression for the sensitivity parameter, e.g.,

$$\exp(\gamma) = \frac{\text{odds}(R_t = 0|Y_t = y+1, \overline{\mathbf{O}}_{t-1})}{\text{odds}(R_t = 0|Y_t = y, \overline{\mathbf{O}}_{t-1})},$$

by standardized probabilities

$$\frac{1}{n}\sum_{i=1}^{n} \text{expit}\left\{h_t(\overline{\mathbf{O}}_{i,t-1}; \widehat{\boldsymbol{\alpha}}) + \gamma y\right\}$$

for a given choice of y.

Acknowledgment

We thank the International Breast Cancer Study Group for sharing data from Trial VI, and IAP research network grant nr. P7/06 from the Belgian government (Belgian Science Policy).

References

Birmingham, J., Rotnitzky, A., and Fitzmaurice, G.M. (2003). Pattern-mixture and selection models for analysing longitudinal data with monotone missing patterns. *Journal of the Royal Statistical Society, Series B* **65**, 275–297.

Buzkova, P. and Lumley, T. (2009). Semiparametric modeling of repeated measurements under outcome-dependent follow-up. *Statistics in Medicine* **28**, 987–1003.

Diggle, P.J. and Kenward, M.G. (1994). Informative drop-out in longitudinal data analysis (with discussion). *Applied Statistics* **43**, 49–93.

Egleston, B.L. and Wong, Y.-N. (2009). Sensitivity analysis to investigate the impact of a missing covariate on survival analyses using cancer registry data. *Statistics in Medicine* **28**, 1498–1511.

Gill, R.D., van der Laan, M.J., and Robins, J.M. (1997). Coarsening at random: Characterizations, conjectures and counterexamples. *Proceedings of the First Seattle Symposium on Survival Analysis*, pp. 255–294.

Gore, S.M. (1987). *Biostatistics and the Medical Research Council.* Medical Research Council News.

Greenland, S., Robins, J.M., and Pearl, J. (1999). Confounding and collapsibility in causal inference. *Statistical Science* **14**, 29–46.

Horowitz, J.L. and Manski, C.F. (2000). Non-parametric analysis of randomized experiments with missing covariate and outcome data. *Journal of the American Statistical Association* **95**, 77–88.

Hürny, C., Bernhard, J., Gelber, R.D., Coates, A., Castiglione, M., Isley, M., Dreher, D., Peterson, H., Goldhirsch, A., and Senn, H.J. (1992). Quality-of-life measures for patients receiving adjuvant therapy for breast-cancer: An international trial. *European Journal of Cancer* **28A**, 118–124.

Ibrahim, J.G., Chen, M.H., and Lipsitz, S.R. (2001). Missing responses in generalised linear mixed models when the missing data mechanism is nonignorable. *Biometrika* **88**, 551–564.

Imbens, G.W. and Manski, C.F. (2004). Confidence intervals for partially identified parameters. *Econometrica* **72**, 1845–1857.

Kenward, M.G., Molenberghs, G., and Goetghebeur, E. (2001). Sensitivity analysis for incomplete categorical data. *Statistical Modeling* **1**, 31–48.

Lin, H., Scharfstein, D.O., and Rosenheck, R.A. (2003). Analysis of longitudinal data with irregular, informative follow-up. *Journal of the Royal Statistical Society, Series B* **66**, 791–813.

Manski, C.F. (1990). Non-parametric bounds on treatment effects. *American Economic Review, Papers and Proceedings* **80**, 319–323.

Manski, C.F. (2003). *Partial Identification of Probability Distributions.* New York: Springer.

Molenberghs, G., Kenward, M.G., and Goetghebeur, E. (2001). Sensitivity analysis for incomplete contingency tables: the Slovenian plebiscite case. *Journal of the Royal Statistical Society, Series C* **50**, 15-29.

Robins, J.M. (1989). The analysis of randomized and non-randomized AIDS treatment trials using a new approach to causal inference in longitudinal studies. in *Health Service Research Methodology: A Focus on AIDS.* Eds: Sechrest L., Freeman H., Mulley A., Washington, D.C.: U.S. Public Health Service, National Center for Health Services Research, pp. 113–159.

Robins, J.M. (1997). Non-response models for the analysis of non-monotone non-ignorable missing data. *Statistics in Medicine* **16**, 21–37.

Robins, J.M. and Gill, R. (1997). Non-response models for the analysis of non-monotone ignorable missing data. *Statistics in Medicine* **16**, 39–56.

Robins, J.M., Rotnitzky, A., and Scharfstein, D. (1999). Sensitivity Analysis for Selection Bias and Unmeasured Confounding in Missing Data and Causal Inference Models. In: *Statistical Models in Epidemiology: The Environment and Clinical Trials.* M.E. Halloran and D. Berry, D. (Eds.). IMA Volume 116, New York: Springer, pp. 1–92.

Robins, J.M. and Rotnitzky, A. (2001). Comment on the Bickel and Kwon article, "Inference for semiparametric models: Some questions and an answer." *Statistica Sinica* **11**, 920–936.

Romano, J.P. and Shaikh, A.M. (2008). Inference for identifiable parameters in partially identified econometric models. *Journal of Statistical Planning and Inference* **138**, 2786–2807.

Rotnitzky, A., Bergesio, A., and Farall, A. (2009). Analysis of quality-of-life adjusted failure time data in the presence of competing, possibly informative, censoring mechanisms. *Lifetime Data Analysis* **15**, 1–23.

Rotnitzky, A., Faraggi, D. and Schisterman, E. (2006). Doubly robust estimation of the area under the receiver-operating characteristic curve in the presence of verification bias. *Journal of the American Statistical Association* **101**, 1276–1288.

Rotnitzky, A., Robins, J.M., and Scharfstein, D.O. (1998). Semiparametric regression for repeated outcomes with nonignorable nonresponse. *Journal of the American Statistical Association* **93**, 1321–1339.

Rotnitzky, A., Scharfstein, D., Su, T.L., and Robins, J.M. (2001). Methods for conducting sensitivity analysis of trials with potentially nonignorable competing causes of censoring. *Biometrics* **57**, 103–113.

Rubin. D.B. (1976). Inference and missing data. *Biometrika* **63**, 581–592.

Scharfstein, D.O., Rotnitzky, A., and Robins, J.M. (1999a). Adjusting for non-ignorable drop-out using semiparametric non-response models. *Journal of the American Statistical Association* **94**, 1096–1120.

Scharfstein, D.O., Rotnitzky, A., and Robins, J.M. (1999b). Adjusting for non-ignorable drop-out using semiparametric non-response models: Rejoinder. *Journal of the American Statistical Association* **94**, 1135–1146.

Scharfstein, D., Robins, J.M., Eddings, W., and Rotnitzky, A. (2001). Inference in randomized studies with informative censoring and discrete time-to-event endpoints. *Biometrics* **57**, 404–413.

Scharfstein, D.O. and Robins, J.M. (2002). Estimation of the failure time distribution in the presence of informative censoring. *Biometrika* **89**, 617–634.

Scharfstein, D.O., Daniels, M.J. and Robins, J.M. (2003). Incorporating prior beliefs about selection bias into the analysis of randomized trials with missing outcomes.

Biostatistics **4**, 495–512.

Shardell, M., Scharfstein, D.O., and Bozzette, S.A. (2007). Survival curve estimation for informatively coarsened discrete event-time data. *Statistics in Medicine* **26**, 2184–2202.

Shardell, M., Scharfstein, D.O., Vlahov, D., and Galai, N. (2008a). Inference for cumulative incidence functions with informatively coarsened discrete event-time data. *Statistics in Medicine* **27**, 5861–5879.

Shardell, M., Scharfstein, D.O., Vlahov, D., and Galai, N. (2008b). Sensitivity analysis using elicited expert information for inference with coarsened data: Illustration of censored discrete event times in the AIDS Link to Intravenous Experience (ALIVE) study. *American Journal of Epidemiology* **168**, 1460–1469.

Tan, Z. (2008). Understanding OR, PS, and DR. *Statistical Science* **22**, 560–568.

Todem, D., Fine, J., and Peng, L. (2010). A global sensitivity test for evaluating statistical hypotheses with non-identifiable models. *Biometrics* **66**, 558–566.

Tsiatis, A.A. (2006). *Semiparametric Theory and Missing Data.* New York: Springer.

Vansteelandt, S., Goetghebeur, E., Kenward, M. G., and Molenberghs, G. (2006). Ignorance and uncertainty regions as inferential tools in a sensitivity analysis. *Statistica Sinica* **16**, 953–979.

Vansteelandt, S., Rotnitzky, A., and Robins, J.M. (2007). Estimation of regression models for the mean of repeated outcomes under non-ignorable non-monotone non-response. *Biometrika* **94**, 841-860.

Vansteelandt, S. (2009). Discussion on "Identifiability and Estimation of Causal Effects in Randomized Trials with Noncompliance and Completely Non-Ignorable Missing-Data." *Biometrics* **65**, 686–689.

Vansteelandt, S., Carpenter, J., and Kenward, M.G. (2010). Analysis of incomplete data using inverse probability weighting and doubly robust estimators. *Methodology* **6**, 37–48.

18

A Bayesian Perspective on Assessing Sensitivity to Assumptions about Unobserved Data

Joseph W. Hogan

Brown University, Providence, RI

Michael J. Daniels

University of Texas at Austin, TX

Liangyuan Hu

Brown University, Providence, RI

CONTENTS

18.1 Introduction

This chapter provides a Bayesian perspective on how inference might proceed in settings where data that are intended to be collected are missing. Assessment of model sensitivity is a broad topic, encompassing many aspects of inference that might include distributional assumptions, parametric structure, sensitivity to outliers, and assessment of influence of individual data points. When the intended sample is completely observed, many of these modeling assumptions can be checked empirically; our ability to refute the assumptions with any degree of confidence is limited only by sample size, so in some sense these assumptions can be subjected to empirical critique. Assumptions required for fitting models to incomplete data are different because they apply to data that cannot be observed and are therefore inherently untestable. Put simply, they are *subjective.*

The need for subjectivity in analyses of incomplete data is not frequentist or Bayesian, parametric or non-parametric; it is just a feature of the problem. But this feature makes analysis of incomplete data fertile ground for the Bayesian approach, in which subjective components of a model are formally represented in terms of prior distributions.

Approaches to inference from incomplete data are essentially based on one of three approaches: (i) making an assumption such as missing at random (MAR), and assuming it holds without further critique; (ii) fitting and comparing models under several different assumptions about the missing data distribution; (iii) reporting bounds rather than point estimates to reflect the incompleteness of information in the data (as in Manski 2007).

The use of bounds for frequentist inference is usually advocated on the grounds that lack of information about untestable assumptions must be properly reflected in inferences. The philosophy of the Bayesian approach is essentially the same. In this chapter, we demonstrate how models can be parameterized in ways that make clear where purely subjective input is being used. Degree of uncertainty about untestable assumptions is encoded in a prior and is ultimately reflected in posterior distributions. In a broad sense, then, a Bayesian sensitivity analysis is simply an assessment of the impact of prior distributions used to represent the structure of and uncertainty about purely subjective assumptions.

The process of assessing sensitivity to untestable assumptions requires careful consideration of model parameterization and prior specification. The objective of this chapter is to describe how Bayesian modeling—and in particular, appropriately formulated prior distributions —can be used to represent these untestable assumptions, and to illustrate using (stylized) data analyses. Section 18.2 of this chapter lays out notation and definitions, and reviews common missing data assumptions. In Section 18.3, we provide an overview of model identification, a key concept when dealing with incomplete data. Section 18.4 contains the basis of our recommended approach: we show how to parameterize models for incomplete data so that untestable assumptions are represented strictly in terms of non-identifiable parameters. Although this may seem like an intuitive or even obvious idea, we also show that not all models of incomplete data can be decomposed this way; the parametric selection model is an example. Section 18.5 illustrates the ideas through several analyses applied to data from the Commit to Quit Study, a randomized trial examining the effect of vigorous exercise on smoking cessation among women (Marcus et al. 1999).

18.2 Notation and Definitions

18.2.1 Outcome variables, covariates, and response indicators

We are interested to draw inference about features of a distribution of a dependent variable Y conditional on a $1 \times K$ vector of independent variables $\boldsymbol{X} = (X_1, \ldots, X_K)$. For univariate Y, the density or mass function for the distribution of interest is denoted $f(y \mid \boldsymbol{x})$. For multivariate Y, let $\boldsymbol{Y} = (Y_1, \ldots, Y_J)'$ denote the $J \times 1$ vector of outcomes that are intended to be observed, where J may vary over individuals. For each component Y_j of $\boldsymbol{Y}$, there is an associated $1 \times K$ vector of covariates $\boldsymbol{X}_j = (X_{j1}, \ldots, X_{jK})$, some of which may vary over j and some of which may be constant over j. The distribution of a multivariate outcome is therefore written as $f(\boldsymbol{y} \mid \boldsymbol{x})$ (note that $\boldsymbol{x}$ will be a $1 \times K$ vector when y is scalar, and a $J \times K$ matrix when $\boldsymbol{y}$ is vector-valued). For simplicity, we assume all covariates are *exogenous*, including time-varying covariates in settings where $\boldsymbol{Y}$ represents longitudinal measurements. There may be available, in addition, information on auxiliary covariates $\boldsymbol{W}$; these are indexed in the same fashion as $\boldsymbol{X}$. The vector $\boldsymbol{R} = (R_1, \ldots, R_J)'$ of binary indicators is defined such that $R_j = 1$ when Y_j is observed, and $R_j = 0$ when it is not. Throughout this chapter, we assume all elements of $\boldsymbol{X}$ and $\boldsymbol{W}$ are fully observed.

18.2.2 Distributions and models

The notation $f(\boldsymbol{y} \mid \boldsymbol{x})$ refers generically to *density* or *mass function* as appropriate. A *model* of $f(\boldsymbol{y} \mid \boldsymbol{x})$ is written as $p(\boldsymbol{y} \mid \boldsymbol{x}, \boldsymbol{\theta})$, where $\boldsymbol{\theta}$ is a finite-dimensional parameter indexing the model p. Throughout this chapter, it frequently will be necessary to draw distinctions between models, which carry assumptions and structure, from the distributions they are intended to represent or capture.

In the Bayesian context, $\boldsymbol{\theta}$ is a random variable; hence $p(\boldsymbol{y} \mid \boldsymbol{x}, \boldsymbol{\theta})$ is the data model, specified conditionally on $\boldsymbol{\theta}$, while $p(\boldsymbol{\theta})$ is the a-priori model, or *prior*, for the $\boldsymbol{\theta}$. The *posterior*

$$p(\boldsymbol{\theta} \mid \boldsymbol{y}, \boldsymbol{x}) = \frac{p(\boldsymbol{y} \mid \boldsymbol{x}, \boldsymbol{\theta})\, p(\boldsymbol{\theta})}{\int p(\boldsymbol{y} \mid \boldsymbol{x}, \boldsymbol{\theta})\, p(\boldsymbol{\theta})\, d\boldsymbol{\theta}}$$

is the implied parameter model, conditional on the data. The *likelihood function* is any function $L(\boldsymbol{\theta} \mid \boldsymbol{y}, \boldsymbol{x})$ that is proportional in $\boldsymbol{\theta}$ to the data model. It follows that $p(\boldsymbol{\theta} \mid \boldsymbol{y}, \boldsymbol{x}) \propto L(\boldsymbol{\theta} \mid \boldsymbol{y}, \boldsymbol{x})\, p(\boldsymbol{\theta})$.

The key feature of drawing inference from incomplete data is the need to characterize the joint distribution of the outcomes $\boldsymbol{Y}$ and the response indicators $\boldsymbol{R}$, usually conditionally on covariates of interest $\boldsymbol{X}$ and sometimes also conditionally on auxiliary covariates $\boldsymbol{W}$. Missing data assumptions in particular (MCAR, MAR) can be operationalized in terms of distributions governing $(\boldsymbol{Y}, \boldsymbol{R}, \boldsymbol{X}, \boldsymbol{W})$. Consequently, we need to define *full data*, *observed data*, and their associated distributions.

The *full data* $(\boldsymbol{Y}, \boldsymbol{R}, \boldsymbol{X})$ for an individual comprise the data that were intended to be collected, plus the response indicators. The *full data distribution* is defined as $f(\boldsymbol{y}, \boldsymbol{r} \mid \boldsymbol{x})$. With auxiliary covariates, it is necessary to expand the definition of the full data and full-data distribution to be $(\boldsymbol{Y}, \boldsymbol{R}, \boldsymbol{X}, \boldsymbol{W})$ and $f(\boldsymbol{y}, \boldsymbol{r} \mid \boldsymbol{x}, \boldsymbol{w})$, respectively.

Because data on $\boldsymbol{Y}$ are only partially observed, we must define the observed data and its corresponding distribution. The *observed data* for an individual are denoted by $(\boldsymbol{Y}_{\text{obs}}, \boldsymbol{R}, \boldsymbol{X})$, where $\boldsymbol{Y}_{\text{obs}} = \{Y_j : R_j = 1\}$. The full data can therefore be partitioned as $(\boldsymbol{Y}_{\text{obs}}, \boldsymbol{Y}_{\text{mis}}, \boldsymbol{R}, \boldsymbol{X})$; the *observed-data distribution* is obtained by averaging the full-data distribution over the possible realizations of the missing outcomes,

$$\begin{aligned} f(\boldsymbol{y}_{\text{obs}}, \boldsymbol{r} \mid \boldsymbol{x}) &= \int f(\boldsymbol{y}_{\text{obs}}, \boldsymbol{y}_{\text{mis}}, \boldsymbol{r} \mid \boldsymbol{x})\, d\boldsymbol{y}_{\text{mis}} \\ &= \int f(\boldsymbol{y}_{\text{obs}}, \boldsymbol{r} \mid \boldsymbol{x})\, dF(\boldsymbol{y}_{\text{mis}} \mid \boldsymbol{y}_{\text{obs}}, \boldsymbol{r}, \boldsymbol{x}). \end{aligned}$$

Ultimately, the goal is to draw inference about $f(\boldsymbol{y} \mid \boldsymbol{x})$, which we call the *target distribution*. It is the distribution of full-data outcomes $\boldsymbol{Y}$ conditional on covariates of interest $\boldsymbol{X}$. The target distribution is obtained by integrating the full-data distribution over the response indicators; in our case, where the response indicators are discrete, the integration is carried out via the summation

$$f(\boldsymbol{y} \mid \boldsymbol{x}) = \sum_{\boldsymbol{r} \in \mathscr{R}} f(\boldsymbol{y}, \boldsymbol{r} \mid \boldsymbol{x}),$$

where $\mathscr{R} = \{0,1\}^{\otimes J}$ is the sample space of $\boldsymbol{R}$.

18.2.3 Missing data mechanisms

Inference about the target distribution relies on assumptions about the full-data distribution. This can be seen most easily in the univariate case where the target distribution is expressed as a mixture over observed and missing outcome distributions. Let $\pi(\boldsymbol{x}) = \Pr(R = 1 \mid \boldsymbol{x})$, and for $r \in \{0,1\}$, denote $f(y \mid \boldsymbol{x}, R = r)$ by $f_r(y \mid \boldsymbol{x})$. Set aside auxiliary covariates for the time being. The target distribution can be written as

$$f(y \mid \boldsymbol{x}) = \pi(\boldsymbol{x})\, f_1(y \mid \boldsymbol{x}) + \{1 - \pi(\boldsymbol{x})\} f_0(y \mid \boldsymbol{x}). \tag{18.1}$$

It is clear from this simple case that no amount of observed data can be used to infer $f_0(y \mid \boldsymbol{x})$, which necessitates the use of untestable assumptions in order to infer $f(y \mid \boldsymbol{x})$.

In the statistics literature, these assumptions are typically characterizing a *missing data mechanism*, expressed in terms of the relationship between $\boldsymbol{Y}$ and $\boldsymbol{R}$, possibly conditioned on covariates $\boldsymbol{X}$ and $\boldsymbol{W}$. The missing data mechanism is typically expressed in terms of the distribution $f(\boldsymbol{r} \mid \boldsymbol{y}, \boldsymbol{x}, \boldsymbol{w}) = f(\boldsymbol{r} \mid \boldsymbol{y}_{\text{obs}}, \boldsymbol{y}_{\text{mis}}, \boldsymbol{x}, \boldsymbol{w})$, which captures the probability of (non-)response as a function of covariates, observed responses, and possibly missing responses. More generally it can be expressed in terms of the joint distribution $f(\boldsymbol{y}_{\text{obs}}, \boldsymbol{y}_{\text{mis}}, \boldsymbol{r} \mid \boldsymbol{x}, \boldsymbol{w})$.

The standard taxonomy for missing data mechanisms was formalized by Rubin (1976) and has been generalized to settings involving longitudinal data (Robins, Rotnitzky, and Zhao 1995; Scharfstein, Rotnitzky, and Robins 1999) and event history data (Fleming and Harrington 1991; Scharfstein et al. 2001). Although the core of the taxonomy is well understood, there are small variations in how the assumptions are stated and applied; see Little (1995), Fitzmaurice (2003), Hogan, Roy, and Korkontzelou (2004), Tsiatis (2006), Molenberghs and Kenward (2007), Diggle et al. (2007), and Daniels and Hogan (2008) for a variety of perspectives, particularly as they relate to the incorporation of information on model covariates and auxiliary covariates.

Owing to the small but measurable variation in how the assumptions are stated and used, it is useful to write the assumptions as they will be used in this chapter. We find it helpful to define missing data mechanisms (i) relative to the distribution of interest and (ii) in terms of all data that are available, including auxiliary covariates. What we mean by (i) is that if the distribution of interest conditions on $\boldsymbol{X}$, then the missing data mechanisms are, by default, defined conditionally on $\boldsymbol{X}$. (It does not seem worthwhile to consider missing data mechanisms that do not condition on $\boldsymbol{X}$.) This avoids the need to subdivide the assumptions into cases such as covariate-dependent, auxiliary-variable-dependent, and so forth. We will sometimes assume that observable data are limited, particularly where auxiliary covariates are concerned. Moreover, we will focus only on the missing at random (MAR) assumption and its complement, not missing at random (NMAR). The stronger *ignorability* condition applies to likelihood-based inference from parametric models, and is described in detail in Chapter 5.

First consider the case where Y is scalar, covariate information is available in the form of $(\boldsymbol{X}, \boldsymbol{W})$, and the target of inference is $f(y \mid \boldsymbol{x})$. The outcome Y is said to be *missing at random (MAR)* if Y is conditionally independent of R within distinct levels of $(\boldsymbol{X}, \boldsymbol{W})$, denoted by

$$Y \perp\!\!\!\perp R \,|\, (\boldsymbol{X}, \boldsymbol{W}).$$

In short, the observed Y are obtained by random sampling within levels of $(\boldsymbol{X}, \boldsymbol{W})$. An immediate consequence of MAR is that $f_0(y \,|\, \boldsymbol{x}, \boldsymbol{w}) = f_1(y \,|\, \boldsymbol{x}, \boldsymbol{w})$. If auxiliary covariates are not available, MAR states that $Y \perp\!\!\!\perp R \,|\, \boldsymbol{X}$, which implies $f_0(y \,|\, \boldsymbol{x}) = f_1(y \,|\, \boldsymbol{x})$. In fact, by replacing the missing data distribution $f_0(y \,|\, \boldsymbol{x})$ with its observed-data counterpart $f_1(y \,|\, \boldsymbol{x})$ in (18.1), we see that $f(y \,|\, \boldsymbol{x}) = f_1(y \,|\, \boldsymbol{x})$ under MAR.

When auxiliaries are available, additional assumptions are needed. Note first that the target distribution is

$$f(y \,|\, \boldsymbol{x}) \quad = \quad \int f(y \,|\, \boldsymbol{x}, \boldsymbol{w}) \, dF(\boldsymbol{w} \,|\, \boldsymbol{x}). \tag{18.2}$$

Furthermore, the integrand $f(y \,|\, \boldsymbol{x}, \boldsymbol{w})$ is itself a mixture over the response indicator,

$$f(y \,|\, \boldsymbol{x}, \boldsymbol{w}) \quad = \quad \pi(\boldsymbol{x}, \boldsymbol{w}) f_1(y \,|\, \boldsymbol{x}, \boldsymbol{w}) + \{1 - \pi(\boldsymbol{x}, \boldsymbol{w})\} f_0(y \,|\, \boldsymbol{x}, \boldsymbol{w}); \tag{18.3}$$

as with (18.1), the observed data offer no information about $f_0(y \,|\, \boldsymbol{x}, \boldsymbol{w})$. Under MAR however, $f_0(y \,|\, \boldsymbol{x}, \boldsymbol{w}) = f_1(y \,|\, \boldsymbol{x}, \boldsymbol{w})$, and $f(y \,|\, \boldsymbol{x})$ can in theory be inferred from the observed data; in practice, however, modeling assumptions are needed for $\pi(\boldsymbol{x}, \boldsymbol{w})$ and for $f(\boldsymbol{w} \,|\, \boldsymbol{x})$, and the integral (18.2) must be calculated. Although it is possible to use other factorizations of the joint distribution $f(y, r \,|\, \boldsymbol{x}, \boldsymbol{w})$, (18.2) and (18.3) convey the degree to which a combination of (testable) modeling assumptions and (untestable) missing data mechanism assumptions must be applied in order to draw inference about $f(y \,|\, \boldsymbol{x})$ from incomplete data.

Missingness is *not missing at random (NMAR)* when MAR fails to hold. Like MAR, it can be defined conditionally on $\boldsymbol{X}$ alone or both $\boldsymbol{X}$ and $\boldsymbol{W}$. We consider the latter case. Let $\boldsymbol{V} = (\boldsymbol{X}, \boldsymbol{W})$. Outcomes are MAR when $Y \perp\!\!\!\perp R \,|\, \boldsymbol{V}$. Outcomes are NMAR when this independence condition does not hold; formally, when there exists at least one value $\boldsymbol{v}^*$ in the support of $\boldsymbol{V}$ such that $f_0(y \,|\, \boldsymbol{v}^*) \neq f_1(y \,|\, \boldsymbol{v}^*)$.

18.3 Inference from Under-Identified Models

18.3.1 Overview

This section contrasts frequentist and Bayesian approaches to inference for distributions or parameters that are not identified. The literature on model identification is substantial, and reaches well beyond situations involving missing data. For accessible accounts, see Manski (2007) (non-parametric models) and Dawid (1979) (parametric Bayesian models). Rather than approach the discussion by laying out conditions under which parameters or distributions are identified, it is sometimes easier to characterize the issues in terms of model *non*-identifiability. Our goal here is to focus only on missing datasettings, and to describe the key issues that are relevant to setting up a Bayesian framework for sensitivity analysis. Rather than describing a general theory for (non)identifiability, we will make our points through the use of concrete examples.

Recall that the target distribution is $f(y \,|\, \boldsymbol{x})$. The target of inference is usually a feature or parameter of this distribution, which we write as $\boldsymbol{\phi} = \boldsymbol{\phi}\{f(y \mid \boldsymbol{x})\}$. We assume that $\boldsymbol{\phi}$ is a real-valued and possibly multivariate quantity (though for the most part we will illustrate using univariate $\boldsymbol{\phi}$). For example, we may be interested in the mean $\phi = E(Y \mid \boldsymbol{X} = \boldsymbol{x}) = \int y \; dF(y \mid \boldsymbol{x})$, or in a vector of important centiles of $f(y \mid \boldsymbol{x})$. For a parametric or semi-parametric model $p(y \mid \boldsymbol{x}, \boldsymbol{\theta})$, $\boldsymbol{\phi}$ will generally be expressible as a function $\boldsymbol{\phi}(\boldsymbol{\theta})$ of the model parameters. For example, if we assume $E(Y \mid \boldsymbol{x}) = \boldsymbol{x}'\boldsymbol{\beta}$, then $\boldsymbol{\phi} = \boldsymbol{\beta}$.

18.3.2 Assessing veracity of modeling assumptions

Statistical models and methods can rely on many types of assumptions. In ideal settings, where complete samples of data are drawn randomly from target populations, and sample sizes are sufficiently large, assumptions about the distribution of observed data can be checked empirically. When empirical checks are not conclusive (e.g., due to finite samples), it is possible to assess sensitivity of inferences to modeling assumptions by fitting several different models and comparing lack of fit.

For example, given a random sample $(Y_1, \ldots, Y_N)$, we can assess empirically whether the sample departs from a normal distribution. From a sample $(Y_1, \boldsymbol{X}_1), \ldots, (Y_N, \boldsymbol{X}_N)$, we can assess evidence against the linearity assumption $E(Y \mid \boldsymbol{X} = \boldsymbol{x}) = \boldsymbol{x}'\boldsymbol{\beta}$. In the frequentist context, these assessments can be formulated in terms of hypothesis tests. For Bayesian models, posterior predictive distributions can be used (e.g., Gelman, Meng, and Stern 1996). As $n \to \infty$, the observed data theoretically would allow us to accept or reject modeling assumptions with certainty; indeed, as $n \to \infty$, we can infer features of $f(y \mid \boldsymbol{x})$, or indeed the entire distribution, up to sampling variation. If we assume $f(y \mid \boldsymbol{x})$ follows a model $p(y \mid \boldsymbol{x}, \boldsymbol{\theta})$, we can evaluate the parametric and distributional assumptions underlying $p(y \mid \boldsymbol{x}, \boldsymbol{\theta})$.

When a sample (or more generally, intended information) is incomplete, it is no longer possible to empirically check assumptions underlying a model or method. To lend some formality, we return to the simple setting where $f(y \mid \boldsymbol{x})$ is the target distribution and there are no auxiliary covariates. Recall that $f(y \mid \boldsymbol{x})$ can be factored using the mixture distribution (18.1). When Y is missing for some individuals, we observe $(Y, \boldsymbol{X})$ when $R = 1$ and observe only $\boldsymbol{X}$ when $R = 0$.

Clearly the observed data provide information about $f_1(y \mid \boldsymbol{x})$ and $\pi(\boldsymbol{x})$, but provide no information about $f_0(y \mid \boldsymbol{x})$, even as $n \to \infty$. In this simple setting, *purely subjective assumptions* are needed to draw inference about features of $f(y \mid \boldsymbol{x})$. That subjectivity is required has nothing to do with whether inferences are Bayesian or frequentist; it simply means that the data alone, even in infinite samples, do not provide sufficient information to infer $f(y \mid \boldsymbol{x})$.

Another common objective of statistical modeling is estimation of a causal treatment effect, which frequently is formulated in terms of potential outcomes. To illustrate, consider the setting of binary treatment, where Y_1 is the outcome when treatment is received and Y_0 when not received. Let $A = 1$ if treatment is received, $A = 0$ if not, so that the observed data for each individual is

$$(Y, A) \quad = \quad \begin{cases} (Y_1, 1) & \text{if} \quad A = 1, \\ (Y_0, 0) & \text{if} \quad A = 0. \end{cases}$$

In general, the causal effect of treatment can be summarized using features of the joint distribution $f(y_1, y_0)$; more commonly, the marginal distributions $f(y_1)$ and $f(y_0)$ are used. Consider drawing inference about $f(y_1)$. The issue with identification mirrors that for the missing datasetting, where now Y_1 is observed when $A = 1$ and is missing when $A = 0$. Without further assumptions, $f(y_1)$ is not identifiable (the same holds for $f(y_0)$). Assumptions analogous to MAR can be invoked to identify $f(y_1)$ and $f(y_0)$ (cf. Rosenbaum and Rubin 1983); in randomized trials, they can be expected to hold, but in observational studies, they are completely subjective in the sense that even an infinite sample of the observable data will not allow empirical assessment (Rubin 1978; Robins 1999; Dawid 2000; Robins and Greenland 2000).

18.3.3 Non-parametric (non-)identifiability

A simple and nontechnical way to think about whether a distribution is identifiable without reliance on distributional or parametric assumptions is to consider cases where infinite amounts of observable data are available. If $\pi(\boldsymbol{x}) < 1$ for all possible realizations of $\boldsymbol{x}$, then observations on Y may be missing. Referring back to (18.1), it is clear that no amount of observable data can be used to identify $f(y \mid \boldsymbol{x})$: an infinite sample of data yields perfect knowledge of $f_1(y \mid \boldsymbol{x})$ and $\pi(\boldsymbol{x})$, but yields no information about $f_0(y \mid \boldsymbol{x})$. The distribution $f(y \mid \boldsymbol{x})$ is therefore *non-parametrically non-identifiable*: in the absence of parametric assumptions, $f(y \mid \boldsymbol{x})$ cannot be determined from observed data.

Some progress can be made if we are willing to focus on specific aspects of $f(y \mid \boldsymbol{x})$ instead of the entire distribution. Consider the case where Y is binary and we are interested in $\mu(\boldsymbol{x}) = E(Y \mid \boldsymbol{x}) = \Pr(Y = 1 \mid \boldsymbol{x})$. Let $\mu_r(\boldsymbol{x}) = E(Y \mid \boldsymbol{x}, R = r)$. Then

$$\mu(\boldsymbol{x}) \quad = \quad \pi(\boldsymbol{x})\mu_1(\boldsymbol{x}) + \{1 - \pi(\boldsymbol{x})\}\mu_0(\boldsymbol{x}). \tag{18.4}$$

Although no amount of data will provide information about $\mu_0(\boldsymbol{x})$, we know it is bounded between 0 and 1. Hence the possible values of $\mu(\boldsymbol{x})$ also are bounded, with

$$\pi(\boldsymbol{x})\mu_1(\boldsymbol{x}) \ \leq \ \mu(\boldsymbol{x}) \ \leq \ 1 - \pi(\boldsymbol{x})\{1 - \mu_1(\boldsymbol{x})\}. \tag{18.5}$$

In this case we say that $\mu(\boldsymbol{x})$ is *partially identified*, because its possible values are constrained by quantities that can be estimated from data. Similar arguments can be used to show that for continuous Y, quantiles of $f(y \mid \boldsymbol{x})$ are non-parametrically partially identified (Manski 2007).

A main focus of this paper is on drawing inference about means. Generally, mean parameters are not identifiable, or even partially identifiable, from incomplete data unless there are natural restrictions on the support of Y (e.g., body weights that are bounded below by zero). This is evident by examining the analogue of (18.4) for the case of continuous and unbounded Y. If $\mu_0(\boldsymbol{x})$ is unbounded and $\pi(\boldsymbol{x}) < 1$, then $\mu(\boldsymbol{x})$ also is unbounded.

It is possible to place bounds on $\mu_0(\boldsymbol{x})$, in which case $\mu(\boldsymbol{x})$ also will be bounded. For example, if we impose the constraint $a \leq \mu_0(\boldsymbol{x}) \leq b$, then $\mu(\boldsymbol{x})$ is partially identified and bounded (the bounds on $\mu_0(\boldsymbol{x})$ can depend on $\boldsymbol{x}$ in general). But unless $\mu_0(\boldsymbol{x})$ has a natural bound, such as with measured quantities like weight and blood pressure, the choices for a and b are entirely subjective.

At the same time, it is unusual in practice to encounter settings where there is no prior knowledge to inform potential values for the bounds; e.g., body weight within specific adult populations is likely to have some pragmatic bounds, even though experts may disagree on specific values for these. As we illustrate in the data analysis examples of Section 18.5, the Bayesian framework provides a formal way to set "soft" boundaries, in the sense that they do not have to take fixed values for the analysis, and to have the final inferences reflect uncertainty about where the boundaries should be set.

18.3.4 Parametric identification

In this section we describe identifiability conditions for parametric models of a distribution $f(y \mid \boldsymbol{x})$. Parameter identification for frequentist inference has been written about extensively. The Bayesian framework provides a useful and elegant characterization of parameter identification in terms of conditional independence (Dawid 1979). In a Bayesian model, we say that a parameter is not identified by data when, conditional on other parameters in a model, it is conditionally independent of the observed data. Specifically, consider a model $p(y \mid \theta_1, \theta_2)$ that characterizes $f(y)$ in terms of $\boldsymbol{\theta} = (\theta_1, \theta_2)$. The parameter θ_2 is *not identifiable* if

$$p(\theta_2 \mid \boldsymbol{y}, \theta_1) \quad = \quad p(\theta_2 \mid \theta_1).$$

To illustrate identifiability in the context of missing data problems, we use several examples. To simplify the exposition we omit covariates, but the examples here are easily extended to accommodate them. We also confine attention to scalar Y, so that the full-data distribution is $f(y, r)$ and the target distribution is $f(y)$. Unless stated otherwise, we are interested in drawing inference about $\mu = E(Y)$.

We begin with a simple example where the full-data model is parameterized as a mixture model, and non-identifiability of μ is clear. We then move to a selection model parameterization to illustrate how parametric and structural assumptions can be used to identify μ. The assumptions used to identify μ in the selection model cannot, however, be assessed empirically, even with an infinite sample of observable data.

Example 5 (Mixture of normal distributions)**.** The full-data model is

$$p(y, r \mid \boldsymbol{\theta}) \quad = \quad \pi\, p_1(y \mid \boldsymbol{\theta}_1) + (1 - \pi) p_0(y \mid \boldsymbol{\theta}_0),$$

where $\pi = \Pr(R = 1)$ and $p_r(y \mid \boldsymbol{\theta}_r) = p(y \mid R = r, \boldsymbol{\theta}_r)$ for $r \in \{0, 1\}$. Assume $p_r(y \mid \boldsymbol{\theta}_r)$ is a normal density function indexed by mean and variance given by $\boldsymbol{\theta}_r = (\mu_r, \sigma^2)$; hence $\boldsymbol{\theta} = (\mu_0, \mu_1, \sigma, \pi)$. Further assume the joint prior for $\boldsymbol{\theta}$ can be factored so that μ_0 is a-priori independent of the other parameters,

$$p(\mu_0, \mu_1, \sigma, \pi) \quad = \quad p(\mu_0)\, p(\mu_1, \sigma, \pi).$$

For this prior, it is straightforward to show that $p(\mu_0 \mid \text{Data}, \mu_1, \sigma, \pi) = p(\mu_0 \mid \mu_1, \sigma, \pi) = p(\mu_0)$, and hence that observed data contribute no information about μ_0. Consequently, any functions of $\boldsymbol{\theta}$ that depend on μ_0, such as $E(Y \mid \boldsymbol{\theta}) = \pi\mu_1 + (1 - \pi)\mu_0$, cannot be identified by data alone. □

Example 6 (Parametric selection model)**.** In this example we consider a different parameterization of the joint distribution using a selection model (Heckman 1979). The joint distribution $f(y, r \mid \boldsymbol{w})$ is modeled via

$$\begin{pmatrix} Y \\ U \end{pmatrix} \sim N\left[\begin{pmatrix} \mu \\ 0 \end{pmatrix}, \begin{pmatrix} \sigma^2 & \rho\sigma \\ \rho\sigma & 1 \end{pmatrix}\right],$$

where U is a latent variable and $R = \quad (U > 0)$. This is a standard formulation of the parametric selection model typically used in econometrics, and implemented in Stata Version 11.1 (covariates typically would be used in the distributions of Y and U, but are omitted here for simplicity).

The implied missing data mechanism for this model can be written in terms of a probit regression of R on Y,

$$\Phi^{-1}\{\Pr(R = 1 \mid Y = y)\} \quad = \quad \frac{\rho}{\sigma\sqrt{1 - \rho^2}}(y - \mu), \tag{18.6}$$

where $\Phi(\cdot)$ is the cumulative distribution function of the standard normal distribution. The coefficient of y, which characterizes the missing data mechanism, is a function of ρ. We have MAR if and only if $\rho = 0$, and NMAR otherwise. The model also implies $E(Y \mid R = 1) = \mu + \rho\sigma_1\phi(0)/\Phi(0)$, where $\phi(\cdot)$ is the density function of the standard normal. □

Although this model is commonly used in practice, it has some features that can present difficulties in assessing sensitivity to underlying assumptions.

Remark 1. The parameter ρ governs the dependence between Y and R. The parameter ρ is actually identifiable even when its prior is independent of other model parameters. It can be shown that under a prior of the form $p(\rho, \mu, \sigma, \boldsymbol{\alpha}) = p(\rho)p(\mu, \sigma)$, the conditional posterior $p(\rho \mid \text{Data}, \mu, \sigma)$ *does* depend on the other model parameters. We argue in Section 18.4 that ρ is not a suitable sensitivity parameter.

Remark 2. The distribution $f_0(y)$ cannot be non-parametrically identified, and in the mixture model formulation $p_0(y \mid \mu_0, \sigma)$ is not identifiable because μ_0 is not. In the selection model, however, because all components of $\boldsymbol{\theta}$ are identified, so is the implied model $p_0(y \mid \boldsymbol{\theta})$. Moreover, the implied model $p_1(y \mid \boldsymbol{\theta})$ for the observed Ys is an explicit function of ρ; the structural and parametric assumptions underlying this model translate directly into restrictions on $f_1(y)$ and hence $E(Y \mid R = 1)$.

This latter property can lead to some unintended consequences. Specifically, one type of sensitivity analysis recommended for these models is to fit the model under several different fixed values for the coefficient of y in the selection model (18.6), which amounts to fitting the model for several fixed values of ρ. But because the observed-data mean $E(Y \mid R = 1)$ also is a function of ρ, this type of sensitivity analysis may place unnecessary restrictions on the observed data distribution.

Remark 3. Parametric selection models are sometimes seen as preferable because when covariate effects are of primary interest, the model $p(y \mid \boldsymbol{\theta})$ can be specified directly. Mixture models do not generally have this property, and indeed in some cases it is difficult to calculate covariate effects from mixtures (but see Fitzmaurice, Laird, and Shneyer 2001 and Roy and Daniels 2008 for workable approaches). However it is important to keep in mind that all parameters of the model, including effects of covariates that we may add to the model, are identified by the joint normality assumption between Y and U (which also implies that $\Pr(R = 1 \mid Y = y)$ is monotone in y). Even though no aspect of the normality assumption is testable, the model has been shown in practice to be extremely sensitive to it (Kenward 1998). Chapter 5 of this volume has further discussion of the properties of parametric selection models.

18.3.5 Summary

The distribution $f(y \mid \boldsymbol{x})$ is non-parametrically non-identifiable when observations on Y are missing. When parametric models are used, the model parameters may be identifiable under certain assumptions, but these are always subjective. In the mixture model of Example 5, the model $p_0(y \mid \mu_0, \sigma)$, and hence the target model $p(y \mid \boldsymbol{\theta})$, can only be identified if we make subjective assumptions about the value of μ_0. In the selection model of Example 6, the subjective assumption of joint normality between the partially observed Y and the latent response variable U drives model identification.

In the next section, we describe Bayesian model formulations that permit incorporation of subjective assumptions and assessment of sensitivity of inferences about target distributions to these assumptions. A key feature of our approach is to formulate the models in terms of parameters that, in the Bayesian sense, are not identified by data. This approach clearly differentiates assumptions that are purely subjective from those that can be critiqued using observed data. It also differentiates models that are well suited to sensitivity analysis from those that are not.

18.4 Sensitivity Analysis from a Bayesian Perspective

This section is concerned with formalizing the notion of sensitivity analysis in a Bayesian framework. The approach outlined here relies on a specification of a model (likelihood) $p(\boldsymbol{y}, \boldsymbol{r} \mid \boldsymbol{x}, \boldsymbol{\theta})$ for the full data, and a prior $p(\boldsymbol{\theta})$ for the vector of parameters indexing this model. In the next sections we (i) describe a factorization of the likelihood that separates the (identified) observed-data distribution from the (non-identified) missing data distribution; (ii) provide a formal definition of "sensitivity parameter"; and (iii) discuss implications for prior specification when the set of parameters includes identified and non-identified parameters.

A principled inference should reflect all sources of uncertainty, including uncertainty about the missing data mechanism. In the frequentist domain, one approach to reflecting uncertainty about untestable assumptions is to report bounds instead of point estimates (as in (18.5)), and to construct confidence intervals that incorporate sampling variability (see Vansteelandt et al. 2006 for example). Frequently-cited advantages to using bounds are that they remove or alleviate reliance on subjective assumptions such as MAR and normality, which are frequently motivated by familiarity or convenience rather than scientific or contextual considerations; and that they communicate information content about the parameter of interest, no more and no less.

On the other hand, there are several disadvantages to using bounds in practice. First, bounds implicitly assign equal weight to all possible versions of the untestable assumptions, including the extreme cases. For example, in a smoking cessation study where the outcome is a binary indicator of cessation status, equal weight is given to the two extreme assumptions (i) all missing observations are "smokers" and (ii) all missing observations are "non-smokers." This simple example amplifies the importance of context: it can be almost universally agreed by those who study smoking cessation that the boundary associated with "all those with missing outcomes are non-smokers" has no credibility. Hence, in practice, subjective but contextually appropriate restrictions can be placed on the distribution of missing observations.

Second, in their pure form, resorting to bounds for inference about means yields no information at all unless the measurement scale for the outcome is itself naturally bounded. When it is not, subjectivity is required to fix upper and lower boundaries for the measurement scale; even among subject-matter experts, variability in opinion about where these should be fixed is inevitable.

18.4.1 Likelihood parameterization

Recall that the full-data distribution is $f(\boldsymbol{y}, \boldsymbol{r} \mid \boldsymbol{x})$ and that the target distribution is $f(\boldsymbol{y} \mid \boldsymbol{x})$. The full-data distribution can be factored as

$$f(\boldsymbol{y}, \boldsymbol{r} \mid \boldsymbol{x}) \quad = \quad f(\boldsymbol{y}_{\text{mis}} \mid \boldsymbol{y}_{\text{obs}}, \boldsymbol{r}, \boldsymbol{x}) f(\boldsymbol{y}_{\text{obs}}, \boldsymbol{r} \mid \boldsymbol{x}).$$

The first part we call the *extrapolation distribution* because it describes the rule for extrapolating missing data from the observed data.

Example 7 (Bivariate data with no covariates). Assume the full data are (Y_1, Y_2, R), where $R = 1$ if Y_2 is observed and $R = 0$ otherwise. The full-data distribution can be written as

$$\begin{aligned} f(y_1, y_2, r) &= f(y_1, y_2, r=0)^{1-r} f(y_1, y_2, r=1)^r \\ &= \{f(y_2 \mid y_1, r=0) f(y_1, r=0)\}^{1-r} f(y_1, y_2, r=1)^r \\ &= \underbrace{f(y_2 \mid y_1, r=0)^{1-r}}_{} \times \underbrace{f(y_1, r=0)^{1-r} f(y_1, y_2, r=1)^r}_{} \\ &= \text{extrapolation dist.} \times \text{observed-data dist.} \end{aligned}$$

We have written the factorization in a very general way; unless there are sufficient data to estimate the observed-data distribution non-parametrically, choices will have to be made about imposing structure. The data analyst can compare different assumptions about the observed-data distribution using the posterior predictive distribution. Chapter 5 in this volume provides more details. □

Example 8 (Bivariate normal data with no covariates.). If the data are continuous, it may be useful to use a bivariate normal model for the observed-data distribution. First, write

$$\begin{aligned} f(y_1, r=0) &= f(y_1 \mid r=0) \Pr(r=0) \\ f(y_1, y_2, r=1) &= f(y_2 \mid y_1, r=1) f(y_1 \mid r=1) \Pr(r=1). \end{aligned}$$

Next, make parametric assumptions about the observed data distribution,

$$\begin{aligned} Y_1 \mid R=0 &\sim N(\mu_0, \sigma_0^2), \\ Y_1 \mid R=1 &\sim N(\mu_1, \sigma_1^2), \\ Y_2 \mid Y_1 = y_1, R=1 &\sim N(\alpha_1 + \beta_1 y_1, \ \tau_1^2), \\ R &\sim \text{Ber}(\pi). \end{aligned} \tag{18.7}$$

To complete the model, assumptions are needed about the non-identified distribution $f(y_2 \mid y_1, r=0)$. In what follows, we recommend specifying this model in a way that links it to identified parts of the distribution using sensitivity parameters. □

18.4.2 Sensitivity parameters

A sensitivity parameter indexes the extrapolation model using parameters that cannot be identified by observed data. An additional property of a sensitivity parameter is that once its value is fixed, the full data model is identified. We adopt a somewhat modified but equivalent version of the definition given by Daniels and Hogan (2008).

Definition: Sensitivity parameter. Let $f(\boldsymbol{y}, \boldsymbol{r})$ denote a full-data distribution; the target of inference is $f(\boldsymbol{y})$. The associated full-data model $p(\boldsymbol{y}, \boldsymbol{r} \mid \boldsymbol{\theta})$ has factorization

$$p(\boldsymbol{y}, \boldsymbol{r} \mid \boldsymbol{\theta}) = p(\boldsymbol{y}_{\text{mis}} \mid \boldsymbol{y}_{\text{obs}}, \boldsymbol{r}, \boldsymbol{\theta}_E)\, p(\boldsymbol{y}_{\text{obs}}, \boldsymbol{r} \mid \boldsymbol{\theta}_D).$$

Let Θ_D and Θ_E define the parameter spaces for $\boldsymbol{\theta}_D$ and $\boldsymbol{\theta}_E$, respectively. Let $\boldsymbol{\Delta}$ be a $d \times 1$ parameter vector with support on $\mathscr{D}$. Finally, define $g : \{\Theta_D \times \mathscr{D}\} \to \Theta_E$ such that $g(\boldsymbol{\theta}_D, \boldsymbol{\Delta}) = \boldsymbol{\theta}_E$. If

(i) the parameter $\boldsymbol{\Delta}$ is non-identifiable in the sense that

$$p(\boldsymbol{\Delta} \mid \boldsymbol{\theta}_D, \boldsymbol{y}_{\text{obs}}, \boldsymbol{r}) = p(\boldsymbol{\Delta} \mid \boldsymbol{\theta}_D),$$

(ii) the parameter $\boldsymbol{\theta}_D$ is identifiable, and

(iii) the function $g(\boldsymbol{\theta}_D, \boldsymbol{\Delta})$ is non-constant in $\boldsymbol{\Delta}$,

then $\boldsymbol{\Delta}$ is a sensitivity parameter. The first condition states that the observed data contribute no information about the sensitivity parameter; the second and third conditions imply that fixing $\boldsymbol{\Delta}$ identifies the full-data model. □

A parameter $\boldsymbol{\Delta}$ is therefore a sensitivity parameter if it indexes the extrapolation distribution and is not identifiable. In many cases, models can be parameterized so that $\boldsymbol{\Delta}$ captures assumptions about the missing data mechanism. The following two examples illustrate models that incorporate a sensitivity parameter satisfying our definition.

Example 9 (Location shift for bivariate normal distribution)**.** We continue with Example 8. Recall that we are interested in $f(y_1, y_2)$. Following the parameterization (18.7), the extrapolation model is $p_0(y_2 \mid y_1, \boldsymbol{\theta}_E)$, and the full-data model is

$$\begin{aligned} p(y_1, y_2, r \mid \boldsymbol{\theta}) &= p_0(y_2 \mid y_1, \boldsymbol{\theta}_E)^{1-r} \, \{(1-\pi)\, p_0(y_1 \mid \mu_0, \sigma_0)\}^{1-r} \\ &\quad \times \{\pi \, p_1(y_2 \mid y_1, \alpha_1, \beta_1, \tau_1)\, p_1(y_1 \mid \mu_1, \sigma_1)\}^r . \end{aligned}$$

Hence $\boldsymbol{\theta}_D = (\pi, \alpha_1, \beta_1, \mu_0, \mu_1, \sigma_0, \sigma_1, \tau_1)$ are the (identifiable) parameters of the observed-data model.

Now consider specifying the non-identified distribution $f_0(y_2 \mid y_1)$ in terms of a location shift model, relative to its identified counterpart $f_1(y_2 \mid y_1) = p_1(y_2 \mid y_1, \alpha_1, \beta_1, \tau_1)$. Specifically, assume

$$\begin{aligned} f_0(y_2 \mid y_1) &= f_1(y_2 + \Delta \mid y_1) \\ &= p_1(y_2 + \Delta \mid y_1, \alpha_1, \beta_1, \tau_1). \end{aligned}$$

Recall that $p_1(y_2 \mid y_1, \alpha_1, \beta_1, \tau_1)$ is a normal distribution of the form

$$Y_2 \mid Y_1 = y_1, R = 1 \sim N(\alpha_1 + \beta_1 y_1,\ \tau_1^2).$$

Hence the model $p_0(y_2 \mid y_1)$ is $N(\Delta + \alpha_1 + \beta_1 y_1,\ \tau_1^2)$. In the context of a more general pattern-mixture formulation where $[Y_2 \mid Y_1 = y_1, R = 0] \sim N(\alpha_0 + \beta_0 y_1,\ \tau_0^2)$, the location shift assumption is induced by

$$\boldsymbol{\theta}_E = \begin{pmatrix} \alpha_0 \\ \beta_0 \\ \tau_0 \end{pmatrix} = \begin{pmatrix} \alpha_1 + \Delta \\ \beta_1 \\ \tau_1 \end{pmatrix} = g(\boldsymbol{\theta}_D, \Delta),$$

where

$$\begin{aligned} \boldsymbol{\theta}_D &= (\alpha_1, \beta_1, \tau_1, \mu_1, \mu_0, \sigma_1, \sigma_0, \pi) \\ \boldsymbol{\theta}_E &= (\alpha_0, \beta_0, \tau_0). \end{aligned} \tag{18.8}$$

This parameterization is "centered" at MAR in this sense that setting $\Delta = 0$ yields the equality $f_0(y_2 \mid y_1) = f_1(y_2 \mid y_1)$ and thus implies MAR. Setting $\Delta \neq 0$ yields NMAR mechanisms. The normality assumption on $f_0(y_2 \mid y_1)$ is used here for illustration, but it is neither testable nor necessary. It is possible to introduce additional sensitivity parameters that, in addition to the location shift, perturb $f_0(y_2 \mid y_1)$ away from normality. □

Example 10 (Exponential tilt for bivariate normal distribution). An alternative parameterization of the bivariate model uses exponential tilting (Birmingham, Rotnitzky, and Fitzmaurice 2003), which links $f_0(y_2 \mid y_1)$ to $f_1(y_2 \mid y_1)$ via

$$f_0(y_2 \mid y_1) \quad = \quad \exp(\lambda y_2)\, f_1(y_2 \mid y_1) \,/\, K(y_1, \lambda), \qquad (18.9)$$

where $K(y_1, \lambda) = \int \exp(\lambda y_2)\, f_1(y_2 \mid y_1)\, dy_2$ is a normalizing constant that ensures $f_0(y_2 \mid y_1)$ is a proper density, and λ satisfies the definition for sensitivity parameter. One attractive property of the exponential tilt model is that it can be represented either as a pattern-mixture or selection model.

Continuing with Example 8, replace $f_1(y_2 \mid y_1)$ with the normal model $p_1(y_2 \mid y_1, \alpha_1, \beta_1, \tau_1)$. With some algebra, it can be shown that the exponential tilt model (18.9) implies

$$Y_2 \mid Y_1 = y_1, R = 0 \quad \sim \quad N(\lambda \tau_1^2 + \alpha_1 + \beta_1 y_1, \tau_1^2),$$

with

$$\boldsymbol{\theta}_E \;=\; \begin{pmatrix} \alpha_0 \\ \beta_0 \\ \tau_0 \end{pmatrix} \;=\; \begin{pmatrix} \alpha_1 + \lambda \tau_1^2 \\ \beta_1 \\ \tau_1 \end{pmatrix} \;=\; g(\boldsymbol{\theta}_D, \lambda).$$

Hence $f(y_2 \mid y_1)$ is a mixture of normals with a location shift whose scale depends on τ_1^2. The selection model representation can be derived using Bayes' rule, and yields

$$\text{logit}\{\Pr(R = 1 \mid Y_1 = y_1, Y_2 = y_2)\} \quad = \quad \text{logit}(\pi) + \log\{K(y_1, \lambda)\} - \lambda y_2,$$

so that λ is a log odds ratio capturing the association between y_2 and missingness probability. In this case, because K involves an expectation taken over $f_1(y_2 \mid y_1) = p_1(y_2 \mid y_1, \boldsymbol{\theta}_D)$, it also is a function of $\boldsymbol{\theta}_D$. □

18.4.3 Priors

In addition to the likelihood, we must specify priors for the full-data model parameters, including sensitivity parameters. To facilitate sensitivity analysis, or more generally to facilitate clear representation of untestable modeling assumptions, we will typically formulate the prior as

$$p(\boldsymbol{\Delta}, \boldsymbol{\theta}_E, \boldsymbol{\theta}_D) \quad = \quad p(\boldsymbol{\theta}_E \mid \boldsymbol{\Delta}, \boldsymbol{\theta}_D)\, p(\boldsymbol{\Delta} \mid \boldsymbol{\theta}_D)\, p(\boldsymbol{\theta}_D),$$

where $p(\boldsymbol{\theta}_E \mid \boldsymbol{\Delta}, \boldsymbol{\theta}_D) = \{\boldsymbol{\theta}_E = g(\boldsymbol{\theta}_D, \boldsymbol{\Delta})\}$ is simply the reparameterization of the extrapolation model parameters $\boldsymbol{\theta}_E$ in terms of $(\boldsymbol{\theta}_D, \boldsymbol{\Delta})$. The key specification is of course $p(\boldsymbol{\Delta} \mid \boldsymbol{\theta}_D)$, which captures those assumptions about the full-data model that cannot be informed or critiqued using observed data. It is generally possible to parameterize $g(\boldsymbol{\theta}_D, \boldsymbol{\Delta})$ so that the prior $p(\boldsymbol{\Delta} \mid \boldsymbol{\theta}_D)$ captures assumptions and associated uncertainty about the missing data mechanism, and opens up several possibilities for prior specification:

1. Setting $p(\boldsymbol{\Delta} \mid \boldsymbol{\theta}_D)$ to a point mass conveys absolute prior certainty about a specific assumption regarding the missing data mechanism. For example, if $\Delta = 0$ corresponds to missing at random, assigning point mass at 0 yields the standard MAR analysis. If diffuse or flat priors are assigned to $p(\boldsymbol{\theta}_D)$, one can expect roughly similar inferences between Bayesian and frequentist inference about $\boldsymbol{\theta}$ based on the observed-data likelihood.

2. The prior $p(\boldsymbol{\Delta} \mid \boldsymbol{\theta}_D)$ can be used to convey degree of belief or degree of uncertainty about the missing data mechanism. For example, if Δ is a scalar, we can assume $\Delta \mid \boldsymbol{\theta}_D \sim N(0, \xi^2)$ to convey the prior belief that missingness is MAR, with uncertainty captured in terms of ξ^2. The prior also can be skewed to place more weight on directional departures from MAR. We will show in the examples that choices for "degree of belief" parameters such as ξ^2 must be calibrated in the context of the specific application, and should not be assigned arbitrarily.

3. The prior can be constructed using information elicited from historical data or experts, and then incorporated directly and formally into the final inferences.

4. The prior allows degree of departure from MAR to be *calibrated* in terms of identified parameters. Recall in the bivariate normal distributions given above that $\tau_1 = \text{SD}(Y_2 \mid Y_1, R = 1)$. We can write $p(\Delta \mid \boldsymbol{\theta}_D) = p(\Delta \mid \tau_1)$, where $\Delta \mid \tau_1 \sim N(\kappa\tau_1, \tau_1^2)$ and κ is a fixed constant. To allow for uncertainty in κ, we specify $p(\Delta \mid \tau_1)$ as

$$p(\Delta \mid \tau_1) \quad = \quad \int \phi\left(\frac{\Delta - \kappa\tau_1}{\tau_1}\right) p(\kappa)\, d\kappa, \tag{18.10}$$

where $\phi(\cdot)$ is the standard normal density function, and $p(\kappa)$ is a hyperprior on κ (e.g., a Unif(a, b)).

The frequentist approach of summarizing inferences at different fixed values of a sensitivity parameter can, in this context, be characterized as an examination of sensitivity to prior belief. If the prior on $\boldsymbol{\theta}_D$ is flat or sufficiently vague, then maximum likelihood inference at a fixed value of $\boldsymbol{\Delta}$ is essentially equivalent to posterior inference based on the same observed-data likelihood with a point mass assigned to $p(\boldsymbol{\Delta} \mid \boldsymbol{\theta}_D)$ (as in item 1 above).

Frequentist methods of constructing bounds that reflect both sampling variability and uncertainty about the missing data mechanism essentially emerge as special cases of the approach described in item 2 above. When uncertainty about the missing data distribution is represented using a prior distribution, the Bayesian approach allows us to formally weight possible values of $\boldsymbol{\Delta}$ within a (possibly pre-specified) set of bounds. When the choice of boundary values for $\boldsymbol{\Delta}$ is not obvious, it is possible to attach uncertainty to these as well. In short, a summary inference using an appropriately constructed prior for $\boldsymbol{\Delta}$ correctly propagates this uncertainty from all relevant sources to inferences about the parameter(s) of interest.

18.4.4 Summary

In this section we have shown how to parameterize a full-data distribution in terms of the observed-data model parameters and a separate set of parameters that cannot be identified by data. Ideally, sensitivity parameters should represent assumptions about the missing data mechanism. In many cases it will be helpful to use a default assumption such as MAR as a centering point. Parameterization of the priors is then separated into priors for the data model in one component, and priors for the sensitivity parameters, conditional on the data parameters. Sensitivity in the Bayesian context can then be formulated in terms of comparison of prior beliefs about subjective components of the model.

This framework also allows us to characterize many commonly-used methods, such as random effects models and Heckman-style selection models, and indeed to understand their suitability (or lack thereof) for separating subjective assumptions from testable ones. This

TABLE 18.1
Summary of smoking cessation outcomes and missing status for Commit to Quit study.

Treatment	Wk. 12 Status	Quit Smoking? Yes	No	Total
Control	Observed	30	66	96
	Missing	?	?	51
Exercise	Observed	40	53	93
	Missing	?	?	41

framework is readily extendible to settings that can be characterized as missing data problems, such as those encountered in some measurement error models and in causal inference.

18.5 Empirical Illustrations

18.5.1 Overview

This section uses data analytic illustrations to make the foregoing ideas more concrete. We rely on data from the Commit to Quit (CTQ) study, which was a randomized clinical trial of vigorous supervised exercise as an intervention to promote smoking cessation among women (Marcus et al. 1999). This study randomized 281 women to two arms (exercise and control), and ascertained smoking status and other variables weekly over a 12-week follow up period.

The example analyses here focus on two outcomes, smoking cessation status at week 12 (yes/no) and weight at week 12 (in pounds) (weight is an important variable in smoking cessation studies because weight gain is a risk factor for relapse to smoking). Nonresponse rates are relatively high: at week 12, of the 281 women randomized, only 189 women had smoking status ascertained and 161 had weight measured (nonresponse rates of 32 and 44 percent, respectively).

The first part of our analyses concerns rate of smoking cessation at week 12. We focus on the control arm, using additive and multiplicative model formulations to incorporate prior information about missing responses. Although the example is simple, comparison of inferences between additive and multiplicative models sheds light on how sensitivity analyses can be carried out in more complex settings (such as logistic regression). In the second part of our analyses we focus on weight at week 12. We start with the simple case of drawing inference about mean weight in the control arm, and then elaborate the model to handle treatment effects. For more in-depth analyses of these data, see Roy, Hogan, and Marcus (2008), Daniels and Hogan (2008), and Liu, Daniels, and Marcus (2009).

18.5.2 Inference about proportions

Let Y denote actual cessation status (1 if quit smoking, 0 if not), and let R denote whether cessation status is observed (1 if yes, 0 if no). The objective here is to draw inference about the smoking cessation rate at week 12 among all individuals who initiated the study on the control arm; this is denoted by $\mu = E(Y)$. The sample mean among those with $R = 1$ is $30/96 = .31$. A summary of the observed outcomes appears in Table 18.1.

Analysis via additive model. In our first analysis we set up an additive model, where

$$\begin{aligned} Y \mid R = r &\sim \text{Ber}(\mu_r), \quad r = 0, 1 \\ R &\sim \text{Ber}(\pi). \end{aligned}$$

The full-data model is therefore

$$f(y, r \mid \mu_0, \mu_1, \pi) = \underbrace{\pi^r (1-\pi)^{1-r} \left\{\mu_1^y (1-\mu_1)^{1-y}\right\}^r}_{\text{observed-data dist.}} \underbrace{\left\{\mu_0^y (1-\mu_0)^{1-y}\right\}^{1-r}}_{\text{extrapolation dist.}}$$

where $\boldsymbol{\theta}_D = (\mu_1, \pi)$ and $\theta_E = \mu_0$. The parameter μ_0 cannot be identified by observed data. We introduce a sensitivity parameter Δ to capture departures from MAR via a location shift

$$\begin{aligned} \mu_0 &= g(\boldsymbol{\theta}_D, \Delta) \\ &= \mu_1 + \Delta. \end{aligned}$$

Hence the distribution of missing observations follows the model $Y \mid R = 0 \sim \text{Ber}(\mu_1 + \Delta)$, and that the overall mean can be written as $\mu = \mu_1 + (1 - \pi)\Delta$.

Because $0 < \mu_0 < 1$, the sensitivity parameter must satisfy $-\mu_1 < \Delta < 1 - \mu_1$. We use priors of the form

$$p(\mu_1, \pi, \Delta) = p(\Delta \mid \mu_1)\, p(\mu_1)\, p(\pi),$$

where $\mu_1 \sim \text{Unif}(0, 1)$ and $\pi \sim \text{Unif}(0, 1)$. We then consider several different choices for $p(\Delta \mid \mu_1)$:

1. Point mass at $\Delta = 0$, which corresponds to MAR.
2. The uniform prior $\text{Unif}(-\mu_1, 1-\mu_1)$, which assigns equal weight to all possible values of Δ. It should be noted that this prior is not centered at MAR.
3. The uniform prior $\text{Unif}(-\mu, 0)$, which imposes the constraint that cessation rate among those with missing values is systematically lower than those whose cessation status is observed, and assigns equal weight to all possible values satisfying this condition.

Table 18.2 summarizes posteriors for the overall mean $\mu = E(Y)$ and for Δ. The posterior for $\mu_1 = E(Y \mid R = 1)$ has mean .32 with 95% credible interval (.23, .42). For the point mass prior at $\Delta = 0$, which implies MAR, the posterior of μ replicates the posterior for μ_1, as expected. Allowing Δ to range uniformly over its possible values yields a posterior for μ with higher mean (.38) and substantially increased variation (SD = .106) relative to MAR. Conditionally on the sample mean $\widehat{\mu}_1 = .32$, the prior is $p(\Delta \mid \mu_1 = .32)$ is $\text{Unif}(-.32, .68)$ and centered at .18; the marginal posterior $p(\Delta \mid \text{Data})$ for this case has mean .19 and 95% credible interval with limits (−.31, .67); these correspond almost exactly to the posterior means of $-\mu_1$ and $1 - \mu_1$. (In fact, as shown in Moon (2012), the limiting posterior of Δ in this case is uniform with boundaries $(-\widehat{\mu}_1, 1 - \widehat{\mu}_1)$).

When using the third prior for Δ, which conveys the (still relatively vague) prior belief that those with missing response must have, on average, lower cessation rates, the posterior mean $E(\mu \mid \text{Data})$ reflects a downward adjustment relative to the first two priors, to .26.

TABLE 18.2
Summary of posterior mean, standard deviation, and 95% credible intervals for $\mu = E(Y)$ and Δ for additive model having different prior specifications for $p(\Delta \mid \mu_1)$.

	$p(\mu \mid \text{Data})$		$p(\Delta \mid \text{Data})$	
$p(\Delta \mid \mu_1)$	Mean (SD)	95% CI	Mean (SD)	95% CI
$(\Delta = 0)$	.32 (.047)	(.23, .41)		
$\text{Unif}(-\mu_1, 1-\mu_1)$	.38 (.106)	(.20, .57)	.18 (.29)	(–.31, .67)
$\text{Unif}(-\mu_1, 0)$	.26 (.051)	(.17, .37)	–.16 (.10)	(–.34, –.01)

In this simple example, there is relatively little loss in posterior variation relative to MAR, with SD = .047 under the point mass prior for MAR and SD = .051 under this NMAR prior.

The full posterior distributions $p(\mu \mid \text{Data})$ and $p(\Delta \mid \text{Data})$ are shown in Figure 18.1, and demonstrate the importance of using graphical summaries to augment numerical ones. Referring to the left panel of Figure 18.1, when the prior on Δ allows all possible values in $(-\mu_1, 1-\mu_1)$, there is not enough posterior information either in the data or the prior to distinguish the relative likelihood for values of μ that are between .3 and .5. This is seen as a desirable property because lack of plausible prior information about Δ is reflected in the posterior.

On the other hand, for the relatively diffuse NMAR prior $\Delta \mid \mu_1 \sim \text{Unif}(-\mu_1, 0)$, the posterior $p(\mu \mid \text{Data})$ has a distinguishable mode and variation that is roughly equal to the case where an MAR point mass prior is used. This demonstrates the advantage that can be gained by narrowing the assumptions on the missing data mechanism to reflect contextual realities while still retaining prior uncertainty about its precise nature.

Plots of the marginal posterior $p(\Delta \mid \text{Data})$ appear in the right panel. Although Δ is non-identifiable in the sense that $p(\Delta \mid \mu_1, \text{Data}) = p(\Delta \mid \mu_1)$, the marginal posterior $p(\Delta \mid \text{Data}) = \int p(\Delta \mid \mu_1) p(\mu_1 \mid \text{Data}) d\mu_1$ does depend on observed data through the marginal posterior of μ_1, leading to smooth rather than sharp boundaries in the tails.

Analysis with multiplicative model. A potentially more natural parameterization, especially if regression effects are of interest, uses a transformation from the probability scale to the real line, such as logit or probit. Although working in the transformed scale does add flexibility for accommodating covariates, more care must be taken in the parameterization of the model, the specification of priors, and interpretation of posteriors for parameters of interest.

In this example we use the logit transformation. Let $\mu_r = P(Y = 1 \mid R = r)$, and let

$$\text{logit}(\mu_r) = \alpha_1 r + \alpha_0(1-r).$$

We have $\boldsymbol{\theta}_D = (\alpha_1, \pi)$ and $\theta_E = \alpha_0$. A sensitivity parameter λ can be introduced via the mapping $\alpha_0 = g(\alpha_1, \pi, \lambda) = \alpha_1 + \lambda$, which in turn implies

$$\text{logit}(\mu_r) = \alpha_1 + \lambda(1-r). \tag{18.11}$$

It is straightforward to show that λ satisfies the definition of "sensitivity parameter" given in Section 18.4.2.

We take the same general approach for this analysis as with the additive model. There are two features of this parameterization that differ from the previous one. First, like the

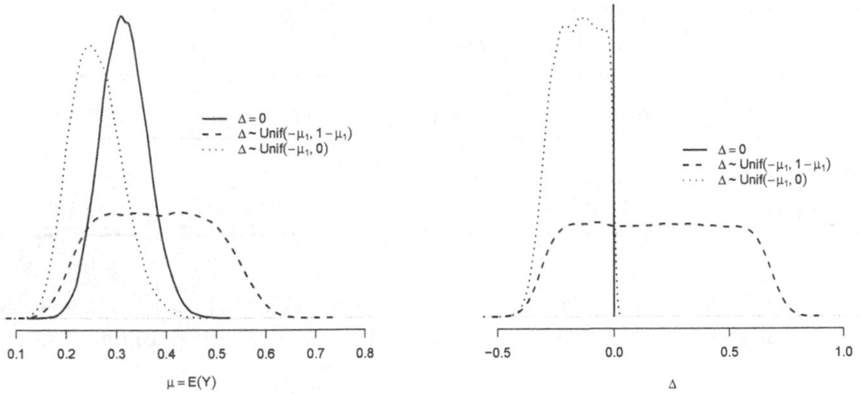

FIGURE 18.1
Posterior distributions $p(\mu \mid Data)$ (left panel) and $p(\Delta \mid Data)$ (right panel) corresponding to different priors $p(\Delta \mid \mu_1)$ used in the additive model.

additive model, $\lambda = 0$ corresponds to MAR. But unlike the additive model, the lack of constraints on λ makes it straightforward to center this model at MAR and add uncertainty, for example by adding prior variance to the distribution of λ. Second, mean cessation rate μ is a nonlinear function of λ, which means special attention must be paid to specification of priors. As a function of λ, the overall mean is

$$\begin{aligned}\mu(\alpha_1, \pi, \lambda) &= \pi\left\{\frac{\exp(\alpha_1)}{1+\exp(\alpha_1)}\right\} + (1-\pi)\left\{\frac{\exp(\alpha_1+\lambda)}{1+\exp(\alpha_1+\lambda)}\right\} \\ &= \pi\mu_1(\alpha_1) + (1-\pi)\mu_0(\alpha_1, \lambda).\end{aligned}$$

It is straightforward to see that $\mu_0(\alpha_1, \lambda) \to 0$ as $\lambda \to -\infty$ and $\mu_0(\alpha_1, \lambda) \to 1$ as $\lambda \to \infty$. The overall mean therefore has lower limit $\pi\mu_1$ and upper limit $1 - \pi(1 - \mu_1)$.

In practice, point mass priors for λ that are far from zero will effectively place point mass for μ_0 at 0 or 1. Diffuse priors may have similar effects. For example, a normal prior with large variance, such as $\lambda \sim N(0, 100)$ places most prior mass on the endpoints $\mu_0 = 0$ and $\mu_0 = 1$. A plot of the posterior of μ for different point mass priors on λ can be used to calibrate priors for λ that use more general distributions, and is illustrated below.

We use priors of the form $p(\alpha_1, \pi, \lambda) = p(\lambda)p(\pi)p(\alpha_1)$, with $\pi \sim \text{Unif}(0, 1)$ and $\alpha_1 \sim N(0, 10^6)$, a diffuse but proper prior. We then use three types of priors for λ: point mass at different values, priors centered at 0, and priors centered away from zero.

Point mass priors. Table 18.3 summarizes posterior inferences about μ for both MAR and NMAR point mass priors. The MAR specification $\lambda = 0$ yields the same inference as the additive model with $\Delta = 0$. NMAR specifications $\lambda = -1$ and $\lambda = -10$ yield lower posterior means but also lower posterior SD relative to MAR. Figure 18.2 shows posterior means and 95% CI for μ associated with point mass priors on λ across a range of values. The posterior mean of μ reaches its lower asymptote at around $\lambda = -4$, implying that point mass priors with $\lambda < -4$ are essentially equivalent to placing prior point mass on μ_0 at zero, and will yield equivalent posterior inferences for μ.

Normal priors centered at MAR. Rows 2 through 5 of Table 18.3 show inferences under normal priors centered at $\lambda = 0$, where the variance term quantifies uncertainty about

TABLE 18.3
Summary of posterior distribution of $\mu = E(Y)$ for the multiplicative model, with several different choices of prior for λ.

Prior $p(\lambda)$		Posterior $p(\mu \mid \text{Data})$		
Type	Specification	Mean	S.D.	95% CI
MAR	$(\lambda = 0)$	.32	.046	(.24, .42)
Centered at MAR	$N(0, 1)$	.33	.081	(.20, .50)
	$N(0, 2^2)$	.34	.115	(.18, .57)
	$N(0, 3^2)$	.35	.133	(.17, .59)
	$N(0, 10^2)$	.37	.164	(.16, .62)
NMAR	$(\lambda = -1)$	.26	.040	(.19, .34)
	$N(0, 1)$ $(\lambda < 0)$	.27	.049	(.18, .38)
	$N(0, 2^2)$ $(\lambda < 0)$	.26	.050	(.17, .36)
	$N(0, 3^2)$ $(\lambda < 0)$	.24	.050	(.16, .35)
	$N(0, 10^2)$ $(\lambda < 0)$	.24	.049	(.16, .35)
	$(\lambda = -10)$	.21	.033	(.15, .28)

MAR. Both the posterior mean and SD of μ increase as a function of the prior variance for λ, although the posterior means are generally close to .32, the MLE under MAR.

Figure 18.3 provides some additional insight about the relationship between priors on λ and posteriors on μ. From (18.11), we see that $\mu_0 = e^{\alpha_0+\lambda}/(1 + e^{\alpha_0+\lambda})$. The MLE of α_0 is $\widehat{\alpha}_0 = \text{logit}(\widehat{\mu}_1) = \text{logit}(.32) = -.75$. Plots on the left panel show conditional priors $p(\mu_0 \mid \widehat{\alpha}_0)$ implied by different MAR specifications of $p(\lambda)$. These are generated using 10^6 simulated draws of $\mu_0^* \sim p(\mu_0 \mid \widehat{\alpha}_0)$, computed as follows: first, draw $\lambda^* \sim p(\lambda)$; then, $\mu_0^* = e^{\widehat{\alpha}_0+\lambda^*}/(1+e^{\widehat{\alpha}_0+\lambda^*})$. Plots on the right panel show corresponding posteriors $p(\mu \mid \text{Data})$. The plots in Figure 18.3 show that the prior variance of λ governs the distribution of prior weight across values of μ_0, with higher variance pushing prior weight toward the boundary values 0 and 1. As $\text{var}(\lambda) \to \infty$, the posterior for μ becomes bimodal. The bottom right panel shows that when $\text{var}(\lambda) = 100$, which in practical terms is a limiting case, the posterior modes are near the frequentist bounds $[\widehat{\pi}\widehat{\mu}_1,\, 1 - \widehat{\pi}(1 - \widehat{\mu}_1)] = [.21, .56]$ for μ.

Priors representing NMAR. The final set of summaries uses priors on λ that are designed to convey the prior belief that those with missing outcomes have a strictly lower mean cessation rate than those with observed outcomes. Posterior summaries for μ based on point mass and half-normal priors for λ having mass strictly less than zero are given in the bottom part of Table 18.3. The NMAR prior $\lambda = -10$ essentially places *a priori* point mass at $\mu_0 = 0$, and the corresponding posterior mean of μ, .21, differs substantially from the posterior mean .32 under MAR. This discrepancy suggests that inferences certainly are sensitive to the missing data assumption (see also Figure 18.2). However, within the class of NMAR mechanisms characterized by half-normal priors, inferences seem relatively insensitive to prior specification, even though these distribute prior weight on μ_0 quite differently (see Figure 18.4).

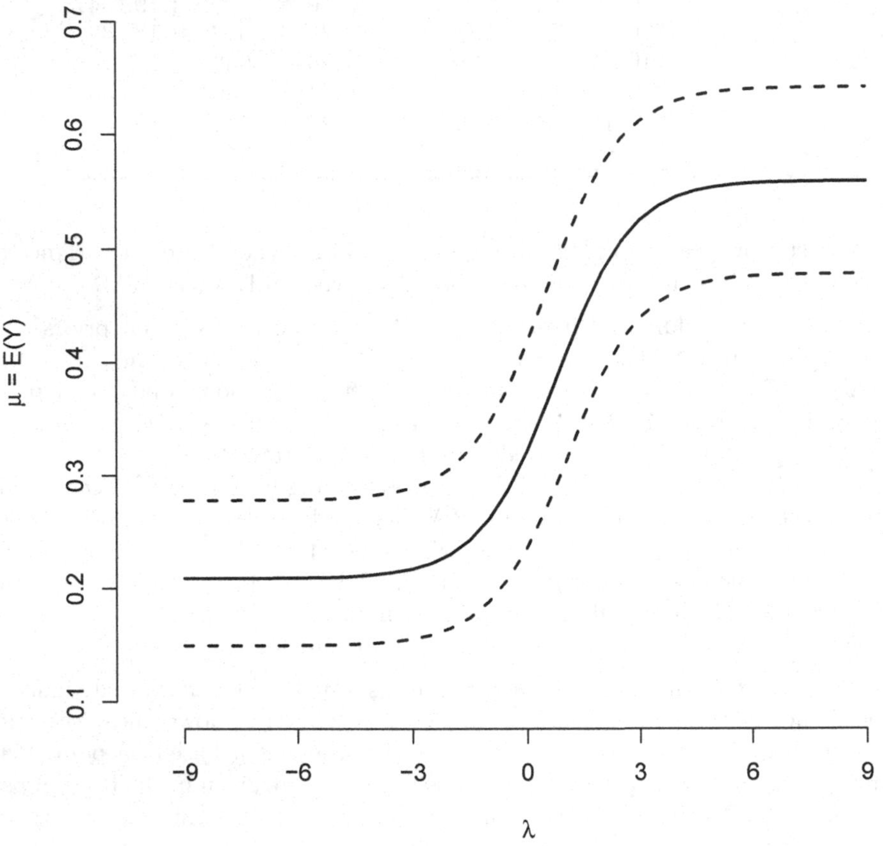

FIGURE 18.2
Posterior means and 95% pointwise credible intervals for for $\mu = E(Y)$ under point mass priors on λ.

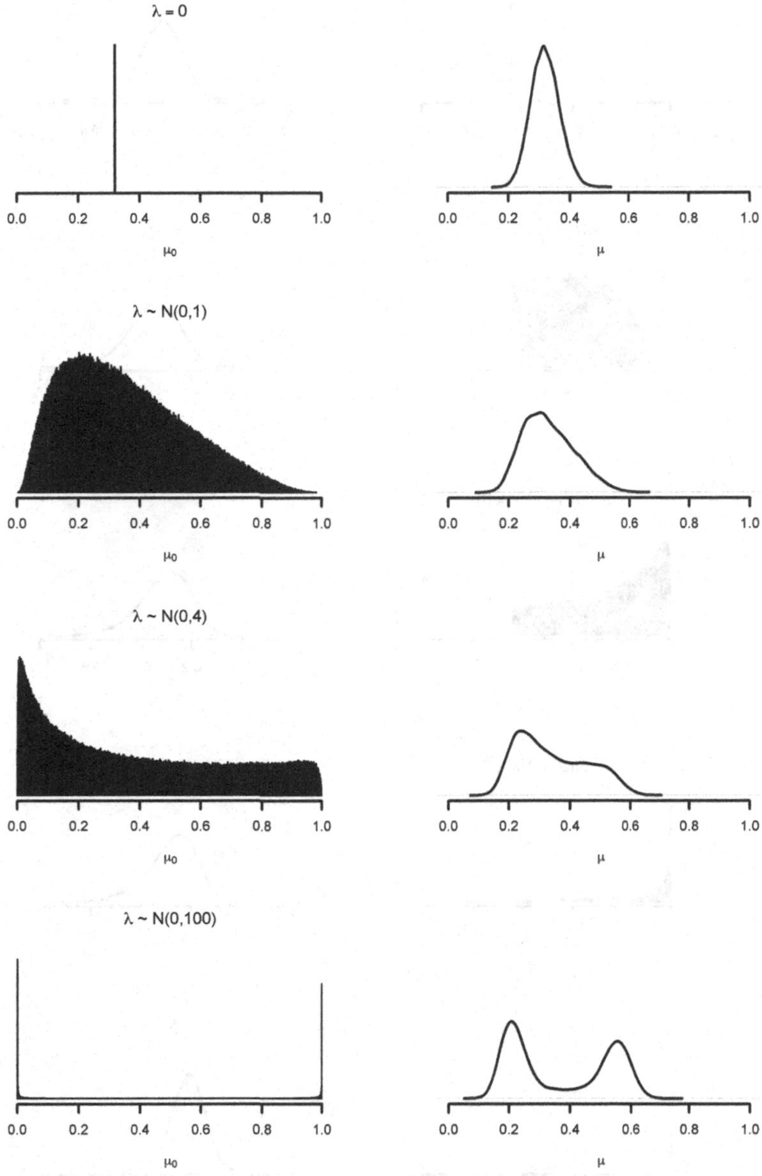

FIGURE 18.3

Summary of MAR priors and resulting posteriors for multiplicative model. Left column shows simulation-based realizations of $p(\mu_0 \mid \widehat{\alpha}_0)$, the conditional prior for μ_0 at the MLE of α_0 under different choices for $p(\lambda)$. Right column shows $p(\mu \mid Data)$, the posterior of $\mu = E(Y)$.

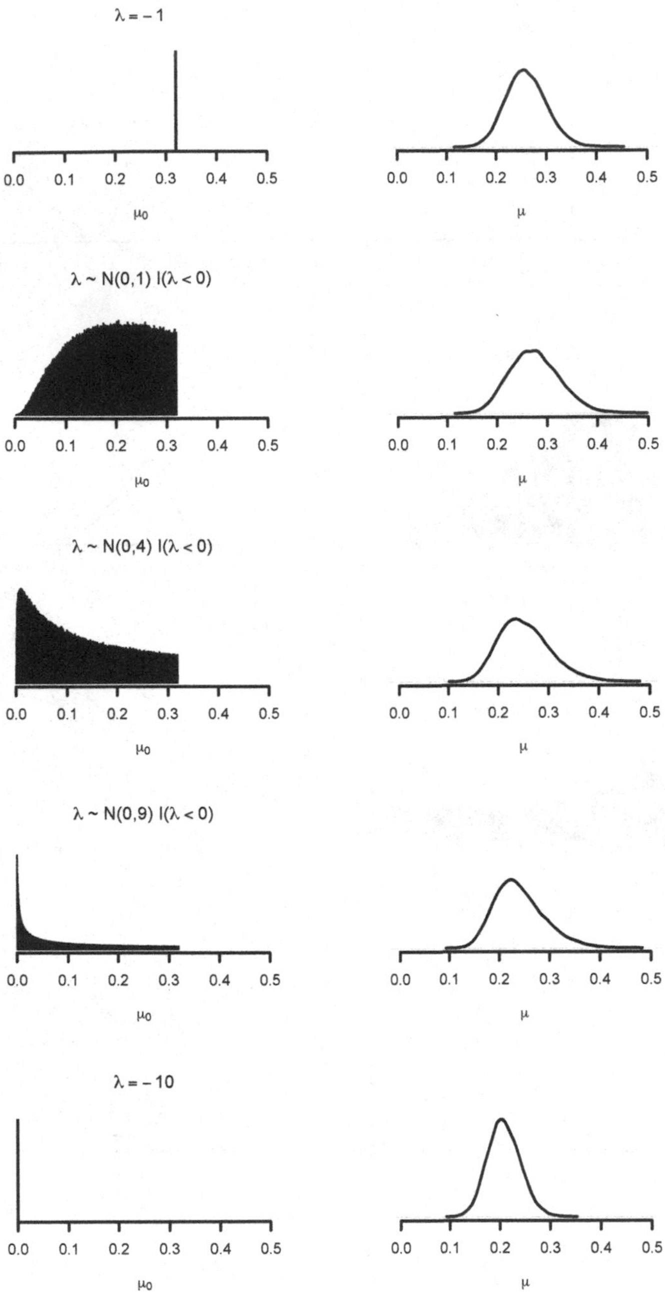

FIGURE 18.4
Summary of NMAR priors and resulting posteriors for multiplicative model. Left column shows simulation-based realizations of $p(\mu_0 \mid \widehat{\alpha}_0)$ *under different choices for* $p(\lambda)$*. Right column shows* $p(\mu \mid Data)$*.*

TABLE 18.4
Summary statistics for weight in the CTQ study, where Y_1 is weight at baseline and Y_2 is weight at week 12. Table entries for Y_1, Y_2 and $Y_2 - Y_1$ are sample mean (sample SD).

Treatm.	Wk 12 Status	n	Y_1	Y_2	$Y_2 - Y_1$	$SD(Y_2 \mid Y_1)$	$Corr(Y_1, Y_2)$
Control	Observed	81	144(28)	148(30)	3.5(5.1)	4.9	.98
	Missing	64	144(31)				
Exercise	Observed	80	151(32)	155 (34)	4.3(6.3)	6.2	.98
	Missing	52	152(30)				
Overall			148(30)				

In smoking cessation research, it is common to make the assumption that all individuals with missing observations are smoking (i.e., have $Y = 0$). This corresponds to setting $\lambda = -\infty$, or equivalently placing prior point mass at $\mu_0 = 0$. This assumption is unlikely to hold in practice; moreover, it assumes zero variation in the unobserved cessation outcomes, and therefore yields *lower* posterior SD for μ than point mass priors which induce positive variability in the missing Ys (e.g., compare posterior SD in Table 18.3 under $\lambda = 0$ and $\lambda = -1$ to that under $\lambda = -10 \approx -\infty$). Using more general and arguably more realistic priors allows less extreme versions of NMAR while still placing considerable prior probability near $\mu_0 = 0$; referring to Figure 18.4, using a half-normal prior such as $\lambda \sim N(0, 3^2)$ $(\lambda < 0)$ accomplishes this goal. It also is possible to calibrate prior variance to obtain a desired fraction of μ_0 values that are essentially equal to zero.

Comparison to frequentist inference about bounds. Using methods described in Section 4.1 of Vansteelandt et al. (2006), we estimated bounds and associated uncertainty regions for μ. The estimated uncertainty region is calculated by expanding the estimated bounds for μ using one-sided critical values from the normal distribution to generate an interval that, in large samples, provides 95% coverage of μ without making any assumptions about the missing data mechanism. The estimated bounds for μ are [.21, .56], with estimated uncertainty region [.15, .62].

Notably, the uncertainty region corresponds closely to the 95% posterior interval corresponding to the prior $\lambda \sim N(0, 10^2)$, which distributes nearly all prior mass for μ_0 near the boundaries of the $(0, 1)$ interval; see Figure 18.3, bottom left. (Note that a "flat" prior on λ distributes mass equally at the points $\mu_0 = 0$ and $\mu_0 = 1$.) In general, priors that assign at least some mass for μ_0 on the interior of $(0, 1)$ will yield Bayesian posterior credible intervals for μ that are more narrow than the frequentist uncertainty regions, and the frequentist bounds are consistent with a prior that puts all prior mass at the extremes. For a more detailed comparison of Bayesian and frequentist inferences for partially identified models, see Moon and Schorfheide (2012) and Gustafson (2012).

18.5.3 Inference about continuous bivariate distribution

Our second example uses a continuous outcome having a bivariate distribution. We are interested to estimate mean weight at week 12. Let Y_1 denote baseline weight, let Y_2 denote weight at week 12, and let $R = 1$ represent the response indicator for Y_2 (the baseline weight is observed on everyone). Treatment group is denoted by Z ($1 =$ exercise, $0 =$ control). Table 18.4 describes the summary statistics for weight, stratified by treatment group.

Our analysis uses a version of the location shift model described in Example 9. The full data model is

$$\begin{aligned} Y_2 \mid Y_1 = y_1, Z = z, R = r &\sim N\{\mu_{zr}(y_1), \sigma^2_{zr}\} \\ Y_1 \mid Z = z, R = r &\sim N(\eta, \tau^2) \\ R \mid Z = z &\sim \text{Ber}(\pi_z). \end{aligned} \qquad (18.12)$$

This model assumes separate distributions for $Y_2 \mid Y_1$ over treatment group and missing data patterns, but a common distribution for Y_1. The conditional mean $\mu_{zr}(y_1) = E(Y_2 \mid Y_1 = y_1, Z = z, R = r)$ follows the linear model

$$\mu_{zr}(y_1) = \alpha_{zr} + \beta_{zr}(y_1 - \overline{y}_0).$$

For this model we have

$$\begin{aligned} \boldsymbol{\theta}_D &= \{(\pi_z, \eta, \tau, \alpha_{z1}, \beta_{z1}, \sigma_{z1}) : z = 0, 1\} \\ \boldsymbol{\theta}_E &= \{(\alpha_{z0}, \beta_{z0}, \sigma_{z0}) : z = 0, 1\}. \end{aligned}$$

We introduce treatment-group-specific sensitivity parameters that shift the conditional means for Y_2 given Y_1 between those with observed and missing values of Y_2, such that

$$\Delta_z(y_1) = E(Y_2 \mid Y_1 = y_1, Z = z, R = 1) - E(Y_2 \mid Y_1 = y_1, Z = z, R = 0). \quad (18.13)$$

In words, this allows the user to specify potential differences in mean weight Y_2 between those with observed and missing Y_2, within groups of individuals having the same baseline weight $Y_1 = y_1$.

Using a slight abuse of notation, we parameterize the sensitivity analysis via the transformation g given by

$$\boldsymbol{\theta}_E = \begin{pmatrix} \alpha_{z0} \\ \beta_{z0} \\ \sigma_{z0} \end{pmatrix} = \begin{pmatrix} \alpha_{z1} + \Delta_z \\ \beta_{z1} \\ \sigma_{z1} \end{pmatrix} = g(\boldsymbol{\theta}_D, \Delta);$$

i.e., it maps the 10×1 vector $\boldsymbol{\theta}_D$ to the 6×1 vector $\boldsymbol{\theta}_E$ (3 parameters for each treatment group). The treatment-specific means $E(Y_2 \mid Z = z)$ are obtained by averaging over the joint distribution $f(y_1, r \mid z)$,

$$\begin{aligned} E(Y_2 \mid Z = z) &= E_{R \mid Z}\left[E_{Y_1 \mid R, Z}\{E(Y_2 \mid Y_1 = y_1, R = r, Z = z)\}\right] \\ &= \sum_{r=0}^{1} \Pr(R = 1 \mid Z = z) \left\{ \int \mu_{zr}(y_1)\, f(y_1)\, dy_1 \right\} \\ &= \pi_z \alpha_{z1} + (1 - \pi_z)(\alpha_{z1} + \Delta_z). \end{aligned}$$

Note that the treatment specific means are being standardized to a single distribution of Y_1 across treatment arms and missing data patterns, as specified by (18.12). Finally, treatment effect is denoted by $\gamma = E(Y_2 \mid Z = 1) - E(Y_2 \mid Z = 0)$.

Although Δ_z as defined by (18.13) is a function of y_1, setting $\beta_{0z} = \beta_{1z}$ eliminates dependence on y_1 (even though Δ_z is still a difference in *conditional* means). Furthermore, centering the conditional mean model for $\mu_{zr}(y_1)$ on $\overline{y}_0$ eliminates dependence on parameters governing $f(y_1)$. As we illustrate below, Δ_z can be calibrated empirically by specifying the differences in units of the residual standard deviation $\sigma_{z1} = \text{SD}(Y_2 \mid Y_1, Z = z, R = 1)$.

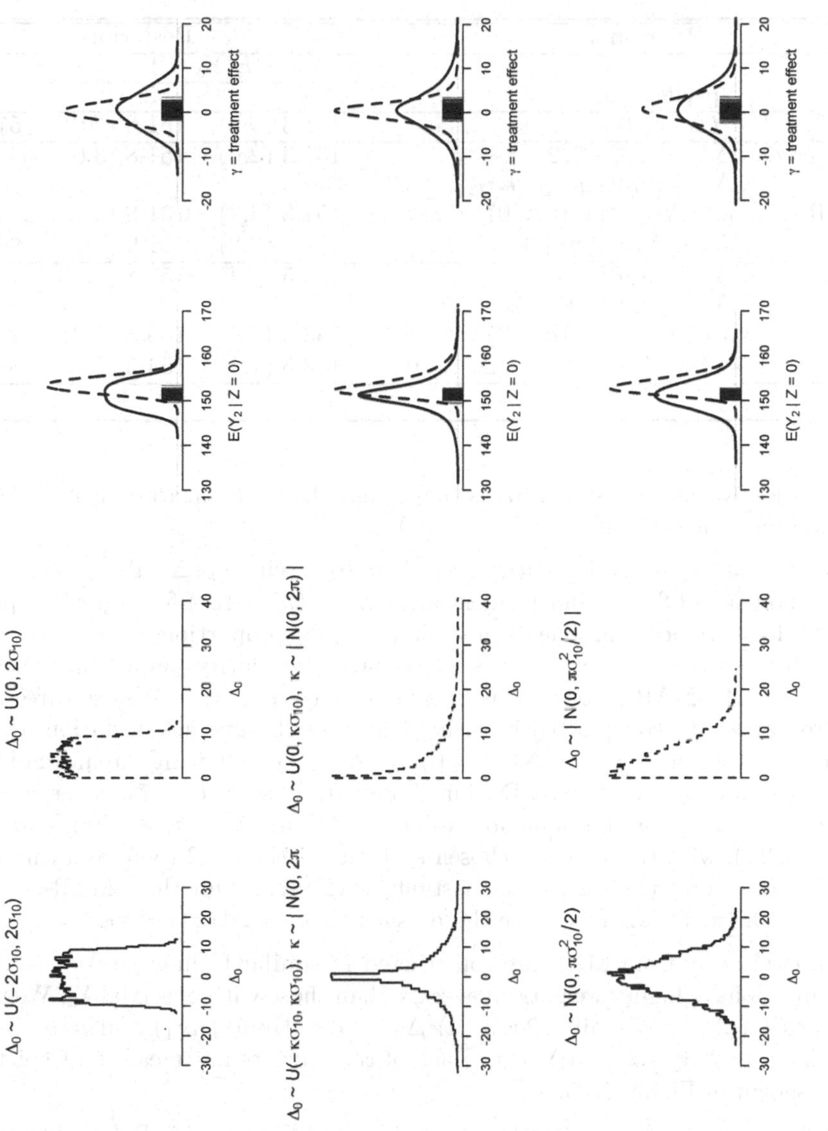

FIGURE 18.5
Posterior distributions for Δ_0, $E(Y_2 \mid Z = 0)$, and γ for analysis of weight data. Each row corresponds to a different set of priors on Δ_z. First column is MAR-centered prior, second column is NMAR-centered prior; third and fourth columns show posteriors for $E(Y_2 \mid Z = 0)$ and γ, respectively. Rug plot shows draws from posterior under MAR point mass prior. Solid lines correspond to posteriors under MAR-centered priors; dashed lines for NMAR-centered priors.

TABLE 18.5
Posterior summaries of treatment-group means $E(Y_2 \mid Z = z)$ and treatment effect γ for different specifications of prior for Δ_z.

Prior on Δ_z		Posteriors		
		$E(Y_2 \mid Z)$		
Type	Specification	$Z = 1$	$Z = 0$	$\gamma =$ Diff.
MAR	$(\Delta_z = 0)$	151.3 (.55)	151.8 (.70)	.51 (.89)
Centered	$\Delta_z \sim \text{Unif}(-2, 2)$	151.3 (2.6)	151.8 (3.0)	.44 (3.9)
at	$\Delta_z \sim \text{Unif}(-\kappa\sigma_{1z},\, \kappa\sigma_{1z})$,			
MAR	$\kappa \sim N(0, 2\pi)$ $(\kappa > 0)$	151.3 (3.3)	151.8 (3.8)	.52 (5.0)
	$\Delta_z \sim N(0,\, \frac{1}{2}\pi\sigma_{1z})$	151.3 (2.9)	151.9 (3.2)	.60 (4.3)
NMAR	$\Delta_z \sim \text{Unif}(0, 2)$	153.5 (1.4)	153.8 (1.4)	.25 (2.0)
	$\Delta_z \sim \text{Unif}(0,\, \kappa\sigma_{1z})$,			
	$\kappa \sim N(0, 2\pi)$ $(\kappa > 0)$	153.5 (2.4)	153.8 (2.2)	.28 (2.4)
	$\Delta_z \sim N(0,\, \frac{1}{2}\pi\sigma_{1z})$ $(\Delta_z > 0)$	153.5 (1.8)	154.3 (2.1)	.81 (2.7)

The motivation for using residual SD is that it has the same measurement scale as Y_2 but already accounts for variability explained by Y_1.

We factor the joint prior as $p(\Delta_0 \mid \boldsymbol{\theta}_D)p(\Delta_1 \mid \boldsymbol{\theta}_D)p(\boldsymbol{\theta}_D)$, where $p(\Delta_z \mid \boldsymbol{\theta}_D) = p(\Delta_z \mid \sigma_{1z})$. Priors for components of $\boldsymbol{\theta}_D$ are independent and diffuse ($N(0, 10^6)$ for regression parameters, Unif$(0, 100)$ for variance components, and Unif$(0, 1)$ for proportions). Priors described here are treatment specific, but we suppress subscripts z for clarity. As a benchmark for comparison, we use the MAR prior that assigns point mass at $\Delta = 0$. We use three priors that are centered at MAR, constructed in such a way that the *average* deviation from MAR is $\sigma_1 = \text{SD}(Y_2 \mid Y_1)$. First we assign $\Delta \mid \sigma_1 \sim \text{Unif}(-2\sigma_1, 2\sigma_1)$, allowing the mean of those with missing Y_2 to differ by up to two SD from those with observed Y_2. Next, as in (18.10), we introduce uncertainty for the boundary values, assigning $\Delta \mid \kappa, \sigma_1 \sim \text{Unif}(-\kappa\sigma_1, \kappa\sigma_1)$ and $\kappa \mid \sigma_1 \sim N(0, 2\pi)$, with the variance chosen such that $E(|\kappa|) = 2$ (here and throughout our specification of priors, π refers to the constant 3.14159...). Our third MAR-centered prior is $\Delta \mid \sigma_1 \sim N(0, \frac{1}{2}\pi\sigma_1^2)$, with variance chosen such that $E(|\Delta| \mid \sigma_1) = \sigma_1$.

Priors centered away from MAR are constructed in similar fashion, and reflect that those with missing Y_2 have higher weights on average than those with observed Y_2. We use NMAR counterparts (i) $\Delta \mid \sigma_1 \sim \text{Unif}(0, 2\sigma_1)$; (ii) $\Delta \mid \kappa, \sigma_1 \sim \text{Unif}(0, \kappa\sigma_1)$ and $\kappa \mid \sigma_1 \sim N(0, 2\pi)$; and (iii) $\Delta \mid \sigma_1 \sim N(0, \frac{1}{2}\pi\sigma_1^2)$ $(\Delta > 0)$. Each of these priors has mean σ_1. Depictions of all priors are shown in Figure 18.5.

Posterior inferences are summarized in Table 18.5 and Figure 18.5. Relative to the analysis assigning point mass at $\Delta = 0$, using priors that convey uncertainty about Δ introduces considerably more variation into posterior inferences about both the treatment-arm-specific means and about the treatment effect itself; posterior SD for treatment effect is inflated by a factor of between 4 and 6. Inferences under NMAR yield upward adjustments in the posterior means of Y_2, as expected, but have relatively little effect on inferences about treatment effect because in both treatment arms, the distribution of Y_1 is similar between those with observed and missing Y_2 (see Table 18.4).

Plots such as the one shown in Figure 18.5 can form the basis of a Bayesian sensitivity analysis. Each row corresponds to a class of prior distributions on Δ for this analysis; the first entries in each row show posterior distributions for Δ_0 for priors centered at MAR and NMAR, respectively. Posteriors for Δ_1 are not shown, but have very similar shape.

The third and fourth entries in each row of Figure 18.5 show posteriors for $E(Y_2 \mid Z = z)$ and γ under the MAR- and NMAR-centered priors. The vertical marks on the bottom of each plot corresponds to draws from the posterior under the point mass MAR prior at $\Delta_z = 0$.

Graphical depictions such as this one are critical for communicating the precise nature of assumptions being made about the missing data mechanism and the sensitivity of posterior inferences to these assumptions. For example, the effect of adding uncertainty to the boundaries of a uniform prior on Δ is easily seen by comparing the second and third rows; these priors admit potential departures from MAR whereby weight for dropouts is up to 40 pounds greater, *on average*, than for completers. This assumption is likely to be too extreme in the context of a 12-week study, and adjustments to the priors can be made.

18.5.4 Summary

We have used two examples to illustrate model parameterization, prior formulation, and the effect of the choice of priors on posterior inferences about parameters of the full-data distribution. Although these examples are simplified, they are intended to illustrate the modeling approach outlined in Section 18.4. The first example, using proportions, shows the importance of the choice of scale in the formulation of priors. The second example, using continuous data, illustrates the importance of choosing appropriate bounds for the sensitivity parameter, and gives specific suggestions for both calibrating the bounds and introducing uncertainty about them. In both cases, the use of priors allows the analyst to make contextually appropriate assumptions about missing data (e.g., dropouts are less likely to quit smoking) while maintaining a substantial degree of uncertainty about the parameter values indexing the missing data mechanism.

18.6 Summary and Discussion

This chapter provides a Bayesian perspective on sensitivity analysis for inferences from incomplete data. Although the focus here is on missing data, the concepts we have outlined apply to a broad range of settings where information is incomplete, such as potential outcomes models for observational data and measurement error models. The Bayesian approach is particularly well suited to analysis of incomplete data because the assumptions governing the distribution of missing observations are purely subjective.

The approach described here parameterizes the full-data model such that the observed-data distribution is modeled directly, using assumptions that can be checked, and the missing data distribution (extrapolation) is indexed by one or more sensitivity parameters that must be non-identifiable in the sense of being conditionally independent of observed data, given identifiable parameters. The sensitivity parameters index the missing data mechanism, and can generally be centered at MAR.

Bayesian inferences allow the analyst to formally represent uncertainty about the missing data mechanism in terms of a prior distribution. If there is consensus about the prior, the Bayesian approach affords the opportunity to have a summary inference that reflects fairly weak, but contextually informed, assumptions about the missing data mechanisms. In other cases, the Bayesian sensitivity analysis compares inferences across different priors for the sensitivity parameters.

Frequentist approaches acknowledge subjectivity as well. A common approach to sensitivity analysis is to summarize inferences across a range of values for the sensitivity parameter. This can be seen as examining sensitivity to different choices of point mass priors, but it cannot be formally translated into a summary inference reflecting a distribution across values of the sensitivity parameter. Another approach is to use bounds in lieu of point estimates, and to calculate confidence intervals that reflect both sampling variability and lack of knowledge about the sensitivity parameter, thereby generating a summary inference. These methods are more well suited to proportions, where the outcome is naturally bounded, than for continuous outcomes, where bounds on the missing data distribution still must be chosen subjectively.

Methods for quantifying and representing uncertainty attributable to untestable assumptions is a high-priority area for statistical research, particularly in health sciences. A recent report by the National Research Council (Little et al. 2012) focuses entirely on handling missing data in analysis of clinical trials used to support regulatory approval of new drugs and devices. The Patient-Centered Outcomes Research Institute (PCORI) recently issued a set of methodological standards for design, analysis and reporting of studies of patient outcomes that will be used to inform clinical and public health practice (Gabriel and Normand 2012). Both reports make the specific recommendation that sensitivity analyses must be included as a routine part of statistical inferences.

There are several potential areas for further research in this area. Bayesian methods typically rely on parametric models for tractability of computations. Non- and semi-parametric modeling approaches provide the flexibility needed for complex applications. Further study of the connections between Bayesian and frequentist approaches in partially identified models, such as that found in Gustafson (2012), can be expected to generate new insights about both types of methods and how they can be most effectively used in practice.

The NRC report raises the important question of decision-making from incomplete data. The decision to approve a new drug or device in the regulatory setting depends on a variety of different factors, but prominent among these are statistical treatment comparisons from randomized clinical trials. In this context, decision-making relies on some form of summary inference. To the extent that Bayesian analyses of incomplete data can be used to generate summary inferences, this suggests the need for rigorous methods of constructing appropriate prior distributions for different settings. This may include elicitation of expert opinion, use of historical data, collection of data from follow-up samples, or some combination of these.

Finally, effective methods of communicating assumptions about missing data mechanisms, and the sensitivity of inferences to these assumptions, are critically needed. New methods for visual or graphical representations of model sensitivity are needed so that end-users of the analyses can understand their limitations and use them to make more fully informed decisions.

Acknowledgments

This work was partially supported by grants RC1-AA-019186, P30-AI-42853 and P01-AA-019072 from the US National Institutes of Health. The authors are grateful to Paul Gustafson and the editors for very helpful comments on an earlier draft.

References

Birmingham, J., Rotnitzky, A., and Fitzmaurice, G.M. (2003). Pattern-mixture and selection models for analysing longitudinal data with monotone missing patterns. *Journal of the Royal Statistical Society, Series B* **65**, 275–297.

Daniels, M.J. and Hogan J.W. (2008). *Missing Data in Longitudinal Studies: Strategies for Bayesian Modeling and Sensitivity Analysis.* Boca Raton: Chapman & Hall.

Dawid, A.P. (1979). Conditional independence in statistical theory. *Journal of the Royal Statistical Society, Series B* **41**, 1–15.

Dawid, A.P. (2000). Causal inference without counterfactuals (with discussion). *Journal of the American Statistical Association* **95**, 407–448.

Diggle, P.J., Farewell, D., and Henderson, R. (2007). Analysis of longitudinal data with drop-out: Objectives, assumptions and a proposal. *Applied Statistics* **56**, 499–550.

Fitzmaurice, G.M. (2003). Methods for handling dropouts in longitudinal clinical trials. *Statistica Neerlandica* **57**, 75–99.

Fitzmaurice, G.M., Laird, N.M., and Shneyer, L. (2001). An alternative parameterization of the general linear mixture model for longitudinal data with non-ignorable drop-outs. *Statistics in Medicine* **20**, 1009–1021.

Fleming, T.R. and Harrington, D.P. (1991). *Counting Processes and Survival Analysis.* New York: John Wiley & Sons.

Gabriel, S.E. and Normand, S.-L.T. (2012). Getting the methods right: The foundation of patient-centered outcomes research. *New England Journal of Medicine* **367**, 787–790.

Gelman, A., Meng, X.-L., and Stern, H. (1996). Posterior predictive assessment of model fitness via realized discrepancies (Disc: P760-807). *Statistica Sinica* **6**, 733–760.

Gustafson, P. (2012). On the behaviour of Bayesian credible intervals in partially identified models. *Electronic Journal of Statistics* **6**, 2107–2124.

Heckman, J.J. (1979). Sample selection bias as a specification error. *Econometrica* **47**, 153–161.

Hogan, J., Roy, J., and Korkontzelou, C. (2004). Handling dropout in longitudinal studies. *Statistics in Medicine* **23**, 1455–1497.

Kenward, M.G. (1998). Selection models for repeated measurements with non-random dropout: An illustration of sensitivity. *Statistics in Medicine* **17**, 2723–2732.

Little, R.J.A. (1995) Modeling the drop-out mechanism in repeated-measures studies. *Journal of the American Statistical Association* **90**, 1112–1121.

Little, R.J.A., D'Agostino, R., Cohen, M.L., Dickersin, K., Emerson, S., Farrar, J.T., Frangakis, C., Hogan, J.W., Molenberghs, G., Murphy, S.A., Neaton, J.D., Rotnitzky, A., Scharfstein, D., Shih, W., Siegel, J.P., and Stern, H. (2012). The prevention and treatment of missing data in clinical studies. *New England Journal of Medicine* **367**, 1355–1360.

Liu, X., Daniels, M.J., and Marcus, B.H. (2009). Joint models for the association of a longitudinal binary and continuous process. *Journal of the American Statistical Association* **104**, 429–439.

Manski, C.F. (2007). *Identification for Prediction and Decision.* Cambridge, MA: Harvard University Press.

Marcus, B.H., Albrecht, A.E., King, T.K., Parisi, A.F., Pinto, B.M., Roberts, M., Niaura, R.S., and Abrams, D.B. (1999). The efficacy of exercise as an aid for smoking cessation in women: a randomized controlled trial. *Archives of Internal Medicine* **159**, 1229–1234.

Molenberghs, G. and Kenward, M.G. (2007). *Missing Data in Clinical Studies.* New York: John Wiley & Sons.

Moon, H.R. and Schorfheide, F. (2012). Bayesian and frequentist inference in partially identified models. *Econometrica*, **80**, 755–782.

Robins, J.M. (1999). Association, causation, and marginal structural models. *Synthese* **121**, 151–179.

Robins, J.M. and Greenland, S. (2000). Comment on "Causal inference without counterfactuals" by A.P. Dawid. *Journal of the American Statistical Association* **95**, 431–435.

Robins J.M., Rotnitzky, A., and Zhao, L.P. (1995). Analysis of semiparametric regression models for repeated outcomes in the presence of missing data. *Journal of the American Statistical Association* **90**, 106–121.

Rosenbaum, R. and Rubin, D.B. (1983). The central role of the propensity score in observational studies for causal effects. *Biometrika* **70**, 41–55.

Roy, J. and Daniels, M.J. (2008). A general class of pattern mixture models for nonignorable dropout with many possible dropout times. *Biometrics* **64**, 538–545.

Roy, J., Hogan, J.W., and Marcus, B.H. (2008). Principal stratification with predictors of compliance for randomized trials with two active treatments. *Biostatistics* **9**, 277–289.

Rubin, D.B. (1976). Inference and missing data. *Biometrika* **63**, 581–590.

Rubin, D.B. (1978). Bayesian inference for causal effects: The role of randomization. *The Annals of Statistics* **6**, 34–58.

Scharfstein, D., Robins, J.M., Eddings, W., and Rotnitzky, A. (2001). Inference in randomized studies with informative censoring and discrete time-to-event endpoints. *Biometrics* **57**, 404–413.

Scharfstein, D.O., Rotnitzky, A., and Robins, J.M. (1999). Adjusting for nonignorable drop-out using semiparametric nonresponse models (with discussion). *Journal of the American Statistical Association* **94**, 1096–1146.

Tsiatis, A.A. (2006). *Semiparametric Theory and Missing Data.* New York: Springer.

Vansteelandt, S., Goetghebeur, E., Kenward, M.G., and Molenberghs, G. (2006). Ignorance and uncertainty regions as inferential tools in a sensitivity analysis. *Statistica Sinica* **16**, 953–979.

19

Sensitivity Analysis with Multiple Imputation

James R. Carpenter

London School of Hygiene and Tropical Medicine and MRC Clinical Trials Unit at UCL, London, UK

Michael G. Kenward

London School of Hygiene and Tropical Medicine, UK

CONTENTS

19.1 Introduction

This chapter describes a range of practical approaches for sensitivity analysis via multiple imputation. Following a general introduction, Section 19.2 briefly reviews the theory underlying the analysis of data when missing values are NMAR, in particular focusing on the contrast between the pattern-mixture and selection model approaches. This has been dealt with in some detail in earlier chapters, especially 4 and 16. Section 19.3 describes a pattern mixture approach, illustrating its application with missing covariate and survival data. Section 19.4 targets specific issues raised by longitudinal data in clinical trials describing the "Δ-method" (Section 19.4.3) and "reference-based imputation" approach (Section 19.4.5). The latter has been recently proposed by Carpenter et al. (2013); see also Mallenckrodt (2013) and O'Kelly and Ratitch (2014, Ch. 7). These approaches are very flexible and practical, with a key advantage being that they avoid direct estimation of an NMAR model. They are contrasted in Section 19.5 with a selection model formulation. Again, direct estimation of an NMAR model is avoided, here through reweighting of the imputations from the MAR model. We conclude with a brief discussion in Section 19.6.

Multiple imputation (MI) has been developed in some detail in the chapters that constitute Part IV of this Handbook. Nearly all the approaches described there made the assumption, implicitly or explicitly, that the missing data mechanism was ignorable. This is often expressed in the imputation setting, rather loosely, as the MAR assumption. MI is certainly convenient to implement under this assumption. Under ignorability, the conditional predictive distribution that determines the imputation model does not depend on $\boldsymbol{R}$, the missing data indicator. One consequence of this is that the observed data is sufficient both to estimate, and assess the fit, of the imputation model. As discussed in Chapters 16–18 however, non-ignorable, non-random (NMAR) missing values mechanisms are often required for constructing sensitivity analyses. In this chapter we approach such use of non-random models through MI. The value of MI for sensitivity analysis was recognised by Rubin at an early stage of its development. For example, Chapter 6 of Rubin (1987) is devoted to the use of MI with non-ignorable response mechanisms. The common thread to the different approaches considered in this chapter is the need to impute under such non-ignorable (or non-random) mechanisms. That is, we need to accommodate the dependence of the imputation model on $\boldsymbol{R}$. It has been seen on various occasions in previous chapters how this can lead to awkward computational issues. We see in the following how, in certain settings, the direct estimation of such NMAR models can be avoided through the use of MI. This is one reason behind its value as a tool for sensitivity analysis.

However, a further, and very important, requirement for a sensitivity analysis to be of practical value is its transparency. This was pointed out by Rubin in the above reference, and by many other authors since; the key requirements for successful sensitivity analyses include a clear expression of the implied assumptions, and relevance and accessibility from the substantive viewpoint. This means that, whichever approach is adopted, a useful sensitivity analysis must frame the assumptions in a way that is accessible to all those who need to make use of the results, so they can in turn identify relevant, plausible forms of these assumptions to explore. In this respect, analysing data under NMAR models is qualitatively different than fitting a model to fully observed data, or examining diagnostics. Further, the correspondence between the pattern-mixture and selection approaches discussed in Section 4.3.2 implies that if we adopt a pattern-mixture approach for framing assumptions, the selection consequences should be plausible, and vice-versa.

The term "sensitivity analysis" has been used loosely above, without a formal definition. There are of course many forms of sensitivity analysis, see for example Saltelli et al. (2008) and other chapters in the current part of this Handbook. For our purposes in the following development we need to make one particular distinction between two different classes of sensitivity analysis, both of which will be developed through multiple imputation. In both classes, an alternative data generating mechanism is postulated that captures a relevant departure from the assumptions underlying the primary analysis. Often this will involve a particular non-random missingness mechanism. The two classes of approach differ in whether this alternative data generating mechanism is reflected in the subsequent analysis model. In the first class, the analysis model is chosen to satisfy the assumptions underlying the chosen mechanism. The subsequent analysis is then a conventional one, albeit typically made under a particular NMAR model, and the frequentist properties of such a procedure can be assessed in the usual way through long-run behaviour. In the second class, the original primary analysis is retained, even though it is *inconsistent* with the postulated data generating mechanism. The aim of such a sensitivity analysis is to assess the impact of departures from the assumptions on the primary analysis as originally used, for example as specified in a trial analysis protocol. Care must then be taken in assessing the frequentist properties of such a procedure because the long-run behaviour must reflect the mismatch of data generating mechanism and analysis model. Both classes of sensitivity analysis have a valuable role, but it is important to be clear about which is being used in any given setting, so that their properties can be assessed appropriately. Multiple imputation is a particularly convenient tool for constructing such analyses because of the separation of the imputation and substantive models, which map naturally onto the data generating mechanism and analysis models, respectively. In the first class of sensitivity analysis the two models are typically congenial, in the second they are uncongenial.

We begin with a review of NMAR modelling.

19.2 Review of NMAR Modelling

In what follows, where the distinction between outcome (Y) and covariate (X) is irrelevant—as is often the case in an imputation model—the generic variable Z will be used. Suppose we have two variables, Z_1 and Z_2, with Z_1 always observed and Z_2 having a non-zero probability of being missing. Let R be the usual indicator of missingness, taking the value 1 if Z_2 is observed and 0 otherwise. Then, as described in Chapter 4, for a generic unit, we have

$$f(Z_1, Z_2 \mid R)f(R) = f(Z_1, Z_2, R) = f(R \mid Z_1, Z_2)f(Z_1, Z_2). \tag{19.1}$$

The central expression is the joint distribution of the data, comprising the variables and the selection indicator. On the right-hand side, this is written as the product of a density for selection given Z_1, Z_2 and a density of Z_1, Z_2, the so-called *selection factorisation*, as discussed in Chapter 4.

By contrast, on the left-hand side we have the *pattern-mixture factorisation*, again as set out in Chapter 4. There is a different distribution of (Z_1, Z_2) depending on whether Z_2 is observed. This is averaged over the probability that Z_2 is observed. In more realistic examples, there will be a number of *patterns* of missing observations, each potentially with a different joint distribution of partially observed and fully observed data, and the overall density as the average over these patterns.

The MAR assumption therefore has two forms: the selection form,

$$f(R \mid Z_1, Z_2) = f(R \mid Z_1),$$

and the pattern-mixture form

$$f(Z_2 \mid Z_1, R) = f(Z_2 \mid Z_1),$$

where it can be seen that the conditional distribution of Z_2 given Z_1 is the same whether Z_2 is observed or not. In this simple case it is obvious that the one implies the other; however this holds true quite generally (Molenberghs et al. 1998). This means that for sensitivity analysis we can either focus on modelling the different patterns, or on modelling the selection process. In a complex analysis we could do both, handling some aspects with a pattern-mixture model and some aspects with a selection model.

If the focus is on pattern-mixture modelling, then for each pattern we need to specify the joint distribution of the partially and fully observed variables. In turn, this implies the conditional distribution of partially observed data given the fully observed data within each pattern. This can take any form commensurate with the data type (continuous, ordinal, categorical). However, the majority of these forms will be extremely implausible, given the scientific context and the observed data. We therefore advocate, in many settings, starting from the conditional distribution implied by MAR, and then changing this to reflect assumptions, which could be based on contextual knowledge or expert belief, about the difference from the observed conditional distribution when the variable, or set of variables, is unobserved. Although the MAR assumption may be implausible in any given setting, it does provide an unambiguous origin from which the impact of departures can be assessed.

Now suppose that there is more than one partially observed variable, i.e., a vector $\boldsymbol{Z}_2$, so that the set of potential missing data patterns becomes non-trivial. Without essential loss of generality we retain a single fully observed variable, Z_1. It follows that a convenient starting point for modelling is to write

$$f(Z_1, \boldsymbol{Z}_2 \mid \boldsymbol{R}) = f(\boldsymbol{Z}_2 \mid Z_1, \boldsymbol{R}) f(Z_1),$$

keeping the marginal model for the fully observed variables the same across patterns, i.e., $f(Z_1 \mid \boldsymbol{R}) = f(Z_1)$, and allowing $f(\boldsymbol{Z}_2 \mid Z_1, \boldsymbol{R})$ to differ with $\boldsymbol{R}$.

Methods for fitting such NMAR models have been developed in many forms, see for example Chapters 4 and 16 of this handbook. They are often computationally awkward, especially in the selection framework. In the following we will see how the use of MI can help both conceptually, in formulating accessible sensitivity analyses, and computationally, through the avoidance of fitting NMAR models directly.

19.3 Pattern-Mixture Modelling with Multiple Imputation

We motivate the development with an example, and then describe and illustrate a generic approach to pattern-mixture modelling using MI. Schroter et al. (2004) report a single-blind randomised controlled trial among reviewers for a general medical journal. The aim was to investigate whether training improved the quality of peer review. The study compared two different types of training (face-to-face training, or a self-taught package) with no training.

Here, attention is restricted to the comparison between those randomised to the self-training package and to no-training. Each participating reviewer was pre-randomised into their intervention group. Prior to any training, each was sent a baseline article to review (termed paper 1). If this was returned, then according to their randomised group, the reviewer was either (i) mailed a self-training package or (ii) received no further intervention.

Two to three months later, participants who had completed their first review were sent a further article to review (paper 2); if this was returned, a third paper was sent three months later (paper 3). The analysis excluded all participants who did not complete their first review: this was not expected to cause bias since these participants were unaware of their randomised allocation.

Reviewers were sent manuscripts in a similar style to the standard *British Medical Journal* request for a review, but were told these articles were part of the study and were not paid. The three articles were based on three previously published papers, with original author names, titles and location changed. In addition, nine major and five minor errors were

TABLE 19.1
Peer Review Trial. Review Quality Index of paper 1 by whether or not paper 2 was reviewed.

		Group		
		Control	Postal	Face-to-Face
Returned review of paper 2	n	162	120	158
	mean	2.65	2.80	2.75
	SD	0.81	0.62	0.70
Did not return review of paper 2	n	11	46	25
	mean	3.02	2.55	2.51
	SD	0.50	0.75	0.73

introduced. The outcome is the quality of the review, as measured by the Review Quality Instrument. This validated instrument contains eight items scored from 1 to 5. Rating was done independently by two editors. The response in our analysis is the mean of the first seven items, averaged over the two editors. This can range between 1 and 5, where a perfect review would score 5.

We restrict attention to the second review. For this, analysis of the observed data showed a statistically significant difference at the 5% level in favour of the self-training package. In Table 19.1 we break down the results of the baseline review (paper 1) by whether reviewers responded to the request to review paper 2. One possible, but unverifiable, missing data mechanism that is suggested by these results is that the improved review quality on paper 2 in the self-taught group may be because poorer reviewers are disproportionately NMAR from this group, even after accounting for baseline. We assume that inference is to be made about the effect of being sent the self-training package, compared with no intervention, regardless of whether it was used. Below, we explore the robustness of inference for this question to different assumptions about the missing review quality.

19.3.1 Modifying the MAR imputation distribution

We now develop the notation introduced earlier for the current setting. For the ith of N units, let $\boldsymbol{Z}_i = (Z_{i1}, \ldots, Z_{ip})'$ and $\boldsymbol{R}_i = (R_{i1}, \ldots, R_{ip})'$ be the vector of observations and response indicators, respectively. Suppose that there are, among all N units, $M \ll N$ distinct response patterns, indexed by $\boldsymbol{R}_m, m \in (1, \ldots, M)$. The missing and observed variables from any unit conform to one of these patterns, say $m(i)$, for the ith unit, and one of the patterns corresponds to complete records on all p variables. Let $\boldsymbol{Z}^o_{m(i)}, \boldsymbol{Z}^m_{m(i)}$ be the observed and missing variables for unit i with response pattern $m(i)$.

For each unit, we denote the distribution of the missing data, given the observed data and missingness pattern, by

$$[\boldsymbol{Z}^m_{m(i)} | \boldsymbol{Z}^o_{m(i)}, m, \boldsymbol{\eta}_m]. \tag{19.2}$$

Here $\boldsymbol{\eta}_m$ are the parameters of this distribution for missingness pattern m, whose values we have to estimate before we can draw missing data from (19.2).

If the missing data are MAR, then (19.2) does not depend on the missingness pattern m; distributions are the same across all patterns, $\boldsymbol{\eta}_m = \boldsymbol{\eta}$, $m \in (1, \ldots, M)$, and we impute missing data from $[\boldsymbol{Z}^m_i | \boldsymbol{Z}^o_i, \boldsymbol{\eta}]$. However, if the data are not missing at random (NMAR), then this distribution will differ across missingness patterns.

TABLE 19.2
Peer Review Trial. Inference for comparison of self-taught package with no training, under MAR and NMAR with $\rho = 0, 0.5$, and 1. Parameter estimates are differences in mean review quality index, on a scale of 0–5. $K = 10,000$.

Analysis	Est.	s.e.	MI d.f.	p	95% CI
Complete records, MAR	0.237	0.070	N/A	<0.001	(0.099, 0.376)
MAR, $K = 10,000$	0.237	0.070	$\approx \infty$	<0.001	(0.099, 0.375)
NMAR, $\rho = 0$, $K = 20$	0.209	0.178	27	0.25	(−0.158, 0.575)
NMAR, $\rho = 0$, $K = 10,000$	0.193	0.151	$\approx \infty$	0.20	(−0.102, 0.488)
NMAR, $\rho = 0.5$, $K = 20$	0.205	0.167	27	0.23	(−0.141, 0.234)
NMAR, $\rho = 1$, $K = 20$	0.213	0.134	34	0.12	(−0.059, 0.486)

Suppose that $\widehat{\theta} = \widehat{\theta}(\boldsymbol{Z}) = \widehat{\theta}(\boldsymbol{Z}^m, \boldsymbol{Z}^o)$ is of interest. For example, θ may be a regression coefficient. With no missing data, we would estimate this from $\boldsymbol{Z}$, using the appropriate regression model. For each missingness pattern m, our approach is to define a form for (19.2) which reflects contextually relevant assumptions. Then we impute K 'complete' datasets using the standard MI procedure:

MI1: take a draw from the Bayesian posterior distribution of $[\boldsymbol{\eta}_m | \boldsymbol{Z}^o, m]$ and then

MI2: impute the missing data from (19.2) using the above draw of $\boldsymbol{\eta}_m$.

Both steps are repeated to create each imputed dataset. The parameter of interest is then estimated from each imputed dataset in turn to give $\widehat{\theta}_k$, with standard error $\widehat{\sigma}_k$, $k = 1, \ldots, K$. These are then combined using Rubin's rules (Chapter 12).

To implement MI1 we need to choose a model for the observed data. To implement MI2 we need to specify (19.2). Taking the former first, the approach we adopt is to estimate a parameter vector $\boldsymbol{\eta}$ from all the observed data assuming MAR. For the latter, specific rules or information used to derive, or draw, $\boldsymbol{\eta}_m$. This is of necessity context-specific. It could involve

1. explicitly specifying the distribution of $\boldsymbol{\eta}_m$ given $\boldsymbol{\eta}$, possibly using opinions elicited from experts, or
2. specifying how $\boldsymbol{\eta}_m$ is constructed from $\boldsymbol{\eta}$, for example in terms of rules across well defined subsets of the data (such as treatment or exposure groups).

The examples considered below illustrate both approaches in which MAR is taken as the starting point for sensitivity analysis.

We now apply the first approach to the Reviewers' Trial. Focusing on the baseline adjusted comparison of the self-taught training package with no training, the substantive model can be written

$$Y_i = \beta_0 + \beta_1 X_{i1} + \beta_2 X_{i2} + e_i, \quad e_i \ i.i.d. \ \mathrm{N}(0, \sigma^2), \tag{19.3}$$

where i indexes participant, Y_i and X_{i1} are the mean review quality index for paper 2 and paper 1, respectively, and X_{i2} is an indicator for the self-training group. Inference for β_2 from fitting (19.3) to the complete records, assuming review 2 is MAR given baseline review and intervention group, is shown in the first row of Table 19.2.

Under the assumption of MAR, missing reviewer scores are imputed using information from participants who completed the second review and complied with the study protocol implying, in particular, that the analysis reflects what would have been seen if those who did not complete the study had instead complied with the protocol. This is in contrast to the inferential goal of the analysis which is to assess the effect of intervention if it were rolled out to all reviewers by the *British Medical Journal*, and such an analysis reflects the fact that non-completers in the study represent those in the population at large who are not expected to comply with the protocol. So, White et al. (2007) devised a questionnaire that was completed by 2 investigators and 20 editors and other staff at the *British Medical Journal*. The questionnaire was designed to elicit the experts' prior belief about the difference between the average missing and average observed review quality index. White et al. (2007) show that it was reasonable to pool information from the experts. The resulting distribution is negatively skewed, with mean −0.21 and SD 0.46 (on the review quality index scale). Suppose that (δ_0, δ_1) denotes a draw from the distribution of the mean difference in review quality between observed and unobserved reviews, in respectively the control and self-training groups. A bivariate normal model is used as an approximation to the prior:

$$\begin{pmatrix} \delta_0 \\ \delta_1 \end{pmatrix} \sim N\left[\begin{pmatrix} -0.21 \\ -0.21 \end{pmatrix}, 0.46^2 \begin{pmatrix} 1 & \rho \\ \rho & 1 \end{pmatrix}\right]. \tag{19.4}$$

Unfortunately, it was not possible to elicit a prior on ρ from the experts; the data are therefore analysed with ρ set to a range of fixed values: 0, 0.5, and 1.

Given a draw (δ_0, δ_1) from this distribution the model is

$$\begin{aligned} Y_i &= \beta_0 + \beta_1 X_{i1} + \beta_2 X_{i2} + e_i && \text{if } Y_i \text{ observed,} \\ Y_i &= (\beta_0 - \delta_0) + \beta_1 X_{i1} + (\beta_2 + \delta_1 - \delta_0) X_{i2} + e_i && \text{if } Y_i \text{ unobserved,} \\ & e_i \; i.i.d. \; \mathrm{N}(0, \sigma^2). \end{aligned} \tag{19.5}$$

The mean review quality, relative to that in the observed data, is therefore reduced by δ_0 in the control arm and δ_1 in the self-taught arm.

Following the general approach for estimating pattern-mixture models via MI described earlier, the procedure is as follows, noting that the substantive and MAR imputation models coincide in this example.

1. Fit model (19.3) to the observed data and draw from the posterior distribution of the parameters $\boldsymbol{\eta} = (\beta_0, \beta_1, \beta_2, \sigma^2)$.
2. Draw (δ_0, δ_1) from (19.4).
3. Using the draws obtained in steps 1 and 2, impute the missing Y_i using (19.5).

Steps 1–3 are repeated to create K imputed datasets. Then, the substantive model (19.3) is fitted to each imputed dataset and Rubin's rules applied.

The results of this are shown in Table 19.2. In line with theory, the results from MI under MAR agree very closely with the complete records analysis. From the NMAR imputations it can be seen that the mean review quality index in the self-training group is no longer statistically significantly different from the no-training group. The standard error is largest when $\rho = 0$, and decreases as $\rho \to 1$. It can also be seen from the NMAR case with $\rho = 0$ that $K = 20$ imputations is enough to clearly show this conclusion to practically relevant precision, and the results with $K = 20$ have therefore been presented for the other cases. The result with 10,000 imputations agrees very closely with both a theoretical approximation and a full Bayesian analysis reported by White et al. (2007), even though the latter allows for uncertainty in estimating the proportion with missing data, which is conditioned on in the pattern-mixture approach.

With regard to the example, we conclude that, taking the experts' prior belief into account, there is no evidence that self-training improves the quality of peer review.

This overall approach can also be applied directly to discrete outcomes. The attractions of interpretability and computational simplicity remain. For example, it could be applied in the setting of Magder (2003), who measured non-random missingness via the response probability ratio. Care must be taken however with the scale on which inferences are made using mixtures of distributions derived from models with non-linear link functions. With a linear link, in particular the identity link as used in a conventional linear model, the marginal expectation under the pattern-mixture model preserves the linear structure of the model, and hence the interpretation of effects. The same is true when a log link is used, for all effects apart from the intercept. However, when a logit or log hazard link is used, the same is not true. For example, in general, a mixture of logistic models will not have a logistic form, and so the analogue of the odds-ratios in the component logistic models does not even exist in the marginalized pattern-mixture model. This property of non-linear models is well-known, and is sometimes termed non-collapsibility, see for example Ford et al. (1995) and Greenland et al. (1999). When using pattern-mixture models in such settings care must be taken in the appropriate choice of scale to avoid such non-coherence.

19.3.2 Missing covariates

Next, consideration is given to sensitivity analysis for partially observed covariates in the substantive model. As above, the coefficients of the imputation model are estimated under MAR, but then modified to reflect a departure from MAR before imputation of the missing values. The approach is illustrated using the Youth Cohort Study of England and Wales (YCS). This is an ongoing UK government funded representative survey of pupils in England and Wales at school leaving age (school year 11, age 16–17) (UK Data Archive 2007). Each year that a new cohort is surveyed, detailed information is collected on each young person's experience of education and their qualifications as well as information on employment and training. A limited amount of information is collected on their personal characteristics, family, home circumstances, and aspirations.

Over the life-cycle of the YCS different organisations have had responsibility for the structure and timings of data collection. Unfortunately, the documentation of older cohorts is poor. Croxford et al. 2007 deposited a harmonised dataset that comprises YCS cohorts from 1984–2002 (UK Data Archive Study Number 5765 dataset). We consider data from pupils attending comprehensive schools from five YCS cohorts (cohorts 5, 7, 8 and 10); these pupils reached the end of Year 11 in 1990, 1993, 1995, 1997, and 1999, respectively.

Relationships between Year 11 educational attainment (the General Certificate of Secondary Education) and key measures of social stratification are of interest. The units are pupils and the items are measurements on these pupils, and a non-trivial number of items are partially observed. The principle missing data pattern, in 11% of the records, is missing parental occupation but observed values for the other variables in our substantive model. The other, more complex patterns, account for $< 3\%$ of the missing records between them. Because imputing with or without the latter records makes no difference to the resulting inference, we restrict this analysis to the 61,609 records with either complete records or only parental occupation missing.

TABLE 19.3
Youth Cohort Study. Model for the differences in GCSE points score (range 0–84) by ethnicity, adjusted for sex, parental occupation and cohort. Analyses used 20 imputations.

Covariate	Estimates (standard errors) from:			
	Complete records	MAR	NMAR ($\delta_I = \delta_W = -2$)	NMAR ($\delta_I = 0, \delta_W = 1.8$)
Black	−5.39	−6.88	−6.92	−6.88
	(0.57)	(0.50)	(0.49)	(0.50)
Indian	3.83	3.25	3.22	3.22
	(0.44)	(0.41)	(0.41)	(0.41)
Pakistani	−1.79	−3.55	−3.59	−3.53
	(0.59)	(0.47)	(0.47)	(0.47)
Bangladeshi	0.69	−2.99	−4.93	−1.44
	(1.05)	(0.72)	(0.79)	(0.70)
Other Asian	5.79	4.90	4.89	4.86
	(0.69)	(0.63)	(0.62)	(0.63)
Other	0.37	−0.72	−0.71	−0.67
	(0.71)	(0.65)	(0.64)	(0.64)
Boys	−3.47	−3.37	−3.37	−3.36
	(0.13)	(0.13)	(0.13)	(0.13)
Intermediate	−7.46	−7.80	−7.73	−7.77
parental occupation	(0.15)	(0.16)	(0.17)	(0.16)
Working	−13.82	−14.33	−14.30	−14.37
parental occupation	(0.17)	(0.17)	(0.17)	(0.18)
1995 cohort	6.30	6.19	6.18	6.19
	(0.18)	(0.17)	(0.17)	(0.17)
1997 cohort	4.96	5.01	5.01	5.01
	(0.18)	(0.17)	(0.17)	(0.17)
1999 cohort	9.57	9.88	9.88	9.88
	(0.19)	(0.18)	(0.18)	(0.18)
constant	4.81	4.11	4.07	4.10
	(0.15)	(0.15)	(0.15)	(0.15)

As an illustration, the substantive model is taken to be the linear regression of GCSE score on ethnic group, adjusted for sex, parental occupation, and cohort. Parental occupation is a three-category variable, with classes: managerial, intermediate and working. A multinomial logistic imputation model is used, with 'managerial' as the reference category. Let $\pi_{iM}, \pi_{iI}, \pi_{iW}$ be the probability that pupil i's parental occupation is classed as 'managerial,' 'intermediate,' or 'working', respectively, where $\pi_{IM} + \pi_{iI} + \pi_{iW} = 1$. The imputation model is

$$\log(\pi_{iI}/\pi_{iM}) = \boldsymbol{X}_i \boldsymbol{\alpha}_I \tag{19.6}$$

$$\log(\pi_{iW}/\pi_{iM}) = \boldsymbol{X}_i \boldsymbol{\alpha}_W, \tag{19.7}$$

where $\boldsymbol{X}_i$ is a $(n \times q)$ matrix with columns corresponding to the constant, pupil's GCSE score centred at 38 points, and indicators for boys, ethnicity (6 variables), and cohort (3 variables).

For MAR imputation, (19.6) and (19.7) are estimated from the complete records and proper imputations are created in the conventional way. Full details are given in Carpenter and Kenward (2013, Ch. 5). The results of the complete records analysis and MAR imputation are shown in the left two columns of Table 19.3. Of particular interest are the coefficients for Pakistani and Bangladeshi ethnicity, which under MAR become substantially more negative and statistically significant. Focusing on Bangladeshi ethnicity, we explore the

TABLE 19.4
Occupation of parents of students of Bangladeshi ethnicity, in complete records and under MAR, NMAR. Twenty imputations were used.

	Complete Records	Imputation under		
		MAR	NMAR ($\delta_I =$ $\delta_W = -2$)	NMAR (δ_I $\delta_W = 1.8$)
Managerial	26 (12%)	48 (9%)	143 (26%)	34 (6%)
Intermediate	91 (41%)	215 (40%)	174 (32%)	126 (23%)
Working	104 (47%)	272 (51%)	221 (41%)	378 (70%)
Total	221	538	538	538

sensitivity of this result to the MAR assumption, which in this setting implies that the relationship between parental occupation and the other variables in the imputation model is the same among those for whom it is observed and those for whom it is not. Two NMAR scenarios are considered as a form of sensitivity analysis. For pupils of Bangladeshi ethnicity (i) missing parental occupations are predominantly 'managerial,' and (ii) missing parental occupation scores are predominantly 'working.' In the former case we add to the Bangladeshi coefficients in (19.6), (19.7), respectively ($\delta_I = -2, \delta_W - 2$), and in the latter ($\delta_I = 0, \delta_W = 1.8$). In the absence of prior information, in this example we do not follow (19.4) but instead set $\text{Var}(\boldsymbol{\delta}) = 0$. The resulting imputed parental occupations are then, respectively, predominantly working, and predominantly managerial; when combined with the observed occupations we obtain the proportions shown in Table 19.4. The rightmost columns of Table 19.3 show the corresponding effect on the coefficient for Bangladeshi ethnicity in the model of inference: under both scenarios, inference is remarkably robust. We note that the NMAR scenarios chosen are quite extreme. In practice, we would also include auxiliary covariates in the imputation model; here these have been omitted to keep the development relatively simple.

We conclude that on average Bangladeshi pupils have significantly lower GCSE scores than the complete records analysis would suggest, and this result is robust to plausible NMAR mechanisms.

19.3.3 Application to survival analysis

We now consider the application of this approach to sensitivity analysis for non-random censoring, or loss to follow-up, in a survival analysis. As an example, a study for the evaluation of antiretroviral treatment (ART) programmes in sub-Saharan Africa is used. HIV infection is a major cause of mortality and morbidity in sub-Saharan Africa, and between 2007–2010 the number of patients starting ART increased steeply (Boulle et al. 2008), along with consequent interest in evaluating the efficacy of ART in this setting. However, during the same period there has been increasing concern about loss to follow up in these programmes (Brinkhof et al. 2010) especially as those lost to follow-up may have substantially worse mortality than those who are followed up, which will bias the evaluation (Bartlett and Shao 2009, Brinkhof et al. 2009). Accordingly, Brinkhof et al. (2010) decided to use a pattern-mixture approach, through multiple imputation, to explore the robustness of inferences about mortality to non-random censoring.

To see how the pattern-mixture approach can be applied directly to survival data, suppose that T is the random variable representing the time to event and C the random variable representing loss to follow-up (i.e., censoring). From each unit, or patient, we observe baseline covariates $\boldsymbol{X}_i$, and $Y_i = \min(T_i, C_i)$.

The Censoring at Random (CAR) assumption is the analogue of the covariate dependent MAR assumption for univariate outcomes in the time-to-event setting. Under CAR it is assumed that T_i and C_i are conditionally independent given the covariates $\boldsymbol{X}_i$. Under CAR, valid estimates for coefficients $\boldsymbol{\beta}$ relating covariates to survival time are obtained from modelling the observed data, where censored units contribute $\Pr(T_i > C_i|\boldsymbol{X}_i, \boldsymbol{\beta})$ to the likelihood.

Similarly, the Not Censoring at Random (NCAR) assumption is the NMAR assumption for survival data. Under NCAR, T_i and C_i are not independent, even given the covariates $\boldsymbol{X}_i$.

To obtain parameter estimates and inferences under NCAR, we adapt the strategy outlined earlier to the survival setting. Specifically, we obtain an estimate of the hazard assuming CAR, and then introduce a sensitivity parameter defining the ratio of the hazard of censored individuals to those that are not censored. The hazard under NCAR can be written

$$h(t_i|C_i, \boldsymbol{X}_i) = \begin{cases} h_{CAR}(t_i|\boldsymbol{X}_i) & \text{if } t_i < C_i \\ \delta h_{CAR}(t_i|\boldsymbol{X}_i) & \text{if } t_i \geq C_i. \end{cases} \tag{19.8}$$

If $\delta = 1$, then given $\boldsymbol{X}_i$, the hazard is the same, irrespective of censoring; in other words CAR holds. Otherwise, the hazards are different. The parameter δ is once again the sensitivity parameter, but this time it represents the hazard ratio of uncensored and censored units. As usual, the data at hand give no information on the distribution of δ. However information on this may be elicited from experts or other data sources; alternatively the data may be re-analysed with δ successively further from 1 and then assess the plausibility of the value of δ at which our conclusions change substantively.

For now it is assumed that the distribution of the log hazard-ratio is

$$\log(\delta) \sim N(\mu_\delta, \sigma_\delta^2), \tag{19.9}$$

although in seeking prior information it will usually be better to work on the hazard-ratio scale itself. A parametric model is assumed for survival. Then, given (19.9) and (19.8), the approach described in Section 19.3.1 can be applied directly, giving the following algorithm:

1. Assuming CAR, fit the survival model to the data, obtaining parameter estimates $\widehat{\boldsymbol{\beta}}$ relating the log-hazard to the covariates, with associated estimated covariance matrix $\widehat{\Sigma}$, typically obtained from the observed information (Kenward and Molenberghs 1998).

2. Draw δ from (19.9) and $\boldsymbol{\beta} \sim N(\widehat{\boldsymbol{\beta}}, \widehat{\Sigma})$.

3. With these values, for each censored unit i impute the time to event from the hazard $\delta h_{CAR}(t_i, \beta, \boldsymbol{X}_i)$, conditional on the time exceeding C_i. The observed and imputed event times together make up the imputed dataset under NCAR.

Repeating steps 2–3, K imputed datasets are generated, to each of which the substantive model is fitted. This could be a parametric or semi-parametric survival model; at its simplest it could be the Kaplan–Meier estimate of survival at a specific time. This gives K point estimates and their standard errors, which can be combined for final inference using Rubin's rules.

In practice, to draw the survival times, we require the cumulative distribution function of the survival time, $F(t; \boldsymbol{X}, \boldsymbol{\beta}, \delta)$, that corresponds to the hazard $\delta h_{CAR}(t; \boldsymbol{X}, \boldsymbol{\beta})$. Because

$$F(T; \boldsymbol{Y}, \boldsymbol{\beta}, \delta) \sim U[0, 1],$$

a survival time for censored unit i can be imputed by drawing u_i from the uniform distribution on $[0, 1]$ and then calculating

$$t_i = F^{-1}(u_i; \boldsymbol{X}_i, \boldsymbol{\beta}, \delta),$$

with appropriate rejection sampling to ensure that the imputed value is greater that C_i. This approach is convenient under a parametric survival model. If a Cox proportional-hazards model is used, then generating proper imputations is harder, as uncertainty in estimating both the coefficients in the relative hazard as well as the baseline hazard need to be taken into account. We do not view this as a major limitation, because, within the rich families of parametric survival models it is likely that an appropriate choice can be found for imputation. Of course, the proportional hazards model can be fitted to the imputed data. If there are concerns about inconsistencies between the imputation model and substantive model, a natural first check is to impute the censored data assuming CAR (i.e., $\delta = 1$), fit the substantive model to each imputed dataset and combine the results using Rubin's rules. Just as in Table 19.2, the resulting inferences should be very close to those from the observed data.

Again, exactly as in the application to the peer review trial above, it may well be appropriate to have different means and variances for $\log(\delta)$ in different treatment or other groups. In that case, values for the correlation between them also need to be selected. In trials, it is also possible to envisage piecing post-censoring hazards together by drawing on other trial arms, in a similar spirit to the method developed for the longitudinal trials in Section 19.4.

Brinkhof et al. (2010) apply this approach to explore the sensitivity of inference about mortality of patients receiving ART in studies conducted in sub-Saharan Africa to NCAR. Their substantive model was the Kaplan–Meier estimate of survival 1-year following initiation of ART. They chose a Weibull model for imputation. Under this model, (19.8) becomes

$$h(t; \boldsymbol{X}, \boldsymbol{\beta}, C) = \begin{cases} \exp(\boldsymbol{X}'\boldsymbol{\beta})\gamma t^{\gamma-1} & \text{if } t < C \\ \delta \exp(\boldsymbol{X}'\boldsymbol{\beta})\gamma t^{\gamma-1} & \text{if } t \geq C, \end{cases} \tag{19.10}$$

where $t > 0, \gamma > 1, \delta > 0$ and the covariate matrix $\boldsymbol{X}$ includes the intercept.

To implement the algorithm described above, the Weibull model is first fitted to the observed data in the usual way. Then using the observed information, to create each imputed dataset draws are made from the distribution of the parameters, $\boldsymbol{\beta}, \gamma$, and from these, sample survival times for censored observations from $S(t \mid t > C, \boldsymbol{X}, \boldsymbol{\beta}, \gamma, \delta)$. Under the Weibull distribution, this is

$$S(t \mid t > C, \boldsymbol{X}, \boldsymbol{\beta}, \gamma, \delta) = \frac{S(t; \boldsymbol{X}, \boldsymbol{\beta}, \gamma, \delta)}{S(C; \boldsymbol{X}, \boldsymbol{\beta}, \gamma, \delta)} = \exp\{-\delta(t^\gamma - C^\gamma)\exp(\boldsymbol{X}'\boldsymbol{\beta})\}.$$

In this application, the substantive model is simply the Kaplan–Meier estimate of 1 year survival. The results for two of the five studies analysed by Brinkhof et al. (2010) are shown in Table 19.5. Notice that the estimated 1-year survival in the observed data, and after CAR imputation ($\delta = 0, K = 10$ imputations) agrees closely. This is what theory predicts: imputation of censored survival times under CAR should give the same results as the observed data CAR analysis.

TABLE 19.5
Illustration of sensitivity analysis for increased mortality among patients lost to follow-up, from Brinkhof et al. (2010).

Study	$\widehat{\delta}$	One Year Mortality			Relative
		In Observed Data	Under (19.10) $\delta = 1$	Under (19.10) $\delta = \widehat{\delta}$	Increase
Lighthouse	6	10.9% (9.6–12.4%)	10.8% (9.4–12.3%)	16.9% (15.0–19.1%)	56%
AMPATH	12	5.7% (4.9–6.5%)	5.9% (5.1–6.9%)	10.2% (8.9–11.6%)	73%

Brinkhof et al. (2010) used meta-regression of other studies available to them to estimate plausible values of δ for these studies, which were then fixed at the estimated values rather than drawn from a distribution as in (19.9). The results show a substantial, practically important, relative increase in mortality in these studies if the hazard ratio relating those lost to follow-up to those remaining in follow-up is consistent with that estimated in the meta-regression.

We conclude this example by noting that while, given $\widehat{\delta}$, the point estimates in Table 19.5 could be approximated analytically, MI also provides estimates of the standard errors, and resulting confidence intervals. The ability to do this when δ has a distribution which is allowed to differ by treatment arm or other subgroups, makes MI a very attractive method for estimation and inference.

19.4 Longitudinal Clinical Trial Data

We now turn to dropout in longitudinal data, and focus on data arising from clinical trials, although the same ideas will typically be applicable to longitudinal data arising in other settings. The methods described here were first suggested by Little and Yau (1996) and subsequently developed in Kenward and Carpenter (2008) and Carpenter et al. (2013). A special case of the general approach is sometimes called *controlled imputation*, see for example, Mallinckrodt et al. (2012), Teshome et al. (2014) and Mallinckrodt (2013, Section 10.5). In this section we will be making use of the second class of sensitivity analysis described in the introduction in which the original substantive model is retained, even when the imputation model is constructed under alternative, NMAR, scenarios.

19.4.1 Estimands

To provide some context for the analyses to be developed here, we first touch on a recent debate in the area of dropout and compliance in the clinical trial setting. As well as the familiar problem of making assumptions about the missing data mechanism, it will be seen that one cannot avoid simultaneously making assumptions about the treatment behaviour of participants who drop out, or have missing data for some other reason. This follows from a crucial issue in the handling of missing data in clinical trials (and in other settings): the definition of the target of the inferences to be made. This has been the subject of much recent discussion, see for example Mallinckrodt and Kenward (2009), Mallinkrodt (2013) and Carpenter et al. (2013) and, in particular, Chapter 22 of this handbook. Following the report by the US National Research Council (2010) we encapsulate this concept in the so-called *estimands.* For the population of eligible patients, as defined by the trial inclusion criteria, let $\tilde{f}_{\text{act}}$ be the joint probability distribution function of baseline and post-randomisation responses for patients randomised to the *active* treatment, who follow the treatment regime exactly, withstanding all the rigours (including adverse events) without departing from the protocol. Define $\tilde{f}_{\text{cont}}$ analogously, for patients receiving the control (or reference) treatment. Following Carpenter et al. (2013) we define two classes of estimand that are of interest in this setting.

(1) The de jure estimand: For response profile $\boldsymbol{Y}$, and any suitable function $g(\,.\,)$, we define the de jure estimand as

$$\mathrm{E}_{\tilde{f}_{\text{act}}}[g(\boldsymbol{Y})] - \mathrm{E}_{\tilde{f}_{\text{cont}}}[g(\boldsymbol{Y})]. \tag{19.11}$$

De jure questions concern the magnitude of this quantity; in other words 'What would the expected treatment effect be in the eligible population if the treatment and control were taken as specified in the protocol,' i.e., 'does the treatment work under the best case scenario?'

The function $g(\,.\,)$ defines the effect. For many choices of g, a more precise estimate of the quantity (19.11) can be obtained by conditioning on baseline, Y_0. Often $g(\boldsymbol{Y})$ is the last scheduled observation, so that if (i) there is no missing data and (ii) the observed data in the trial arms comes respectively from $\tilde{f}_{\text{act}}$, $\tilde{f}_{\text{cont}}$, then (19.11) can be efficiently estimated by regression of the final response on treatment group and baseline.

(2) The de facto estimand: Now further define, for the population of eligible patients as defined by the trial inclusion criteria, f_{act} as the joint probability density function of baseline and post-randomisation responses that would be seen in the context of interest among patients randomised to the active arm. Define f_{cont} analogously for the control arm.

For response profile $\boldsymbol{Y}$, and any suitable function $g(\,.\,)$, we define the de facto estimand as

$$\mathrm{E}_{f_{\text{act}}}[g(\boldsymbol{Y})] - \mathrm{E}_{f_{\text{cont}}}[g(\boldsymbol{Y})]. \tag{19.12}$$

De facto questions concern the magnitude of this quantity; in other words "What would be the effect seen in practice if this treatment were applied to the population defined by the trial inclusion criteria?"

These proposed terms, de facto and de jure, describe estimands in the sense defined by the US National Research Council (2010). By contrast, the commonly used terms terms *effectiveness* and *efficacy*, besides causing confusion by their similarity, actually refer to different, looser concepts about the conditions under which the intervention is being used (National Institute for Health and Care Excellence, 2011).

Because, in some settings, patients may be unable to withstand the rigours of the regime, distributions $\tilde{f}_{\text{act}}$, $\tilde{f}_{\text{cont}}$ may be counter-factual. By contrast, providing patients remain alive, their actual responses, with probability distributions f_{act}, f_{cont}, can almost always be observed given sufficient time and resources.

While the discussion above is focussed on randomised clinical trials, the key issues apply to all analyses. We need to consider carefully the population to which the conclusions are to apply, the reasons why data are missing, and how (if at all) the distribution of these missing data may differ from what is predicted from the observed data under strong assumptions such as MAR. As an example, consider the Youth Cohort Study introduced above. We may be interested in (i) inference for those regularly in school or (ii) inference for the population, including those who have been excluded from school, or whose attendance is erratic. The former we would term a de jure estimand, the latter a de facto estimand.

We emphasise that, whatever the estimand, for inference we will generally wish to estimate the parameters under both MAR and NMAR mechanisms. However, the choices for the latter will typically vary with the choice of estimand.

19.4.2 Deviations

We now focus on the setting of the longitudinal clinical trial. To define estimands unambiguously, and how to develop analyses that target these when data are missing, it is first necessary to be clear about what is meant by a *protocol deviation.* Suppose we intend to observe a response at baseline, and n follow-up times, $Y_{i0}, Y_{i1}, \ldots, Y_{in}$. For the relevant estimand, we define a *deviation from the protocol relevant to the estimand*, which we simply refer to as a *deviation*, as a violation of the protocol, such that post-deviation data (if known) cannot be directly used for inference about the estimand. For now we assume that post-deviation data are missing; we discuss the use of any post-deviation data that may be available in Section 19.6.

Having pre-specified the analysis questions and associated deviations, there remains a dataset in which (i) each patient has longitudinal follow-up data until either they deviate, or reach the scheduled end of the study, and (ii) the nature of each deviation is available. The approach we describe here is that, for each deviation (or—more likely—group of similar deviations occurring for similar reasons) an appropriate post-deviation distribution is constructed that takes account of (i) the patient's pre-deviation observations; (ii) pre-deviation data from other patients in the trial; (iii) the nature of the deviation, and (iv) the reason for the deviation.

19.4.3 Change in slope post-deviation: The 'Δ-method'

Suppose that post-deviation patients have a different, usually poorer response than predicted under MAR. For example in an asthma trial, FEV_1 might improve more slowly (or decline more quickly) after withdrawal. If the change in rate of decline post-deviation in arm a is denoted δ_a, then the MAR conditional mean for the first scheduled post-deviation observation is reduced by δ, the second by 2δ and so on. This is schematically illustrated in Figure 19.1.

Computationally, the outline approach described in Section 19.3.1 is used. The K imputations are created under MAR, then, given two intervention arms, for each imputation, k, the following samples are taken

$$\begin{pmatrix} \delta_{1k} \\ d_{2k} \end{pmatrix} \sim N\left[\begin{pmatrix} \mu_{\delta_1} \\ \mu_{\delta_2} \end{pmatrix}, \begin{pmatrix} \sigma_1^2 & \rho\sigma_1\sigma_2 \\ \rho\sigma_1\sigma_2 & \sigma_2^2 \end{pmatrix}\right].$$

For each patient, in each intervention arm $a = 1, 2$, for each imputation, the first MAR imputed observation is increased/decreased by δ_{ak}, the second by $2\delta_{ak}$ and so on. The resulting datasets are analysed in the conventional MI manner, combining the estimates using Rubin's rules. If the time between observations is not constant, the multipliers of δ from $1, 2, 3, \ldots,$ can be changed accordingly. Interim missing observations can be handled by decreasing them by δ_l, or simply leaving them with their MAR imputed values. The

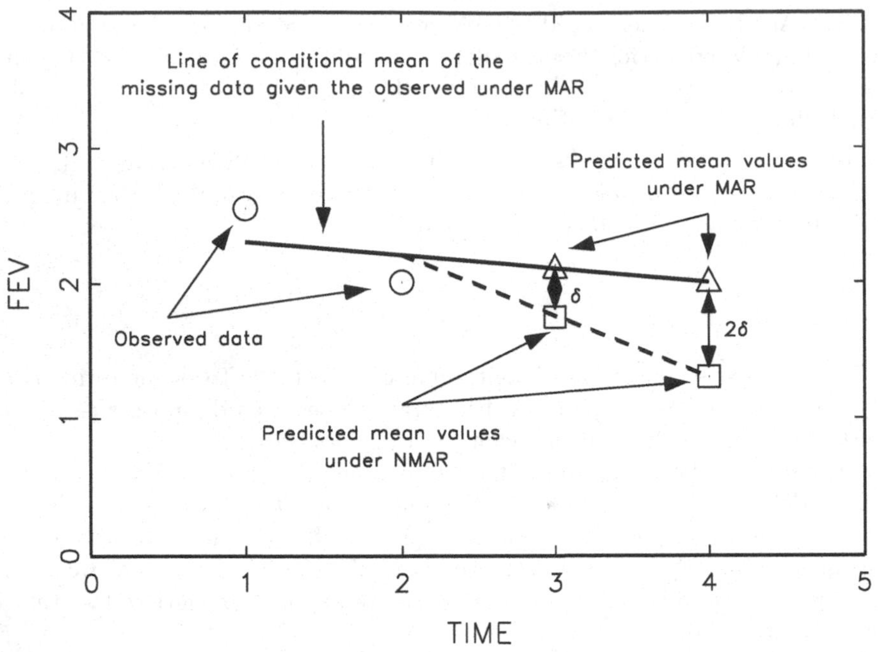

FIGURE 19.1
Schematic illustration of increasing the rate of decline by δ post-deviation.

latter is consistent with a different mechanism driving interim missing data and patient withdrawal.

Some decision must be made about the values chosen for the δ's, in particular whether they should differ among the treatment arms. This might be guided by prior knowledge, or, as is much more likely in practice, in the absence of this $\delta_1 = \delta_2$ might be the most sensible choice. Or if one treatment is placebo, and dropout is associated with withdrawal, then δ might be set to zero for the placebo arm. Depending on the context, one approach may be to increase δ until the treatment effect is no longer clinically relevant. This is sometimes called a 'tipping point analysis,' see for example Yan et al. (2009).

19.4.4 Example: Asthma Study

This approach is now illustrated using data from a 5-arm asthma clinical trial to assess the efficacy and safety of budesonide, a second-generation glucocorticosteroid, on patients with chronic asthma. In this trial, 473 patients with chronic asthma were enrolled in the 12-week randomised, double-blind, multi-centre parallel-group trial, which compared the effect of a daily dose of 200, 400, 800 or 1600 mcg of budesonide with placebo.

Table 19.4.4 shows the withdrawal patterns for the placebo and lowest active dose arms (all the patients are receiving their randomised medication). We have removed three patients with unusual interim missing data from Table 19.4.4 and in all the analyses below. The remaining missingness pattern is monotone in both treatment arms. The outcomes of clinical interest include patient's peak expiratory flow rate (their maximum speed of expiration in litres/minute) and their forced expiratory volume (the volume of air, in litres, the patient

TABLE 19.6
Asthma Study. Withdrawal pattern by treatment arm. Outcome is mean FEV_1 in litres.

	Week						
Dropout Pattern	0	2	4	8	12	#	%
Placebo Arm							
1	√	√	√	√	√	37	40
2	√	√	√	√	·	15	16
3	√	√	√	·	·	22	24
4	√	√	·	·	·	16	17
Lowest Active Arm							
1	√	√	√	√	√	71	78
2	√	√	√	√	·	8	9
3	√	√	√	·	·	8	9
4	√	√	·	·	·	3	3

with fully inflated lungs can breathe out in one second). In summary, the trial found a statistically significant dose-response effect for the mean change from baseline over the study for both morning peak expiratory flow, evening peak expiratory flow and FEV_1 at the 5% level.

Budesonide treated patients also showed reduced asthma symptoms and bronchodilator use compared with placebo, while there were no clinically significant differences in treatment related adverse experiences between the treatment groups. Further details about the conduct of the trial, its conclusions and the variables collected can be found elsewhere (Busse et al., 1998). Here, we focus on FEV_1 and confine our attention to the placebo and lowest active dose arms. FEV_1 was collected at baseline, then 2, 4, 8 and 12 weeks after randomisation. The intention was to compare FEV_1 across treatment arms at 12 weeks. However (excluding 3 patients whose participation in the study was intermittent) only 37 out of 90 patients in the placebo arm, and 71 out of 90 patients in the lowest active dose arm, remained in the trial till twelve weeks.

The substantive model is the regression of 12 week FEV_1 on treatment, adjusted for baseline. It should be noted that this is uncongenial with the imputation model: it does not reflect the different statistical behaviour of those who drop out and those who do not, but does reflect the analysis that would be used in practice, that is, it is essentially the primary analysis of the original study. We focus first on the de jure question, asking what would be the treatment effect if patients in both arms were able to remain in the study under the protocol, despite worsening asthma. In this study patients are followed up systematically until they deviate, typically by discontinuing or unblinding treatment for a variety of reasons. Only limited post-deviation data are available, and these are disregarded for these analyses.

The first row of Table 19.7 shows the analysis of the 108/180 patients who completed the 12 week follow-up. Treatment increases lung function by a clinically relevant 0.25 l on average, and this estimate is significant at the 5% level. This analysis addresses the de jure question under the assumption that data are MAR given baseline and treatment. A more plausible assumption is that data are MAR given baseline, treatment and intermediate measurements. Under this assumption, and imputing separately in the treatment arms, the 12 week effect increases to 0.36 l, with a much reduced p-value. Following Figure 19.1, the third analysis decreases the MAR conditional mean by 0.1 l each post-deviation follow-up visit. Since 53/90 in the placebo, but only 19/90 in the lowest active arm, deviate before 12 weeks, the

TABLE 19.7
Asthma Study. De jure estimates of the treatment effect under MAR, and NMAR with increasing post-deviation increments, δ.

Analysis	Treatment Estimate (l)	s.e.	p
Complete records, de jure	0.247	0.100	0.0155
MI, de jure, randomised arm MAR	0.335	0.107	0.0017
MI, de jure, $\delta = 0.1, \sigma_\delta = 0$	0.361	0.108	0.0008
MI, de jure, $\delta = 0.1, \sigma_\delta = 0.1$	0.362	0.156	0.0207
MI, de jure, $\delta = 0.2, \sigma_\delta = 0.0$	0.388	0.110	0.0004
MI, de jure, $\delta = 0.2, \sigma_\delta = 0.1$	0.392	0.158	0.0132

consequence is to increase the treatment effect. The precision is unchanged if $\sigma_\delta = 0$; with a relatively large $\sigma_\delta = 0.1$, it is increased but the effect remains significant. Analysis with $\delta = 0.2, \sigma_\delta = 0.1$ yields similar results. We conclude that, under this model for sensitivity analysis, the de jure inference is robust to NMAR.

19.4.5 Reference-based imputation

The pattern-mixture sensitivity analyses so far described have all relied on specification of the distribution of a sensitivity parameter vector, $\boldsymbol{\delta}$. We now describe an alternative that is particularly useful for, but not confined to, randomized clinical trials. In this, post-deviation distributions are constructed by making qualitative, not quantitative, reference to other arms—hence the name *reference-based imputation.* This idea was introduced by Little and Yau (1996), and should be contrasted with the use of so-called identifying restrictions in pattern-mixture models in which post-deviation distributions are implicitly constructed using information from the same treatment group. See, for example, Chapter 4 of this Handbook.

We have seen in Section 19.4.1 that in a clinical trial when data are missing or, more strictly, when there are deviations, any analysis that incorporates all randomized patients must, implicitly or explicitly, make assumptions about the treatment regime the patient follows after deviation. In the approach to be developed here this is made explicit. When a patient deviates it is either known, or assumed for the purposes of the sensitivity analysis, that they subsequently follow some particular treatment involved in the trial, or that is assumed to behave like one in the trial. For a de jure estimand this will be the same treatment to which they were originally randomized. For a de facto estimand it will typically be different. This treatment assumption is reflected in the imputation model: the future statistical behaviour is borrowed from other patients who remain in the trial on the relevant treatment. For a de jure estimand these patients come from the same treatment group as the patient for whom the imputations are being made, in which case this is precisely the "borrowing" implied by a likelihood-based analysis under MAR. For a de facto estimand information is borrowed from another group. Such borrowing can be done in many ways, and we shall see some examples below, and suitable choices will depend on the substantive setting, in particular on the nature of the condition being treated and the action of the intervention. A crucial part of the overall procedure, one that makes it relatively easy to implement, is that the component parts of the model that are "borrowed" under the various scenarios are all taken from the fitted MAR model, so no NMAR model needs to be fitted directly to the data.

For a continuous outcome the overall procedure can be summarized as follows.

1. Separately for each treatment arm, fit a multivariate normal linear model with unstructured mean that is, a separate mean for each of the n repeated measurements (baseline plus post-randomisation observations) and a corresponding unstructured covariance matrix. We call this the Multivariate Gaussian Linear Model (MGLM).

2. Separately for each treatment arm, draw a mean vector and covariance matrix from the posterior distribution.

3. For each patient who deviates before the end of the study, use the draws from step 2 to build the joint distribution of their pre- and post-deviation outcome data. Suggested options for constructing this are given below.

4. For each patient who deviates before the end, use their joint distribution in step 3 to construct their conditional distribution of post-deviation given pre-deviation outcome data. Sample their post-deviation data from this conditional distribution, to create a 'completed' dataset.

5. Repeat steps 2–4, K times, resulting in K imputed datasets.

6. Fit the substantive model to each imputed dataset, and combine the resulting parameter estimates and standard errors using Rubin's rules for final inference.

19.4.6 Constructing joint distributions of pre- and post-deviation data

There are many potential ways in which such post-deviation distributions might be constructed. The choice in any particular setting will depend critically on the condition or disease being measured, the action of the treatment, and the reason for the deviation/dropout. The following are examples that may be of value for some situations, and serve to illustrate the basic approach.

Each option represents a difference between de jure and de facto behaviour post-deviation. Many others are possible but these are sufficient to explain the approach and illustrate its flexibility.

Randomised-arm MAR
The joint distribution of the patient's observed and post-deviation outcome data is multivariate normal with mean and covariance matrix from their randomized treatment arm.

Jump to reference
Post-deviation, the patient ceases his/her randomised treatment, and the mean response distribution is now that of a 'reference' group of patients (typically, but not necessarily, control patients). Such a change may be seen as extreme, and choosing the reference group to be the control group might be used as a worst-case scenario in terms of reducing any treatment effect since withdrawn patients on active will lose the effect of their period on treatment.

Post-deviation data in the reference arm are imputed under randomised-arm MAR.

Last mean carried forward
Post-deviation, it is assumed that the patient is expected to neither get worse nor

better. So the mean of the distribution stays constant at the value of the mean for the randomized treatment arm at the last pre-deviation measurement. The covariance matrix remains that for their randomized treatment arm. There are different versions of this depending on whether the group mean profile remains constant, or whether it is the individual conditional mean that is held constant.

It follows that a patient who is well above the mean for his/her arm at the last pre-deviation measurement (giving a large positive pre-deviation residual) will tend to progress back across later visits to within random variation of the mean value for the arm at their pre-deviation measurement visit. The speed and extent of that progression will depend on the strength of the correlation in the covariance matrix.

Copy increments in reference
After the patient deviates, their post-deviation mean increments are taken from the reference group (typically, but not necessarily, control patients). Post-deviation data in the reference arm are imputed under randomised-arm MAR. If the reference is chosen to be the control arm, the patient's mean profile following deviation tracks that of the mean profile in the control arm, but starting from the benefit already obtained. This is what we might be seen, for example, in an Alzheimer's study where treatment halts disease progression but after stopping therapy the disease progression restarts.

Copy reference
Here, for the purpose of imputing the missing response data, a patient's whole distribution, both pre- and post-deviation, is assumed to be the same as the reference (typically, but not necessarily, control) group. Post-deviation data in the reference arm are imputed under randomised-arm MAR. If the reference group is chosen as the control group, this mimics the case where those deviating are in effect non-responders.

Perhaps surprisingly, this may often have a less extreme impact than 'jump to reference' above. This is because if a patient on active treatment is above the control mean then this positive residual will feed through into subsequent draws from the conditional distribution of post-deviation data, to a degree determined by the correlation in the control arm. Thus, the patient's profile will slowly decay back towards the mean for control at later visits.

Note that 'last mean carried forward' and 'copy increments in reference' have an important feature in common. For an active and a placebo patient who deviate at the same time, the difference between their two post-deviation means is maintained at a constant value up to the end of the trial. In the former case the individual group mean profiles are held constant over time, in the latter they are allowed to vary across time. In this sense they both represent ways of implementing in a principled modelling framework the assumptions that might be implied by "last observation carried forward" type methods.

19.4.7 Technical details

As an illustration, we begin by showing for the 'jump to reference' scenario how the joint distribution of the pre- and post-deviation distributions are constructed, that is, step (3) of the earlier summary. It is supposed that there is an active treatment and control arm, and let 'reference' designate the control arm. This leads to a brief presentation of the approach for the other options.

In step 2, the current draw from the posterior for the n reference arm means and covariance matrix are denoted by $\mu_{r,0}, \ldots \mu_{r,n}$, and Σ_r. The subscript a is used for the corresponding draws from the other arm in question.

Under 'jump to reference,' suppose that patient i is not randomised to the reference arm and that their last observation, prior to deviating, is made at time d_i, $d_i \in (1, \ldots, n-2)$. The joint distribution of their observed and post-withdrawal outcomes is multivariate normal with mean

$$\tilde{\boldsymbol{\mu}}_i = (\mu_{a,0}, \ldots, \mu_{a,d_i}, \mu_{r,d_i+1}, \ldots, \mu_{r,n-1})';$$

that is post-deviation they 'jump to reference.'

The new covariance matrix is now constructed for these observations as follows. Denote the covariance matrices from the reference arm (without deviation) and the other arm in question (without deviation), partitioned at time d_i according to the pre- and post-deviation measurements, by:

$$\text{Reference } \Sigma_r = \begin{bmatrix} R_{11} & R_{12} \\ R_{21} & R_{22} \end{bmatrix} \text{ and other arm: } \Sigma_a = \begin{bmatrix} A_{11} & A_{12} \\ A_{21} & A_{22} \end{bmatrix}.$$

These are estimated under MAR using the pre-deviation data.

The new covariance matrix, $\tilde{\Sigma}$ say, must match that from the active arm for the pre-deviation measurements, and the reference arm for the *conditional* components for the post-deviation given the pre-deviation measurements. This also guarantees positive definiteness of the new matrix, because Σ_r and Σ_a are both positive definite. That is,

$$\tilde{\Sigma} = \begin{bmatrix} \Sigma_{11} & \Sigma_{12} \\ \Sigma_{21} & \Sigma_{22} \end{bmatrix},$$

must be subject to the constraints

$$\begin{aligned} \Sigma_{11} &= A_{11}, \\ \Sigma_{21}\Sigma_{11}^{-1} &= R_{21}R_{11}^{-1}, \\ \Sigma_{22} - \Sigma_{21}\Sigma_{11}^{-1}\Sigma_{12} &= R_{22} - R_{21}R_{11}^{-1}R_{12}. \end{aligned}$$

The solution is:

$$\begin{aligned} \Sigma_{11} &= A_{11}, \\ \Sigma_{21} &= R_{21}R_{11}^{-1}A_{11}, \\ \Sigma_{22} &= R_{22} - R_{21}R_{11}^{-1}(R_{11} - A_{11})R_{11}^{-1}R_{12}. \end{aligned}$$

The joint distribution for a patient's pre- and post-deviation outcomes under 'jump to reference' has now been specified, when deviation is at time d_i. This completes the construction.

For 'copy increments in reference' the same $\tilde{\Sigma}$ as for 'jump to reference' is used but now

$$\begin{aligned} \tilde{\boldsymbol{\mu}}_i = \{&\mu_{a,0}, \ldots, \mu_{a,d_i-1}, \mu_{a,d_i}, \mu_{a,d_i} + (\mu_{r,d_i+1} - \mu_{r,d_i}), \\ &\mu_{a,d_i} + (\mu_{r,d_i+2} - \mu_{r,d_i}), \ldots\}.' \end{aligned}$$

For 'last mean carried forward,' $\tilde{\Sigma}$ equals the covariance matrix from the randomization arm. The important change is the way in which $\tilde{\boldsymbol{\mu}}$ is constructed. For patient i in arm a under 'last mean carried forward,'

$$\tilde{\boldsymbol{\mu}}_i = (\mu_{a,0}, \ldots, \mu_{a,d_i-1}, \mu_{a,d_i}, \mu_{a,d_i}, \ldots\ldots); \quad \tilde{\Sigma} = \Sigma_a.$$

Finally for 'copy reference' the mean and covariance both come from the reference (typically, but not necessarily, control) arm, irrespective of deviation time.

19.4.8 Example: Asthma Study

We return to the asthma study, and again focus on the placebo and lowest active dose arms, with the same definition of a deviation as before.

The first row of Table 19.8 (analysis DJ1) shows the complete records analysis (cf Table 19.7) by ANCOVA using the week 12 data, using the 108 patients with data at 12 weeks. It follows from the definition of deviation that the MAR assumption implies that, conditional on treatment and baseline, the distribution of unobserved and observed 12 week responses is the same. Since the latter is estimated from on-treatment, protocol adhering patients, this analysis addresses the de jure question. However, there are only 37 out of 90 patients with 12 week data in the placebo arm, compared with 71 out of 90 in the active.

It follows that analysis DJ2 includes all observed data in a saturated multivariate repeated measures linear model (MGLM) model. This fits a separate mean for each treatment and time, with a full baseline-time interaction and common unstructured covariance matrix. Analysis DJ3 further allows a separate, unstructured covariance matrix in each arm. The results show inference from the primary analysis (DJ1: 0.247l, $p = 0.0155$), which addresses the de jure question, is robust to the different assumptions made by DJ2, DJ3; indeed DJ3 gives both the largest and most significant treatment estimate (0.346l, $p = 0.0013$).

Analysis DJ4 is constructed under the same MAR assumption as DJ3. In line with theory, the result from the MGLM model with separate covariance matrices agrees closely with that from using multiple imputation since the data are well modelled by the normal distribution. The only difference in underlying imputation model for DJ4 from the MGLM model DJ3 is that the multiple imputation approach implicitly allows a full three-way interaction of baseline with treatment and time, whereas the MGLM model has the same regression coefficients for baseline×time in the two arms. Apart from this they are structurally equivalent.

Consider now the de facto question, and analyses DF1–DF7 in Table 19.8. Each of these assumes a different joint distribution for pre- and post-deviation data. For analyses DF2–DF7 the reference arm is always imputed using randomised-arm MAR.

Analysis DF1 corresponds to the underlying mean response remaining static after intervention stops. This would address the de-facto question when after deviation patients took no further relevant medication and their condition was stable. If, allowing for the personal covariates such as baseline, they have a positive residual at their final pre-deviation visit, we expect that post-deviation their residuals will decrease (but continue to randomly vary), so on average they will get closer to their own conditional mean.

In the asthma setting, this flat mean profile post deviation may be plausible for a week or two after deviation, especially in the active treatment arm. In these data, the downward trend in the placebo arm (Table refch20:dpat) suggests that it is likely to yield higher estimates of the latter placebo means than the de jure analyses, while estimates for the lowest active arm will change in the other direction. This leads to a smaller treatment estimate (0.296l), but not as low as that for DJ1.

The 'jump to reference' (DF2,3), 'copy reference' (DF4,5) and 'copy increments in reference' (DF6,7) options are now applied. The implications are considered for the interpretation of

TABLE 19.8
Estimated 12 week treatment effect on FEV_1 (litres), from ANCOVA, MGLM's, multiple imputation and sensitivity analyses. All multiple imputation analyses used 1000 imputations, with a 'burn-in' of 1000 updates and 500 updates between imputations.

Analysis		Est. (l)	s.e.	d.f. (mod)	t	p
De jure						
DJ1	ANCOVA (Compl.), joint var.	0.247	0.101	105.0	2.46	0.0155
DJ2	MGLM†, joint cov.	0.283	0.094	131.0	3.02	0.0030
DJ3	MGLM†, separate cov.	0.346	0.104	72.8	3.34	0.0013
DJ4	Randomised-arm MAR	0.334	0.107	130.6	3.13	0.0022
De facto						
DF1	Macro, last mean carr. fwd.	0.296	0.102	141.6	2.90	0.0043
DF2	Jump to ref. (active)	0.141	0.119	102.7	1.18	0.2390
DF3	Jump to ref. (placebo)	0.264	0.108	135.5	2.46	0.0153
DF4	Copy ref. (active)	0.252	0.087	139.4	2.88	0.0046
DF5	Copy ref. (placebo)	0.295	0.105	146.5	2.82	0.0055
DF6	Copy incr. in ref. (active)	0.295	0.103	139.7	2.87	0.0048
DF7	Copy incr. ref. (placebo)	0.323	0.104	139.6	3.12	0.0022

† MGLM: Multivariate General Linear Model.

choice of the 'reference' arm under these two approaches (in this example it could either be the placebo, or the lowest active dose arm).

Suppose that the de facto question to be addressed corresponds to the assumption that, post-deviation, (i) patients on placebo obtain a treatment equivalent to the active, and (ii) the active treatment patients continue on treatment and adhere to the protocol, so that their post-deviation data can be imputed assuming randomised-arm MAR. In this case, we specify the *active arm* as 'reference.' The early part of the study suggests that 2–3 weeks are needed for a treatment to take effect. Further, the patients have chronic asthma, so it is likely that they will seek an active treatment on withdrawal from the placebo arm. For the de facto question, this assumption is more plausible than considering placebo as 'reference.' Nevertheless, for discussion the results where the placebo is 'reference' are also presented; that is where post-deviation (i) patients on the active treatment switch to a placebo equivalent, and (ii) the placebo treatment patients continue on placebo adhering to the protocol, and their post-deviation data can be imputed assuming randomised-arm MAR. This latter assumption might be appropriate where no alternative treatment is generally available.

In analysis DF2 the treatment effect is estimated under 'jump to reference' when the 'reference' is the active arm. Of all the de facto analyses, this is the most extreme in terms of effect on the treatment difference, since the means prior to deviation follow the patients randomized arm and then abruptly switch to that of the specified 'reference' arm. The effect of this is shown in Figure 19.2; we find plots like this very useful tools for conveying the implications of assumptions to stakeholders. This large change in placebo patients' post-deviation means results in a substantially reduced treatment estimate (0.141ℓ, $p = 0.24$). We conclude that, if post-deviation medication has comparable effect to the lowest active

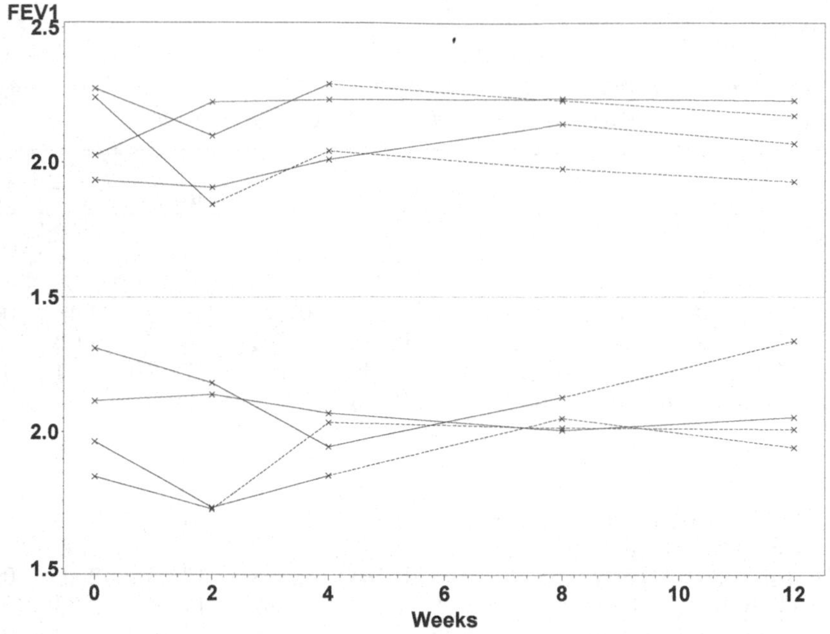

FIGURE 19.2
Mean FEV_1 (litres) against time for the four different deviation patterns. Solid lines join observed means (before deviation) and dotted lines join the means of the imputed data for that pattern. Top panel: lowest active dose, imputed under randomised-arm MAR; bottom panel: placebo arm imputed under 'jump to reference' (where 'reference' is lowest active dose).

dose, patients from both arms will have comparable lung function at the end of the study. This analysis mimics what we might expect for a *retrieved dropout analysis* (Committee for Medicinal Products for Human Use, 2010) where placebo patients are allowed to return to active treatment.

Analysis DF3 corresponds to the less plausible reverse option where after withdrawal the active patients now 'jump to reference' and the reference is the placebo. Because far fewer patients deviate in the active arm, the change from the de jure MAR analysis is much smaller: the treatment difference remains significant ($0.264\,\ell$, $p = 0.015$).

Two further sets of assumptions for addressing the de facto question are now considered: 'copy reference' and 'copy increments in reference.' The former replaces both pre-deviation and post-deviation means with those of the specified 'reference,' when constructing the joint distribution of pre- and post-deviation data in step 3. The latter, 'copy increments in reference,' has the randomized profile prior to deviation, but then the incremental changes in mean FEV_1 from visit to visit follow those in the 'reference' arm.

When the active arm is specified as the 'reference,' 'copy reference' (DF4) and 'copy increments in reference' (DF6) give 12 week treatment estimates of 0.252ℓ and 0.295ℓ, respectively. In this case, 'copy reference' has a larger treatment effect than 'jump to reference' because under 'copy reference,' pre-deviation placebo patients have relatively larger residuals (differenced from the mean for the active arm), which implies that after deviation they track to the active means more slowly; with 'jump to reference' pre-deviation placebo patients have smaller residuals (differenced from the mean for the placebo arm), so post-deviation they track to the active means more rapidly.

When placebo is specified as the 'reference,' 'copy reference' (DF5) and 'copy increments in reference' (DF7) give smaller changes from the randomised-arm MAR estimate, for the reasons discussed in the previous paragraph.

In summary, the primary analysis that addresses the de jure question assuming MAR is consistent with a significant beneficial effect of treatment relative to placebo. Addressing the de facto question with a conservative assumption about the effect of post-deviation (withdrawal) switching to active ('copy reference' or 'copy increments in reference') continues to show a significant improvement. However, if instead 'jump to reference' ('reference' is active treatment) were used then the treatment benefit is reduced by over 50% relative to randomised-arm MAR, and is no longer statistically significant. If the many placebo patients who deviate early switch to the active treatment, this should be expected, and is a consequence of a de facto analysis with this particular type of treatment switching. It is the reason some have argued against de facto (or ITT) analyses in such settings (Keene, 2011).

19.4.9 Distinguishing two types of sensitivity analysis

In closing this section, it is useful to return to the distinction made at the beginning of this chapter concerning the two classes of sensitivity analysis. It is important to be clear about the nature of the sensitivity analysis that has been used here, especially when assessing the expected behaviour of the procedures. The data generating mechanism is assumed to follow the NMAR assumptions but the original substantive analysis has been retained. So, this belongs to the second class of sensitivity analysis, and it attempts to answer the question, "What might be seen under the original analysis if some other sampling behaviour is actually operating?" Expressed in MI terms, the substantive model is retained from the original (primary) analysis, while the imputation model follows the NMAR assumptions, and these two models are uncongenial.

We can contrast this approach with the analogous sensitivity analysis of the first class in which both the imputation model *and substantive model* follow the same NMAR mechanism. In this particular setting, in contrast to other sections in this chapter, and other chapters in this part of the Handbook, such a sensitivity analysis is not very interesting from an inferential perspective. This is because it reflects the same *information* on the treatment effect as that produced by the original MAR-based analysis. For example, in simple settings, the treatment effect may well be scaled down under the NMAR assumption but the accompanying measure of precision will be scaled in a similar way, and the resulting inference in terms of statistical significance will be very similar. If such an analysis is of interest then MI is not needed: the estimator is a prediction from the MAR model and the associated precision can be obtained in a conventional way from the information.

Asymptotically both approaches lead to the same treatment estimator, and so it is very tempting to expect the corresponding precision estimators to also have the same long-run sampling behaviour. But this is not what is required for the uncongenial sensitivity analysis developed in this section. Because the original substantive analysis is being used which does *not* acknowledge the potential heterogeneity of means within treatment groups under NMAR, the estimate of precision must be inflated appropriately for this. For example, in a simple final time point analysis, the residual variance must include mean heterogeneity within each randomized group. At the same time, loss of information due to missing data needs to be acknowledged. The MI variance estimator should, at least approximately, achieve both aims, but it is important that the performance of the overall procedure is *not* assessed through the conventional long-run sampling distribution of the treatment effects estimators, in contrast to the congenial approach of the first class.

19.5 Approximating a Selection Model by Importance Weighting

Up to this point sensitivity analyses have been constructed through the pattern-mixture formulation. In particular, departures from MAR have been expressed in terms of the conditional behaviour of the missing data given the observed. This is very natural in the MI context as this maps directly to the construction of the imputation model.

Although arguably not as natural a route under MI, we now describe an approximate method that can be used for constructing selection model-based sensitivity analyses within this framework. Now, exploration of sensitivity of conclusions to NMAR mechanisms is described directly in terms of the missing data mechanism, but in a way that avoids the need to impute under the NMAR mechanism concerned. The basic idea, introduced in Carpenter et al. (2007), is simple: having imputed K datasets under MAR, and fitted the substantive model to each, instead of averaging the results for the imputation estimate, we perform a weighted average, up-weighting imputations that are more likely under the NMAR mechanism under consideration.

This approach rests on the idea of importance sampling. Suppose that

$$Z^1, \ldots, Z^K \ i.i.d.\ g,$$

are drawn from some distribution g, but the expectation of some function, $h(Z)$ is required when Z is drawn from a different distribution f. Provided that the support of f is contained in that of g, and f/g is bounded,

$$\mathrm{E}_f[h(Z)] \approx \frac{\sum_{k=1}^K w_k h(Z^k)}{\sum_{k=1}^K w_k}, \tag{19.13}$$

for $w_k = f(Z^k)/g(Z^k)$, with the approximation tending to equality as $K \to \infty$.

Here, this result is applied to the imputations from MI. We make the following identifications and definitions:

1. g with the imputation distribution under MAR;
2. f with the imputation distribution under NMAR;
3. k indexes the imputations;
4. Z^k is the kth imputation under MAR;
5. $h(Z^k)$ is the estimate of the parameter in the substantive model obtained from imputation k.

To illustrate how this works, consider the very simple setting of two variables, Z_1, Z_2, from a single unit. Let the parameter of interest $\widehat{\theta} = \widehat{\theta}(Z_1)$ be the mean of Z_1. Suppose that there is a single pair of observations (Z_1, Z_2) with (Z_1) missing. As usual, denote the missingness indicator by R, so that $R = 0$. Two imputations, Z_1^1, Z_1^2, are drawn, under MAR, from the imputation model

$$f(Z_1 \mid Z_2, R = 1). \tag{19.14}$$

However, imputations are needed under NMAR, i.e., from the distribution

$$f(Z_1|Z_2, R = 0). \tag{19.15}$$

In the pattern-mixture approach (19.15) was defined with reference to (19.14) via sensitivity parameters, or other treatment groups. Here a selection model is used, that is, the relationship is defined implicitly through the logistic regression of R on Z_1, Z_2 :

$$\operatorname{logit}\Pr(R = 1) = \alpha_0 + \alpha_1 Z_1 + \alpha_2 Z_2. \tag{19.16}$$

In this case, the weight is the ratio

$$\begin{aligned}\frac{f(Z_1|Z_2, R = 0)}{f(Z_1|Z_2, R = 1)} &= \frac{f(Z_1, Z_2, R = 0)f(Z_2, R = 1)}{f(Z_1, Z_2, R = 1)f(Z_2, R = 0)} \\ &= \frac{f(R = 0|Z_1, Z_2)f(Z_2, R = 1)}{f(R = 1|Z_1, Z_2)f(Z_2, R = 0)}.\end{aligned} \tag{19.17}$$

Now consider $f(R = 0|Z_1, Z_2)$. Under model (19.16), this is $\{1+\exp(\alpha_0+\alpha_1 Z_1+\alpha_2 Z_2)\}^{-1}$, and $f(R = 1|Z_1, Z_2) = 1 - f(R = 0|Z_1, Z_2)$. So (19.17) is

$$\exp\{-(\alpha_0 + \alpha_1 Z_1 + \alpha_2 Z_2)\}\frac{f(Z_2, R = 1)}{f(Z_2, R = 0)}.$$

Therefore the weights for the two imputations, Z_1^1 and Z_1^2 are

$$\begin{aligned}\tilde{w}_1 &= \exp(-\alpha_1 Z_1^1)\left\{\exp[-(\alpha_0 + \alpha_2 Z_2)]\frac{f(Z_2, R = 1)}{f(Z_2, R = 0)}\right\} \text{ and} \\ \tilde{w}_2 &= \exp(-\alpha_1 Z_1^2)\left\{\exp[-(\alpha_0 + \alpha_2 Z_2)]\frac{f(Z_2, R = 1)}{f(Z_2, R = 0)}\right\}.\end{aligned} \tag{19.18}$$

Since the terms between the curly braces are common to both weights, the normalised weights, $w_1 = \tilde{w}_1/(\tilde{w}_1 + \tilde{w}_2)$ and $w_2 = \tilde{w}_2/(\tilde{w}_1 + \tilde{w}_2)$ are proportional to $\exp(-\alpha_1 Z^1)$ and $\exp(-\alpha_1 Z^2)$, respectively. In other words, estimates of α_0, α_2 are not required to estimate the weights; all that is needed are the imputed values under MAR, and α_1. As with pattern-mixture modelling, we do not estimate this parameter (Kenward 1998). Instead, the sensitivity of the results is explored as α_1 varies from 0.

In fact, in the linear predictor '$\alpha_0 + \alpha_1 Z_1 + \alpha_2 Z_2$' can be any function of the observed data because it will cancel out of the weights. This means that, in contrast to full joint modelling approaches, the inference is robust to possible mis-specification of the relationship between the probability of observing Z_i and the observed part of the data, *provided* the relationship of the withdrawal process on the unseen data is correct.

So far the setting with only $N = 1$ units/individuals and 2 imputations has been considered. Suppose now that there are $i = 1, \ldots, N_1$ units (out of a total of N) with Z_1 missing, and, for unit i, K values imputed assuming MAR, $Z_{i1}^1, \ldots Z_{i1}^K$. Because, given $Z_{i1}^k, Z_{i'1}^k, Z_{i2}, Z_{i'2}$ and $i' \neq i$, the probability of observing $Z_{i'}^k$ is independent of the probability of observing Z_i^k, it follows from (19.18) that the weight for imputation k is

$$w_k \propto \exp\left(-\alpha_1 \sum_{i=1}^{N_1} Z_{i1}^k\right). \tag{19.19}$$

In other words, the log-weight for imputation k is proportional to a linear combination of the imputed data. Lastly, $\theta(\boldsymbol{Z}_1) = \mathrm{E}((\boldsymbol{Z}_1)$, can be replaced with any general $\theta(\boldsymbol{Z}_1, \boldsymbol{Z}_2)$. So, in general, the procedure is as described in the following section.

19.5.1 Algorithm for approximate sensitivity analysis by re-weighting

Let the data vector for unit i be $\boldsymbol{Z}_i^o, \boldsymbol{Z}_i^m$, and θ be the scalar quantity of interest, such as a treatment effect. Assume, as is usual with multiple imputation, that if there were no missing data the estimator of θ would be approximately normally distributed. Given data, denote the estimated value of θ by $\widehat{\theta}$. Assuming MAR, suppose that K versions of the missing data, $\boldsymbol{Z}_1^m, \ldots \boldsymbol{Z}_K^m$ have been imputed and K estimates of the parameter of interest $\widehat{\theta}_1, \ldots, \widehat{\theta}_K$ and their corresponding variances, $\widehat{\sigma}_1^2, \ldots, \widehat{\sigma}_K^2$ calculated from thee completed datasets.

Suppose that sensitivity to NMAR on a partially observed variable $\boldsymbol{Z}_1$, is to be explored, with corresponding missingness indicator $\boldsymbol{R}_1$. Consider the selection model

$$\operatorname{logit} \Pr(R_{i1} = 1) = \alpha_1 Z_{i1} + g(\boldsymbol{Z}_i^{\text{full}}), \tag{19.20}$$

where $\boldsymbol{Z}_i^{\text{full}}$ are those variables fully observed on all individuals.

To obtain an estimate of θ when data are NMAR, the analyst first chooses a plausible value of α_1. Suppose we re-order the dataset so that units $i = 1, \ldots, n_1$ have missing Z_{i1}. Let Z_{i1}^k denote the kth MAR imputation of Z_{i1}. Then, for each imputation, k, compute

$$\tilde{w}_k = \exp\left(\sum_{i=1}^{n_1} -\alpha_1 Z_{i1}^k\right), \quad \text{and} \quad w_k = \tilde{w}_k \Big/ \sum_{k=1}^{K} \tilde{w}_k. \tag{19.21}$$

Then, under the NMAR model implied by the analyst's choice of α_1, in (19.20), the estimate of θ and its variance are

$$\widehat{\theta}_{\text{NMAR}} = \sum_{k=1}^{K} w_k \widehat{\theta}_k, \tag{19.22}$$

with variance

$$V_{\text{NMAR}} \approx \tilde{V}_W + (1 + 1/K)\tilde{V}_B, \tag{19.23}$$

where now

$$\tilde{V}_W = \sum_{k=1}^{K} w_m \widehat{\sigma}_m^2, \quad \tilde{V}_B = \sum_{k=1}^{K} w_m (\widehat{\theta}_k - \widehat{\theta}_{\text{NMAR}})^2. \tag{19.24}$$

$\widehat{\theta}_{\text{NMAR}}$ and $\widehat{V}_{\text{NMAR}}$ can then be used for inference.

19.5.2 Reliability of the approximation

The accuracy of the approximation justifying $\widehat{\theta}_{\text{NMAR}}$ and $\widehat{V}_{\text{NMAR}}$ improves as the number of imputations, K increases, provided the two importance sampling conditions hold:

I1 the support of the NMAR distribution of the missing given the observed data is contained within the support of the MAR distribution, and

I2 the ratio of the NMAR distribution to the MAR distribution is bounded.

This has a number of practical implications.

First, for the full generality of NMAR models, I1 cannot hold. Thus, the method is suitable for exploring local departures from MAR. Some graphical diagnostics that can be used to assess 'local' in applications are described in Carpenter and Kenward (2013, Sec. 10.6.1).

Second, even locally, it may be that the ratio of the distributions is unbounded. For example, if the MAR imputation distribution is $N(1,1)$ and the NMAR imputation distribution is $N(0.5,1)$, this occurs in the left tail. Where this is a concern, the ratio can be bounded. Typically this means defining the NMAR distribution to be zero outside a given range.

Interestingly, in terms of under/over estimating the difference between NMAR and MAR inference, I1 and I2 work in opposite directions. If I1 is violated, as we move from the locality of the MAR model, the difference will be underestimated; if I2 is violated, then as the number of imputations increases $\widehat{\theta}_{\text{NMAR}}$ will remain unstable. Despite this, simulation studies (Carpenter et al. 2007, Carpenter et al. 2011) have shown good performance in practically relevant situations, especially if care is taken to choose an appropriate model for the observed data when the sample size is small.

Taken together, these points suggest that when using this approach, often $K \geq 50$ will be needed; but this is not unduly burdensome for many problems. However, even for relatively large values of K, V_{NMAR} tends to slightly underestimate the variance, although the resulting confidence interval coverage is acceptable (Carpenter et al. 2011). An explanation for this is that after re-weighting the effective number of imputations is often considerably smaller than K. One possible correction is to replace K in (19.23) by a measure of the effective sample size, such as the number of $\{Kw_k\}_{k=1}^{K}$ which are ≥ 1.

Finally, it is noted that while the approach can be applied to longitudinal data, more than one sensitivity parameter will typically be needed. For example, let Z_{i1}, Z_{i2}, and Z_{i3} be three longitudinal measurements on subject i. Let $R_i = 1$ if Z_{i3} is observed. A general model for this is

$$\operatorname{logit}\Pr(R_i = 1) = \alpha_0 + \alpha_1 Z_{i1} + \alpha_2 Z_{i2} + \alpha_3 Z_{i3}.$$

If some subjects have Z_{i2} and/or Z_{i3} missing, then we need to specify both sensitivity parameters α_2, α_3, since no parameter in the selection model involving missing data cancels out when we normalise the weights. One possible approach to this would be to specify say α_3, and use an EM type approach to estimate other parameters; however given that typically there is relatively little information on these in the data, both a large number of imputations and a large number of observations are likely to be needed.

More subtly, the justification for the weights relies on (19.17), which has the true MAR distribution, in our notation $f(Y|X, R = 1)$, in the denominator. In practice, of course, this is estimated from the observed data. If the dataset is very small, or there are a lot of missing data, this will cause re-weighted estimates to overshoot the NMAR value. This is because missing data are sampled from a distribution that is an estimate of $f(Y|X, R = 1)$ with a considerably heavier tail, which gets disproportionately up-weighted. So, when datasets are small, the imputation distribution must therefore be chosen carefully. When datasets are much larger than $n = 20$, the simulation results of Carpenter et al. (2011) indicate that this error is of secondary importance.

As an illustration of the approach we revisit the peer review example, considered earlier in Section 19.3 and perform a second sensitivity analyses for these data by imputing under MAR and re-weighting to explore sensitivity to NMAR. Specifically, we investigate the sensitivity of the results to the possibility that the review quality index from paper 2, Y_i, is not missing at random. Let $R_i = 1$ if the Y_i is observed and 0 otherwise. We use the

TABLE 19.9
Parameter estimates (log odds ratios) for logistic regression for probability of paper 2 being reviewed, model (19.25), with $\alpha_3 = 0$.

	Parameter		
	α_0,	α_1	α_2
Estimate	2.14	0.21	−1.75
(s.e.)	(0.64)	(0.22)	(0.35)

following selection model

$$\text{logit}\,\Pr(R_i = 1) = \alpha_0 + \alpha_1 X_{i1} + \alpha_2 X_{i2} + \alpha_3 Y_i, \tag{19.25}$$

where $\boldsymbol{X}_1$ is the review of the baseline paper and $\boldsymbol{X}_2$ is an indicator for the self-training group. If $\alpha_3 = 0$, then Y_i is MAR. As α_3 increases from 0 the probability of Y_i being observed increases with Y_i (i.e., increases with the review quality). If $\alpha_3 = 0$ we can fit this model using logistic regression. This gives the results in Table 19.9. Overall the probability of withdrawing decreases as baseline review quality increases, and is much higher in the self-training arm.

The estimate $\widehat{\alpha}_1 = 0.21$ suggests that each rise in the baseline average RQI of one point increases the odds ratio of response by 1.23. In the light of this we carry out sensitivity analyses with $\alpha_4 = 0.3$ and $\alpha_4 = 0.5$ in (19.25). These correspond, on the odds-scale, to roughly 10% and 35% stronger adjusted association between the chance of seeing the second review and its quality.

We obtain estimates of the effect of the postal intervention *versus* control using a fully Bayesian procedure through `WinBUGS` by fitting models (19.3) and (19.25) jointly. We also obtain estimates and standard errors by re-weighting the imputations obtained under MAR, using (19.22) and (19.24). We do this using $K = 50$, 150, 250, and 1000 imputations. The results presented below are based on analyses including the face-to-face training group, with corresponding indicator variables added to (19.3) and (19.25).

Table 19.10 shows the results. As expected, the estimated effect of the postal intervention is reduced, but it remains significant. The Bayesian estimates agree within at most 0.01 with the estimates obtained by re-weighting. The estimated standard errors are also similar, although for small K they appear to be underestimated.

In applications, it is important to know whether we have enough imputations for a reasonably reliable answer, and also whether the range of parameter estimates from the MAR model is sufficiently wide to give acceptable support to the NMAR distribution—the key assumption for the method. In Carpenter and Kenward (2013, Sec. 10.6.1) there is a description of some graphical approaches for exploring this.

In conclusion, this analysis, by reweighting after MAR imputation, confirms the results for the postal arm are sensitive to poorer reviewers not returning the second paper and hence withdrawing. Specifically, if the adjusted log odds of returning the second paper increases by 0.5 (OR 1.65) for every point increase in the RQI, the confidence interval for the postal intervention comes close to the null value. Analysis using `WinBUGS` and weighting give similar conclusions. However, the latter is far quicker, even taking into account the time taken to draw the MAR imputations, and requires no specialist programming. In comparing the results of this analysis with those presented in Table 19.2, the principle difference is in the standard errors. The reason for this is that in the selection modelling, we have fixed the

TABLE 19.10
Peer review trial. Estimated effects of the postal intervention versus the control on the mean review quality index (RQI). All models adjusted for baseline RQI. Uncertainty in parameter estimates from `WinBUGS` *due to Monte Carlo estimation is less than ±0.001.*

Method	Est.	s.e.
Withdrawal at Random		
regression	0.236	0.070
multiple imputation, $M = 1000$	0.237	0.070
regression using `winBUGS`	0.236	0.070
Withdrawal Not at Random, $\alpha_4 = 0.3$		
Bayesian	0.215	0.071
weighting, $M = 50$	0.209	0.066
weighting, $M = 150$	0.201	0.069
weighting, $M = 250$	0.205	0.068
weighting, $M = 1000$	0.205	0.069
Withdrawal Not at Random, $\alpha_4 = 0.5$		
Bayesian	0.202	0.071
weighting, $M = 50$	0.205	0.063
weighting, $M = 150$	0.197	0.067
weighting, $M = 250$	0.199	0.066
weighting, $M = 1000$	0.195	0.067

sensitivity parameter α_4 at certain values, where as in the pattern-mixture modelling, the corresponding sensitivity parameter, δ, had a distribution.

19.5.3 Further developments

Carpenter et al. (2011) apply this re-weighting approach to publication bias in meta-analysis. Here, careful choice of the imputation distribution is important, especially if there is marked asymmetry in the funnel plot and only a few studies. They report encouraging results from a simulation study and good agreement with a full Bayesian model where the latter is fit using standard software. The advantage of the re-weighting approach in this context is its ability to rapidly handle otherwise computationally awkward (though practically relevant) selection mechanisms.

Bousquet et al. (2011) apply the method to sensitivity analysis for missing covariates. Here, after imputation under MAR, the selection model allows investigation of dependence of the missingness mechanism on the covariate. A separate sensitivity analysis performed for each covariate. The substantive model is a logistic regression. In line with the theoretical results noted at the end of Carpenter and Kenward (2013, Ch. 1), where the complete records analysis suggests missingness does not involve the response, corresponding log-odds ratios are relatively unchanged by MAR imputation or sensitivity analysis. Where selection does involve the response, coefficients change after MAR analysis, and the method allows rapid exploration of sensitivity to NMAR.

19.6 Discussion

The essential ambiguity that accompanies statistical analyses when data are missing implies an important role for sensitivity analysis. We have seen, as set out in Chapters 16, 17, 18, and 20, that there are many alternative ways in which such sensitivity analyses can be constructed. As well there are different classes of sensitivity analysis that have rather different aims. Two important examples were described at the start of this chapter. In the first, the analysis model matches the postulated data generating mechanism. In the second, an original (or primary) analysis is retained under one or more alternative but, at the same time incompatible data generating mechanisms. Both classes have an important role to play, but do correspond to different aims.

In this chapter, a range of sensitivity analyses has been explored in which multiple imputation (MI) plays a central role. In the first part, pattern-mixture-based approaches were described. Given the relationship between assessing departures from MAR through the behaviour of the unobserved data and the form of the imputation model, this can be regarded as a very natural way of using MI for sensitivity analyses. It has also been shown how, with approximation, MI can be used to construct sensitivity analyses through the selection model formulation. Up to this point this latter approach has proved more limited in scope than the former. At one point a relationship with a particular likelihood sensitivity analysis was noted for the pattern-mixture approach1 (Section 19.4).

One strong advantage of the pattern-mixture approach is that the assumptions under which the NMAR scenarios are constructed can be readily understood by non-statistically trained experts. Specifically, distributions estimated from the complete records, together with imputed values under MAR, can be plotted as a starting point for discussion about NMAR distributions. By contrast, in the selection model framework, the alternative assumptions are typically expressed in terms of the adjusted log-odds ratio, or a function of these, that relates the chance of observing a variable to its underlying (possibly unseen) value. This is typically less directly linked to the underlying substantive setting that the statistical behaviour of the missing data. A further advantage of the MI approach is that it is relatively easy to link the two frameworks, if this is required, through an exploration of the implied selection model following a pattern-mixture-based analysis, and vice versa. For example, having created imputed datasets using the pattern-mixture approach under NMAR, we can fit selection models to the imputed data, combine the results using Rubin's rules, and check with experts they are contextually plausible. In the opposite direction, Carpenter et al. (2011) illustrate the calculation of weighted averages of the imputed data to graph the difference in means between the observed and imputed data as selection increases.

A feature of the MI-based pattern-mixture approach for longitudinal data described in this chapter is the avoidance of explicit specification of sensitivity parameters by specifying instead patterns of profiles. This seems most natural in the clinical trials setting, where there are data from both the active and control arms and the appropriate approach to imputation depends on whether interest lies in de jure or de facto questions. As we argued above, in this setting, post-deviation distributions can be accessibly framed with reference to the various imputation arms. We note however, that although motivated by, and developed within, a clinical trial framework, this approach can in principle be applied much wore widely.

In conclusion, the MAR assumption provides a natural starting point for sensitivity analysis, with the implication that conditional distributions of partially observed given fully observed data do not differ by missingness pattern. Moving to NMAR mechanisms, the potential flexibility, and hence broad utility, of an MI-based approaches in constructing NMAR sensitivity analyses has been demonstrated. The approach is a highly practical one, and has the great advantage of allowing the user to focus on the key assumptions that are linked directly to the substantive setting, and on how these should be varied for different sensitivity scenarios. For these reasons the broad approach deserves to be widely adopted.

Acknowledgments

We are grateful to Astra-Zeneca for permission to use the data from the clinical trial to assess the efficacy and safety of budesonide on patients with chronic asthma. We are also grateful to Vernon Gayle for introducing us to the Youth Cohort Time Series for England, Wales and Scotland, 1984–2002. This is published by, and freely available from, the UK Data Archive.

References

Bartlett, J.A. and Shao, J.F. (2009). Successes, challenges, and limitations of current antiretroviral therapy in low-income and middle-income countries. *Lancet Infections Diseases* **9**, 637–649.

Boulle, A., Bock, P., Osler, M., Cohen, K., Channing, L., Hilderbrand, K., Mothibi, E., Zweigenthal, Z., Slingers, N., Cloete, K., and Abdullah, F. (2008). Antiretroviral therapy and early mortality in South Africa. *Bulletin of the World Health Organization*, **86**, 657–736.

Bousquet, A.H., Desenclos, J.C., Larsen, C., Le Strat, Y., and Carpenter, J.R. (2011). Practical considerations for sensitivity analysis after multiple imputation applied to epidemiological studies with incomplete data. *BMC Medical Research Methodology* **12**, 73.

Brinkhof, M.W.G., Dabis, F., Myer, L., Bangsberg, D.R., Boulle, A., Nash, D., Schechter, M., Laurent, C., Keiser, O., May, M., Sprinz, E., Egger, M., and Anglaretb, X. (for the ART-LINC of IeDEA collaboration) (2008). Early loss of HIV-infected patients on potent antiretroviral therapy programmes in lower-income countries. *Bulletin of the World Health Organization* **86**, 497–576.

Brinkhof, M.W.G., Spycher, B.D., Yiannoutsos, C., Weigel, R., Wood, R., Messou, E., Boulle, A., Egger, M., and Sterne, J.A.C. (2010). Adjusting mortality for loss to follow-up: analysis of five ART programmes in Sub-Saharan Africa. *PLOS ONE* **5**, e14149.

Busse, W.W., Chervinsky, P., Condemi, J., Lumry, W.R., Petty, T.L., Rennard, S., and Townley, R.G. (1998). Budesonide delivered by Turbuhaler is effective in a dose-dependent fashion when used in the treatment of adult patients with chronic asthma. *Journal of Allergy and Clinical Immunology* **101**, 457–463.

Carpenter, J.R. and Kenward, M.G. (2013). *Multiple Imputation and Its Application.* Chichester: John Wiley & Sons.

Carpenter, J.R., Kenward, M.G., and White, I.R. (2007). Sensitivity analysis after multiple imputation under missing at random: A weighting approach. *Statistical Methods in Medical Research* **16**, 259–275.

Carpenter, J.R., Roger, J.H., and Kenward, M.G. (2013). Analysis of longitudinal trials with protocol deviations: A framework for relevant, accessible assumptions and inference via multiple imputation. *Journal of Biopharmaceutical Statistics* **23**, 1352–1371.

Carpenter, J.R., Rücker, G., and Schwarzer, G. (2011). Assessing the sensitivity of meta analysis to selection bias: A multiple imputation approach. *Biometrics* **67**, 1066–1072.

Committee for Medicinal Products for Human Use (2010). *Guideline on Missing Data in Confirmatory Clinical Trials.* London: European Medicines Agency.

Croxford, L., Ianelli, C., and Shapira, M. (2007). Documentation of the Youth Cohort Time-Series Datasets, UK Data Archive Study Number 5765, Economic and Social Data Service.

Ford, I., Norrie, J., and Ahmadi, S. (1995). Model inconsistency, illustrated by the Cox proportional hazards model. *Statistics in Medicine* **14**, 735–746.

Greenland, S., Robins, J.M., and Pearl, J. (1999). Confounding and collapsibility in causal inference. *Statistical Science* **14**, 29–46.

Keene, O. (2011). Intent-to-treat analyses in the presence of off-treatment or missing data. *Pharmaceutical Statistics* **1**, 191–195.

Kenward, M.G. (1998). Selection models for repeated measurements with nonrandom dropout: an illustration of sensitivity. *Statistics in Medicine* **17**, 2723–2732.

Kenward, M.G. and Carpenter, J.R. (2008). Multiple Imputation. In: *Longitudinal Data Analysis: A Handbook of Modern Statistical Methods.* M. Davidian, G. Fitzmaurice, G. Verbeke, and G. Molenberghs (Eds.) Boca Raton: Chapman & Hall/CRC, pp. 477–500.

Kenward, M.G. and Molenberghs, G. (1998). Likelihood based frequentist inference when data are missing at random. *Statistical Science* **13**, 236–247.

Little, R.J.A. and Yau, L. (1996). Intent-to-treat analysis for longitudinal studies with drop-outs. *Biometrics* **52**, 1324–1333.

Magder, L.S. (2003). Simple approaches to assess the possible impact of missing outcome information on estimates of risk ratios, odds ratios, and risk differences. *Controlled Clinical Trials* **24**,411–421.

Mallinckrodt, C.H. (2013). *Preventing and Treating Missing Data in Longitudinal Clinical Trials.* Cambridge University Press.

Mallinckrodt, C.H. and Kenward, M.G. (2009). Conceptual considerations regarding choice of endpoints, hypotheses, and analyses in longitudinal clinical trials. *Drug Information Journal* **43**, 449–458.

Mallinckrodt, C.H., Lin, Q., and Molenberghs, G. (2012). A structured framework for assessing sensitivity to missing data assumptions in longitudinal clinical trials. *Pharmaceutical Statistics* **12**, 1–6.

Molenberghs, G., Michiels, B., Kenward, M.G., and Diggle, P.J. (1998). Missing data mechanisms and pattern-mixture models. *Statistica Nederlandica* **52**, 153–161.

National Institute for Health and Care Excellence (NICE) (2011). *Glossary accessed September 4th 2013 at http://www.nice.org.uk/website/glossary/glossary.jsp.*

National Research Council (2010). *The prevention and Treatment of Missing Data in Clinical Trials.* Panel on Handling Missing Data in Clinical Trials. Committee on National Statistics, Division of Behavioral and Social Sciences and Education. Washington, DC: The National Academies Press.

O'Kelly, M. and Ratitch, B. (2014). *Clinical Trials with Missing Data: A Guide for Practitioners.* Chichester: John Wiley & Sons.

Rubin, D.B. (1987). *Multiple Imputation for Nonresponse in Surveys.* New York: John Wiley & Sons.

Saltelli, A., Chan, K., and Scott, E.M. (2008). *Sensitivity Analysis.* Chichester: John Wiley & Sons.

Schroter, S., Black, N., Evans, S., Carpenter, J.R., Godlee, F., and Smith, R. (2004). Effects of training on quality of peer review: Randomised controlled trial. *British Medical Journal* **32**, 673–675.

Teshome, B., Lipkovich, I., Molenberghs, G., and Mallinckrodt, C. (2014). A multiple imputation based approach to sensitivity analysis and effectiveness assessments in longitudinal clinical trials. *Journal of Biopharmaceutical Statistics* **24**, 211–228.

UK Data Archive (2007). *Youth Cohort Time Series for England, Wales and Scotland, 1984-2002 [computer file]. First Edition.* Colchester, Essex: UK Data Archive [distributor], November 2007. SN 5765.

White, I., Carpenter, J.R., Evans, S., and Schroter, S. (2007). Eliciting and using expert opinions about non-response bias in randomised controlled trials. *Clinical Trials* **4**, 125–139.

Yan, X., Lee, S., and Li, N. (2009). Missing data handling methods in medical device clinical trials. *Journal of Biopharmaceutical Statistics* **19**, 1085–1098.

20

Sensitivity Analysis: The Elicitation and Use of Expert Opinion

Ian R. White

MRC Biostatistics Unit, Cambridge, UK

CONTENTS

20.1 Introduction

Chapters 15–19 have shown that analyses with missing data generally rest on untestable assumptions and that sensitivity analysis is required to explore sensitivity of results to departures from these untestable assumptions. We have also seen that a principled approach to sensitivity analysis involves fitting a model containing one or more sensitivity parameters which express the extent of departure from the assumption made in the main analysis. These sensitivity parameters are typically unidentified by the data and must be specified by the data analyst.

In some datasets, a sensitivity analysis over the whole range of sensitivity parameters (a best case-worst case analysis) may give useful information. However, usually the interval of ignorance (Chapter 16) is too wide to be useful and accommodates assumptions too extreme to be plausible. It is therefore necessary to restrict sensitivity parameters to a plausible range.

In this chapter, we will see how this plausible range of values of the sensitivity parameters may be chosen. The methods will elicit a probability distribution for the sensitivity parameters, so either frequentist methods (fixing the sensitivity parameters at various plausible values) or Bayesian methods (using the elicited distribution as a prior) may be used.

Crucially, choice of a plausible range of values of the sensitivity parameters is not just a statistical decision. It requires both statistical understanding and understanding of the scientific context, so it usually requires close collaboration between statisticians and their scientific colleagues. This process will be called "elicitation," and the elicited scientist's opinion an "expert opinion." The methods described in this chapter apply very widely, but they will be illustrated in the context of randomised clinical trials where interest lies in estimating the difference in outcome between two or more randomised groups.

Section 20.2 starts by describing the psychological literature on judgement under uncertainty, which pinpoints some biases to be avoided in elicitation. Section 20.3 proposes a pattern-mixture model which is well-suited to sensitivity analysis because the sensitivity parameter is easily interpreted. Sections 20.4 and 20.5 describe the elicitation process itself. Section 20.6 discusses practicalities and Section 20.7 considers possible extensions.

20.2 Background on Elicitation

There is a very large statistical and psychological literature on eliciting expert opinions about unknown quantities, summarised by O'Hagan et al. (2006). This section briefly introduces one key issue relevant to eliciting expert opinions about the sensitivity parameters.

The "heuristics and biases" research programme of the psychologists Amos Tversky and Daniel Kahneman showed that human judgements can easily be distorted (Tversky and Kahneman 1974). Three main *heuristics* (ways in which humans make judgements under uncertainty) can lead to *biases* (errors in those judgements):

Availability, in which the probability of an event A is over-estimated if examples of A come readily to mind;

Representativeness, in which the conditional probability of event A given another event B is overestimated if B is perceived as representative of A; and

Adjustment and anchoring, in which a numerical quantity is estimated by adjusting from an easily available value, and bias occurs because adjustment is typically insufficient.

This chapter is concerned with eliciting a range of plausible values or a probability distribution for one or more sensitivity parameters, so adjustment and anchoring is perhaps the largest concern.

For example, the author's first attempt at elicitation for sensitivity parameters (White et al. 2007) was based on a previously published questionnaire designed to elicit prior opinions about the treatment effect in a clinical trial (Parmar, Spiegelhalter, and Freedman 1994). It asked the expert to assess the probability that the sensitivity parameter fell into nine categories ranging from < -1 to $> +1$. These bounds of ± 1 were arbitrarily chosen and may have acted as anchors which discouraged experts from reporting either the view that the sensitivity parameter could be much larger than ± 1, or the view that the sensitivity parameter is very likely to be much smaller than ± 1. In later work, the author has started by asking the experts themselves to suggest a suitable extreme value, a say, and then asked them to assess categories ranging from $< -a$ to $> +a$.

20.3 How to Parameterise a Model to Elicit Expert Opinion

Given that statisticians must discuss the values of the sensitivity parameters in detail with non-statistical colleagues, it is important that sensitivity parameters are easily interpretable. This has implications for the form of the statistical model used. Here we consider the modelling options for a randomised controlled trial in which a quantitative outcome Y is measured at a single time point: other types of outcome are considered in the discussion.

Sensitivity analysis may be conducted around any assumption, but for simplicity, we consider sensitivity analyses around a missing at random (MAR) assumption, so that the sensitivity parameters describe the degree of departure from MAR. Because the sensitivity analysis does not assume MAR, it requires a joint model of the incomplete data and the missing data mechanism. The two main classes of joint model are the selection model and the pattern-mixture model (Chapter 4). For a continuous outcome Y, the sensitivity parameter has different interpretations in these two models: in a selection model (here assumed to be a logistic regression), it is the increase in the log odds of missingness for a one-unit increase in Y; in a pattern-mixture model it is the difference between the mean of the missing data and the mean of the observed data. The pattern-mixture parameter is far easier to interpret by non-statisticians, and is adopted here.

20.3.1 Pattern-mixture model

Suppose n_I individuals are randomised to the intervention arm. Let Y_{Ii} be the outcome for the ith individual in the intervention arm, $i = 1, \ldots, n_I$. Let $R_{Ii} = 0$ if this individual drops out, so that Y_{Ii} is unseen, and $R_{Ii} = 1$ if this individual completes the trial, so that Y_{Ii} is observed. Denote by $\pi_I = \Pr(R_{Ii} = 0)$ the probability of dropout in the intervention arm. For the control arm, define n_C, Y_{Ci}, R_{Ci}, and π_C similarly.

For the intervention arm, our model is that observed outcomes come from a distribution with mean μ_I and variance σ^2, while unobserved outcomes come from a different distribution with mean $(\mu_I + \delta_I)$ and variance σ_M^2. This can be written

$$\begin{aligned} Y_{Ii}|R_{Ii} = 1 &\sim (\mu_I, \sigma^2), \\ Y_{Ii}|R_{Ii} = 0 &\sim (\mu_I + \delta_I, \sigma_M^2). \end{aligned} \tag{20.1}$$

For the control arm, define μ_C and δ_C analogously, and assume the variances are equal to those in the control arm (although this can readily be relaxed if desired). Thus δ_I and δ_C are the two sensitivity parameters governing the degree of non-random missingness; σ_M is not a sensitivity parameter because we will see below that it does not contribute to the analysis. The MAR assumption corresponds to the case $\delta_C = \delta_I = 0$. It is important to allow δ_I and δ_C to differ: for example, missingness may well be more informative among individuals who have been encouraged to change their behaviour than among controls.

Under model (20.1), the average outcome in the control arm is $\mu_C + \pi_C\delta_C$. Likewise, the average outcome in the intervention arm is $\mu_I + \pi_I\delta_I$. The true effect of the intervention is then

$$\begin{aligned} \Delta &= (\mu_I + \delta_I\pi_I) - (\mu_C + \delta_C\pi_C) \\ &= \Delta^{CC} + (\delta_I\pi_I - \delta_C\pi_C), \end{aligned} \tag{20.2}$$

where $\Delta^{CC} = (\mu_I - \mu_C)$ is the apparent treatment effect amongst complete cases. Thus the true treatment effect equals the apparent treatment effect in complete cases plus bias due to informative dropout.

An extension to handle baseline covariates is reasonably straightforward and is described in White et al. (2007).

20.3.2 Prior

Assume normal priors $\delta_I \sim N(m_I, s_I^2)$ and $\delta_C \sim N(m_C, s_C^2)$. Since knowing the value of δ_I is likely to help experts to predict δ_C, we allow δ_I and δ_C to have correlation r. A bivariate normal prior is therefore assumed:

$$\begin{pmatrix} \delta_I \\ \delta_C \end{pmatrix} \sim N\left(\begin{pmatrix} m_I \\ m_C \end{pmatrix}, \begin{pmatrix} s_I^2 & rs_Is_C \\ rs_Is_C & s_C^2 \end{pmatrix}\right). \tag{20.3}$$

Here, m_I, s_I, m_C, s_C, and r are numbers which are elicited from experts using the methods described in Sections 20.4 and 20.5 below.

20.3.3 Estimation: Sensitivity analysis

In a sensitivity analysis, (δ_I, δ_C) are treated as known parameters and varied over a range of values consistent with the prior (20.3). Δ in equation (20.2) can be estimated by correcting the usual complete-cases estimate $\widehat{\Delta}^{CC}$ (the difference between the observed means in the two groups) to take account of informative dropout:

$$\widehat{\Delta} = \widehat{\Delta}^{CC} + B, \tag{20.4}$$

where

$$B = \delta_I\widehat{\pi}_I - \delta_C\widehat{\pi}_C \tag{20.5}$$

and $\widehat{\pi}_I$ and $\widehat{\pi}_C$ are the observed dropout fractions in the intervention and control arms.

Similarly, the squared standard error $\widehat{\text{var}}\left(\widehat{\Delta}\right)$ is computed by correcting the squared standard error from the complete-cases analysis $\widehat{\text{var}}\left(\widehat{\Delta}^{CC}\right)$:

$$\widehat{\text{var}}\left(\widehat{\Delta}\right) = \widehat{\text{var}}\left(\widehat{\Delta}^{CC}\right) + V, \tag{20.6}$$

where

$$V = \frac{\delta_I^2\widehat{\pi}_I(1-\widehat{\pi}_I)}{n_I} + \frac{\delta_C^2\widehat{\pi}_C(1-\widehat{\pi}_C)}{n_C}. \tag{20.7}$$

Sensitivity analysis is illustrated in Section 20.5.

20.3.4 Estimation: Bayesian analysis

A fully Bayesian analysis requires priors for all the parameters $(\delta_I, \delta_C, \mu_I, \mu_C, \pi_I, \pi_C, \sigma^2)$. There is a strong case for restricting the use of prior information to the unidentified parameters (δ_I, δ_C), so that non-informative priors are used for $(\mu_I, \mu_C, \pi_I, \pi_C, \sigma^2)$, independent of δ_I and δ_C (White et al. 2007). σ_M does not appear in expressions for the posterior mean and variance of the overall treatment effect, and therefore requires no prior.

The treatment effect is estimated using equation (20.2). Using the prior distribution (20.3), the corrections required to compute the posterior mean and variance of the treatment effect for non-random dropout are

$$\text{Posterior mean} \quad = \quad \widehat{\Delta}^{CC} + B, \qquad (20.8)$$

$$\text{Posterior variance} \quad = \quad \widehat{\text{var}}\left(\widehat{\Delta}^{CC}\right) + V_1 + V_2. \qquad (20.9)$$

From (20.2), the correction term for the point estimate in (20.8) is

$$B = m_I \widehat{\pi}_I - m_C \widehat{\pi}_C, \qquad (20.10)$$

which uses the experts' best estimates of the values of δ_I and δ_C, together with the observed dropout proportions. The variance correction terms in (20.9) are derived in White et al. (2007) as

$$V_1 \quad = \quad s_I^2 \widehat{\pi}_I^2 - 2r s_I s_C \widehat{\pi}_I \widehat{\pi}_C + s_C^2 \widehat{\pi}_C^2 \qquad (20.11)$$

$$V_2 \quad = \quad \frac{(m_I^2 + s_I^2)\widehat{\pi}_I(1 - \widehat{\pi}_I)}{n_I} + \frac{(m_C^2 + s_C^2)\widehat{\pi}_C(1 - \widehat{\pi}_C)}{n_C}. \qquad (20.12)$$

V_1 allows for uncertainty about δ_I and δ_C using the prior variances s_I^2, s_C^2, and their correlation r, and hence does not depend on sample size. V_2 allows for uncertainty about $\widehat{\pi}_I$ and $\widehat{\pi}_C$ and hence decreases with sample size: in practice, V_2 is often negligible compared with V_1. By approximating the posterior distribution as a normal distribution, approximate 95% credible intervals can be calculated as posterior mean $\pm$ 1.96 $\times\sqrt{\text{posterior variance}}$.

We will see below that r can be hard to elicit. To explore its importance, we rewrite $V_1 = (s_I \widehat{\pi}_I - s_C \widehat{\pi}_C)^2 + 2(1-r) s_I s_C \widehat{\pi}_I \widehat{\pi}_C$. Thus the value of r strongly affects V_1 when $s_I \widehat{\pi}_I \approx s_C \widehat{\pi}_C$ but is less important when $s_I \widehat{\pi}_I \gg s_C \widehat{\pi}_C$ or $s_I \widehat{\pi}_I \ll s_C \widehat{\pi}_C$. The value of r does not affect the point estimate correction B or the other variance correction term V_2.

Bayesian analysis is illustrated in Section 20.4.

20.4 Eliciting Expert Opinion about a Single Sensitivity Parameter

20.4.1 The peer review trial

The peer review trial was a single blind randomised controlled trial comparing two different types of training versus no training for reviewers for the *British Medical Journal (BMJ)* (Schroter et al. 2004a). Each participating reviewer was pre-randomised into one of three groups. However, prior to any training, each was sent a baseline article to review (termed paper 1). If this was returned, then according to their randomised group, the reviewer either (i) was invited to participate in a full day's face to face training, (ii) was mailed a

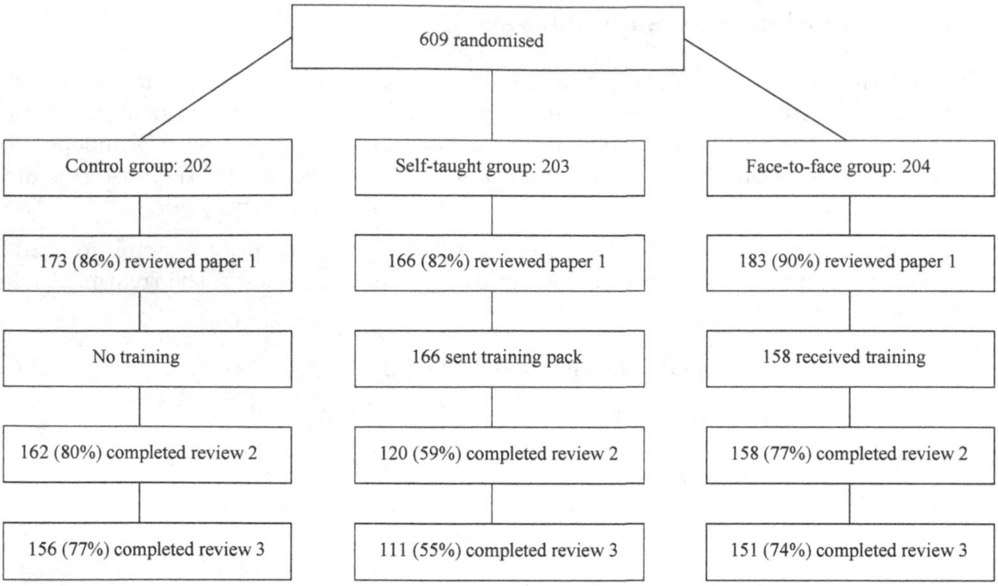

FIGURE 20.1
Progress of participants through the peer review trial. All percentages relate to the total number randomised in a given group.

self-taught training package, or (iii) received no further intervention. Two to three months later, participants who had completed their first review were sent a further article to review (paper 2); if this was returned, a third paper was sent three months later (paper 3). The analysis excluded all participants who did not complete their first review: this was not expected to cause bias since these participants were unaware of their randomised allocation.

Reviewers were sent manuscripts in a similar style to the standard *BMJ* request for a review, but were told these articles were part of the study and were not paid. The three articles were based on three previously published papers, with original author names, titles and location changed. In addition, 9 major and 5 minor errors were introduced (Schroter et al. 2004b).

The outcome considered here is the quality of the review, measured by the Review Quality Instrument (RQI) (van Rooyen et al. 1998, 1999; Walsh et al. 2000). This validated instrument contains 8 items scored from 1 to 5 (van Rooyen, Black, and Godlee 1999) and was rated independently by two editors. The primary results paper used the mean of the first seven items, averaged over the two editors. This mean is approximately normally distributed.

The progress of participants through the trial is summarised in Figure 20.1. A substantially higher proportion of reviewers dropped out in the self-taught arm. The published analysis, based on analysis of covariance of complete cases, showed statistically significant differences between groups for paper 2 but not for paper 3. Here we focus on the results for paper 2, following a previous analysis (White et al. 2007).

Table 20.1 shows the mean RQI at paper 1 for those who did, and did not, complete the review of paper 2. It suggests that the missing observations may not be missing completely at random; furthermore, in contrast to the control group, in the intervention groups reviewers who dropped out tended to have worse RQI scores for paper 1. This is particularly

TABLE 20.1
Peer Review Trial. Review Quality Index of paper 1 by whether or not paper 2 was completed.

		Control Group	Self-Taught Group	Face-to-Face Group
Completed review of paper 2	n	162	120	158
	mean	2.65	2.80	2.75
	SD	0.81	0.62	0.70
Did not complete review of paper 2	n	11	46	25
	mean	3.02	2.55	2.51
	SD	0.50	0.75	0.73

important in the self-taught arm, because the number of reviewers dropping out in this arm is substantially greater.

When the investigators saw these results, they approached *BMJ* editorial staff to elicit prior information on the difference between missing and observed outcomes in this study, with a view to performing a Bayesian analysis to assess the impact of dropout.

20.4.2 Elicitation

The elicitation questionnaire (Table 20.2) considered a single sensitivity parameter δ, without distinguishing δ_I from δ_C. Focusing on a common δ in this way simplified the elicitation process, but complicated the analysis discussed below which allows δ_I and δ_C to differ. Section 20.5 discusses the elicitation of δ_I and δ_C separately.

The elicitation questionnaire was completed by 22 individuals: 2 investigators, 12 *BMJ* editors and editorial staff, and 8 other *BMJ* staff. Unlike the *BMJ* staff, the two investigators had seen the data before completing the questionnaire: this could have influenced their assessments, but they gave priors broadly in line with the *BMJ* staff, so they were retained in the sample. 95% of the 22 prior means were less than or equal to 0, and 50% of these were between -0.3 and -0.05.

It was considered reasonable to pool the experts by averaging their prior distributions (Genest and Zidek 1986). The resulting pooled prior distribution is shown in Figure 20.2: it has mean -0.21 and standard deviation 0.46. The average within-questionnaire standard deviation was 0.4, so an approximate intra-class correlation was 0.76, suggesting reasonable consensus among the experts.

20.4.3 Analysis

Two adaptations were needed to allow δ_I and δ_C to differ despite the elicitation of a single sensitivity parameter δ. Firstly, we assumed that the elicited distribution applied both to δ_C and δ_I, so that $m_C = m_I$ and $s_C = s_I$. Results below would be sensitive to departures from both of these assumptions, especially the former. Secondly, because information about the correlation r between δ_C and δ_I was not elicited, values $r = 0$, 0.5, and 1 were assumed.

The Bayesian analysis described in Section 20.3 required a normal approximation to the prior. A second analysis used a fully Bayesian Monte Carlo Markov Chain (MCMC) analysis

TABLE 20.2
Peer Review Trial. Questionnaire eliciting sensitivity parameter, before completion.

Suppose the mean review quality for reviewers who respond to the second and third (i.e., final) paper is 3, with standard deviation 0.5, so that about 95% of these responders have values between 2 and 4.
What is your expectation for the mean review quality for those who do not respond to either the second or the third paper?
To help you, the hypothetical example shows a rather idiosyncratic statistician who is convinced that non-responders will differ on average from responders by 1/2 or 3/4 points, but is not sure whether they will be better or worse than responder.

Mean review quality in reviewers who do not respond to either the 2nd or the 3rd paper (minimum 0, maximum 5)

	Non-responders compared with responders									
	worse				same	better				Total
	≥ 1	0.75	0.5	0.25		0.25	0.5	0.75	≥ 1	
Hypoth. example	0	25	25	0	0	0	25	25	0	100
Your answers										

based on the actual elicited prior in Figure 20.2, but it was only possible to fit the cases $r = 0$ (corresponding to independent δ_C and δ_I) and $r = 1$ (corresponding to $\delta_C = \delta_I = \delta$) (White et al. 2007). Table 20.3 shows the results. The normal approximation and MCMC analyses gave very similar results. Adjustment for non-random dropout reduced the magnitude of the estimated intervention effect (because missing data were likely to be worse than observed data and were commoner in the intervention arms) and increased the uncertainty (because the magnitude of departure from MAR is not known with any certainty). Lower values of the prior correlation r increased the uncertainty still more. Results were only moderately sensitive to the value of r, presumably because of the large imbalance in missing data fractions between arms ($s_I\widehat{\pi}_I \gg s_C\widehat{\pi}_C$ as discussed in Section 20.3.4).

20.5 A Spreadsheet to Elicit Expert Opinion about Two Sensitivity Parameters

We now discuss a trial with a substantial missing data problem, in which the author attempted to elicit the full prior in equation (20.3).

20.5.1 The Down Your Drink (DYD) Trial

Hazardous drinking in the general population is an important public health problem (Prime Minister's Strategy Unit 2004). Brief interventions are effective (Moyer et al. 2002) but hard

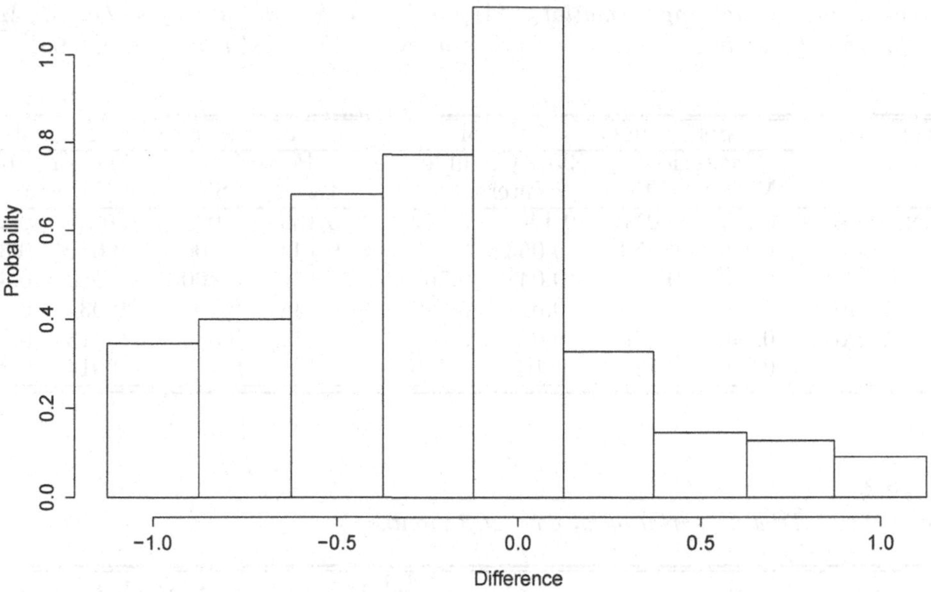

FIGURE 20.2
Peer Review Trial. Experts' prior distribution for δ, the difference between mean missing and observed RQI.

to implement. The Internet is increasingly used to deliver behaviour change interventions (Murray 2008), and a new "Down Your Drink" website was developed, building on psychological theories and aiming to engage users by providing interactive tools (Linke et al. 2008).

The "Down Your Drink" (DYD) trial was a randomised evaluation of the "Down Your Drink" website compared with a non-interactive control website providing information only (Murray et al. 2007). All stages of the trial—recruitment, randomisation, intervention and data collection—were conducted online. This presented a number of challenges (Murray et al. 2009), in particular whether sufficient numbers of participants would use the intervention website and would provide follow-up data (White, Kalaitzaki, and Thompson 2011).

The primary trial outcome, the TOT-AL, was the number of units of alcohol consumed in the past week, measured 3 months after randomisation. It was highly skewed but with some zeroes, so analyses used the transformation log(TOT-AL+1). The results shown here refer to the pilot phase and are unadjusted for baseline covariates; greater detail, including covariate adjustment, is given in (White, Kalaitzaki, and Thompson 2011).

20.5.2 Spreadsheet for each arm

The sensitivity parameter δ here corresponds to the difference between missing and observed individuals in transformed mean weekly alcohol consumption, and is expressed as

TABLE 20.3
Peer Review Trial. Posterior mean intervention effect, standard deviation and 95% credible intervals using normal approximation ("Approx") and MCMC analyses. For MCMC methods, the uncertainty in posterior means due to Monte Carlo estimation is < 0.001.

Analysis		Self-Taught vs. Control				Face-to-Face vs. Control			
		Posterior		95% Credible		Posterior		95% Credible	
		Mean	SD	Interval		Mean	SD	Interval	
Complete cases		0.291	0.077	0.140	0.442	0.160	0.071	0.021	0.299
r=0	Approx	0.246	0.153	−0.053	0.545	0.144	0.100	−0.052	0.341
	MCMC	0.246	0.151	−0.042	0.564	0.144	0.100	−0.050	0.344
r=0.5	Approx	0.246	0.140	−0.028	0.520	0.144	0.091	−0.033	0.322
r=1	Approx	0.246	0.126	−0.001	0.493	0.144	0.080	−0.013	0.301
	MCMC	0.246	0.126	0.004	0.505	0.145	0.080	−0.014	0.302

TABLE 20.4
Down Your Drink Trial. Description of data at 3 months

		Intervention ($n = 1880$)		Control ($n = 1866$)	
Responded	%	38%		46%	
TOT-AL (units/week)	mean (SD)	38.6	(32.6)	37.0	(32.5)
log(TOT-AL + 1)	mean (SD)	3.25	(1.12)	3.18	(1.18)

a *proportional* change in alcohol consumption (ignoring the 1 unit/week added). Thus if non-responders drank 10% more than responders, $\delta = \log 1.1$.

The spreadsheet in Table 20.5 was used to elicit five investigators' views about the sensitivity parameter and shows the results for one particular investigator (White, Kalaitzaki, and Thompson 2011). The spreadsheet requires the user to start by setting the group width in order to avoid the "adjustment and anchoring" bias described in Section 20.2: in Table 20.5 the group width has been set to 20% after discussion with this particular expert. From the values in Table 20.5, the lower quartile, median, and upper quartile of the prior distribution for δ_I for this expert were taken as −15%, 3% and 28% respectively.

A similar spreadsheet was used for the control arm, yielding for this expert a slightly different marginal prior for δ_C with lower quartile, median and upper quartile −19%, 0%, and 19%, respectively.

20.5.3 Spreadsheet for the correlation

The next task was to elicit a joint prior. This was done using the spreadsheet in Table 20.6, which simplifies the problem by considering just three possible values—the median, lower, and upper quartiles—for each of δ_I and δ_C. The expert was asked to give their prior probability for each of the resulting nine combinations of δ_I and δ_C. If, for example, the expert had $r = 0$, then the prior probabilities would show no association, while if the expert had

TABLE 20.5
Down Your Drink Trial. Spreadsheet to elicit δ_I, the sensitivity parameter in the intervention arm. Expert's responses are given in bold.

	Mean TOT-AL in Non-Responders Compared with Responders								
	Less than −70%	−70% to −50%	−50% to −30%	−30% to −10%	−10% to 10%	10% to 30%	30% to 50%	50% to 70%	More than 70%
wt.	**2**	**5**	**10**	**20**	**40**	**20**	**15**	**10**	**5**
	non-resp. drink less...				equal	... non-resp. drink more			

$r = 1$, then the off-diagonal cells would have zero prior probabilities. The prior correlation r was estimated as the weighted correlation in the table.

Table 20.6 shows the opinions of the same expert whose prior for δ_I was given in Table 20.5. The expert's weights do not sum to 100, but only the ratios of weights matter. They show a moderate preference for the diagonal cells, and the correlation r was estimated as +0.27.

20.5.4 Use of the elicited values

Figure 20.3 shows contours plots of the prior distribution elicited from 5 investigators: our previous expert was expert 3. The contours are asymmetrical because the parameters (δ_I, δ_C) are plotted on an exponentiated scale.

Using these contour plots, "extreme" sensitivity analyses were identified using points with $\exp(\delta_I)$ and $\exp(\delta_C)$ in multiples of 25% that lay within the 95% contour for 3 or more investigators and handling (δ_I, δ_C) symmetrically; similarly, "moderate" sensitivity analyses included points that lay within the 50% contour for 3 or more investigators. The extreme and moderate sensitivity analyses are indicated by black dots and open circles, respectively, in Figure 20.3.

The primary publication of the DYD trial reported results for each sensitivity analysis in a web appendix (Wallace et al. 2011). The results in Table 20.7, taken from a methods paper (White, Kalaitzaki, and Thompson 2011), show that results are reasonably robust to NMAR models with $\delta_C = \delta_I$, but they are much less robust to moderate departures from MAR in one arm only (especially with $\delta_C = 0, \delta_I \neq 0$), and they are hugely affected if extreme departures from MAR in one arm only are considered.

20.6 Practicalities

20.6.1 Choice of experts

It is important to choose appropriate experts: this may depend on the question being asked. In the peer review study, the authors wanted to quantify for the general reader how journal editors would interpret the results of the trial. In the DYD trial, the authors used the

TABLE 20.6
Down Your Drink Trial. Spreadsheet to elicit r, the prior correlation between the sensitivity parameters in the two arms. Expert's responses are given in bold.

Just focussing on your median, lower and upper quartiles, these were:

	Lower Quartile	Median	Upper Quartile
Intervention arm	Non-resp. drink 15% less than resp.	Non-resp. drink 3% more than resp.	Non-resp. drink 28% more than resp.
Control Arm	Non-resp. drink 19% less than resp.	Non-resp. drink the same as resp.	Non-resp. drink 19% more than resp.

Now please give your weights for the following 9 combinations:

	Control Arm		
Intervention arm	Non-resp. drink 19% less than resp.	Non-resp. drink the same as resp.	Non-resp. drink 19% more than resp.
Non-resp. drink 15% less than resp.	**5**	**2**	**5**
Non-resp. drink 3% more than resp.	**2**	**10**	**2**
Non-resp. drink 28% more than resp.	**2**	**5**	**10**

prior views of the investigators, as they know most about the new treatment; however investigators may be overly enthusiastic for the intervention.

20.6.2 Format of elicitation

The elicitations described in this chapter have been implemented in written questionnaires and in interactive spreadsheets, using face-to-face discussions, telephone discussions and by experts on their own, and with a single expert or a group. Some form of interactivity seems highly desirable, because the task is unfamiliar to scientists, so that careful questioning can avoid anchoring biases, and in order to check that elicited opinions do represent experts' views. The face-to-face format with the use of a computer to process elicited views seems ideal, but may be impractical in some circumstances.

20.6.3 Feedback

Feedback on elicited opinions is essential. For example, once expert 3 had given the results in Figure 20.5, a possible line of feedback would be based on the fitted median and quartiles, asking if the expert was happy that the following four statements were equally likely: the

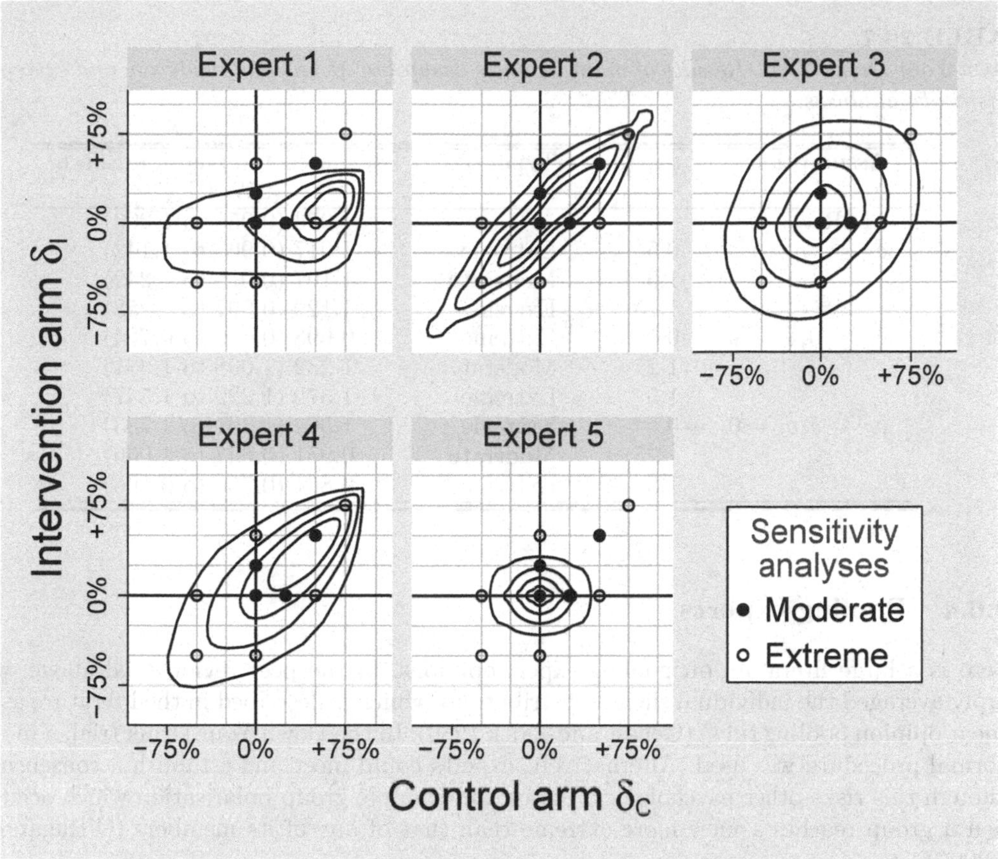

FIGURE 20.3
Down Your Drink Trial. Contour plot of expert priors for (δ_I, δ_C)*, showing 25%, 50%, 75% and 95% prior regions and the values chosen for moderate and extreme sensitivity analyses.*

average non-responder drinks more than 15% less; between 15% less and 3% more; between 3% more and 28% more; and more than 28% more than responders.

TABLE 20.7
Down Your Drink Trial. Results of main analysis assuming MAR, and moderate and extreme sensitivity analyses.

Assumption	exp (δ)	Type	Estimated Intervention Effect (95% CI)
MAR			1.073 (0.956 to 1.203)
$\delta_C = \delta_I = \delta$	0.5	Extreme	1.017 (0.905 to 1.142)
	1.5	Moderate	1.107 (0.986 to 1.242)
	1.75	Extreme	1.120 (0.997 to 1.258)
$\delta_C = 0, \delta_I = \delta$	0.5	Extreme	0.698 (0.622 to 0.784)
	1.25	Moderate	1.232 (1.098 to 1.381)
	1.5	Extreme	1.379 (1.229 to 1.547)
$\delta_C = \delta, \delta_I = 0$	0.5	Extreme	1.562 (1.391 to 1.753)
	1.25	Moderate	0.951 (0.847 to 1.066)
	1.5	Extreme	0.861 (0.768 to 0.966)

20.6.4 Pooling experts

There is a large literature on pooling expert opinions. In the peer review trial above, we simply averaged the individual elicited distributions, which is described in the literature as a "linear opinion pooling rule" (Genest and Zidek 1986). In the Down Your Drink trial, a more informal procedure was used. Alternatively, experts could meet and establish a consensus, although this risks other psychological biases, for example group polarisation which occurs when a group reaches a view more extreme than that of any of its members (O'Hagan et al. 2006).

20.6.5 When to elicit

Quantifying the characteristics of individuals with missing outcomes may be difficult for investigators at the start of the trial and may become easier as the trial progresses and knowledge develops, perhaps including personal knowledge of particular individuals who have dropped out of the study.

However, to avoid bias, it is desirable for all analyses in clinical trials to be specified at the design stage, and sensitivity analyses for missing data are no exception. Thus it is recommended that elicitation exercises such as those described here are conducted at the start of a study, before outcome data are collected. Prior opinions should be unaffected by the extent of missing data. If experts feel they need other information then it may be reasonable to elicit prior opinions about multiple parameters: for example, if experts want to know the distribution of reasons for missing data, then it may be reasonable to elicit prior opinions about reason-specific departures from MAR. Such pre-specification is especially important from a regulatory perspective.

Tackling the issue at the design stage also helps to focus attention on the issue of incomplete data and may help investigators to identify other ways to improve follow-up. For example, in the QUATRO trial (Jackson, White, and Leese 2010), when the investigators were asked their views about the missing data, one response was that instead we should collect carers' assessments of the patient's quality of life when patient reports were unavailable. We agreed to collect carers' assessments for all patients (not just those with missing outcomes), thus enabling statistical evaluation of how the carers' assessments should be used in the analysis, as well as eliciting views about the missing data. (Unfortunately, the carers' assessments turned out to be so weakly associated with the patients' assessments—Pearson correlation 0.31—that they were unhelpful for the analysis; Jackson, White, and Leese 2010.)

20.7 Discussion

Allowing different departures from MAR in the two randomised groups (that is, allowing $\delta_I \neq \delta_C$) is crucial: results are often very sensitive to this, as was demonstrated in both the peer review trial and the Down Your Drink trial. Eliciting a single δ, as in the peer review trial, is not recommended, and explicit efforts to elicit the prior correlation $r = \text{corr}(\delta_I, \delta_C)$ are needed.

It is tempting to allow a richer model than that considered above: for example, in the intervention arm, instead of applying a single sensitivity parameter $\delta_I \sim N(m_I, s_I^2)$ one might allow separate sensitivity parameters δ_{IM}, δ_{IW} for men and women. There is, however, a pitfall here. If δ_{IM} and δ_{IW} are both given a $N(m_I, s_I^2)$ prior with correlation $r_{MW} < 1$, then the overall difference between unobserved and observed data in the intervention arm has prior variance less than s_I^2, underestimating the impact of possible departures from MAR. Systematic departures from MAR should therefore be expressed assuming perfect correlations within trial arms ($r_{MW} = 1$).

This chapter has described methods for quantitative outcomes, but they can also be applied to binary outcomes. Various parameters can be used to describe departures from MAR, such as the "response probability ratio" (the ratio of the probability of response in successes to the probability of response in failures) (Magder 2003), or the ratio of response probability ratios (Greenland 2005). It is simpler and more natural to quantify departures from MAR through the odds ratio between outcome and missingness, the "informatively missing odds ratio" (IMOR) (Higgins, White, and Wood 2008). Model (20.3) may then be used with δ_I, δ_C defined as the IMORs in the two arms. Expressions corresponding to those in sections 20.3.3 and 20.3.4 can then be derived (Higgins, White, and Wood 2008; White, Wood, and Higgins 2008). Related methods have been applied to categorical data (Resseguier, Giorgi, and Paoletti 2011) and to time-to-event data (Jackson et al. 2013a).

We have considered an outcome measured at a single time. For repeated measures, the difficulty is in choosing suitable priors for all the different response patterns: for example, it seems likely that individuals who drop out immediately after baseline have a different degree of departure from MAR from individuals who drop out only at the final observation time. One suggestion is to elicit a prior for δ in individuals with no follow-up measurements, and then to assume that departure from MAR in individuals with more nearly complete data is some specified fraction of δ, possibly proportional to the variance of the outcome given the observed intermediate data (White et al. 2007).

We have focussed on analyses where MAR is the main analysis, but sensitivity analyses can be performed in other settings. For example, smoking cessation trials with successful quit as outcome commonly make a "missing=smoking" assumption, where the main analysis assumes the IMOR is 0. Sensitivity analyses would then vary the IMOR away from zero in one or both arms (Jackson et al. 2013b).

In contrast to our pattern-mixture model approach, Mason (2009) adopted a selection model approach (that is, a logistic regression for response on the outcome and covariates) and elicited the parameters of the selection model using ELICITOR software (Kynn 2005). Other software available to aid elicitation is the Sheffield Elicitation Framework (O'Hagan 2013).

The examples presented here have been in clinical trials. Other examples of the use of elicited opinions about missing data in medical research (including observational studies) are given in the literature (Shardell et al. 2008; Paddock and Ebener 2009, Wang et al. 2010; Jackson, White, and Leese 2010). However, the ideas presented here should apply equally well outside medicine, whenever scientific interest lies in the association between a particular covariate X and outcome Y. In this general setting, it remains important to consider both the likely magnitude of differences between unobserved and observed values of Y, and how much such differences may vary with X.

Finally, these methods are no substitute for collecting data as completely as possible. On the contrary, encouraging investigators to think about the potential impact of dropout at the planning stage, through eliciting a prior or otherwise, is a useful exercise: it raises awareness of the ways in which missing data can seriously undermine the definitiveness of a trial's conclusions, and encourages investigators to work to minimise dropout.

Acknowledgments

This work was supported by the Medical Research Council [Unit Programme number U105260558]. The author is grateful to his co-authors in previous papers, in particular James Carpenter; to the investigators of the peer review trial and the Down Your Drink trial for permission to use the data; and especially to those investigators who worked hard to quantify their views about missing data.

References

Genest, C. and Zidek, J.V. (1986). Combining probability distributions: A critique and an annotated bibliography. *Statistical Science* **1**, 114–148.

Greenland, S. (2005). Multiple-bias modelling for analysis of observational data. *Journal of the Royal Statistical Society, Series A* **168**, 267–306.

Higgins, J.P.T., White, I.R., and Wood, A.M. (2008). Imputation methods for missing outcome data in meta-analysis of clinical trials. *Clinical Trials* **5**, 225–239

Jackson, D., White, I.R., Carpenter, J., Baisley, K,, Evans, H., and Seaman, S. (2013a). Non-ignorable censoring in the proportional hazards regression model. *Submitted for publication.*

Jackson, D., White, I.R., Mason, D., and Sutton, S. (2013b). An improved method for handling missing binary outcome data in randomised controlled trials. *Submitted for publication.*

Jackson, D., White, I.R., and Leese, M. (2010). How much can we learn about missing data? An exploration of a clinical trial in psychiatry. *Journal of the Royal Statistical Society, Series A* **173**, 593–612.

Kynn, M. (2005). *Eliciting Expert Knowledge for Bayesian Logistic Regression in Species Habitat Modelling.* PhD Thesis, Queensland University of Technology. http://eprints.qut.edu.au/16041/, accessed 23 August 2013.

Linke, S., McCambridge, J., Khadjesari, Z., Wallace, P., and Murray, E. (2008). Development of a psychologically enhanced interactive online intervention for hazardous drinking. *Alcohol and Alcoholism* **43**, 669–674, 2008.

Magder, L.S. (2003). Simple approaches to assess the possible impact of missing outcome information on estimates of risk ratios, odds ratios, and risk differences. *Controlled Clinical Trials* **24**, 411–421.

Mason, A.J. (2009). *Bayesian Methods for Modelling Non-random Missing Data Mechanisms in Longitudinal Studies.* PhD thesis. Imperial College London. http://www.bias-project.org.uk/papers/AJMasonPhDThesis2009.pdf, accessed 23 August 2013.

Moyer, A., Finney, J.W., Swearingen, C.E., and Vergun, P. (2002). Brief interventions for alcohol problems: A meta-analytic review of controlled investigations in treatment-seeking and non-treatment-seeking populations. *Addiction* **97**, 279–292.

Murray, E. (2008). Internet-delivered treatments for long-term conditions: strategies, efficiency and cost-effectiveness. *Expert Review of Pharmacoeconomics and Outcomes Research* **8**, 261–272.

Murray, E., Khadjesari, Z., White, I.R., Kalaitzaki, E., Godfrey, C., McCambridge, J., Thompson, S., and Wallace, P. (2009). Methodological challenges in on-line trials. *Journal of Medical Internet Research* **11**, e9.

Murray, E., McCambridge, J., Khadjesari, Z., White, I.R., Thompson, S.G., Godfrey, C., Linke, S., and Wallace, P. (2007). The DYD-RCT protocol: An on-line randomised controlled trial of an interactive computer-based intervention compared with a standard information website to reduce alcohol consumption among hazardous drinkers. *BMC Public Health* **7**, 306.

O'Hagan, A., Buck, C.E., Daneshkhah, A., Eiser, R., and Gartwaite, P.H. (2006). *Uncertain Judgements: Eliciting Expert Probabilities.* Chichester: John Wiley & Sons.

O'Hagan, T. (2013). SHELF: the Sheffield Elicitation Framework. http://www.tonyohagan.co.uk/shelf/, accessed 23 August 2013.

Paddock, S.M. and Ebener, P. (2009). Subjective prior distributions for modeling longitudinal continuous outcomes with non-ignorable dropout. *Statistics in Medicine* **28**, 659–678.

Parmar, M.K., Spiegelhalter, D.J., and Freedman, L.S. (1994). The CHART trials: Bayesian design and monitoring in practice. *Statistics in Medicine* **13**, 1297–1312.

Prime Minister's Strategy Unit. (2004). *Alcohol Harm Reduction Strategy for England.* UK: Cabinet Office.
http://www.referrersguide.com/Guidelines/Alcohol Harm Reduction Strategy.pdf, accessed 23 August 2013.

Resseguier, N., Giorgi, R., and Paoletti, X. (2011). Sensitivity analysis when data are missing not-at-random. *Epidemiology* **22**, 282–283.

Schroter, S., Black, N., Evans, S., Carpenter, J., Godlee, F., and Smith, R. (2004a). Effects of training on the quality of peer review: A randomised controlled trial. *British Medical Journal* **328**, 673–675.

Schroter, S., Black, N., Evans, S., Carpenter, J., Godlee, F., and Smith, R. (2004b). Test papers, descriptions of errors and review quality instrument.
http://www.bmj.com/content/suppl/2004/03/23/328.7441.673.DC1, accessed 23 August 2013.

Shardell, M., Scharfstein, D.O., Vlahov, D., and Galai, N. (2008). Sensitivity analysis using elicited expert information for inference with coarsened data: Illustration of censored discrete event times in the AIDS link to intravenous experience (ALIVE) study. *American Journal of Epidemiology* **168**, 1460–1469.

Tvesky, A., and Kahneman, A., (1974). Judgment under uncertainty: Heuristics and biases. *Science* **185**, 1124–1131.

van Rooyen, S., Black, N., and Godlee, F. (1999). Development of the review quality instrument (RQI) for assessing peer reviews of manuscripts. *Journal of Clinical Epidemiology* **52**, 625–629.

van Rooyen, S., Godlee, F., Smith, R., Evans, S., and Black, N. (1998). The effect of blinding and unmasking on the quality of peer review: a randomised trial. *Journal of the American Medical Association* **280**, 234–237.

van Rooyen, S., Godlee, F., Evans, S., Black, N., and Smith, R. (1999). Effect of open peer review on quality of review and on reviewers' recommendations; A randomised trial. *British Medical Journal* **318**, 23–27.

Wallace, P., Murray, E., McCambridge, J., Kahdjesari, Z., White, I.R., Thompson, S.G., Kalaitzaki, E., Godfrey, C., and Linke, S. (2011). On-line randomized controlled trial of an Internet based interactive intervention for members of the general public with alcohol use disorders. *PLoS ONE* **6**, e14740.

Walsh, E., Rooney, M., Appleby, L., and Wilkinson, G. (2000). Open peer review: a randomised controlled trial. *British Journal of Psychiatry* **176**, 47–51.

Wang, C., Daniels, M.J., Scharfstein, D.O., and Land, S. (2010). A Bayesian shrinkage model for incomplete longitudinal binary data with application to the breast cancer prevention trial. *Journal of the American Statistical Association* **105**, 1333–1346.

White, I.R., Carpenter, J., Evans, S., and Schroter, S. (2007). Eliciting and using expert opinions about dropout bias in randomised controlled trials. *Clinical Trials* **4**, 125–139.

White, I.R., Kalaitzaki, R., and Thompson, S.G. (2011). Allowing for missing outcome data and incomplete uptake of randomised interventions, with application to an Internet-based alcohol trial. *Statistics in Medicine* **30**, 3192–3207.

White, I.R., Wood, A.M., and Higgins, J.P.T. (2008). Allowing for uncertainty due to missing data in meta-analysis — part 1: Two-stage methods. *Statistics in Medicine* **27**, 711–727.

Part VI

Special Topics

21

Special Topics: Introduction and Overview

Geert Molenberghs

Universiteit Hasselt & KU Leuven, Belgium

CONTENTS

21.1 Introduction

Earlier parts of the book were structured around inferential paradigms. In Part II, the focus was on likelihood and Bayesian methods. Because of their shared properties under ignorability, bringing these together is sensible, as explained in Chapters 1 and 3. Semiparametric methods were considered in Part III, while multiple imputation was treated as a proper subject in Part IV.

Part V was devoted to sensitivity analysis. The relevance of sensitivity analysis is widely recognized, precisely because the analysis of incomplete data, to a greater or lesser degree, rests upon non-verifiable assumptions about the distribution of the missing data, conditional upon the observed data. Assumptions made for the primary analysis differ qualitatively across inferential paradigms, which is why the aforementioned Parts II–III each have their own chapter in Part V.

These non-verifiable assumptions, and hence the implied sensitivity of ensuing inferences, bring non-trivial challenges, from a substantive as well as from a methodological standpoint. This is why we focus, in this part, on two substantive areas, clinical trials and sample surveys, as well as on the important topic of model diagnostics.

21.2 Missing Data in Clinical Trials

Apart from methodological issues, incomplete data in clinical trials also entail ethical considerations. The Helsinki Convention properly protects the rights of human study subjects, preventing them from being subjected to unethical experiments, ensuring that the risks do not outweigh the benefits. This is a delicate balance, continually monitored by regulators, data monitoring committees, and institutional review boards. The Convention allows both patients and healthy volunteer to withdraw from a study to which they have given consent. At the same time, physicians and other health professionals, as well as monitoring committees, may take the decision to remove some or all of the subjects from a study. As such, issues stemming from missing data in clinical studies will remain unavoidable.

Researchers have been confronted with incomplete trial data ever since the inception of the clinical trial itself. Standard texts, such as Little and Rubin (2002) as well as recent, targeted texts (Mallinckrodt 2013, Molenberghs and Kenward 2007, Little et al. 2010, Carpenter and Kenward 2013, O'Kelly and Ratich 2014) deal to a greater or lesser extent with the issue.

Early work focused on the practical issues implied by loss of balance stemming from incomplete data, and the computational consequences thereof. Also, the loss of precision induced by incomplete data was of concern. Near the end of the twentieth century, focus shifted towards induced biases. This awareness has been instrumental in the development of a variety of methods known and used, across the various inferential paradigms reviewed in the previous section, although motivation came from other areas as well. For over half a century, it has been of concern to reconcile the intention-to-treat principle with the analysis of incomplete sets of data.

A paradigm shift has taken place, moving away to a large extent from methods that are overly simplistic, such as complete case analysis, last observation carried forward, and other single imputation methods, replacing them with the methodology presented in Parts II–IV. Having a sufficiently wide and flexible array of methodological tools available is necessary but not sufficient. One also needs to carefully reflect on whether the proper question is asked and answered. This means that the estimand needs to be defined adequately and accurately. At the same time, regulatory requirements need to be met, such as specifying the analysis beforehand in study protocols and statistical analysis plans. This is compounded when sensitivity analysis is added to the analysis plan.

It is against this background that the National Research Council Report (Little et al. 2010, 2012) came into being and indeed was very welcome. The report carefully addresses both prevention as well as handling of incomplete data. The report clearly states the issues at stake and provides a taxonomy within which both the problems as well as solutions can be placed. Methodology from the various schools, organized along the methodological classification in this book's Parts II–IV, as well as in terms of sensitivity analysis (Part V), is discussed, with advantages and drawbacks highlighted. Arguably, the report takes a large step in a direction away from the simpler methods, focusing largely on methodology valid under MAR and even NMAR for primary analysis, thereby not losing sight of sensitivity analysis.

A very important component is the proper definition of estimands. This forces all involved to adopt the perspective of what one wants to learn about the population, rather than starting from what tools happen to be available. Carpenter, Roger, and Kenward (2013) make relevant contributions in this sense, allowing researchers to consider plausible scenarios for what could have happened after dropout, thereby making use of pattern-mixture modeling combined with multiple imputation.

Mallinckrodt (2013) and O'Kelly and Ratich (2014) operationalize a number of concepts laid out in the National Research Council Report. Craig Mallinckrodt also adopts this operational perspective in Chapter 22.

21.3 Missing Data in Surveys

Sample surveys, as well as other empirical studies in the social, behavioral, and educational areas, have in common with data from clinical and epidemiological studies that they virtually always involve human subjects. While incomplete data occur in virtually every area of research, both prevalence and severity are obviously exacerbated with studies in humans.

Sample surveys, whether conducted from a governmental, academic, or commercial perspective, are liable to sometimes extremely high fractions of incomplete data. It is therefore unsurprising that the handling of incomplete data in sample surveys has received considerable attention for at least half a century. The area has developed its own terminology, such as "list-wise deletion" and "case-wise deletion."

Whereas the older texts predominantly focus on design-based survey analysis, where simple estimators for means and totals are adjusted for the survey design (Cochran 1977, Kish 1965), more recently a partial paradigm shift has taken place towards model-based methodology (Skinner, Holt, and Smith 1989). The same is true for the handling of incomplete data. In Chapter 23, this is nicely described by Thomas Belin and Juwon Song.

As in the clinical-trials settings, methodology is now also available under various inferential paradigms. Belin and Song describe likelihood and Bayesian methodology, imputation-based strategies, and inverse probability weighted methods. Moreover, both multiple imputation and inverse probability weighting have their roots in survey sampling and censuses. Donald Rubin developed multiple imputation with, among others, the US Census in mind. The use of sample weights is as old as the occurrence of differential selection probabilities in survey samples. Such differential selection probabilities can result from stratification, multi-stage sampling, or on their own terms. Of course, one should not lose sight of fundamental differences between design weights, which are unambiguously known once the design is chosen, and weights used to address incompleteness in the data, unknown and to be estimated from the data based on only partially verifiable assumptions.

21.4 Model Diagnostics

Routinely, statistical analysis is preceded by data exploration and topped off with and assessment of goodness-of-fit and applying model diagnostics. Broadly, model diagnostics are concerned with how well a model fits the data as a whole on the one hand, and whether individual study subjects, individual observations, or small groups of these do not have undue influence on the inferences drawn on the other. Therefore, most diagnostics are based on comparing the data with the fitted model, be it parametric, semi-parametric, or non-parametric.

This is not straightforward when data are incomplete, as virtually all models describe more than just the data, but also "predict" the unobserved data given the observed ones. The implied sensitivities have been discussed at length in Part V. In a similar manner incompleteness complicates the construction of model diagnostics. This is already the case for (longitudinal) data with missingness only. An account of issues that occur in this setting is given in Verbeke, Molenberghs, and Beunckens (2008). Gelman *et al* (2005) proposed a multiple-imputation-based method to assess goodness-of-fit with incomplete data.

The matter becomes more complex when incomplete longitudinal data and censored time-to-event data are simultaneously collected. This topic was discussed by Rizopoulos, Verbeke, and Molenberghs (2010). Evidently, the topic of model diagnostics with incomplete data of some form is still in full development and does need additional work. A current state of the art is given by Dimitris Rizopoulos in Chapter 24.

References

Carpenter, J.R. and Kenward, M.G. (2013). *Multiple Imputation and Its Application.* New York: John Wiley & Sons.

Carpenter, J.R., Roger, J., and Kenward, M.G. (2013). Analysis of longitudinal trials with protocol deviation: A framework for relevant, accessible assumptions, and inference via multiple imputation. *Journal of Biopharmaceutical Statistics* **23**, 1352–1371.

Cochran, W.G. (1977). *Sampling Techniques.* New York: John Wiley & Sons.

Gelman, A., Van Mechelen, I., Verbeke, G., Heitjan, D., and Meulders, M. (2005). Multiple imputation for model checking: Completed-data plots with missing and latent data. *Biometrics* **61**, 74–85.

Kish, L. (1965). *Survey Sampling.* New York: John Wiley & Sons.

Little, R.J.A., D'Agostino, R., Dickersin, K., Emerson, S.S., Farrar, J.T., Frangakis, C., Hogan, J.W., Molenberghs, G., Murphy, S.A., Neaton, J.D., Rotnitzky, A., Scharfstein, D., Shih, W., Siegel, J.P., and Stern, H. National Research Council (2010). *The Prevention and Treatment of Missing Data in Clinical Trials. Panel on Handling Missing Data in Clinical Trials.* Committee on National Statistics, Division of Behavioral and Social Sciences and Education. Washington, D.C.: The National Academies Press.

Little, R., D'Agostino, R., Cohen, M.L., Dickersin, K., Emerson, S., Farrar, J.T., Frangakis, C., Hogan, J.W., Molenberghs, G., Murphy, S.A., Neaton, J.D., Rotnitzky, A., Scharfstein, D., Shih, W., Siegel, J.P., and Stern, H. (2012). The Prevention and Treatment of Missing Data in Clinical Studies. *New England Journal of Medicine* **367**, 1355–1360.

Little, R.J.A. and Rubin, D.B. (2002) *Statistical Analysis with Missing Data (Second Edition).* Chichester: Wiley.

Mallinckrodt, C.H. (2013). *Preventing and Treating Missing Data in Longitudinal Clinical Trials: A Practical Guide.* New York: Cambridge University Press.

Molenberghs, G. and Kenward, M.G. (2007). *Missing Data in Clinical Studies.* Chichester: John Wiley & Sons.

O'Kelly, M. and Ratich, B. (2014). *Clinical Trials with Missing Data: A Guide for Practitioners.* New York: John Wiley & Sons.

Rizopoulos, D., Verbeke, G., and Molenberghs, G. (2010). Multiple-imputation-based residuals and diagnostic plots for joint models of longitudinal and survival outcomes. *Biometrics* **66**, 20–29.

Skinner, C.J., Holt, D., and Smith, T.M.F. (1989). *Analysis of Complex Surveys.* New York: John Wiley & Sons.

Verbeke, G., Molenberghs, G., and Beunckens, C. (2008). Formal and informal model selection with incomplete data. *Statistical Science* **23**, 201–218.

22

Missing Data in Clinical Trials

Craig Mallinckrodt
Eli Lilly & Company, Indianapolis, IN

CONTENTS

22.1 Introduction

This chapter focuses on issues concerning missing data in clinical trials intended to support regulatory applications for drugs, medical devices, and biologics. Although many of the issues regarding missing data are similar in many contexts, several aspects of the regulatory setting have particular bearing on how missing data are handled. Therefore, this chapter begins with a brief review of clinical trials and the factors specific to the regulatory setting that influence the handling of missing data.

22.1.1 Clinical trials

The evidence to support new medicines, devices, or other medical interventions is based primarily on randomized clinical trials. Many of these trials involve assessments taken at the start of treatment (baseline), followed by assessments taken repeatedly during, and in some scenarios, after the treatment period. In some cases, such as cancer trials, the primary post-baseline assessments are whether or not some important event occurred during the assessment interval(s). These outcomes can be summarized by expressing the multiple post-baseline outcomes as a time to the event or as a percentage of patients experiencing the event at or before some landmark time point. Alternatively, the multiple post-baseline assessments can all be used in a longitudinal, repeated measures analysis, which can either focus on a landmark time point or consider outcomes across time points (Mallinckrodt 2013a).

Regardless of the scenario, randomization facilitates fair comparisons between treatment and control groups by balancing known and unknown factors across the groups. The intent of randomization in particular, and the design of clinical trials in general, is that differences observed between the treatment and control groups are attributable to causal differences in the treatments and not to other factors (Mallinckrodt 2013a).

Missing data is an ever-present problem in clinical trials and has been the subject of considerable debate and research. The fundamental problem caused by missing data is that the balance provided by randomization is lost if, as is usually the case, the patients who discontinue the study differ in regards to the outcome of interest from those who complete the study. This imbalance can lead to bias in treatment group comparisons. As the proportion of missing data increases, the potential for greater bias increases. The bias cannot be overcome by larger sample sizes. In fact, biased results from larger studies can be even more problematic because the larger studies engender greater confidence–in the wrong result (Mallinckrodt 2013a).

Regulators render yes or no decisions regarding the approvability of a drug rather than just describing the data and possible interpretations. Clinical trial sponsors, who make substantial investments in pursuit of regulatory approval, seek predictability regarding what findings would support a favorable decision; and regulators, eager to ensure common requirements across all sponsors and to enable quality development, also value predictability. Regulators generally require a high level of confidence before concluding a drug is safe and effective, preferring in close or ambiguous cases to err on the side of withholding approval. It is necessary to specify primary analytic methods and sensitivity analyses prior to a trial in order to preserve the type I error rate and to improve the predictability of the decision process (Little et al. 2010).

The ICH E9 guideline (www.ich.org/cache/compo/276254-1.html), which provides the fundamental principles that guide researchers and regulators in medical research, states that despite missing data, a trial may still be valid provided the statistical methods used are sensible. Carpenter and Kenward (2007) define a sensible analysis as one where:

1. The variation between the intervention effect estimated from the trial and that in the population is random. In other words, trial results are not systematically biased.

2. As the sample size increases, the variation between the intervention effect estimated from the trial and that in the population gets smaller and smaller. In other words, the estimates are consistent: as the size of the trial increases, the estimated intervention effect hones in on the true value in the population.

3. The estimate of the variability between the trial intervention effect and the true effect in the population (i.e., the standard error) correctly reflects the uncertainty in the data.

If all these conditions hold, then valid inference can be drawn despite the missing data. However, the analyses required to meet these conditions may be different from the analyses that satisfy these conditions when no data are missing. Regardless, whenever data intended to be collected are missing, information is lost and estimates are less precise than if data were complete (Carpenter and Kenward 2007).

The extent to which useful information can be gleaned from trials with missing data depends on the amount of missing data, how well the reasons or mechanisms driving the missingness are understood, and how robust conclusions are across the plausible reasons (mechanisms). Although it is impossible to be certain what mechanism(s) gave rise to the missing data, the extent to which it is understood why data are missing narrows the possibilities. Results can be compared across these various possibilities. Of course, all else equal, the more complete the data the more interpretable the findings (Mallinckrodt 2013a).

Modern statistical analyses can reduce the potential for bias from missing data. However, all means of handling missing data rely on untestable assumptions about the missing values and the mechanism giving rise to them (Verbeke and Molenberghs 2000). The conundrum inherent to analyses of incomplete data is that data about which the missing data assumptions are made are missing. Hence, the assumptions cannot be tested from data, and the appropriateness of analyses and inferences cannot be assured. The greater the rate of missing data, the greater the potential for increased bias. Therefore, minimizing missing data is the best way of dealing with it (Molenberghs and Kenward 2007; CHMP 2010; Little et al. 2010; Fleming 2011; Mallinckrodt 2013a).

Sensitivity analyses are a series of analyses with differing assumptions. The aim is that by comparing results across sensitivity analyses it becomes apparent how much inference about the treatment effect relies on the assumptions. In fact, many of the newer statistical approaches are finding their best application as sensitivity analyses rather than as primary analyses (Molenberghs and Kenward 2007; Mallinckrodt et al. 2008; Little et al. 2010). Reasonable measures to reduce missing data combined with appropriate analytic plans that include sensitivity analyses can markedly reduce the uncertainty in results and increase the information gained from medical research (Mallinckrodt 2013a).

22.1.2 History

Until recently, guidelines for analyzing clinical trial data provided limited advice on how to handle missing data, and analyses tended to be simple and *ad hoc*. The calculations required to estimate parameters from a balanced dataset are far easier than the calculations required with unbalanced data, such as when patients drop out. Hence, the initial motivation for dealing with missing data may have been as much to foster computational feasibility in an era of limited computing power as to counteract the potential bias from the missing values (Verbeke and Molenberghs 2000; Molenberghs and Kenward 2007).

One such simple method, complete case analysis, includes only those cases for which all measurements were recorded. This method yields a data structure that would have resulted with no missing data. Therefore, standard software and simple statistical analyses can be used. Unfortunately, the loss of information is usually substantial and severe bias can result when the outcomes for patients who discontinue differ from those who complete (Verbeke and Molenberghs 2000; Molenberghs and Kenward 2007).

Alternative means to obtain complete datasets are based on imputing the missing data. However, simple imputation strategies such as baseline and last observation carried forward (BOCF, LOCF) that were used widely in clinical trials also have serious drawbacks. These methods entail restrictive assumptions that are unlikely to hold and the uncertainty of imputation is not taken into account because imputed values are not distinguished from observed values. Therefore, biased estimates of treatment effects and inflated rates of false positive and false negative results are likely (Verbeke and Molenberghs 2000; Molenberghs and Kenward 2007; Mallinckrodt et al. 2008; Little et al. 2010).

Initial widespread use of simple methods set historical precedent that when combined with the desire to compare current results with historical findings fostered continued use of the simple methods even as advances in statistical theory and implementation might have otherwise relegated these methods to the museum of statistics (Mallinckrodt 2013a). Continued acceptance of LOCF and BOCF was also fostered by the belief that they yielded conservative estimates of treatment effects; thereby providing additional protection against erroneous approval of ineffective interventions (Mallinckrodt et al. 2008).

However, analytic proof showed that the direction and magnitude of bias in LOCF (and BOCF) depended on factors not known at the start of a trial (Molenberghs et al. 2004). A large volume of empirical research showed that in common clinical trial scenarios the bias from LOCF and BOCF could favor the treatment group and inflate the rate of false positive results, while some of the newer analytic methods were either not biased in these settings or the magnitude of the bias was smaller (Mallinckrodt et al. 2008; Lane 2008; Siddiqui et al. 2009). Not surprisingly, recent guidance almost universally favors the newer methods over LOCF, BOCF, and complete case analyses. Some of the more commonly used newer methods are discussed in subsequent sections.

22.1.3 National Research Council recommendations

At the request of and with funding from the Food and Drug Administration (FDA), an expert panel was created by the National Research Council's Committee on National Statistics. The panel interviewed prominent statisticians in the pharmaceutical industry and FDA. Information from these interviews and the panels experience formed the basis of their report that was published in 2010 (Little et al. 2010). This report is clearly influencing practice. Regulators are asking drug development teams to utilize the recommendations when proposing and implementing plans to deal with missing data.

Although the panel focused on phase III trials, the recommendations are useful regardless of stage of development. The recommendations set forth an overarching framework for tackling the problem of missing data. Key pillars of this framework include: 1) trial design and conduct features to maximize patient retention; 2) precise and clear specification of trial objectives; 3) and, choice of, and reporting results from, a sensible primary analysis and sensitivity analyses that support the research question and assess robustness of the primary result to missing data assumptions (Mallinckrodt et al. 2014). Each of these pillars is examined in subsequent sections.

22.2 Preventing Missing Data

An important conundrum inherent to all analyses of incomplete data is that they require assumptions about the missing data, but the data about which the assumptions are made are missing. Hence, the assumptions cannot be tested from the data and the appropriateness of analyses and inference cannot be assured (Verbeke and Molenberghs, 2000). The greater the proportion of missing data, the greater the potential for increased bias. Therefore, agreement is universal that minimizing missing data is the best way of dealing with it (Molenberghs and Kenward, 2007; CHMP, 2010; Little et al. 2010; Fleming, 2011; Mallinckrodt, 2013a).

However, the merits of trial design and conduct features to prevent missing data are difficult to evaluate. Clinical trials are not designed to assess factors that influence retention. Therefore, confounding factors can mask or exaggerate differences in rates of missing data due to trial conduct or methods. Nevertheless, recent guidance includes a number of specific suggestions, with the NRC recommendations (Little et al. 2010) being the most comprehensive and specific.

For example, some of the trial design options noted in the NRC guidance (Little et al. 2010) included enrolling a target subpopulation for whom the risk-benefit ratio of the drug is more favorable, or to identify such subgroups during the course of the trial via enrichment or run-in designs. Other design options in the NRC guidance included use of add-on designs and flexible dosing. These design features generally influence only discontinuation due to lack of efficacy or adverse events (primarily in non placebo groups) and also entail limitations and trade-offs. For example, enrichment and run-in designs require that a subset with more favorable risk-benefit can be readily and rapidly identified in a trial; and, the inferential focus is on the enriched subset, not all patients. Flexible dosing cannot be used in trials where inference about specific doses is required, such as dose-response studies (Mallinckrodt 2013a).

Moreover, lack of efficacy and adverse events typically do not account for an overwhelming majority of early discontinuation, limiting the degree to which designs that foster more favorable drug response can reduce dropout. Flexible dosing and enriching the sample for more favorable drug response would have little impact on reducing dropout in placebo groups.

However, further reduction in dropout may be achieved via trial conduct and procedures that encourage maximizing the number of patients retained on the randomized medications. These approaches may be most useful in reducing dropout for reasons other than adverse events and lack of efficacy, such as patient decision, physician decision, protocol violation, and loss to follow up (Mallinckrodt 2013a). Specific guidance on trial conduct from the NRC panel included minimizing patient burden, efficient data capture procedures, education on the importance of complete data, along with monitoring and incentives for complete data (Little et al. 2010).

Simply put, lowering rates of dropout can be as much about behavior as design and conduct. If completion rates received as much attention as enrollment rates considerable progress might be possible. Importantly, changing attitudes and behaviors regarding missing data would likely help increase retention in all arms, whereas design features may have greater impact on drug groups than on placebo groups (Mallinckrodt 2013a).

Minimizing loss to follow-up has particularly important consequences for validity of analyses. To appreciate this importance consider the taxonomy of missing data (Little and Rubin 2002). Another useful way to think about NMAR is that if, conditioning on observed outcomes, the statistical behavior (means, variances, etc) of the unobserved data is equal to the behavior had the data been observed, then the missingness is MAR, if not, then NMAR (Mallinckrodt et al. 2013b).

To illustrate, consider a clinical trial for an anti-depressant where a patient had meaningful improvement during the first six weeks of the eight-week study. At Week-6 the patient had a marked worsening in symptom severity and discontinued study medication. If the patient was lost to follow up and there was no Week-6 observation to reflect the worsened condition the missingness was NMAR. If the Week-6 observation was obtained before the patient discontinued it is possible the missingness was MAR (when conditioning on previous outcomes).

Trials should therefore aim to maximize retention, minimize loss to follow-up, and capture reasons for discontinuing study medication. Patient or physician decision does not explain the cause of discontinuation, only who made the decision. Success in these efforts would result in completion rates that are as high as possible given the drug(s) being studied, and what missing data exists would be more readily understood, thereby fostering formulation of sensible analyses (Mallinckrodt et al. 2013b).

22.3 Estimands

An important evolution in the discussions on missing data has been the focus on clarity of objectives. In fact, the first recommendation from the recent National Research Council (Little et al. 2010) recommendations on the prevention and treatment of missing data was that the objectives be clearly specified. The need for clarity in objectives is driven by the differences in, and the ambiguities that arise from, the missing data.

For example, data may be intermittently missing or missing due to dropout. Patients may or may not be given rescue medications. Assessments after withdrawal from the initially randomized study medication or after the addition of rescue medications may or may not be taken. Whether or not—and if so, how—these follow-up data should be used in analyses and inference is critically important (Mallinckrodt and Kenward 2009).

Conceptually, an estimand is simply what is being estimated. Components of estimands for longitudinal trials may include the parameter (e.g., difference between treatments in mean change), time point or duration of exposure (e.g., at Week 8), outcome measure (e.g., diastolic blood pressure), population (e.g., in patients diagnosed with hypertension), and inclusion/exclusion of follow-up data after discontinuation of the originally assigned study medication and/or initiation of rescue medication (Mallinckrodt 2013a).

During the development of an intervention the primary goals of the research evolve and therefore the primary estimand may also evolve. Therefore, one estimand cannot be advocated as universally most relevant. Rather, the need is to clarify the strengths and limitations of the estimands in order to choose the best one for various situations (Mallinckrodt 2013a).

An important consideration in choosing appropriate estimands is whether the focus is on efficacy or effectiveness. Efficacy may be viewed as the effects of the drug if taken as directed, and effectiveness as the effects of the drug as actually taken. However, referring to estimands in the efficacy-versus-effectiveness context ignores that many safety parameters need to be analyzed. It does not make sense to test an efficacy estimand for a safety outcome. A more general terminology is to refer to hypotheses about efficacy and effectiveness as the de-jure (if taken as directed) and de-facto (as actually taken) hypotheses, respectively (Carpenter, Roger, and Kenward 2013).

The NRC guidance (Little et al. 2010) lists five example estimands for longitudinal trials with continuous endpoints, such as depression, pain, diabetes, etc. Mallinckrodt (2013a) summarized those five estimands and a sixth estimand as follows.

1. Difference in outcome improvement at the planned endpoint for all randomized participants. This estimand compares the mean outcome for treatment versus control regardless of what treatment participants actually received. Follow-up data (after withdrawal of initially randomized medication and/or initiation of rescue medication) are included in the analysis.

In the ITT framework, where inference is based on the originally assigned treatment, including follow-up data when rescue medications are allowed can mask or exaggerate both the efficacy and safety effects of the initially assigned treatments, thereby invalidating causal inferences for the originally assigned medication (Mallinckrodt and Kenward 2009). Therefore, inference for estimand 1 is on de-facto hypotheses about the effectiveness of treatment regimens or policies.

However, in regulatory settings the most relevant research questions are often in regards to the causal effects of the investigational drugs, not treatment policies. As O'Neill and Temple (2012) noted, including follow-up data as part of the primary estimand is more customary in outcomes trials, whereas in symptomatic treatment trials, such as depression or pain, follow-up data are usually not included in the primary estimand, for the previously noted reasons.

2. Difference in outcome improvement in tolerators. This estimand compares the mean outcomes for treatment versus control in the subset of the population who initially tolerated the treatment. An open label run-in phase is used to identify patients that meet outcome criteria to continue. Patients that continue are randomized (usually double-blind) to either continue on the investigational drug or switch to control. Including only patients with initially favorable outcomes for the drug should reduce dropouts, thereby providing a better opportunity to assess de-jure (efficacy) hypotheses. However, estimand 2 focuses on a patient subset and would not be applicable when inference to all patients was desired. Moreover, in most situations it is not known who will tolerate, and thus all patients must be exposed to the safety risks of the drug, whereas efficacy inferences apply only to the tolerators.

3. Difference in outcome improvement if all patients adhered. This estimand addresses the expected change if all patients remained in the study. Estimand 3 addresses de-jure hypotheses about the causal effects of the initially randomized drug, if taken as directed–an efficacy estimand. Although knowing what to expect if a patient takes the drug as directed is important, on a group basis it is hypothetical because there will always be some patients who do not adhere (Little et al. 2010).

4. Difference in area under the outcome curve during adherence to treatment.

5. Difference in outcome improvement during adherence to treatment.

Estimands 4 and 5 assess de-facto hypotheses regarding the initially randomized drug. These estimands are based on all patients and simultaneously quantify treatment effects on the outcome measure and the duration of adherence. As such, there is no missing data attributable to dropout. However, assessing a drug for effectiveness only during adherence ignores the fact that if patients cannot continue to take the drug they will in many instances have no lasting benefit from it (Permutt and Pinheiro 2009; Kim 2011). In such situations, estimands 4 and 5 overestimate the effectiveness of the drug at the planned endpoint of the trial.

Another estimand, referred to as estimand 6 for convenience, has been proposed (Mallinckrodt et al. 2012) that may be particularly relevant in the early evaluations and initial regulatory approvals of new medications.

6. Difference in outcome improvement in all randomized patients at the planned endpoint of the trial attributable to the initially randomized medication. Estimand 6 assesses effectiveness at the planned endpoint, focusing on the causal effects of the initially randomized medications. Conceptually, estimand 1 and estimand 6 require follow-up data. Unlike estimand 1, estimand 6 must be free of the confounding effects of rescue medications. However, ethical considerations often mandate that rescue medications be allowed after patients discontinue randomized study medication (Mallinckrodt 2013a).

Conceptually, estimands 3 and 6 both focus on causal effects of the initially randomized medications, in all randomized patients, at the planned endpoint of the trial. However, estimand 3 focuses on what would have happened if patients adhered to treatment and estimand 6 focuses on what was actually observed. Estimand 3 addresses de-jure (efficacy) hypotheses and estimand 6 addresses de-facto (effectiveness) hypotheses (Mallinckrodt 2013a).

Given the confounding effects of rescue medications and the ethical need to allow them, one approach to testing de-facto hypotheses is to impute the data after discontinuation of the initially randomized study medication under the assumption that initially randomized active medications have no effect (or a diminished effect) after they are discontinued. Historically, this has been done by imputing values using baseline observation carried forward (BOCF). However, BOCF entails assumptions that are unlikely to hold and it underestimates the uncertainty of imputation (Mallinckrodt 2013a).

Another approach to testing de-facto hypotheses is to eliminate missing data by making explicit use of dropout in defining a single endpoint outcome for each patient. For example, treatment differences in response rates can be compared where response is considered successful if symptom severity (efficacy) has improved compared with baseline by a certain absolute or relative amount, and if the patient completed the planned assessment interval. All patients who discontinue early are considered treatment failures (Mallinckrodt 2013a). However, Fleming (2011) warned against changing primary endpoints to reduce missing data if it meaningfully compromises the endpoints clinical relevance.

The treatment success/failure approach to assessing effectiveness assumes that doctors and patients make similar decisions regarding continuation of therapy in the clinical trial as they would in clinical practice. This assumption will not always be valid. For example, in contrast to clinical practice, in a double-blind trial patients and clinicians are unsure whether the patient is taking drug or placebo, and the properties of the experimental drug are not known. Nevertheless, the success/failure approach becomes more meaningful as the proportion of patients who discontinue early due so for reasons causally related to the medications; that is, when most of the discontinuation are due to adverse events or lack of efficacy.

22.4 Analyses

Technical specifications for many of the analyses noted in this section can be found elsewhere in this Handbook. The intent here is to cover the basic principles of certain analyses and to develop an overall analytic approach for longitudinal clinical trials in regulatory settings.

22.4.1 Primary analysis

Despite all efforts to minimize missing data, anticipating complete data is not realistic and analysis plans must include means of dealing with incomplete data. In order to develop an appropriate analysis plan the mechanism(s) leading to the missingness must be considered. In longitudinal clinical trials, MCAR is unlikely to be valid; MAR is often plausible but never provable; and, going beyond MAR to NMAR requires assumptions that are not testable.

Therefore, no single NMAR analysis can be definitive, and consensus is emerging that a primary analysis based on MAR is often reasonable. Complete case and single imputation methods that require MCAR and/or other restrictive assumptions are generally not reasonable (Verbeke and Molenberghs 2000; Molenberghs and Kenward 2007; Mallinckrodt et al. 2008; Little et al. 2010). The CHMP guidance takes a somewhat more favorable view of one simple method, LOCF, noting that even though it has suboptimal statistical properties, LOCF may provide a conservative estimate of treatment effects when patients in the experimental group discontinue more frequently and/or earlier. Clinical trials in depression are given as an example where LOCF yields conservative results. However, it is unclear how trialists can know if these conditions will hold, especially when testing novel therapies.

Moreover, in summarizing over 200 outcomes from an entire new drug application for an antidepressant Mallinckrodt et al. (2004) reported that LOCF yielded a p-value lower than the corresponding p-value from an MAR likelihood-based analysis for over 1/3 of the outcomes. Molenberghs et al. (2004) showed that in addition to the proportion and timing of withdrawal, the bias from LOCF was also influenced by the magnitude of the true difference between treatments, which is of course unknown.

Primary analyses based on MAR may be especially reasonable when combined with rigorous efforts to maximize retention on the initially randomized medications. Methods common in the statistical literature based on MAR include likelihood-based analyses, multiple imputation (MI) and weighted generalized estimating equations (WGEE) (Molenberghs and Kenward 2007). The specific attributes of each method can be used to tailor an analysis to the situation at hand.

With an MAR primary analysis several aspects of the model fit can be addressed by standard model-checking diagnostics. For example, correlation and distributional assumptions for the observed data can be objectively evaluated. Of course, assumptions for the unobserved data cannot be assessed. The influence of individual or clusters of outlying observations on various aspects of model fit can also be evaluated, along with identifying observations that may be driving weakly identified parts of an NMAR model (Molenberghs and Kenward 2007; Mallinckrodt 2013a).

22.4.2 Model diagnostics

Important assumptions required for valid regression-type analyses of continuous outcomes include linearity, normality, and independence (Wonnacott and Wonnacott 1981). These assumptions are in regards to the residuals, not on the actual observations and entail no special considerations for missing data, except that departures from assumed distributions can be more problematic in NMAR models. The focus is not as much on whether or not deviations from the assumptions existed, but rather on how much (if at all) departures from assumptions influenced results. For example, if the residuals were not normally distributed, but this lack of normality had a trivial impact on the primary treatment contrast, then the result would be useful, even if not entirely valid in the strictest sense, because inferences were not contingent on this assumption (Mallinckrodt 2013a).

Although methods to test for the existence and impact of outlier (influential) observations have been around for decades, new methods have been developed for use in NMAR analyses. To this end, interest has grown in local influence approaches (Thijs, Molenberghs, and Verbeke 2000; Verbeke et al. 2001; Zhu and Lee 2001; Molenberghs et al. 2001; Troxel, Ma, and Heitjan 2004; Ma, Troxel and Heitjan, 2005; Shen et al. 2006), which are often associated with selection models. Local influence provides an objective approach to identifying and examining the impact of influential observations and clusters of observations on various aspects of the analysis, including the missing-data mechanisms and treatment effects.

However, given the newness and complexity of such methods, and that key principles do not markedly differ from simpler methods, the key principles are illustrated using simpler methods that can be implementable using standard, commercially available software. The basic idea is again not to simply identify if or if not influential observations were present, but rather to assess how the most influential observations influenced the parameters of interest.

The validity of likelihood-based analyses also hinge on correct specification of the correlation structure. Standard practice has moved to use of an unstructured correlation matrix for modeling the within-patient errors as a means of avoiding misspecification. However, this approach does not guarantee validity. For example, although failure to converge is often attributable to data preparation issues, pre-specification of the primary analysis should include provisions for failure to converge (Mallinckrodt et al. 2008). If unstructured fails to converge, a more parsimonious model is likely necessary, but not necessarily correct. Moreover, structures more general than unstructured are possible, such as separate unstructured matrices by treatment (Mallinckrodt 2013a).

Pre-specification of the primary analysis can include an ever-more parsimonious set of plausible correlation structures to be tested if unstructured fails to converge. The primary analysis could be considered the first structure in the ever-more parsimonious set to converge, or the structure yielding the best fit as measured by common model-fitting criteria (e.g., Akaikes information criterion). For confirmatory trials it has been more common to use the first-to-converge approach because it avoids model building and hypothesis testing from the same data, and it ensures the most general plausible structure is used (Mallinckrodt 2013a).

Use of the sandwich estimator for standard errors rather than model-based estimates provides valid inference when the correlation structure is misspecified (Verbeke and Molenberghs 2000). Therefore, use of the sandwich estimator as the default approach would protect against correlation misspecification. However, when the sandwich estimator is used in SAS PROC MIXED, only the between-within method for estimating denominator degrees-of-freedom is available, an approach that is known to be biased, especially in small samples. Moreover, the sandwich estimator assumes MCAR, which is difficult to justify a priori.

22.4.3 Sensitivity analyses

Sensitivity analyses can be defined as analyses in which several statistical models are considered simultaneously or in which a statistical model is further scrutinized using specialized tools, such as diagnostic measures (Little et al. 2010). In clinical trial data it is not possible to prove the validity of MAR or have a singly definitive NMAR analysis. Therefore, a natural analytic framework is that of sensitivity analyses (Verbeke and Molenberghs 2000).

To facilitate decision-making--for sponsors and regulators—the standard approach is to pre-specify a plausible primary analysis and assess robustness of inferences to departures from assumptions via sensitivity analyses. Although need for additional sensitivity analyses inspired by trial results may arise, a parsimonious set of plausible sensitivity analyses should be pre-specified and reported (Little et al. 2010; Mallinckrodt 2013a). A straight-forward approach is to fit a selected number of plausible models. The degree to which conclusions (inferences) are stable across the analyses provides an indication of the confidence that can be placed in them. With a primary analysis that assumes MAR, departures from MAR must be assessed using NMAR methods.

The focus should be on comparing the magnitude of the primary treatment contrasts from the various sensitivity analyses with that from the primary analysis. Emphasis should not be placed heavily on p-values because this can be misleading. For example, assume the primary result from an MAR analysis yielded a p-value of 0.049, and that MAR was valid.

Sensitivity analyses could yield p-values distributed randomly around the primary p-value. That is, half the sensitivity analyses would yield non-significant results, thereby potentially suggesting results were not robust to departures from MAR.

Three common families of NMAR analyses are shared-parameter models, pattern-mixture models, and selection models. Selection models are multivariate models for repeated measures where one variable is the efficacy outcome from the primary analysis and the second is the repeated binary outcome for dropout that is modeled via logistic regression. The two parts of the selection model are often referred to as the measurement model and the dropout model. The two models are linked in that the dependent variable from the measurement model is a predictor (independent) variable in the dropout model (Molenberghs and Kenward 2007; Mallinckrodt 2013a).

Selection models can be implemented in which an estimate of the association between the present, possibly missing efficacy outcome and the probability of dropout is obtained from the dropout model. However, it is hard to have confidence in the estimate of this parameter that describes the "NMAR" part of the model because the estimate is driven by assumption, not data. Therefore, subsequent selection models can be run wherein plausible values of the "NMAR" parameters are input and fixed in the analysis, thereby facilitating assessment of changes in the magnitude of the primary treatment contrast across plausible NMAR models.

Shared-parameter models can also be thought of as multivariate models, where one variable is again the continuous efficacy outcome from the primary analysis and the second variable is time to dropout. The measurement and dropout models are linked by a set of latent variables, latent classes, and/or random effects that are assumed to influence both the outcome variable and time to dropout (Molenberghs and Kenward 2007; Mallinckrodt 2013a). These models are fitted either using maximum likelihood or as Bayesian models using data augmentation and Markov Chain Monte Carlo (MCMC) techniques (Mallinckrodt et al. 2013b).

Pattern-mixture models fit separate response models for each pattern of missing values, weighted by their respective probabilities, with an overall outcome derived across patterns. Patterns are often defined by time of dropout, but could be defined by reason for discontinuation or other means. Pattern-mixture models are by construction under-identified, i.e., over-specified. For example, assume the goal is to estimate the difference between treatments at endpoint and three dropout patterns are used: early dropouts, late dropouts, and completers. Missing endpoint values for the early and late dropout groups must be imputed; however, information must be borrowed from other groups because there are no endpoint values in the early and late dropout patterns (Molenberghs and Kenward 2007; Mallinckrodt 2013a).

This problem can be resolved through the use of identifying restrictions where inestimable parameters of the incomplete patterns are set equal to (functions of) those of other patterns. Three common identifying restrictions are: 1) Complete Case Missing Values (CCMV) where information is borrowed from the completers; 2) Neighboring Case Missing Values (NCMV) where information is borrowed from the nearest identified pattern; 3) Available Case Missing Values (ACMV) where information is borrowed from all patterns where the information is available (Molenberghs and Kenward 2007; Mallinckrodt et al. 2014).

In addition, a general family of restrictions can be defined as non-future dependent missing value restrictions (NFD) where one conditional distribution per incomplete pattern is left unidentified. In other words, the distribution of the "current" unobserved measurement given the previous measurements is unconstrained. In practice, this can be accomplished by using data only up to and including the time point being imputed as the basis for estimating parameters for the imputation. When information is borrowed in the NFD family via CCMV or NCMV the mechanism is NMAR. The ACMV restriction is assumes MAR. Therefore, comparing results from ACMV with NFD_CCMV or NFD_NCMV assesses sensitivity of results to departures from MAR (Molenberghs and Kenward 2007; Mallinckrodt et al. 2014).

Recently, another family of methods referred to as controlled imputation has seen increasing discussion in the literature and use in practice. Controlled imputation approaches such as those discussed by Little and Yao (1996), Carpenter, and Kenward (2007), Ratitch and O'Kelly (2011), Carpenter, Roger, and Kenward (2013), can be thought of as specific versions of multiple-imputation-based pattern-mixture models. The aim is to construct a principled set of imputations that exhibit a specific departure from MAR in order to assess either sensitivity of de-jure estimands or as a primary means to assess de-facto estimands (Carpenter, Roger, and Kenward 2013).

Multiple imputation has most commonly been implemented in the MAR setting with separate imputation models for the drug and placebo (control) arms (in a two-arm study). For NMAR analyses, one sub-family of approaches within controlled imputation, referred to as reference-based imputation, uses one imputation model derived from the reference (e.g., placebo, or standard of care) group but then applies that model to both the drug and placebo arms. Alternatively, a single imputation model can be developed from all the data and applied to both arms (Mallinckrodt 2013a, 2014; Carpenter, Roger, and Kenward 2013; Ratitch and O'Kelly 2011; Teshome et al. 2013).

Using one imputation model for both treatment arms generally diminishes the difference between the arms compared with MAR approaches that use separate imputation models for each arm. The intent is to generate a plausibly conservative, or worst plausible estimate of efficacy. If inferences agree with the MAR primary result the findings are robust to the plausible departures from MAR. Alternatively, the same result can be interpreted as an estimate of effectiveness that reflects a change in or discontinuation of treatment (Mallinckrodt 2013a).

Reference-based imputations include several sub-families which are useful in tailoring analyses to specific scenarios. In the jump to reference (J2R) method imputed values for patients who discontinue the active arm immediately take on the attributes of the reference arm (placebo). That is, the treatment benefit in patients who discontinue the active arm disappears immediately upon discontinuation. The J2R approach is therefore useful for symptomatic treatments with short duration of action. In the copy reference (CR) method the imputations result in a treatment effect that diminishes after dropout in accordance with the correlation structure implied by the imputation model. The CR approach is therefore useful for symptomatic treatments with long duration of action. The copy increment from reference (CIR) method maintains the treatment effect after discontinuation by matching changes after withdrawal to changes in the reference arm, and would therefore be appropriate for treatments thought to alter the underlying disease process (disease modification) (Carpenter, Roger, and Kenward 2013; Mallinckrodt et al. 2014).

Controlled imputation can also be used to assess sensitivity by repeatedly adjusting the imputations to provide a progressively more severe stress test to assess how extreme departures from MAR must be to overturn the primary result. Typically, only imputed values for the experimental arm are adjusted while the control arm is handled using an MAR-based approach (Ratitch and O'Kelly 2011; Ratitch, O'Kelly, and Tosiello 2013).

The basic idea is to impute the missing values and subtract a value (delta) from the imputed values of the experimental arm. Two distinct implementations exist. One method uses the sequential regression approach to MI and adds delta to the imputed value so that it feeds through in the imputation process to later visits. In the second approach, delta is added to the imputed value for that visit only (Mallinckrodt et al. 2014).

The primary analysis is applied to the delta-adjusted dataset to see if the conclusion of the primary analysis is overturned. If not, a larger delta is chosen and the process repeated until the primary result is overturned. If the delta required to overturn the primary result is not a plausible departure from MAR then the primary result is robust to plausible departures from MAR (Mallinckrodt et al. 2014).

Two additional approaches to sensitivity analyses include doubly robust methods and using MAR methods with inclusive models. A restrictive model is one which typically contains only the design factors of the experiment, a parsimonious set of baseline covariates, and usually the interactions between baseline covariates and time. Inclusive models add on ancillary variables to improve the performance of the missing data procedure (Mallinckrodt et al. 2008; Little et al. 2010).

Inclusive models are most easily implemented by including the ancillary variables in the dropout model for WGEE or the imputation model in MI. Ancillary variables need not be included in the analysis models (Molenberghs and Kenward 2007). This is particularly useful in avoiding confounding with treatment for those covariates that are related to both missingness and treatment. It is possible to implement inclusive models with likelihood-based analyses, but avoiding the aforementioned confounding can complicate or limit the analysis (Mallinckrodt 2013a).

The genesis of doubly robust methods can be seen in the following. Although GEE is valid only under MCAR, inverse probability weighting (IPW) can correct for MAR, provided an appropriate model for the missingness process (dropout) is used whereby missingness depends on observed outcomes but not further on unobserved outcomes (Molenberghs and Kenward 2007).

The WGEE yield semi-parametric estimators because they do not model the entire distribution. These semi-parametric estimates are generally not as efficient as maximum likelihood estimators obtained using the correct model, but they remain consistent when maximum likelihood estimators from a misspecified parametric model are inconsistent (Molenberghs and Kenward 2007).

The efficiency of WGEE can be improved by augmenting the estimating equations with the predicted distribution of the unobserved data given the observed data (Molenberghs and Kenward 2007). Augmentation also introduces the property of double robustness. To understand double robustness, consider that efficient IPW estimators require three models: 1) The substantive (analysis) model which relates the outcome to explanatory variables and/or covariates of interest; 2) A model for the probability of observing the data (usually a logistic model of some form); and, 3) A model for the joint distribution of the partially and fully observed data, which is compatible with the substantive model in (1) (Molenberghs and Kenward 2007; Mallinckrodt et al. 2014).

If model (1) is wrong, e.g., because a key confounder is omitted, then estimates of all parameters will typically be inconsistent. The intriguing property of augmented WGEE is that if either model (2) or model (3) is wrong, but not both, the estimators in model (1) are still consistent (Molenberghs and Kenward, 2007). However, doubly robust methods are fairly new with few rigorous simulation studies or real data applications in the literature. Readers can refer to Carpenter, Kenward, and Vansteelandt (2006), Tsiatis (2006), and Daniel and Kenward (2012) for further background on doubly robust methods.

22.5 Example

The following example is an extension of an abbreviated example of sensitivity analyses previously reported in Mallinckrodt et al. (2014). The example is expanded here to illustrate a thorough approach to sensitivity analyses. The example illustrates the three pillars of dealing with missing data: Clearly stating objectives and estimands, preventing missing data, and a reasonable primary analysis that is supported by sensitivity analyses.

22.5.1 Data

The data used in this example was partially contrived to avoid implications for marketed drugs. However, key features of the original data were preserved. The original data were from two nearly identically designed antidepressant clinical trials that were originally reported by Goldstein et al. (2004) and Detke et al. (2004). Each trial had four treatment arms with approximately 90 patients each that included two doses of an experimental medication (subsequently granted marketing authorization), an approved medication, and placebo. Assessments on the Hamilton 17-item rating scale for depression (HAMD17) (Hamilton 1960) were taken at baseline and weeks 1, 2, 4, 6, and 8 in each trial. All patients from the original placebo arm were included along with a drug arm that was created by randomly selecting 100 patients from the non-placebo arms. In addition to including all the original placebo-treated patients, additional placebo-treated patients were randomly re-selected so that there were also 100 patients in the contrived placebo arms. For these re-selected placebo-treated patients a new patient identification number was assigned and outcomes were altered to create new observations. These trials are referred to as the low and high dropout datasets. In the high dropout dataset completion rates were 70% for drug and 60% for placebo. In the low dropout dataset completion rates were 92% in both the drug and placebo arms. The dropout rates in the contrived datasets closely mirrored those in the corresponding original studies. The design differences that may explain the difference in dropout rates between these two otherwise similar trials were that the low dropout dataset came from a study conducted in Eastern Europe that included a 6-month extension treatment period after the 8-week acute treatment phase, and used titration dosing. The high dropout dataset came from a study conducted in the US that did not have the extension treatment period and used fixed dosing (Mallinckrodt et al. 2014).

The number of patients who discontinued is summarized by study week in Table 22.1.

22.5.2 Primary analyses

The primary analytic objective for each dataset was to compare the efficacy (benefit if taken as directed) of drug versus placebo. The primary estimand was the difference between treatments in mean change from baseline to week 8.

The primary analysis used a restricted maximum likelihood (REML)-based repeated measures approach. The analyses included the fixed, categorical effects of treatment, investigative site, visit, treatment-by-visit interaction, site-by-visit interaction, and the continuous, fixed covariates of baseline score and baseline score-by-visit-interaction. An unstructured (co)variance structure shared across treatment groups was used to model the within-patient errors. The Kenward–Roger approximation was used to estimate denominator degrees of

TABLE 22.1
Week of last visit (Number of subjects).

	Week				
	1	2	4	6	8
		High dropout			
Placebo	8	7	12	13	60
Drug	9	6	10	5	70
		Low dropout			
Placebo	2	0	3	3	92
Drug	2	3	2	1	92

freedom and adjust standard errors. Analyses were implemented using SAS PROC MIXED (SAS 2003). The primary comparison was the contrast (difference in LSMEANS) between treatments at the last Visit (Week-8).

The missing data assumption for the primary analysis that assessed the de-jure (efficacy) estimand was that missing data arose from an MAR mechanism. Sensitivity analyses to assess the consequences of departures from MAR included selection and pattern-mixture models, as well as controlled imputations. An alternative de-facto (effectiveness) estimand where subjects were assumed to revert to placebo following withdrawal was addressed using reference-based imputation.

Results from the primary analyses are summarized in Table 22.2. In the high dropout dataset the advantage of drug over placebo in mean change from baseline to Week 8 was 2.29 (s.e. 1.00, $p = 0.024$). This precision is equivalent that of 74 completers per arm. The corresponding values in the low dropout dataset were 1.82 (s.e. 0.70, $p = 0.010$). The standard error for the difference in LSMEANS at Week 8 in the high dropout dataset was 47% larger than in the low dropout dataset. This was partly due to the lower variability in the low dropout dataset (See Table 22.3). The larger variability in the high dropout group would have led to an increased standard error of 32% on its own. The additional variability in the high dropout set came from the additional quantity of missing data.

22.5.3 Model diagnostics

Unstructured correlation matrices were used for the primary analyses. Correlations and (co)variances from the primary analyses are summarized in Table 22.3.

Treatment contrasts from more general and more parsimonious (co)variance structures, with and without use of the sandwich estimator are summarized in Table 22.4. In the high dropout dataset an unstructured matrix common to both treatment groups that was used as the primary analysis provided the best fit, thereby supporting results from the primary analysis as valid. However, the potential importance of choice of correlation structure with this high rate of dropout can be seen in the comparatively large difference in treatment contrasts, standard errors and p-values across the various correlation structures.

In the low dropout dataset choice of correlation structure had a smaller impact on treatment contrasts and standard errors than in the high dropout dataset, and all structures yielded a significant treatment contrast. Separate unstructured matrices by treatment group yielded the best fit, with the second best fit being from a single unstructured matrix.

TABLE 22.2
Visitwise LSMEANS and contrasts for HAMD17 from the primary analysis.

	Placebo	Drug	Contrast[1]	(s.e.)	p-value
		High Dropout			
Week 1	−1.74	−1.72	−0.02	(0.64)	0.966
Week 2	−3.72	−4.12	0.40	(0.80)	0.621
Week 4	−5.16	−6.25	1.09	(0.84)	0.199
Week 6	−5.96	−8.00	2.04	(0.89)	0.024
Week 8	−5.91	−8.21	2.29	(1.00)	0.024
		Low Dropout			
Week 1	−2.19	−1.74	−0.44	(0.38)	0.246
Week 2	−4.92	−4.89	−0.03	(0.56)	0.955
Week 4	−7.78	−8.26	0.49	(0.61)	0.423
Week 6	−9.38	−10.62	1.24	(0.66)	0.062
Week 8	−10.51	−12.32	1.82	(0.70)	0.010

[1] *Advantage of drug over placebo. Negative values indicate an advantage for placebo.*

The influence option in the model statement of SAS PROC MIXED (SAS 2008) was used to determine which levels of clustering factors had the greatest influence on the endpoint treatment contrast. The clustering factors of interest were investigative site and patient.

The influence procedure as implemented in SAS PROC MIXED (SAS 2003) is similar to a case deletion approach. Therefore, whichever effect is being diagnosed cannot be included in the fixed effects model because the case deletion alters the rank of the X matrix. Hence, when assessing the influence of sites, site was excluded from the model. This was not necessary when assessing the influence of patients as this effect was only in the within-in patient correlation structure, not in the fixed effects.

No specific cutoff was used to identify a site or patient as influential. Instead, natural breaks in the ascending sequence of Cooks D statistic (Cook and Weisberg 1982) were used to determine cutoffs for identifying patients as influential or not. Given the comparatively small number of sites, influence of each site was investigated by removing sites one at a time and repeating the primary analysis on the data subsets. Presence vs. absence of statistical significance was not a useful measure of influence because of the reduced number of patients and observations when a site was deleted. Instead, emphasis was placed on the magnitude of change in the treatment contrasts.

Results for the influence of sites are summarized in Table 22.5. In the high dropout dataset no site had a large influence on the treatment contrast. Deleting sites 003 and 028 decreased the treatment contrasts slightly, whereas deleting sites 005 and 001 increased the treatment contrasts slightly. In the low dropout dataset deleting site 121 reduced the treatment contrast by about 20% (1.85 versus 1.56), with other sites having smaller influences. This heterogeneity was not present in the original dataset and is therefore attributable to the creation of contrived data. However, in real data such a finding could trigger further investigation to better understand the heterogeneity and/or identify possible causes.

Influential patients were identified using a similar process as used to assess influential sites. However, all patients identified as influential were deleted by treatment group rather than one at a time. The primary analysis was repeated with all influential drug-treated patients removed, with all influential placebo-treated patients removed, and with all influential patients removed.

TABLE 22.3
Variance-covariance and correlation matrices from primary analysis.

	High Dropout					Low Dropout				
Wk.	1	2	4	6	8	1	2	4	6	8
					(Co)variances					
1	20.16					7.22				
2	14.05	29.86				4.79	15.50			
4	11.55	19.74	32.09			3.87	11.54	18.07		
6	10.70	18.56	25.27	35.10		3.74	9.96	14.65	20.59	
8	11.51	17.67	22.57	30.88	39.55	2.23	7.03	10.73	16.61	22.82
					Correlations					
1	1.000					1.000				
2	0.573	1.000				0.453	1.000			
4	0.454	0.638	1.000			0.339	0.689	1.000		
6	0.402	0.573	0.753	1.000		0.307	0.558	0.760	1.000	
8	0.408	0.514	0.634	0.829	1.000	0.174	0.374	0.528	0.766	1.000

Results from analyses of the primary outcome excluding influential patients are summarized in Table 22.6. In the high dropout dataset set five patients were identified as influential, two on drug and three on placebo. Dropping placebo-treated influential patients decreased the endpoint contrast slightly, and dropping drug-treated influential patients increased the endpoint contrast slightly, as did dropping all influential patients. Similar trends were seen in the low dropout dataset where six patients were identified as influential, two on drug and four on placebo. Again, dropping placebo-treated influential patients decreased the magnitude of the endpoint contrast, and dropping drug-treated influential patients increased the endpoint contrast, as did dropping all influential patients. Therefore, as expected from confirmatory trials, the treatment effects were not driven by one or a few subjects.

Results from analyses of the primary outcome excluding patients with aberrant residuals are summarized in Table 22.7. Patients were considered to have an aberrant residual if the absolute value of the studentized residual was equal to 2.0. In both the high dropout and low dropout datasets excluding all placebo-treated patients with aberrant residuals decreased the endpoint contrast and excluding drug-treated patients and all patients with aberrant residuals increased the endpoint contrast. In all cases, statistical significance of the treatment effect was preserved. Therefore, non-normality of residuals did not influence inferences.

22.5.4 Sensitivity analyses

Past experience with similar data was used to guide sensitivity analyses for assessing robustness of results to departures from MAR. Sensitivity analyses included a selection model in which the parameters describing the NMAR part of the model were varied across a plausible range. In addition, several controlled imputation approaches were utilized. First, reference-based imputation was used as a worst plausible case NMAR analysis. Although for illustration purposes results were also assessed using jump to reference, copy reference, and copy increment from reference, the copy reference approach most closely matched the characteristics of the scenario at hand. In addition, the delta-adjustment approach was also applied.

TABLE 22.4
Variance-covariance and correlation matrices from primary analysis.

Structure[1]	AIC	Endpoint Contrast	(s.e.)	p-value
	High Dropout			
UN	4679.82	2.29	(1.00)	0.024
UN EMPIRICAL	4679.82	2.29	(0.97)	0.020
TOEPH	4684.44	2.10	(0.91)	0.023
TOEPH EMPIRICAL	4684.44	2.10	(0.92)	0.023
TOEPH GROUP=TRT	4689.88	1.82	(0.91)	0.048
UN GROUP=TRT	4692.05	1.96	(1.00)	0.053
CSH	4735.81	1.86	(0.93)	0.047
CSH EMPIRICAL	4735.81	1.86	(0.91)	0.041
CSH GROUP=TRT	4739.34	1.69	(0.93)	0.070
	Low Dropout			
UN GROUP=TRT	4861.70	1.85	(0.703)	0.009
UN	4867.68	1.82	(0.699)	0.010
UN EMPIRICAL	4867.68	1.82	(0.666)	0.007
TOEPH GROUP=TRT	4888.93	1.82	(0.647)	0.005
TOEPH	4897.89	1.79	(0.649)	0.006
TOEPH EMPIRICAL	4897.89	1.79	(0.662)	0.006
CSH	5030.40	1.76	(0.705)	0.013
CSH EMPIRICAL	5030.40	1.76	(0.667)	0.008
CSH GROUP=TRT	5031.92	1.80	(0.708)	0.011

[1] *UN = unstructured; toeph = heterogeneous toeplitz; CSH = heterogeneous compound symmetric; GROUP = TRT means that separate structures were fit for each treatment group; Empirical means that empirical (sandwich-based) estimators of the standard error were used rather than the model-based standard errors.*

In the selection model the primary outcome was assessed using the repeated measures model as in the primary analysis, and the probability of dropout was modeled using a logistic regression that fit the log odds of dropout as a function of visit, separate intercepts (ψ_1, ψ_2) for each treatment group, and separate linear regression coefficients for previous (ψ_3, ψ_4) and current (possibly unobserved) efficacy outcomes (ψ_5, ψ_6). Hence the dependent variable from the measurement model was an independent variable in the dropout model. Fitting separate missingness models for each treatment allowed for different departures from MAR for drug and placebo groups.

The parameters ψ_5 andψ_6 were of particular interest because they were the "NMAR" part of the model. Whenever possible, sensitivity analysis should be based on a pre-defined, plausible range of values for ψ_5 and ψ_6. These values assessed the increase in log odds for withdrawal per unit increase in the outcome measure. The range of values input for ψ_5 and ψ_6 was plausible based on previous experience from similar data. A value of 0.2 indicated that the odds of withdrawal increased by a factor 1.22 for each change of 1 unit on the HAMD17. However, pairing the largest value for ψ_5 with the smallest value for ψ_6 (and vice verse) resulted in combinations of values that were not plausible. These combinations are noted in the table. Setting $\psi_5 = \psi_6 = 0$ was an important validation because the missingness and outcome models become independent and the model reduces to the MAR model in the primary analysis.

TABLE 22.5
Influence of sites on endpoint contrasts.

Dataset	# Pat.	# Obs.	Endpoint Contrast	*p*-value
	High Dropout			
All Data	200	830	2.37	0.0218
Drop POOLINV 005	175	723	2.96	0.0084
Drop POOLINV 001	165	684	2.78	0.0130
Drop POOLINV 028	159	677	2.18	0.0498
Drop POOLINV 003	132	536	2.27	0.0793
	Low Dropout			
All Data	200	961	1.85	0.008
Drop POOLINV 121	167	801	1.56	0.047
Drop POOLINV 131	145	694	1.82	0.028
Drop POOLINV 141	147	710	2.10	0.005
Drop POOLINV 101	141	678	1.77	0.044

TABLE 22.6
Endpoint contrasts for all data and for data with influential patients removed.

Dataset	# Pat.	# Obs.	Endpoint Contrast	*p*-value
	High Dropout			
All Data	200	830	2.29	0.024
Drop All	195	808	2.50	0.015
Drop Drug	198	823	2.59	0.010
Drop Placebo	197	815	2.21	0.032
	Low Dropout			
All Data	200	961	1.82	0.010
Drop All	194	931	1.97	0.003
Drop Drug	196	941	1.97	0.003
Drop Placebo	198	951	1.80	0.011

Across the range of plausible input values for ψ_5 and ψ_6 in the high dropout dataset, the endpoint contrast ranged from 1.30 to 3.60, which was a deviation of about ± 1 around the MAR-based estimate. The corresponding range in the low dropout dataset was 1.71 to 2.01, a much smaller deviation of ± 0.15 around the MAR estimate. Therefore, selection model results clearly support the existence of a treatment effect in the low dropout dataset; however, the support is less strong in the high dropout dataset.

The reasons for greater variation in results from the high dropout dataset were, in decreasing order of importance: (1) more missing values; (2) greater variance that led to increased leverage for the same value of ψ in the missingness model; (3) stronger correlation between weeks that increased the impact of withdrawals at early weeks; and (4) larger variance leading to increased potential impact of missing values on the mean.

Additional results from selection model analyses of the high dropout dataset are summarized in Table 22.8. The additional results from the low dropout dataset essentially agreed with those from the high dropout dataset and are therefore not included for brevity.

TABLE 22.7
Endpoint contrasts for all data and data with patients having aberrant residuals removed.

Dataset	# Obs.	Endpoint Contrast	p-value
	High Dropout		
All Data	830	2.29	0.024
DropAll	796	2.42	0.009
DropDrug	809	2.57	0.006
DropPlacebo	817	2.13	0.035
	Low Dropout		
All Data	961	1.82	0.010
DropAll	915	1.92	0.002
DropDrug	938	2.16	0.001
DropPlacebo	938	1.59	0.019

As expected, when $\psi_5 = \psi_6 = 0$, results matched results from the primary direct likelihood analysis. With equal negative values for ψ_5 and ψ_6 the within group mean changes were greater than from equivalent MAR results. Conversely, identical positive values for ψ_5 and ψ_6 led to smaller within group mean changes. The impact of a positive ψ value was to make those subjects with negative residuals more likely to withdraw leading to an increased mean in observed data, with the he selection model compensated for this by reducing the LSMEAN.

When there is more dropout in the placebo group and $\psi_5 = \psi_6 < 0$, the within group mean change is increased more in the placebo group than the drug group and the endpoint contrast is reduced. Conversely with $\psi_5 = \psi_6 > 0$ the endpoint contrast was increased. When the input values for ψ_5 and ψ_6 differed, between group differences (endpoint contrasts) followed a consistent pattern dictated by the within group changes. Whenever ψ_6 (the regression coefficient for the drug group) was less than ψ_5 (the regression coefficient for the placebo group) the treatment contrast was greater than from the MAR primary analysis; when ψ_5 was greater than ψ_6 the treatment contrast was smaller than in MAR.

The separation of missingness models for each treatment group means that changing ψ_5 has more impact on the placebo mean, while changing ψ_6 has more impact on the drug group. But some change is carried through to the other group, mostly driven by changes to the regression coefficients for baseline.

Results from pattern-mixture model (PMM) analyses under various identifying restrictions are summarized in Table 22.9. The ACMV restriction assumed MAR, whereas CCMV and NCMV assumed NMAR. Therefore, comparing results from ACMV with those from other restrictions assessed the impact of departures from MAR. Pattern-mixture model analysis require that all parameters are estimable in all patterns of dropouts. For analysis of the high dropout dataset centers had to pooled into two grouped centers. Analysis of the low dropout dataset was not feasible because with so few missing observations the treatment effect was not estimable in several dropout patterns. Compared with ACMV, the endpoint contrast and standard error in the high dropout dataset was slightly smaller in CCMV and slightly larger in NCMV. However, these departures from the MAR results were not meaningful, thereby supporting the primary result as valid.

Reference-based imputations were constructed using a full multivariate repeated measures model for estimating parameters for the imputation model that included treatment, inves-

TABLE 22.8
Results from selection model analyses of high dropout dataset.

Input Values		Week 8 LSMEANS				
ψ_5	ψ_6	Placebo	Drug	Contrast	(s.e.)	2×Bayes p
0.2	0.2	4.87	7.33	2.46	(1.09)	0.023
0.0	0.2	5.60	7.38	1.78	(1.05)	0.091
−0.2	0.2	6.28	7.41	1.18	(1.05)	0.282
−0.42	0.22	6.76	7.42	0.66	(1.06)	0.527
0.2	0.0	4.94	7.97	3.03	(1.07)	0.005
0.03	0.0	5.63	8.00	2.37	(1.04)	0.022
−0.2	0.0	6.29	8.04	1.75	(1.02)	0.087
−0.4	0.0	6.75	8.05	1.30	(1.02)	0.204
0.2	−0.2	4.97	8.57	3.60	(1.06)	0.001
0.0	−0.2	5.67	8.57	2.89	(1.03)	0.004
−0.2	−0.2	6.31	8.59	2.29	(1.01)	0.024
−0.4	−0.2	6.76	8.63	1.86	(1.01)	0.064
0.22	-0.42	4.97	8.97	4.01	(1.07)	<0.001
0.0	-0.4	5.68	8.96	3.28	(1.03)	0.002
-0.2	-0.4	6.33	8.98	2.64	(1.01)	0.009
-0.4	-0.4	6.78	9.01	2.22	(1.01)	0.027

[1] *ψ_5 and ψ_6 are the regression coefficients (placebo and drug, respectively) for the association between the current, possibly missing efficacy scores and the logit for probability of dropout.*
[2] *This combination of values is not plausible based on previous experience but is included for completeness of illustration.*
[3] *Results differ from the primary result because the baseline value by site interaction was not fit in the selection model.*

tigative site, baseline score, and their interactions with week. Missing values for both the drug and placebo groups were imputed using a model developed from only the placebo group data. The analysis model was ANOVA at week 8 with treatment, baseline and pooled investigator in the model. In the copy reference (CR) approach, which was most applicable for the example data, any advantage gained from the treatment during adherence decays after discontinuation, the speed of which is determined by the correlation between the repeated measurements. Results from jump to reference (J2R) and copy increment from reference (CIR) are also included for illustration. Results are summarized in Table 22.10.

In the high dropout dataset the endpoint contrast from CR was 1.75 (s.e. 0.98, $p = 0.075$), which was approximately 76% of the magnitude of the MAR estimate (2.29). In the low dropout dataset the endpoint contrast was 1.72 (s.e. 0.70, $p = 0.015$), which was 95% of the magnitude of the MAR estimate (1.82). Therefore, the difference from the MAR result was much smaller and statistical significance was preserved in the low dropout dataset. As expected, J2R results were more conservative than CR, and CR was more conservative than CIR.

Progressive stress tests were implemented using a delta-adjustment controlled imputation approach based on the sequential regression algorithm. Delta-adjustments were applied to imputed values at all visits, but only for the drug group. Imputations were performed visit-by-visit, with patients' delta-adjusted imputed data contributing to imputed values at subsequent visits. The impact of delta for a subject in the drug group withdrawing at week 2 was therefore the accumulation of an increment of approximately $\Delta \times 1 + \sum_{j=2}^{n} \rho_{2j}$ with

TABLE 22.9
Results from pattern-mixture model analyses of high dropout dataset.

Identifying Restriction[1-3]	Endpoint Contrast	(s.e.)	p-value
ACMV	2.67	(1.17)	0.0224
CCMV	2.51	(1.05)	0.0166
NCMV	2.87	(1.69)	0.0895

[1] *ACMV: available case missing values*
[2] *CCMV: complete case missing values*
[3] *NCMV: neighboring case missing values*

TABLE 22.10
Results from reference-based multiple imputation.

	Week 8 LSMEANS				
	Placebo	Drug	Contrast	(s.e.)	p-value
		High Dropout			
MAR	−5.95	−8.24	2.29	1.00	0.024
CR	−5.96	−7.71	1.75	0.98	0.075
J2R	−5.97	−7.57	1.60	0.99	0.110
CIR	−5.95	−7.78	1.83	0.97	0.004
		Low Dropout			
MAR	−10.56	−12.40	1.84	0.70	0.009
CR	−10.55	−12.27	1.72	0.70	0.015
J2R	−10.55	−12.26	1.71	0.70	0.016
CIR	−10.55	−12.27	1.72	0.70	0.015

ρ_{2j} the correlation between week 2 and week j, and subsequent visits. That is, $\Delta \times 1 + (0.51 + 0.63 + 0.83 + 1) = 2.97$ in the high dropout dataset. An early dropout therefore had a greater accumulation—and therefore larger impact–than a later withdrawal. The "tipping point" was identified by repeating the imputation process with progressively larger deltas. Delta-adjustment stress test results are summarized in Table 22.11.

In the high dropout dataset the Δ had to be a worsening of 0.5 points on the HAMD17 in order to overturn the primary result (produce a non-significant result of $p > 0.05$). The endpoint contrast changed 0.31 with $\Delta = 0.5$.

This result can be approximately explained using the accumulations described above, as applied to each assessment week, and the number of patients discontinuing at each week, divided by the total number of patients: $0.5 \times (9 \times 2.97 + 6 \times 2.46 + 10 \times 1.83 + 5 \times 1)/100 = 0.32$. The corresponding tipping point in the low dropout dataset was 2.5 points with a shift of 0.41, which is approximately explained as $2.5 \times (2 \times 2.66 + 3 \times 2.29 + 2 \times 1.77 + 1 \times 1)/100 = 0.42$.

The preceding examples illustrate some fundamental points in dealing with missing data. The primary analysis focused on a precisely defined efficacy estimand. A sensible primary analysis was specified and sensitivity analyses aided understanding of the degree to which departures from MAR could alter inferences from the primary analysis. The sensitivity analyses were interpreted based on the properties of each approach, rather than seen as

TABLE 22.11
Results from delta-adjustment multiple imputation—delta applied on all visits after discontinuation to active arm only.

	Low Dropout Dataset			High Dropout Dataset		
Δ	Contrast	(s.e.)	p-value	Contrast	(s.e.)	p-value
0	1.85	(0.71)	0.009	2.31	(1.02)	0.024
0.5	1.77	(0.71)	0.013	2.00	(1.03)	0.051
2.0	1.52	(0.73)	0.037			
2.5	1.44	(0.74)	0.051			

simply another alternative approach. Most importantly, the results illustrated the benefit from lower rates of missing data.

Although sensitivity analyses of the high dropout dataset generally supported the existence of a treatment effect, the possibility of plausible departures from MAR overturning the primary result, while unlikely, could not be ruled out. In the low dropout dataset inferences from the primary analysis were robust to even the largest plausible departures from MAR. Importantly, the comparatively larger variability in results from the high dropout dataset was not limited to sensitivity analyses for plausible NMAR scenarios. Results from some of the standard diagnostics, such as choice of correlation structure, also showed greater variability with higher dropout.

22.6 Discussion

This chapter focused on issues concerning missing data in clinical trials intended to support regulatory applications for drugs, medical devices, and biologics. Missing data is an ever-present problem in clinical trials and has been the subject of considerable debate and research. Whether or not follow-up data should be collected and/or included in the primary estimand can be considered on a case-by-case basis. However, given the confounding influences of rescue medications, the role for follow-up data in the analysis of symptomatic treatment trials would usually be secondary.

Consensus has emerged that a primary analysis based on MAR is often reasonable. Likelihood-based methods, MI, and WGEE are all useful MAR approaches whose specific attributes can be considered when tailoring a primary analysis to specific situations. With an MAR-based primary analysis a focal point of sensitivity assessments is the impact of departures from MAR. Model-based NMAR methods such as selection models, pattern-mixture models and shared-parameter models can be considered. Prior experience can guide analytic decisions such as plausible ranges of input values for selection models, and appropriate identifying restrictions for pattern-mixture models.

Controlled-imputation methods can be especially useful in assessing the consequences of departures from MAR in regulatory settings. If a plausibly conservative controlled imputation analysis agrees sufficiently with the primary result the primary result can be declared robust to departures from MAR. Alternatively, a tipping point (progressive stress-testing) format can be used to assess how severe departures from MAR must be in order to overturn conclusions from the primary analysis. If implausible departures from MAR.

References

Carpenter, J.R., Kenward, M.G., and Vansteelandt, S. (2006). A comparison of multiple imputation and doubly robust estimation for analyses with missing data. *Journal of the Royal Statistical Society, Series A* **169**, 571-584.

Carpenter, J.R. and Kenward, M.G. (2007) *Missing Data in Randomised Controlled Trials: A Practical Guide.*
Available at http://missingdata.lshtm.ac.uk/downloads/rm04_jh17_mk.pdf (accessed 23 January 2012).

Carpenter J.R., Roger, J., and Kenward, M.G. (2013). Analysis of longitudinal trials with missing data: A framework for relevant, accessible assumptions, and inference via multiple imputation. *Journal of Biopharmaceutical Statistics* **23**, 000–000.

Committee for Medicinal Products for Human Use (CHMP, 2010). *Guideline on Missing Data in Confirmatory Clinical Trials.* London: EMA/CPMP/EWP/1776/99 Rev. 1.

Cook, R.D. and Weisberg, S. (1982). *Residuals and Influence in Regression.* New York: CRC / Chapman & Hall.

Daniel. R. and Kenward, M.G. (2012) A method for increasing the robustness of multiple imputation. *Computational Statistics and Data Analysis* **56**, 1624–1643.

Detke, M.J., Wiltse, C.G., Mallinckrodt, C.H., McNamara, R.K., Demitrack, M.A., and Bitter, I. (2004) Duloxetine in the acute and long-term treatment of major depressive disorder: A placebo- and paroxetine-controlled trial. *European Neuropsychopharmacology* **14**, 457–470.

Fleming, T.R. (2011). Addressing missing data in clinical trials. *Annals of Internal Medicine* **154**, 113–117.

Goldstein, D.J., Lu, Y., Detke, M.J., Wiltse, C., Mallinckrodt, C., and Demitrack, M.A. (2004). Duloxetine in the treatment of depression: A double-blind placebo-controlled comparison with paroxetine. *Journal of Clinical Psychopharmacology* **24**, 389–399.

Hamilton, M. (1960). A rating scale for depression. *Journal of Neurological and Neurosurgical Psychiatry* **23**, 56–61.

International Conference on Harmonization (ICH) Guidelines. Online at: http://www.ich.org/cache/compo/276254-1.html.

Kim, Y. (2011). Missing data handling in chronic pain trials. *Journal of Biopharmaceutical Statistics* **21**, 311-325.

Lane, P.W. (2008). Handling drop-out in longitudinal clinical trials: A comparison of the LOCF and MMRM approaches. *Pharmaceutical Statistics* **7**, 93–106.

Little, R.J.A., D'Agostino, R., Dickersin, K., Emerson, S.S., Farrar, J.T., Frangakis, C., Hogan, J.W., Molenberghs, G., Murphy, S.A., Neaton, J.D., Rotnitzky, A., Scharfstein, D., Shih, W., Siegel, J.P., and Stern, H. National Research Council (2010). *The Prevention and Treatment of Missing Data in Clinical Trials. Panel on Handling Missing Data in Clinical Trials.* Committee on National Statistics, Division of Behavioral and Social Sciences and Education. Washington, D.C.: The National Academies Press.

Little, R.J.A., and Rubin, D.B. (2002). *Statistical Analysis with Missing Data* (2nd Ed.). New York: John Wiley & Sons.

Little, R., and Yau, L. (1996). Intent-to-treat analysis for longitudinal studies with drop-outs. *Biometrics* **52**, 1324–1333.

Ma, G., Troxel, A.B., and Heitjan, D.F. (2005). An index of local sensitivity to nonignorable drop-out in longitudinal modeling. *Statistics in Medicine* **24**, 2129–2150.

Mallinckrodt, C.H., Raskin, J., Wohlreich, M.M., Watkin, J.G., and Detke, M.J. (2004). The efficacy of duloxetine: A comprehensive summary of results from MMRM and LOCF in eight clinical trials. *BMC Psychiatry* 4, 26.

Mallinckrodt, C.H., Lane, P.W., Schnell, D., Peng, Y., and Mancuso, J.P. (2008). Recommendations for the primary analysis of continuous endpoints in longitudinal clinical trials. *Drug Information Journal* **42**, 305–319.

Mallinckrodt, C.H. and Kenward, M.G. (2009). Conceptual considerations regarding choice of endpoints, hypotheses, and analyses In longitudinal clinical trials. *Drug Information Journal* **43**, 449–458.

Mallinckrodt, C.H, Lin, Q., Lipkovich, I., and Molenberghs, G. (2012). A structured approach to choosing estimands and estimators in longitudinal clinical trials. *Pharmaceutical Statistics* **11**, 456–461.

Mallinckrodt, C.H. (2013a). *Preventing and Treating Missing Data in Longitudinal Clinical Trials: A Practical Guide.* New York: Cambridge University Press.

Mallinckrodt, C.H., Roger, J., Chuang-Stein C, Molenberghs, G., O'Kelly, M., Ratitch, B., Janssens, M., and Bunouf, P. (2014). Recent developments in the prevention and treatment of missing data. *Therapeutic Innovation & Regulatory Science* **48**, 68–80.

Mallinckrodt, C.H., Roger, J., Molenberghs, G., Lane, P.W., O'Kelly, M., Ratitch, B., Xu, L., Gilbert, S., Mehrotra, D., Wolfinger, R., and Thijs, H. (2013b). Missing data: Turning guidance into action. *Statistics in Biopharmaceutical Research* **5**, 369–382.

Molenberghs, G. and Kenward, M.G. (2007). *Missing Data in Clinical Studies.* Chichester: John Wiley & Sons.

Molenberghs, G., Thijs, H., Jansen, I., Beunckens, C., Kenward, M.G., Mallinckrodt, C., and Carroll, R.J. (2004). Analyzing incomplete longitudinal clinical trial data. *Biostatistics* **5**, 445–464.

Molenberghs, G., Verbeke, G., Thijs, H., Lesaffre, E., and Kenward, M. (2001). Mastitis in dairy cattle: Local influence to assess sensitivity of the dropout process. *Computational Statistics & Data Analysis* **37**, 93–113.

O'Neill, R.T. and Temple, R. (2012). The prevention and treatment of missing data in clinical trials: An FDA Perspective on the importance of dealing with it. *Clinical Pharmacology & Therapeutics* **91**, 550–554.

Permutt, T. and Pinheiro J. (2009). Dealing with the missing data challenge in clinical trials. *Drug Information Journal* **43**, 403–408.

Ratitch, B. and O'Kelly, M. (2011). *Implementation of Pattern-Mixture Models Using Standard SAS/STAT Procedures.* PharmaSUG 2011. Available at http://pharmasug.org/proceedings/2011/SP/PharmaSUG-2011-SP04.pdf (accessed October 4, 2011)

Ratitch, B., O'Kelly, M., and Tosiello, R. (2013). Missing data in clinical trials: from clinical assumptions to statistical analysis using pattern mixture models. *Journal of Pharmaceutical Statistics* **12**, 337–347.

SAS Institute, Inc. (2008). *SAS/STAT 9.2. Users Guide.* Cary, NC: SAS Institute, Inc.

Shen, S., Beunckens, C., Mallinckrodt, C., and Molenberghs, G. (2006). A local influence sensitivity analysis for incomplete longitudinal depression data. *Journal of Biopharmacological Statistics* **16**, 365–384.

Siddiqui, O., Hung, H.M., and ONeill, R.O. (2009). MMRM vs. LOCF: A comprehensive comparison based on simulation study and 25 NDA datasets. *Journal of Biopharmaceutical Statistics* **19**, 227–246.

Teshome, B., Lipkovich, I., Molenberghs, G., and Mallinckrodt, C.H. (2013) Placebo multiple imputation: A new approach to sensitivity analyses for incomplete longitudinal clinical trial data. *Submitted for publication.*

Thijs, H., Molenberghs, G., and Verbeke, G. (2000). The milk protein trial: influence analysis of the dropout process. *Biometrical Journal* **42**, 617–646.

Troxel, A.B., Ma, G., and Heitjan, D.F. (2004). An index of local sensitivity to nonignorability. *Statistica Sinica* **14**, 1221–1237.

Tsiatis, A.A. (2006). *Semiparametric Theory and Missing Data.* New York: Springer.

Verbeke, G. and Molenberghs, G. (2000). *Linear Mixed Models for Longitudinal Data.* New York: Springer.

Wonnacott, T.H. and Wonnacott, R.J. (1981). *Regression: A Second Course in Statistics.* New York: John Wiley & Sons.

Zhu, H.T. and Lee, S.Y. (2001). Local influence for incomplete-data models. *Journal of the Royal Statistical Society, Series B* **63**, 111–126.

23

Missing Data in Sample Surveys

Thomas R. Belin

UCLA Jonathan and Karin Fielding School of Public Health, Los Angeles, CA

Juwon Song

Korea University, Seoul, Korea

CONTENTS

Data collected from surveys often have missing values due to nonresponse. For example, some sampled individuals may not answer questions about financial matters or their personal life. In a political election survey, some respondents may not reveal their preferred political party or candidate. Censuses, like surveys, also face challenges from missing data. In an earlier era, statisticians struggled with very limited tools to address potential errors in estimation due to missing data. For example, in his seminal textbook on sample-survey techniques, Cochran (1977, Ch. 13) showed that there is potential for substantial bias in estimates of quantities of interest due to survey nonresponse but offered little in the way of guidance unless the investigator was prepared to pursue call-backs to try to convert initial nonrespondents into respondents. For binary outcomes, Cochran considered intervals incorporating bounds reflecting the possibilities that all of the missing items either took on the value 0 or took on the value 1, but he acknowledged that such intervals might be so wide as to be of little practical use.

This chapter aims to provide an overview of strategies for addressing missing data in sample surveys as well as to discuss different (and to some extent divergent) perspectives on how best to approach the challenges presented by missing data in sample surveys.

23.1 Design-Based versus Model-Based Inference

We would place the discussion of these perspectives in the context of broader consideration that has been given in the literature to the relative merits of design-based versus model-based inference from survey data. The paper by Hansen, Madow, and Tepping (1983a) and companion discussion by Royall (1983), Little (1983a), Dalenius (1983), Smith (1983), and Rubin (1983) along with the rejoinder by Hansen, Madow, and Tepping (1983b) provide an excellent springboard for broader consideration of the relevant philosophical perspectives. Relevant subtleties at the interface between design-based and model-based inference are also presented in Rubin's (1987) text on multiple imputation, which embraces a finite-population inference perspective but intertwines model-based strategies in an effort to develop interval-estimation procedures with good coverage properties, as well as in Little (2004), which reviews strengths and weaknesses of both frameworks.

In particular, we discuss the implications of different perspectives on missing data in surveys in a setting where one of us (TRB) had first-hand involvement (Belin, et al. 1993a) and where diverging perspectives fueled a lively debate (reflected in Rubin 1996 and Fay 1996 among a number of other contributions to the literature). The applied context, involving unresolved cases in the 1990 Post-Enumeration Survey (PES) that followed the decennial U.S. census, was part of a broader decision analysis on whether to incorporate statistical adjustments into official census counts to address inaccuracies in the census. The setting also featured complex model-based estimation procedures, and when it was recognized that meaningful distinctions in variance estimates could arise depending on whether a model-based or design-based inference approach was adopted, discussions about the procedures used in the 1990 PES morphed into a broader methodological debate about the role of model-based procedures in survey-sampling settings. In addition to influencing scholarship in this area, these exchanges also helped clarify the foundation of multiple-imputation inference, particularly with respect to subtleties that were considered in the original development of the topic (Rubin, 1978; Rubin, 1987) but that did not emerge in sharp relief until the debate between Fay (1996) and Rubin (1996).

23.2 Design-Based versus Model-Based Perspectives on Missing-Data Uncertainty in Estimation of Census Undercount Rates

A brief sketch of the procedure that drew scrutiny in the context of census undercount estimation is as follows, with further detail available in Hogan (1993) and Belin, et al. (1993a). Undercount estimation in the 1990 census was based on "dual-system estimation" comparing the individuals enumerated by the census in a sample of blocks and the individuals enumerated in the PES. Computer-matching and clerical-matching operations identified individuals included in both samples; where there was a discrepancy, further field operations sought to resolve whether PES cases not found in the census were real (thereby providing evidence about omissions in the census) and to resolve whether census cases not found in the PES were correct enumerations (or, alternatively, erroneous enumerations, which could be viewed as balancing omissions). Inevitably, a fraction of the discrepant cases came back unresolved, typically because a follow-up interview was not obtained.

Rather than attempting further field follow-up, logistic regression models were fit to data on resolved cases to estimate the proportion of unresolved cases that would have been deemed omissions and erroneous enumerations. As summarized in Belin et al. (1993a), these logistic regression models controlled for main effects and low-order interactions of all of the factors defining sampling strata in the PES (primarily geographic and demographic characteristics) as well as numerous other quantities including a key covariate characterizing process information on the nature of the match or non-match (e.g., whole-household non-match with an address match, whole-household non-match without an address match, possible match found in the other data source, etc.) Building on a model-based perspective, a component of uncertainty due to unresolved match/enumeration status was added to the sampling covariance matrix to be incorporated in broader loss-function analyses; this component of uncertainty incorporated contributions due to sampling variation, uncertainty in parameter estimation, and uncertainty about the specific choice of estimation model, although all of the alternative approaches undertaken to estimate this component of model uncertainty made a missing-at-random (MAR) assumption for the unresolved cases given observed covariates.

An initial concern, elaborated in Fay (1991), had to do with underestimation of uncertainty associated with unresolved cases (whose match/enumeration statuses were being viewed as missing data) due to the possibility that the underlying models were wrong. Fay offered illustrative examples from a cluster sample with correlated outcomes within clusters where imputation assuming simple random sampling systematically underestimated uncertainty. Fay further argued that a design-based perspective on missing-data uncertainty offered greater robustness against model misspecification.

In follow-up discussions that included both Fay and Rubin (and that included TRB as a participant), Rubin commented that the multiple-imputation framework cannot be expected to work in settings where the imputation model is wrong in an important way. With no consensus emerging from those discussions, additional methodological investigation followed.

Fay developed examples, presented in Fay (1992) and Fay (1993), where the imputation model is correct, yet the traditional multiple-imputation variance estimate is not consistent, providing an overestimate of the target quantity. A key feature of these scenarios is that the "imputer," in charge of producing multiple imputations for downstream use, knows something that downstream analysts do not, namely that there is a characteristic (e.g., gender) where there is truly no difference between subgroups, so when analysts includes gender in their complete-data analyses, the estimated variance is systematically too large and interval estimates can have greater than the nominal level of 95% coverage. Meanwhile, Fay noted that a jackknife resampling strategy due to Rao and Shao (1992) produced intervals with very close to the nominal coverage level. In the context of the 1990 PES, the Rao/Shao approach would have involved leaving out one of the 5,280 block clusters and fitting the logistic regression model to the remaining clusters, then going through the same steps for each of the 5,280 clusters, with perturbations in the estimates from the logistic-regression models providing information for estimating missing-data variance.

We return to this applied context in the discussion section and turn now to more general considerations in handling missing data in surveys.

23.3 Weighting and Imputation as Strategies for Unit and Item Nonresponse

A useful distinction can be drawn between two types of nonresponse in surveys: unit nonresponse and item nonresponse (Levy and Lemeshow 1999; Dillman et al. 2002). Unit nonresponse occurs when an individual (or other unit of analysis) fails to provide any information in response to the survey despite having been selected into the sample. Item nonresponse refers to missing values on specific questions or items for an individual who does provide information on other questionnaire items. Although it is possible to address unit nonresponse using an imputation-based analysis strategy (e.g., Marker, Judkins, and Winglee, 2002; Pfefferman and Nathan, 2002), a common approach in survey-sampling contexts is to handle unit nonresponse by weighting and item nonresponse by imputation (Lohr, 1999; Little, 2003). There is a substantial literature on weighting techniques in surveys; in this chapter, we focus primarily on the issues that arise with item nonresponse and refer interested readers to the broader literature on weighting techniques (see., e.g., Little 1983b, 1991; Lazzeroni and Little 1998; Elliott and Little 2000; Bethlehem 2002; Little and Rubin 2002, Ch. 3; Little and Vartivarian 2003; Little and Vartivarian 2005; Elliott, 2007, 2009; and the references in each of these works).

Imputation refers to a statistical technique of filling in missing values with plausible values so that the filled-in, or imputed, datasets can be analyzed with standard analysis techniques for completely observed data. Many imputation strategies have been used to handle missing values in surveys, some based on explicit models and others based on implicit models (Little and Rubin 2002, Ch. 4 and 5). It is worth noting that aside from weighting strategies, essentially every alternative procedure involves some variation on the idea of imputation; for example, Efron (1994) uses a bootstrap procedure to draw inference about the largest eigenvalue in a covariance matrix, but an intermediate step of the procedure involves filling in missing values using a fitted least-squares regression.

We proceed from the perspective that the central purpose of surveys is to make inferences about population quantities. In that sense, the crucial consideration is not the accuracy associated with guesses or predictions of individual missing values but rather the statistical properties of procedures for drawing inferences about population quantities. In that spirit, Rubin (1987, pp. 118–132) develops a framework built around a definition of "proper" imputation methods that can be expected to support valid inferences for population quantities of interest. The framework, which is further elaborated in Rubin (1996), is built around the notion that if the first two moments of repeated-imputation estimates are asymptotically unbiased for the corresponding population quantities under the posited response mechanism, then the repeated-imputation inference will be "randomization valid" in the sense that confidence intervals will achieve the nominal level of coverage.

As a practical matter, it cannot be directly determined that an imputation procedure is proper, but as discussed in Rubin (1996), experience in applied settings underscores the importance of incorporating a wide array of variables that are apt to be predictive of missing values for the sake of avoiding bias in predictions. Relatedly, Collins, Schafer, and Kam (2001) show the value of including not just variables that will be included in the complete-data analysis but available auxiliary variables as well. See also David et al. (1986), Belin et al. (1993a), and Rubin, Stern, and Vehovar (1995) regarding the role of the assumption that the missing-data mechanism is "missing at random" (MAR) or the alternative that the missing-data mechanism is "not missing at random" (NMAR), in which case the missing-data mechanism is nonignorable for likelihood-based inference (for example, Greenlees, Reece, and Zieschang 1982; Rubin and Zanutto 2002; Yuan and Little 2008). Chapters 16–19 of this volume discuss sensitivity analysis under the NMAR assumption, which was also a recurrent theme in a recent National Academy of Sciences report on missing data in clinical trials (Little et al. 2010). Uncertainty about whether the missing-data mechanism is MAR or NMAR translates into greater uncertainty in inferences, and it can be challenging to be precise about this source of uncertainty; see, for example, Wachter (1993) and the rejoinder by Belin et al. (1993b).

23.4 Strategies for Producing Imputations

Once it is decided to pursue an imputation strategy, questions arise regarding how best to produce imputed values. A general principle is to make use of observed patterns of association between variables in the dataset, but this principle still leaves room for a wide range of possible alternative imputation methods. Imputation techniques can be classified according to whether they are based on explicit or implicit models (Little and Rubin, 2002). In explicit modeling, a statistical model is posited for multivariate data, and missing data are imputed by draws from the predictive distribution of the missing values given observed values. This framework includes imputation methods based on Bayesian iterative simulation procedures such as data augmentation (Tanner and Wong 1987; Gelman et al. 2004; Su et al. 2011). In contrast, algorithms that fill in missing values based on seemingly "common-sense" ideas, such as borrowing values from an individual whose values were observed and using them to impute for missing values on an individual who lives in the same neighborhood, can be thought of as being based on implicit models that have certain exchangeability assumptions embedded within them. Hot-deck imputation methods, where individuals with observed data serve as donors of imputed values to individuals with incomplete data, can be thought of as being based on implicit models. We consider these alternatives in turn.

23.5 Imputation Based on an Explicit Bayesian Model

When Y includes missing values, let Y_{obs} and Y_{mis} denote the observed and missing components of Y, respectively. Multiple imputation can be thought of as a form of numerical integration, and as such, it is possible to produce multiple imputations for either ignorable or non-ignorable models. In practice, since fitting non-ignorable models requires information from a source outside the observed data, it is typical for implementations of imputation procedures to assume an ignorable missing data mechanism, where the process consists of a modeling step and an imputation step.

The modeling step specifies the distribution of data, $f(Y|\theta)$, where θ denotes the unknown parameters (Little 1992; Little and Zheng 2007). An appropriate model for multivariate incomplete data Y is selected in this step. For continuous data, a multivariate normal distribution is a common choice given the availability of software for this model, and, if necessary, some variables may be transformed to achieve normality. For categorical data, a multinomial distribution can be assumed, or a log-linear model can be chosen for model parsimony (Schafer 1997). In many cases, survey data may include different types of variables. For data with both continuous and categorical variables, a general location model can be chosen (Schafer 1997; Raghunathan and Grizzle 1995). When data include various different types of variables, Raghunathan et al. (2001) proposed a "sequential regression multiple imputation" method using a set of overlapping regression models as conditional distributions, which might be thought of as an approximation to the set of conditional distributions that would define the steps of an iterative simulation procedure for a fully specified joint multivariate model. For Bayesian modeling, a non-informative prior distribution for θ, $p(\theta)$, is often considered, but an informative prior might be needed to obtain stable parameter estimates in complex models.

Typically, the imputation step consists of iteratively taking draws from the following distributions. At the tth iteration,

(1) draw a value of θ, $\theta^{(t)}$, from its posterior distribution, $p\left(\theta \middle| Y_{obs}, Y_{mis}^{(t-1)}\right)$;

(2) draw a value of Y_{mis}, $Y_{mis}^{(t)}$, from its predictive distribution, $p\left(Y_{mis} \middle| Y_{obs}, \theta^{(t)}\right)$.

In a fully specified joint model, iterating such a sequence gives rise to a Markov chain (hence the name "Markov chain Monte Carlo," or MCMC, being used to describe such approaches), and theory suggests that under mild conditions, such a chain converges in distribution to the Bayesian posterior distribution that can be used to support inferences about the target quantities. Put differently, imputed values of Y_{mis} can be obtained from $Y_{mis}^{(t)}$ by running an MCMC algorithm for a number of iterations and taking the values from the last draw to serve as imputed values. When using sequential regression multiple imputation, the justification rests on an approximation, but the idea is the same.

Imputation procedures can avoid the need to model the mechanism giving rise to missing values when the missing data mechanism is missing at random (MAR) and the parameters of the data model and the parameters of the missingness mechanism model are distinct (Rubin 1976, 1987; Little and Rubin 2002). Typically, since information about the parameters in the data model would not ordinarily be expected to provide information about the parameters in the missingness mechanism model, the MAR assumption is more important than the distinctness assumption for the sake of justifying an assumption of an ignorable missing data mechanism.

Schafer (1997) points out that any MAR assumption has to be understood as a relative assumption, that is, relative to a set of other variables in the model. For example, in the context of imputation of wage and salary data in the U.S. Current Population Survey, Greenlees et al. (1982) controlled for 19 variables yet still found evidence of systematic differences between model predictions and observed values in a matched dataset, thus violating MAR. However, working in the same context, David et al. (1986) controlled for 90 variables including indicators for nominal categorical variables, interactions, and quadratic terms and, consistent with MAR, found little evidence of any systematic difference between model predictions and observed values in a matched dataset. As a practical matter, this suggests that an imputation model should include as many related variables as possible to satisfy MAR and avoid bias in survey inference (Rubin 1996; Schafer and Graham 2002). Recent work in the area of missing-data methodology has considered settings where part of the missing-data mechanism can be considered MAR and part not, giving rise to notions of partial or latent ignorability that can be thought of as expanding the scope of models that might be useful in applications (Harel and Schafer 2009).

With survey data, nonresponse could also be handled by predicting missing values with an imputation model that contains inclusion probabilities as covariates (Little, 2004). Model-based inference provides consistent and efficient estimators when the model is correctly specified but can give rise to bias when the model is misspecified. Semi-parametric approaches are also possible; Paik (1997) describes generalized-estimating-equation approaches that involve modeling the response mechanism, and Chen, Elliott, and Little (2010, 2012) propose spline-based inference in a Bayesian framework and discuss its advantages over design-based inference and linear model-based inference. In line with comparisons in the literature on design-based versus model-based inference, the extent to which there is an advantage to either parametric or semi-parametric modeling approaches will depend on the nature of model misspecification (e.g., Liu, Taylor, and Belin 2000).

23.6 Hot-Deck Imputation

Hot-deck imputation has considerable appeal as an imputation method for survey data due to its flexibility to handle various types of variables as well as the face validity of imputing values that have been observed in the same dataset on other individuals. In a simple scenario, assume that a sample survey has n targeted participants, with r individuals who provided responses and $n-r$ individuals who did not. In simple random hot-deck imputation, the r respondents can be thought of as donors, with $n-r$ values randomly chosen from the r respondents used to impute missing values to the $n-r$ nonrespondents. One can envision variations on this idea. For example, it would be possible to obtain matched pairs between $n-r$ donors and $n-r$ nonrespondents by sequential selection (e.g., borrowing values from a nearest neighbor based on geographic proximity).

Unless the missing-data mechanism is missing completely at random, the simple random hot-deck can be expected to give rise to biased estimates. In practice, various alternative hot-deck imputation methods seek to condition on salient characteristics of individuals to yield values that are more plausible and in better alignment with observed characteristics. Because hot-deck methods are widely used in sample-survey settings, we offer additional more detailed background on alternative hot-deck techniques.

23.6.1 Hot-deck imputation within adjustment cells

Suppose missing values occasionally occur on one of the variables, Y_1, in a dataset comprised of a broader set of variables $Y = (Y_1, \ldots, Y_p)$ where at least some of the other variables, $Y_2, \cdots, Y_p$, may be closely related to Y_1. If all participants, both respondents and nonrespondents, can be classified into groups with similar values of related variables, then it seems reasonable to expect that respondents and nonrespondents within the same group have similar Y_1 values. The groups with similar characteristics are called adjustment cells (also called imputation cells or imputation classes). In hot-deck imputation within adjustment cells, each nonrespondent is linked to a donor from among the respondents in the same adjustment cell.

Clearly, it would be desirable to include as many related variables as possible to build adjustment cells, thereby conditioning on as much available information as possible. The idea that the imputation method should include all highly related variables suggests that variables characterizing a survey design, such as stratum or cluster indicators, should be reflected in the procedure. As discussed in the previous section, the formation of hot-deck cells might similarly be done with an eye toward ensuring that an MAR assumption is satisfied. Andridge and Little (2010) discuss two key properties of a set of variables, X, used to form adjustment cells in hot-deck imputation. When the first variable Y_1 of data Y includes missing values, the first property is whether or not X is associated with incomplete variable Y_1, and the second one is whether or not X is associated with a missingness indicator vector, M_1, where M_1 indicates the first column of the missingness indicator matrix M. Following Little and Vartivarian (2005), they suggest including a set of variables X if they are highly related with both Y_1 and M_1 to reduce bias and increase precision in the estimation. On the other hand, when X is highly related with Y_1 but not with M_1, inclusion of X may increase precision, but it does not reduce bias.

However, cross-classification on a large number of characteristics gives rise to a combinatorial explosion of covariate patterns and inevitably leads to sparse representation in adjustment cells. Hot-deck imputation within adjustment cells is successful when the number of respondents is much larger than the number of nonrespondents in each adjustment cell, but if many related variables are used to build adjustment cells, it can be anticipated that some nonrespondents would be in adjustment cells with no available donors. If this difficulty occurs, a possible approach is to drop variables in defining adjustment cells until all nonrespondents have donors within their adjustment cells, but discarding available information in this manner can be unappealing. Another possible approach starts with adjustment cells generated by a large set of related variables, yielding as many donors as possible, after which adjustment cells are coarsened by dropping variables in a sequence until a donor is found for each non-respondent. Such a hot-deck procedure has been used in the Current Population Survey (CPS) in the U.S. (David *et al.* 1986), with the method now sometimes known as the "CPS hot-deck."

23.6.2 Hot-deck imputation using distance metrics

When related variables are continuous variables, hot-deck within adjustment cells requires categorizing continuous variables to build the adjustment cells, which can have undesirable consequences. As an alternative, distance metrics can be chosen to measure similarity between respondents and nonrespondents. When one of variables Y_1 in data Y include missing values, other related variables, $X = (Y_2, \cdots, Y_p)$, are used to calculate a distance between ith nonrespondent and jth donor by using various distance metrics. Examples of commonly used distance metrics include

(1) Mahalanobis distance,

$$d(i,j) = (X_i - X_j)' S_x^{-1} (X_i - X_j),$$

where X_i and X_j indicate a $(p-1)$ vector of $Y_2, \cdots, Y_p$ for the ith nonrespondent and jth respondent, and S_x indicates an estimated variance-covariance matrix of X_i;

(2) the maximum deviation,

$$d(i,j) = \max_k |X_{ik} - X_{jk}|,$$

for $k = 1, \ldots, p-1$, where X_{ik} and X_{jk} indicates the kth variable of X_i, that is, Y_{k+1}, for ith nonrespondent and jth respondent, respectively; and

(3) a distance between the predictive mean for ith nonrespondent and jth respondent (Little, 1988),

$$d(i,j) = \left(\widehat{Y}_{1i} - \widehat{Y}_{1j}\right)^2,$$

where $\widehat{Y}_{1i}$ and $\widehat{Y}_{1j}$ are the predicted value of Y_1 for the ith nonrespondent and for the jth respondent from the regression of Y_1 on $X = (Y_2, \cdots, Y_p)$ obtained from the respondents, respectively.

Siddique and Belin (2008) present a procedure that involves randomly selecting a donor with probability inversely proportional to the donor's calculated distance from a given nonrespondent. The nearest neighbor hot-deck selects donors from respondents with the smallest $d(i,j)$. Instead of choosing the closest donor, adjustment cells can be defined by a range of $d(i,j)$. For example, a donor pool for ith nonrespondent is defined by respondents whose $d(i,j) \leq c$, for a pre-specified c.

Instead of relying on a calculated distance based on a predictive mean, the predictive mean can be used to form adjustment cells (Bell 1999). With this approach, the predictive mean $\widehat{Y}_{1i}$ is sorted for all participants, and adjustment cells can be formed based on the predictive mean. Hot-deck imputation within adjustment cells is a special case of hot-deck using the predictive mean, where all related variables are categorical and all interactions are included in the prediction.

23.6.3 Maintaining relationships between variables in multivariate data

Survey data often include many variables that are inter-related. When missing values occur on more than one variable, imputation should be conducted in a way that maintains consistency among related variables. Bernaards, Belin, and Schafer (2007) offer an illustration of how problems can arise when procedures do not respect associations between variables. Exploring the performance of adapting multivariate normal imputations for missing binary data, they considered alternative rounding procedures as well as a "coin-flipping" procedure that treated imputed values between 0 and 1 as probabilities that the binary variable should take on the value 1. Although the procedure performed well for estimating marginal properties, it gave rise to severe under-coverage for estimating an odds ratio. The issue was that the true odds ratio in their simulation evaluation was close to 8, but performing coin flips separately for different variables implicitly assumed an odds ratio of 1.

Hot-deck procedures have been developed with an eye toward addressing similar concerns. Single-partition common-donor hot-deck (Marker et al. 2002) selects a donor for each nonrespondent using a multivariate version of the predictive mean metric. For incomplete data Y, assume that more than one variable, $Y_1^* = (Y_1, \cdots, Y_q)$ include missing values, and other fully observed related variables, $X = (Y_{q+1}, \cdots, Y_p)$, are used to calculate the distance between the ith nonrespondent and jth donor by using distance metrics as follows:

$$d(i,j) = \left(\widehat{Y}_{1i}^* - \widehat{Y}_{1j}^*\right)' S_{Y_1^*}^{-1} \left(\widehat{Y}_{1i}^* - \widehat{Y}_{1j}^*\right),$$

where $\widehat{Y}_{1i}^*$ is a $(q \times 1)$ vector of predicted values of Y_1^* for the ith nonrespondent from the multivariate regression of Y_1^* on $X = (Y_2, \cdots, Y_p)$ obtained from the respondents and $S_{Y_1^*}$ is an estimated residual variance-covariance matrix of the multivariate regression of Y_1^* on X.

Another approach, called the n-partition hot-deck (Marker et al. 2002), sequentially conducts hot-deck imputation within adjustment cells. To impute the first variables Y_1, adjustment cells are formed by X. To impute the next variable Y_2, the imputed Y_1 and X are used to build adjustment cells. Similarly, imputation of the third variable Y_3 uses the imputed Y_1, imputed Y_2, and X to form adjustment cells. Finally, the variable Y_q is imputed within adjustment cells that have been created by using the imputed Y_1 to Y_{q-1}, and X. Since the previously imputed values are used to form adjustment cells for later variables, the imputation order may be important in this hot-deck. Decisions about the imputation order can make use of background information and knowledge, and the percentage of missing values and importance of the variables may be also considered.

The single-partition common-donor hot-deck and n-partition hot-deck can be combined by dividing the variables Y_1^* into sets. In each set, the single-partition common donor hot-deck is implemented, and the earlier imputed values are used to form adjustment cells in imputation of later sets. Another approach (Grau et al. 2004) implements hot-deck using a distance metric within each set in that hot-deck imputation in later sets uses imputed values from earlier sets to calculate the distance metric.

The full-information common-donor hot-deck (Marker et al. 2002) applies a separate single-partition common donor hot-deck to each subset of data with different missingness patterns. At this point, comparisons of the performance of alternative methods are limited.

23.7 Sampling Weights

As reflected in Hansen, Madow, and Tepping (1983a) and the discussions of that work (see list in Section 23.1), there have been long-running debates about the merits of incorporating sample-design weights in analyses of survey data. A more recent review that clarifies the key issues is offered by Korn and Graubard (1999).

In a complex survey with stratification and/or clustering, participants might have different design weights, and imputation ignoring these weights can give rise to bias in estimating the variance of the parameters (Kott 1995; Wang and Robins 1998; Kim et al. 2006). In the context of estimating regression parameters for multiple-phase samples using sampling and nonresponse weights along with hot-deck imputation, Fuller (2003) is among the authors advocating that imputation of data with a complicated sampling scheme should incorporate the sampling weights. Rao and Shao (1992) and Rao (1996) suggest selecting donors with probability proportional to respondents' sampling weight. Under the assumption that the response probability is constant within an adjustment cell, this method yields an asymptotically unbiased estimator of variances of key target quantities. In sequential hot-deck imputation, sampling weights have been used to restrict the number of times a respondent serve as donors (Williams and Folsom 1981).

In contrast, Ezzati-Rice et al. (1995) suggest incorporating design weights as well as clustering and stratification indicators as covariates in imputation models. Pfefferman and LaVange (1989) point out that sampling weights can become redundant, providing little or no additional protection against model misspecification, when design characteristics are used as predictors in a model-based inference framework. Little and Vartivarian (2003) clarify the tradeoffs involved in such discussions, suggesting that it might not be advisable to offer blanket advice such as that it is always best to perform weighted analyses when analyzing incomplete survey data.

23.8 Multiple Imputation

Multiple imputation (Rubin 1978, 1987), where each missing value is filled in more than one time, has been a successful, widely-used method for representing uncertainty due to missingness. In multiple imputation, each imputed dataset is analyzed using a standard complete-data technique, with inference based on combining individual analysis results in a way that reflects both between-imputation variability and average within-imputation variability in target quantities. Unless the fraction of missing information is high, simulation studies suggest that even just a few imputations are sufficient to provide a solid foundation for inference in incomplete-datasettings (Rubin 1987). Chapters 11–14 of this volume cover a range of topics pertaining to multiple imputation in greater detail.

Rubin (1987) discusses the statistical validity of multiple imputation from a frequentist viewpoint, developing the notion of "proper" imputation as a way to link imputation procedures (including model-based procedures) with frequency-based evaluations of interval-estimate coverage. In brief, the idea requires that point estimation (including the second moments of quantities of interest) must be approximately unbiased, and confidence intervals and tests must be valid in the context of repeated sampling under the posited nonresponse mechanism. Rubin (1987) shows that if complete-data inference is randomization-valid and multiple imputation is proper, inference based on an infinite number of imputations is randomization-valid under the posited response mechanism.

A key conceptual insight is that when multiple imputation is based on drawing from the posterior distribution of Y_{mis} under the posited response mechanism and an appropriate data model, the imputation procedure can be expected to be proper if the model is successful in representing key patterns of association and key sources of uncertainty in the data (Rubin 1987, pp. 118–132). Imputation under a Bayesian model typically achieves these goals through two steps: (1) draw the parameters of the model from their posterior distribution, and (2) draw missing values from their posterior distribution conditional on the drawn values of the parameters in step (1).

Another key conceptual insight relevant to hot-deck imputation is that multiple hot-deck imputation, even when sampling with replacement, is not proper unless another stage is incorporated into the procedure. Hot-deck imputation can become proper by using either a Bayesian bootstrap (Rubin 1981) or approximate Bayesian bootstrap (Rubin and Schenker 1986), where bootstrap replicates are generated from the original data values and then hot-deck imputation with replacement is performed. Intuitively, the resampling step implicitly represents estimation uncertainty for the underlying model parameters, akin to step (1) in the Bayesian-model procedure.

When the sample size is small and the response rate is very low, Kim (2002) shows that bias in the variance estimator of the population mean may occur under multiple imputation using ABB. When the sample size gets larger, the bias becomes negligible.

23.9 When the Imputer's Model Differs from the Analyst's Model

Large-scale surveys typically include many items that are related to each other, potentially in a complicated manner. An appropriate imputation of missing values in large-scale surveys can require substantial time and effort, since many proper imputation techniques require statistical and computing expertise. Therefore, it is often recommended that multiple imputations should be produced by a person or an organization having extensive relevant knowledge. In this case, the "imputer" and the eventual data analyst might not be the same. Moreover, the available data might include some confidential information that is useful for imputing missing values but cannot be released to public, and this could give rise to discrepancies between an imputation model chosen by the imputer and an analysis model chosen by the analyst. On the other hand, different analysts may use different models for their analyses, and some of variables they consider in their models might not be included in the imputation model. Discrepancies between the imputer's model and the analyst's model gives rise to what has been termed uncongeniality (Meng 1994), and the validity of multiple imputation inference has been debated for the case where the imputer's model and the analyst's model are different (Fay 1991, 1992, 1993; Kott 1992; Meng 1994, 2002; Rubin 1996).

23.9.1 When the analyst considers fewer variables than the imputer

When imputation is conducted by an expert or an organization that collects data, the imputer might have additional variables that the analyst cannot access. Moreover, the analyst may focus on a model that includes only certain variables of interest. In these cases, the imputation model may include more variables than the analysis model.

For a continuous data Y and a variable X having two values, 0 and 1, suppose that Y has r_0 responses among n_0 observations where $X = 0$ and has r_1 responses among n_1 observations where $X = 1$, where $n_0 + n_1 = n$. Consider a situation where the imputer uses a variable X in the imputation, but the analyst does not include X in the analysis. In the explicit imputation modeling, the variable X is included as a covariate in the imputation model. In hot-deck imputation, two adjustment cells are formed with X and hot-deck imputation is conducted within adjustment cells. When the variable X is not related to Y, the imputer's model is still valid, although it might be somewhat conservative due to reflecting an extra degree of uncertainty by considering X in the imputation model. On the other hand, when the variable X is related to Y, the analyst's estimated mean of Y can be biased due to ignoring X. This problem occurs since the analyst's model is misspecified; there is nothing inherently wrong with the imputed dataset in this instance.

23.9.2 When the imputer assumes more than the analyst

It is also possible that the imputer would assume that certain variables are not related to missing data, while some analysts might include them in their model.

When the variable X is not related to Y, the estimated variance of $\widehat{\theta}_\infty$, T_∞, can be smaller than the variance of $\widehat{\theta}$, $\widehat{T}$, reflecting that the estimated $\widehat{\theta}_\infty$ is more efficient than an observed-data estimate $\widehat{\theta}$ ignoring imputation, because it incorporates the imputer's superior knowledge about the variable X. Rubin (1996) related this result to the mathematical-statistics idea of super-efficiency, noting that the coverage of an interval estimate in such a setting can be greater than the nominal $100(1-\alpha)\%$ level. Alluding to the way Neyman defined confidence intervals, Rubin (1996) further argues that the appropriate standard for evaluation purposes is not how close the coverage of a procedure is to the nominal level but rather is the precision of inference (e.g., narrowness of interval estimates) among those procedures that do achieve the nominal level of coverage. In simulations presented in Rubin (1996), both the Rao and Shao (1992) procedure and multiple imputation succeed in achieving the nominal coverage level, but even though the coverage levels of the multiple imputation intervals are the same or greater than the coverage of Rao-Shao intervals, the widths of the multiple-imputation intervals are the same or narrower than those with the Rao-Shao procedure. Rubin (1996) does not view super-efficiency as a serious problem because it is beneficial to have narrower intervals. On the other hand, if the variable X is related to Y but is not included in imputation, the imputation model might give rise to biased estimates and invalid inferences. To avoid this bias, Rubin (1996) suggests that the imputation model should be chosen to include as many variables as possible, which also might be helpful in terms of satisfying a missing-at-random assumption.

Mislevy et al. (1992) evaluates the effect of omitting variables in the imputation model. Specifically, they evaluate the magnitude of the bias using data from the National Assessment of Educational Progress (NAEP), noting that it depends on the extent to which the variables are related to missing data and the degree to which omitted variables are explained by variables in the imputation model. Interestingly, it is not essential for the imputation model to be perfect in order for it to be useful. Mislevy et al. (1992) offered an example where only half of the variables that had important associations relevant to predicting missing values were included in the imputation model, yet the bias in analyses remained negligible.

23.10 Variance Estimation with Imputed Data

Imputed data can be analyzed by using standard analysis techniques as if they have no missing values. However, the imputed portion of data is not known to be real, so treating imputed values as real can be expected to exaggerate the information available in the data, resulting in underestimation of the variance of the parameters. There are three main approaches to obtain valid variance estimates from imputed data.

23.10.1 Applying explicit variance formulae

Explicit variance formulae have been suggested under a few simple survey settings. For example, under MCAR, the explicit variance can be easily calculated for hot-deck imputation within adjustment cells (Little and Rubin 2002). Särndal (1992) decomposed the variance of an estimator in the model-based imputation under MAR, and Brick et al. (2004) extended this method to handle hot-deck imputed data. Chen and Shao (2000) derived an asymptotic variance formula for the mean in nearest-neighbor hot-deck imputation, but their formula is restricted to handle one missing variable Y and one auxiliary variable X, and one has to assume a conditional distribution of Y on X. For complex survey data, it is often hard to calculate explicit variance formulae.

23.10.2 Resampling methods

Resampling methods such as the bootstrap or jackknife can be applied to assess parameter uncertainty. One approach proposes modifying formulae for the estimated variance when single imputation is conducted; another approach calculates the estimated variance of the parameters using resampling-based multiple imputations.

Rao and Shao (1992) propose a modification of the variance estimate of the population mean in single hot-deck imputation using the jackknife method in complex survey data, leaving out each unit one at a time, re-estimating target quantities, and taking account of parameter uncertainty due to perturbing the dataset. Suppose that r participants respond among n sampled individuals and that simple random sampling is chosen to recruit participants. Hot-deck imputation can be applied to create an imputed dataset, and the estimator of the mean is calculated by

$$\bar{y}_{HD} = \frac{1}{n}\left(r\bar{y}_r + (n-r)\,\bar{y}^*_{nr}\right),$$

where $\bar{y}_r$ is the mean of y among r respondents and $\bar{y}^*_{nr}$ is the mean of the imputed values. For $i = 1,\ldots, n$, jackknife replicates are obtained by removing the ith observation from the imputed dataset and recalculating the mean, $\bar{y}^{(-i)}_{HD}$, using the remaining $n-1$ observations. The jackknife variance estimate is then calculated by

$$v_J = \frac{n-1}{n}\sum_{i=1}^{n}\left(\bar{y}^{(-i)}_{HD} - \bar{y}_{HD}\right)^2.$$

This can be generalized to handle hot-deck imputation within adjustment cells and to incorporate stratified multistage sampling. Rao (1996) also extends this single-imputation hot-deck approach to handle ratio imputation and regression imputation. The main limitation of this approach is that users of the imputed data need to calculate v_J by themselves, and computation requires knowledge of precise details about the chosen hot-deck imputations.

For multiply imputed data, jackknife methods can be applied to impute missing values and compute the estimated variance of the parameters (Little and Rubin 2002). Suppose that r participants respond among n samples in data Y. For $i = 1, \ldots, n$, the ith observation is removed from Y, yielding $Y^{(-i)}$. Missing values in $Y^{(-i)}$ are imputed and $\widehat{\theta}^{(-i)}$ is estimated. The jackknife variance estimate is calculated by

$$v_{J,MI} = \frac{n-1}{n}\sum_{i=1}^{n}\left(\widehat{\theta}^{(-i)} - \bar{\theta}\right)^2,$$

where $\bar{\theta} = \sum_{i=1}^{n}\frac{\widehat{\theta}^{(-i)}}{n}$.

This method can also be generalized to calculate the bootstrap estimated variance (Little and Rubin, 2002). The main limitation of this approach is that it can require a much larger number of imputations than is traditional with multiple imputation because the jackknife requires n imputations and the bootstrap is usually implemented with at least several hundred replicates.

Fay (1996) proposed fractionally weighted imputation that builds on m multiple imputations and assigns weights of $1/m$ to each imputed values. The estimator of the mean is calculated by

$$\bar{y}_{FWI} = \frac{1}{n}\left(\sum_{j=1}^{r} y_j + \sum_{j\in A_{nr}}\frac{1}{m}\sum_{l=1}^{n-r} y^*_{(FWI)jl}\right),$$

where A_{nr} indicates a set of nonrespondents and $y^*_{(FWI)jl}$ is the imputed value for jth nonrespondent in lth imputation. The jackknife variance is then estimated by

$$v_{J,FWI} = \frac{n-1}{n}\sum_{i=1}^{n}\left(\bar{y}_{FWI}^{(-i)} - \bar{y}_{FWI}\right)^2.$$

Kim and Fuller (2004) describe applications of the jackknife variance estimator in fractional hot-deck imputation and Kim et al. (2006) extends it to parametric fractional imputation. A feature of this method is that it requires multiply imputed datasets and computation of $v_{J,FWI}$ by analysts.

23.10.3 Multiple imputation

Multiple imputation of incomplete data Y generates M sets of multiply imputed datasets, $Y_{imp,1}, \ldots, Y_{imp,M}$. For each imputed data $Y_{imp,i}$, where $i = 1, \ldots, M$, parameter estimates $\widehat{\theta}_1, \ldots, \widehat{\theta}_M$ and their estimated variances $W_1, \ldots, W_M$ are obtained. The estimate of the parameter θ is calculated by

$$\bar{\theta} = \sum_{i=1}^{M}\widehat{\theta}_i$$

and the associated variance is

$$T_M = W_M + \frac{M+1}{M}B_M,$$

where

$$W_M = \sum_{i=1}^{M}\frac{W_i}{M}$$

represents within-imputation variability and

$$B_M = \frac{1}{M-1}\sum_{i=1}^{M}\left(\widehat{\theta}_i - \bar{\theta}\right)^2$$

indicates between-imputation variability.

This formula can be extended to handle a multivariate parameter θ. Rubin (1996) discusses advantages of multiple imputation in detail, including that data storage is no longer a serious constraint, computation time is typically not a problem, and available software has made it much easier to implement the method.

23.11 Discussion

In this chapter, we have reviewed some of the relevant considerations when confronting missing data in sample surveys. Many of the issues that arise when handling incomplete data in other settings carry over naturally to the sample-survey setting, but the sample-survey setting gives rise to additional issues stemming from subtle differences between design-based and model-based inference. The need to deal with unknown probabilities of nonresponse has motivated a range of strategies for statistical inference, where the basis for preferring one strategy over another, already blurry with complete data, is even more so with incomplete data. We have sought here to illuminate the underlying issues both through a review of the literature and consideration of a motivating example involving census undercount estimation where the model-based approach that was used accommodated certain features of the survey design but did not specifically accommodate other features of the design. And with an imperfect model, it is natural to wonder about the validity of estimates of uncertainty for population quantities of interest. In Sec. 5.5 of the second edition of their *Statistical Analysis with Missing Data* book, Little and Rubin (2002) provide what we regard as a very helpful summary of subtleties relating to the estimation of imputation uncertainty, which we reproduce here in its entirety:

1. None of the methods is model-free, in the sense that they all make assumptions about the predictive distribution of the missing values in order to generate estimates based on the filled-in data that are consistent for population parameters.
2. In large samples where asymptotic arguments apply, resampling methods yield consistent estimates of variance with minimal modeling assumptions, whereas MI estimates of variance tend to be more closely tied to a particular model for the data and missing-data mechanism. Thus MI standard errors may be more appropriate for the particular dataset when the model is sound, whereas resampling standard errors are more generic (or less conditioned on features of the observed data) but are potentially less vulnerable to model misspecification. The issue of standard errors under misspecified models is discussed further in Chapter 6 [of Little and Rubin (2002)].
3. Resampling methods are based on large-sample theory, and their properties in small samples are questionable. The theory underlying MI is Bayesian and can provide useful inferences in small samples.
4. Some survey samplers have tended to be suspicious of MI because of its model-based, Bayesian etiology, and have favored sample reuse methods since they are apparently less dependent on parametric modeling assumptions. This may be more a question of complete-data analysis paradigms than of the method for dealing with imputation uncertainty. The relatively simple models of regression and ratio estimation with known covariates can form the basis for MI methods, and conversely resampling standard errors can be computed for more complex parametric models when these are deemed appropriate. The right way to assess the relative merits of the methods, from a frequentist perspective, is through comparison of their repeated-sampling operating characteristics in realistic settings, not their theoretical etiologies.
5. Some survey samplers have questioned the ability of MI to incorporate features of the sample design in propagating imputation uncertainty (e.g., Fay 1996). However, imputation models can incorporate stratification by including strata indicators as covariates, and clustering by multilevel models that include random cluster effects. In contrast, the complete-data inference can be design-based to incorporate these features, and can be based on a model that takes into account these features (Skinner, Smith, and Holt 1989).
6. MI is more useful than resampling methods for multiple-user data base construction, since a dataset with a relatively small set of MIs (say 10 or less) can allow users to derive excellent inferences for a broad range of

> estimands with complete-data methods, provided the MIs are based on a sound model (e.g., Ezzati-Rice et al. 1995). In contrast, resampling methods require 200 or more different imputed datasets, with imputations based on each resampled dataset, and transmitting this large set of resampled and imputed datasets to users may not be practical. Thus in practice the user needs software to implement a resampling imputation scheme on each replication.

Even as carefully drawn as these comments are, our expectation is that professionals in the field might not be entirely clear about the underlying message without additional context. In the census-undercount estimation setting, imputation models were pre-specified to include dozens of predictor variables reflecting main effects and some interactions among factors defining the stratified sample design, but the models did not include all possible interactions among these factors. Furthermore, no formal steps were incorporated into estimation procedures to account for clustering in the sample design. But by incorporating design characteristics into the model, the approach was consistent with the guidance of Pfefferman and LaVange (1989) for making the sampling weights redundant and providing protection against model misspecification in a model-based inference framework. As also documented in Belin *et al.* (1993a), investigation using evaluation follow-up data suggested little evidence of bias in model predictions, although questions might remain about the extent to which interval estimates should be wider to account for sources of uncertainty. Relevant to the decision about whether to adjust the official 1990 census counts (where it was ultimately decided not to do so), a model-based estimate of imputation uncertainty (including a component of uncertainty due to treating each of several variations on the underlying modeling assumptions as a priori equally likely) was incorporated in formal loss-function analyses. But as seems inherent in complicated applications, uncertainty about underlying assumptions is destined to force decision-makers to address some issues informally rather than formally, and subtleties having to do with diverging perspectives on design-based versus model-based inference fall into that category.

Efforts to bridge design-based and model-based perspectives on missing data can be expected to remain important in other settings as well. We hope that the material we have reviewed in this chapter provides additional context that can help guide future practice in this area.

References

Andridge, R.R. and Little, R.J.A. (2010). A review of hot deck imputation for survey non-response. *International Statistical Review* **78**, 40–64.

Belin, T.R., Diffendal, G.J., Mack, S., Rubin, D.B., Schafer, J.L., and Zaslavsky, A.M. (1993a). Hierarchical logistic regression models for imputation of unresolved enumeration status in undercount estimation (with discussion). *Journal of the American Statistical Association* **88**, 1149–1166.

Belin, T.R., Diffendal, G.J., Mack, S., Rubin, D.B., Schafer, J.L., and Zaslavsky, A.M. (1993b). Rejoinder on: Hierarchical logistic regression models for imputation of unresolved enumeration status in undercount estimation. *Journal of the American Statistical Association* **88**, 1163–1166.

Bell, R. (1999). *Depression PORT Methods Workshop (I).* Santa Monica, CA: RAND.

Bernaards, C., Belin, T.R., Schafer, J.L. (2007). Robustness of a multivariate normal approximation for imputation of incomplete binary data. *Statistics in Medicine* **26**, 1368–1382.

Bethlehem, J. (2002). Weighting nonresponse adjustments based on auxiliary information. In: *Survey Nonresponse.* R.M. Groves, D.A. Dillman, J.L. Eltinge, and R.J.A. Little (Eds.). New York: John Wiley & Sons, pp. 275–288.

Brick, J.M., Kalton, G., and Kim, J.K. (2004). Variance estimation with hot deck imputation using a model. *Survey Methodology* **30**, 57–66.

Chen, J. and Shao, J. (2000). Nearest neighbor imputation for survey data. *Journal of Official Statistics* **16**, 113–141.

Chen, Q. Elliott, M.R., and Little, R.J.A. (2010). Bayesian penalized spline model-based inference for finite population proportion in unequal probability sampling. *Survey Methodology* **36**, 23–34.

Chen, Q. Elliott, M.R., and Little, R.J.A. (2012). Bayesian inference for finite population quantiles from unequal probability samples. *Survey Methodology* **38**, 203–214.

Cochran, W.G. (1977). *Sampling Techniques,* (3rd Ed.). New York: John Wiley & Sons.

Collins, L.M., Schafer, J.L., and Kam, C.M. (2001). A comparison of inclusive and restrictive strategies in modern missing data procedures. *Psychological Methods* **6**, 330–351.

Dalenius, T. (1983). Comment on "An evaluation of model-dependent and probability-sampling inferences in sample surveys." *Journal of the American Statistical Association* **78**, 799–800.

David, M.H., Little, R.J.A., Samuhel, M.E., and Triest, R.K. (1986). Alternative methods for CPS income imputation. *Journal of the American Statistical Association* **81**, 29–41.

Dillman, D.A., Eltinge, J.L., Groves, R.M., and Little, R.J.A. (2002). Survey nonresponse in design, data collection, and analysis. In: *Survey Nonresponse.* R.M. Groves, D.A. Dillman, J.L. Eltinge, and R.J.A. Little (Eds.). New York: John Wiley & Sons, pp. 3–26.

Efron, B. (1994). Missing data, imputation, and the bootstrap (with discussion by D.B. Rubin). *Journal of the American Statistical Association* **89**, 463–479.

Elliott, M.R. (2007). Bayesian weight trimming for generalized linear regression models. *Survey Methodology* **33**, 23–34.

Elliott, M.R. (2009). Model averaging methods for weight trimming in generalized linear regression models. *Journal of Official Statistics* **25**, 1–20.

Elliott, M.R. and Little, R.J.A. (2000). Model-based alternatives to trimming survey weights. *Journal of Official Statistics* **16**, 191–209.

Ezzati-Rice, T.M., Johnson, W., Khare, M., Little, R.J.A., Rubin, D.B., and Schafer, J.L. (1995). A simulation study to evaluate the performance of model-based multiple imputation in NCHS Health Examination Survey. *Proceedings of the Bureau of the Census Eleventh Annual Research Conference*, pp. 257–266.

Fay, R.E. (1991). A design-based perspective on missing data variance. *Proceedings of the 1991 Annual Research Conference*, U.S. Bureau of the Census, 429-440.

Fay, R.E. (1992). When are inferences from multiple imputation valid? *ASA Proceedings in the Survey Research Methods Section*, pp. 227–232.

Fay, R.E. (1993). *Valid Inferences From Imputed Survey Data.* Paper presented at the Annual Meeting of the American Statistical Association, San Francisco, CA.

Fay, R.E. (1996). Alternative paradigms for the analysis of imputed survey data. *Journal of the American Statistical Association* **91**, 490–498.

Fuller, W.A. (2003). Estimation for multiple phase samples. In: *Analysis of Survey Data.* R.L. Chambers and C.J. Skinner (Eds.). New York: John Wiley & Sons, pp. 307–322.

Gelman, A., Carlin, J.B., Stern, H.S., and Rubin, D.B. (2004). *Bayesian Data Analysis*, (2nd Ed.). London: CRC / Chapman & Hall.

Greenlees, J.S., Reece, W.S., and Zieschang, K.D. (1982). Imputation of missing values when the probability of response depends on the variable being imputed. *Journal of the American Statistical Association* **82**, 251–261.

Grau, E.A., Frechtel, P.A., and Odom, D.M. (2004). A simple evaluation of the imputation procedures used in HSDUH. *ASA Proceedings in Section on Survey Research Methods*, pp. 3588–3595.

Hansen, M.H., Madow, W.G., and Tepping, B.J. (1983a). An evaluation of model-dependent and probability-sampling inferences in sample surveys (with discussion). *Journal of the American Statistical Association* **78**, 776–807.

Hansen, M.H., Madow, W.G., and Tepping, B.J. (1983b). Rejoinder on: An evaluation of model-dependent and probability-sampling inferences in sample surveys. *Journal of the American Statistical Association* **78**, 805–807.

Harel, O. and Schafer, J.L. (2009). Partial and latent ignorability in missing-data problems. *Biometrika* **96**, 37–50.

Hogan, H. (1993). The 1990 post-enumeration survey: Operations and results. *Journal of the American Statistical Association* **88**, 1047–1061.

Kim, J.K. (2002). A note on approximate Bayesian bootstrap. *Biometrika*, **89**, 470–477.

Kim, J.K., Brick, J.M., Fuller, W.A., and Kalton, G. (2006). On the bias of the multiple-imputation variance estimator in survey sampling. *Journal of the Royal Statistical Society, Series B* **68**, 509–521.

Kim, J.K. and Fuller, W. (2004). Fractional hot deck imputation. *Biometrika* **91**, 559–578.

Korn, E.L. and Graubard, B.I. (1999). *Analysis of Health Surveys.* New York: John Wiley & Sons.

Kott, P.S. (1992). *A Note on a Counter-Example to Variance Estimation Using Multiple Imputation.* Technical Report. U.S. National Agriculture Service.

Kott, P.S. (1995). A paradox of multiple imputation. *ASA Proceedings in Section on Survey Research Methods*, pp. 380–383.

Lazzeroni, L.C. and Little, R.J.A. (1998). Random-effects models for smoothing post-stratification weights. *Journal of Official Statistics* **14**, 61–78.

Levy, P.S. and Lemeshow, S. (1999). *Sampling of Populations: Methods and Applications.* New York: John Wiley & Sons.

Little, R.J.A. (1983a). Comment on: An evaluation of model-dependent and probability-sampling inferences in sample surveys. *Journal of the American Statistical Association* **78**, 797–799.

Little, R.J.A. (1983b). Estimating a finite population mean from unequal probability samples. *Journal of the American Statistical Association* **78**, 596–604.

Little, R.J.A. (1988). Missing data adjustments in large surveys. *Journal of Business and Economic Statistics* **6**, 287–301.

Little, R.J.A. (1991). Inference with survey weights. *Journal of Official Statistics* **7**, 405–424.

Little, R.J.A. (1992). Regression with missing X's: a review. *Journal of the American Statistical Association* **87**, 1227–1237.

Little, R.J.A. (2003). Bayesian methods for unit and item nonresponse. In: *Analysis of Survey Data.* R.L. Chambers and C.J. Skinner (Eds.). New York: John Wiley & Sons, pp. 289–306.

Little, R.J.A. (2004). To model or not to model? Competing modes of inference for finite population sampling. *Journal of the American Statistical Association* **99**, 546–556.

Little, R.J.A., D'Agostino, R., Dickersin, K., Emerson, S.S., Farrar, J.T., Frangakis, C., Hogan, J.W., Molenberghs, G., Murphy, S.A., Neaton, J.D., Rotnitzky, A., Scharfstein, D., Shih, W., Siegel, J.P., and Stern, H. National Research Council (2010). *The Prevention and Treatment of Missing Data in Clinical Trials. Panel on Handling Missing Data in Clinical Trials.* Committee on National Statistics, Division of Behavioral and Social Sciences and Education. Washington, D.C.: The National Academies Press.

Little, R.J.A. and Rubin, D.B. (2002). *Statistical Analysis with Missing Data.* (2nd ed.). New York: John Wiley & Sons.

Little, R.J.A. and Vartivarian, S. (2003). On weighting the rates in non-response weights. *Statistics in Medicine* **22**, 1589–1599.

Little, R.J.A. and Vartivarian, S. (2005). Does weighting for nonresponse increase the variance of survey means? *Survey Methodology* **31**, 161–168.

Little, R.J.A. and Zheng, H. (2007). The Bayesian approach to the analysis of finite population surveys. In: *Bayesian Statistics 8.* J.M. Bernardo, M.J. Bayarri, J.O. Berger, A.P. Dawid, D. Heckerman, A.F.M. Smith, and M. West (Eds.) (with discussion and rejoinder). Oxford University Press, pp. 283–302.

Liu, M., Taylor, J.M.G., and Belin, T.R. (2000). Multiple imputation and posterior simulation for multivariate missing data in longitudinal studies. *Biometrics* **56**, 1157–1163.

Lohr, S.L. (1999). *Sampling: Design and Analysis.* Boston: Duxbury Press.

Marker, D.A., Judkins, D.R., and Winglee, M. (2002). Large-scale imputation for complex surveys. In: *Survey Nonresponse.* R.M. Groves, D.A. Dillman, J.L. Eltinge, R.J.A. Little (Eds.). New York: John Wiley & Sons, pp. 329–341.

Meng, X.L. (1994). Multiple imputation with uncongenial sources of input (with discussion). *Statistical Science* **9**, 538–574.

Meng, X.L. (2002). A congenial overview and investigation of multiple imputation inferences under uncongeniality. In: *Survey Nonresponse.* R.M. Groves, D.A. Dillman, J.L. Eltinge, and R.J.A. Little (Eds.). New York: John Wiley & Sons, pp. 343–356.

Mislevy, R.J., Johnson, E.G., and Muraki E. (1992). Scaling procedures in NAEP. *Journal of Educational Statistics* **17**, 131–154.

Paik, M.C. (1997). The generalized estimating equation approach when data are not missing completely at random. *Journal of the American Statistical Association* **92**, 1320–1329.

Pfefferman, D. and LaVange, L. (1989). Regression models for stratified multi-stage cluster samples. In: *Analysis of Complex Surveys.* C.J. Skinner, D. Holt, and T.M.F. Smith (Eds.). New York: John Wiley & Sons, pp. 237–260.

Pfefferman, D. and Nathan, G. (2002). Imputation for wave nonresponse: Existing methods and a time series approach. In: *Survey Nonresponse.* R.M. Groves, D.A. Dillman, J.L. Eltinge, and R.J.A. Little (Eds.). New York: John Wiley & Sons, pp. 417–430.

Raghunathan, T.E. and Grizzle, J.E. (1995). A split questionnaire survey design. *Journal of the American Statistical Association* **90**, 55–63.

Raghunathan, T.E., Lepkowski, J.M., Hoewyk, J.V., and Solenberger, P. (2001). A Multivariate technique for multiply imputing missing values using a sequence of regression models. *Survey Methodology* **27**, 85–95.

Rao, J.N.K. (1996). On variance estimation with imputed survey data. *Journal of the American Statistical Association* **91**, 499–506.

Rao, J.N.K. and Shao, J. (1992). Jackknife variance estimation with survey data under hot deck imputation. *Biometrika* **79**, 811–822.

Royall, R.M. (1983). Comment on: An evaluation of model-dependent and probability-sampling inferences in sample surveys. *Journal of the American Statistical Association* **78**, 794–796.

Rubin, D.B. (1976). Inference and missing data (with discussion). *Biometrika* **63**, 581–592.

Rubin, D.B. (1978). Multiple imputation in sample surveys. *Proceedings of the American Statistical Association Survey Research Methods Section*, pp. 20–34.

Rubin, D.B. (1981). The Bayesian bootstrap. *The Annals of Statistics* **9**, 130–134.

Rubin, D.B. (1983). Comment on: An evaluation of model-dependent and probability-sampling inferences in sample surveys. *Journal of the American Statistical Association* **78**, 803–805.

Rubin, D.B. (1987). *Multiple Imputation for Nonresponse in Surveys.* New York: John Wiley & Sons.

Rubin, D.B. (1996). Multiple imputation after 18+ years (with discussion). *Journal of the American Statistical Association* **91**, 473–489.

Rubin, D.B. and Schenker, N. (1986). Multiple imputation for interval estimation from simple random samples with ignorable nonresponse. *Journal of the American Statistical Association* **81**, 366–374.

Rubin, D.B., Stern, H.S. and Vehovar, V. (1995). Handling "Don't Know" survey responses: the case of the Slovenian plebiscite. *Journal of the American Statistical Association* **90**, 822–828.

Rubin, D.B. and Zanutto, E. (2002). Using matched substitutes to adjust for nonignorable nonresponse through multiple imputations. In: *Survey Nonresponse.* R.M. Groves, D.A. Dillman, R.J.A. Little, and J. Eltinge (Eds.). New York: John Wiley & Sons, pp. 389–402.

Särndal, C.E. (1992). Methods for estimating the precision of survey estimates when imputation has been used. *Survey Methodology* **18**, 241–252.

Schafer, J.L. (1997). *Analysis of Incomplete Multivariate Data.* New York: CRC / Chapman & Hall.

Schafer, J.L. and Graham, J.W. (2002). Missing data: Our view of the state of the art. *Psychological Methods* **7**, 147–177.

Siddique, J. and Belin, T.R. (2008). Multiple imputation using an iterative hot-deck with distance-based donor selection. *Statistics in Medicine* **27**, 83–102.

Skinner, C.J., Smith, T.M.F., and Holt, D. (Eds.) (1989). *Analysis of Complex Surveys.* New York: John Wiley & Sons.

Smith, T.M.F. (1983). Comment on: An evaluation of model-dependent and probability-sampling inferences in sample surveys. *Journal of the American Statistical Association* **78**, 801–802.

Su, Y.S., Gelman, A., Hill, J., and Yajima, M. (2011). Multiple imputation with diagnostics (MI) in R: Opening windows into the black box. *Journal of Statistical Software* **45**, http://www.jstatsoft.org/v45/i02/.

Tanner, M.A. and Wong, W.H. (1987). The calculation of posterior distributions by data augmentation (with discussion). *Journal of the American Statistical Association* **82**, 528–550.

Wachter, K.W. (1993). Comment on: Hierarchical logistic regression models for imputation of unresolved enumeration status in undercount estimation. *Journal of the American Statistical Association* **88**, 1161–1163.

Wang, N. and Robins, J.M. (1998). Large-sample theory for parametric multiple imputation procedures. *Biometrika* **85**, 935–948.

Williams, R.Lh and Folsom, R.E. (1981). Weighted hot-deck imputation of medical expenditures based on a record check subsample. *ASA Proceedings in Section on Survey Research Methods*, pp. 406-411.

Yuan, Y. and Little, R.J.A. (2008). Model-based inference for two-stage cluster samples subject to nonignorable item nonresponse. *Journal of Official Statistics* **24**, 193–211.

24

Model Diagnostics

Dimitris Rizopoulos

Erasmus University Medical Center, the Netherlands

Geert Molenberghs

Universiteit Hasselt & KU Leuven, Belgium

Geert Verbeke

KU Leuven & Universiteit Hasselt, Belgium

CONTENTS

24.1 Introduction

There is a growing literature on methods for statistical analysis of datasets with missing values, with much of this research focused on the development of several modeling frameworks to handle missing data. These include the classes of selection, pattern-mixture and shared-parameter models under a likelihood perspective, and the inverse probability weighted and doubly robust estimators under the generalized estimating equations framework. The interested reader is referred to Chapters 4 and 8 of this volume. The most important lesson that has been learned from all this work is that in missing datasettings, inferences can be strongly influenced by modeling assumptions. Therefore, there has been much interest in sensitivity analysis and the investigation of how results change when key components of the model are altered.

However, much less attention has been given to the related problem of diagnostic and model-assessment tools for regression models in missing datasettings. In particular, even though the properties of residuals of several statistical models have been extensively studied in the literature, in the presence of missing data these basic properties do not always carry over, making the use of residuals challenging. In this chapter, we focus on this problem, and in particular on the analysis of residuals for longitudinal outcomes under attrition. We start with a brief review of marginal and mixed-effects models and we present basic types of residuals for each one of these modeling frameworks. Following this, we provide a thorough discussion of the implications of dropout, and we present a method for calculating residuals and producing diagnostic plots for multivariate models that are expected to have the same properties as residuals based on the complete data (i.e., had there been no dropout). The basic idea behind this approach is to multiply impute the missing longitudinal responses under the fitted model, thus creating random versions of the completed dataset. These completed datasets can then be used to extract conclusions regarding the modeling assumptions, and how these assumptions are affected by dropout. This method is based on an adaptation of the multiply-imputed residuals for joint models for longitudinal and time-to-event data of Rizopoulos et al. (2010), and also shares similarities with the approach of Gelman et al. (2005) who used multiple imputation for posterior predictive checks in missing data and latent variable contexts. These ideas are applied to the data from two randomized longitudinal studies.

The first study concerns 467 HIV infected patients who had failed or were intolerant of zidovudine therapy. The aim of this study was to compare the efficacy and safety of two alternative antiretroviral drugs, namely didanosine (ddI) and zalcitabine (ddC). Patients were randomly assigned to receive either ddI or ddC, and CD4 cell counts were recorded at study entry, where randomization took place, as well as 2, 6, 12, and 18 months thereafter. Of 467 patients, 279 completed the study, resulting in 40.4% dropout. More details about this dataset can be found in Abrams et al. (1994). In the second example, we consider the primary biliary cirrhosis (PBC) data collected by the Mayo Clinic from 1974 to 1984 (Murtaugh et al. 1994). PBC is a chronic, fatal, but rare liver disease characterized by inflammatory destruction of the small bile ducts within the liver, which eventually leads to cirrhosis of the liver. Patients with PBC have abnormalities in several blood tests, such as elevated levels of serum bilirubin. In this study, 312 patients are considered of whom 158 were randomly assigned to receive D-penicillamine and 154 placebo. In our analysis we are interested in testing for a treatment effect on the average longitudinal evolution of bilirubin. Patients did not return to the study centers at prespecified time points to provide serum bilirubin measurements, and thus we observe great variability among their visiting patterns. In particular, patients made on average 6.23 visits (standard deviation 3.77 visits), resulting in a total of 1945 observations. By the end of the study, 169 either received a transplant or died, leading to 54.2% dropout. Dropout due to death raises subtle issues and is distinguished by many from other forms of dropout. Dufouil et al. (2004), Kurland and Heagerty (2005), and Frangakis et al. (2007) offer various strategies to treat this particular form of dropout.

24.2 Multivariate Models

24.2.1 Marginal and mixed-effects models

We start with a short review of multivariate regression models for repeated measurements data. For the sake of illustration and for simplicity, we focus on normally distributed continuous responses and present linear regression models, but our ideas can be easily extended to more complex settings. Let y_{ij} denote the response of subject i, $i = 1, \ldots, N$ at time t_{ij}, $j = 1, \ldots, n_i$, and we collect all measurements of this subject in a vector $\boldsymbol{y}_i$. The statistical models that are often used for the analysis of such data can be broadly classified as marginal models or mixed-effects models. The former class constitutes a direct extension of the simple linear regression model to the multivariate setting. In particular, the multivariate normal model has the form:

$$\boldsymbol{y}_i = \boldsymbol{X}_i\boldsymbol{\beta} + \boldsymbol{\varepsilon}_i^*, \quad \boldsymbol{\varepsilon}_i^* \sim \mathcal{N}(\mathbf{0}, \boldsymbol{V}_i), \tag{24.1}$$

where $\boldsymbol{X}_i$ denotes the design matrix of the covariates of each subject we wish to incorporate in the analysis. In a longitudinal study this almost always will include a time effect and possibly its interactions with other baseline covariates. The interpretation of the regression coefficients $\boldsymbol{\beta}$ is exactly the same as in a simple linear regression. The correlation between the repeated measurements of each subject are explicitly modeled using the variance-covariance matrix $\boldsymbol{V}_i$. There are several covariance structures available, that can capture the features of the data. In the majority of cases, the parameters in this covariance matrix are not of primary interest, and are regarded as nuisance. With complete data, a misspecification of the structure of $\boldsymbol{V}_i$ does not affect consistency but it does affect the efficiency of $\widehat{\boldsymbol{\beta}}$. Nevertheless, when some of the longitudinal responses are missing and depending on the nature of the missing data mechanism, appropriate modeling of $\boldsymbol{V}_i$ is required to ensure unbiased estimates for $\boldsymbol{\beta}$ (Fitzmaurice, Laird, and Ware 2004). In the literature there have been several proposals of different models for $\boldsymbol{V}_i$ that lead to different types of serial correlation functions. Some of the most frequently used are autoregressive processes, and exponential and Gaussian spatial correlation structures, but standard statistical software for fitting multivariate normal models provide many more options (Verbeke and Molenberghs 2000; Pinheiro and Bate, 2000).

Another approach for the analysis of longitudinal responses is based on the linear mixed model. Important early work in the model is by Harville (1977) and Laird and Ware (1982). Verbeke and Molenberghs (2000) and Fitzmaurice, Laird, and Ware (2004), among others, offer detailed treatments. In its general form, it can be posited as:

$$\left\{\begin{array}{lcl} \boldsymbol{y}_i & = & \boldsymbol{X}_i\boldsymbol{\beta} + \boldsymbol{Z}_i\boldsymbol{b}_i + \boldsymbol{\varepsilon}_i, \\ \boldsymbol{b}_i & \sim & \mathcal{N}(\mathbf{0}, \boldsymbol{D}), \\ \boldsymbol{\varepsilon}_i & \sim & \mathcal{N}(\mathbf{0}, \sigma^2\mathbf{I}_{n_i}), \end{array}\right. \tag{24.2}$$

where $\boldsymbol{X}_i$ and $\boldsymbol{Z}_i$ are known design matrices, for the fixed-effects regression coefficients $\boldsymbol{\beta}$, and the random-effects regression coefficients $\boldsymbol{b}_i$, respectively, and $\mathbf{I}_{n_i}$ denotes the n_i-dimensional identity matrix. The random effects are assumed to be normally distributed with mean zero and variance-covariance matrix $\boldsymbol{D}$, and are assumed independent of the error terms $\boldsymbol{\varepsilon}_i$, i.e., $\text{cov}(\boldsymbol{b}_i, \boldsymbol{\varepsilon}_i) = \mathbf{0}$. The interpretation of the fixed effects $\boldsymbol{\beta}$ is exactly the same as in the multivariate regression model (24.1). Similarly, the random effects $\boldsymbol{b}_i$ have interpretation in terms of how a subset of the regression parameters for the ith subject deviates from those in the population. An advantageous characteristic of mixed models is that it is not only possible to estimate parameters that describe how the mean response changes in the population of interest, but it is also possible to predict how individual response trajectories change over time. Moreover, the random effects account for the correlation between the repeated measurements of each subject in a relatively parsimonious way. In particular, the outcomes of the ith subject will be marginally correlated because they share the same random-effect vector $\boldsymbol{b}_i$. In other words, we assume that the longitudinal responses of a subject are independent conditionally on her random effect, i.e.,

$$p(\boldsymbol{y}_i \mid \boldsymbol{b}_i; \boldsymbol{\theta}) = \prod_{j=1}^{n_i} p(y_{ij} \mid \boldsymbol{b}_i; \boldsymbol{\theta}),$$

where $\boldsymbol{\theta}$ denotes the full parameter vector including the fixed effects $\boldsymbol{\beta}$, the measurement error variance σ^2 and the unique element of the covariance matrix $\boldsymbol{D}$, and $p(\cdot)$ denotes an appropriate probability density function. When the chosen random-effects structure is not sufficient to capture the correlation in the data (especially for models with few random effects), we can extend the linear mixed model defined above and allow for an appropriate, more general, covariance matrix for the subject-specific error components, i.e., $\varepsilon_i \sim \mathcal{N}(\mathbf{0}, \Sigma_i)$, with Σ_i depending on i only through its dimension n_i. Similarly to the multivariate normal model (24.1), there are several available covariance structures in standard software for linear mixed models.

Furthermore, taking advantage of the fact that the random effects enter linearly in the specification of the conditional mean $E(\boldsymbol{y}_i \mid \boldsymbol{b}_i)$, and that both the conditional distribution of the longitudinal responses given the random effects $\{\boldsymbol{y}_i \mid \boldsymbol{b}_i\}$ and the distribution of the random effects $\{\boldsymbol{b}_i\}$ are normal, we can derive in closed-form the marginal distribution of the responses $\{\boldsymbol{y}_i\}$, which is an n_i-dimensional normal distribution with mean $\boldsymbol{X}_i\boldsymbol{\beta}$ and variance-covariance matrix $\boldsymbol{V}_i = \boldsymbol{Z}_i\boldsymbol{D}\boldsymbol{Z}_i' + \sigma^2\mathbf{I}_{n_i}$. Thus, the marginal model for $\boldsymbol{y}_i$ implied by the hierarchical representation takes the form:

$$\left\{\begin{array}{lcl} \boldsymbol{y}_i & = & \boldsymbol{X}_i\boldsymbol{\beta} + \varepsilon_i^*, \\ \\ \varepsilon_i^* & \sim & \mathcal{N}(\mathbf{0}, \boldsymbol{Z}_i\boldsymbol{D}\boldsymbol{Z}_i' + \sigma^2\mathbf{I}_{n_i}). \end{array}\right. \tag{24.3}$$

24.2.2 Analysis of the AIDS and PBC datasets

We return to the AIDS and PBC datasets, introduced in Section 24.1, and we perform analyses based on the modeling frameworks presented in the previous section. Taking advantage of the randomization set-up in both studies, we only allow for different average longitudinal evolutions per treatment group and do not control for any other covariates. In particular, for the AIDS dataset we fit the multivariate marginal normal model

$$y_i(t) = \beta_0 + \beta_1 t + \beta_2\{t \times \texttt{ddI}_i\} + \varepsilon_i^*(t),$$

where $y_i(t)$ denotes the square root CD4 cell count at any time point t, $\texttt{ddI}_i$ is the dummy variable for the ddI treatment group, and the covariance matrix for the error terms ε_i^* is assumed to be of the form $\boldsymbol{V}_i = \sigma^2\boldsymbol{H}_i(\phi)$, with $\boldsymbol{H}_i(\phi)$ denoting the first-order autoregressive correlation structure (Pinheiro and Bates 2000, Sec. 5.3). Moreover, we have not included the main effect of treatment because we do not expect any differences in the average square root CD4 cell counts at baseline ($t = 0$), where randomization takes place. The results are presented in Table 24.1. The magnitude of the p-value for parameter β_2 suggests that there is very little evidence of a difference in the average longitudinal evolutions between the two treatments.

For the PBC dataset we fit a linear mixed-effects model and we include in the fixed-effects part the linear and quadratic effects of time and their interaction with treatment, and in the random-effects part random intercepts, random slopes and quadratic random slopes. As in the AIDS dataset, due to randomization we do not control for treatment differences at baseline ($t = 0$). The mixed model takes the form:

$$\begin{aligned} y_i(t) \quad = \quad & \beta_0 + \beta_1 t + \beta_2 t^2 + \beta_3\{\texttt{D-pnc}_i \times t\} + \beta_4\{\texttt{D-pnc}_i \times t^2\} \\ & + b_{i0} + b_{i1}t + b_{i2}t^2 + \varepsilon_i(t), \quad \varepsilon_i(t) \sim \mathcal{N}(0, \sigma^2), \end{aligned}$$

with $y_i(t)$ denoting the logarithm of the observed levels of serum bilirubin, and $\texttt{D-pnc}_i$ the

TABLE 24.1
Parameter estimates, standard errors and p-values under the marginal model for the AIDS dataset. ϕ denotes the parameter of the first-order autoregressive correlation structure.

	Est.	(s.e.)	p-value
β_0	7.148	0.222	< 0.0001
β_1	−0.138	0.024	< 0.0001
β_2	0.012	0.033	0.7173
σ	4.869	0.143	
ϕ	0.861	0.009	

TABLE 24.2
Parameter estimates, standard errors and p-values under the linear mixed model for the PBC dataset. σ_{b_k}, $k = 0, 1, 2$ denotes the standard deviation of the random effect b_k, and ρ_{b_k, b'_k}, $k, k' = 0, 1, 2$ denotes the correlation between random effect b_k and b'_k.

	Est.	(s.e.)	p-value		Est.	(s.e.)
β_0	0.513	0.058	< 0.0001	σ_{b_0}	1.000	0.046
β_1	0.161	0.031	< 0.0001	σ_{b_1}	0.304	0.022
β_2	0.001	0.003	0.6870	σ_{b_2}	0.025	0.003
β_3	0.010	0.043	0.8217	ρ_{b_0, b_1}	0.171	0.115
β_4	−0.002	0.004	0.6745	ρ_{b_0, b_2}	0.011	0.210
σ	0.305	0.006		ρ_{b_1, b_2}	−0.882	0.053

dummy variable for the D-penicillamine group. The results are presented in Table 24.2. Under the posited model we can test whether D-penicillamine stabilizes the average longitudinal evolutions of serum bilirubin (increasing levels of this marker are indicative of worsening of a patient's condition) by statistically testing the null hypothesis $H_0 : \beta_3 = \beta_4 = 0$. The corresponding likelihood ratio test shows that there is no evidence in the observed data for a difference in the average longitudinal evolutions of log serum bilirubin between the placebo and D-penicillamine groups (LRT = 0.234, df = 2, $p = 0.8895$).

24.3 Residuals for Mixed-Effects and Marginal Models

24.3.1 Definitions

A prerequisite before making any final statements based on the analyses of the AIDS and PBC datasets presented in the previous section, is the evaluation of the model assumptions. The statistical tool that is typically used in regression analysis to assess model assumptions is residual plots. Starting from the marginal model (24.1), we can observe that an estimate for the marginal errors $\boldsymbol{\varepsilon}_i^*$ is provided by the residual terms

$$\boldsymbol{r}_i^m = \boldsymbol{y}_i - \boldsymbol{X}_i\widehat{\boldsymbol{\beta}}, \tag{24.4}$$

with corresponding standardized version

$$\boldsymbol{r}_i^{ms} = \widehat{\boldsymbol{V}}_i^{-1/2}(\boldsymbol{y}_i - \boldsymbol{X}_i\widehat{\boldsymbol{\beta}}),$$

where $\widehat{\boldsymbol{V}}_i$ denotes the estimated marginal covariance matrix of $\boldsymbol{y}_i$, and $\boldsymbol{H}^{-1/2}$ denotes the Choleski factor of $\boldsymbol{H}^{-1}$. These residuals can be used to investigate misspecification of the mean structure $\boldsymbol{X}_i\boldsymbol{\beta}$ as well as to validate the assumptions for the within-subjects covariance structure $\boldsymbol{V}_i$.

Similarly for the linear mixed-effects model, we can define two types of residuals, namely the subject-specific (conditional) residuals, and marginal (population averaged) residuals (Nobre and Singer 2007). The subject-specific residuals can be used to validate the assumptions of the hierarchical version of the model (24.2) and are defined as:

$$\boldsymbol{r}_i^{s} = \boldsymbol{y}_i - \boldsymbol{X}_i\widehat{\boldsymbol{\beta}} - \boldsymbol{Z}_i\widehat{\boldsymbol{b}}_i, \qquad (24.5)$$

with corresponding standardized version

$$\boldsymbol{r}_i^{ss} = \{\boldsymbol{y}_i - \boldsymbol{X}_i\widehat{\boldsymbol{\beta}} - \boldsymbol{Z}_i\widehat{\boldsymbol{b}}_i\}/\widehat{\sigma},$$

where, as before, $\widehat{\boldsymbol{\beta}}$ and $\widehat{\sigma}$ denote the MLEs, and $\widehat{\boldsymbol{b}}_i$ the empirical Bayes estimates for the random effects. These residuals predict the conditional errors $\boldsymbol{\varepsilon}_i$, and can be used for checking the homoscedasticity and normality assumptions. On the other hand, the marginal residuals focus on the implied marginal model (24.3) and have exactly the same definition as the residuals for a marginal model (24.4); however, for their standardized version:

$$\boldsymbol{r}_i^{ms} = \widehat{\boldsymbol{V}}_i^{-1/2}(\boldsymbol{y}_i - \boldsymbol{X}_i\widehat{\boldsymbol{\beta}}), \qquad (24.6)$$

the matrix $\widehat{\boldsymbol{V}}_i = \boldsymbol{Z}_i\widehat{\boldsymbol{D}}\boldsymbol{Z}_i' + \widehat{\sigma}^2\mathbf{I}_{n_i}$ denotes the estimated marginal covariance matrix of $\boldsymbol{y}_i$ based on the mixed model formulation. Similarly to the residuals of marginal models defined above, the marginal residuals for a mixed model predict the marginal errors $\boldsymbol{y}_i - \boldsymbol{X}_i\boldsymbol{\beta} = \boldsymbol{Z}_i\boldsymbol{b}_i + \boldsymbol{\varepsilon}_i$, and can be used to investigate misspecification of the mean structure as well as to validate the assumptions for the specific structure of $\boldsymbol{V}_i$ implied by the hierarchical model.

24.3.2 Residuals for the AIDS and PBC datasets

Based on the fitted multivariate normal models for the AIDS and PBC datasets presented in Section 24.2.2 we produce the scatter plots of the marginal residuals $\boldsymbol{r}_i^m$ versus their corresponding fitted values $\boldsymbol{X}_i\widehat{\boldsymbol{\beta}}$, and the scatterplot of the subject-specific residuals $\boldsymbol{r}_i^s$ versus the subject-specific fitted values $\boldsymbol{X}_i\widehat{\boldsymbol{\beta}} + \boldsymbol{Z}_i\widehat{\boldsymbol{b}}_i$. These are depicted in Figures 24.1 and 24.2 for the AIDS and PBC datasets, respectively. In the scatter plots of marginal residuals versus the fitted values for both datasets (though more evidently for the PBC dataset), we observe that the fitted loess curves show a systematic trend with more positive residuals for small fitted values for the AIDS dataset, and more negative residuals for large fitted values for the PBC dataset. At first glance, this is an alarming signal indicating that the form of the design matrix of the fixed effects $\boldsymbol{X}_i$ is not the appropriate one and/or that there is a misspecification of the within errors covariance structure $\boldsymbol{V}_i$.

24.3.3 Dropout and residuals

When there are missing data in the longitudinal responses, extra care is required in drawing conclusions from residual plots. This is mainly due to the fact that the residuals presented

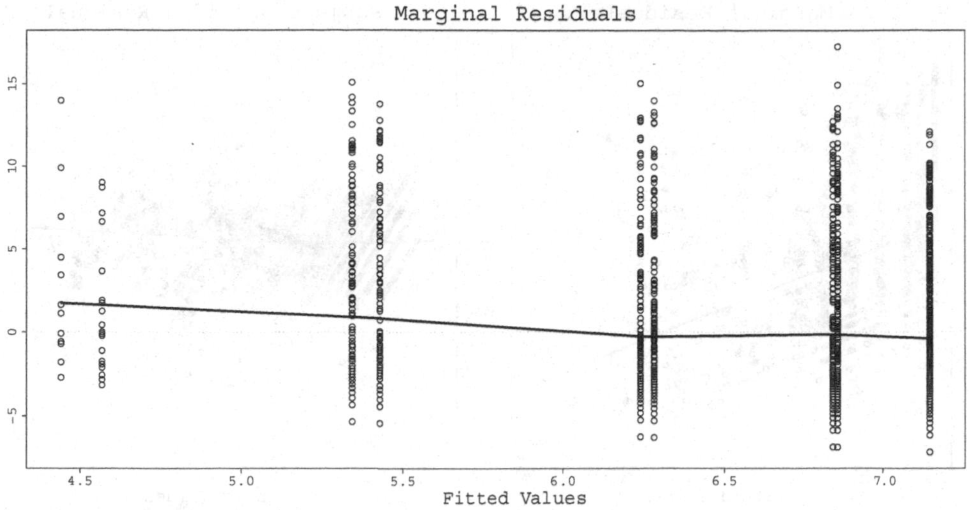

FIGURE 24.1
Scatterplot of the marginal residuals versus marginal fitted values for the AIDS dataset based on the marginal normal model with first-order autoregressive errors. The solid line denotes the fit of the loess smoother.

in Section 24.3.1 are calculated based on the observed longitudinal responses, implying that these residuals will inherit the statistical properties of the observed data, which may be different from the properties of the complete data. To clarify this, we introduce the standard missing data notation utilized in longitudinal settings. In general, we assume that each subject i in the study is designed to be measured at occasions $t_{i1}, \ldots, t_{in_i}$, meaning that, for this subject, we expect to collect the vector of measurements $\boldsymbol{y}_i = (y_{i1}, \ldots, y_{in_i})'$. To distinguish between the response measurements we actually collected from the ones we planned to collect, we introduce the missing data indicator defined as:

$$r_{ij} = \begin{cases} 1 & \text{if } y_{ij} \text{ is observed,} \\ 0 & \text{otherwise.} \end{cases}$$

Therefore, we obtain a partition of the *complete* response vector y_i into two sub-vectors, the observed data sub-vector $\boldsymbol{y}_i^o$ containing those y_{ij} for which $r_{ij} = 1$, and the missing data sub-vector $\boldsymbol{y}_i^m$ containing the remaining components. The vector $\boldsymbol{r}_i = (r_{i1}, \ldots, r_{in_i})'$ and the process generating r_i are referred to as the *missing data process.* When missingness is restricted to dropout or attrition, the missing data indicator $\boldsymbol{r}_i$ is always of the form $(1, \ldots, 1, 0, \ldots, 0)$, and therefore can be replaced by the scalar variable d_i, defined as

$$d_i = 1 + \sum_{j=1}^{n_i} r_{ij}.$$

For an incomplete sequence, d_i denotes the occasion at which dropout occurs, whereas for a complete sequence, $d_i = n_i + 1$. In both cases, d_i equals one plus the length of the observed measurement sequence, whether this is complete or incomplete.

The appropriateness of the use of different types of residuals for incomplete longitudinal data is determined by the *missing data mechanism.* As it has been introduced earlier in this book, this can be thought of as the probability model describing the relation between the

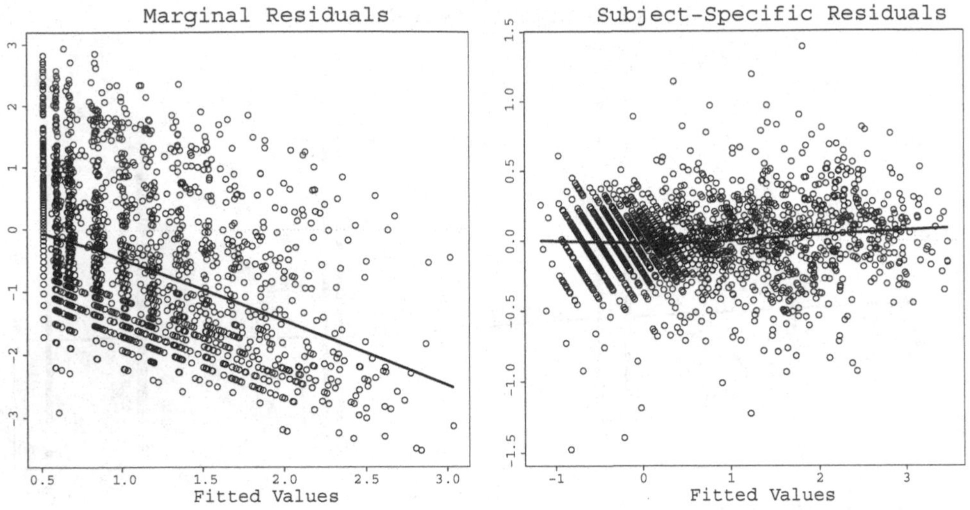

FIGURE 24.2
Scatter plots of the marginal residuals versus marginal fitted values (left panel), and of the subject-specific residuals versus the subject-specific fitted value (right panel) for the PBC dataset based on the linear mixed-effects model. The solid lines denote the fit of the loess smoother.

missing data ($\boldsymbol{r}_i$) and response data ($\boldsymbol{y}_i$) processes. Following the taxonomy of missing data mechanisms, first proposed by Rubin (1976), and further developed in Little and Rubin (2002), we distinguish between the three types of mechanisms as in Chapters 1 and 4, i.e., missing completely at random (MCAR), missing at random (MAR), and not missing at random (NMAR).

It is useful to consider the implications of each of the mechanisms for residual analysis. Under MCAR, we can obtain valid conclusions from residual plots, while ignoring the process(es) generating the missing values. Matters are more subtle under MAR. Then, owing to the fact that the missing data mechanism depends on $\boldsymbol{y}_i^o$, the distribution of $\boldsymbol{y}_i^o$ does not coincide with the distribution of $\boldsymbol{y}_i$, and therefore the observed data cannot be considered a random sample from the target population (Verbeke and Molenberghs 2000; Fitzmaurice, Laird, and Ware 2004, Sec. 14.2). Only the distribution of each subject's missing values $\boldsymbol{y}_i^m$, conditioned on her observed values $\boldsymbol{y}_i^o$, is the same as the distribution of the corresponding observations in the target population. Thus, missing values can be validly predicted using the observed data under a model for the joint distribution $\{\boldsymbol{y}_i^o, \boldsymbol{y}_i^m\}$. The important implication of this feature of MAR is that sample moments are not unbiased estimates of the same moments in the target population. Thus, statistics based on the sample moments without accounting for MAR, such as loess plots of the sample average longitudinal evolutions or scatter plots of residuals calculated on the observed data alone, may prove misleading. Under NMAR, even more so, the observed data do not constitute a random sample from the target population (Verbeke et al. 2008; Molenberghs and Kenward 2007). However, contrary to MAR, the predictive distribution of $\boldsymbol{y}_i^m$ conditional on $\boldsymbol{y}_i^o$ is not the same as in the target population, but rather depends on both $\boldsymbol{y}_i^o$ and on $p(\boldsymbol{r}_i \mid \boldsymbol{y}_i)$. Thus, again plots of residuals based on the observed data will not be expected to exhibit the same properties (i.e., zero mean and independence) as the residuals we would have observed had there been no missing values.

Revisiting the analysis of residuals for the AIDS and PBC datasets presented in Section 24.3.2, we can hypothesize about the origin of the systematic trends in the loess curves in the marginal residual plots. In particular, in Figure 24.1 we observed that for small fitted values we have more positive than negative residuals. However, small fitted values correspond to lower levels of square root CD4 cell count, which in turn corresponds to a worsening of the patient's condition and therefore to higher chance of dropout. Similarly for the PBC data (Figure 24.2) high levels of serum bilirubin indicate a worsening of a patient's condition resulting in higher death rates (i.e., dropout). Thus, for both datasets we cannot discern whether the systematic trends seen in the residual plots are truly attributed to a misspecification of the design matrix $\boldsymbol{X}_i$ of the fixed effects or to dropout.

24.4 Multiple Imputation Residuals

24.4.1 Fixed visit times

In this section, we introduce an approach that is relatively easy to implement. It produces residuals that can be readily used in diagnostic plots. The main idea is to augment the observed data with randomly imputed longitudinal responses under the complete data model, corresponding to the longitudinal outcomes that would have been observed had the patients not dropped out. Using these augmented longitudinal responses, residuals are then calculated for the complete data, and a multiple imputation approach is used to properly account for the uncertainty in the imputed values due to missingness (Rizopoulos et al. 2010; Gelman et al. 2005).

For simplicity, we focus on the missing at random setting, and in particular random dropout. We assume that the multivariate model of interest has been fitted to the dataset at hand, and that we have obtained the maximum likelihood estimates $\widehat{\boldsymbol{\theta}}$ and an estimate of their asymptotic covariance matrix $\widehat{\text{var}}(\widehat{\boldsymbol{\theta}})$. In addition, we assume that every subject in the study is planned to provide longitudinal measurements at the same time points $t_0, t_1, \ldots, t_{max}$, that every subject has a baseline measurement at $t = 0$, and that, for the ith subject, measurements are available up to the last pre-specified visit time before d_i. For this setting, we aim to construct complete datasets using multiple imputation. As introduced in Part IV of this handbook, multiple imputation has a Bayesian foundation and is based on repeated sampling from the posterior distribution of $\boldsymbol{y}_i^m$ given the observed data, averaged over the posterior distribution of the parameters. In our context, this distribution has the form

$$p(\boldsymbol{y}_i^m \mid \boldsymbol{y}_i^o, d_i) = \int p(\boldsymbol{y}_i^m \mid \boldsymbol{y}_i^o, d_i; \boldsymbol{\theta})\, p(\boldsymbol{\theta} \mid \boldsymbol{y}_i^o, d_i)\, d\boldsymbol{\theta}. \tag{24.7}$$

For the first term of the integrand in (24.7), and under the random dropout mechanism and the multivariate normal model (24.1) we obtain that:

$$p(\boldsymbol{y}_i^m \mid \boldsymbol{y}_i^o, d_i; \boldsymbol{\theta}) = p(\boldsymbol{y}_i^m \mid \boldsymbol{y}_i^o; \boldsymbol{\theta}_y),$$

which is a multivariate normal distribution with mean vector and covariance matrix given by the standard formulas used to derive the conditional distributions of a multivariate normal distribution. For model (24.2), the posterior predictive distribution $p(\boldsymbol{y}_i^m \mid \boldsymbol{y}_i^o; \boldsymbol{\theta}_y)$ can be further simplified as:

$$p(\boldsymbol{y}_i^m \mid \boldsymbol{y}_i^o; \boldsymbol{\theta}_y) = \int p(\boldsymbol{y}_i^m \mid \boldsymbol{b}_i; \boldsymbol{\theta}_y)\, p(\boldsymbol{b}_i \mid \boldsymbol{y}_i^o; \boldsymbol{\theta}_y)\, d\boldsymbol{b}_i,$$

where both terms in the integrand are densities of the multivariate normal distribution. The second term in the integrand is the posterior distribution of the parameters given the observed data. Since we have not actually fitted the model under the Bayesian approach but instead under maximum likelihood, we can approximate this posterior distribution using arguments of standard asymptotic Bayesian theory (Cox and Hinkley 1974, Sec. 10.6). More specifically, assuming that the sample size n is sufficiently large, $\{\boldsymbol{\theta} \mid \boldsymbol{y}_i^o\}$ can be well approximated by $\mathcal{N}\{\widehat{\boldsymbol{\theta}}, \widehat{\text{var}}(\widehat{\boldsymbol{\theta}})\}$. This assumption, combined with the previous statements, suggests the following simulation scheme:

S1: Draw $\widehat{\boldsymbol{\theta}}^{(l)} \sim \mathcal{N}\{\widehat{\boldsymbol{\theta}}, \widehat{\text{var}}(\widehat{\boldsymbol{\theta}})\}$.

S2: For random-effects models, draw $\widehat{\boldsymbol{b}}_i^{(l)} \sim \{\boldsymbol{b}_i \mid \boldsymbol{y}_i^o, \widehat{\boldsymbol{\theta}}^{(l)}\}$.

S3: Obtain realizations of $\boldsymbol{y}_i^m$

S3a: For marginal models, draw $[\boldsymbol{y}_i^m(t_{ij})]^{(l)} \sim \mathcal{N}(\boldsymbol{X}_i\widehat{\boldsymbol{\beta}}^{(l)}, \widehat{\boldsymbol{V}}_i^{(l)})$, for the pre-specified visit times t_{ij}, $j = 1, \ldots, n_i'$ that were not observed for the ith subject.

S3b: For random-effects models, draw $[\boldsymbol{y}_i^m(t_{ij})]^{(l)} \sim \mathcal{N}\Big(\boldsymbol{X}_i\widehat{\boldsymbol{\beta}}^{(l)} + \boldsymbol{Z}_i\widehat{\boldsymbol{b}}_i^{(l)}, [\widehat{\sigma}^2]^{(l)}\Big)$ for the pre-specified visit times t_{ij}, $j = 1, \ldots, n_i'$ that were not observed for the ith subject.

Steps 1–3 are repeated for each subject, $l = 1, \ldots, L$ times, where L denotes the number of imputations. Steps 1 and 2 account for the uncertainties in the parameter and empirical Bayes estimates, respectively, whereas Step 3 imputes the missing longitudinal responses. Evidently, Step 2 is only required for random effects models. All steps are straightforward to perform since they simply require sampling from a multivariate normal distribution. The simulated $[\boldsymbol{y}_i^m(t_{ij})]^{(l)}$ values together with $\boldsymbol{y}_i^o$ can now be used to create complete datasets based on which we can calculate the residuals presented in Section 24.3.1. A key advantage of these multiply-imputed residuals is that they inherit the properties of the complete data model. This greatly facilitates common graphical model checks without requiring formal derivation of the reference distribution of the observed data residuals. We should note, however, that in some clinical studies in which the terminating event is death, such as in the AIDS and PBC datasets, it may be conceptually unreasonable to consider potential values of the longitudinal outcome after the event time; for instance, see Kurland and Heagerty (2005). Nonetheless, the multiply-imputed residuals are merely used as a mechanism to help us investigate the fit of the model, and we are not actually interested in inferences after the event time. Finally, regarding the choice of L, we expect that a moderate number of multiple imputations, say between 5 and 10, will be sufficient for the propagation of uncertainty.

We illustrate the use of the multiply-imputed residuals for the AIDS dataset. Following the residual analysis presented in Section 24.3.2, we imputed each of the missing square root CD4 cell counts five times. The resulting scatterplot of the observed and the multiply-imputed marginal residuals versus the fitted values is shown in Figure 24.3. As in Figure 24.1, the black solid lines denotes the fit of the loess smoother based on the observed data. To produce the grey loess curve, which describes the relationship between the complete residuals (i.e., the multiply-imputed residuals together with the observed residuals) versus their corresponding fitted values, some extra steps are required. In particular, we need to take into account that, for each of the time points the ith subject did not appear in the study center, we have $L = 5$ multiply-imputed residuals, whereas for the times that she did appear we only have one (Little and Rubin 2002). Thus, in the calculation of the loess curve

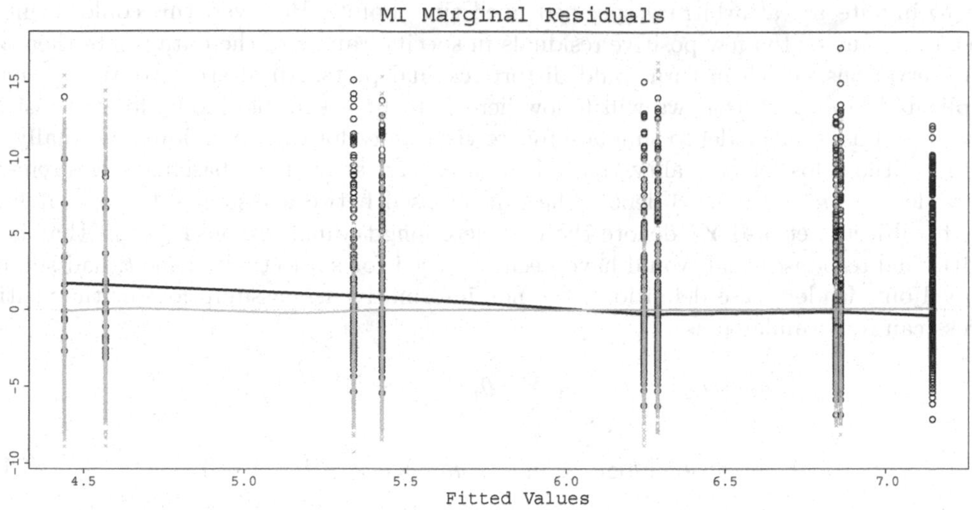

FIGURE 24.3
Observed standardized marginal residuals (black points), augmented with all the multiply-imputed residuals produced by the $L = 5$ imputations (grey points) for the AIDS dataset. The superimposed solid lines represent a loess fit based only on the observed residuals (black solid line), and a weighted loess fit based on all residuals (grey solid line).

we will use case weights with the value 1 for the observed residuals and $1/L = 1/5 = 0.2$ for the multiply-imputed ones. A comparison between the two loess smoothers reveals that the systematic trends that were present in the residual plots based on the observed data alone are probably attributed to the dropout and not to a model lack-of-fit.

24.4.2 Random visit times

The multiple imputation scheme presented in Section 24.4.1 assumes that visit times were fixed and the same for all patients. However, in observational studies the time points at which the longitudinal measurements are taken are not fixed by design but rather determined by the physician or even the patients themselves. This may even occur in randomized studies, which, per protocol, have prespecified visit times. For instance, for the PBC dataset and during the first two years of follow-up, measurements of serum bilirubin were taken at baseline, 0.5, 1, and 2 years, with little variability; however, in later years the variability in the visit times increased considerably. In these settings and provided that the multivariate model is correctly specified, we can obtain valid inferences for the parameters of interest, while ignoring the visiting process (i.e., the mechanism (stochastic or deterministic) that generates the time points at which longitudinal measurements are collected (Lipsitz et al. 2002)). This non-informativeness assumption is, in fact, analogous to the MAR assumption for the missingness process.

However, the possibility of random visit times considerably complicates the computation of multiple imputation residuals, presented in Section 24.4.1. The problem is that time points at which the ith subject was supposed to provide measurements after the dropout time are not known, and thus the corresponding rows $\boldsymbol{x}_i'(t_{ij})$ and $\boldsymbol{z}_i'(t_{ij})$, for $t_{ij} \geq d_i$, of the design matrices $\boldsymbol{X}_i$ and $\boldsymbol{Z}_i$, respectively, cannot be specified. A potential easy solution could have been to impute $\boldsymbol{y}_i^m$ at arbitrary specified fixed time points. However, this could result in either too many or too few positive residuals in specific ranges of the data where there are few observations, which in turn could distort residual plots. An alternative, though more complicated, approach that we will follow here is to fit a suitable model for the visiting process, and use this model to simulate future visit times for each individual. Formally, we assume, without loss of generality, that all subjects have at least the baseline measurement, and we let u_{ik} $(k = 2, \ldots, n_i)$ denote the time elapsed between visit $k-1$ and visit k for the ith subject. Let also $\boldsymbol{Y}_i^*$ denote the complete longitudinal response vector, that is all longitudinal responses that would have been observed for subject i by time t, had she not dropped out. Under these definitions, the non-informativeness assumption for the visiting process can be formulated as

$$\begin{aligned} & p(u_{ik} \mid u_{i2}, \ldots, u_{i,k-1}, \boldsymbol{Y}_i^*; \boldsymbol{\theta}_v) \\ & \quad = \quad p\{u_{ik} \mid u_{i2}, \ldots, u_{i,k-1}, y_i(t_1), \ldots, y_i(t_{k-1}); \boldsymbol{\theta}_v\}, \end{aligned} \tag{24.8}$$

where $\boldsymbol{\theta}_v$ is the vector parameterizing the visiting process density, and $\boldsymbol{\theta}$ and $\boldsymbol{\theta}_v$ have disjoint parameter spaces. To formulate a model for the visiting process, we need to account for the fact that the visit times $\boldsymbol{u}_i' = (u_{i2}, \ldots, u_{in_i})$ of each subject are correlated. There are two main classes of model for multivariate survival data, namely, marginal and conditional/frailty models (Hougaard 2000; Therneau and Grambsch 2000). Marginal models are based on an idea similar to the generalized estimating equations approach (Liang and Zeger 1986). This entails fitting an ordinary Cox model to the multivariate data treating events from the same subjects as independent (i.e., ignoring the correlation), and then adjust the estimated standard errors using a sandwich-type of estimator. On the other hand, conditional models explicitly model the correlation using latent variables. As with mixed models, we make a conditional independence assumption that states that the multivariate survival responses are independent given the frailty term. Therefore, frailty models require the specification of both a model for the multivariate survival responses conditional on the frailty term, and appropriate distributional assumptions for the frailty term itself (Duchateau and Janssen 2008).

For our purposes, and because we want to simulate visit times for each subject, we need to provide a full specification of the conditional distribution $p\{u_{ik} \mid u_{i2}, \ldots, u_{i,k-1}, y_i(t_1), \ldots, y_i(t_{k-1}); \boldsymbol{\theta}_v\}$, and therefore a conditional model is more appropriate. In particular, we use a Weibull model with a multiplicative Gamma frailty, defined as:

$$\lambda(u_{ik} \mid \boldsymbol{x}_{vi}, \omega_i) = \lambda_0(u_{ik})\omega_i \exp(\boldsymbol{x}_{vi}'\gamma_v), \quad \omega_i \sim \text{Gamma}(\sigma_\omega, \sigma_\omega), \tag{24.9}$$

where $\lambda(\cdot)$ is the risk function conditional on the frailty term ω_i, $\boldsymbol{x}_{vi}$ denotes the covariate vector that may contain a functional form of the observed longitudinal responses $y_i(t_{i1}), \ldots, y_i(t_{i,k-1})$, $\boldsymbol{\gamma}_v$ is the vector of regression coefficients, and σ_ω^{-1} is the unknown variance of ω_i's. The Weibull baseline risk function is given by $\lambda_0(u_{ik}) = \phi\psi u_{ik}^{\psi-1}$, with $\psi, \phi > 0$. Our choice for this model is motivated, not only by its flexibility and simplicity, but also by the fact that the posterior distribution of the frailty term, given the observed data, is of standard form (Sahu et al. 1997), which as will be shown below, facilitates simulation.

As in Section 24.4.1, we assume that both the joint model and the visiting process model (24.9) have been fitted to the data at hand, and that the maximum likelihood estimates $\widehat{\boldsymbol{\theta}}$ and $\widehat{\boldsymbol{\theta}}_v$, and their corresponding asymptotic covariance matrices, $\widehat{\text{var}}(\widehat{\boldsymbol{\theta}})$ and $\widehat{\text{var}}(\widehat{\boldsymbol{\theta}}_v)$, respectively, have been obtained. Let also t_{max} denote the end of the study, and $\delta_{v,ik}$

the event indicator corresponding to u_{ik}. Furthermore, taking into consideration the non-informativeness assumption (24.8), the future elapsed visit time u_{i,n_i+1} can be simulated independently from $y_i^m(t_{i,n_i+1})$. Thus, the simulation scheme under the random visit times setting takes the following form:

S1: Parameter Values

a. Draw $\widehat{\boldsymbol{\theta}}_v^{(l)} \sim \mathcal{N}\{\widehat{\boldsymbol{\theta}}_v, \widehat{\text{var}}(\widehat{\boldsymbol{\theta}}_v)\}$.

b. Draw $\widehat{\boldsymbol{\theta}}^{(l)} \sim \mathcal{N}\{\widehat{\boldsymbol{\theta}}, \widehat{\text{var}}(\widehat{\boldsymbol{\theta}})\}$.

S2: Frailties and Random Effects

a. Draw

$$\widehat{\omega}_i^{(l)} \sim \text{Gamma}(A, B),$$

with $A = \widehat{\sigma}_\omega^{(l)} + \sum_{k=2}^{n_i} \delta_{v,ik}$, and $B = \widehat{\sigma}_\omega^{(l)} + \widehat{\phi}^{(l)} \sum_{k=2}^{n_i} u_{ik}^{\widehat{\psi}^{(l)}} \exp(\boldsymbol{x}_{vi}' \widehat{\boldsymbol{\gamma}}_v^{(l)})$ for subjects with two or more visits, and

$$\widehat{\omega}_i^{(l)} \sim \text{Gamma}(\widehat{\sigma}_\omega^{(l)}, \widehat{\sigma}_\omega^{(l)}),$$

for subjects with one visit.

b. Draw $\widehat{\boldsymbol{b}}_i^{(l)} \sim \{\boldsymbol{b}_i \mid \boldsymbol{y}_i^o, \widehat{\boldsymbol{\theta}}^{(l)}\}$.

S3: Outcomes

a. Draw $\widehat{u}_i^{(l)} \sim \text{Weibull}\Big\{\widehat{\psi}^{(l)}, \widehat{\phi}^{(l)} \widehat{\omega}_i^{(l)} \exp(\boldsymbol{x}_{vi}' \widehat{\boldsymbol{\gamma}}_v^{(l)})\Big\}$.

b. Set $\tilde{t}_i = u_i^{(l)} + t_{in_i}$, where t_{in_i} denotes the last observed visit time for the ith subject. If $\tilde{t}_i > t_{max}$, no $\boldsymbol{y}_i^m$ need to be imputed for this subject; otherwise for a mixed model draw $y_i^{m(l)}(\tilde{t}_i) \sim \mathcal{N}\Big\{\boldsymbol{x}_i'(\tilde{t}_i)\widehat{\boldsymbol{\beta}}^{(l)} + \boldsymbol{z}_i'(\tilde{t}_i)\widehat{\boldsymbol{b}}_i^{(l)}, \widehat{\sigma}^{2,(l)}\Big\}$, whereas if a marginal model has been considered instead, then draw $y_i^{m(l)}(\tilde{t}_i) \sim \mathcal{N}\Big\{\boldsymbol{x}_i'(\tilde{t}_i)\widehat{\boldsymbol{\beta}}^{(l)}, \widehat{\boldsymbol{V}}_i^{(l)}\Big\}$.

c. Set $t_{in_i} = \tilde{t}_i$, and repeat a–b until $t_{in_i} > t_{max}$ for all i.

Steps 1–3 are repeated $l = 1, \ldots, L$ times. As in Section 24.4.1, Steps 1–3 simultaneously account for uncertainties in both the joint and visiting process models. Note that subjects who have only one longitudinal measurement provide no information to the visiting process model. For these cases, in Step 3a, we can only simulate future elapsed visit times using a simulated frailty value from the Gamma prior distribution (Step 2a).

An important feature exploited in the above simulation scheme is the form of the linear predictor of the visiting process model. More specifically, we should note that assumption (24.8) is the weakest assumption under which the joint model provides valid inferences even if the visiting process is ignored, but a model satisfying (24.8) involves many parameters, and thus it may be unstable. A set of stronger but perhaps more plausible assumptions is

$$\begin{aligned} & p(u_{ik} \mid u_{i2}, \ldots, u_{i,k-1}, \boldsymbol{Y}_i^*; \boldsymbol{\theta}_v) \\ & \quad = \quad p\{u_{ik} \mid u_{i2}, \ldots, u_{i,k-1}, y_i(t_{k-1}); \boldsymbol{\theta}_v\}, \end{aligned} \qquad (24.10)$$

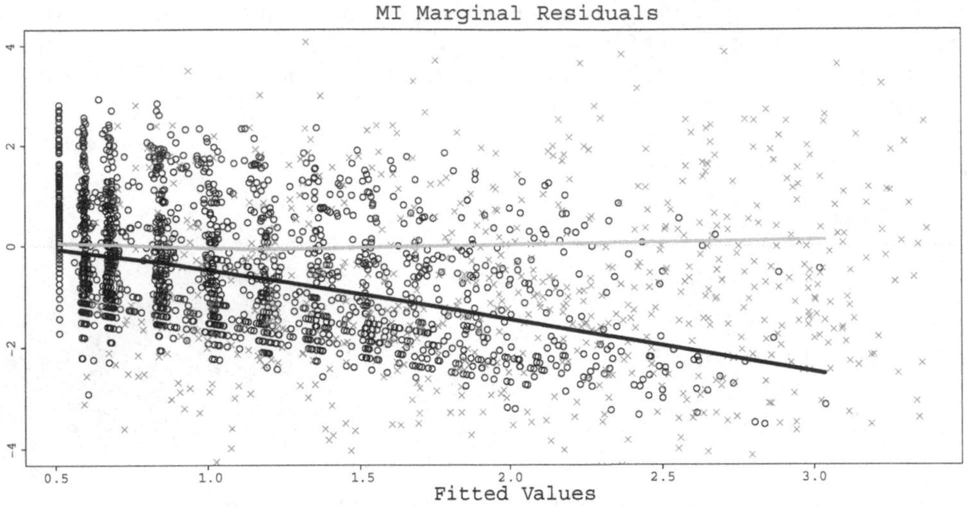

FIGURE 24.4
Observed standardized marginal residuals (black points), augmented with all the multiply-imputed residuals produced by the $L = 5$ imputations (grey points) for the PBC dataset. The superimposed solid lines denote a loess fit based only on the observed residuals (black solid line), and a weighted loess fit based on all residuals (grey solid line).

and

$$p(u_{ik} \mid u_{i2}, \dots, u_{i,k-1}, \boldsymbol{Y}_i^*; \boldsymbol{\theta}_v) = p\{u_{ik} \mid y_i(t_{k-1}); \boldsymbol{\theta}_v\}. \qquad (24.11)$$

Equation (24.10) posits that the time elapsed between visit $k-1$ and visit k depends on the previous elapsed times and the last observed longitudinal measurement, whereas under (24.11) it depends only on the last observed longitudinal measurement. These assumptions describe the situation in which physicians base their decision for a future visit for a patient on the last observed outcome and possibly the past visiting pattern.

We illustrate the practical use of the multiple imputation approach to augment the standardized residuals for the longitudinal process for the linear mixed model fitted to the PBC dataset in Section 24.3.2. As a first step, we need to specify the visiting process model. For the sake of illustration, we assume here a simple model that postulates that the elapsed time u_{ik} between visits $k-1$ and k depends on the previous visit times, and the current value of serum bilirubin. This model corresponds to assumption (24.10), and it takes the following form:

$$\lambda(u_{ik} \mid y_i^*(t), \omega_i) = \lambda_0(u_{ik})\omega_i \exp\{\gamma_v y_i(t_{k-1})\}, \quad \omega_i \sim \text{Gamma}(\sigma_\omega, \sigma_\omega).$$

Based on this model for the visiting process and the linear mixed model for the PBC dataset presented in Section 24.2.2, we implemented the above Monte Carlo scheme with five imputations. To help us spot any systematic trends in the residuals versus fitted values scatter plot, we calculate the loess smoother. Again when calculating the loess smoother for the completed data, we need to account for the fact that residuals after the dropout time are over-represented. As in Section 24.4.1, we achieve that using the weighted loess smoother, with weights $1/L = 1/5 = 0.2$ for the residuals after dropout. Similarly to the AIDS dataset, we observe from Figure 24.4 that the weighted loess, based on the completed data, does not exhibit the systematic trend that was clearly seen in the loess curve based on the observed data alone. This is a strong indication that this systematic trend was primarily due to dropout and not because of model misfit.

24.5 Discussion and Extensions

The take home message of this chapter is two-fold. First, missing data do not only complicate statistical analyses, but they also complicate checking the assumptions behind these analyses. Even though standard plots and diagnostics can be performed and are frequently reported by statistical software for a variety of regression models, these were designed for the setting of complete data. When they are computed for incomplete data, their statistical properties are linked to the nature of the missing data mechanism. Under the convenient but strict scenario of MCAR, all these tools can be used without requiring any further adjustments. Under the more commonly assumed MAR assumption however, the use of such tools becomes problematic because they may show alarming signals due to dropout and not because of model misfit. This may be the case even if the model-fitting procedure we have chosen provides valid parameter estimates and standard errors under MAR (e.g., using maximum likelihood). Our second goal was to introduce a relatively easy-to-implement method that allows the use of regression diagnostics in the presence of missing data. This approach is based on the idea of multiple imputation, i.e., to create random complete versions of our incomplete dataset. Residuals or other diagnostics can be then computed in these completed datasets and utilized in the usual manner because they will exhibit the same properties as if there were no missing data. However, it needs to be stressed that by using the multiple imputation approach, we fill-in the missing data with values that "agree," in a sense, with the postulated model. At first glance, this could be regarded as a vicious circle, because residuals based on the imputed values will never show a model misfit. Nonetheless, the aim is not actually to check the fit in the missing part; this cannot be done because we do not get to see $\boldsymbol{y}_i^m$. The aim is to check the fit based on the observed part, but without getting distracted by the dropout.

Even though we have discussed in this chapter the use of the multiply-imputed residuals for multivariate linear models, several extensions can be considered. For example, these residuals are directly applicable to other types of hierarchical data (e.g., multilevel designs). In addition, for non-normal responses and by suitably changing the definitions of (24.4) and (24.5), the same ideas carry over for checking the assumptions behind generalized linear mixed models (Breslow and Clayton 1993) and generalized estimating equations (Liang and Zeger 1986). Moreover, in our developments we have concentrated on random dropout. Nonetheless, using a small adaptation of the methodology presented in Section 24.4, these residuals also can be used in nonrandom dropout settings and in conjunction with selection, pattern-mixture or shared-parameter models. For the latter modeling framework the interested reader is referred to Rizopoulos et al. (2010).

Acknowledgments

Geert Verbeke and Geert Molenberghs acknowledge support from IAP research network grant P07/06 from the Belgian government (Belgian Science Policy).

References

Abrams, D., Goldman, A., Launer, C., et al. (1994). Comparative trial of didanosine and zalcitabine in patients with human immunodeficiency virus infection who are intolerant of or have failed zidovudine therapy. *New England Journal of Medicine*, **330**, 657–662.

Breslow, N. and Clayton, D. (1993). Approximate inference in generalized linear mixed models. *Journal of the American Statistical Association*, **88**, 9–25.

Cox, D.R. and Hinkley, D.V. (1974). *Theoretical Statistics.* London: Chapman & Hall.

Duchateau, L. and Janssen, P. (2008). *The Frailty Model.* New York: Springer.

Dufouil, C., Brayne, D., and Clayton, D. (2004). Analysis of longitudinal studies with death and drop-out: a case study. *Statistics in Medicine* **23**, 2215–2226.

Fitzmaurice, G., Laird, N., and Ware, J. (2004). *Applied Longitudinal Analysis.* Hoboken: John Wiley & Sons.

Frangakis, C.E., Rubin, D.B., An, M.-W., and MacKenzie, E. (2007). Principal stratification designs to estimate input data missing due to death. *Biometrics* **63**, 641–649.

Gelman, A., Van Mechelen, I., Verbeke, G., Heitjan, D., and Meulders, M. (2005). Multiple imputation for model checking: Completed-data plots with missing and latent data. *Biometrics*, **61**, 74–85.

Harville, D. (1977). Maximum likelihood approaches to variance component estimation and to related problems. *Journal of the American Statistical Association* **72**, 320–340.

Hougaard, P. (2000). *Analysis of Multivariate Survival Data.* New York: Springer.

Kurland, B. and Heagerty, P. (2005). Directly parameterized regression conditioning on being alive: Analysis of longitudinal data truncated by deaths. *Biostatistics* **6**, 241–258.

Laird, N. and Ware, J. (1982). Random-effects models for longitudinal data. *Biometrics*, **38**, 963–974.

Liang, K.-Y. and Zeger, S. (1986). Longitudinal data analysis using generalized linear models. *Biometrika*, **73**, 13–22.

Lipsitz, S., Fitzmaurice, G., Ibrahim, J., Gelber, R., and Lipshultz, S. (2002). Parameter estimation in longitudinal studies with outcome-dependent follow-up. *Biometrics* **58**, 621–630.

Little, R. and Rubin, D. (2002). *Statistical Analysis with Missing Data* (2nd Ed.) New York: John Wiley & Sons.

Molenberghs, G. and Kenward, M. (2007). *Missing Data in Clinical Studies.* New York: John Wiley & Sons.

Murtaugh, P., Dickson, E., Van Dam, G., Malincho, M., Grambsch, P., Langworthy, A., and Gips, C. (1994). Primary biliary cirrhosis: Prediction of short-term survival based on repeated patient visits. *Hepatology* **20**, 126–134.

Nobre, J. and Singer, J. (2007). Residuals analysis for linear mixed models. *Biometrical Journal* **6**, 863–875.

Pinheiro, J. and Bates, D. (2000). *Mixed-Effects Models in S and S-PLUS.* New York: Springer.

Rizopoulos, D., Verbeke, G., and Molenberghs, G. (2010). Multiple-imputation-based residuals and diagnostic plots for joint models of longitudinal and survival outcomes. *Biometrics* **66**, 20–29.

Rubin, D. (1976). Inference and missing data. *Biometrika* **63**, 581-592.

Sahu, S., Dey, D., Aslanidou, H., and Sinha, D. (1997). A Weibull regression model with gamma frailties for multivariate survival data. *Lifetime Data Analysis* **3**, 123–137.

Therneau, T. and Grambsch, P. (2000). *Modeling Survival Data: Extending the Cox Model.* New York: Springer.

Verbeke, G. and Molenberghs, G. (2000). *Linear Mixed Models for Longitudinal Data.* New York: Springer.

Verbeke, G., Molenberghs, G., and Beunckens, C. (2008). Formal and informal model selection with incomplete data. *Statistical Science* **23**, 201–218.

Index

Printed in the United States
by Baker & Taylor Publisher Services

Printed in the United States
by Baker & Taylor Publisher Services